Student Solutions Manual

Intermediate Algebra

FIFTH EDITION

Ron Larson
The Pennsylvania State University
The Behrend College

With the assistance of
Kimberly Nolting
Hillsborough Community College

Prepared by

Gerry C. Fitch
Louisiana State University

BROOKS/COLE
CENGAGE Learning

Australia • Brazil • Japan • Korea • Mexico • Singapore • Spain • United Kingdom • United States

BROOKS/COLE
CENGAGE Learning

For product information and technology assistance, contact us at
Cengage Learning Academic Resource Center,
1-800-423-0563

For permission to use material from this text or product, submit all requests online at **www.cengage.com/permissions**
Further permissions questions can be emailed to
permissionrequest@cengage.com

ISBN-13: 978-0-547-14019-3
ISBN-10: 0-547-14019-3

Brooks/Cole
20 Channel Center Street
Boston, MA 02210
USA

Cengage Learning products are represented in Canada by Nelson Education, Ltd.

For your course and learning solutions, visit **academic.cengage.com**

Purchase any of our products at your local college store or at our preferred online store **www.ichapters.com**

Printed in the United States of America
2 3 4 5 6 7 13 12 11 10 09

PREFACE

This *Student Solutions Manual* is a supplement to *Intermediate Algebra*, Fifth Edition, by Ron Larson and Kimberly Nolting. The *Student Solutions Manual* includes solutions to the odd-numbered exercises in the text, including chapter reviews. In addition, this book contains the solutions to all exercises, even-numbered as well as odd-numbered, from the Review: Concepts, Skills, and Problem Solving, mid-chapter quizzes, chapter tests, and cumulative tests.

These solutions give step-by-step details of each exercise. There are usually several "correct" ways to arrive at a solution to a mathematics problem. Therefore, you should not be concerned if you have approached problems differently than I have. Several accuracy checks have been made to ensure that these solutions are correct. Corrections to the solutions or suggestions for improvement are welcome.

Producing this manual has been quite a challenge and a learning experience for me. I would like to thank the editors of Cengage Learning for allowing me this experience and Larson Texts, Inc. And finally, a special word of thanks goes to my husband, Chuck, for his support and patience during the writing and proofing of the manuscript.

I hope you will find the *Student Solutions Manual* a helpful supplement as you study algebra.

Gerry C. Fitch
Louisiana State University
Baton Rouge, Louisiana 70803

CONTENTS

CHAPTER 1
Fundamentals of Algebra

CHAPTER 1
Fundamentals of Algebra

Section 1.1 The Real Number System

1. $\left\{-6, -\sqrt{6}, -\frac{4}{3}, 0, \frac{5}{8}, 1, \sqrt{2}, 2, \pi, 6\right\}$

(a) Natural numbers: $\{1, 2, 6\}$

(b) Integers: $\{-6, 0, 1, 2, 6\}$

(c) Rational numbers: $\left\{-6, -\frac{4}{3}, 0, \frac{5}{8}, 1, 2, 6\right\}$

(d) Irrational numbers: $\left\{-\sqrt{6}, \sqrt{2}, \pi\right\}$

3. $\left\{-4.2, \sqrt{4}, -\frac{1}{9}, 0, \frac{3}{11}, \sqrt{11}, 5.\bar{5}, 5.543\right\}$

(a) Natural numbers: $\left\{\sqrt{4}\right\}$

(b) Integers: $\left\{\sqrt{4}, 0\right\}$

(c) Rational numbers: $\left\{4.2, \sqrt{4}, -\frac{1}{9}, 0, \frac{3}{11}, 5.\bar{5}, 5.543\right\}$

(d) Irrational numbers: $\left\{\sqrt{11}\right\}$

5. $0.2222\ldots = 0.\bar{2}$

7. $2.121212\ldots = 2.\overline{12}$

9. $\{-5, -4, -3, -2, -1, 0, 1, 2, 3\}$

11. $\{5, 7, 9\}$

13. (a) The point representing the real number 3 lies between 2 and 4.

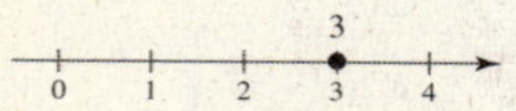

(b) The point representing the real number $\frac{5}{2}$ lies between 2 and 3.

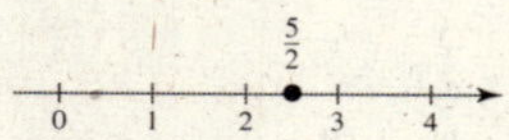

(c) The point representing the real number $-\frac{7}{2}$ lies between -4 and -3.

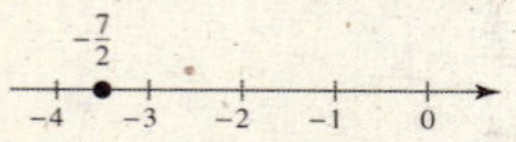

(d) The point representing the real number -5.2 lies between -6 and -5, but closer to -5.

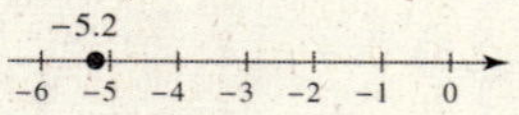

15. $a = -1, b = \frac{1}{2}$

$-1 < \frac{1}{2}$

17. $a = -\frac{9}{2}, b = -2,$

$-\frac{9}{2} < -2$

19. $\frac{4}{5} < 1$ because $\frac{4}{5}$ is to the left of 1 on the real number line.

21. $-5 < 2$ because -5 is to the left of 2 on the real number line.

23. $-5 < -2$ because -5 is to the left of -2 on the real number line.

25. $\frac{5}{8} > \frac{1}{2}$ because $\frac{5}{8}$ is to the right of $\frac{1}{2}$ on the real number line.

27. $-\frac{2}{3} > -\frac{10}{3}$ because $-\frac{2}{3}$ is to the right of $-\frac{10}{3}$ on the real number line.

29. Distance $= 10 - 4 = 6$

31. Distance $= 7 - (-12) = 7 + 12 = 19$

33. Distance $= 18 - (-32) = 18 + 32 = 50$

35. Distance $= 0 - (-8) = 0 + 8 = 8$

37. Distance $= 35 - 0 = 35$

39. Distance $= (-6) - (-9) = (-6) + 9 = 3$

41. $|10| = 10$

43. $|-225| = 225$

45. $-|-85| = -85$

47. $-|16| = -16$

49. $-\left|-\frac{3}{4}\right| = -\frac{3}{4}$

51. $-|3.5| = -3.5$

53. $|-\pi| = \pi$

55. $|-6| > |2|$ because $|-6| = 6$ and $|2| = 2$, and 6 is greater than 2.

57. $|47| > |-27|$ because $|47| = 47$ and $|-27| = 27$, and 47 is greater than 27.

59. $|-1.8| = |1.8|$ because $|-1.8| = 1.8$ and $|1.8| = 1.8$.

61. $\left|-\frac{3}{4}\right| > -\left|\frac{4}{5}\right|$ because $\left|-\frac{3}{4}\right| = \frac{3}{4}$ and $-\left|\frac{4}{5}\right| = -\frac{4}{5}$, and $\frac{3}{4}$ is greater than $-\frac{4}{5}$.

63. Opposite: -34
Absolute value: 34

65. Opposite: 160
Absolute value: 160

67. Opposite: $\frac{3}{11}$
Absolute value: $\frac{3}{11}$

69. Opposite: $-\frac{5}{4}$
Absolute value: $\frac{5}{4}$

71. Opposite: -4.7
Absolute value: 4.7

73. The opposite of -7 is 7.

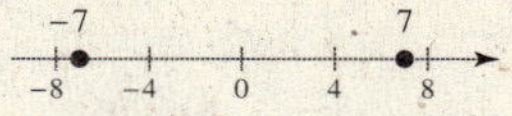

The distance of both -7 and 7 from 0 is 7.

75. The opposite of 5 is -5.

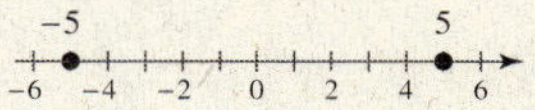

The distance of both -5 and 5 from 0 is 5.

77. The opposite of $-\frac{3}{5}$ is $\frac{3}{5}$.

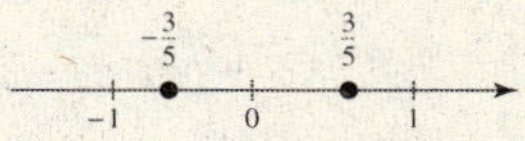

The distance of both $-\frac{3}{5}$ and $\frac{3}{5}$ from 0 is $\frac{3}{5}$.

79. The opposite of $\frac{5}{3}$ is $-\frac{5}{3}$.

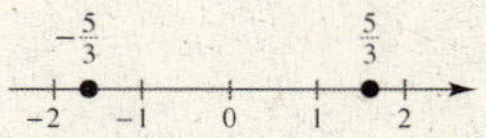

The distance of both $\frac{5}{3}$ and $-\frac{5}{3}$ from 0 is $\frac{5}{3}$.

81. The opposite of -4.25 is 4.25.

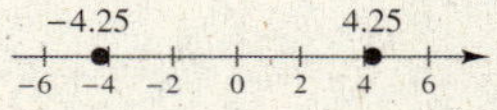

The distance of both -4.25 and 4.25 from 0 is 4.25.

83. $x < 0$

85. $u \geq 16$

87. $16 \leq s \leq 28$

89. $p < 225$

91. Because $|-4| = 4$ and $|4| = 4$, the two possible values of a are -4 and 4.

93. Because $|-2 - 3| = |-5| = 5$ and $|8 - 3| = |5| = 5$, the two possible values of a are -2 and 8.

95. True. If a number can be written as ratio of two integers, it is rational. If not, the number is irrational.

97. $0.15 = \frac{15}{100}$ and $0.\overline{15} = 0.151515\ldots = \frac{15}{99}$

Section 1.2 Operations with Real Numbers

1. $13 + 32 = 45$

3. $-8 + 12 = +(12 - 8) = 4$

5. $-6.4 + 3.7 = -(6.4 - 3.7) = -2.7$

7. $13 + (-6) = +(13 - 6) = 7$

9. $12.6 + (-38.5) = -(38.5 - 12.6) = -25.9$

11. $-8 - 12 = -8 + (-12) = -(8 + 12) = -20$

13. $-21.5 - (-6.3) = -21.5 + 6.3 = -(21.5 - 6.3) = -15.2$

15. $4 - (-11) + 9 = 4 + 11 + 9 = 24$

17. $5.3 - 2.2 - 6.9 = 5.3 + (-2.2) + (-6.9)$
$= 5.3 - (2.2 + 6.9)$
$= 5.3 - 9.1$
$= -3.8$

19. $15 - 6 + 31 + (-18) = 9 + 31 + (-18)$
$= 40 + (-18)$
$= +(40 - 18)$
$= 22$

21. $\frac{3}{8} + \frac{7}{8} = \frac{3 + 7}{8} = \frac{10}{8} = \frac{5}{4}$

23. $\frac{3}{4} - \frac{1}{4} = \frac{3 - 1}{4} = \frac{2}{4} = \frac{1}{2}$

25. $\frac{3}{5} + \left(-\frac{1}{2}\right) = \frac{3(2)}{5(2)} - \frac{1(5)}{2(5)}$
$= \frac{6}{10} - \frac{5}{10}$
$= \frac{6 - 5}{10}$
$= \frac{1}{10}$

27. $\frac{5}{8} + \frac{1}{4} - \frac{5}{6} = \frac{5(3)}{8(3)} + \frac{1(6)}{4(6)} - \frac{5(4)}{6(4)}$
$= \frac{15}{24} + \frac{6}{24} - \frac{20}{24}$
$= \frac{15 + 6 - 20}{24} = \frac{1}{24}$

29. $3\frac{1}{2} + 4\frac{3}{8} = \frac{7}{2} + \frac{35}{8}$
$= \frac{7(4)}{2(4)} + \frac{35}{8}$
$= \frac{28}{8} + \frac{35}{8}$
$= \frac{28 + 35}{8} = \frac{63}{8}$

31. $10\frac{5}{8} - 6\frac{1}{4} = \frac{85}{8} - \frac{25}{4}$
$= \frac{85}{8} - \frac{25(2)}{4(2)}$
$= \frac{85}{8} - \frac{50}{8}$
$= \frac{85 - 50}{8} = \frac{35}{8}$

33. $85 - |-25| = 85 - 25 = 60$

35. $-(-11.325) + |34.625| = 11.325 + 34.625 = 45.95$

37. $-\left|-6\frac{7}{8}\right| - 8\frac{1}{4} = -6\frac{7}{8} - 8\frac{1}{4}$
$= -\frac{55}{8} - \frac{33(2)}{4(2)}$
$= -\frac{55}{8} - \frac{66}{8}$
$= \frac{-55 - 66}{8}$
$= -\frac{121}{8}$

39. $9 + 9 + 9 + 9 = 4 \cdot 9$

41. $\frac{1}{4} + \frac{1}{4} + \frac{1}{4} + \frac{1}{4} + \frac{1}{4} + \frac{1}{4} = 6\left(\frac{1}{4}\right)$

43. $\left(-\frac{1}{5}\right) + \left(-\frac{1}{5}\right) + \left(-\frac{1}{5}\right) + \left(-\frac{1}{5}\right) = 4\left(-\frac{1}{5}\right)$

45. $5(-6) = -30$

47. $(-8)(-6) = 48$

49. $2(4)(-5) = 8(-5) = -40$

51. $(-1)(12)(-3) = (-12)(-3) = 36$

53. $\left(-\frac{5}{8}\right)\left(-\frac{4}{5}\right) = \frac{1}{2}$

55. $-\frac{3}{2}\left(\frac{8}{5}\right) = -\frac{24}{10} = -\frac{12}{5}$

57. $\frac{1}{2}\left(\frac{1}{6}\right) = \frac{1}{12}$

59. $-\frac{9}{8}\left(\frac{16}{27}\right)\left(\frac{1}{2}\right) = \frac{-9 \cdot \cancel{2} \cdot \cancel{8} \cdot 1}{\cancel{8} \cdot \cancel{9} \cdot 3 \cdot \cancel{2}} = -\frac{1}{3}$

61. $\frac{1}{3}\left(-\frac{3}{4}\right)(2) = -\frac{1}{4}(2) = -\frac{2}{4} = -\frac{1}{2}$

63. The reciprocal of 6 is $\frac{1}{6}$ because $6\left(\frac{1}{6}\right) = 1$.

65. The reciprocal of $\frac{2}{3}$ is $\frac{3}{2}$ because $\frac{2}{3}\left(\frac{3}{2}\right) = 1$.

67. The reciprocal of $-\frac{9}{7}$ is $-\frac{7}{9}$ because $-\frac{9}{7}\left(-\frac{7}{9}\right) = 1$.

69. $\dfrac{-18}{-3} = \dfrac{-6 \cdot -3}{-3} = 6$

71. $\dfrac{-48}{16} = \dfrac{-3 \cdot 16}{16} = -3$

73. $63 \div (-7) = \dfrac{63}{-7} = \dfrac{9 \cdot 7}{-7} = -9$

75. $-\dfrac{4}{5} \div \dfrac{8}{25} = -\dfrac{4}{5} \cdot \dfrac{25}{8} = \dfrac{(-4)(25)}{(5)(8)} = -\dfrac{5}{2}$

77. $\left(-\dfrac{1}{3}\right) \div \left(-\dfrac{5}{6}\right) = \left(-\dfrac{1}{3} \div -\dfrac{5}{6}\right)$

$= \left(\dfrac{-1}{3} \cdot \dfrac{-6}{5}\right) = \dfrac{(-1)(-6)}{(3)(5)} = \dfrac{2}{5}$

79. $-4\dfrac{1}{4} \div \left(-5\dfrac{5}{8}\right) = -\dfrac{17}{4} \div \left(-\dfrac{45}{8}\right)$

$= -\dfrac{17}{4} \cdot \left(-\dfrac{8}{45}\right) = \dfrac{17(8)}{4(45)} = \dfrac{34}{45}$

81. $4\dfrac{1}{8} \div 4\dfrac{1}{2} = \dfrac{33}{8} \div \dfrac{9}{2} = \dfrac{33}{8} \cdot \dfrac{2}{9} = \dfrac{(33)(2)}{(8)(9)} = \dfrac{11}{12}$

83. $(-7) \cdot (-7) \cdot (-7) = (-7)^3$

85. $\left(\frac{1}{4}\right) \cdot \left(\frac{1}{4}\right) \cdot \left(\frac{1}{4}\right) \cdot \left(\frac{1}{4}\right) = \left(\frac{1}{4}\right)^4$

87. $-(7 \cdot 7 \cdot 7) = -7^3$

89. $2^5 = (2)(2)(2)(2)(2) = 32$

91. $(-2)^4 = (-2)(-2)(-2)(-2) = 16$

93. $-4^3 = -(4)(4)(4) = -64$

95. $\left(\frac{4}{5}\right)^3 = \left(\frac{4}{5}\right)\left(\frac{4}{5}\right)\left(\frac{4}{5}\right) = \frac{64}{125}$

97. $-\left(-\frac{1}{2}\right)^5 = -\left(-\frac{1}{2}\right)\left(-\frac{1}{2}\right)\left(-\frac{1}{2}\right)\left(-\frac{1}{2}\right)\left(-\frac{1}{2}\right) = -\left(-\frac{1}{32}\right) = \frac{1}{32}$

99. $(0.3)^3 = (0.3)(0.3)(0.3) = 0.027$

101. $5(-0.4)^3 = 5(-0.4)(-0.4)(-0.4) = 5(-0.064) = -0.32$

103. $16 - 6 - 10 = (16 - 6) - 10 = 10 - 10 = 0$

105. $24 - 5 \cdot 2^2 = 24 - 5 \cdot 4$

$= 24 - (5 \cdot 4) = 24 - 20 = 4$

107. $28 \div 4 + 3 \cdot 5 = (28 \div 4) + (3 \cdot 5)$

$= 7 + 15$

$= 22$

109. $14 - 2(8 - 4) = 14 - 2(4)$

$= 14 - 8$

$= 6$

111. $17 - 5(16 \div 4^2) = 17 - 5(16 \div 16)$

$= 17 - 5(1)$

$= 17 - 5$

$= 12$

113. $5^2 - 2[9 - (18 - 8)] = 25 - 2[9 - 10]$

$= 25 - 2[-1]$

$= 25 + 2$

$= 27$

115. $5^3 + |-14 + 4| = 125 + |-10|$

$= 125 + 10$

$= 135$

117. $\dfrac{6 + 8(3)}{7 - 12} = [6 + 8(3)] \div (7 - 12)$

$= (6 + 24) \div (7 - 12)$

$= 30 \div (-5)$

$= -6$

119. $\dfrac{4^2 - 5}{11} - 7 = [(4^2 - 5) \div 11] - 7$

$= [(16 - 5) \div 11] - 7$

$= (11 \div 11) - 7$

$= 1 - 7$

$= -6$

121. $\dfrac{6 \cdot 2^2 - 12}{3^2 + 3} = [(6 \cdot 2^2) - 12] \div (3^2 + 3)$

$= (24 - 12) \div (9 + 3)$

$= 12 \div 12$

$= 1$

123. $\dfrac{3+\frac{3}{4}}{\frac{1}{8}} = \left(3+\frac{3}{4}\right) \div \frac{1}{8}$

$= \left(\frac{12}{4}+\frac{3}{4}\right) \div \frac{1}{8}$

$= \frac{15}{4} \div \frac{1}{8}$

$= \frac{15}{4} \cdot \frac{8}{1}$

$= \frac{15(8)}{4} = 30$

125. $5.6[13 - 2.5(-6.3)] = 5.6[13 + 15.75]$

$= 5.6[28.75]$

$= 161$

127. $5^6 - 3(400) = 15{,}625 - 1200 = 14{,}425$

129. $\dfrac{500}{(1.055)^{20}} = \dfrac{500}{2.9177575} = 171.36448 \approx 171.36$

131. $\frac{1}{4} + \frac{2}{9} + \frac{1}{10} + x + \frac{1}{3} = 1$

So,

$x = 1 - \left(\frac{1}{4} + \frac{2}{9} + \frac{1}{10} + \frac{1}{3}\right)$

$= 1 - \left(\frac{45}{180} + \frac{40}{180} + \frac{18}{180} + \frac{60}{180}\right)$

$= 1 - \left(\frac{45 + 40 + 18 + 60}{180}\right)$

$= 1 - \frac{163}{180} = \frac{180}{180} - \frac{163}{180} = \frac{17}{180}.$

133. $2618.68 + $1236.45 − $25.62 − $455.00 − $125.00 − $715.95 = $2533.56

The balance at the end of the month was $2533.56.

135. (a)

Day	Daily Gain or Loss
Tuesday	+5
Wednesday	+8
Thursday	−5
Friday	+16

(b) $(+5) + (+8) + (-5) + (+16) = +24$ = the sum of the daily gains and losses. The sum of the daily gains and losses is equal to the difference of the value of the stock on Friday and the value of the stock on Monday. This sum could be determined from the graph by $524 (value on Friday) − $500 (value on Monday) = $24.

137. (a) $50(12)(18) = $10,800

(b) The account would have $15,832.22.

(c) $15,832.22 − $10,800 = $5032.22

$5032.22 is earnings from interest.

139. $l = 5$ meters, $w = 3$ meters

$A = lw$

$A = 5 \cdot 3 = 15$ square meters

141. $b = 8$ inches, $h = 5$ inches

$A = \frac{1}{2}bh$

$A = \frac{1}{2}(8)(5) = 20$ square inches

143. $V = l \cdot w \cdot h$

= 14 inches · 18 inches · 42 inches

= 10,584 cubic inches ÷ 1728 cubic inches

= 6.125 cubic feet

145. True. A nonzero rational number is an integer divided by an integer. The reciprocal of such a number is still an integer divided by an integer, and so it is a rational number.

147. True. Any negative real number raised to an even numbered power will be a positive real number.

149. $a \div b = b \div a$

False. Division is not commutative.

151. If the numbers have like signs, the product or quotient is positive. If the numbers have unlike signs, the product or quotient is negative.

153. (a) $40 - 10 + 3 = 30 + 3 = 33 \neq 27$

Insert parentheses: $40 - (10 + 3) = 40 - 13 = 27$

(b) $5^2 + \frac{1}{2} \cdot 4 = 25 + \frac{1}{2} \cdot 4 = 25 + 2 = 27$

(c) $8 \cdot 3 + 30 \div 2 = 24 + 15 = 39 + 27$

Insert parentheses:

$(8 \cdot 3 + 30) \div 2 = (24 + 30) \div 2$

$= 54 \div 2 = 27$

(d) $75 \div 2 + 1 + 2 = 37.5 + 1 + 2$

$= 38.5 + 2 = 40.5 \neq 27$

Insert parentheses:

$75 \div (2 + 1) + 2 = 75 \div 3 + 2 = 25 + 2 = 27$

155. Only common factors (not terms) of the numerator and denominator can be divided out.

$\frac{5 + 12}{5} = \frac{17}{5}$

Section 1.3 Properties of Real Numbers

1. $18 - 18 = 0$

Additive Inverse Property

3. $\frac{1}{12} \cdot 12 = 1$

Multiplicative Inverse Property

5. $13 + 12 = 12 + 13$

Commutative Property of Addition

7. $3 + (12 - 9) = (3 + 12) - 9$

Associative Property of Addition

9. $(8 - 5)(10) = 8 \cdot 10 - 5 \cdot 10$

Distributive Property

11. $10(2x) = (10 \cdot 2)x$

Associative Property of Multiplication

13. $10x \cdot \frac{1}{10x} = 1$

Multiplicative Inverse Property

15. $2x - 2x = 0$

Additive Inverse Property

17. $3(2 + x) = 3 \cdot 2 + 3x$

Distributive Property

19. $(x + 1) - (x + 1) = 0$

Additive Inverse Property

21. $15(-3) = (-3)15$

23. $5(6 + z) = 5 \cdot 6 + 5 \cdot z$

25. $25 + (-x) = -x + 25$

27. $(x + 8) \cdot 1 = x + 8$

29. (a) Additive Inverse: -10

(b) Multiplicative Inverse: $\frac{1}{10}$

31. (a) Additive Inverse: 19

(b) Multiplicative Inverse: $-\frac{1}{19}$

33. (a) Additive Inverse: $-\frac{1}{2}$

(b) Multiplicative Inverse: 2

35. (a) Additive Inverse: $\frac{5}{8}$

(b) Multiplicative Inverse: $-\frac{8}{5}$

37. (a) Additive Inverse: $-6z$

(b) Multiplicative Inverse: $\frac{1}{6z}$

39. (a) Additive Inverse: $-(x - 2)$ or $-x + 2$

(b) Multiplicative Inverse: $\frac{1}{x - 2}$

41. $32 + (4 + y) = (32 + 4) + y$

43. $9(6M) = (9 \cdot 6)M$

45. $20(2 + 5) = 20 \cdot 2 + 20 \cdot 5$

47. $(x + 6)(-2) = x \cdot (-2) + 6 \cdot (-2)$ or $-2x - 12$

49. $-6(2y - 5) = -6(2y) + (-6)(-5)$ or $-12y + 30$

51. $7x + 2x = (7 + 2)x = 9x$

53. $\frac{7x}{8} - \frac{5x}{8} = (7 - 5)\left(\frac{x}{8}\right) = \frac{2x}{8} = \frac{x}{4}$

55. $3(x + 5) = 3x + 15$

57. $-2(x + 8) = -2x - 16$

59.

$ac = bc,\ c \neq 0$	Write original equation.
$\frac{1}{c}(ac) = \frac{1}{c}(bc)$	Multiplication Property of Equality
$\frac{1}{c}(ca) = \frac{1}{c}(cb)$	Commutative Property of Multiplication
$\left(\frac{1}{c} \cdot c\right)a = \left(\frac{1}{c} \cdot c\right)b$	Associative Property of Multiplication
$1 \cdot a = 1 \cdot b$	Multiplicative Inverse Property
$a = b$	Multiplicative Identity Property

61.

$a = (a + b) + (-b)$	Write original equation.
$a = a + [b + (-b)]$	Associative Property of Addition
$a = a + 0$	Additive Inverse Property
$a = a$	Additive Identity Property

63.

$x + 5 = 3$	Write original equation.
$(x + 5) + (-5) = 3 + (-5)$	Addition Property of Equality
$x + (5 + (-5)) = 3 - 5$	Associative Property of Addition
$x + 0 = -2$	Additive Inverse Property
$x = -2$	Additive Identity Property

65.

$2x - 5 = 6$	Write original equation.
$(2x - 5) + 5 = 6 + 5$	Addition Property of Equality
$2x + (-5 + 5) = 11$	Associative Property of Addition
$2x + 0 = 11$	Additive Inverse Property
$2x = 11$	Additive Identity Property
$\frac{1}{2}(2x) = \frac{1}{2}(11)$	Multiplication Property of Equality
$\left(\frac{1}{2} \cdot 2\right)x = \frac{11}{2}$	Associative Property of Multiplication
$1 \cdot x = \frac{11}{2}$	Multiplicative Inverse Property
$x = \frac{11}{2}$	Multiplicative Identity Property

67. $16(1.75) = 16\left(2 - \frac{1}{4}\right) = 16(2) - 16\left(\frac{1}{4}\right) = 32 - 4 = 28$

69. $7(62) = 7(60 + 2) = 7(60) + 7(2) = 420 + 14 = 434$

71.
$$\begin{aligned} 9(6.98) &= 9(7 - 0.02) \\ &= 9(7) - 9(0.02) \\ &= 63 - 0.18 \\ &= 62.82 \end{aligned}$$

73. $a(b + c) = ab + ac$

75.
$$\begin{aligned} 4 + (x + 5) + (3x + 2) &= 4 + (5 + x) + (3x + 2) \\ &= (4 + 5) + x + (3x + 2) \\ &= 9 + (x + 3x) + 2 \\ &= 9 + 4x + 2 \\ &= 4x + 9 + 2 \\ &= 4x + 11 \end{aligned}$$

77. (a)
$$\begin{aligned} 2(x + 6) + 2(2x) &= 2x + 12 + 4x \\ &= 2x + 4x + 12 \\ &= 6x + 12 \end{aligned}$$

(b) $(x + 6)(2x) = x(2x) + 6(2x) = 2x^2 + 12x$

79. The additive inverse of a real number a is the number $-a$. The sum of a number and its additive inverse is the additive identity 0. For example, $8 + (-8) = 0$.

81. Given two real numbers a and b, the sum a plus b is the same as the sum b plus a.

83. Sample answer: $4 \odot 7 = 2 \cdot 4 + 7 = 8 + 7 = 15$

$$7 \odot 4 = 2 \cdot 7 + 4 = 14 + 4 = 18$$

Because $15 \neq 18$, $4 \odot 7 \neq 7 \odot 4$. So, the operation is not commutative.

$$\begin{aligned} 3 \odot (4 \odot 7) &= 3 \odot (2 \cdot 4 + 7) \\ &= 3 \odot 15 \\ &= 2 \cdot 3 + 15 \\ &= 6 + 15 \\ &= 21 \end{aligned}$$

$$\begin{aligned} (3 \odot 4) \odot 7 &= (2 \cdot 3 + 4) \odot 7 \\ &= 10 \odot 7 \\ &= 2 \cdot 10 + 7 \\ &= 20 + 7 \\ &= 27 \end{aligned}$$

Because $21 \neq 27$, $3 \odot (4 \odot 7) \neq (3 \odot 4) \odot 7$. So, the operation is not associative.

Mid-Chapter Quiz for Chapter 1

1. $-4.5 > -6$

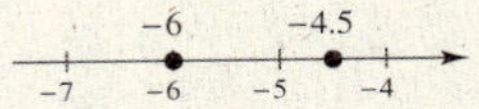

2. $\frac{3}{4} < \frac{3}{2}$

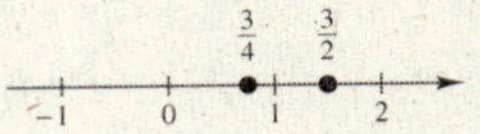

3. $|-15 - 7| = |-22| = 22$

4. $|-8.75 - (-2.25)| = |-8.75 + 2.25| = |-6.5| = 6.5$

5. $|-7.6| = 7.6$

6. $-|9.8| = -9.8$

7. $32 + (-18) = 14$

8. $-12 - (-17) = -12 + 17 = 5$

9. $\frac{3}{4} + \frac{7}{4} = \frac{3+7}{4} = \frac{10}{4} = \frac{5}{2}$

10. $\frac{2}{3} - \frac{1}{6} = \frac{4}{6} - \frac{1}{6} = \frac{4-1}{6} = \frac{3}{6} = \frac{1}{2}$

11. $(-3)(2)(-10) = (-6)(-10) = 60$

12. $\left(-\frac{4}{5}\right)\left(\frac{15}{32}\right) = \frac{(-4)(15)}{(5)(32)} = -\frac{3}{8}$

13. $\frac{7}{12} \div \frac{5}{6} = \frac{7}{12} \cdot \frac{6}{5} = \frac{(7)(6)}{(12)(5)} = \frac{7}{10}$

14. $\left(-\frac{3}{2}\right)^3 = \left(-\frac{3}{2}\right)\left(-\frac{3}{2}\right)\left(-\frac{3}{2}\right) = -\frac{27}{8}$

15. $3 - 2^2 + 25 \div 5 = 3 - 4 + 25 \div 5$

$$= 3 - 4 + 5 = -1 + 5 = 4$$

16. $\frac{18 - 2(3+4)}{6^2 - (12 \cdot 2 + 10)} = \left[18 - 2(3+4)\right] \div \left[6^2 - (12 \cdot 2 + 10)\right] = (18 - 14) \div (36 - 34) = 4 \div 2 = 2$

17. (a) $8(u - 5) = 8 \cdot u - 8 \cdot 5$ Distributive Property

(b) $10x - 10x = 0$ Additive Inverse Property

18. (a) $(7 + y) - z = 7 + (y - z)$ Associative Property of Addition

(b) $2x \cdot 1 = 2x$ Multiplicative Identity Property

19. $1406.98 − $375.03 − $59.20 − $225.00 + $320.45 = $1068.20

20. $\$45(2)(12)(8) = \$8640$

21.
$$1 = \tfrac{1}{3} + \tfrac{1}{4} + \tfrac{1}{8} + x$$
$$1 - \tfrac{1}{3} - \tfrac{1}{4} - \tfrac{1}{8} = x$$
$$\tfrac{24}{24} - \tfrac{8}{24} - \tfrac{6}{24} - \tfrac{3}{24} = x$$
$$\tfrac{7}{24} = x$$

The sum of the parts of a circle is equal to 1.

Section 1.4 Algebraic Expressions

1. Terms: $10x$, 5

Coefficients: 10, 5

3. Terms: $-6x^2$, 12

Coefficients: -6, 12

5. Terms: $-3y^2$, $2y$, -8

Coefficients: -3, 2, -8

7. Terms: $-4a^3$, $1.2a$

Coefficients: -4, 1.2

9. Terms: $4x^2$, $-3y^2$, $-5x$, 21

Coefficients: 4, -3, -5, 21

11. Terms: $-5x^2y$, $2y^2$, xy

Coefficients: -5, 2, 1

13. Terms: $\frac{1}{4}x^2$, $-\frac{3}{8}x$, 5

Coefficients: $\frac{1}{4}$, $-\frac{3}{8}$, 5

15. $4 - 3x = -3x + 4$ illustrates the Commutative Property of Addition.

17. $-5(2x) = (-5 \cdot 2)x$ illustrates the Associative Property of Multiplication.

19. $(5 - 2)x = 5x - 2x$ illustrates the Distributive Property.

21. $5(x + 6) = 5x + 30$ or $5x + 5 \cdot 6$

23. $5(x + 6) = (x + 6) \cdot 5$

25. $3x + 4x = (3 + 4)x = 7x$

27. $-2x^2 + 4x^2 = (-2 + 4)x^2 = 2x^2$

29. $7x - 11x = (7 - 11)x = -4x$

31. $9y - 5y + 4y = (9 - 5 + 4)y = 8y$

33.
$$\begin{aligned} 3x - 2y + 5x + 20y &= (3x + 5x) + (-2y + 20y) \\ &= (3 + 5)x + (-2 + 20)y \\ &= 8x + 18y \end{aligned}$$

35.
$$\begin{aligned} 7x^2 - 2x - x^2 &= 7x^2 - x^2 - 2x \\ &= (7 - 1)x^2 - 2x \\ &= 6x^2 - 2x \end{aligned}$$

37. $-3z^4 + 6z - z + 8 + z^4 - 4z^2 = (-3z^4 + z^4) - 4z^2 + (6z - z) + 8 = -2z^4 - 4z^2 + 5z + 8$

39. $x^2 + 2xy - 2x^2 + xy + y = x^2 - 2x^2 + 2xy + xy + y = (1 - 2)x^2 + (2 + 1)xy + y = -x^2 + 3xy + y$

41. $4(2x^2 + x - 3) = 8x^2 + 4x - 12$

43. $-3(6y^2 - y - 2) = -18y^2 + 3y + 6$

45. $-(3x^2 - 2x + 4) = -3x^2 + 2x - 4$

47. $x(5x + 2) = 5x^2 + 2x$

49. $3x(17 - 4x) = 51x - 12x^2 = -12x^2 + 51x$

51. $-5t(7 - 2t) = -35t + 10t^2 = 10t^2 - 35t$

53.
$$\begin{aligned} 10(x - 3) + 2x - 5 &= 10x - 30 + 2x - 5 \\ &= (10x + 2x) + (-30 - 5) \\ &= (10 + 2)x + (-30 - 5) \\ &= 12x - 35 \end{aligned}$$

55. $x - (5x + 9) = x - 5x - 9 = (1 - 5)x - 9 = -4x - 9$

57.
$$\begin{aligned} 5a - (4a - 3) &= 5a - 4a + 3 \\ &= (5 - 4)a + 3 \\ &= a + 3 \end{aligned}$$

59. $-3(3y - 1) + 2(y - 5) = -9y + 3 + 2y - 10$
$= -9y + 2y + 3 - 10$
$= (-9 + 2)y - 7$
$= -7y - 7$

61. $-3(y^2 - 2) + y^2(y + 3) = -3y^2 + 6 + y^3 + 3y^2$
$= (-3 + 3)y^2 + 6 + y^3$
$= 6 + y^3$

63. $x(x^2 + 3) - 3(x + 4) = x^3 + 3x - 3x - 12$
$= x^3 + (3 - 3)x - 12$
$= x^3 - 12$

65. $9a - [7 - 5(7a - 3)] = 9a - [7 - 35a + 15]$
$= 9a - [-35a + 22]$
$= 9a + 35a - 22$
$= (9 + 35)a - 22$
$= 44a - 22$

67. $3[2x - 4(x - 8)] = 3[2x - 4x + 32]$
$= 3[-2x + 32]$
$= -6x + 96$

69. $8x + 3x[10 - 4(3 - x)] = 8x + 3x[10 - 12 + 4x]$
$= 8x + 3x[-2 + 4x]$
$= 8x - 6x + 12x^2$
$= 2x + 12x^2$

71. $2[3(b - 5) - (b^2 + b + 3)] = 2[3b - 15 - b^2 - b - 3]$
$= 6b - 30 - 2b^2 - 2b - 6$
$= (-2b^2) + (6b - 2b) + (-30 - 6)$
$= -2b^2 + 4b - 36$

73. (a) When $x = \frac{2}{3}$, the expression $5 - 3x$ has a value of $5 - 3\left(\frac{2}{3}\right) = 5 - 2 = 3.$

(b) When $x = 5$, the expression $5 - 3x$ has a value of $5 - 3(5) = 5 - 15 = -10.$

75. (a) When $x = -1$, the expression $10 - 4x^2$ has a value of $10 - 4(-1)^2 = 10 - 4 = 6.$

(b) When $x = \frac{1}{2}$, the expression $10 - 4x^2$ has a value of $10 - 4\left(\frac{1}{2}\right)^2 = 10 - 1 = 9.$

77. (a) When $y = 2$, the expression $y^2 - y + 5$ has a value of $(2)^2 - 2 + 5 = 4 - 2 + 5 = 7.$

(b) When $y = -2$, the expression $y^2 - y + 5$ has a value of $(-2)^2 - (-2) + 5 = 4 + 2 + 5 = 11.$

79. (a) When $x = 0$, the expression $\frac{1}{x^2} + 3 = \frac{1}{0^2} + 3$ is undefined.

(b) When $x = 3$, the expression $\frac{1}{x^2} + 3$ has a value of $\frac{1}{3^2} + 3 = \frac{1}{9} + 3 = \frac{1}{9} + \frac{27}{9} = \frac{28}{9}.$

81. (a) When $x = 1$ and $y = 5$, the expression $3x + 2y$ has a value of $3(1) + 2(5) = 3 + 10 = 13.$

(b) When $x = -6$, and $y = -9$, the expression $3x + 2y$ has a value of $3(-6) + 2(-9) = -18 + -18 = -36.$

83. (a) When $x = 2$ and $y = -1$, the expression $x^2 - xy + y^2$ has a value of $(2)^2 - (2)(-1) + (-1)^2 = 4 + 2 + 1 = 7.$

(b) When $x = -3$ and $y = -2$, the expression $x^2 - xy + y^2$ has a value of $(-3)^2 - (-3)(-2) + (-2)^2 = 9 - 6 + 4 = 7.$

85. (a) When $x = 4$ and $y = 2$, the expression $\frac{x}{y^2 - x} = \frac{4}{2^2 - 4} = \frac{4}{4 - 4} = \frac{4}{0}$ is undefined.

(b) When $x = 3$ and $y = 3$, the expression $\frac{x}{y^2 - x}$ has a value of $\frac{3}{3^2 - 3} = \frac{3}{9 - 3} = \frac{3}{6} = \frac{1}{2}.$

87. (a) When $x = 2$ and $y = 5$, the expression $|y - x|$ has a value of $|5 - 2| = |3| = 3.$

(b) When $x = -2$ and $y = -2$, the expression $|y - x|$ has a value of $|-2 - (-2)| = |0| = 0.$

89. (a) When $r = 40$ and $t = 5\frac{1}{4}$, the expression rt has a value of $(40)\left(5\frac{1}{4}\right) = (40)\left(\frac{21}{4}\right) = 210$.

(b) When $r = 35$ and $t = 4$, the expression rt has a value of $(35)(4) = 140$.

91. $lwh = 6(6)(7) = 252$

The volume is 252 cubic feet.

93. $lwh = 27(18)(8) = 3888$

The volume is 3888 cubic inches.

95. When $p = 11$, $n = 7$, $d = 0$, and $q = 3$, the expression $0.01p + 0.05n + 0.10d + 0.25q$ has a value of $0.01(11) + 0.05(7) + 0.10(0) + 0.25(3) = 0.11 + 0.35 + 0 + 0.75 = \$1.21$.

97. When $p = 43$, $n = 27$, $d = 17$, and $q = 15$, the expression $0.01p + 0.05n + 0.10d + 0.25q$ has a value of $0.01(43) + 0.05(27) + 0.10(17) + 0.25(15) = 0.43 + 1.35 + 1.70 + 3.75 = \$7.23$.

99. $A = \frac{1}{2}b(b - 3) = \frac{1}{2}b^2 - \frac{3}{2}b$

$A = \frac{1}{2}(15)(15 - 3)$

$= \frac{1}{2}(15)(12)$

$= 90$

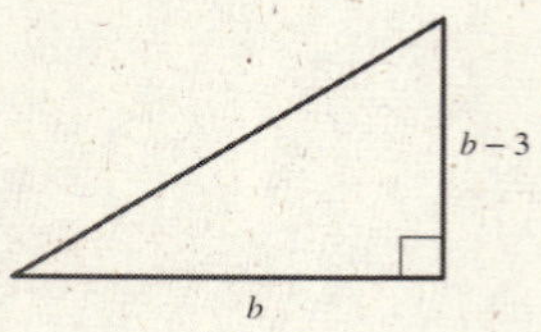

100. $A = h\left(\frac{5}{4}h + 10\right) = \frac{5}{4}h^2 + 10h$

$A = 12\left[\frac{5}{4}(12) + 10\right] = 12[15 + 10] = 12[25] = 300$

101. Graphically, the sales in 2005 is approximately \$23,500 million. Let $t = 5$.

Sales $= 607.6(5) + 20{,}737 = 3038 + 20{,}737 = \$23{,}775$ million

103. Graphically, the total amount of FFEL disbursements in 2000 is approximately \$22 billion. Let $t = 10$.

Disbursements $= 0.322(10)^2 - 3.75(10) + 27.6 = 32.2 - 37.5 + 27.6 = \$22.3$ billion

105. [Total area] = 2 · [Area of Trapezoid] + 2 · [Area of triangle]

Area $= 2\left[\frac{1}{2} \cdot h(b_1 + b_2)\right] + 2\left[\frac{1}{2} \cdot b \cdot h\right]$

Area $= 12(60 + 40) + 20 \cdot 12 = 12(100) + 240 = 1200 + 240 = 1440$ square feet

(b) For any natural number n, $n(n - 3)$ is a product of an even and an odd natural number. So, the product is even and $\dfrac{n(n-3)}{2}$ is a natural number.

107. It is not possible to evaluate $\dfrac{x + 2}{y - 3}$ when $x = 5$ and $y = 3$ because $\dfrac{7}{0}$ is undefined.

109. To remove a set of parentheses preceded by a minus sign, distribute -1 to each term inside the parentheses. For example, $13 - (-10 + 5) = 13 + 10 - 5 = 18$.

111. A factor can consist of a sum of terms. The term x is part of the sum $x + y$, which is a factor of $(x + y) \cdot z$.

113. No. There are an infinite number of values of x and y that would satisfy $8y - 5x = 14$. For example, $x = 10$ and $y = 8$ would be a solution and so would $x = 2$ and $y = 3$.

Section 1.5 Constructing Algebraic Expressions

1. The sum of 23 and a number n is translated into the algebraic expression $23 + n$.

3. The sum of 12 and twice a number n is translated into the algebraic expression $12 + 2n$.

5. Six less than a number n is translated into the algebraic expression $n - 6$.

7. Four times a number n minus 10 is translated into the algebraic expression $4n - 10$.

9. Half of a number n is translated into the algebraic expression $\frac{1}{2}n$.

11. The quotient of a number x and 6 is translated into the algebraic expression $\frac{x}{6}$.

13. Eight times the ratio of N and 5 is translated into the algebraic expression $8 \cdot \frac{N}{5}$.

15. The number c is quadrupled and the product is increased by 10 is translated into the algebraic expression $4c + 10$.

17. Thirty percent of the list price L is translated into the algebraic expression $0.30L$.

19. The sum of a number n and 5 divided by 10 is translated into the algebraic expression $\frac{n+5}{10}$.

21. The absolute value of the difference between a number and 8 is translated into the algebraic expression $|n - 8|$.

23. The product of 3 and the square of a number decreased by 4 is translated into the algebraic expression $3x^2 - 4$.

25. A verbal description of $t - 2$ is a number decreased by 2.

27. A verbal description of $y + 50$ is the sum of a number and 50 or a number increased by 50.

29. A verbal description of $2 - 3x$ is 2 decreased by 3 times a number.

31. A verbal description of $\frac{z}{2}$ is the ratio of a number and 2.

33. A verbal description of $\frac{4}{5}x$ is four-fifths of a number.

35. A verbal description of $8(x - 5)$ is 8 times the difference of a number and 5.

37. A verbal description of $\frac{x+10}{3}$ is the sum of a number and 10, divided by 3.

39. A verbal description of $y^2 - 3$ is the square of a number, decreased by 3.

41. *Verbal Description:* The amount of money (in dollars) represented by n quarters
Label: n = number of nickels
Algebraic Description: $0.25n$ = amount of money (in dollars)

43. *Verbal Description:* The amount of money (in dollars) represented by m dimes
Label: m = number of dimes
Algebraic Description: $0.10m$ = amount of money (in dollars)

45. *Verbal Description:* The amount of money (in cents) represented by m nickels and n dimes
Labels: m = number of nickels
n = number of dimes
Algebraic Description: $5m + 10n$ = amount of money (in cents)

47. *Verbal Description:* The distance traveled in t hours at an average speed of 55 miles per hour
Label: t = number of hours
Algebraic Description: $55t$ = distance

49. *Verbal Description:* The time to travel 320 miles at an average speed of r miles per hour
Label: r = average speed
Algebraic Description: $\frac{320}{r}$ = time

51. *Verbal Description:* The amount of antifreeze in a cooling system containing y gallons of coolant that is 45% antifreeze
Label: y = number of gallons
Algebraic Description: $0.45y$ = amount of antifreeze

53. *Verbal Description:* The amount of wage tax due for a taxable income of I dollars that is taxed at the rate of 1.25%
Label: I = number of dollars
Algebraic Description: $0.0125I$ = amount of wage tax

55. *Verbal Description:* The sale price of a coat that has a list price of L dollars if the sale is a "20% off" sale
Label: L = number of dollars
Algebraic Description: $0.80L$ = sale price

57. *Verbal Description:* The total hourly wage for an employee when the base pay is \$8.25 per hour plus 60 cents for each of q units produced per hour

Label: q = number of units produced

Algebraic Description: $8.25 + 0.60q$ = total hourly wage

59. *Verbal Description:* The sum of a number n and five times the number

Labels: n = the number

$5n$ = five times the number

Algebraic Description: $n + 5n = 6n$ = sum

61. *Verbal Description:* The sum of three consecutive odd integers, the first of which is $2n + 1$

Labels: $2n + 1$ = first odd integer

$2n + 3$ = second odd integer

$2n + 5$ = third odd integer

Algebraic Description: $(2n + 1) + (2n + 3) + (2n + 5) = 6n + 9$ = sum

63. *Verbal Description:* The product of two consecutive even integers, divided by 4

Labels: $2n$ = first even integer

$2n + 2$ = second even integer

Algebraic Description: $\dfrac{2n(2n+2)}{4} = \dfrac{4n(n+1)}{4} = n(n+1) = n^2 + n$ = product

65. Area = side · side = $s \cdot s = s^2$

67. Area = $\frac{1}{2}$(base)(height) = $\frac{1}{2}(b)(0.75b) = 0.375b^2$

69. Perimeter = $2(2w) + 2(w) = 4w + 2w = 6w$

Area = $2w \cdot w = 2w^2$

71. Perimeter = $3 + 2x + 6 + x + 3 + x = 4x + 12$

Area = $(x \cdot 3) + (3 \cdot 2x) = 3x + 6x = 9x$

73. Area = length · width = $b(b - 50) = b^2 - 50b$

The unit measure for the area is square meters.

75.

n	0	1	2	3	4	5
$5n - 3$	−3	2	7	12	17	22
Differences	5	5	5	5	5	

The differences are constant.

77. The third row difference for the algebraic expression $an + b$ would be a.

79. $4x$ is the equivalent to (a) x multiplied by 4 and (c) the product of x and 4.

81. Using a specific case may make it easier to see the form of the expression for the general case.

Review Exercises for Chapter 1

1. (a) Natural numbers: $\{52, \sqrt{9}\}$

(b) Integers: $\{-4, 0, \sqrt{9}, 52\}$

(c) Rational numbers: $\{-4, -\frac{1}{8}, 0, \frac{3}{5}, \sqrt{9}, 52\}$

(d) Irrational numbers: $\{\sqrt{2}\}$

3. $\{1, 2, 3, 4, 5, 6\}$

5. (a)

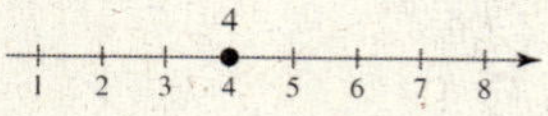

(b)

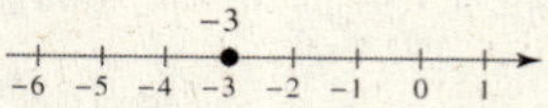

(c)

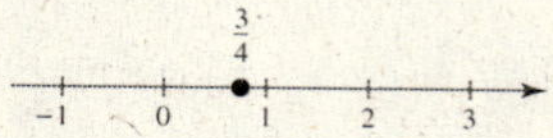

(d)

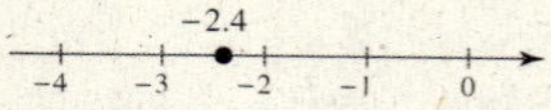

7. $-5 < 3$

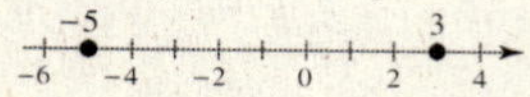

9. $-\frac{8}{5} < -\frac{2}{5}$

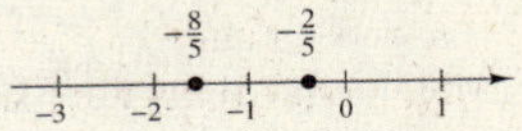

11. $d = |11 - (-3)| = |11 + 3| = |14| = 14$

13. $d = |-13.5 - (-6.2)| = |-13.5 + 6.2| = |-7.3| = 7.3$

15. $|-5| = 5$

17. $-|-7.2| = -7.2$

19. $15 + (-4) = 11$

21. $340 - 115 + 5 = 225 + 5 = 230$

23. $-63.5 + 21.7 = -41.8$

25. $\frac{4}{21} + \frac{7}{21} = \frac{11}{21}$

27. $-\frac{5}{6} + 1 = -\frac{5}{6} + \frac{6}{6} = \frac{1}{6}$

29. $8\frac{3}{4} - 6\frac{5}{8} = \frac{35}{4} - \frac{53}{8} = \frac{70}{8} - \frac{53}{8} = \frac{17}{8}$

31. $-7 \cdot 4 = -28$

33. $120(-5)(7) = -4200$

35. $\frac{3}{8} \cdot \left(-\frac{2}{15}\right) = -\frac{6}{120} = -\frac{1}{20}$

37. $\frac{-56}{-4} = 14$

39. $-\frac{7}{15} \div -\frac{7}{30} = -\frac{7}{15} \cdot \frac{30}{-7} = 2$

41. $7(-3) = (-3) + (-3) + (-3) + (-3) + (-3) + (-3) + (-3)$

43. $8 + 8 + 8 + 8 + 8 + 8 + 8 + 8 = 8(8)$

45. $6 \cdot 6 \cdot 6 \cdot 6 \cdot 6 \cdot 6 \cdot 6 = 6^7$

47. $(-6)^4 = (-6)(-6)(-6)(-6) = 1296$

49. $-4^2 = (-1)(4)(4) = -16$

51. $-\left(-\frac{1}{2}\right)^3 = -1 \cdot \left(-\frac{1}{2}\right)\left(-\frac{1}{2}\right)\left(-\frac{1}{2}\right) = \frac{1}{8}$

53. $120 - (5^2 \cdot 4) = 120 - (25 \cdot 4) = 120 - 100 = 20$

55.
$$\begin{aligned} 8 + 3[6^2 - 2(7 - 4)] &= 8 + 3[36 - 2(3)] \\ &= 8 + 3[36 - 6] \\ &= 8 + 3[30] \\ &= 8 + 90 = 98 \end{aligned}$$

57.
$$\begin{aligned} 7(408.2^2 - 39.5 \div 0.3) &= 7(166{,}627.24 - 39.5 \div 0.3) \\ &= 7(166{,}627.24 - 131.\overline{6}) \\ &= 7(166{,}495.57\overline{3}) \\ &= 1{,}165{,}469.01\overline{3} \\ &\approx 1{,}165{,}469.01 \end{aligned}$$

59. $395 + 9(45) = 395 + 405 = 800$

You paid \$800 for the entertainment system.

61. Additive Inverse Property

63. Distributive Property

65. Associative Property of Addition

67. Commutative Property of Multiplication

69. Distributive Property

71. $-(-u + 3v) = u - 3v$

73. $-a(8 - 3a) = -a(8) + (-a)(-3a) = -8a + 3a^2$

75. Terms: $4y^3, -y^2, \frac{17}{2}y$

Coefficients: $4, -1, \frac{17}{2}$

77. Terms: $-1.2x^3, \frac{1}{x}, 52$

Coefficients: $-1.2, 1, 52$

79. $6x + 3x = (6 + 3)x = 9x$

81. $3u - 2v + 7v - 3u = (3u - 3u) + (-2v + 7v) = 5v$

83. $5(x - 4) + 10 = 5x - 20 + 10 = 5x - 10$

85. $3x - (y - 2x) = 3x - y + 2x = 5x - y$

87.
$$\begin{aligned} 3[b + 5(b - a)] &= 3[b + 5b - 5a] \\ &= 3b + 15b - 15a \\ &= 18b - 15a \end{aligned}$$

89. (a) When $x = 3$, the expression $x^2 - 2x - 3$ has a value of $(3)^2 - 2(3) - 3 = 9 - 6 - 3 = 0$.

(b) When $x = 0$, the expression $x^2 - 2x - 3$ has a value of $(0)^2 - 2(0) - 3 = 0 - 0 - 3 = -3$.

91. (a) When $x = 4$ and $y = -1$, the expression $y^2 - 2y + 4x$ has a value of $(-1)^2 - 2(-1) + 4(4) = 1 + 2 + 16 = 19$.

(b) When $x = -2$ and $y = 2$, the expression $y^2 - 2y + 4x$ has a value of $2^2 - 2(2) + 4(-2) = 4 - 4 - 8 = -8$.

93. Twelve decreased by twice the number n is translated into the algebraic expression $12 - 2n$.

95. The sum of the square of a number y and 49 is translated into the algebraic expression $y^2 + 49$.

97. The sum of twice a number and 7

99. The difference of a number and 5, all divided by 4

101. $0.18I$ = tax on I dollars at 18%

103. $l \cdot (l - 5) = l^2 - 5l$ = area of rectangle with length l and width $(l - 5)$

Chapter Test for Chapter 1

1. (a) $+\frac{5}{2} < |-3|$

(b) $-\frac{2}{3} > -\frac{3}{2}$

2. $d = |-4.4 - 6.9| = |-11.3| = 11.3$

3. $-14 + 9 - 15 = (-14 + 9) - 15 = -5 - 15 = -20$

4. $\frac{2}{3} + \left(-\frac{7}{6}\right) = \frac{4}{6} + \left(-\frac{7}{6}\right) = -\frac{3}{6} = -\frac{1}{2}$

5. $-2(225 - 150) = -2(75) = -150$

6. $(-3)(4)(-5) = (-12)(-5) = 60$

7. $\left(-\frac{7}{16}\right)\left(-\frac{8}{21}\right) = \frac{1}{6}$

8. $\frac{5}{18} \div \frac{15}{8} = \frac{5}{18} \cdot \frac{8}{15} = \frac{4}{27}$

9. $\left(-\frac{3}{5}\right)^3 = -\frac{27}{125}$

10. $\frac{4^2 - 6}{5} + 13 = \frac{16 - 6}{5} + 13 = \frac{10}{5} + 13 = 2 + 13 = 15$

11. (a) Associative Property of Multiplication

(b) Multiplicative Inverse Property

12. $-6(2x - 1) = -6(2x) + -6 \cdot (-1) = -12x + 6$

13. $3x^2 - 2x - 5x^2 + 7x - 1 = -2x^2 + 5x - 1$

14. $x(x + 2) - 2(x^2 + x - 13) = x^2 + 2x - 2x^2 - 2x + 26 = (x^2 - 2x^2) + (2x - 2x) + 26 = -x^2 + 26$

15. $a(5a - 4) - 2(2a^2 - 2a) = 5a^2 - 4a - 4a^2 + 4a = a^2$

16. $4t - [3t - (10t + 7)] = 4t - [3t - 10t - 7] = 4t - [-7t - 7] = 4t + 7t + 7 = 11t + 7$

17. Evaluating an expression is solving the expression when values are provided for its variables.

(a) When $x = -1$:

$7 + (x - 3)^2 = 7 + (-1 - 3)^2 = 7 + (-4)^2 = 7 + 16 = 23$

(b) When $x = 3$:

$7 + (x - 3)^2 = 7 + (3 - 3)^2 = 7 + 0^2 = 7 + 0 = 7$

18. *Verbal Model:* $17 \cdot \boxed{\text{Length of each piece}} = \boxed{\text{Total length}}$

Equation: $17 \cdot n = 102$

$n = 6$

Each piece should be 6 inches.

19. *Verbal Model:* $\boxed{\text{Volume of 1 cord}} = \boxed{\text{Length}} \cdot \boxed{\text{Width}} \cdot \boxed{\text{Height}}$

Equation: $V = 4 \cdot 4 \cdot 8$

$V = 128$ cubic feet

Verbal Model: $\boxed{\text{Volume of 5 cords}} = \boxed{5} \cdot \boxed{\text{Volume of 1 cord}}$

Equation: $V = 5 \cdot 128 = 640$

There are 640 cubic feet in 5 cords of wood.

20. The product of a number n and 5 is decreased by 8 is translated into the algebraic expression $5n - 8$.

21. *Verbal Description:* The sum of two consecutive even integers, the first of which is $2n$

Labels: $2n$ = first even integer

$2n + 2$ = second even integer

Algebraic Description: $2n + (2n + 2) = 4n + 2$

22. Perimeter = $2l + 2(0.6l) = 2l + 1.2l = 3.2l$

Area = $l(0.6l) = 0.6l^2$

When $l = 45$:

Perimeter = $3.2(45) = 144$

Area = $0.6(45)^2 = 1215$

CHAPTER 2
Linear Equations and Inequalities

CHAPTER 2
Linear Equations and Inequalities

Section 2.1 Linear Equations

1. (a) $x = 0$

$3(0) - 7 \stackrel{?}{=} 2$

$-7 \neq 2$

Not a solution

(b) $x = 3$

$3(3) - 7 \stackrel{?}{=} 2$

$9 - 7 = 2$

$2 = 2$

Solution

3. (a) $x = 4$

$4 + 8 \stackrel{?}{=} 3(4)$

$12 = 12$

Solution

(b) $x = -4$

$-4 + 8 \stackrel{?}{=} 3(-4)$

$4 \neq -12$

Not a solution

5. (a) $x = -4$

$\frac{1}{4}(-4) \stackrel{?}{=} 3$

$-1 \neq 3$

Not a solution

(b) $x = 12$

$\frac{1}{4}(12) \stackrel{?}{=} 3$

$3 = 3$

Solution

7. $6(x + 3) = 6x + 3$

$6x + 18 \neq 6x + 3$

No solution

9. $\frac{2}{3}x + 4 = \frac{1}{3}x + 12$

$\frac{2}{3}x - \frac{1}{3}x + 4 = \frac{1}{3}x - \frac{1}{3}x + 12$

$\frac{1}{3}x + 4 = 12$

$\frac{1}{3}x + 4 - 4 = 12 - 4$

$\frac{1}{3}x = 8$

$3\left(\frac{1}{3}x\right) = 3(8)$

$x = 24$

Conditional equation

11.

$3x + 15 = 0$	Original equation
$3x + 15 - 15 = 0 - 15$	Subtract 15 from each side.
$3x = -15$	Combine like terms.
$\frac{3x}{3} = \frac{-15}{3}$	Divide each side by 3.
$x = -5$	Simplify.

13. $4x = x + 10$

$4x - x = x - x + 10$

$3x = 10$

Equivalent

15. $x + 5 = 12$

$2(x + 5) = 2(12)$

$2x + 10 = 24$

Not equivalent

17. $3(4 - 2t) = 5$

$12 - 6t = 5$

Equivalent

19. $2x - 7 = 3$

$2x - 7 + 7 = 3 + 7$

$2x = 10$

$\frac{2x}{2} = \frac{10}{2}$

$x = 5$

Not equivalent

21.
$$\begin{aligned} x - 3 &= 0 \\ x - 3 + 3 &= 0 + 3 \\ x &= 3 \end{aligned}$$

Check: $3 - 3 \stackrel{?}{=} 0$

$0 = 0$

23.
$$\begin{aligned} 3x - 12 &= 0 \\ 3x &= 12 \\ \frac{3x}{3} &= \frac{12}{3} \\ x &= 4 \end{aligned}$$

Check: $3(4) \stackrel{?}{=} 12$

$12 = 12$

25.
$$\begin{aligned} 6x + 4 &= 0 \\ 6x + 4 - 4 &= 0 - 4 \\ 6x &= -4 \\ \frac{6x}{6} &= \frac{-4}{6} \\ x &= -\frac{4}{6} \\ x &= -\frac{2}{3} \end{aligned}$$

Check: $6\left(-\frac{2}{3}\right) + 4 \stackrel{?}{=} 0$

$-4 + 4 \stackrel{?}{=} 0$

$0 = 0$

27.
$$\begin{aligned} 3t + 8 &= -2 \\ 3t + 8 - 8 &= -2 - 8 \\ 3t &= -10 \\ \frac{3t}{3} &= \frac{-10}{3} \\ t &= -\frac{10}{3} \end{aligned}$$

Check: $3\left(-\frac{10}{3}\right) + 8 \stackrel{?}{=} -2$

$-10 + 8 \stackrel{?}{=} -2$

$-2 = -2$

29.
$$\begin{aligned} 4y - 3 &= 4y \\ 4y - 3 + 3 &= 4y + 3 \\ 4y &= 4y + 3 \\ 4y - 4y &= 4y + 3 - 4y \\ 0 &= 3 \\ 0 &\neq 3 \end{aligned}$$

No solution

31.
$$\begin{aligned} -9y - 4 &= -9y \\ -9y + 9y - 4 &= -9y + 9y \\ -4 &= 0 \\ -4 &\neq 0 \end{aligned}$$

No solution

33.
$$\begin{aligned} 7 - 8x &= 13x \\ 7 - 8x + 8x &= 13x + 8x \\ 7 &= 21x \\ \frac{7}{21} &= \frac{21x}{21} \\ \frac{1}{3} &= x \end{aligned}$$

Check: $7 - 8\left(\frac{1}{3}\right) \stackrel{?}{=} 13\left(\frac{1}{3}\right)$

$7 - \frac{8}{3} \stackrel{?}{=} \frac{13}{3}$

$\frac{21}{3} - \frac{8}{3} \stackrel{?}{=} \frac{13}{3}$

$\frac{13}{3} = \frac{13}{3}$

35.
$$\begin{aligned} 3x - 1 &= 2x + 14 \\ 3x - 2x - 1 &= 2x + 14 - 2x \\ x - 1 &= 14 \\ x - 1 + 1 &= 14 + 1 \\ x &= 15 \end{aligned}$$

Check: $3(15) - 1 \stackrel{?}{=} 2(15) + 14$

$45 - 1 \stackrel{?}{=} 30 + 14$

$44 = 44$

37.
$$\begin{aligned} 8(x - 8) &= 24 \\ 8x - 64 &= 24 \\ 8x - 64 + 64 &= 24 + 64 \\ 8x &= 88 \\ \frac{8x}{8} &= \frac{88}{8} \\ x &= 11 \end{aligned}$$

Check: $8(11 - 8) \stackrel{?}{=} 24$

$8(3) \stackrel{?}{=} 24$

$24 = 24$

39.
$$\begin{aligned}3(x-4) &= 7x+6\\ 3x-12 &= 7x+6\\ 3x-7x-12 &= 7x-7x+6\\ -4x-12 &= 6\\ -4x-12+12 &= 6+12\\ -4x &= 18\\ \frac{-4x}{-4} &= \frac{18}{-4}\\ x &= -\frac{9}{2}\end{aligned}$$

Check:
$$\begin{aligned}3\left(-\frac{9}{2}-4\right) &\stackrel{?}{=} 7\left(-\frac{9}{2}\right)+6\\ 3\left(-\frac{9}{2}-\frac{8}{2}\right) &\stackrel{?}{=} -\frac{63}{2}+\frac{12}{2}\\ 3\left(-\frac{17}{2}\right) &\stackrel{?}{=} -\frac{51}{2}\\ -\frac{51}{2} &= -\frac{51}{2}\end{aligned}$$

41.
$$\begin{aligned}4(2x-3) &= 8x-12\\ 8x-12 &= 8x-12\\ 8x-12-8x &= 8x-12-8x\\ -12 &= -12\end{aligned}$$

Infinitely many solutions

43.
$$\begin{aligned}12(x+3) &= 7(x+3)\\ 12x+36 &= 7x+21\\ 12x+36-7x &= 7x+21-7x\\ 5x+36 &= 21\\ 5x+36-36 &= 21-36\\ 5x &= -15\\ \frac{5x}{5} &= \frac{-15}{5}\\ x &= -3\end{aligned}$$

Check:
$$\begin{aligned}12[(-3)+3] &\stackrel{?}{=} 7[(-3)+3]\\ 12[0] &\stackrel{?}{=} 7[0]\\ 0 &= 0\end{aligned}$$

45.
$$\begin{aligned}7(x+6) &= 3(2x+14)+x\\ 7x+42 &= 6x+42+x\\ 7x+42 &= 7x+42\\ 7x+42-7x &= 7x+42-7x\\ 42 &= 42\end{aligned}$$

Infinitely many solutions

47.
$$\begin{aligned}t-\tfrac{2}{5} &= \tfrac{3}{2}\\ t-\tfrac{2}{5}+\tfrac{2}{5} &= \tfrac{3}{2}+\tfrac{2}{5}\\ t &= \tfrac{19}{10}\end{aligned}$$

Check:
$$\begin{aligned}\tfrac{19}{10}-\tfrac{2}{5} &\stackrel{?}{=} \tfrac{3}{2}\\ \tfrac{19}{10}-\tfrac{4}{10} &\stackrel{?}{=} \tfrac{3}{2}\\ \tfrac{15}{10} &\stackrel{?}{=} \tfrac{3}{2}\\ \tfrac{3}{2} &= \tfrac{3}{2}\end{aligned}$$

49.
$$\begin{aligned}\frac{t}{5}-\frac{t}{2} &= 1\\ 10\left(\frac{t}{5}-\frac{t}{2}\right) &= (1)10\\ 2t-5t &= 10\\ -3t &= 10\\ \frac{-3t}{-3} &= \frac{10}{-3}\\ t &= -\frac{10}{3}\end{aligned}$$

Check:
$$\begin{aligned}\frac{-\frac{10}{3}}{5}-\frac{-\frac{10}{3}}{2} &\stackrel{?}{=} 1\\ \frac{10}{-15}+\frac{10}{6} &\stackrel{?}{=} 1\\ -\frac{2}{3}+\frac{5}{3} &\stackrel{?}{=} 1\\ \frac{3}{3} &\stackrel{?}{=} 1\\ 1 &= 1\end{aligned}$$

51.
$$\begin{aligned}\frac{8x}{5}-\frac{x}{4} &= -3\\ 20\left(\frac{8x}{5}-\frac{x}{4}\right) &= 20(-3)\\ 32x-5x &= -60\\ 27x &= -60\\ \frac{27x}{27} &= \frac{-60}{27}\\ x &= -\frac{20}{9}\end{aligned}$$

Check:
$$\begin{aligned}\frac{8\left(-\frac{20}{9}\right)}{5}-\frac{\left(-\frac{20}{9}\right)}{4} &\stackrel{?}{=} -3\\ \frac{\left(-\frac{160}{9}\right)}{5}-\frac{\left(-\frac{20}{9}\right)}{4} &\stackrel{?}{=} -3\\ -\frac{32}{9}+\frac{5}{9} &\stackrel{?}{=} -3\\ -\frac{27}{9} &\stackrel{?}{=} -3\\ -3 &= -3\end{aligned}$$

53.
$$\begin{aligned}0.3x + 1.5 &= 8.4\\ 10(0.3x + 1.5) &= (8.4)10\\ 3x + 15 &= 84\\ 3x + 15 - 15 &= 84 - 15\\ 3x &= 69\\ \frac{3x}{3} &= \frac{69}{3}\\ x &= 23\end{aligned}$$

Check:
$$\begin{aligned}0.3(23) + 1.5 &\stackrel{?}{=} 8.4\\ 6.9 + 1.5 &\stackrel{?}{=} 8.4\\ 8.4 &= 8.4\end{aligned}$$

55.
$$\begin{aligned}1.2(x - 3) &= 10.8\\ 1.2x - 3.6 &= 10.8\\ 10(1.2x - 3.6) &= (10.8)10\\ 12x - 36 &= 108\\ 12x - 36 + 36 &= 108 + 36\\ 12x &= 144\\ \frac{12x}{12} &= \frac{144}{12}\\ x &= 12\end{aligned}$$

Check:
$$\begin{aligned}1.2(12 - 3) &\stackrel{?}{=} 10.8\\ 1.2(9) &\stackrel{?}{=} 10.8\\ 10.8 &= 10.8\end{aligned}$$

57.
$$\begin{aligned}\frac{2}{3}(2x - 4) &= \frac{1}{2}(x + 3) - 4\\ 6\left[\frac{2}{3}(2x - 4)\right] &= \left[\frac{1}{2}(x + 3) - 4\right]6\\ 4(2x - 4) &= 3(x + 3) - 24\\ 8x - 16 &= 3x + 9 - 24\\ 8x - 16 &= 3x - 15\\ 8x - 3x - 16 &= 3x - 3x - 15\\ 5x - 16 &= -15\\ 5x - 16 + 16 &= -15 + 16\\ 5x &= 1\\ \frac{5x}{5} &= \frac{1}{5}\\ x &= \frac{1}{5}\end{aligned}$$

Check:
$$\begin{aligned}\frac{2}{3}\left[2\left(\frac{1}{5}\right) - 4\right] &\stackrel{?}{=} \frac{1}{2}\left(\frac{1}{5} + 3\right) - 4\\ \frac{2}{3}\left(\frac{2}{5} - \frac{20}{5}\right) &\stackrel{?}{=} \frac{1}{2}\left(\frac{1}{5} + \frac{15}{5}\right) - 4\\ \frac{2}{3}\left(-\frac{18}{5}\right) &\stackrel{?}{=} \frac{1}{2}\left(\frac{16}{5}\right) - 4\\ -\frac{12}{5} &\stackrel{?}{=} \frac{8}{5} - \frac{20}{5}\\ -\frac{12}{5} &= -\frac{12}{5}\end{aligned}$$

59.
$$\begin{aligned}n + (n + 1) &= 251\\ 2n + 1 &= 251\\ 2n + 1 - 1 &= 251 - 1\\ 2n &= 250\\ \frac{2n}{2} &= \frac{250}{2}\\ n &= 125\\ n + 1 &= 126\end{aligned}$$

61.
$$\begin{aligned}38h + 162 &= 257\\ 38h + 162 - 162 &= 257 - 162\\ 38h &= 95\\ \frac{38h}{38} &= \frac{95}{38}\\ h &= \frac{5}{2} = 2.5 \text{ hours}\end{aligned}$$

So, the repair work took 2.5 hours.

63.
$$\begin{aligned}\frac{t}{10} + \frac{t}{15} &= 1\\ \frac{3t}{30} + \frac{2t}{30} &= 1\\ \frac{5t}{30} &= 1\\ \frac{t}{6} &= 1\\ 6\left(\frac{t}{6}\right) &= 6(1)\\ t &= 6 \text{ hours}\end{aligned}$$

65. The fountain reaches its maximum height when the velocity of the stream of water is zero.

$$0 = 48 - 32t$$
$$0 + 32t = 48 - 32t + 32t$$
$$32t = 48$$
$$\frac{32t}{32} = \frac{48}{32}$$
$$t = \frac{3}{2} \text{ seconds} = 1.5 \text{ seconds}$$

67. (a)

t	1	1.5	2	3	4	5
Width	250	200	166.7	125	100	83.3
Length	250	300	333.4	375	400	416.5
Area	62,500	60,000	55,577.8	46,875	40,000	34,694.5

(b) Because the perimeter is fixed, as t increases the length increases and the width and area decrease. The maximum area occurs when the length and width are equal.

69. Graphically: Graph $y = 355.3t + 3725$, and zoom in to find t when $y = 7633.30$; $t = 11$.

Numerically: Substitute different values for t until you find the correct value of y; $t = 11$.

Algebraically: Let $7633.30 = 355.3t + 3725$ and solve for t; $t = 11$.

71. False. Multiplying both sides of an equation by zero does not yield an equivalent equation because this does not follow the Multiplication Property of Equality.

73. No. An identity in standard form would be written as $ax + b = 0$, with $a = b = 0$. But, the equation would not be in standard form.

75.
$$x + 0.20x = 50.16$$
$$1.20x = 50.16$$
$$x = 41.8$$

The equation is a conditional equation. The equation could represent a 20% tip on a restaurant bill.

77.
$$0.25(40 + x) = 10 + 0.25x$$
$$10 + 0.25x = 10 + 0.25x$$
$$10 - 10 + 0.25x = 10 - 10 + 0.25x$$
$$0.25x = 0.25x$$
$$\frac{0.25x}{0.25} = \frac{0.25x}{0.25}$$
$$x = x$$

The equation is an identity. Let x be the number of quarters in a dollar.

79. $\frac{2}{5} + \frac{4}{5} = \frac{6}{5}$

81. $-5 - (-3) = -5 + 3 = -2$

83. $8 + 7(2) = 8 + 14 = 22$

$8 + 7(3) = 8 + 21 = 29$

85. $(-4)^2 - 1 = 16 - 1 = 15$

$(3)^2 - 1 = 9 - 1 = 8$

87. $n - 8$

89. $2(n + 3)$

Section 2.2 Linear Equations and Problem Solving

1. *Verbal Model:* [Number] + 24 = 68

Label: A number $= x$

Equation: $x + 24 = 68$

$x = 44$

3. *Verbal Model:*

26 · [Amount of each paycheck] + [Bonus] = [Income for year]

Labels: Amount of each paycheck $= x$

Bonus $= 2800$

Income for year $= 37{,}120$

Equation: $26x + 2800 = 37{,}120$

$26x = 34{,}320$

$x = 1320$

Each paycheck will be $1320.

5. Percent: 30%

Parts out of 100: 30

Decimal: 0.30

Fraction: $\frac{30}{100} = \frac{3}{10}$

7. Percent: 7.5%

Parts out of 100: 7.5

Decimal: 0.075

Fraction: $\frac{75}{1000} = \frac{3}{40}$

9. Percent: $66\frac{2}{3}\%$

Parts out of 100: $66\frac{2}{3}$

Decimal: 0.66. . .

Fraction: $\frac{2}{3}$

11. Percent: 100%

Parts out of 100: 100

Decimal: 1.00

Fraction: 1

13. *Verbal Model:* [Compared number] = [Percent] · [Base number]

Labels: Compared number $= a$

Percent $= p$

Base number $= b$

Equation: $a = p \cdot b$

$a = (0.35)(250)$

$a = 87.5$

So, 35% of 250 is 87.5.

15. *Verbal Model:*

[Compared number] = [Percent (decimal form)] · [Base number]

Labels: Compared number $= a$

Percent $= 0.425$

Base number $= 816$

Equation: $a = 0.425(816)$

$a = 346.8$

So, 346.8 is 42.5% of 816.

17. *Verbal Model:* [Compared number] = [Percent] · [Base number]

Labels: Compared number $= a$

Percent $= p$

Base number $= b$

Equation: $a = p \cdot b$

$a = (0.125)(1024)$

$a = 128$

So, 128 is 12.5% of 1024.

19. *Verbal Model:* [Compared number] = [Percent] · [Base number]

Labels: Compared number $= a$

Percent $= p$

Base number $= b$

Equation: $a = p \cdot b$

$a = (0.004)(150{,}000)$

$a = 600$

So, 600 is 0.4% of 150,000.

21. *Verbal Model:* [Compared number] = [Percent] · [Base number]

Labels: Compared number $= a$

Percent $= p$

Base number $= b$

Equation: $a = p \cdot b$

$a = (2.50)(32)$

$a = 80$

So, 80 is 250% of 32.

23. *Verbal Model:* $\boxed{\text{Compared number}} = \boxed{\text{Percent}} \cdot \boxed{\text{Base number}}$

Labels: Compared number $= a$

Percent $= p$

Base number $= b$

Equation:
$$\begin{aligned} a &= p \cdot b \\ 84 &= (0.24)(b) \\ \frac{0}{0.24} &= b \\ 350 &= b \end{aligned}$$

So, 84 is 24% of 350.

25. *Verbal Model:* $\boxed{\text{Compared number}} = \boxed{\text{Percent}} \cdot \boxed{\text{Base number}}$

Labels: Compared number $= a$

Percent $= p$

Base number $= b$

Equation:
$$\begin{aligned} a &= p \cdot b \\ 42 &= (1.2)(b) \\ \frac{42}{1.2} &= b \\ 35 &= b \end{aligned}$$

So, 42 is 120% of 35.

27. *Verbal Model:*

$\boxed{\text{Compared number}} = \boxed{\text{Percent (decimal form)}} \cdot \boxed{\text{Base number}}$

Labels: Compared number $= 22$

Percent $= 0.008$

Base number $= b$

Equation:
$$\begin{aligned} 22 &= 0.008b \\ \frac{22}{0.008} &= b \\ 2750 &= b \end{aligned}$$

So, 22 is 0.8% of 2750.

29. *Verbal Model:* $\boxed{\text{Compared number}} = \boxed{\text{Percent}} \cdot \boxed{\text{Base number}}$

Labels: Compared number $= a$

Percent $= p$

Base number $= b$

Equation:
$$\begin{aligned} a &= p \cdot b \\ 496 &= (p)(800) \\ \tfrac{496}{800} &= p \\ 0.62 &= p \\ p &= 62\% \end{aligned}$$

So, 496 is 62% of 800.

31. *Verbal Model:* $\boxed{\text{Compared number}} = \boxed{\text{Percent}} \cdot \boxed{\text{Base number}}$

Labels: Compared number $= a$

Percent $= p$

Base number $= b$

Equation:
$$\begin{aligned} a &= p \cdot b \\ 2.4 &= (p)(480) \\ \frac{2.4}{480} &= p \\ 0.005 &= p \\ p &= 0.5\% \end{aligned}$$

So, 2.4 is 0.5% of 480.

33. *Verbal Model:* $\boxed{\text{Compared number}} = \boxed{\text{Percent}} \cdot \boxed{\text{Base number}}$

Labels: Compared number $= a$

Percent $= p$

Base number $= b$

Equation:
$$\begin{aligned} a &= p \cdot b \\ 2100 &= (p)(1200) \\ \tfrac{2100}{1200} &= p \\ 175\% &= p \end{aligned}$$

So, 2100 is 175% of 1200.

35. $\dfrac{120 \text{ meters}}{180 \text{ meters}} = \dfrac{12}{18} = \dfrac{2}{3}$

37. $\dfrac{36 \text{ inches}}{48 \text{ inches}} = \dfrac{36}{48} = \dfrac{3}{4}$

39. $\dfrac{40 \text{ milliliters}}{1 \text{ liter}} = \dfrac{0.04 \text{ liter}}{1} = \dfrac{4}{100} = \dfrac{1}{25}$

41. $\dfrac{5 \text{ pounds}}{24 \text{ ounces}} = \dfrac{80 \text{ ounces}}{24 \text{ ounces}} = \dfrac{10}{3}$

43.
$$\begin{aligned} \frac{x}{6} &= \frac{2}{3} \\ 3 \cdot x &= 6 \cdot 2 \\ 3x &= 12 \\ x &= 4 \end{aligned}$$

45.
$$\begin{aligned} \frac{y}{36} &= \frac{6}{7} \\ 7 \cdot y &= 36 \cdot 6 \\ 7y &= 216 \\ y &= \frac{216}{7} \end{aligned}$$

47. $\frac{5}{4} = \frac{t}{6}$

$$6 \cdot 5 = 4 \cdot t$$
$$30 = 4t$$
$$\frac{30}{4} = t$$
$$\frac{15}{2} = t$$

49. $\frac{y}{6} = \frac{y-2}{4}$

$$4y = 6(y - 2)$$
$$4y = 6y - 12$$
$$12 = 2y$$
$$6 = y$$

51. $\frac{z-3}{3} = \frac{z+8}{12}$

$$12(z - 3) = 3(z + 8)$$
$$\frac{12(z-3)}{3} = \frac{3(z+8)}{3}$$
$$4(z - 3) = z + 8$$
$$4z - 12 = z + 8$$
$$3z - 12 = 8$$
$$3z = 20$$
$$z = \frac{20}{3}$$

53. *Verbal Model:*

[Total enrollment] = [Percent (decimal form)] · [Total admitted]

Labels: Total enrollment $= a$

Percent $= 0.40$

Total admitted $= 20{,}181$

Equation: $a = 0.40(20{,}181)$

$a = 8072.4$

So, 8072 students were enrolled.

55. *Verbal Model:* [Students failing test] = [Percent] · [Total Students]

Labels: Students failing test $= a$

Percent $= p$

Total students $= b$

Equation: $a = p \cdot b$

$a = (1 - 0.95)(40)$

$a = (0.05)(40)$

$a = 2$

So, 2 students failed the test.

57. *Verbal Model:* [Number laid off] = [Percent] · [Number of employees]

Labels: Number laid off $= a$

Percent $= p$

Number of employees $= b$

Equation: $a = p \cdot b$

$$25 = (p)(160)$$
$$\frac{25}{160} = \frac{(p)(160)}{160}$$
$$\frac{25}{160} = p$$
$$15.625 = p$$

So, 15.625% of the workforce was laid off.

59. *Verbal Model:* [Tip] = [Percent (decimal form)] · [Cost of meal]

Labels: Tip $= a$

Percent $= 0.15$

Cost of meal $= 32.60$

Equation: $a = 0.15(32.60)$

$a = 4.89$

So, you should leave a \$4.89 tip.

61. *Verbal Model:* [Tip] = [Percent] · [Cost of meal]

Labels: Tip $= a$

Percent $= p$

Cost of meal $= b$

Equation: $25 - 20.66 = p \cdot 20.66$

$$4.34 = p \cdot 20.66$$
$$\frac{4.34}{20.66} = \frac{p \cdot 20.66}{20.66}$$
$$0.21 \approx p$$

So, it is about a 21% tip.

63. *Verbal Model:* [Tip] = [Percent] · [Cost of meal]

Labels: Tip $= a$

Percent $= p$

Cost of meal $= b$

Equation: $9 - 8.20 = p \cdot 8.20$

$$0.8 = p \cdot 8.20$$
$$\frac{0.8}{8.20} = \frac{p \cdot 8.20}{8.20}$$
$$0.098 \approx p$$

So, it is about a 9.8% tip.

65. *Verbal Model:* $\boxed{\text{Commission}} = \boxed{\text{Percent}} \cdot \boxed{\text{Price of home}}$

Labels: Commission $= a$

Percent $= p$

Price of home $= b$

Equation:

$$a = p \cdot b$$
$$12{,}250 = p \cdot 175{,}000$$
$$\frac{12{,}250}{175{,}000} = \frac{p \cdot 175{,}000}{175{,}000}$$
$$0.07 = p$$

So, it is a 7% commission.

67. *Verbal Model:* $\boxed{\text{Defective parts}} = \boxed{\text{Percent}} \cdot \boxed{\text{Total parts}}$

Labels: Defective parts $= a$

Percent $= p$

Total parts $= b$

Equation:

$$a = p \cdot b$$
$$3 = (0.015)(b)$$
$$\frac{3}{0.015} = \frac{(0.015)(b)}{0.015}$$
$$\frac{3}{0.015} = b$$
$$200 = b$$

So, the sample contained 200 parts.

69. (a) *Verbal Model:*

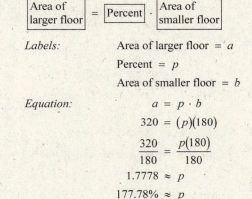

$\boxed{\text{Area of larger floor}} = \boxed{\text{Percent}} \cdot \boxed{\text{Area of smaller floor}}$

Labels: Area of larger floor $= a$

Percent $= p$

Area of smaller floor $= b$

Equation:

$$a = p \cdot b$$
$$320 = (p)(180)$$
$$\frac{320}{180} = \frac{p(180)}{180}$$
$$1.7778 \approx p$$
$$177.78\% \approx p$$

(b) *Verbal Model:*

$\boxed{\text{Area of smaller floor}} = \boxed{\text{Percent}} \cdot \boxed{\text{Area of larger floor}}$

Labels: Area of smaller floor $= a$

Percent $= p$

Area of larger floor $= b$

Equation:

$$a = p \cdot b$$
$$180 = (p)(320)$$
$$\frac{180}{320} = \frac{p(320)}{320}$$
$$0.5625 \approx p$$
$$56.25\% \approx p$$

71. *Verbal Model:*

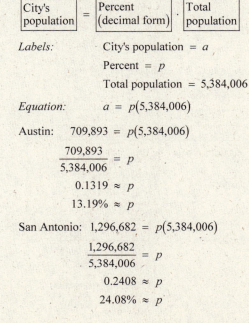

$\boxed{\text{City's population}} = \boxed{\text{Percent (decimal form)}} \cdot \boxed{\text{Total population}}$

Labels: City's population $= a$

Percent $= p$

Total population $= 5{,}384{,}006$

Equation: $a = p(5{,}384{,}006)$

Austin: $709{,}893 = p(5{,}384{,}006)$

$$\frac{709{,}893}{5{,}384{,}006} = p$$
$$0.1319 \approx p$$
$$13.19\% \approx p$$

Dallas: $1{,}232{,}940 = p(5{,}384{,}006)$

$$\frac{1{,}232{,}940}{5{,}384{,}006} = p$$
$$0.229 = p$$
$$22.9\% = p$$

San Antonio: $1{,}296{,}682 = p(5{,}384{,}006)$

$$\frac{1{,}296{,}682}{5{,}384{,}006} = p$$
$$0.2408 \approx p$$
$$24.08\% \approx p$$

Houston: $2{,}144{,}491 = p(5{,}384{,}006)$

$$\frac{2{,}144{,}491}{5{,}384{,}006} = p$$
$$0.3983 \approx p$$
$$39.83\% \approx p$$

73. The change is about \$37 per 100 pounds.

Verbal Model: $\boxed{\text{Percent increase}} = \boxed{\dfrac{\text{Change}}{\text{Original amount}}}$

Labels: Percent increase $= p$

Change $= 37$

Original amount $= 112$

Equation: $p = \dfrac{37}{112}$

$p \approx 33\%$

The percent increase is about 33%.

75. *Verbal model:* $\boxed{\text{Price of ham}} = \boxed{\text{Percent}} \cdot \boxed{\text{Price of lamb}}$

Labels: Price of ham $= 50$

Percent $= p$

Price of lamb $= 160$

Equation: $50 = p(160)$

$\frac{50}{160} = p$

$0.3125 \approx p$

So, the price of ham was about 31.25% of the price of lamb.

77. $\dfrac{\text{Tax}}{\text{Pay}} = \dfrac{\$12.50}{\$625} = \dfrac{125}{6250} = \dfrac{1}{50}$

79. $\dfrac{\text{Expanded volume}}{\text{Compressed volume}} = \dfrac{425 \text{ cu cm}}{20 \text{ cu cm}} = \dfrac{85}{4}$

81. $\dfrac{\text{Area 1}}{\text{Area 2}} = \dfrac{\pi(4)^2}{\pi(6)^2} = \dfrac{16\pi}{36\pi} = \dfrac{4}{9}$

83. Unit price $= \dfrac{\text{Total price}}{\text{Total units}}$

$= \dfrac{\$1.10}{20 \text{ ounces}} \approx \$0.06$ per ounce

85. Unit price $= \dfrac{\text{Total price}}{\text{Total units}}$

$= \dfrac{\$2.29}{20 \text{ ounces}} \approx \$0.11$ per ounce

87. (a) Unit price $= \dfrac{2.32}{14.5} = \$0.16$ per ounce

(b) Unit price $= \dfrac{0.99}{5.5} = \$0.18$ per ounce

The $14\frac{1}{2}$-ounce bag is a better buy.

89. (a) Unit price $= \dfrac{1.69}{4} = \$0.4225$ per ounce

(b) Unit price $= \dfrac{2.39}{6} = \$0.3983$ per ounce

The 6-ounce tube is a better buy.

91. $\dfrac{x}{7} = \dfrac{4}{5.5}$

$5.5 \cdot x = 7 \cdot 4$

$5.5x = 28$

$x = 5.\overline{09}$

93. $\dfrac{x}{6} = \dfrac{2}{4}$

$4 \cdot x = 6 \cdot 2$

$4x = 12$

$x = 3$

95. $\dfrac{h}{86} = \dfrac{6}{11}$

$11 \cdot h = 86 \cdot 6$

$11h = 516$

$h = \dfrac{516}{11}$

$h \approx 46.9$

So, the height of the tree is about 46.9 feet.

97. $\dfrac{5}{105} = \dfrac{x}{360}$

$360 \cdot 5 = 105 \cdot x$

$1800 = 105x$

$\dfrac{1800}{105} = x$

$17.1 \approx x$

So, about 17.1 gallons of fuel are used.

99. *Verbal Model:*

$\boxed{\dfrac{\text{Tax 1}}{\text{Assessed value 1}}} = \boxed{\dfrac{\text{Tax 2}}{\text{Assessed value 2}}}$

Labels: Tax 1 $= x$

Assessed value 1 $= 160{,}000$

Tax 2 $= 1650$

Assessed value 2 $= 110{,}000$

Proportion: $\dfrac{x}{160{,}000} = \dfrac{1650}{110{,}000}$

$110{,}000 \cdot x = 1650 \cdot 160{,}000$

$110{,}000x = 264{,}000{,}000$

$x = 2400$

So, the tax is \$2400.

101. *Verbal Model:*

$$\frac{\boxed{\text{Total defective units}}}{\boxed{\text{Total units}}} = \frac{\boxed{\text{Defective units}}}{\boxed{\text{Sample}}}$$

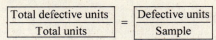

Labels: Total defective units $= x$

Total units $= 200{,}000$

Defective units $= 1$

Sample $= 75$

Proportion: $\dfrac{x}{200{,}000} = \dfrac{1}{75}$

$$75 \cdot x = 200{,}000 \cdot 1$$
$$75x = 200{,}000$$
$$x = 2667$$

So, the expected number of defective units is 2667.

103. *Verbal Model:*

$$\frac{\boxed{\text{Total color defects}}}{\boxed{\text{Total units}}} = \frac{\boxed{\text{Color defects}}}{\boxed{\text{Sample}}}$$

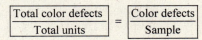

Labels: Total color defects $= x$

Total units $= 235$

Color defects $= 1$

Sample $= 40$

Proportion: $\dfrac{x}{235} = \dfrac{1}{40}$

$$40 \cdot x = 235 \cdot 1$$
$$40x = 235$$
$$x \approx 6$$

So, the expected number of color defects is about 6.

105. *Verbal Model:* $\dfrac{\boxed{\text{Current in favor}}}{\boxed{\text{Current poll}}} = \dfrac{\boxed{\text{Total in favor}}}{\boxed{\text{Total vote}}}$

Labels: Current in favor $= 870$

Current poll $= 1500$

Total in favor $= x$

Total vote $= 80{,}000$

Proportion: $\dfrac{870}{1500} = \dfrac{x}{80{,}000}$

$$80{,}000 \cdot 870 = 1500 \cdot x$$
$$69{,}600{,}000 = 1500x$$
$$46{,}400 = x$$

So, the candidate can expect 46,400 votes.

107. To change percents to decimals divide by 100. To change decimals to percents multiply by 100.

Examples: $42\% = \frac{42}{100} = 0.42$

$0.38 = (0.38)(100)\% = 38\%$

109. Mathematical modeling is the use of mathematics to solve problems that occur in real-life situations. For examples review the real-life problems in the exercise set.

111. $-\frac{4}{15} \cdot \frac{15}{16} = -\frac{60}{240} = -\frac{1}{4}$

113. $(12 - 15)^3 = (-3)^3 = -27$

115. Commutative Property of Addition

117. Distributive Property

119. $2x - 5 = x + 9$

$$2x = x + 14$$
$$x = 14$$

121. $2x + \frac{3}{2} = \frac{3}{2}$

$$2x = 0$$
$$x = 0$$

123. $-0.35x = 70$

$$x = -200$$

Section 2.3 Business and Scientific Problems

1. *Verbal Model:* [Selling price] = [Cost] + [Markup]

Labels: Selling price = 64.33
Cost = 45.97
Markup = x

Equation:
$$64.33 = 45.97 + x$$
$$x = 64.33 - 45.97$$
$$x = 18.36$$

The markup is $18.36.

Verbal Model: [Markup] = [Markup rate] · [Cost]

Labels: Markup = 18.36
Markup rate = x
Cost = 45.97

Equation:
$$18.36 = x \cdot 45.97$$
$$\frac{18.36}{45.97} = x$$
$$0.40 \approx x$$

The markup rate is 40%.

3. *Verbal Model:* [Selling price] = [Cost] + [Markup]

Labels: Selling price = 250.80
Cost = x
Markup = 98.80

Equation:
$$250.80 = x + 98.80$$
$$250.80 - 98.80 = x$$
$$152.00 = x$$

The cost is $152.00.

Verbal Model: [Markup] = [Markup rate] · [Cost]

Labels: Markup = 98.80
Markup rate = x
Cost = 152.00

Equation:
$$98.80 = x \cdot 152.00$$
$$\frac{98.80}{152.00} = x$$
$$0.65 = x$$

The markup rate is 65%.

5. *Verbal Model:* [Selling price] = [Cost] + [Markup]

Labels: Selling price = 26,922.50
Cost = x
Markup = 4672.50

Equation:
$$26{,}922.50 = x + 4672.50$$
$$26{,}922.50 - 4672.50 = x$$
$$22{,}250.00 = x$$

The cost is $22,250.00.

Verbal Model: [Markup] = [Markup rate] · [Cost]

Labels: Markup = 4672.50
Markup rate = x
Cost = 22,250.00

Equation:
$$4672.50 = x \cdot 22{,}250.00$$
$$\frac{4672.50}{22{,}250.00} = x$$
$$0.21 = x$$

The markup rate is 21%.

7. *Verbal Model:* [Markup] = [Markup rate] · [Cost]

Labels: Markup = x
Markup rate = 85.2%
Cost = 225.00

Equation:
$$x = 0.852 \cdot 225.00$$
$$x = 191.70$$

The markup is $191.70.

Verbal Model: [Selling price] = [Cost] + [Markup]

Labels: Selling price = x
Cost = 225.00
Markup = 191.70

Equation:
$$x = 225.00 + 191.70$$
$$x = 416.70$$

The selling price is $416.70.

9. *Verbal Model:* $\boxed{\text{Sale price}} = \boxed{\text{List price}} - \boxed{\text{Discount}}$

Labels: Sale price = 25.74

List price = 49.95

Discount = x

Equation:

$$25.74 = 49.95 - x$$
$$x = 49.95 - 25.74$$
$$x = 24.21$$

The discount is $24.21.

Verbal Model: $\boxed{\text{Discount}} = \boxed{\text{Discount rate}} \cdot \boxed{\text{List price}}$

Labels: Discount = 24.21

Discount rate = x

List price = 49.95

Equation:

$$24.21 = x \cdot 49.95$$
$$\frac{24.21}{49.95} = x$$
$$0.485 \approx x$$

The discount rate is 48.5%.

11. *Verbal Model:* $\boxed{\text{Sale price}} = \boxed{\text{List price}} - \boxed{\text{Discount}}$

Labels: Sale price = x

List price = 300.00

Discount = 189.00

Equation:

$$x = 300.00 - 189.00$$
$$x = 111.00$$

The sale price is $111.00.

Verbal Model: $\boxed{\text{Discount}} = \boxed{\text{Discount rate}} \cdot \boxed{\text{List price}}$

Labels: Discount = 189.00

Discount rate = x

List price = 300.00

Equation:

$$189.00 = x \cdot 300.00$$
$$\frac{189.00}{300.00} = x$$
$$0.63 = x$$

The discount rate is 63%.

13. *Verbal Model:* $\boxed{\text{Sale price}} = \boxed{\text{Percent}} \cdot \boxed{\text{List price}}$

Labels: Sale price = 27.00

Percent = 0.60

List price = x

Equation:

$$27.00 = 0.60x$$
$$\frac{27.00}{0.60} = x$$
$$45.00 = x$$

The list price is $45.00.

Verbal Model: $\boxed{\text{Discount}} = \boxed{\text{Discount rate}} \cdot \boxed{\text{List price}}$

Labels: Discount = x

Discount rate = 0.40

List price = 45.00

Equation:

$$x = 0.40 \cdot 45.00$$
$$x = 18.00$$

The discount is $18.00.

15. *Verbal Model:* $\boxed{\text{Sale price}} = \boxed{\text{List price}} - \boxed{\text{Discount}}$

Labels: Sale price = 831.96

List price = x

Discount = 323.54

Equation:

$$831.96 = x - 323.54$$
$$831.96 + 323.54 = x$$
$$1155.50 = x$$

The list price is $1155.50.

Verbal Model: $\boxed{\text{Discount}} = \boxed{\text{Discount rate}} \cdot \boxed{\text{List price}}$

Labels: Discount = 323.54

Discount rate = p

List price = 1155.50

Equation:

$$323.54 = p \cdot 1155.50$$
$$\frac{323.54}{1155.50} = p$$
$$0.28 = p$$

The discount rate is 28%.

17. *Verbal Model:* $\boxed{\text{Selling price}} = \boxed{\text{Cost}} + \boxed{\text{Markup}}$

Labels: Selling price $= 157.14$

Cost $= 130.95$

Markup $= x$

Equation:
$$157.14 = 130.95 + x$$
$$x = 157.14 - 130.95$$
$$x = 26.19$$

The markup is \$26.19.

19. *Verbal Model:* $\boxed{\text{Markup}} = \boxed{\text{Markup rate}} \cdot \boxed{\text{Cost}}$

Labels: Markup $= 37.33$

Markup rate $= p$

Cost $= 46.67$

Equation:
$$37.33 = p \cdot 46.67$$
$$\frac{37.33}{46.67} = p$$
$$0.80 \approx p$$

The markup rate is 80%.

21. *Verbal Model:* $\boxed{\text{Sale price}} = \boxed{\text{List price}} - \boxed{\text{Discount}}$

Labels: Sale price $= 45$

List price $= 75$

Discount $= x$

Equation:
$$45 = 75 - x$$
$$45 - 75 = -x$$
$$30 = x$$

The discount is \$30.

23. *Verbal Model:* $\boxed{\text{Sale price}} = \boxed{\text{List price}} - \boxed{\text{Discount}}$

Labels: Sale price $= 16$

List price $= 20$

Discount $= x$

Equation:
$$16 = 20 - x$$
$$x = 20 - 16$$
$$x = 4$$

Verbal Model: $\boxed{\text{Discount}} = \boxed{\text{Discount rate}} \cdot \boxed{\text{List price}}$

Labels: Discount $= 4$

Discount rate $= x$

List price $= 20$

Equation:
$$4 = x \cdot 20$$
$$\tfrac{4}{20} = x$$
$$0.20 = x$$

The discount rate is 20%.

25. *Verbal Model:*

$\boxed{\text{Total cost}} = \boxed{\text{Cost of first minute}} + \boxed{\text{Cost of additional minutes}}$

Labels: Total cost $= 5.15$

Cost of first minute $= 0.75$

Cost of additional minutes $= 0.55x$

Equation:
$$5.15 = 0.75 + 0.55x$$
$$4.40 = 0.55x$$
$$8 = x$$

The length of call is 9 minutes.

Verbal Model: $\boxed{\text{Discount}} = \boxed{\text{Discount rate}} \cdot \boxed{\text{List price}}$

Labels: Discount $= x$

Discount rate $= 60\%$

List price $= 5.15$

Equation:
$$x = 0.60 \cdot 5.15$$
$$x = 3.09$$

Verbal Model: $\boxed{\text{Selling price}} = \boxed{\text{List price}} - \boxed{\text{Discount}}$

Labels: Selling price $= x$

List price $= 5.15$

Discount $= 3.09$

Equation:
$$x = 5.15 - 3.09$$
$$x = 2.06$$

The call would have cost \$2.06.

27. *Verbal Model:* $\boxed{\text{Cost}} + \boxed{\text{Markup}} = \boxed{\text{Selling price}}$

Labels: Cost $= x$

Markup $= 0.10x$

Selling price $= 59.565$

Four tires cost: $3(\$79.42) = \$238.26$

Each tire costs: $\dfrac{\$238.26}{4} = \$59.565$

Equation:
$$x + 0.10x = 59.565$$
$$1.10x = 59.565$$
$$x = 54.15$$

The cost to the store for each tire is \$54.15.

29. *Verbal Model:* [Total bill] = [Bill for parts] + [Bill for labor]

Labels: Total bill = 216.37

Bill for parts = 136.37

Bill for labor = $32x$

Number of hours of labor = x

Equation:

$$216.37 = 136.37 + 32x$$
$$80 = 32x$$
$$\frac{80}{32} = x$$
$$2.5 = x$$

The repairs took 2.5 hours.

31. *Verbal Model:* [Total bill] = [Parts charge] + [Labor charge]

Labels: Total bill = 648

Parts charge = 315

Charge per hour = x

Labor charge = $9x$

Equation:

$$648 = 315 + 9x$$
$$333 = 9x$$
$$37 = x$$

The charge per hour was $37.

33. *Verbal Model:*

[Amount of solution 1] + [Amount of solution 2] = [Amount of final solution]

Labels: Percent of solution 1 = 20%

Gallons of solution 1 = x

Percent of solution 2 = 60%

Gallons of solution 2 = $100 - x$

Percent of final solution = 40%

Gallons of final solution = 100

Equation:

$$0.20x + 0.60(100 - x) = 0.40(100)$$
$$0.20x + 60 - 0.60x = 40$$
$$-0.40x = -20$$
$$x = 50$$
$$100 - x = 50$$

50 gallons of solution 1 and 50 gallons of solution 2 are needed.

35. *Verbal Model:*

[Amount of solution 1] + [Amount of solution 2] = [Amount of final solution]

Labels: Percent of solution 1 = 15%

Quarts of solution 1 = x

Percent of solution 2 = 60%

Quarts of solution 2 = $24 - x$

Percent of final solution = 45%

Quarts of final solution = 24

Equation:

$$0.15x + 0.60(24 - x) = 0.45(24)$$
$$0.15x + 14.4 - 0.60x = 10.8$$
$$-0.45x = -3.6$$
$$x = 8$$
$$24 - x = 16$$

8 quarts of solution 1 and 16 quarts of solution 2 are needed.

37. *Verbal Model:* [Cost of seed 1] + [Cost of seed 2] = [Cost of final seed mix]

Labels:

Number of pounds of seed 1 = x

Cost per pound of seed 1 = 12

Number of pounds of seed 2 = $100 - x$

Cost per pound of seed 2 = 20

Number of pounds of final seed mix = 100

Cost per pound of final seed mix = 14

Equation:

$$12x + 20(100 - x) = 14(100)$$
$$12x + 2000 - 20x = 1400$$
$$-8x = -600$$
$$x = 75$$

75 pounds of seed 1 and 25 pounds of seed 2 are needed.

39. *Verbal Model:* [Total sales] = [Adult sales] + [Children sales]

Labels: Total sales = 2200

Number of adult tickets = $3x$

Price of adult tickets = 6

Number of children's tickets = x

Price of children's tickets = 4

Equation:

$$2200 = 6(3x) + 4x$$
$$2200 = 18x + 4x$$
$$2200 = 22x$$
$$100 = x$$

100 children's tickets are sold.

41. *Verbal Model:*

$\boxed{\text{Original antifreeze solution}} - \boxed{\text{Some antifreeze solution}} + \boxed{\text{Pure antifreeze}} = \boxed{\text{Final antifreeze solution}}$

Labels:

Number of gallons of original antifreeze = 5

Percent of antifreeze in original mix = 40%

Number of gallons antifreeze withdrawn = x

Number of gallons of pure antifreeze = x

Percent of pure antifreeze = 100%

Number of gallons of final solution = 5

Percent of antifreeze in final solution = 50%

Equation:

$$0.40(5) - 0.40x + 1.00x = 0.50(5)$$
$$2 - 0.40x + 1.00x = 2.5$$
$$0.60x = 0.5$$
$$x = \frac{5}{6}$$

$\frac{5}{6}$ gallon must be withdrawn and replaced.

43. *Verbal Model:* $\boxed{\text{Distance}} = \boxed{\text{Rate}} \cdot \boxed{\text{Time}}$

Labels: Distance = d

Rate = 650

Time = 3.5

Equation: $d = 650 \cdot 3.5$

$d = 2275$

The distance is 2275 miles.

45. *Verbal Model:* $\boxed{\text{Distance}} = \boxed{\text{Rate}} \cdot \boxed{\text{Time}}$

Labels: Distance = 1000

Rate = 110

Time = t

Equation: $1000 = 110 \cdot t$

$\frac{1000}{110} = t$

$\frac{100}{11} = t$

The time is $\frac{100}{11}$ hours.

47. *Verbal Model:* $\boxed{\text{Distance}} = \boxed{\text{Rate}} \cdot \boxed{\text{Time}}$

Labels: Distance = 385

Rate = r

Time = 7

Equation: $385 = r \cdot 7$

$\frac{385}{7} = r$

$55 = r$

The rate is 55 miles/hour.

49. *Verbal Model:* $\boxed{\text{Distance}} = \boxed{\text{Rate}} \cdot \boxed{\text{Time}}$

Labels: Distance = d

Rate = 10

Time = 2.5

Equation: $d = 10 \cdot 2.5$

$d = 25$

You ride your bike 25 miles.

51. *Verbal Model:* $\boxed{\text{Distance}} = \boxed{\text{Rate}} \cdot \boxed{\text{Time}}$

Labels: Distance = 20

Rate = 16

Time = x

Equation: $20 = 16 \cdot x$

$20 = 16x$

$\frac{20}{16} = x$

$\frac{5}{4} = x$

$1\frac{1}{4} = x$

It will take you $1\frac{1}{4}$ hours.

53. *Verbal Model:* $\boxed{\text{Distance}} = \boxed{\text{Rate}} \cdot \boxed{\text{Time}}$

Labels: Distance = 4

Rate = r

Time = $\frac{2}{3}$

Equation: $4 = r \cdot \frac{2}{3}$

$6 = r$

Your jogging rate is 6 miles/hour.

55. *Verbal Model:* $\boxed{\text{Distance}} = \boxed{\text{Rate}} \cdot \boxed{\text{Time}}$

Labels: Distance = x

Rates = 480 and 600

Time = $\frac{4}{3}$

Equation: $x = 480\left(\frac{4}{3}\right) + 600\left(\frac{4}{3}\right)$

$x = 1440$

The planes are 1440 miles apart after $1\frac{1}{3}$ hours.

57. *Verbal Model:* [Distance] = [Rate] · [Time]

Labels: Distance = 5000

Rate = 17,500

Time = t

Equation:

$$5000 = 17{,}500 \cdot t$$

$$\frac{5000}{17{,}500} = t$$

$$\frac{2}{7} \text{ hour} = t$$

$$17.14 \approx t$$

About 17.14 minutes are required.

59. *Verbal Model:* [Distance] = [Rate] · [Time]

Labels: Distance = 317

Rate for first part of trip = 58

Time for first part of trip = x

Rate for second part of trip = 52

Time for second part of trip = $5\frac{3}{4} - x$

Equation:

$$58x = 58 \cdot x \text{(1st part of trip)}$$

$$52\left(5\tfrac{3}{4} - x\right) = 52 \cdot \left(5\tfrac{3}{4} - x\right)\text{(2nd part of trip)}$$

$$317 = 58x + 52\left(5\tfrac{3}{4} - x\right)$$

$$317 = 58x + 299 - 52x$$

$$18 = 6x$$

$$3 = x$$

The first part of the trip took 3 hours, and the second part took $5\frac{3}{4} - 3 = 2\frac{3}{4}$ hours.

61. (a) Your rate = $\frac{1}{5}$ job per hour

Friend's rate = $\frac{1}{8}$ job per hour

(b) *Verbal Model:*

[Word done] = [Work done by first person] + [Work done by second person]

Labels: Work done = 1

Rate for you = $\frac{1}{5}$

Time for you = t

Rate for friend = $\frac{1}{8}$

Time for friend = t

Equation:

$$1 = \left(\tfrac{1}{5}\right)(t) + \left(\tfrac{1}{8}\right)(t)$$

$$1 = \left(\tfrac{1}{5} + \tfrac{1}{8}\right)t$$

$$1 = \left(\tfrac{13}{40}\right)t$$

$$\frac{1}{13/40} = t$$

$$3\tfrac{1}{13} = \tfrac{40}{13} = t$$

It will take $3\frac{1}{13}$ hours.

63. *Verbal Model:*

[Word done] = [Work done by smaller pump] + [Work done by larger pump]

Labels: Work done = 1

Rate of smaller pump = $\frac{1}{30}$

Rate of larger pump = $\frac{1}{15}$

Time for each pump = t

Equation:

$$1 = \left(\tfrac{1}{30}\right)(t) + \left(\tfrac{1}{15}\right)(t)$$

$$1 = \left(\tfrac{1}{30} + \tfrac{1}{15}\right)t$$

$$1 = \tfrac{3}{30}t$$

$$\frac{1}{3/30} = t$$

$$10 = t$$

It will take 10 minutes.

65.

$$E = IR$$

$$\frac{E}{I} = R$$

67.

$$S = L - rL$$

$$S = L(1 - r)$$

$$\frac{S}{1 - r} = L$$

69. $h = 48t + \frac{1}{2}at^2$

$h - 48t = \frac{1}{2}at^2$

$2(h - 48t) = at^2$

$2h - 96t = at^2$

$\frac{2h - 96t}{t^2} = a$

71. $h = 36t + \frac{1}{2}at^2 + 50$

$h - 36t - 50 = \frac{1}{2}at^2$

$2(h - 36t - 50) = at^2$

$2h - 72t - 100 = at^2$

$\frac{2h - 72t - 100}{t^2} = a$

73. $S = 2\pi r^2 + 2\pi rh$

$S - 2\pi r^2 = 2\pi rh$

$\frac{(S - 2\pi r^2)}{2\pi r} = h$

$\frac{S}{2\pi r} - r = h$

75. *Common formula:* $V = \pi r^2 h$

Equation: $V = \pi\left(3\frac{1}{2}\right)^2 12$

$V = 147\pi$

$V \approx 461.8$ cubic centimeters

77. *Verbal Model:* Perimeter = 2 [Width] + 2 [Height]

Labels: Perimeter = 3

Height = x

Width = $0.62x$

Equation: $3 = 2(0.62x) + 2(x)$

$3 = 1.24x + 2x$

$3 = 3.24x$

$0.926 \approx x$

The height is about 0.926 foot.

79. *Verbal Model:* [Perimeter] = [Side] + [Side] + [Side]

Equation: $129 = x + x + x$

$129 = 3x$

$43 = x$

$x = 43$ centimeters

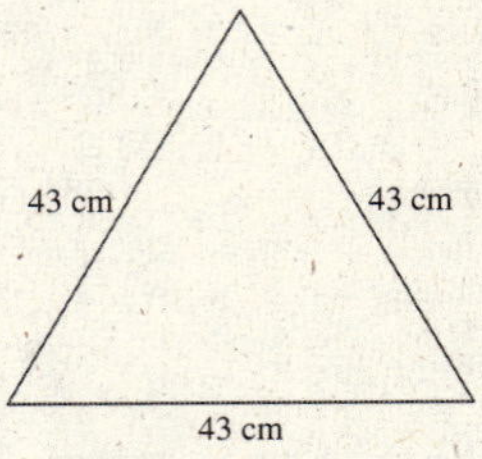

81. *Verbal Model:*

[Fahrenheit temperature] = $\frac{9}{3}$ [Celsius temperature] + 32

Labels: Fahrenheit temperature = 86

Celsius temperature = C

Equation: $86 = \frac{9}{5}C + 32$

$54 = \frac{9}{5}C$

$30 = C$

The daily high temperature is 30°C.

83. *Verbal Model:* [Interest] = [Principal] · [Rate] · [Time]

Labels: Interest = I

Principal = 5000

Rate = 6.5%

Time = 6

Equation: $I = (5000)(0.065)(6)$

$I = 1950$

The interest is $1950.

85. *Verbal Model:*

[Interest] = [Principal] · [Rate] · [Time]

Labels: Interest = 500

Principal = P

Rate = 7%

Time = 2

Equation: $500 = (P)(0.07)(2)$

$500 = P(0.14)$

$\frac{500}{0.14} = P$

$3571.43 \approx P$

The principal required is $3571.43.

87. (a) $y = 0.264t + 11.49, 0 \le t \le 5$

From the graph, the average hourly rate was \$11.96 in 2002.

$$\begin{aligned} 11.96 &= 0.264t + 11.49 \\ 0.47 &= 0.264t \\ 1.78 &\approx t \\ 2 &\approx t \end{aligned}$$

Yes, the result is the same.

(b) The average annual hourly raise for custodians is \$0.26. Determine the average hourly wage for each year. The difference between each two consecutive years is \$0.26.

89. The sale price of an item is the list price minus the discount rate times the list price.

91. No, it quadruples. The area of a square of side s is s^2. If the length of the sides is $2s$, the area is $(2s)^2 = 4s^2$.

93. (a) $21 + (-21) = 0$, so it is -21.

(b) $21\left(\frac{1}{21}\right) = 1$, so it is $\frac{1}{21}$.

95. (a) $-5x + 5x = 0$, so it is $5x$.

(b) $-5x\left(-\frac{1}{5x}\right) = 0$, so it is $-\frac{1}{5x}$.

97. $2x(x - 4) + 3 = 2x^2 - 8x + 3$

99. $x^2(x - 4) - 2x^2 = x^3 - 4x^2 - 2x^2 = x^3 - 6x^2$

101.
$$\begin{aligned} 52 &= 0.40x \\ 130 &= x \end{aligned}$$

103.
$$\begin{aligned} 117 &= p(900) \\ 13\% &= p \end{aligned}$$

Mid-Chapter Quiz for Chapter 2

1.
$$\begin{aligned} 4x - 8 &= 0 \\ 4x - 8 + 8 &= 0 + 8 \\ 4x &= 8 \\ \frac{4x}{4} &= \frac{8}{4} \\ x &= 2 \end{aligned}$$

Check:

$$\begin{aligned} 4(2) - 8 &\stackrel{?}{=} 0 \\ 8 - 8 &= 0 \\ 0 &= 0 \end{aligned}$$

2.
$$\begin{aligned} -3(z - 2) &= 0 \\ \frac{-3(z - 2)}{-3} &= \frac{0}{-3} \\ z - 2 &= 0 \\ z - 2 + 2 &= 0 + 2 \\ z &= 2 \end{aligned}$$

Check:

$$\begin{aligned} -3(2 - 2) &\stackrel{?}{=} 0 \\ -3(0) &\stackrel{?}{=} 0 \\ 0 &= 0 \end{aligned}$$

3.
$$\begin{aligned} 2(y + 3) &= 18 - 4y \\ 2y + 6 &= 18 - 4y \\ 2y + 4y + 6 &= 18 - 4y + 4y \\ 6y + 6 - 6 &= 18 - 6 \\ 6y &= 12 \\ \frac{6y}{6} &= \frac{12}{6} \\ y &= 2 \end{aligned}$$

Check:

$$\begin{aligned} 2(2 + 3) &\stackrel{?}{=} 18 - 4(2) \\ 2(5) &\stackrel{?}{=} 18 - 8 \\ 10 &= 10 \end{aligned}$$

4.
$$\begin{aligned} 5t + 7 &= 7(t + 1) - 2t \\ 5t + 7 &= 7t + 7 - 2t \\ 5t + 7 &= 5t + 7 \end{aligned}$$

Identity

5.
$$\frac{1}{4}x + 6 = \frac{3}{2}x - 1$$
$$4\left(\frac{1}{4}x + 6\right) = 4\left(\frac{3}{2}x - 1\right)$$
$$x + 24 = 6x - 4$$
$$x - x + 24 = 6x - 4 - x$$
$$24 = 5x - 4$$
$$24 + 4 = 5x - 4 + 4$$
$$28 = 5x$$
$$\frac{28}{5} = \frac{5x}{5}$$
$$\frac{28}{5} = x$$

Check:
$$\frac{1}{4}\left(\frac{28}{5}\right) + 6 \stackrel{?}{=} \frac{3}{2}\left(\frac{28}{5}\right) - 1$$
$$\frac{7}{5} + \frac{30}{5} \stackrel{?}{=} \frac{42}{5} - \frac{5}{5}$$
$$\frac{37}{5} = \frac{37}{5}$$

6.
$$\frac{2b}{5} + \frac{b}{2} = 3$$
$$\frac{4b}{10} + \frac{5b}{10} = 3$$
$$\frac{9b}{10} = 3$$
$$9b = 30$$
$$b = \frac{30}{9} = \frac{10}{3}$$

Check:
$$\frac{2\left(\frac{10}{3}\right)}{5} + \frac{\left(\frac{10}{3}\right)}{2} \stackrel{?}{=} 3$$
$$\frac{\left(\frac{20}{3}\right)}{5} + \frac{\left(\frac{10}{3}\right)}{2} \stackrel{?}{=} 3$$
$$\frac{20}{15} + \frac{10}{6} \stackrel{?}{=} 3$$
$$\frac{4}{3} + \frac{5}{3} \stackrel{?}{=} 3$$
$$\frac{9}{3} \stackrel{?}{=} 3$$
$$3 = 3$$

7.
$$\frac{4 - x}{5} + 5 = \frac{5}{2}$$
$$10\left(\frac{4 - x}{5} + 5\right) = 10\left(\frac{5}{2}\right)$$
$$2(4 - x) + 50 = 25$$
$$8 - 2x + 50 = 25$$
$$-2x + 58 = 25$$
$$-2x + 58 - 58 = 25 - 58$$
$$-2x = -33$$
$$\frac{-2x}{-2} = \frac{-33}{-2}$$
$$x = \frac{33}{2}$$

Check:
$$\frac{4 - \frac{33}{2}}{5} + 5 \stackrel{?}{=} \frac{5}{2}$$
$$\frac{\frac{8}{2} - \frac{33}{2}}{5} + 5 \stackrel{?}{=} \frac{5}{2}$$
$$-\frac{25}{2} \cdot \frac{1}{5} + 5 \stackrel{?}{=} \frac{5}{2}$$
$$-\frac{5}{2} + \frac{10}{2} \stackrel{?}{=} \frac{5}{2}$$
$$\frac{5}{2} = \frac{5}{2}$$

8.
$$3x + \frac{11}{12} = \frac{5}{16}$$
$$3x + \frac{11}{12} - \frac{11}{12} = \frac{5}{16} - \frac{11}{12}$$
$$3x = \frac{15}{48} - \frac{44}{48}$$
$$3x = -\frac{29}{48}$$
$$\frac{3x}{3} = -\frac{29}{48} \div 3$$
$$x = -\frac{29}{48} \cdot \frac{1}{3}$$
$$x = -\frac{29}{144}$$

Check:
$$3\left(-\frac{29}{144}\right) + \frac{11}{12} \stackrel{?}{=} \frac{5}{16}$$
$$-\frac{29}{48} + \frac{44}{48} \stackrel{?}{=} \frac{5}{16}$$
$$\frac{15}{48} \stackrel{?}{=} \frac{5}{16}$$
$$\frac{5}{16} = \frac{5}{16}$$

9.
$$\begin{aligned} 0.25x + 6.2 &= 4.45x + 3.9 \\ 0.25x - 0.25x + 6.2 &= 4.45x + 3.9 - 0.25x \\ 6.2 &= 4.2x + 3.9 \\ 6.2 - 3.9 &= 4.2x - 3.9 \\ 2.3 &= 4.2x \\ \frac{2.3}{4.2} &= \frac{4.2x}{4.2} \\ 0.55 &\approx x \end{aligned}$$

10.
$$\begin{aligned} 0.42x + 6 &= 5.25x - 0.80 \\ 0.42x + 6 - 5.25x &= 5.25x - 0.80 - 5.25x \\ -4.83x + 6 &= -0.80 \\ -4.83x + 6 - 6 &= -0.80 - 6 \\ -4.83x &= -6.80 \\ \frac{-4.83x}{-4.83} &= \frac{-6.80}{-4.83} \\ x &\approx 1.41 \end{aligned}$$

11. 0.45 is 45 hundredths, so $0.45 = \frac{45}{100}$ which reduces to $\frac{9}{20}$ and because percent means hundredths, $0.45 = 45\%$.

12. *Verbal Model:* [Compared number] = [Percent] · [Base number]

Labels: Compared number $= a$

Percent $= p$

Base number $= b$

Equation:
$$\begin{aligned} a &= p \cdot b \\ 500 &= (2.50)(b) \\ \frac{500}{2.50} &= b \\ 200 &= b \end{aligned}$$

500 is 250% of 200.

13. $\text{Unit price} = \dfrac{\text{Total price}}{\text{Total units}} = \dfrac{\$4.85}{12 \text{ ounces}} = \$0.40 \text{ per ounce}$

14. *Verbal Model:* $\dfrac{\text{Number defective}}{\text{Sample}} = \dfrac{\text{Total defective}}{\text{Shipment}}$

Labels: Number defective $= 1$

Sample $= 150$

Total defective $= x$

Shipment $= 750{,}000$

Equation:
$$\begin{aligned} \frac{1}{150} &= \frac{x}{750{,}000} \\ 750{,}000 &= 150x \\ 5000 &= x \end{aligned}$$

The expected number of defective units is 5000.

15. Store computer:

Verbal Model: [Discount] = [Discount rate] · [List price]

Labels: Discount $= x$

Discount rate $= 0.25$

List price $= 1080$

Equation:
$$\begin{aligned} x &= (0.25)(1080) \\ x &= 270.00 \end{aligned}$$

Verbal Model: [Selling price] = [List price] − [Discount]

Labels: Selling price $= x$

List price $= 1080$

Discount $= 270$

Equation:
$$\begin{aligned} x &= 1080 - 270 \\ x &= 810 \end{aligned}$$

Mail-order catalog computer:

Verbal Model: [Selling price] = [List price] + [Shipping]

Labels: Selling price $= x$

List price $= 799$

Shipping $= 14.95$

Equation:
$$\begin{aligned} x &= 799 + 14.95 \\ x &= 813.95 \end{aligned}$$

The store computer is the better buy.

16. *Verbal Model:* $\boxed{\text{Total wages}} = \boxed{\text{Regular wages}} + \boxed{\text{Overtime wages}}$

Labels: Total wages $= 616$

Regular wages $= 40(12.25)$

Overtime wages $= x(18)$

Number of hours $= x$

Equation:

$$616 = 40(12.25) + x(18)$$
$$616 = 490 + 18x$$
$$126 = 18x$$
$$7 = x$$

You worked 7 hours of overtime.

17. *Verbal Model:*

$\boxed{\text{Amount of solution 1}} + \boxed{\text{Amount of solution 2}} = \boxed{\text{Amount of final solution}}$

Labels: Percent of solution 1 $= 25\%$

Gallons of solution 1 $= x$

Percent of solution 2 $= 50\%$

Gallons of solution 2 $= 50 - x$

Percent of final solution $= 30\%$

Gallons of final solution $= 50$

Equation:

$$0.25x + 0.50(50 - x) = 0.30(50)$$
$$0.25x + 25 - 0.50x = 15$$
$$25 - 0.25x = 15$$
$$-0.25x = -10$$
$$x = 40$$
$$50 - x = 10$$

40 gallons of solution 1 and 10 gallons of solution 2 are required.

18. *Verbal Model:* $\boxed{\text{Distance}} = \boxed{\text{Rate}} \cdot \boxed{\text{Time}}$

Labels: Distance $= 300$

Rate of first part $= 62$

Time for first part $= x$

Rate of second part $= 46$

Time for seond part $= 6 - x$

Equation:

$$300 = 62x + 46(6 - x)$$
$$300 = 62x + 276 - 46x$$
$$24 = 16x$$

1.5 hours $= x$ (first part of trip at 62 miles/hour)

4.5 hours $= 6 - x$ (second part of trip at 46 miles/hour)

19. *Verbal Model:* $\boxed{\text{Work done}} = \boxed{\text{Part time by you}} + \boxed{\text{Part done by friend}}$

Labels: Work done $= 1$

Time for each portion $= t$

Per hour work rate for you $= \frac{1}{3}$

Per hour work rate for friend $= \frac{1}{5}$

Equation:

$$1 = \left(\tfrac{1}{3}\right)t + \left(\tfrac{1}{5}\right)t$$
$$1 = \tfrac{8}{15}t$$
$$\tfrac{15}{8} = t$$

It will take $\frac{15}{8}$ hours, or 1.875 hours.

20. Perimeter of square I $= 20$

$$4s = 20$$
$$s = 5$$

Perimeter of square II $= 32$

$$4s = 32$$
$$s = 8$$

Length of side of square III $= 5 + 8 = 13$

Area $= s^2 = 13^2 = 169$

The area of square III is 169 square inches.

Section 2.4 Linear Inequalities

1. (a) $7(3) - 10 > 0$
$$21 - 10 > 0$$
$$11 > 0$$
Yes

(b) $7(-2) - 10 > 0$
$$-14 - 10 > 0$$
$$-24 > 0$$
No

(c) $7\left(\frac{5}{2}\right) - 10 > 0$
$$\tfrac{35}{2} - 10 > 0$$
$$\tfrac{35}{2} - \tfrac{20}{2} > 0$$
$$\tfrac{15}{2} > 0$$
Yes

(d) $7\left(\frac{1}{2}\right) - 10 > 0$
$$\tfrac{7}{2} - 10 > 0$$
$$\tfrac{7}{2} - \tfrac{20}{2} > 0$$
$$-\tfrac{13}{2} > 0$$
No

3. (a) $0 < \dfrac{x+4}{5} < 2$

$0 < \dfrac{10+4}{5} < 2$

$0 < \dfrac{14}{5} < 2$

No

(b) $0 < \dfrac{x+4}{5} < 2$

$0 < \dfrac{-4+4}{5} < 2$

$0 < 0 < 2$

No

(c) $0 < \dfrac{x+4}{5} < 2$

$0 < \dfrac{0+4}{5} < 2$

$0 < \dfrac{4}{5} < 2$

Yes

(d) $0 < \dfrac{x+4}{5} < 2$

$0 < \dfrac{6+4}{5} < 2$

$0 < 2 < 2$

No

5. Matches graph (a).

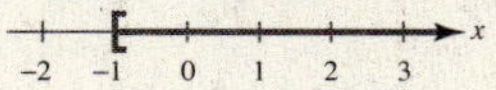

7. Matches graph (d).

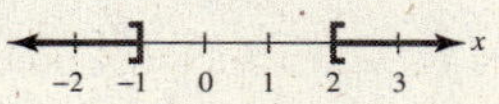

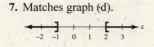

9. Matches graph (f).

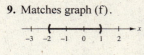

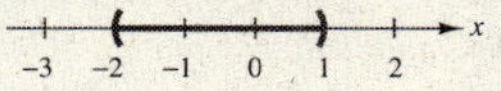

11.

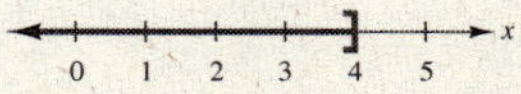

13. $x > 3.5$

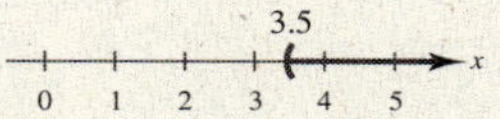

15. $-5 < x \le 3$

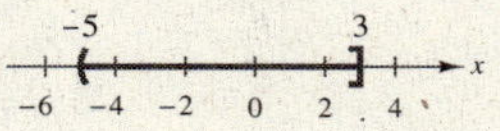

17. $4 > x \ge 1$

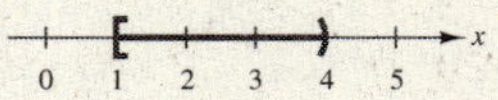

19. $\frac{3}{2} \ge x > 0$

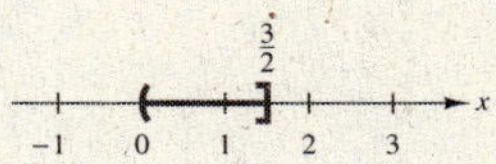

21. $x < -5$ or $x \ge -1$

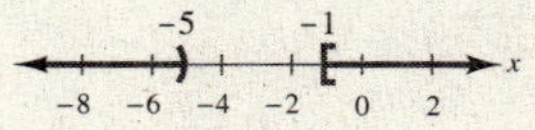

23. $x \le 3$ or $x > 7$

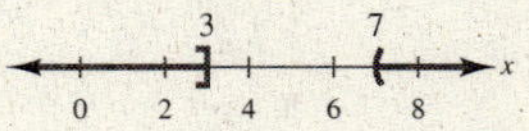

25.
$$5 - \tfrac{1}{3}x > 8$$
$$-3\left(5 - \tfrac{1}{3}x\right) < (8) - 3$$
$$-15 + x < -24$$

27.
$$3x - 2 < 12$$
$$3x - 2 + 2 < 12 + 2$$
$$3x < 14$$
$$\frac{3x}{3} < \frac{14}{3}$$
$$x < \frac{14}{3}$$

$$3x < 10$$
$$\frac{3x}{3} < \frac{10}{3}$$
$$x < \frac{10}{3}$$

The inequalities are not equivalent.

29.
$$-5(x+12) > 25$$
$$\frac{-5(x+12)}{-5} < \frac{25}{-5}$$
$$x + 12 < -5$$
$$x + 12 - 12 < -5 - 12$$
$$x < -17$$

$$x + 12 > -5$$
$$x + 12 - 12 > -5 - 12$$
$$x > -17$$

The inequalities are not equivalent.

31.
$$7x - 6 \le 3x + 12$$
$$7x - 6 + 6 \le 3x + 12 + 6$$
$$7x \le 3x + 18$$
$$7x - 3x \le 3x + 18 - 3x$$
$$4x \le 18$$
$$\frac{4x}{4} \le \frac{18}{4}$$
$$x \le \frac{9}{2}$$

$$4x \le 18$$
$$\frac{4x}{4} \le \frac{18}{4}$$
$$x \le \frac{9}{2}$$

The inequalities are equivalent.

33.
$$3x > 5x$$
$$3x - 3x > 5x - 3x$$
$$0 > 2x$$
$$\frac{0}{2} > \frac{2x}{2}$$
$$0 > x$$

$$3 > 5$$

The inequalities are not equivalent.

35.
$$x - 4 \ge 0$$
$$x - 4 + 4 \ge 0 + 4$$
$$x \ge 4$$

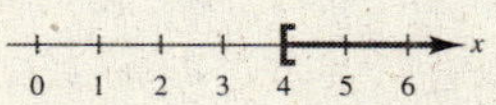

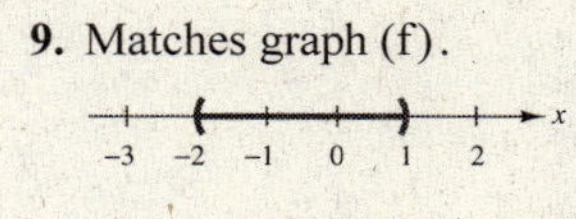

37. $x + 7 \le 9$

$x + 7 - 7 \le 9 - 7$

$x \le 2$

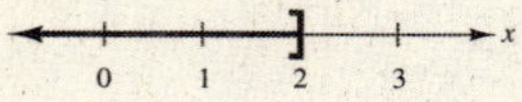

39. $2x < 8$

$\frac{2x}{2} < \frac{8}{2}$

$x < 4$

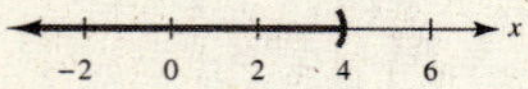

41. $-9x \ge 36$

$\frac{-9x}{-9} \le \frac{36}{-9}$

$x \le -4$

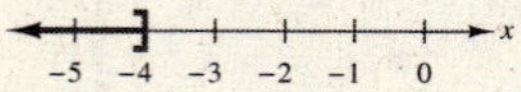

43. $-\frac{3}{4}x < -6$

$-\frac{4}{3} \cdot -\frac{3}{4}x > -6 \cdot -\frac{4}{3}$

$x > 8$

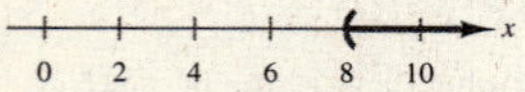

45. $5 - x \le -2$

$5 - x - 5 \le -2 - 5$

$-x \le -7$

$-1 \cdot x \ge -7 \cdot -1$

$x \ge 7$

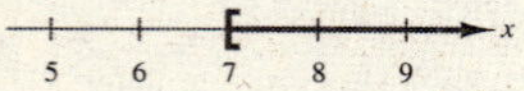

47. $2x - 5.3 > 9.8$

$2x - 5.3 + 5.3 > 9.8 + 5.3$

$2x > 15.1$

$\frac{2x}{2} > \frac{15.1}{2}$

$x > 7.55$

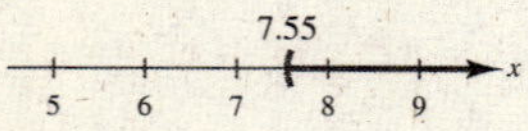

49. $5 - 3x < 7$

$5 - 3x - 5 < 7 - 5$

$-3x < 2$

$\frac{-3x}{-3} > \frac{2}{-3}$

$x > -\frac{2}{3}$

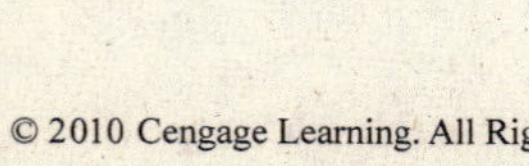

51. $3x - 11 > -x + 7$

$3x - 11 + x > -x + 7 + x$

$4x - 11 > 7$

$4x - 11 + 11 > 7 + 11$

$4x > 18$

$\frac{4x}{4} > \frac{18}{4}$

$x > \frac{9}{2}$

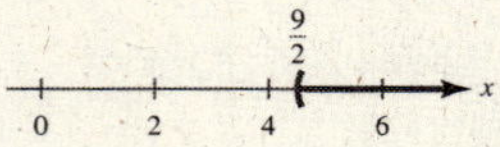

53. $-3x + 7 < 8x - 13$

$-3x - 8x + 7 < 8x - 8x - 13$

$-11x + 7 < -13$

$-11x + 7 - 7 < -13 - 7$

$-11x < -20$

$\frac{-11x}{-11} > \frac{-20}{-11}$

$x > \frac{20}{11}$

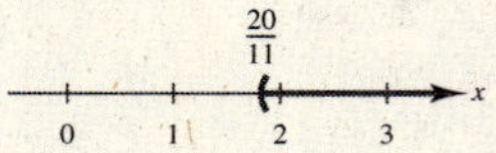

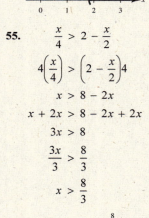

55. $\frac{x}{4} > 2 - \frac{x}{2}$

$4\left(\frac{x}{4}\right) > \left(2 - \frac{x}{2}\right)4$

$x > 8 - 2x$

$x + 2x > 8 - 2x + 2x$

$3x > 8$

$\frac{3x}{3} > \frac{8}{3}$

$x > \frac{8}{3}$

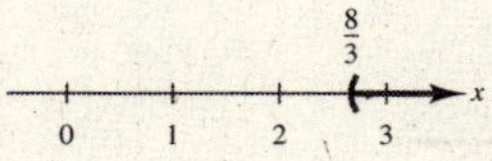

57.
$$\begin{aligned}
\frac{x-4}{3} + 3 &\le \frac{x}{8}\\
24\left(\frac{x-4}{3} + 3\right) &\le \left(\frac{x}{8}\right)24\\
8(x-4) + 72 &\le 3x\\
8x - 32 + 72 &\le 3x\\
8x + 40 &\le 3x\\
8x - 3x + 40 &\le 3x - 3x\\
5x + 40 &\le 0\\
5x + 40 - 40 &\le 0 - 40\\
5x &\le -40\\
\frac{5x}{5} &\le -\frac{40}{5}\\
x &\le -8
\end{aligned}$$

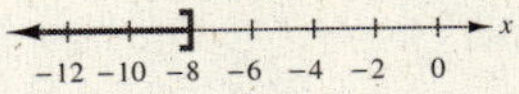

59.
$$\begin{aligned}
\frac{3x}{5} - 4 &< \frac{2x}{3} - 3\\
15\left(\frac{3x}{5} - 4\right) &< \left(\frac{2x}{3} - 3\right)15\\
9x - 60 &< 10x - 45\\
9x - 10x - 60 &< 10x - 45 - 10x\\
-x - 60 &< -45\\
-x - 60 + 60 &< -45 + 60\\
-x &< 15\\
x &> -15
\end{aligned}$$

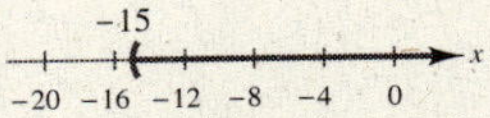

61.
$$\begin{aligned}
0 < 2x &- 5 < 9\\
0 + 5 < 2x &- 5 + 5 < 9 + 5\\
5 &< 2x < 14\\
\frac{5}{2} &< \frac{2x}{2} < \frac{14}{2}\\
\frac{5}{2} &< x < 7
\end{aligned}$$

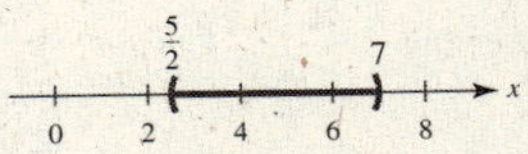

63.
$$\begin{aligned}
8 &< 6 - 2x \le 12\\
8 - 6 &< 6 - 6 - 2x \le 12 - 6\\
2 &< -2x \le 6\\
\frac{2}{-2} &> \frac{-2x}{-2} \ge \frac{6}{-2}\\
-1 &> x \ge -3\\
-3 &\le x < -1
\end{aligned}$$

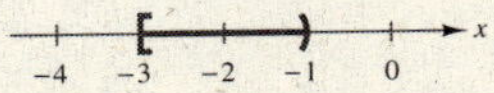

65.
$$\begin{aligned}
-1 &< -0.2x < 1\\
\frac{-1}{-0.2} &> \frac{-0.2x}{-0.2} > \frac{1}{-0.2}\\
5 &> x > -5\\
-5 &< x < 5
\end{aligned}$$

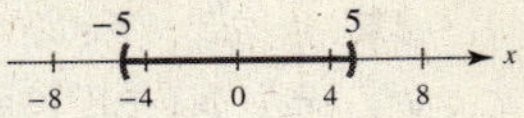

67.
$$\begin{aligned}
-3 &< \frac{2x-3}{2} < 3\\
-6 &< 2x - 3 < 6\\
-6 + 3 &< 2x - 3 + 3 < 6 + 3\\
-3 &< 2x < 9\\
\frac{-3}{2} &< \frac{2x}{2} < \frac{9}{2}\\
-\frac{3}{2} &< x < \frac{9}{2}
\end{aligned}$$

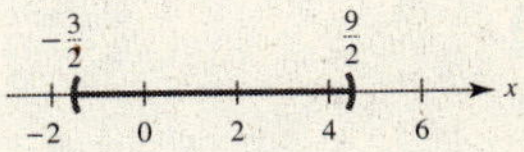

69.
$$\begin{aligned}
1 &> \frac{x-4}{-3} > -2\\
-3 &< x - 4 < 6\\
-3 + 4 &< x - 4 + 4 < 6 + 4\\
1 &< x < 10
\end{aligned}$$

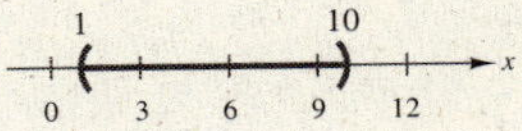

71.
$$\begin{aligned}
2x - 4 &\le 4 & &\text{and} & 2x + 8 &> 6\\
2x - 4 + 4 &\le 4 + 4 & &\text{and} & 2x + 8 - 8 &> 6 - 8\\
2x &\le 8 & &\text{and} & 2x &> -2\\
\frac{2x}{2} &\le \frac{8}{2} & &\text{and} & \frac{2x}{2} &> \frac{-2}{2}\\
x &\le 4 & &\text{and} & x &> -1
\end{aligned}$$

$-1 < x \le 4$

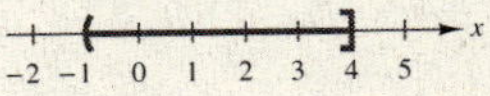

73.
$$8 - 3x > 5 \quad \text{and} \quad x - 5 \ge 10$$
$$8 - 3x - 8 > 5 - 8 \qquad x - 5 + 5 \ge 10 + 5$$
$$-3x > -3 \qquad x \ge 15$$
$$\frac{-3x}{-3} < \frac{-3}{-3}$$
$$x < 1$$

There is no solution.

75.
$$7.2 - 1.1x > 1 \quad \text{or} \quad 1.2x - 4 > 2.7$$
$$7.2 - 1.1x - 7.2 > 1 - 7.2 \qquad 1.2x - 4 + 4 > 2.7 + 4$$
$$-1.1x > -6.2 \qquad 1.2x > 6.7$$
$$\frac{-1.1x}{-1.1} < \frac{-6.2}{-1.1} \qquad \frac{1.2x}{1.2} > \frac{6.7}{1.2}$$
$$x < \frac{62}{11} \qquad x > \frac{67}{12}$$
$$-\infty < x < \infty$$

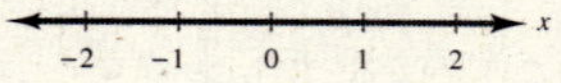

77.
$$7x + 11 < 3 + 4x \quad \text{or} \quad \frac{5}{2}x - 1 \ge 9 - \frac{3}{2}x$$
$$7x - 4x + 11 < 3 + 4x - 4x \qquad \frac{5}{2}x + \frac{3}{2}x - 1 \ge 9 - \frac{3}{2}x + \frac{3}{2}x$$
$$3x + 11 < 3 \qquad 4x - 1 \ge 9$$
$$3x + 11 - 11 < 3 - 11 \qquad 4x - 1 + 1 \ge 9 + 1$$
$$3x < -8 \qquad 4x \ge 10$$
$$\frac{3x}{3} < -\frac{8}{3} \qquad \frac{4x}{4} \ge \frac{10}{4}$$
$$x < -\frac{8}{3} \qquad x \ge \frac{5}{2}$$

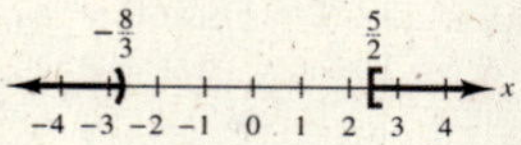

79.
$$-3(y + 10) \ge 4(y + 10)$$
$$-3y - 30 \ge 4y + 40$$
$$3y - 3y - 30 \ge 3y + 4y + 40$$
$$-30 \ge 7y + 40$$
$$-40 - 30 \ge 7y + 40 - 40$$
$$-70 \ge 7y$$
$$-10 \ge y$$

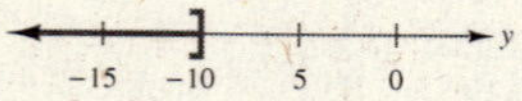

81.
$$-4 \le 2 - 3(x + 2) < 11$$
$$-4 \le 2 - 3x - 6 < 11$$
$$-4 \le -4 - 3x < 11$$
$$-4 + 4 \le -4 - 3x + 4 < 11 + 4$$
$$0 \le -3x < 15$$
$$\frac{0}{-3} \ge \frac{-3x}{-3} > \frac{15}{-3}$$
$$0 \ge x > -5$$

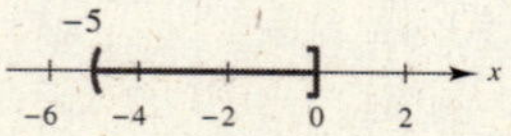

83. $x < -3$ or $x \ge 2$
$\{x|x < -3\} \cup \{x|x \ge 2\}$

85. $-5 \le x < 4$
$\{x|x \ge -5\} \cap \{x|x < 4\}$

87. $x \le -2.5$ or $x \ge -0.5$
$\{x|x \le -2.5\} \cup \{x|x \ge -0.5\}$

89. $\{x|x \ge -7\} \cap \{x|x < 0\}$

91. $\{x|x > -\frac{9}{2}\} \cap \{x|x \le -\frac{3}{2}\}$

93. $\{x|x < 0\} \cup \{x|x \ge \frac{2}{3}\}$

95. $x \ge 0$

97. $z \ge 8$

99. $10 \le n \le 16$

101. x is at least $\frac{5}{2}$.

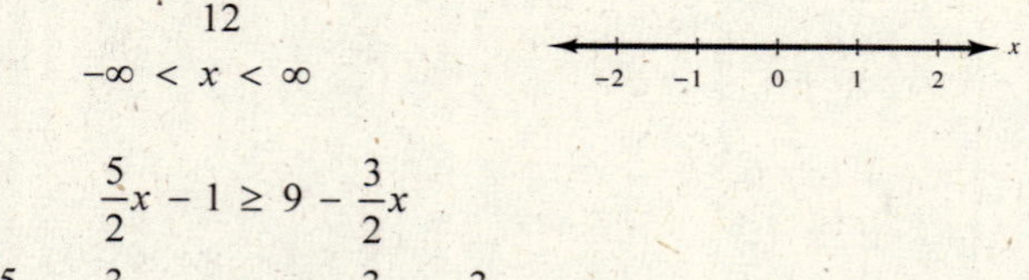

103. y is at least 3 and less than 5.

105. *Verbal Model:*

$$\boxed{\text{Transportation costs}} + \boxed{\text{Other costs}} \le \boxed{\text{Total money for trip}}$$

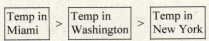

Labels: Transportation costs $= 1900$

Other costs $= C$

Total money $= 4500$

Inequality:

$$1900 + C \le 4500$$
$$1900 + C - 1900 \le 4500 - 1900$$
$$C \le 2600$$

C must be no more than \$2600.

107. *Verbal Model:*

$$\boxed{\text{Temp in Miami}} > \boxed{\text{Temp in Washington}} > \boxed{\text{Temp in New York}}$$

The average temperature in Miami, therefore, is greater than (>) the average temperature in New York.

109. *Verbal Model:* $\boxed{\text{Operating cost}} < \boxed{\$12{,}000}$

Label: Operating cost $= 0.35m + 2900$

Inequality:

$$0.35m + 2900 < 12{,}000$$
$$0.35m + 2900 - 2900 < 12{,}000 - 2900$$
$$0.35m < 9100$$
$$\frac{0.35m}{0.35} < \frac{9100}{0.35}$$
$$m < 26{,}000$$

The maximum number of miles is 26,000.

111. *Verbal Model:* $\boxed{\text{Revenue}} > \boxed{\text{Cost}}$

Labels: Revenue $= 89.95x$

Cost $= 61x + 875$

Inequality:

$$89.95x > 61x + 875$$
$$89.95x - 61x > 61x + 875 - 61x$$
$$28.95x > 875$$
$$\frac{28.95x}{28.95} > \frac{875}{28.95}$$
$$x > 30.224525$$
$$x \ge 31$$

113. *Verbal Model:* $90 \le \boxed{\text{Perimeter}} \le \boxed{120}$

Label: Perimeter $= 2(x + 22)$

Inequality:

$$90 \le 2(x + 22) \le 120$$
$$\frac{90}{2} \le \frac{2(x + 22)}{2} \le \frac{120}{2}$$
$$45 \le x + 22 \le 60$$
$$45 - 22 \le x + 22 - 22 \le 60 - 22$$
$$23 \le x \le 38$$

115.

$$12 \le 4n \le 30$$
$$\frac{12}{4} \le \frac{4n}{4} \le \frac{30}{4}$$
$$3 \le n \le \frac{15}{2}$$

117. *Verbal Model:* $\boxed{\text{Second plan}} > \boxed{\text{First plan}}$

Labels: First plan: \$12.50 per hour

Second plan: \$8 + \$0.75n per hour where n represents the number of units produced.

Inequality:

$$8 + 0.75n > 12.5$$
$$0.75n > 4.5$$
$$n > 6$$

If more than 6 units are produced per hour, the second payment plan yields the greater hourly wage.

119.

$$100 > 21.8t - 160$$
$$100 + 160 > 21.8t - 160 + 160$$
$$260 > 21.8t$$
$$\frac{260}{21.8} > \frac{21.8t}{21.8}$$
$$\frac{260}{21.8} > t$$

The consumption of energy produced by wind was less than 100 trillion Btu in 1999, 2000, and 2001.

121. The multiplication and division properties differ. The inequality symbol is reversed if both sides of the inequality are multiplied or divided by a negative real number.

123. The solution set of a linear inequality is a bounded interval if all x-values are contained by two endpoints of its graph. The solution set of a linear inequality is an unbounded interval if it is not bounded.

125. $a < x < b$; A double inequality is always bounded.

127. $x > a$ or $x < b$; The solution set includes all values on the real number line.

129. $|4| \stackrel{?}{=} |-5|$

$4 \leq 5$

131. $|-7| \stackrel{?}{=} |7|$

$7 = 7$

133. $3(6) \stackrel{?}{=} 27$ $\qquad$ $3(9) \stackrel{?}{=} 27$

$18 \neq 27$ $\qquad$ $27 = 27$

Not a solution $\qquad$ Solution

135. $7(2) - 5 \stackrel{?}{=} 7 + 2$ $\qquad$ $7(6) - 5 \stackrel{?}{=} 7 + 6$

$14 - 5 \stackrel{?}{=} 9$ $\qquad$ $42 - 5 \stackrel{?}{=} 14$

$9 = 9$ $\qquad$ $37 \neq 14$

Solution $\qquad$ Not a solution

137.
$$\begin{aligned} 2x - 17 &= 0 \\ 2x - 17 + 17 &= 0 + 17 \\ 2x &= 17 \\ \frac{2x}{2} &= \frac{17}{2} \\ x &= \frac{17}{2} \end{aligned}$$

139.
$$\begin{aligned} 32x &= -8 \\ \frac{32x}{32} &= \frac{-8}{32} \\ x &= -\frac{1}{4} \end{aligned}$$

Section 2.5 Absolute Value Equations and Inequalities

1. $|4x + 5| = 10, x = -3$

$$\begin{aligned} |4(-3) + 5| &\stackrel{?}{=} 10 \\ |-12 + 5| &\stackrel{?}{=} 10 \\ |-7| &\stackrel{?}{=} 10 \\ 7 &\neq 10 \end{aligned}$$

Not a solution

3. $|6 - 2w| = 2, w = 4$

$$\begin{aligned} |6 - 2(4)| &\stackrel{?}{=} 2 \\ |6 - 8| &\stackrel{?}{=} 2 \\ |-2| &\stackrel{?}{=} 2 \\ 2 &= 2 \end{aligned}$$

Solution

5. $x - 10 = 17$ or $x - 10 = -17$

7. $4x + 1 = \frac{1}{2}$ or $4x + 1 = -\frac{1}{2}$

9.
$$\begin{aligned} |3x| + 7 &= 8 \\ |3x| + 7 - 7 &= 8 - 7 \\ |3x| &= 1 \end{aligned}$$

11.
$$\begin{aligned} 3|2x| - 1 &= 5 \\ 3|2x| - 1 + 1 &= 5 + 1 \\ 3|2x| &= 6 \\ \frac{3|2x|}{3} &= \frac{6}{3} \\ |2x| &= 2 \end{aligned}$$

13. $|x| = 4$

$x = 4$ or $x = -4$

15. $|t| = -45$

No solution

17. $|h| = 0$

$h = 0$

19. $|5x| = 15$

$5x = 15$ $\quad$ or $\quad$ $5x = -15$

$x = 3$ $\qquad$ $x = -3$

21. $|x + 1| = 5$

$x + 1 = 5$ $\quad$ or $\quad$ $x + 1 = -5$

$x = 4$ $\qquad$ $x = -6$

23. $\left|\frac{2s+3}{5}\right| = 5$

$\frac{2s+3}{5} = 5$ or $\frac{2s+3}{5} = -5$

$2s + 3 = 25$ $\quad$ $2s + 3 = -25$

$2s = 22$ $\quad$ $2s = -28$

$s = 11$ $\quad$ $s = -14$

25. $|4 - 3x| = 0$

$4 - 3x = 0$

$-3x = -4$

$x = \frac{4}{3}$

27. $|5x - 3| + 8 = 22$

$|5x - 3| = 14$

$5x - 3 = 14$ or $5x - 3 = -14$

$5x = 17$ $\quad$ $5x = -11$

$x = \frac{17}{5}$ $\quad$ $x = -\frac{11}{5}$

29. $\left|\frac{x-2}{3}\right| + 6 = 6$

$\left|\frac{x-2}{3}\right| = 0$

$\frac{x-2}{3} = 0$

$x - 2 = 0$

$x = 2$

31. $-2|7 - 4x| = -16$

$|7 - 4x| = 8$

$7 - 4x = 8$ or $7 - 4x = -8$

$-4x = 1$ $\quad$ $-4x = -15$

$x = -\frac{1}{4}$ $\quad$ $x = \frac{15}{4}$

33. $3|2x - 5| + 4 = 7$

$3|2x - 5| = 3$

$|2x - 5| = 1$

$2x - 5 = 1$ or $2x - 5 = -1$

$2x = 6$ $\quad$ $2x = 4$

$x = 3$ $\quad$ $x = 2$

35. $|x + 8| = |2x + 1|$

$x + 8 = 2x + 1$ or $x + 8 = -(2x + 1)$

$8 = x + 1$ $\quad$ $x + 8 = -2x - 1$

$7 = x$ $\quad$ $3x + 8 = -1$

$3x = -9$

$x = -3$

37. $|3x + 1| = |3x - 3|$

$3x + 1 = 3x - 3$ or $3x + 1 = -(3x - 3)$

$1 \neq -3$ $\quad$ $3x + 1 = -3x + 3$

$6x + 1 = 3$

$6x = 2$

$x = \frac{1}{3}$

The only solution is $x = \frac{1}{3}$.

39. $|4x - 10| = 2|2x + 3|$

$4x - 10 = 2(2x + 3)$ or $4x - 10 = -2(2x + 3)$

$4x - 10 = 4x + 6$ $\quad$ $4x - 10 = -4x - 6$

$-10 \neq 6$ $\quad$ $8x - 10 = -6$

$8x = 4$

$x = \frac{4}{8} = \frac{1}{2}$

The only solution is $x = \frac{1}{2}$.

41. $|x - 4| = 9$

43. $x = 2$

$|2| < 3$

$2 < 3$

Solution

45. $x = 9$

$|9 - 7| \geq 3$

$|2| \geq 3$

$2 \geq 3$

Not a solution

47. $|y + 5| < 3$

$-3 < y + 5 < 3$

49. $|7 - 2h| \geq 9$

$7 - 2h \geq 9$ or $7 - 2h \leq -9$

51. $|y| < 4$

$-4 < y < 4$

53. $|x| \geq 6$

$x \geq 6$ or $x \leq -6$

55. $|2x| < 14$

$-14 < 2x < 14$

$-7 < x < 7$

57. $\left|\frac{y}{3}\right| \leq \frac{1}{3}$

$-\frac{1}{3} \leq \frac{y}{3} \leq \frac{1}{3}$

$-1 \leq y \leq 1$

59. $|x + 6| > 10$

$x + 6 > 10$ or $x + 6 < -10$

$x > 4$ $\qquad x < -16$

61. $|2x - 1| \leq 7$

$-7 \leq 2x - 1 \leq 7$

$-6 \leq 2x \leq 8$

$-3 \leq x \leq 4$

63. $|3x + 10| < -1$

No solution

Absolute value is never negative.

65. $\frac{|y - 16|}{4} < 30$

$|y - 16| < 120$

$-120 < y - 16 < 120$

$-104 < y < 136$

67. $|0.2x - 3| < 4$

$-4 < 0.2x - 3 < 4$

$-1 < 0.2x < 7$

$\frac{-1}{0.2} < x < \frac{7}{0.2}$

$-5 < x < 35$

69. $\left|\frac{3x - 2}{4}\right| + 5 \geq 5$

$\left|\frac{3x - 2}{4}\right| \geq 0$

$-\infty < x < \infty$

Absolute value is always positive.

71. $|3x + 2| < 4$

$-2 < x < \frac{2}{3}$

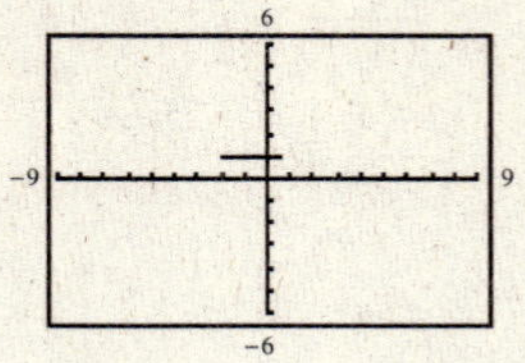

73. $|2x + 3| > 9$

$x < -6$ or $x > 3$

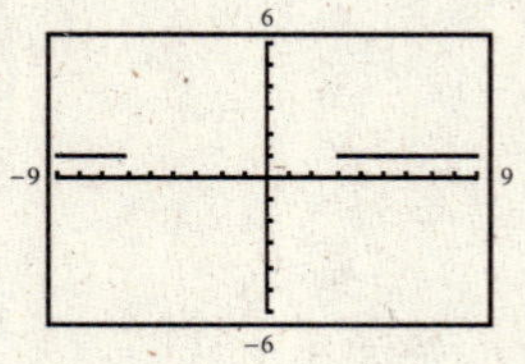

75. $|x - 5| + 3 \leq 5$

$3 \leq x \leq 7$

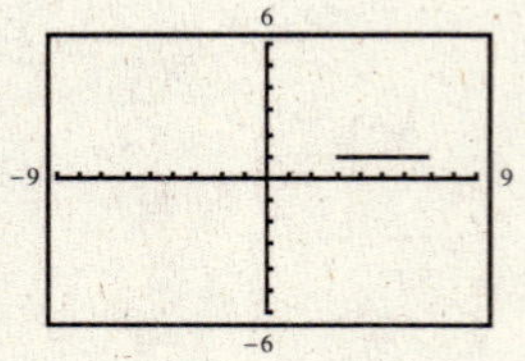

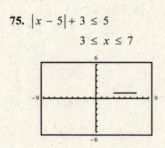

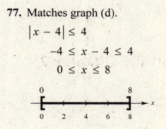

77. Matches graph (d).

$|x - 4| \leq 4$

$-4 \leq x - 4 \leq 4$

$0 \leq x \leq 8$

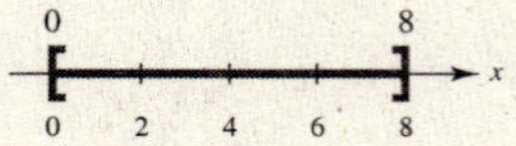

79. Matches graph (b).

$\frac{1}{2}|x - 4| > 4$

$|x - 4| > 8$

$x - 4 > 8$ or $x - 4 < -8$

$x > 12$ $\qquad x < -4$

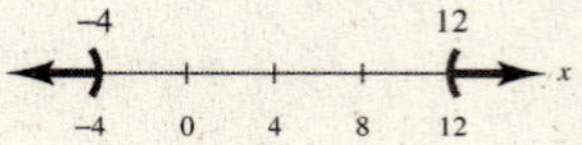

81. $|x| \leq 2$

83. $|x - 19| > 2$

85. $|x| < 3$

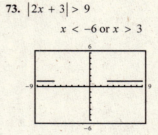

87. $|2x - 3| > 5$

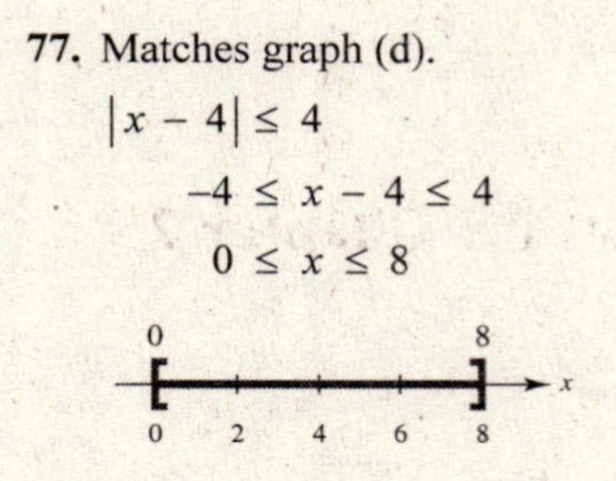

89.

$$|t - 42.238| \le 0.412$$

$$-0.412 \le t - 42.238 \le 0.412$$

$$-0.412 + 42.238 \le t - 42.238 + 42.238 \le 0.412 + 42.238$$

$$41.826 \le t \le 42.65$$

The fastest time is 41.826 seconds and the slowest time is 42.65 seconds.

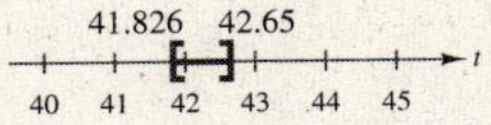

91. (a) $|s - x| \le \frac{3}{16}$

(b) $\left|5\frac{1}{8} - x\right| \le \frac{3}{16}$

$$-\frac{3}{16} \le 5\frac{1}{8} - x \le \frac{3}{16}$$

$$-\frac{3}{16} \le \frac{41}{8} - x \le \frac{3}{16}$$

$$-\frac{85}{16} \le -x \le -\frac{79}{16}$$

$$\frac{85}{16} \ge x \ge \frac{79}{16}$$

$$4\frac{15}{16} \le x \le 5\frac{5}{16}$$

93. The graph of $|x - 4| < 1$ can be described as all real numbers that are within one unit of four.

95. $|2x - 6| \le 6$ because:

$$-6 \le 2x - 6 \le 6$$

$$0 \le 2x \le 12$$

$$0 \le x \le 6$$

97. $4(n + 3)$

99. *Verbal Model:* [Markup] = [Cost] · [Markup rate]

Labels: Markup $= x$

Cost $= 80.00$

Markup rate $= 0.40$

Equation: $x = 80.00 \cdot 0.40$

$x = 32.00$

Verbal Model: [Selling price] = [Cost] + [Markup]

Labels: Selling price $= x$

Cost $= 80.00$

Markup $= 32.00$

Equation: $x = 80.00 + 32.00$

$x = 112.00$

The selling price is \$112.00.

101. $x - 7 > 13$

$x > 20$

103. $4x + 11 \ge 27$

$4x \ge 16$

$x \ge 4$

Review Exercises for Chapter 2

1. (a) $45 - 7(3) \stackrel{?}{=} 3$

$45 - 21 \stackrel{?}{=} 3$

$24 \ne 3$

Not a solution

(b) $45 - 7(6) \stackrel{?}{=} 3$

$45 - 42 \stackrel{?}{=} 3$

$3 = 3$

Solution

3. (a) $\frac{28}{7} + \frac{28}{5} \stackrel{?}{=} 12$

$\frac{140}{35} + \frac{196}{35} \stackrel{?}{=} 12$

$\frac{336}{35} \stackrel{?}{=} 12$

$\frac{48}{5} \ne 12$

Not a solution

(b) $\frac{35}{7} + \frac{35}{5} \stackrel{?}{=} 12$

$5 + 7 = 12$

Solution

5.

$$\begin{aligned} 3x + 21 &= 0 \\ 3x + 21 - 21 &= 0 - 21 \\ 3x &= -21 \\ \frac{3x}{3} &= \frac{-21}{3} \\ x &= -7 \end{aligned}$$

Check:

$$\begin{aligned} 3(-7) + 21 &\stackrel{?}{=} 0 \\ -21 + 21 &= 0 \end{aligned}$$

7.

$$\begin{aligned} 5x - 120 &= 0 \\ 5x - 120 + 120 &= 0 + 120 \\ 5x &= 120 \\ \frac{5x}{5} &= \frac{120}{5} \\ x &= 24 \end{aligned}$$

Check:

$$\begin{aligned} 5(24) - 120 &\stackrel{?}{=} 0 \\ 120 - 120 &= 0 \end{aligned}$$

9.

$$\begin{aligned} x + 4 &= 9 \\ x - 4 + 4 &= 9 - 4 \\ x &= 5 \end{aligned}$$

Check:

$$\begin{aligned} 5 + 4 &\stackrel{?}{=} 9 \\ 9 &= 9 \end{aligned}$$

11.

$$\begin{aligned} -3x &= 36 \\ \frac{-3x}{-3} &= \frac{36}{-3} \\ x &= -12 \end{aligned}$$

Check:

$$\begin{aligned} -3(-12) &\stackrel{?}{=} 36 \\ 36 &= 36 \end{aligned}$$

13.

$$\begin{aligned} -\tfrac{1}{8}x &= 3 \\ (-8)\left(-\tfrac{1}{8}x\right) &= (3)(-8) \\ x &= -24 \end{aligned}$$

Check:

$$\begin{aligned} -\tfrac{1}{8}(-24) &\stackrel{?}{=} 3 \\ 3 &= 3 \end{aligned}$$

15.

$$\begin{aligned} 5x + 4 &= 19 \\ 5x + 4 - 4 &= 19 - 4 \\ 5x &= 15 \\ \frac{5x}{5} &= \frac{15}{5} \\ x &= 3 \end{aligned}$$

Check:

$$\begin{aligned} 5(3) + 4 &\stackrel{?}{=} 19 \\ 15 + 4 &\stackrel{?}{=} 19 \\ 19 &= 19 \end{aligned}$$

17.

$$\begin{aligned} 17 - 7x &= 3 \\ 17 - 7x - 17 &= 3 - 17 \\ -7x &= -14 \\ \frac{-7x}{-7} &= \frac{-14}{-7} \\ x &= 2 \end{aligned}$$

Check:

$$\begin{aligned} 17 - 7(2) &\stackrel{?}{=} 3 \\ 17 - 14 &\stackrel{?}{=} 3 \\ 3 &= 3 \end{aligned}$$

19.

$$\begin{aligned} 7x - 5 &= 3x + 11 \\ 7x - 3x - 5 &= 3x - 3x + 11 \\ 4x - 5 &= 11 \\ 4x - 5 + 5 &= 11 + 5 \\ 4x &= 16 \\ \frac{4x}{4} &= \frac{16}{4} \\ x &= 4 \end{aligned}$$

Check:

$$\begin{aligned} 7(4) - 5 &\stackrel{?}{=} 3(4) + 11 \\ 28 - 5 &\stackrel{?}{=} 12 + 11 \\ 23 &= 23 \end{aligned}$$

21.
$$3(2y-1)=9+3y$$
$$6y-3=9+3y$$
$$6y-3y-3=9+3y-3y$$
$$3y-3=9$$
$$3y-3+3=9+3$$
$$3y=12$$
$$\frac{3y}{3}=\frac{12}{3}$$
$$y=4$$

Check:
$$3(2(4)-1)\stackrel{?}{=}9+3(4)$$
$$3(7)\stackrel{?}{=}9+12$$
$$21=21$$

23.
$$4y-6(y-5)=2$$
$$4y-6y+30=2$$
$$-2y+30=2$$
$$-2y+30-30=2-30$$
$$-2y=-28$$
$$\frac{-2y}{-2}=\frac{-28}{-2}$$
$$y=14$$

Check:
$$4(14)-6(14-5)\stackrel{?}{=}2$$
$$56-6(9)\stackrel{?}{=}2$$
$$56-54\stackrel{?}{=}2$$
$$2=2$$

25.
$$4(3x-5)=6(2x+3)$$
$$12x-20=12x+18$$
$$12x-12x-20=12x-12x+18$$
$$-20=18$$

No solution

27.
$$\frac{4}{5}x-\frac{1}{10}=\frac{3}{2}$$
$$10\left[\frac{4}{5}x-\frac{1}{10}\right]=\left[\frac{3}{2}\right]10$$
$$8x-1=15$$
$$8x-1+1=15+1$$
$$8x=16$$
$$\frac{8x}{8}=\frac{16}{8}$$
$$x=2$$

Check:
$$\frac{4}{5}(2)-\frac{1}{10}\stackrel{?}{=}\frac{3}{2}$$
$$\frac{8}{5}-\frac{1}{10}\stackrel{?}{=}\frac{3}{2}$$
$$\frac{16}{10}-\frac{1}{10}\stackrel{?}{=}\frac{3}{2}$$
$$\frac{15}{10}\stackrel{?}{=}\frac{3}{2}$$
$$\frac{3}{2}=\frac{3}{2}$$

29.
$$1.4t+2.1=0.9t$$
$$1.4t+2.1-0.9t=0.9t-0.9t$$
$$0.5t+2.1=0$$
$$0.5t+2.1-2.1=0-2.1$$
$$0.5t=-2.1$$
$$\frac{0.5t}{0.5}=\frac{-2.1}{0.5}$$
$$t=-4.2$$

Check:
$$1.4(-4.2)+2.1\stackrel{?}{=}0.9(-4.2)$$
$$-5.88+2.1\stackrel{?}{=}-3.78$$
$$-3.78=-3.78$$

31.
$$x+(x+1)=115$$
$$2x+1=115$$
$$2x=114$$
$$x=57$$

The two integers are 57 and 58.

33. *Verbal Model:* $\boxed{\text{Total pay}} = \boxed{\text{Pay per week}} \cdot x + \boxed{\text{Pay for training}}$

Labels: Total pay $= 2635$

Pay per week $= 320$

Pay for training $= 75$

Equation: $2635 = 320x + 75$

$2560 = 320x$

$8 = x$

The internship is 8 weeks long.

35.

Percent	Parts out of 100	Decimal	Fraction
68%	68	0.68	$\frac{17}{25}$

37.

Percent	Parts out of 100	Decimal	Fraction
60%	60	0.6	$\frac{3}{5}$

39. *Verbal Model:* $\boxed{\text{Compared number}} = \boxed{\text{Percent}} \cdot \boxed{\text{Base number}}$

Labels: Compared number $= a$

Percent $= 1.30$

Base number $= 50$

Equation: $a = p \cdot b$

$a = 1.30 \cdot 50$

$a = 65$

So, 65 is 130% of 50.

41. *Verbal Model:* $\boxed{\text{Compared number}} = \boxed{\text{Percent}} \cdot \boxed{\text{Base number}}$

Labels: Compared number $= 645$

Percent $= 0.215$

Base number $= b$

Equation: $645 = 0.215 \cdot b$

$\frac{645}{0.215} = b$

$3000 = b$

So, 645 is $21\frac{1}{2}\%$ of 3000.

43. *Verbal Model:* $\boxed{\text{Compared number}} = \boxed{\text{Percent}} \cdot \boxed{\text{Base number}}$

Labels: Compared number $= 250$

Percent $= p$

Base number $= 200$

Equation: $250 = p \cdot 200$

$\frac{250}{200} = p$

$1.25 = p$

So, 250 is 125% of 200.

45. *Verbal Model:* $\boxed{\text{Commission}} = \boxed{\text{Percent rate}} \cdot \boxed{\text{Sales}}$

Labels: Commission $= 9000$

Percent rate $= x$

Sales $= 150{,}000$

Equation: $9000 = x \cdot 150{,}000$

$\frac{9000}{150{,}000} = x$

$0.06 = x$

It is a 6% commission.

47. *Verbal Model:* $\boxed{\text{Defective parts}} = \boxed{\text{Percent rate}} \cdot \boxed{\text{Sample}}$

Labels: Defective parts $= 6$

Percent rate $= 1.6\%$

Sample $= x$

Equation: $6 = 0.016 \cdot x$

$\frac{6}{0.016} = x$

$375 = x$

The sample contained 375 parts.

49. (a) $\text{Unit price} = \frac{\text{Total price}}{\text{Total units}}$

$= \frac{\$9.79}{39 \text{ ounces}} = \$0.25 \text{ per ounce}$

(b) $\text{Unit price} = \frac{\text{Total price}}{\text{Total units}}$

$= \frac{\$1.79}{8 \text{ ounces}} = \$0.22 \text{ per ounce}$

The 8-ounce can is the better buy.

51. $\frac{\text{Tax}}{\text{Pay}} = \frac{9.90}{396} = \frac{1.10}{44} = \frac{0.1}{4} = \frac{1}{40}$

53. $\frac{7}{8} = \frac{y}{4}$

$8y = 28$

$y = \frac{28}{8}$

$y = \frac{7}{2}$

55. $\frac{b}{6} = \frac{5 + b}{15}$

$15b = 6(5 + b)$

$15b = 30 + 6b$

$9b = 30$

$b = \frac{30}{9}$

$b = \frac{10}{3}$

57. *Verbal Model:* $\boxed{\frac{\text{Leg 1}}{\text{Leg 2}}} = \boxed{\frac{\text{Leg 1}}{\text{Leg 2}}}$

Proportion: $\frac{2}{6} = \frac{x}{9}$

$6x = 18$

$x = 3$

59. *Verbal Model:*

$\boxed{\frac{\text{Tax 1}}{\text{Assessed value 1}}} = \boxed{\frac{\text{Tax 2}}{\text{Assessed value 2}}}$

Labels: Tax 1 = 1680

Assessed value 1 = 105,000

Tax 2 = x

Assessed value 2 = 125,000

Proportion: $\frac{1680}{105{,}000} = \frac{x}{125{,}000}$

$125{,}000 \cdot 1680 = 105{,}000 \cdot x$

$210{,}000{,}000 = 105{,}000x$

$2000 = x$

The tax is $2000.

61. *Verbal Model:* $\boxed{\frac{\text{Flagpole's height}}{\text{Length of flagpole's shadow}}} = \boxed{\frac{\text{Lamp post's height}}{\text{Length of lamp post's shadow}}}$

Labels: Flagpole's height = h

Length of flagpole's shadow = 30

Lamp post's height = 5

Length of lamp post's shadow = 3

Proportion: $\frac{h}{30} = \frac{5}{3}$

$3 \cdot h = 30 \cdot 5$

$3h = 150$

$h = 50$

The flagpole's height is 50 feet.

63. *Verbal Model:* $\boxed{\text{Selling price}} = \boxed{\text{Cost}} + \boxed{\text{Markup}}$

Labels: Selling price = 149.93

Cost = 99.95

Markup = x

Equation:

$$149.93 = 99.95 + x$$
$$149.93 - 99.95 = x$$
$$49.98 = x$$

The markup is \$49.98.

Verbal Model: $\boxed{\text{Markup}} = \boxed{\text{Markup rate}} \cdot \boxed{\text{Cost}}$

Labels: Markup = 49.98

Markup rate = x

Cost = 99.95

Equation:

$$49.98 = x \cdot 99.95$$
$$\frac{49.98}{99.95} = x$$
$$0.50 \approx x$$

The markup rate is about 50%.

65. *Verbal Model:* $\boxed{\text{Sale price}} = \boxed{\text{List price}} - \boxed{\text{Discount}}$

Labels: Sale price = 53.96

List price = 71.95

Discount = x

Equation:

$$53.96 = 71.95 - x$$
$$x = 71.95 - 53.96$$
$$x = 17.99$$

The discount is \$17.99.

Verbal Model: $\boxed{\text{Discount}} = \boxed{\text{Discount rate}} \cdot \boxed{\text{List price}}$

Labels: Discount = 17.99

Discount rate = x

List price = 71.95

Equation:

$$17.99 = x \cdot 71.95$$
$$\frac{17.99}{71.95} = x$$
$$0.25 \approx x$$

The discount rate is about 25%.

67. *Verbal Model:* $\boxed{\text{Sales tax}} = \boxed{\text{Tax rate}} \cdot \boxed{\text{Cost}}$

Labels: Sales tax = x

Tax rate = 0.06

Cost = 2795

Equation:

$$x = 0.06 \cdot 2795$$
$$x = 167.7$$

The sales tax is \$167.70.

Verbal Model: $\boxed{\text{Total bill}} = \boxed{\text{Sales tax}} + \boxed{\text{Cost}}$

Labels: Total bill = x

Sales tax = 167.70

Cost = 2795

Equation:

$$x = 167.70 + 2795$$
$$x = 2962.70$$

The total bill is \$2962.70.

Verbal Model:

$\boxed{\text{Amount financed}} = \boxed{\text{Total bill}} - \boxed{\text{Down payment}}$

Labels: Amount financed = x

Total bill = 2962.70

Downpayment = 800

Equation:

$$x = 2962.70 - 800$$
$$x = 2162.70$$

The amount financed is \$2162.70.

69. *Verbal Model:*

$\boxed{\text{Amount of solution 1}} + \boxed{\text{Amount of solution 2}} = \boxed{\text{Amount of final solution}}$

Labels: Percent of solution 1 = 30%

Liters of solution 1 = x

Percent of solution 2 = 60%

Liters of solution 2 = $10 - x$

Percent of final solution = 50%

Liters of final solution = 10

Equation:

$$0.30x + 0.60(10 - x) = 0.50(10)$$
$$0.30x + 6 - 0.60x = 5$$
$$-0.30x = -1$$
$$x = 3\tfrac{1}{3}$$
$$10 - x = 6\tfrac{2}{3}$$

$3\frac{1}{3}$ liters of solution 1 and $6\frac{2}{3}$ liters of solution 2 are required.

71. *Verbal Model:* $\boxed{\text{Distance}} = \boxed{\text{Rate}} \cdot \boxed{\text{Time}}$

Labels: Distance $= d$

Rate $= 1500$ mph

Time $= 2\frac{1}{3}$ hours

Equation: $d = 1500 \cdot 2\frac{1}{3}$

$d = 3500$

The distance is 3500 miles.

73. *Verbal Model:* $\boxed{\text{Distance}} = \boxed{\text{Rate}} \cdot \boxed{\text{Time}}$

Labels: Distance $= 100$ miles

Rates $= 48$ mph and 40 mph

Time $= t$

Equation: $d = rt$

$$t = \frac{d}{r}$$

$$t = \frac{100}{48} + \frac{100}{40}$$

$$t = 4.58\overline{3} \text{ or } \frac{55}{12}$$

Verbal Model: $\boxed{\text{Average speed}} = \boxed{\text{Total distance}} \div \boxed{\text{Total time}}$

Labels: Average speed $= r$

Total distance $= 200$ miles

Total time $= 4.58\overline{3}$ hours

Equation: $r = 200 \div 4.58\overline{3}$

$r \approx 43.6$

The average speed is 43.6 miles per hour.

75. *Verbal Model:* $\boxed{\text{Work done}} = \boxed{\text{Work done by person 1}} + \boxed{\text{Work done by person 2}}$

Labels: Work done $= 1$

Rate of person 1 $= \frac{1}{4.5}$

Rate of person 2 $= \frac{1}{6}$

Time $= t$

Equation:

$$1 = \frac{t}{4.5} + \frac{t}{6}$$

$$27 = 6t + 4.5t$$

$$27 = 10.5t$$

$$\frac{27}{10.5} = t$$

$$2.57 \approx t$$

The time required is about 2.57 hours.

77. *Verbal Model:* $\boxed{\text{Interest}} = \boxed{\text{Principal}} \cdot \boxed{\text{Rate}} \cdot \boxed{\text{Time}}$

Labels: Interest $= i$

Principal $= \$1000$

Rate $= 0.085$

Time $= 4$

Equation: $i = 1000 \cdot 0.085 \cdot 4$

$i = 340$

The interest is \$340.

79. *Verbal Model:* $\boxed{\text{Interest}} = \boxed{\text{Principal}} \cdot \boxed{\text{Rate}} \cdot \boxed{\text{Time}}$

Labels: Interest $= \$20{,}000$

Principal $= p$

Rate $= 0.095$

Time $= 4$

Equation:

$$20{,}000 = p \cdot 0.095 \cdot 4$$

$$\frac{20{,}000}{0.38} = p$$

$$52{,}631.58 \approx p$$

The principal required is about \$52,631.58.

81. *Verbal Model:* $\boxed{\text{Interest}} = \boxed{\text{Principal}} \cdot \boxed{\text{Rate}} \cdot \boxed{\text{Time}}$

Labels: Interest $= 4700$

Principal 1 $= p$

Rate 1 $= 0.085$

Principal 2 $= 50{,}000 - p$

Rate 2 $= 0.10$

Time $= 1$

Equation:

$$4700 = 0.085p + 0.10(50{,}000 - p)$$

$$4700 = 0.085p + 5000 - 0.10p$$

$$-300 = -0.015p$$

$$\frac{-300}{-0.015} = p$$

$$20{,}000 = p$$

$$30{,}000 = 50{,}000 - p$$

The smallest amount you can invest is \$30,000.

83. *Verbal Model:*

$\boxed{\text{Perimeter}} = 2 \cdot \boxed{\text{Width}} + 2 \cdot \boxed{\text{Length}}$

Labels: Perimeter $= 64$

Width $= \frac{3}{5}l$

Length $= l$

Equation:

$$64 = 2\left(\tfrac{3}{5}l\right) + 2l$$
$$64 = \tfrac{6}{5}l + 2l$$
$$64 = \tfrac{16}{5}l$$
$$20 = l$$

The dimensions are 20 feet × 12 feet.

85. *Verbal Model:*

$\boxed{\text{Fahrenheit temperature}} = \frac{9}{5} \cdot \boxed{\text{Celsius temperature}} + 32$

Labels: Fahrenheit temperature $= x$

Celsius temperature $= 10.5$

Equation:

$$x = \tfrac{9}{5} \cdot 10.5 + 32$$
$$x = 18.9 + 32$$
$$x = 50.9 \text{ degrees Fahrenheit}$$

87. 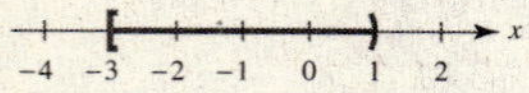

89. 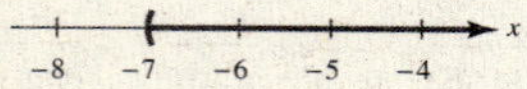

91.
$$x - 5 \le -1$$
$$x - 5 + 5 \le -1 + 5$$
$$x \le 4$$

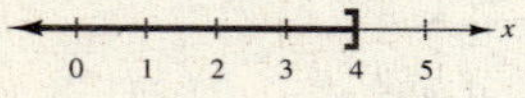

93.
$$-6x < -24$$
$$\frac{-6x}{-6} > \frac{-24}{-6}$$
$$x > 4$$

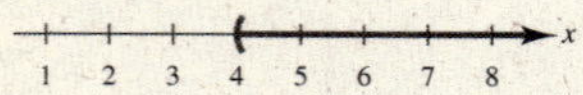

95.
$$5x + 3 > 18$$
$$5x > 15$$
$$x > 3$$

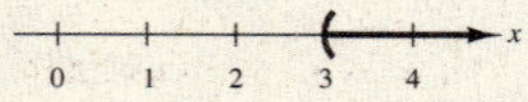

97.
$$8x + 1 \ge 10x - 11$$
$$8x - 10x + 1 \ge 10x - 10x - 11$$
$$-2x + 1 \ge -11$$
$$-2x + 1 - 1 \ge -11 - 1$$
$$-2x \ge -12$$
$$\frac{-2x}{-2} \le \frac{-12}{-2}$$
$$x \le 6$$

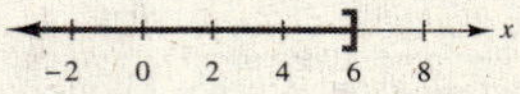

99.
$$\frac{1}{3} - \frac{1}{2}y < 12$$
$$2 - 3y < 72$$
$$-3y < 70$$
$$y > -\frac{70}{3}$$

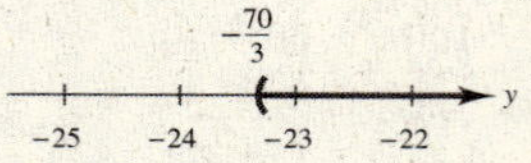

101.
$$-4(3 - 2x) \le 3(2x - 6)$$
$$-12 + 8x \le 6x - 18$$
$$-12 + 8x - 6x \le 6x - 6x - 18$$
$$-12 + 2x \le -18$$
$$-12 + 12 + 2x \le -18 + 12$$
$$2x \le -6$$
$$\frac{2x}{2} \le \frac{-6}{2}$$
$$x \le -3$$

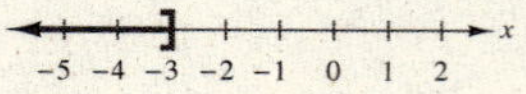

103.
$$-6 \le 2x + 8 < 4$$
$$-6 - 8 \le 2x + 8 - 8 < 4 - 8$$
$$-14 \le 2x < -4$$
$$\frac{-14}{2} \le \frac{2x}{2} < \frac{-4}{2}$$
$$-7 \le x < -2$$

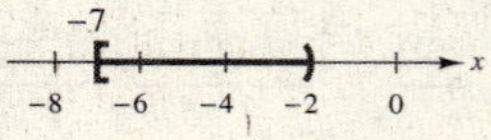

105.
$$5 > \frac{x + 1}{-3} > 0$$
$$-15 < x + 1 < 0$$
$$-16 < x < -1$$

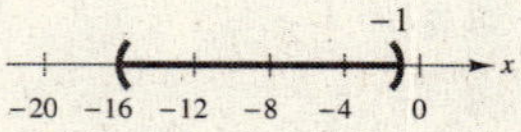

107. $5x - 4 < 6$ and $3x + 1 > -8$

$$5x - 4 + 4 < 6 + 4 \qquad 3x + 1 - 1 > -8 - 1$$
$$5x < 10 \qquad 3x > -9$$
$$\frac{5x}{5} < \frac{10}{5} \qquad \frac{3x}{3} > \frac{-9}{3}$$
$$x < 2 \qquad x > -3$$

$-3 < x < 2$

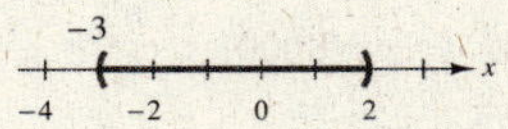

109. *Verbal Model:* [Rate per hour] + [Tips] ≥ [Total]

Labels: Rate per hour $= 6$

Tab $= x$

Tip rate $= 0.15$

Total $= 150$

Equation: $6(5) + 0.15x \geq 150$

$$0.15x \geq 120$$
$$x \geq 800$$

The tab must be at least \$800.

111. $|x| = 6$

$x = 6$ or $x = -6$

113. $|4 - 3x| = 8$

$4 - 3x = 8$ or $4 - 3x = -8$

$$4 - 4 - 3x = 8 - 4 \qquad 4 - 4 - 3x = -8 - 4$$
$$-3x = 4 \qquad -3x = -12$$
$$\frac{-3x}{-3} = \frac{4}{-3} \qquad \frac{-3x}{-3} = \frac{-12}{-3}$$
$$x = -\frac{4}{3} \qquad x = 4$$

115. $|5x + 4| - 10 = -6$

$$|5x + 4| = 4$$

$5x + 4 = 4$ or $5x + 4 = -4$

$$5x + 4 - 4 = 4 - 4 \qquad 5x + 4 - 4 = -4 - 4$$
$$5x = 0 \qquad 5x = -8$$
$$\frac{5x}{5} = \frac{0}{5} \qquad \frac{5x}{5} = \frac{-8}{5}$$
$$x = 0 \qquad x = -\frac{8}{5}$$

117. $|3x - 4| = |x + 2|$

$3x - 4 = x + 2$ or $3x - 4 = -(x + 2)$

$$2x = 6 \qquad 3x - 4 = -x - 2$$
$$x = 3 \qquad 4x = 2$$
$$x = \tfrac{2}{4} = \tfrac{1}{2}$$

119. $|x - 4| > 3$

$x - 4 < -3$ or $x - 4 > 3$

$$x < 1 \qquad x > 7$$

121. $|3x| < 12$

$$-12 < 3x < 12$$
$$-4 < x < 4$$

122. $\left|\frac{t}{3}\right| < 1$

$$-1 < \frac{t}{3} < 1$$
$$-3 < t < 3$$

123. $|2x - 7| < 15$

$$-15 < 2x - 7 < 15$$
$$-8 < 2x < 22$$
$$-4 < x < 11$$

125. $|b + 2| - 6 > 1$

$$|b + 2| > 7$$

$b + 2 < -7$ or $b + 2 > 7$

$$b < -9 \qquad b > 5$$

127. $|4(x - 3)| \geq 8$

$x \leq 1$ or $x \geq 5$

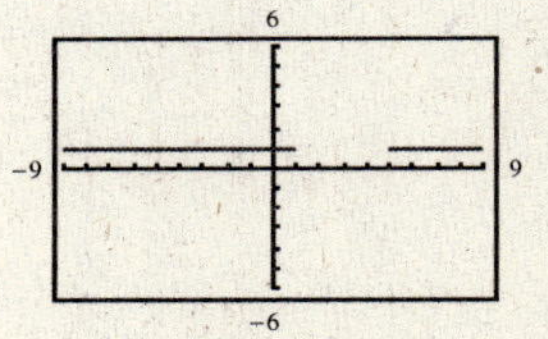

129. $(1, 5)$

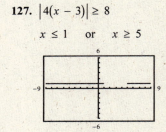

$$1 < x < 5$$
$$1 - 3 < x - 3 < 5 - 3$$
$$-2 < x - 3 < 2$$
$$|x - 3| < 2$$

131. $|t - 78.3| \leq 38.3$

$$-38.3 \leq t - 78.3 \leq 38.3$$
$$40 \leq t \leq 116.6$$

The minimum temperature is 40 degrees Fahrenheit and the maximum temperature is 116.6 degrees Fahrenheit.

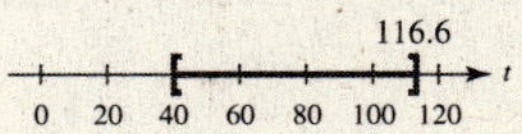

Chapter Test for Chapter 2

1.
$$6x - 5 = 19$$
$$6x - 5 + 5 = 19 + 5$$
$$6x = 24$$
$$\frac{6x}{6} = \frac{24}{6}$$
$$x = 4$$

2.
$$5x - 6 = 7x - 12$$
$$5x - 7x - 6 = 7x - 7x - 12$$
$$-2x - 6 = -12$$
$$-2x - 6 + 6 = -12 + 6$$
$$-2x = -6$$
$$\frac{-2x}{-2} = \frac{-6}{-2}$$
$$x = 3$$

3.
$$15 - 7(1 - x) = 3(x + 8)$$
$$15 - 7 + 7x = 3x + 24$$
$$8 + 7x = 3x + 24$$
$$8 + 7x - 3x = 3x + 24 - 3x$$
$$8 - 8 + 4x = 24 - 8$$
$$4x = 16$$
$$\frac{4x}{4} = \frac{16}{4}$$
$$x = 4$$

4.
$$\frac{2x}{3} = \frac{x}{2} + 4$$
$$6\left(\frac{2x}{3}\right) = \left(\frac{x}{2} + 4\right)6$$
$$4x = 3x + 24$$
$$4x - 3x = 3x + 24 - 3x$$
$$x = 24$$

5. *Verbal Model:* [Compared number] = [Percent] · [Base number]

Labels: Compared number = a

Percent = 1.25

Base number = 3200

Equation: $a = 1.25 \cdot 3200$

$a = 4000$

So, 125% of 3200 is 4000.

6. *Verbal Model:* [Compared number] = [Percent] · [Base number]

Labels: Compared number = a

Percent = p

Base number = b

Equation: $32 = p \cdot 8000$

$\frac{32}{8000} = p$

$0.004 = p$

So, 32 is 0.4% of 8000.

7. *Verbal Model:* [List price] − [Discount] = [Sale price]

Labels: List price = x

Discount = $0.20x$

Sale price = 8900

Equation: $x - 0.20x = 8900$

$0.80x = 8900$

$x = 11{,}125$

The list price is \$11,125.

8. $\frac{\text{Total price}}{\text{Total units}} = \frac{\$2.49}{12 \text{ ounces}} = \frac{249}{1200} = \$0.2075$ per ounce

$\frac{\text{Total price}}{\text{Total units}} = \frac{\$2.99}{15 \text{ ounces}} = \frac{299}{1500} = \$0.199\overline{3}$ per ounce

The 15-ounce can is the better buy. It has a lower unit price.

9. *Verbal Model:* [Tax] = [Tax rate] · [Assessed value]

Labels: Tax = 1650

Tax rate = p

Assessed value = 110,000

Equation: $1650 = p \cdot 110{,}000$

$0.015 = p$

Verbal Model: [Tax] = [Tax rate] · [Assessed value]

Labels: Tax = x

Tax rate = 0.015

Assessed value = 145,000

Equation: $x = 0.015 \cdot 145{,}000$

$x = 2175$

The tax is \$2175.

10. *Verbal Model:* $\boxed{\text{Total bill}} = \boxed{\text{Cost of parts}} + \boxed{\text{Cost of labor}}$

Labels: Total bill $= 165$

Cost of parts $= 85$

Number of half hours of labor $= x$

Cost of labor $= 16x$

Equation:

$$165 = 85 + 16x$$
$$80 = 16x$$
$$5 \text{ half hours} = x$$

The repairs took $2\frac{1}{2}$ hours.

11. *Verbal Model:*

$\boxed{\text{Amount of food 1}} + \boxed{\text{Amount of food 2}} = \boxed{\text{Mixture}}$

Labels: Cost for food 1 $= 2.60$

Pounds of food 1 $= x$

Cost for food 2 $= 3.80$

Pounds of food 2 $= 40 - x$

Cost for mixture $= 3.35$

Pounds of mixture $= 40$

Equation:

$$2.60x + 3.80(40 - x) = 3.35(40)$$
$$-1.20x + 152 = 134$$
$$-1.20x = -18$$
$$x = 15$$

There are 15 pounds at \$2.60 and 25 pounds at \$3.80.

12. *Verbal Model:*

$\boxed{\text{Distance of car 1}} + 10 \text{ miles} = \boxed{\text{Distance of car 2}}$

Labels: Time $= x$

Distance of car 1 $= 40x$

Distance of car 2 $= 55x$

Equation:

$$40x + 10 = 55x$$
$$10 = 15x$$
$$\frac{10}{15} = x$$
$$\frac{2}{3} = x$$

$\frac{2}{3}$ hour, or 40 minutes, must elapse.

13. *Verbal Model:* $\boxed{\text{Interest}} = \boxed{\text{Principal}} \cdot \boxed{\text{Rate}} \cdot \boxed{\text{Time}}$

Labels: Interest $= 300$

Principal $= p$

Rate $= 0.075$

Time $= 2$

Equation:

$$300 = p \cdot 0.075 \cdot 2$$
$$2000 = p$$

The principal required is \$2000.

14. (a) $|3x - 6| = 9$

$3x - 6 = 9$ and $-3x + 6 = 9$

$3x = 15 \qquad -3x = 3$

$\frac{3x}{3} = \frac{15}{3} \qquad \frac{-3x}{-3} = \frac{3}{-3}$

$x = 5 \qquad x = -1$

(b) $|3x - 5| = |6x - 1|$

$3x - 5 = 6x - 1$ and $-3x + 5 = 6x - 1$

$-4 = 3x \qquad 6 = 9x$

$\frac{-4}{3} = \frac{3x}{3} \qquad \frac{6}{9} = x$

$-\frac{4}{3} = x \qquad \frac{2}{3} = x$

(c) $|9 - 4x| + 4 = 1$

$|9 - 4x| = -3$

There is no solution.

15. (a) $3x + 12 \geq -6$

$3x \geq -18$

$x \geq -6$

(b) $9 - 5x < 5 - 3x$

$-2x < -4$

$x > 2$

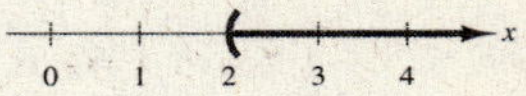

(c) $0 \leq \frac{1 - x}{4} < 2$

$0 \leq 1 - x < 8$

$-1 \leq -x < 7$

$1 \geq x > -7$

$-7 < x \leq 1$

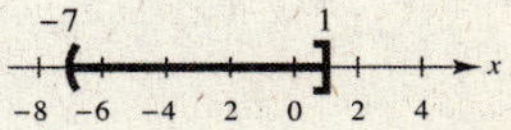

(d) $-7 < 4(2 - 3x) \leq 20$

$-7 < 8 - 12x \leq 20$

$-15 < -12x \leq 12$

$1 \leq x < \frac{5}{4}$

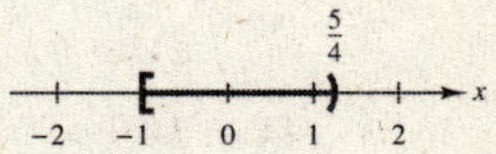

16. $t \geq 8$

17. (a) $|x - 3| \le 2$

$$-2 \le x - 3 \le 2$$
$$1 \le x \le 5$$

(b) $|5x - 3| > 12$

$$5x - 3 > 12 \quad \text{or} \quad 5x - 3 < -12$$
$$5x > 15 \qquad\qquad 5x < -9$$
$$x > 3 \qquad\qquad x < -\frac{9}{5}$$

(c) $\left|\frac{x}{4} + 2\right| < 0.2$

$$-0.2 < \frac{x}{4} + 2 < 0.2$$
$$-0.8 < x + 8 < 0.8$$
$$-8.8 < x < -7.2$$
$$-\frac{44}{5} < x < -\frac{36}{5}$$

18. *Verbal Model:* $\boxed{\text{Operating cost}} \le 11{,}950$

Label: Number of miles $= m$

Equation:
$$0.37m + 2700 \le 11{,}950$$
$$0.37m \le 9250$$
$$m \le 25{,}000$$

The maximum number of miles is 25,000.

CHAPTER 3
Graphs and Functions

CHAPTER 3
Graphs and Functions

Section 3.1 The Rectangular Coordinate System

1.

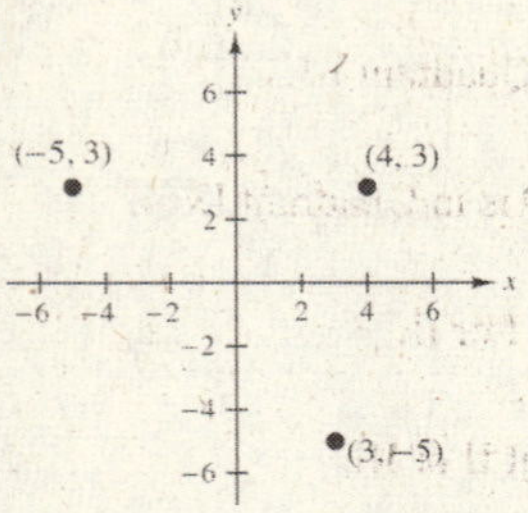

$(4, 3)$ is 4 units to the right of the vertical axis and 3 units above the horizontal axis.

$(-5, 3)$ is 5 units to the left of the vertical axis and 3 units above the horizontal axis.

$(3, -5)$ is 3 units to the right of the vertical axis and 5 units below the horizontal axis.

3.

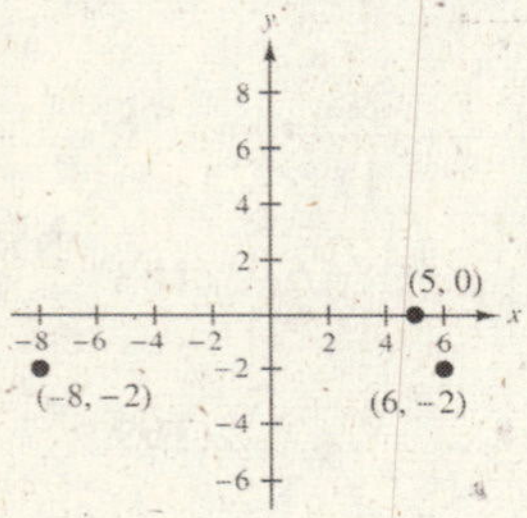

$(-8, -2)$ is 8 units to the left of the vertical axis and 2 units below the horizontal axis.

$(6, -2)$ is 6 units to the right of the vertical axis and 2 units below the horizontal axis.

$(5, 0)$ is 5 units to the right of the vertical axis and 0 units above or below the horizontal axis.

5.

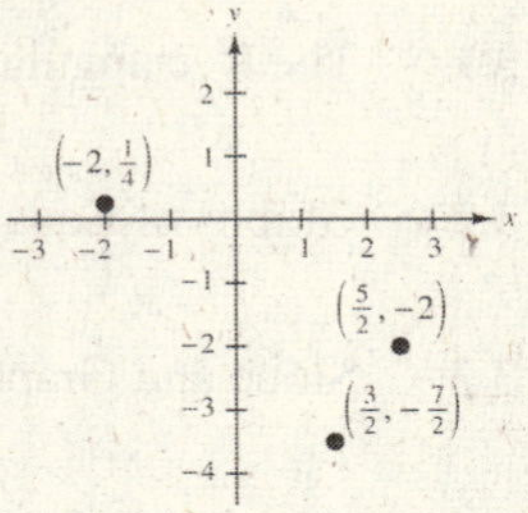

$\left(\frac{5}{2}, -2\right)$ is $\frac{5}{2}$ units to the right of the vertical axis and 2 units below the horizontal axis.

$\left(-2, \frac{1}{4}\right)$ is 2 units to the left of the vertical axis and $\frac{1}{4}$ unit above the horizontal axis.

$\left(\frac{3}{2}, -\frac{7}{2}\right)$ is $\frac{3}{2}$ units to the right of the vertical axis and $\frac{7}{2}$ units below the horizontal axis.

7.

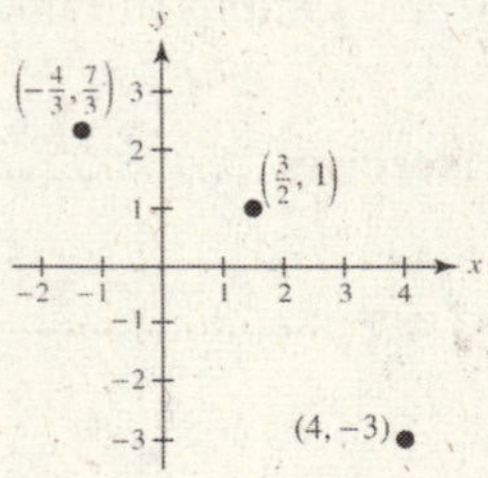

$\left(\frac{3}{2}, 1\right)$ is $\frac{3}{2}$ units to the right of the vertical axis and 1 unit above the horizontal axis.

$(4, -3)$ is 4 units to the right of the vertical axis and 3 units below the horizontal axis.

$\left(-\frac{4}{3}, \frac{7}{3}\right)$ is $\frac{4}{3}$ units to the left of the vertical axis and $\frac{7}{3}$ units above the horizontal axis.

9.

Point	Position	Coordinates
A	2 units left, 4 units up	$(-2, 4)$
B	0 units right or left, 2 units down	$(0, -2)$
C	4 units right, 2 units down	$(4, -2)$

11.

Point	Position	Coordinates
A	4 units right, 2 units down	$(4, -2)$
B	3 units left, $\frac{5}{2}$ units down	$\left(-3, -\frac{5}{2}\right)$
C	3 units right, 0 units above or below	$(3, 0)$

13.

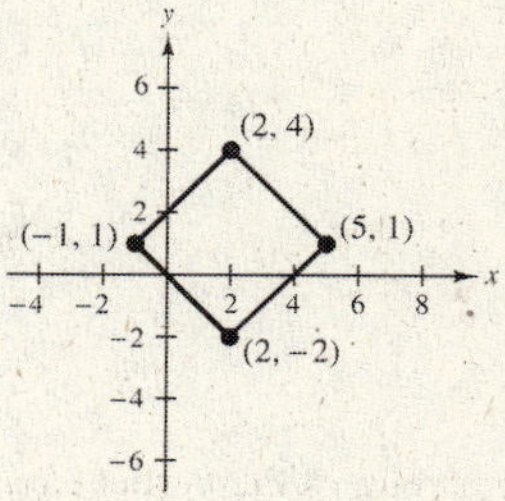

15.

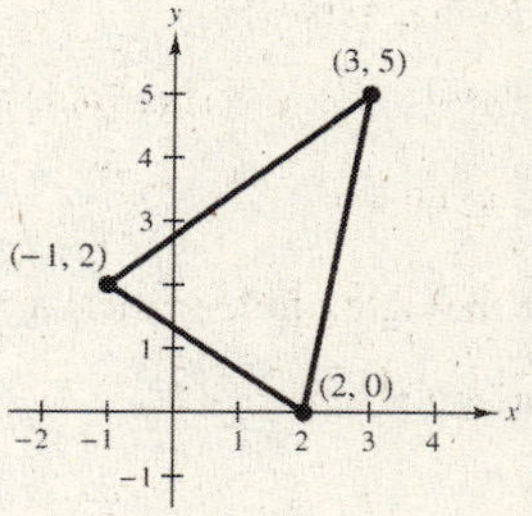

17.

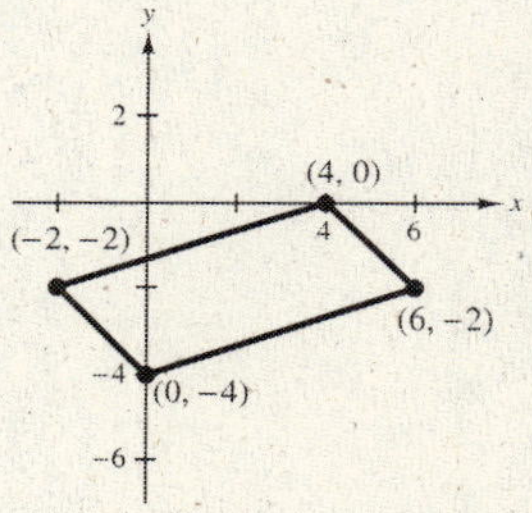

19.

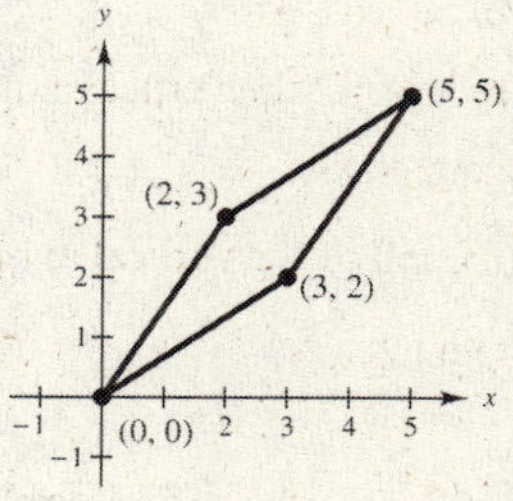

21. Point 1 unit right of y-axis and 4 units above x-axis $= (1, 4)$.

23. Point 10 units right of y-axis and 4 units below x-axis $= (10, -4)$.

25. Point on positive x-axis 10 units from the origin $= (10, 0)$.

27. The coordinates of the point are equal and located in Quadrant III, 8 units left of y-axis $= (-8, -8)$.

29. $(-3, -5)$ is in Quadrant III.

31. $\left(-\frac{8}{9}, \frac{3}{4}\right)$ is in Quadrant II.

33. $(-9.5, -12.13)$ is in Quadrant III.

35. (x, y), $x > 0$, $y < 0$ is in Quadrant IV.

37. $(x, 4)$ is in Quadrant I or II.

39. $(-3, y)$ is in Quadrant II or III.

41. (x, y), $xy > 0$ is in Quadrant I or III.

43.

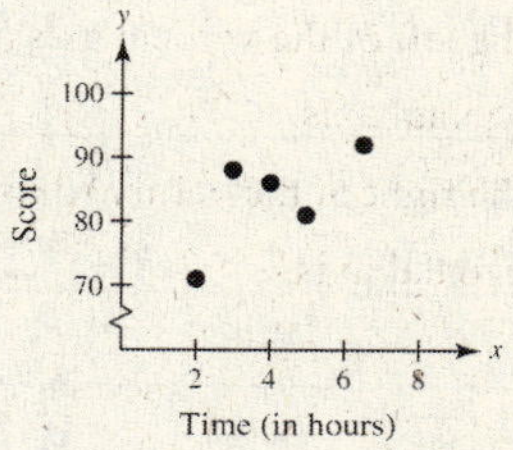

45.

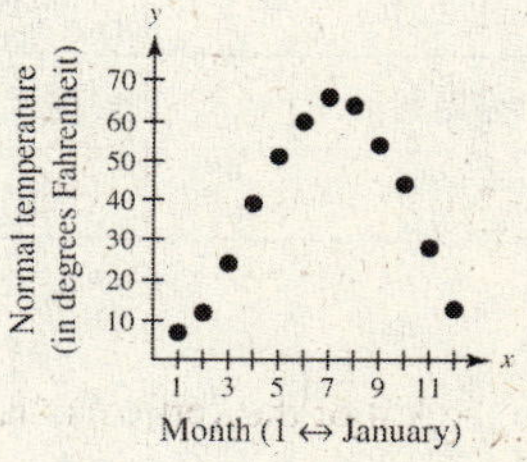

The relationship between x and y is as x increases from 1 to 7, y also increases, but as x increases from 7 to 12, y decreases.

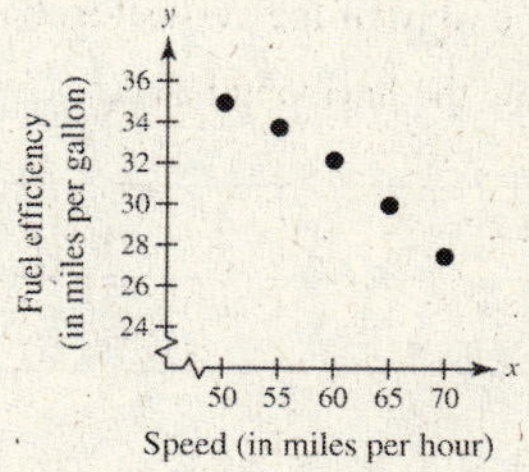

47. $(-2, -1)$ shifted 2 units right and 5 units up $= (0, 4)$

$(-3, -4)$ shifted 2 units right and 5 units up $= (-1, 1)$

$(1, -3)$ shifted 2 units right and 5 units up $= (3, 2)$

49.

Choose x	Calculate y from $y = 5x + 3$	Solution point
-2	$y = 5(-2) + 3 = -7$	$(-2, -7)$
0	$y = 5(0) + 3$	$(0, 3)$
2	$y = 5(2) + 3$	$(2, 13)$
4	$y = 5(4) + 3$	$(4, 23)$
6	$y = 5(6) + 3$	$(6, 33)$

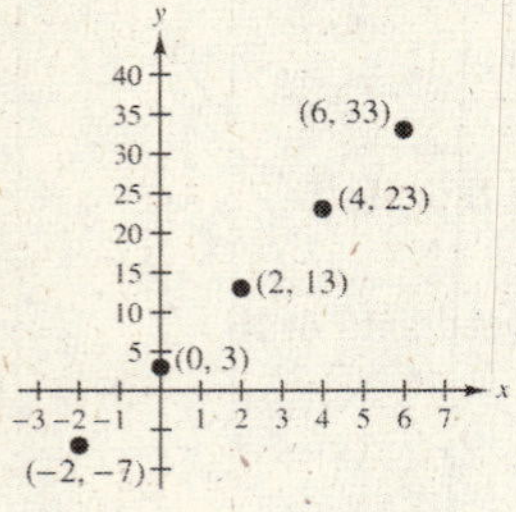

51.

Choose x	Calculate y from $y = \lvert 2x - 7 \rvert + 2$	Solution point
-4	$y = \lvert 2(-4) - 7 \rvert + 2$	$(-4, 17)$
0	$y = \lvert 2(0) - 7 \rvert + 2$	$(0, 9)$
3	$y = \lvert 2(3) - 7 \rvert + 2$	$(3, 3)$
5	$y = \lvert 2(5) - 7 \rvert + 2$	$(5, 5)$
10	$y = \lvert 2(10) - 7 \rvert + 2$	$(10, 15)$

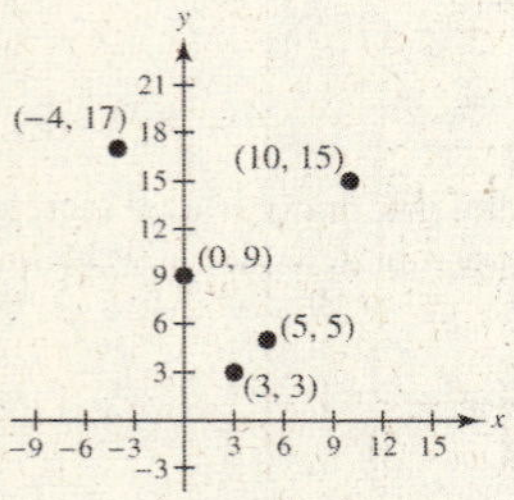

53.

x	-2	0	2	4	6
$y = x^2 + 2x + 5$	5	5	13	29	53

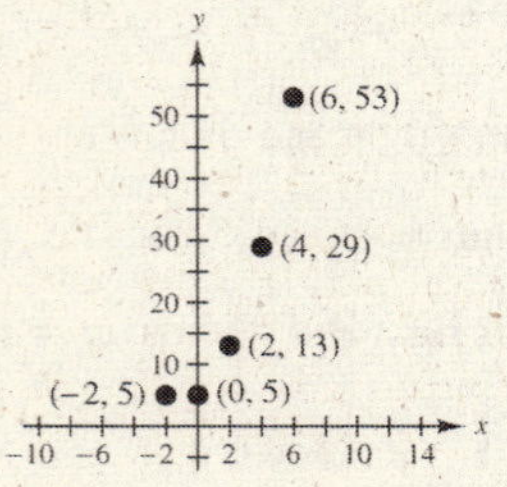

55. $4y - 2x + 1 = 0$

(a) $4(0) - 2(0) + 1 \stackrel{?}{=} 0$

$1 \neq 0$

Not a solution

(b) $4(0) - 2\left(\frac{1}{2}\right) + 1 \stackrel{?}{=} 0$

$0 - 1 + 1 \stackrel{?}{=} 0$

$0 = 0$

Solution

(c) $4\left(-\frac{7}{4}\right) - 2(-3) + 1 \stackrel{?}{=} 0$

$-7 + 6 + 1 \stackrel{?}{=} 0$

$0 = 0$

Solution

(d) $4\left(-\frac{3}{4}\right) - 2(1) + 1 \stackrel{?}{=} 0$

$-3 - 2 + 1 \stackrel{?}{=} 0$

$-4 \neq 0$

Not a solution

57. $y = \frac{7}{8}x + 3$

(a) $4 \stackrel{?}{=} \frac{7}{8}\left(\frac{8}{7}\right) + 3$

$4 \stackrel{?}{=} 1 + 3$

$4 = 4$

Solution

(b) $10 \stackrel{?}{=} \frac{7}{8}(8) + 3$

$10 \stackrel{?}{=} 7 + 3$

$10 = 10$

Solution

(c) $0 \stackrel{?}{=} \frac{7}{8}(0) + 3$

$0 \stackrel{?}{=} 0 + 3$

$0 \neq 3$

Not a solution

(d) $14 \stackrel{?}{=} \frac{7}{8}(-16) + 3$

$14 \stackrel{?}{=} -14 + 3$

$14 \neq -11$

Not a solution

59. $x^2 + 3y = -5$

(a) $3^2 + 3(-2) \stackrel{?}{=} -5$

$9 - 6 \stackrel{?}{=} -5$

$3 \neq -5$

Not a solution

(b) $(-2)^2 + 3(-3) \stackrel{?}{=} -5$

$4 - 9 \stackrel{?}{=} -5$

$-5 = -5$

Solution

(c) $3^2 + 3(-5) \stackrel{?}{=} -5$

$9 - 15 \stackrel{?}{=} -5$

$-6 \neq -5$

Not a solution

(d) $4^2 + 3(-7) \stackrel{?}{=} -5$

$16 - 21 \stackrel{?}{=} -5$

$-5 = -5$

Solution

61. $d = |5 - (-2)|$

$= |7|$

$= 7$

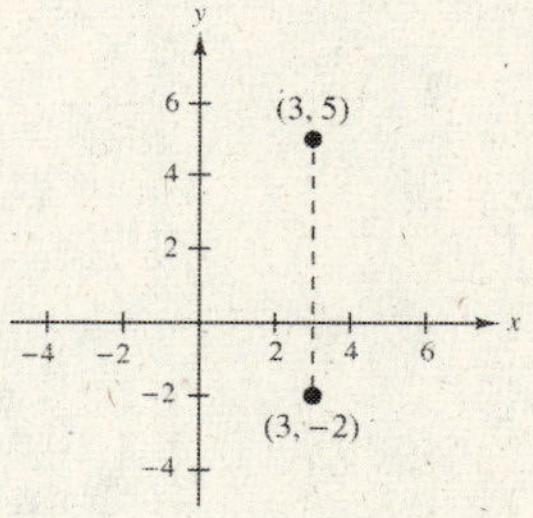

Vertical line

63. $d = |10 - 3|$

$= |7|$

$= 7$

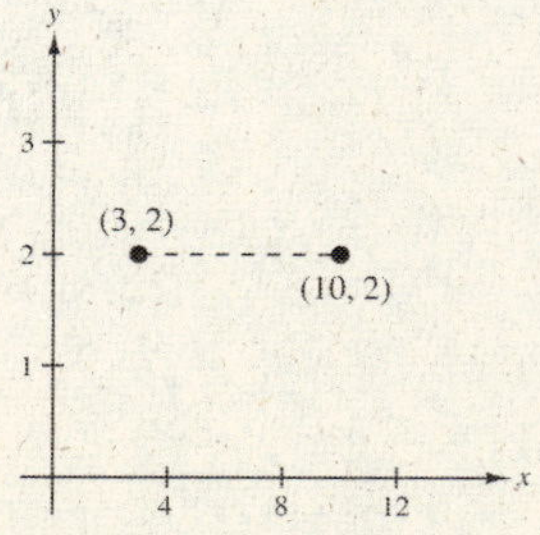

Horizontal line

65. $d = \left|\frac{3}{2} - \frac{9}{4}\right|$

$= \left|\frac{6}{4} - \frac{9}{4}\right|$

$= \frac{3}{4}$

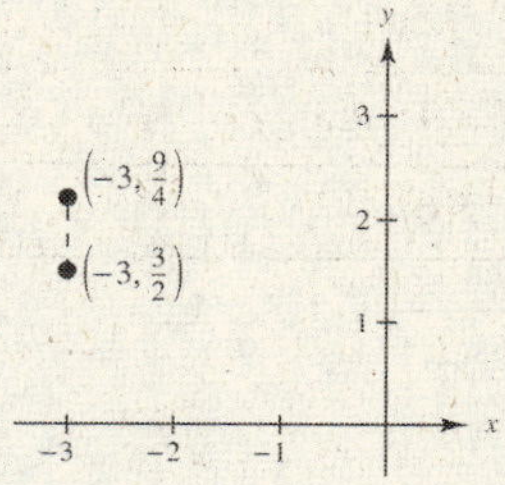

Vertical line

67. $d = \left|\frac{5}{2} - (-4)\right|$

$= \left|\frac{5}{2} + \frac{8}{2}\right|$

$= \left|\frac{13}{2}\right|$

$= \frac{13}{2}$

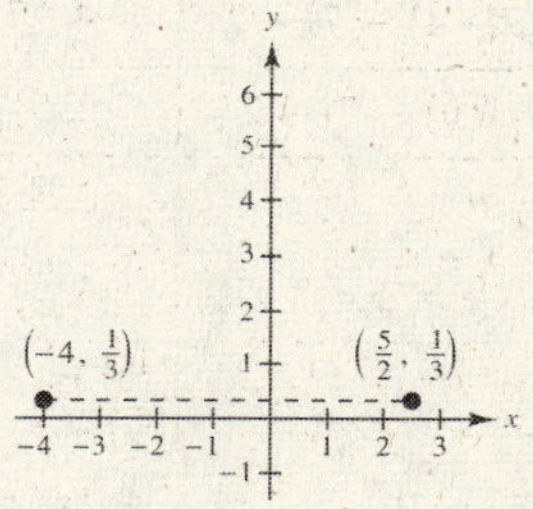

Horizontal line

69. $d = \sqrt{(1-5)^2 + (3-6)^2}$

$= \sqrt{(-4)^2 + (-3)^2} = \sqrt{16 + 9} = \sqrt{25} = 5$

71. $d = \sqrt{(3-4)^2 + (7-5)^2}$

$= \sqrt{(-1)^2 + (2)^2} = \sqrt{1 + 4} = \sqrt{5}$

73. $d = \sqrt{(-3-4)^2 + [0 - (-3)]^2}$

$= \sqrt{(-7)^2 + (3)^2} = \sqrt{49 + 9} = \sqrt{58}$

75. $d = \sqrt{(-2-4)^2 + (-3-2)^2}$

$= \sqrt{(-6)^2 + (-5)^2} = \sqrt{36 + 25} = \sqrt{61}$

77. $d = \sqrt{(3-7)^2 + \left[\frac{3}{4} - \left(-\frac{1}{4}\right)\right]^2}$

$= \sqrt{(-4)^2 + (1)^2} = \sqrt{16 + 1} = \sqrt{17}$

79.

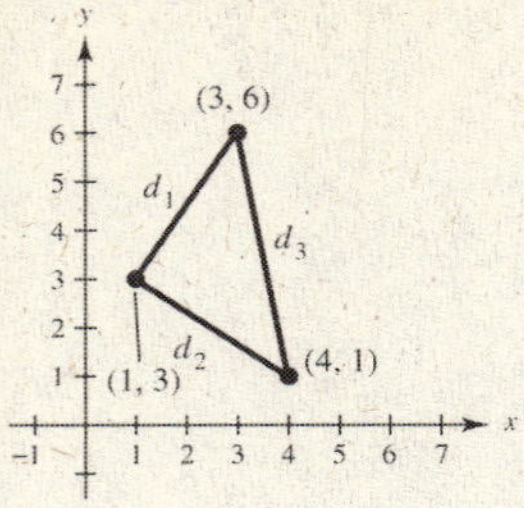

$$d_1 = \sqrt{(1-3)^2 + (3-6)^2} = \sqrt{4+9} = \sqrt{13}$$
$$d_2 = \sqrt{(1-4)^2 + (3-1)^2} = \sqrt{9+4} = \sqrt{13}$$
$$d_3 = \sqrt{(3-4)^2 + (6-1)^2} = \sqrt{1+25} = \sqrt{26}$$
$$\left(\sqrt{13}\right)^2 + \left(\sqrt{13}\right)^2 \stackrel{?}{=} \left(\sqrt{26}\right)^2$$
$$13 + 13 \stackrel{?}{=} 26$$
$$26 = 26$$

By the converse of the Pythagorean Theorem, it is a right triangle.

81.

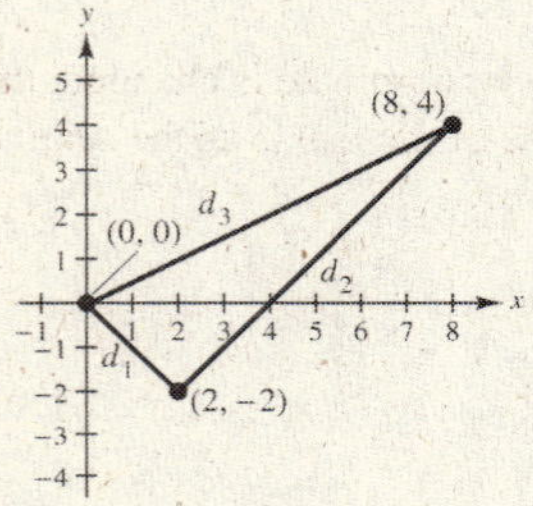

$$d_1 = \sqrt{(2-0)^2 + (-2-0)^2} = \sqrt{4+4} = \sqrt{8}$$
$$d_2 = \sqrt{(8-2)^2 + \left[4-(-2)\right]^2} = \sqrt{36+36} = \sqrt{72}$$
$$d_3 = \sqrt{(8-0)^2 + (4-0)^2} = \sqrt{64+16} = \sqrt{80}$$
$$\left(\sqrt{8}\right)^2 + \left(\sqrt{72}\right)^2 \stackrel{?}{=} \left(\sqrt{80}\right)^2$$
$$8 + 72 \stackrel{?}{=} 80$$
$$80 = 80$$

By the converse of the Pythagorean Theorem, it is a right triangle.

83. $d_1 = \sqrt{(2-2)^2 + (3-6)^2} = \sqrt{0+9} = 3$

$d_2 = \sqrt{(2-6)^2 + (3-3)^2} = \sqrt{16+0} = 4$

$d_3 = \sqrt{(2-6)^2 + (6-3)^2} = \sqrt{16+9} = 5$

$3 + 4 \neq 5$

Not collinear

85. $d_1 = \sqrt{(8-5)^2 + (3-2)^2} = \sqrt{9+1} = \sqrt{10}$

$d_2 = \sqrt{(8-2)^2 + (3-1)^2}$

$= \sqrt{36+4} = \sqrt{40} = 2\sqrt{10}$

$d_3 = \sqrt{(5-2)^2 + (2-1)^2} = \sqrt{9+1} = \sqrt{10}$

$\sqrt{10} + \sqrt{10} = 2\sqrt{10}$

Collinear

87. $d_1 = \sqrt{(-2-0)^2 + (0-5)^2} = \sqrt{4+25} = \sqrt{29}$

$d_2 = \sqrt{(0-1)^2 + (5-0)^2} = \sqrt{1+25} = \sqrt{26}$

$d_3 = \sqrt{(-2-1)^2 + (0-0)^2} = \sqrt{9} = 3$

$P = \sqrt{29} + \sqrt{26} + 3 \approx 13.48$

89. $M = \left(\dfrac{-2+4}{2}, \dfrac{0+8}{2}\right) = (1, 4)$

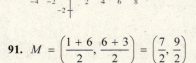

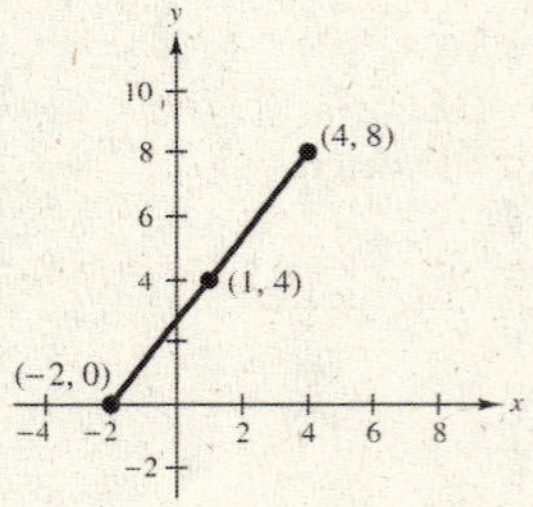

91. $M = \left(\dfrac{1+6}{2}, \dfrac{6+3}{2}\right) = \left(\dfrac{7}{2}, \dfrac{9}{2}\right)$

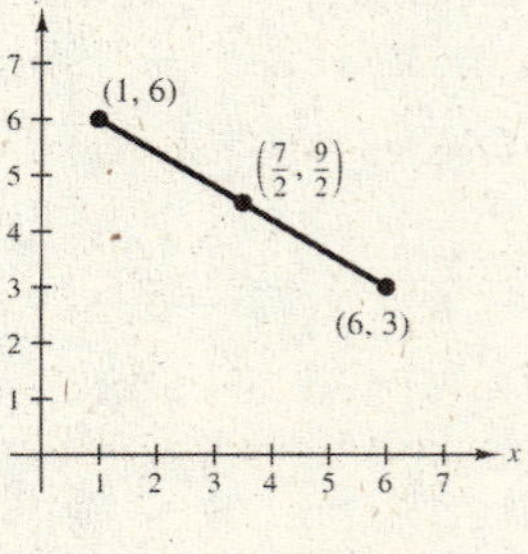

93.

x	1	2	3	4	5
$y = 150x + 425$	575	725	875	1025	1175

$$y = 150(1) + 425 = 150 + 425 = 575$$
$$y = 150(2) + 425 = 300 + 425 = 725$$
$$y = 150(3) + 425 = 450 + 425 = 875$$
$$y = 150(4) + 425 = 600 + 425 = 1025$$
$$y = 150(5) + 425 = 750 + 425 = 1175$$

The cost of installation is \$425, plus \$150 for every window installed.

95. $d = \sqrt{(x_2 - x_1)^2 + (y_2 - y_1)^2}$

Let $(x, y) = (10, 10)$ and $(x_2, y_2) = (35, 40)$.

$$d = \sqrt{(35 - 10)^2 + (40 - 10)^2}$$
$$= \sqrt{25^2 + 30^2}$$
$$= \sqrt{625 + 900}$$
$$= \sqrt{1525}$$
$$= 5\sqrt{61} \approx 39.05$$

The pass is about 39.05 yards long.

97. The faster the car travels, up to 60 kilometers per hour, the less gas it uses. As the speed increases past 60 kilometers per hour, the car uses progressively more gas.

10 liters (per 100 kilometers traveled) is used when the car is traveling at 120 kilometers per hour.

99. $M = \left(\frac{2004 + 2006}{2}, \frac{8279 + 19{,}315}{2}\right)$

$$= \left(\frac{4010}{2}, \frac{27{,}594}{2}\right)$$
$$= (2005, 13{,}797)$$

The net sales in 2005 is estimated to be \$13,797 million.

101. The Pythagorean Theorem states that, for a right triangle with hypotenuse c and sides a and b, $a^2 + b^2 = c^2$.

If a right triangle has leg lengths 3 and 4, then the hypotenuse has length
$\sqrt{3^2 + 4^2} = \sqrt{9 + 16} = \sqrt{25} = 5$.

103. No. The scales on the x-and y-axes are determined by the magnitudes of the quantities being measured by x and y.

105.

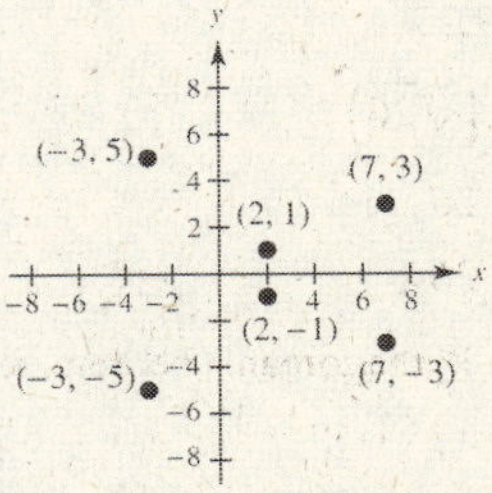

When the sign of the y-coordinate is changed, the point is on the opposite side of the x-axis as the original point.

107.
$$\frac{5}{6} = \frac{y}{36}$$
$$5 \cdot 36 = 6y$$
$$180 = 6y$$
$$30 = y$$

109.
$$\frac{m}{49} = \frac{5}{7}$$
$$7m = 5 \cdot 49$$
$$7m = 245$$
$$m = 35$$

111.
$$\frac{n}{16} = \frac{n - 3}{8}$$
$$8n = 16(n - 3)$$
$$8n = 16n - 48$$
$$-8n = -48$$
$$n = 6$$

113. Verbal model: $\boxed{\text{Discount}} = \boxed{\text{Discount rate}} \cdot \boxed{\text{List price}}$

Labels: Discount $= x$

Discount rate $= 0.35$

List price $= 55$

Equation: $x = 0.35(55)$

$x = 19.25$

The discount is \$19.25.

Verbal model: $\boxed{\text{Sale price}} = \boxed{\text{List price}} - \boxed{\text{Discount}}$

Labels: Sale price $= x$

List price $= 55$

Discount $= 19.25$

Equation: $x = 55 - 19.25$

$x = 35.75$

The sale price is \$35.75.

115. Verbal model: $\boxed{\text{List price}} = \boxed{\text{Sale price}} + \boxed{\text{Discount}}$

Labels: List price $= x$

Sale price $= 134.42$

Discount $= 124.08$

Equation: $x = 134.42 + 124.08$

$x = 258.50$

The list price is \$258.50.

Verbal model: $\boxed{\text{Discount}} = \boxed{\text{Discount rate}} \cdot \boxed{\text{List price}}$

Labels: Discount $= 124.08$

Discount rate $= p$

List price $= 258.50$

Equation: $124.08 = p(258.50)$

$$\frac{124.08}{258.50} = p$$

$0.48 = p$

The discount rate is 48%.

Section 3.2 Graphs of Equations

1. $y = 2$ matches graph (e).

3. $y = 2 - x$ matches graph (f).

5. $y = x^2 - 4$ matches graph (d).

7.

x	-2	-1	0	1	2
$y = 3x$	-6	-3	0	3	6
Solution point	$(-2, -6)$	$(-1, -3)$	$(0, 0)$	$(1, 3)$	$(2, 6)$

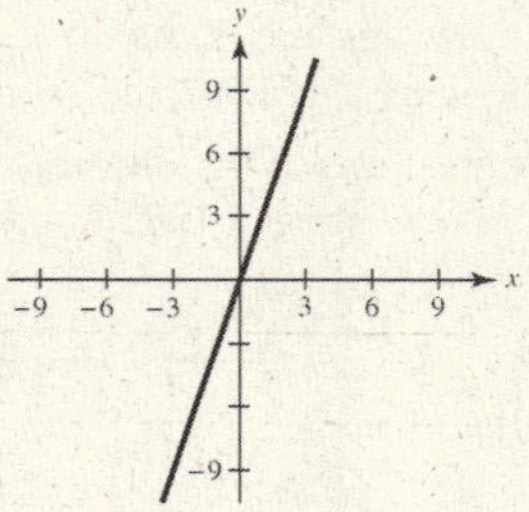

9.

x	-2	-1	0	1	2
$y = 4 - x$	6	5	4	3	2
Solution point	$(-2, 6)$	$(-1, 5)$	$(0, 4)$	$(1, 3)$	$(2, 2)$

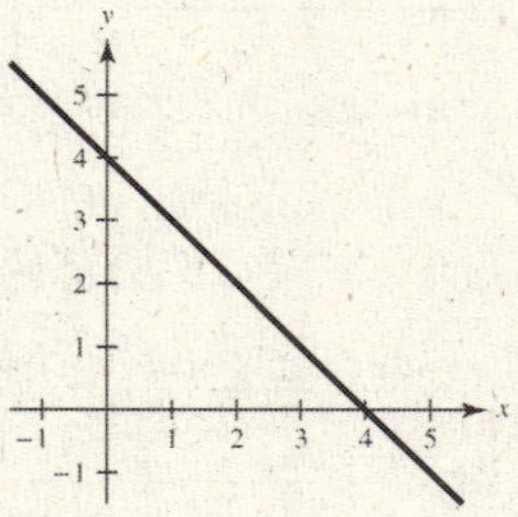

11. $2x - y = 3$

$-y = -2x + 3$

$y = 2x - 3$

x	-2	-1	0	1	2
$y = 2x - 3$	-7	-5	-3	-1	1
Solution point	$(-2, -7)$	$(-1, -5)$	$(0, -3)$	$(1, -1)$	$(2, 1)$

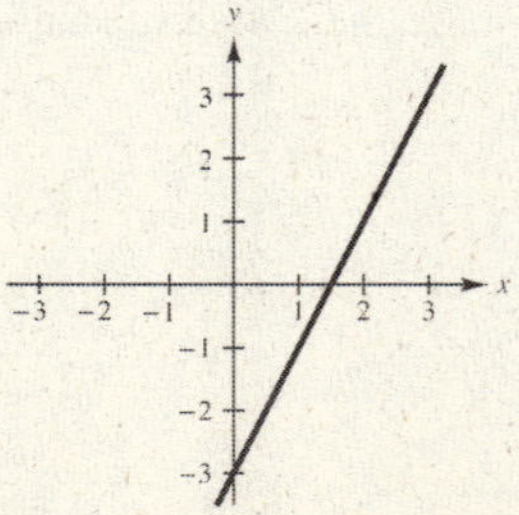

13. $3x + 2y = 2$

$2y = -3x + 2$

$y = -\frac{3}{2}x + 1$

x	-2	-1	0	1	2
$y = -\frac{3}{2}x + 1$	4	$\frac{5}{2}$	1	$-\frac{1}{2}$	-2
Solution point	$(-2, 4)$	$\left(-1, \frac{5}{2}\right)$	$(0, 1)$	$\left(1, -\frac{1}{2}\right)$	$(2, -2)$

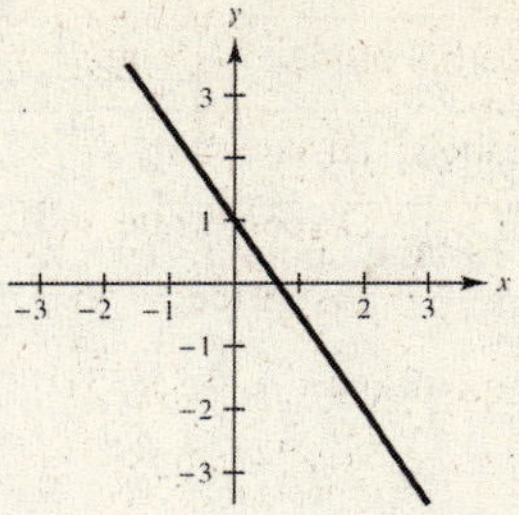

15.

x	-2	-1	0	1	2
$y = -x^2$	-4	-1	0	-1	-4
Solution point	$(-2, -4)$	$(-1, -1)$	$(0, 0)$	$(1, -1)$	$(2, -4)$

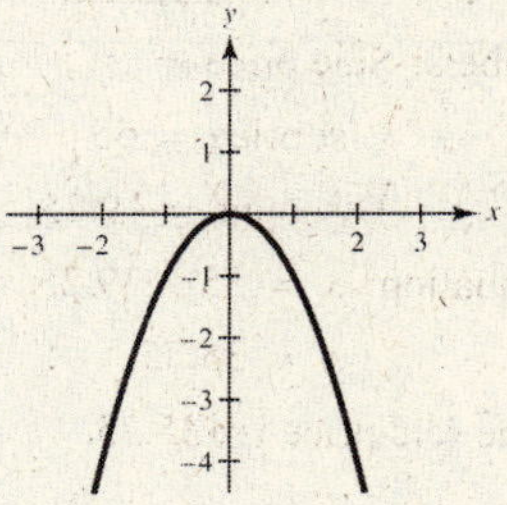

17.

x	-2	-1	0	1	2
$y = x^2 - 3$	1	-2	-3	-2	1
Solution point	$(-2, 1)$	$(-1, -2)$	$(0, -3)$	$(1, -2)$	$(2, 1)$

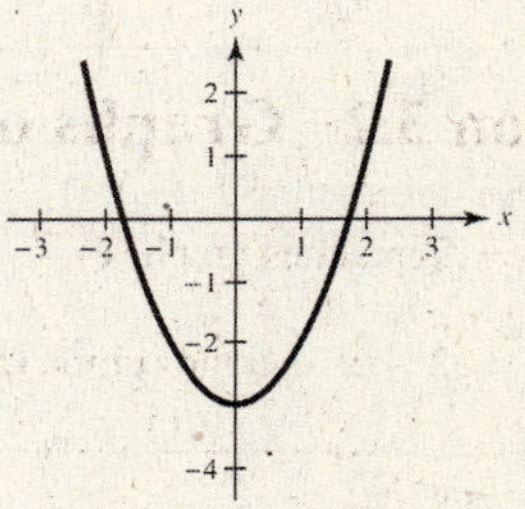

19. $-x^2 - 3x + y = 0$

$y = x^2 + 3x$

x	-2	-1	0	1	2
$y = x^2 + 3x$	-2	-2	0	4	10
Solution point	$(-2, -2)$	$(-1, -2)$	$(0, 0)$	$(1, 4)$	$(2, 10)$

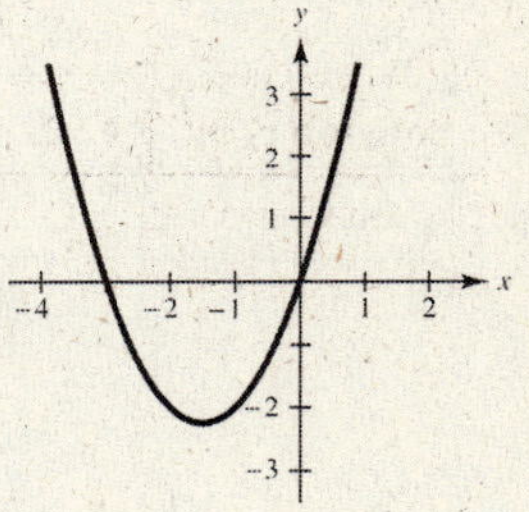

21. $x^2 - 2x - y = 1$

$-y = -x^2 + 2x + 1$

$y = x^2 - 2x - 1$

x	-2	-1	0	1	2
$y = x^2 - 2x - 1$	7	2	-1	-2	-1
Solution point	$(-2, 7)$	$(-1, 2)$	$(0, -1)$	$(1, -2)$	$(2, -1)$

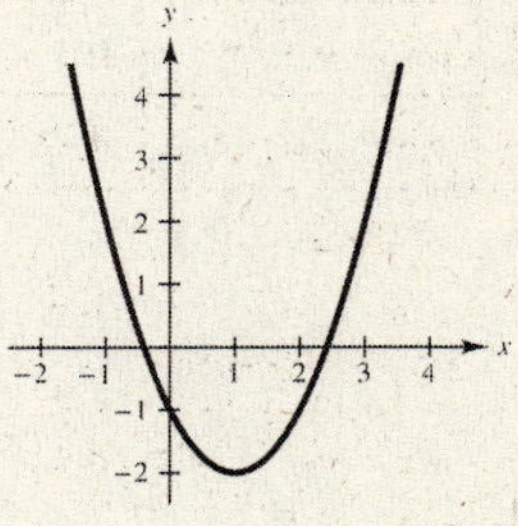

23.

x	-2	-1	0	1	2
$y = \lvert x \rvert$	2	1	0	1	2
Solution point	$(-2, 2)$	$(-1, 1)$	$(0, 0)$	$(1, 1)$	$(2, 2)$

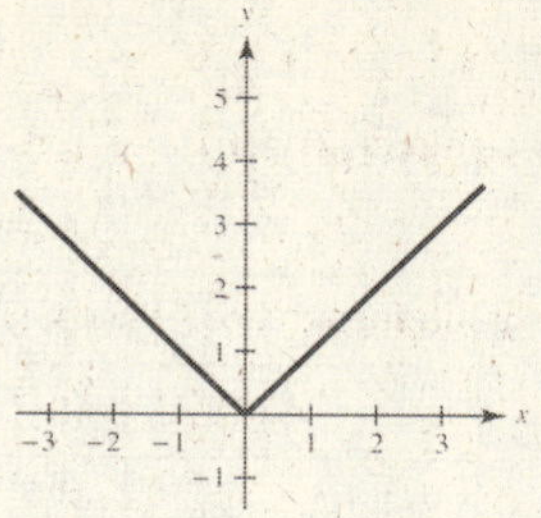

25.

x	-2	-1	0	1	2
$y = \lvert x \rvert + 3$	5	4	3	4	5
Solution point	$(-2, 5)$	$(-1, 4)$	$(0, 3)$	$(1, 4)$	$(2, 5)$

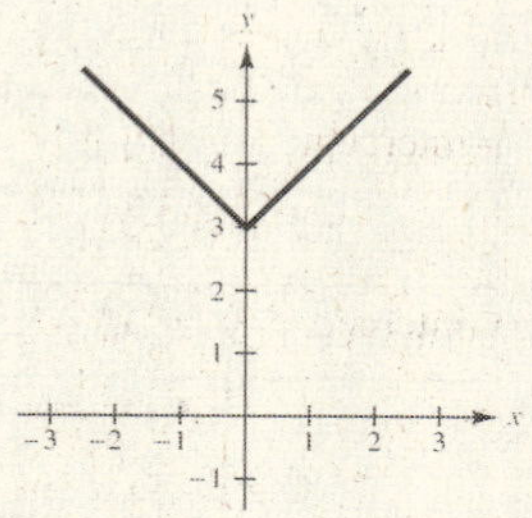

27.

x	-2	-1	0	1	2
$y = \lvert x + 3 \rvert$	1	2	3	4	5
Solution point	$(-2, 1)$	$(-1, 2)$	$(0, 3)$	$(1, 4)$	$(2, 5)$

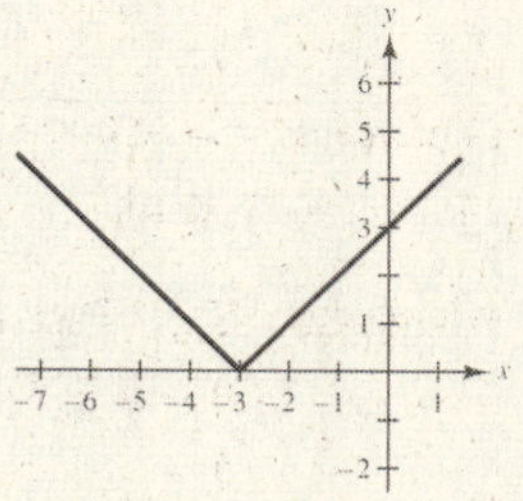

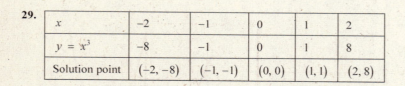

29.

x	-2	-1	0	1	2
$y = x^3$	-8	-1	0	1	8
Solution point	$(-2, -8)$	$(-1, -1)$	$(0, 0)$	$(1, 1)$	$(2, 8)$

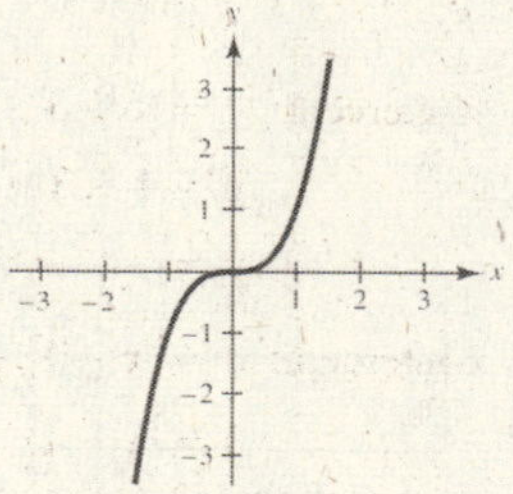

31. $y = 6x - 3$

y-intercept: $y = 6(0) - 3$

$y = -3 \quad (0, -3)$

x-intercept: $0 = 6x - 3$

$3 = 6x$

$\frac{3}{6} = x$

$\frac{1}{2} = x \quad \left(\frac{1}{2}, 0\right)$

33. $y = 12 - \frac{2}{5}x$

y-intercept: $y = 12 - \frac{2}{5}(0)$

$y = 12 \quad (0, 12)$

x-intercept: $0 = 12 - \frac{2}{5}x$

$\frac{2}{5}x = 12$

$x = 30 \quad (30, 0)$

35. $x + 2y = 10$

y-intercept: $0 + 2y = 0$

$y = 5 \quad (0, 5)$

x-intercept: $x + 2(0) = 10$

$x = 10 \quad (10, 0)$

37. $4x - y + 3 = 0$

y-intercept: $4(0) - y + 3 = 0$

$3 = y \qquad (0, 3)$

x-intercept: $4x - 0 + 3 = 0$

$4x = -3$

$x = -\frac{3}{4} \qquad \left(-\frac{3}{4}, 0\right)$

39. $y = |x| - 1$

y-intercept: $y = |0| - 1$

$y = -1 \qquad (0, -1)$

x-intercept: $0 = |x| - 1$

$1 = |x|$

$\pm 1 = x \qquad (1, 0), (-1, 0)$

41. $y = -|x + 5|$

y-intercept: $y = -|0 + 5|$

$y = -5 \qquad (0, -5)$

x-intercept: $0 = -|x + 5|$

$0 = x + 5$

$-5 = x \qquad (-5, 0)$

43. $y = |x - 1| - 3$

y-intercept: $y = |0 - 1| - 3$

$y = 1 - 3$

$y = -2 \qquad (0, -2)$

x-intercept: $0 = |x - 1| - 3$

$3 = |x - 1|$

$3 = x - 1$ or $-3 = x - 1$

$4 = x \qquad -2 = x \qquad (4, 0), (-2, 0)$

45. $2x + 3y = 6$

Estimate: y-intercept: $(0, 2)$

x-intercept: $(3, 0)$

47. $y = x^2 + 3$

Estimate: y-intercept: $(0, 3)$

no x-intercepts

49. $y = -2$

Estimate: y-intercept: $(0, -2)$

no x-intercepts

51. $y = 8 - 4x$

Estimate: y-intercept: $(0, 8)$

x-intercept: $(2, 0)$

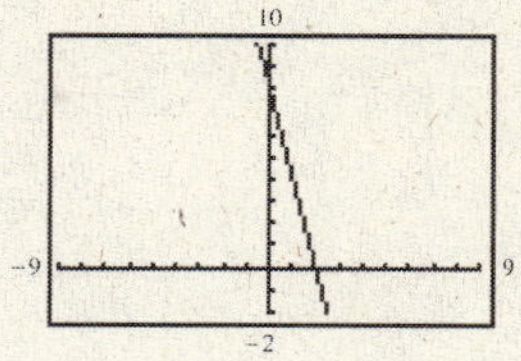

53. $y = (x - 1)(x - 6)$

Estimate: y-intercept: $(0, 6)$

x-intercepts: $(1, 0), (6, 0)$

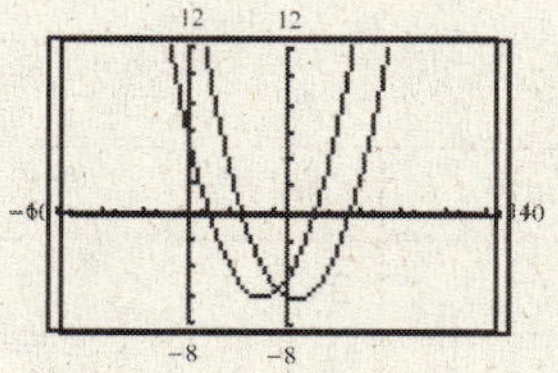

55. $y = |4x + 6| - 2$

Estimate: y-intercept: $(0, 4)$

x-intercepts: $(-1, 0), (-2, 0)$

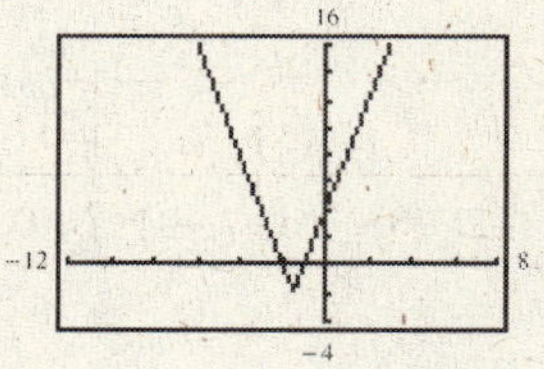

57. $y = 3 - x$

$y = 3 - 0$

$y = 3 \qquad (0, 3)$

$0 = 3 - x$

$x = 3 \qquad (3, 0)$

$y = 3 - 1$

$y = 2 \qquad (1, 2)$

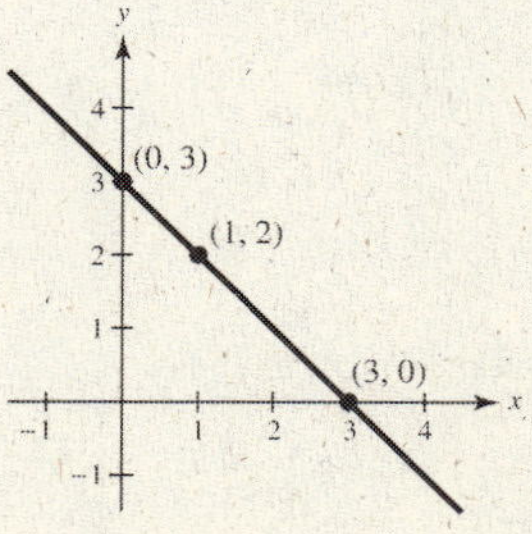

59. $y = 2x - 3$

$y = 2(0) - 3$

$y = -3$ $(0, -3)$

$0 = 2x - 3$

$3 = 2x$

$\frac{3}{2} = x$ $\left(\frac{3}{2}, 0\right)$

$y = 2(3) - 3$

$y = 3$ $(3, 3)$

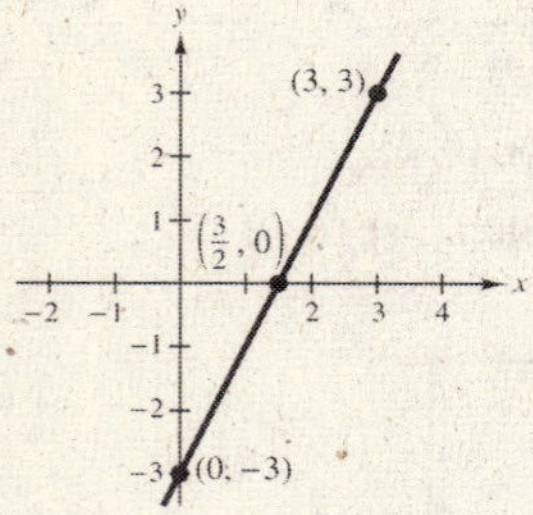

61. $4x + y = 3$

$4(0) + y = 3$

$y = 3$ $(0, 3)$

$4x + 0 = 3$

$4x = 3$

$x = \frac{3}{4}$ $\left(\frac{3}{4}, 0\right)$

$4(1) + y = 3$

$y = -1$ $(1, -1)$

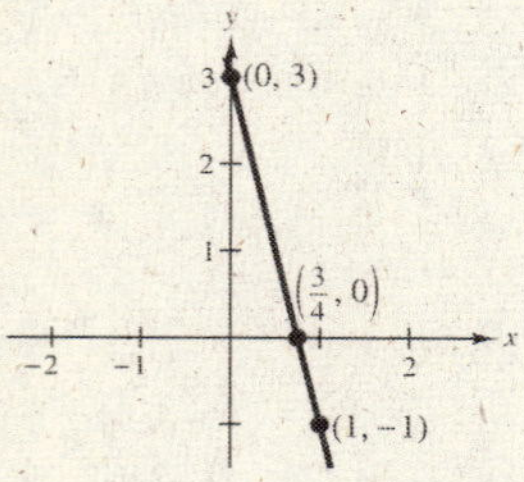

63. $2x - 3y = 6$

$2(0) - 3y = 6$

$-3y = 6$

$y = -2$ $(0, -2)$

$2x - 3(0) = 6$

$2x = 6$

$x = 3$ $(3, 0)$

$2(1) - 3y = 6$

$-3y = 4$

$y = -\frac{4}{3}$ $\left(1, -\frac{4}{3}\right)$

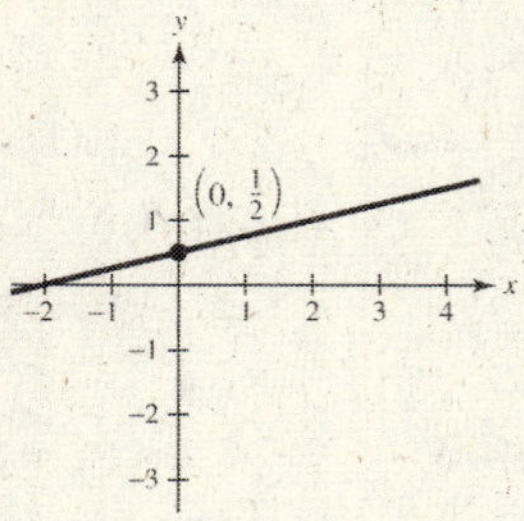

65. $3x + 4y = 12$

$3(0) + 4y = 12$

$0 + 4y = 12$

$4y = 12$

$y = 3$ $(0, 3)$

$3x + 4(0) = 12$

$3x + 0 = 12$

$3x = 12$

$x = 4$ $(4, 0)$

$3(1) + 4y = 12$

$3 + 4y = 12$

$4y = 9$

$y = \frac{9}{4}$ $\left(1, \frac{9}{4}\right)$

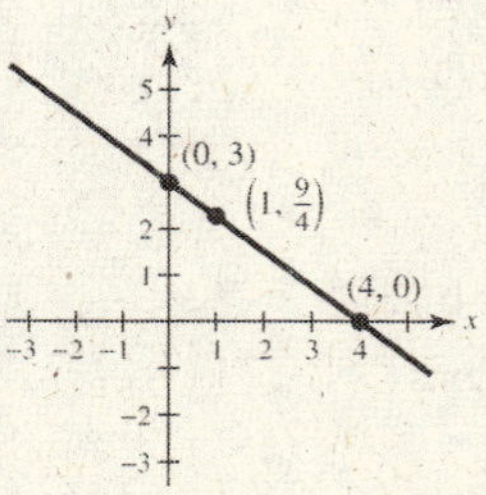

67. $x + 5y = 10$
$0 + 5y = 10$
$y = 2$ $(0, 2)$
$x + 5(0) = 10$
$x = 10$ $(10, 0)$
$5 + 5y = 10$
$5y = 5$
$y = 1$ $(5, 1)$

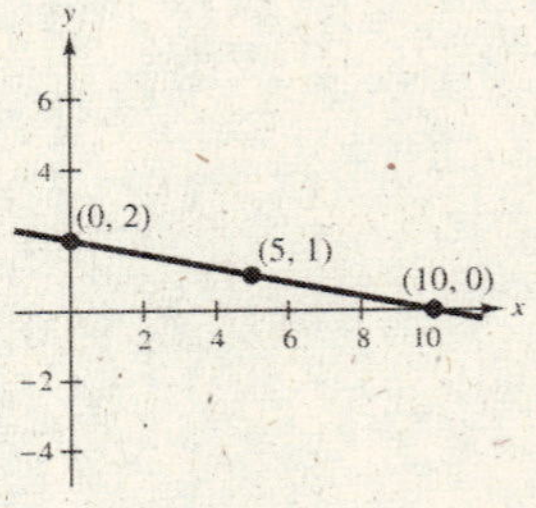

69. $5x - y = 10$
$5(0) - y = 10$
$0 - y = 10$
$-y = 10$
$y = -10$ $(0, -10)$
$5x - (0) = 10$
$5x = 10$
$x = 2$ $(2, 0)$
$5(1) - y = 10$
$5 - y = 10$
$-y = 5$
$y = -5$ $(1, -5)$

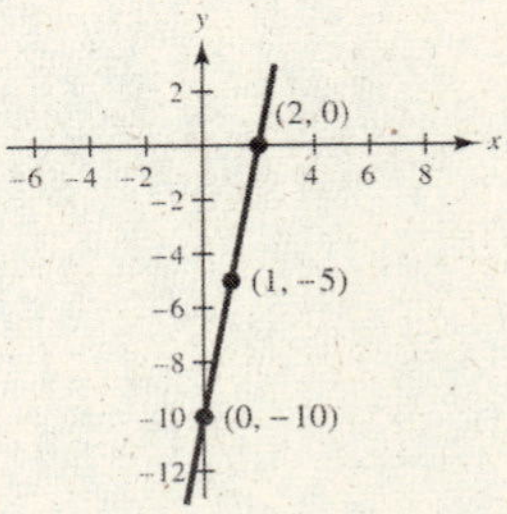

71. $y = x^2 - 9$
$0 = x^2 - 9$
$0 = (x - 3)(x + 3)$
$x = \pm 3$ $(3, 0), (-3, 0)$
$y = 0^2 - 9$
$y = -9$ $(0, -9)$

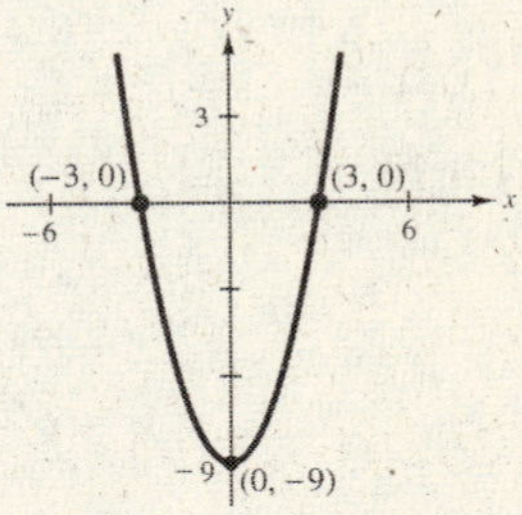

73. $y = 9 - x^2$
$0 = 9 - x^2$
$x^2 = 9$
$x = \pm 3$ $(3, 0), (-3, 0)$
$y = 9 - 0^2$
$y = 9$ $(0, 9)$

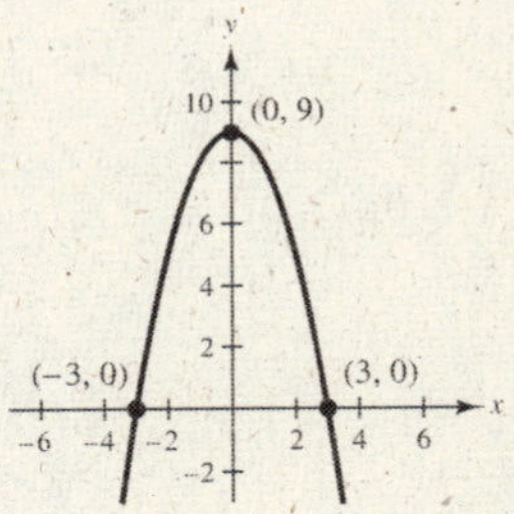

75. $y = 1 - x^2$

$y = 1 - 0$

$y = 1$ $\quad (0, 1)$

$y = 1 - (1)^2$

$y = 1 - 1$

$y = 0$ $\quad (1, 0)$

$y = 1 - (-1)^2$

$y = 1 - 1$

$y = 0$ $\quad (-1, 0)$

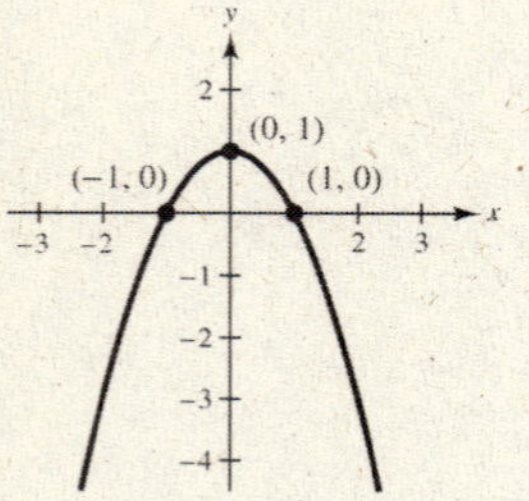

77. $y = x^2 - 4$

$y = 0^2 - 4$

$= -4$ $\quad (0, -4)$

$0 = x^2 - 4$

$4 = x^2$

$\pm 2 = x$ $\quad (-2, 0), (2, 0)$

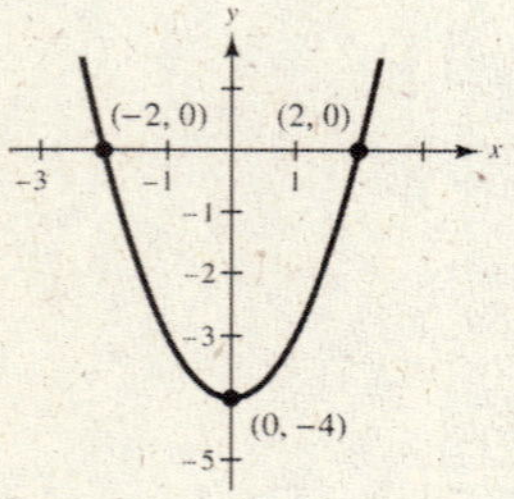

79. $y = x(x - 2)$

$y = 0^2 - 2(0)$

$y = 0$ $\quad (0, 0)$

$0 = x^2 - 2x$

$0 = x(x - 2)$

$x = 0, 2$ $\quad (0, 0), (2, 0)$

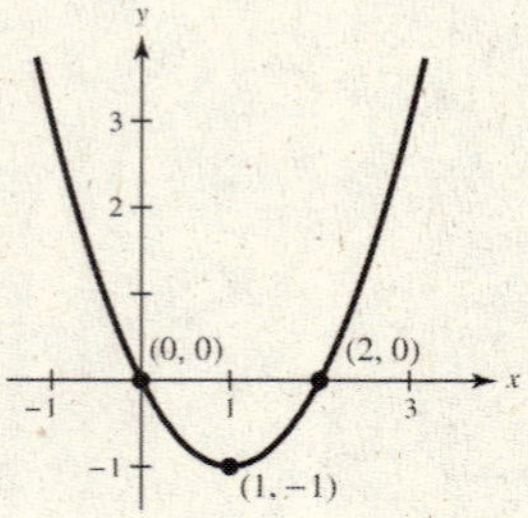

81. $y = -x(x + 4)$

$y = -(0)(0 + 4)$

$y = 0$ $\quad (0, 0)$

$0 = -x(x + 4)$

$x = 0, -4$ $\quad (0, 0), (-4, 0)$

$y = -(-2)(-2 + 4)$

$y = 2(2)$

$y = 4$ $\quad (-2, 4)$

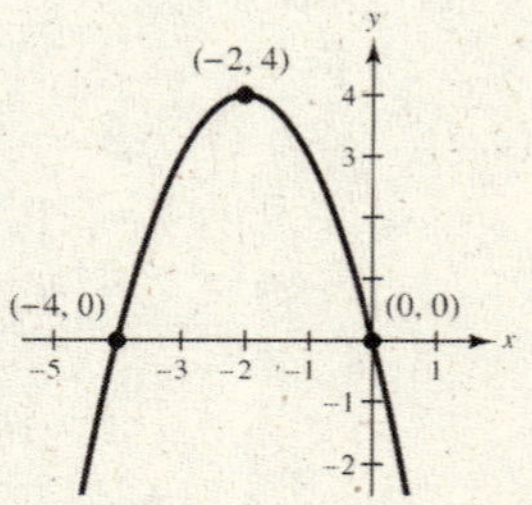

83. $y = |x| - 3$

$y = -3 \qquad (0, -3)$

$0 = |x| - 3$

$3 = |x|$

$x = -3, 3 \qquad (3, 0), (-3, 0)$

85. $y = |x| + 2$

$y = |-1| + 2$

$y = 1 + 2$

$y = 3 \qquad (-1, 3)$

$y = |0| + 2$

$y = 0 + 2$

$y = 2 \qquad (0, 2)$

$y = |1| + 2$

$y = 1 + 2$

$y = 3 \qquad (1, 3)$

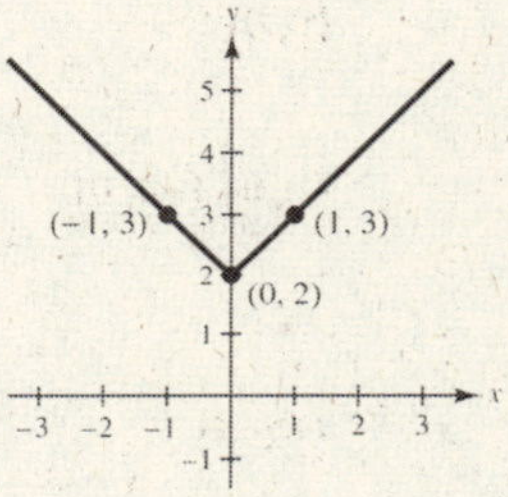

87. $y = |x + 2|$

$= 2 \qquad (0, 2)$

$y = |x + 2|$

$0 = x + 2$

$-2 = x \qquad (-2, 0)$

$y = |-4 + 2|$

$= 2 \qquad (-4, 2)$

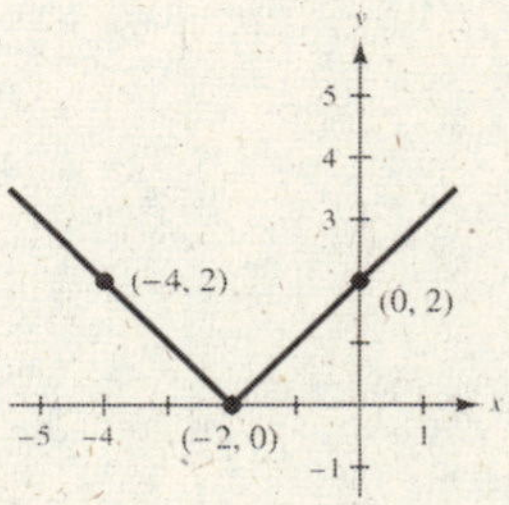

89. $y = |x - 3|$

$0 = |x - 3|$

$x = 3 \qquad (3, 0)$

$y = |0 - 3|$

$y = |-3|$

$y = 3 \qquad (0, 3)$

$y = |6 - 3|$

$y = |3|$

$y = 3 \qquad (6, 3)$

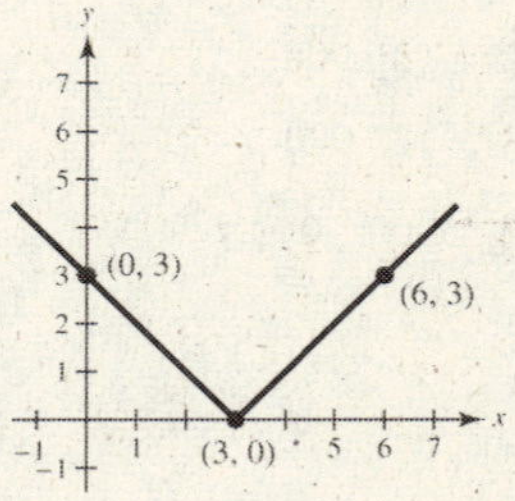

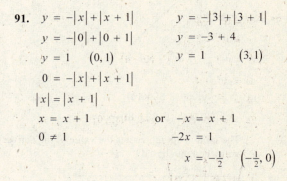

91. $y = -|x| + |x + 1|$

$y = -|0| + |0 + 1|$

$y = 1 \qquad (0, 1)$

$0 = -|x| + |x + 1|$

$|x| = |x + 1|$

$x = x + 1$ or $-x = x + 1$

$0 \neq 1 \qquad -2x = 1$

$x = -\frac{1}{2} \qquad \left(-\frac{1}{2}, 0\right)$

$y = -|3| + |3 + 1|$

$y = -3 + 4$

$y = 1 \qquad (3, 1)$

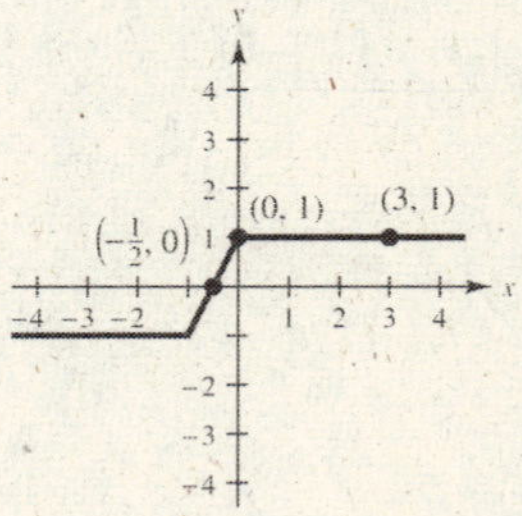

93. $y = 230{,}000 - 25{,}000t$

$y = 230{,}000 - 25{,}000(0)$

$y = 230{,}000 \qquad (0, 230{,}000)$

$y = 230{,}000 - 25{,}000(8)$

$y = 30{,}000 \qquad (0, 30{,}000)$

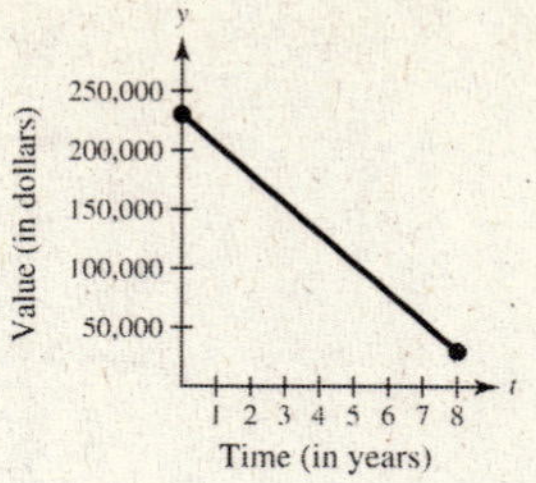

95. (a) The annual depreciation is

$$\frac{40{,}000 - 5000}{7} = 5000.$$

$40{,}000 - 5000(1) = 35{,}000$

$40{,}000 - 5000(2) = 30{,}000$

$40{,}000 - 5000(3) = 25{,}000$

So, $y = 40{,}000 - 5000t,\ 0 \le t \le 7.$

(b)

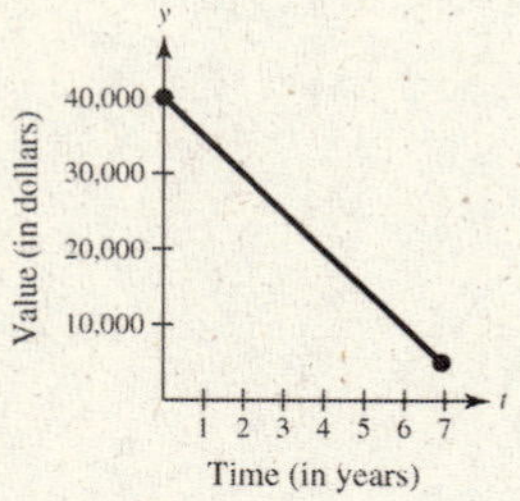

(c) $y = 40{,}000 - 5000(0)$

$y = 40{,}000,\ (0, 40{,}000)$

The y-intercept represents the value of the delivery van when purchased.

97. (a)

x	0	3	6	9	12
$\frac{4}{3}x$	0	4	8	12	16

(b)

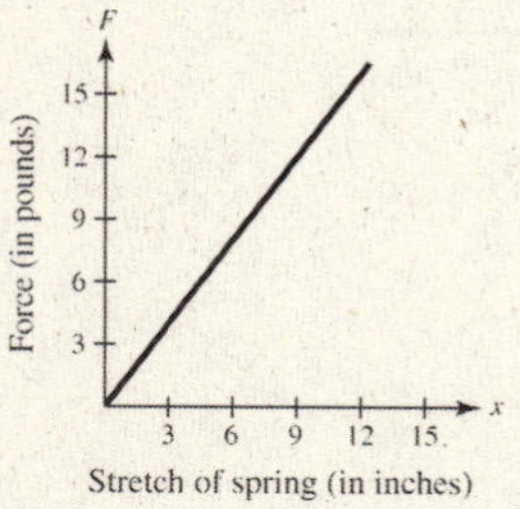

(c) F doubles because F is directly proportional to x.

99. The scales on the y-axes are different. From graph (a) it appears that sales have not increased. From graph (b) it appears that sales have increased dramatically.

101.

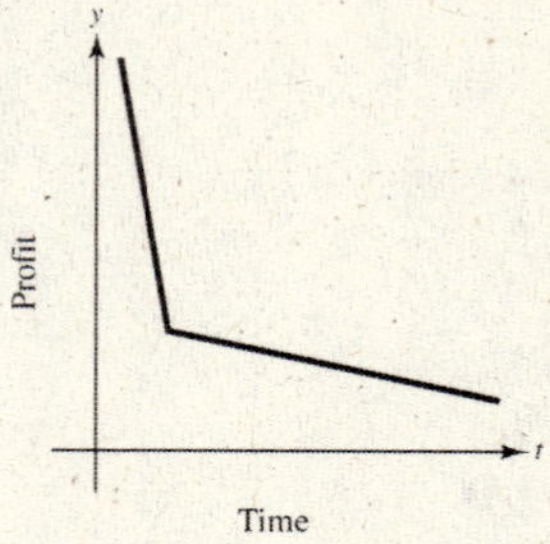

103. A horizontal line has no x-intercepts unless $y = 0$. The y-intercept is $(0, b)$, where b is any real number. So, a horizontal line has one y-intercept.

105. $\frac{1}{7}$

107. $\frac{5}{4}$

109. $x - 8 = 0$

$x = 8$

Check:

$8 - 8 \stackrel{?}{=} 0$

$0 = 0$

111. $4x + 15 = 23$

$4x = 8$

$x = 2$

Check:

$4(2) + 15 \stackrel{?}{=} 23$

$8 + 15 = 23$

113.

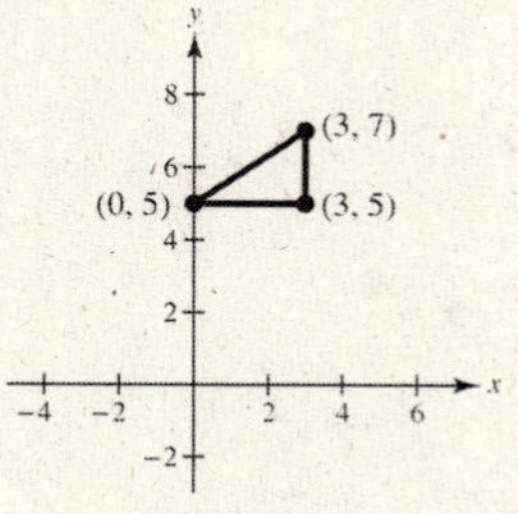

Section 3.3 Slope and Graphs of Linear Equations

1. $(0, 8)$ and $(4, 0)$

$$m = \frac{0 - 8}{4 - 0} = \frac{-8}{4} = -2$$

3. $(0, 0)$ and $(5, 2)$

$$m = \frac{2 - 0}{5 - 0} = \frac{2}{5}$$

5. $(3, 0)$ and $(3, 8)$

$$m = \frac{8 - 0}{3 - 3} = \frac{8}{0} = \text{undefined}$$

7. (a) $m = \frac{3}{4} \Rightarrow L_3$

(b) $m = 0 \Rightarrow L_2$

(c) $m = -3 \Rightarrow L_1$

9. $m = \dfrac{8 - 0}{4 - 0} = \dfrac{8}{4} = 2$

The slope is 2.

11. $m = \dfrac{4 - 4}{7 - 2} = \dfrac{0}{5} = 0$

The slope is 0.

13. $m = \dfrac{3 - 5}{-4 - (-2)} = \dfrac{-2}{-2} = 1$ Line rises.

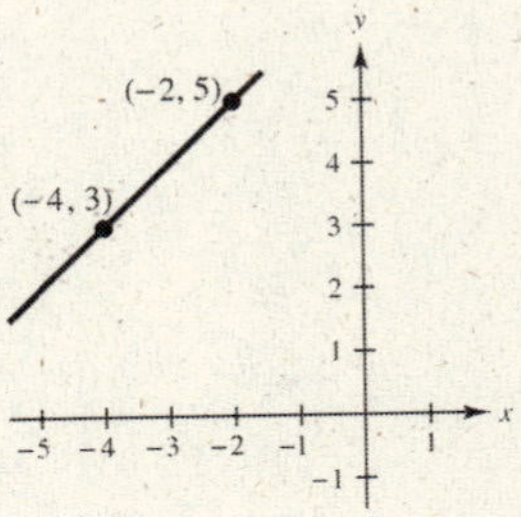

15. $m = \dfrac{4 - (-3)}{-5 - (-5)} = \dfrac{7}{0} = \text{undefined}$ Line is vertical.

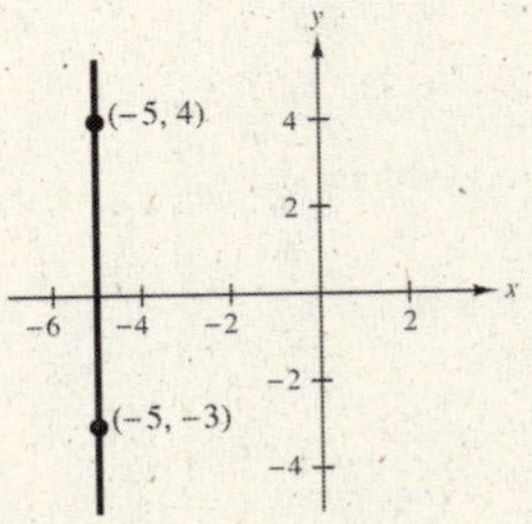

17. $m = \dfrac{-5 - (-5)}{7 - 2} = \dfrac{0}{5} = 0$ Line is horizontal.

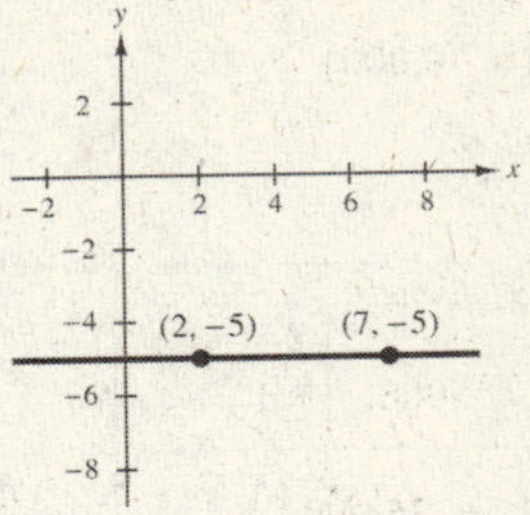

19. $m = \dfrac{2 - \frac{-5}{2}}{\frac{3}{4} - 5} \cdot \dfrac{4}{4} = \dfrac{8 + 10}{3 - 20} = -\dfrac{18}{17}$ Line falls.

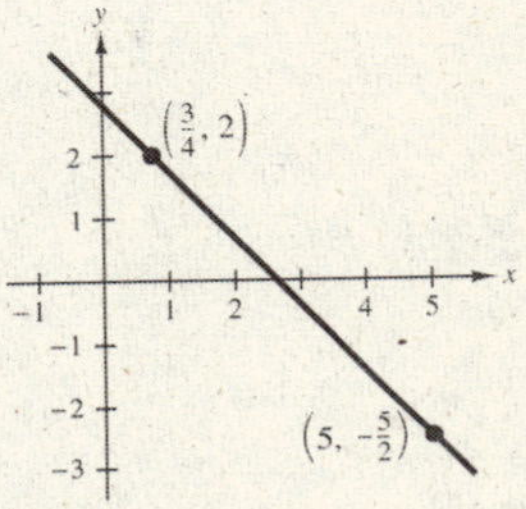

21. $m = \dfrac{\frac{1}{4} - \frac{1}{8}}{\frac{3}{4} - \frac{-3}{2}} \cdot \dfrac{8}{8} = \dfrac{2 - 1}{6 + 12} = \dfrac{1}{18}$ Line rises.

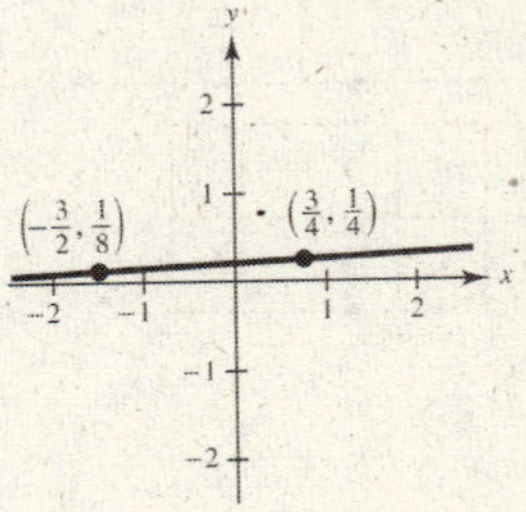

23. $m = \dfrac{6 - (-1)}{-4.2 - 4.2} = \dfrac{7}{-8.4} = -\dfrac{70}{84} = -\dfrac{5}{6}$ Line falls.

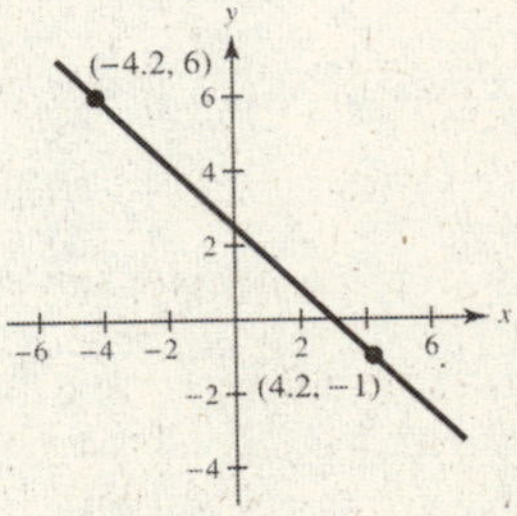

25.

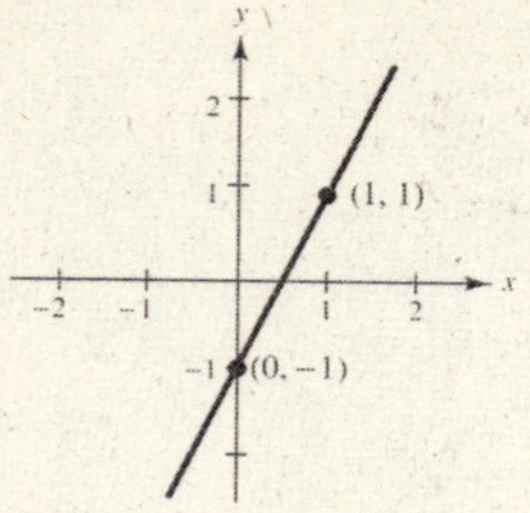

x	-1	0	1
$y = 2x - 1$	-3	-1	1
Solution point	$(-1, -3)$	$(0, -1)$	$(1, 1)$

$$m = \frac{1 - (-1)}{1 - 0} = \frac{2}{1} = 2$$

27.

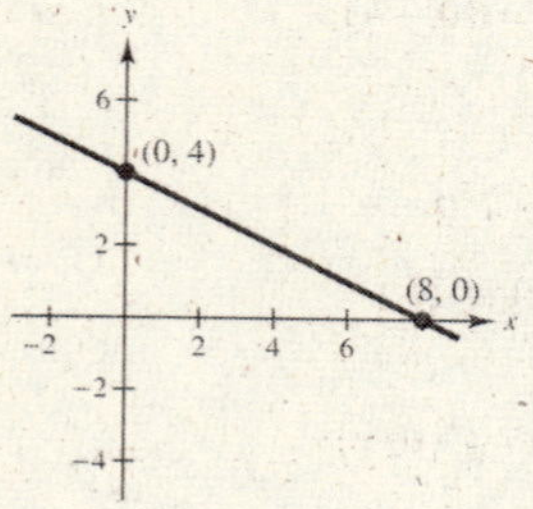

x	-1	0	1
$y = -\frac{1}{2}x + 4$	$\frac{9}{2}$	4	$\frac{7}{2}$
Solution point	$\left(-1, \frac{9}{2}\right)$	$(0, 4)$	$\left(1, \frac{7}{2}\right)$

$$m = \frac{\frac{7}{2} - 4}{1 - 0} = \frac{7}{2} - \frac{8}{2} = -\frac{1}{2}$$

29. $3x - 9y = 18$

$$-9y = -3x + 18$$

$$y = \frac{1}{3}x - 2$$

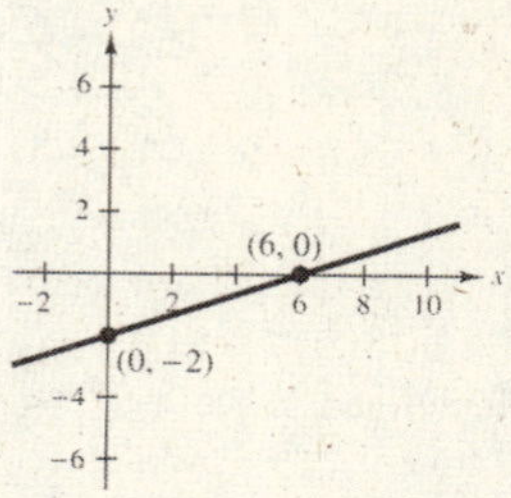

x	0	3	6
$y = \frac{1}{3}x - 2$	-2	-1	0
Solution point	$(0, -2)$	$(3, -1)$	$(6, 0)$

$$m = \frac{0 - (-2)}{6 - 0} = \frac{2}{6} = \frac{1}{3}$$

31. $\frac{-2}{3} = \frac{7 - 5}{x - 4}$

$$-2(x - 4) = 6$$

$$-2x + 8 = 6$$

$$-2x = -2$$

$$x = 1$$

33. $\frac{3}{2} = \frac{3 - y}{9 - (-3)}$

$$3(12) = 2(3 - y)$$

$$36 = 6 - 2y$$

$$30 = -2y$$

$$-15 = y$$

35. Sample answer: $0 = \frac{y - 2}{x - 5}$

Horizontal line: $(1, 2)$, $(0, 2)$, $(3, 2)$

Any points with a y-coordinate of 2

37. Sample answer: $3 = \dfrac{y + 4}{x - 3}$

Let $x = 4$. Solve for y:

$3 = \dfrac{y + 4}{4 - 3}$

$3 = y + 4$

$-1 = y, (4, -1)$

Let $x = 5$. Solve for y:

$3 = \dfrac{y + 4}{5 - 3}$

$6 = y + 4$

$2 = y, (5, 2)$

39. Sample answer: $-1 = \dfrac{y + 3}{x + 2}$

Let $x = -3$. Solve for y:

$-1 = \dfrac{y + 3}{-3 + 2}$

$-1 = -y - 3$

$y = -2, (-3, -2)$

Let $x = -1$. Solve for y:

$-1 = \dfrac{y + 3}{-1 + 2}$

$-1 = y + 3$

$-4 = y, (-1, -4)$

41. Sample answer: $\dfrac{4}{3} = \dfrac{y - 0}{x + 5}$

Let $x = -2$. Solve for y:

$\dfrac{4}{3} = \dfrac{y}{-2 + 5}$

$4 = y, (-2, 4)$

Let $x = 1$. Solve for y:

$\dfrac{4}{3} = \dfrac{y}{1 + 5}$

$8 = y, (1, 8)$

43. $6x - 3y = 9$

$-3y = -6x + 9$

$y = 2x - 3$

45. $4y - x = -4$

$4y = x - 4$

$y = \frac{1}{4}x - 1$

47. $2x + 5y - 3 = 0$

$5y = -2x + 3$

$y = -\frac{2}{5}x + \frac{3}{5}$

49. $x = 2y - 4$

$-2y = -x - 4$

$y = \frac{1}{2}x + 2$

51. $y = 3x - 2$

$m = 3$; y-intercept $= (0, -2)$

53. $4x - 6y = 24$

$-6y = -4x + 24$

$y = \frac{2}{3}x - 4$

$m = \frac{2}{3}$; y-intercept $= (0, -4)$

55. $5x + 3y - 2 = 0$

$3y = -5x + 2$

$y = -\frac{5}{3}x + \frac{2}{3}$

$m = -\frac{5}{3}$; y-intercept $= \left(0, \frac{2}{3}\right)$

57. $x + y = 0$

$y = -x$

slope $= -1$

y-intercept $= 0$

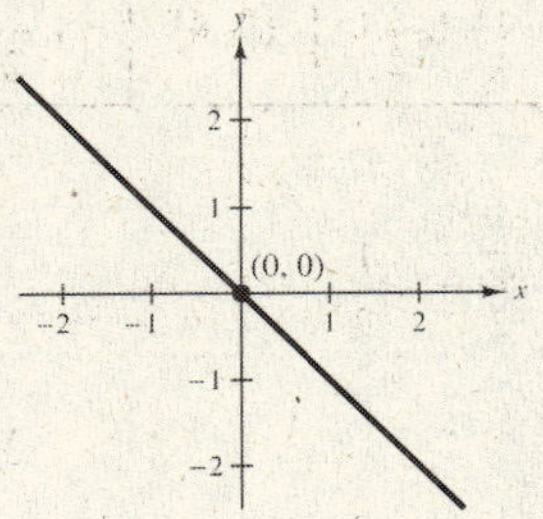

59. $3x - y - 2 = 0$

$-y = -3x + 2$

$y = 3x - 2$

slope $= 3$

y-intercept $= -2$

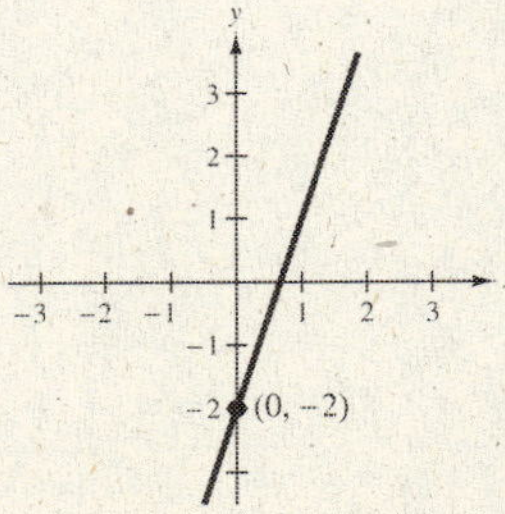

61. $3x + 2y - 2 = 0$

$$2y = -3x + 2$$

$$y = -\frac{3}{2}x + 1$$

slope $= -\frac{3}{2}$

y-intercept $= 1$

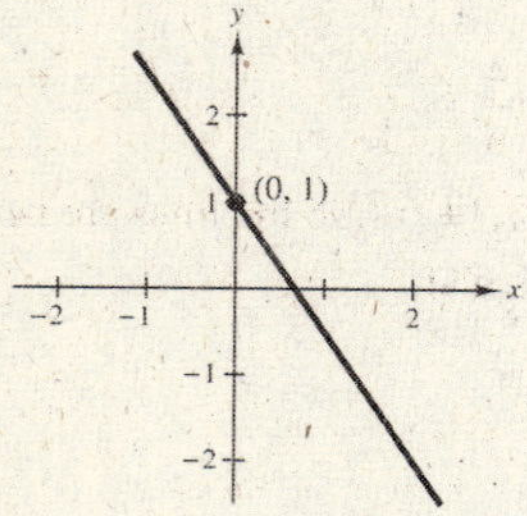

63. $x - 4y + 2 = 0$

$$-4y = -x - 2$$

$$y = \frac{1}{4}x + \frac{1}{2}$$

slope $= \frac{1}{4}$

y-intercept $= \frac{1}{2}$

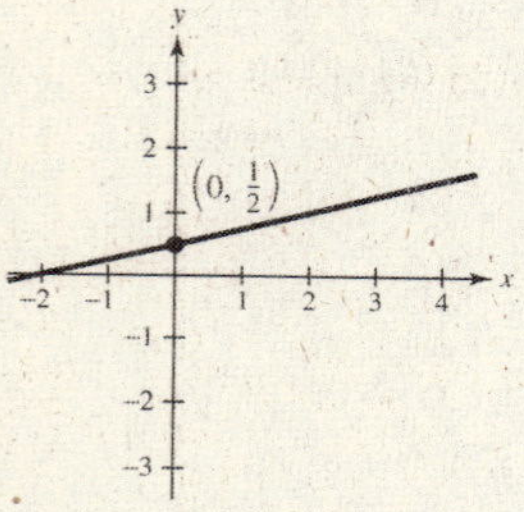

65. $0.2x - 0.8y - 4 = 0$

$$-0.8y = -0.2x + 4$$

$$y = \frac{1}{4}x - 5$$

slope $= \frac{1}{4}$

y-intercept $= (0, -5)$

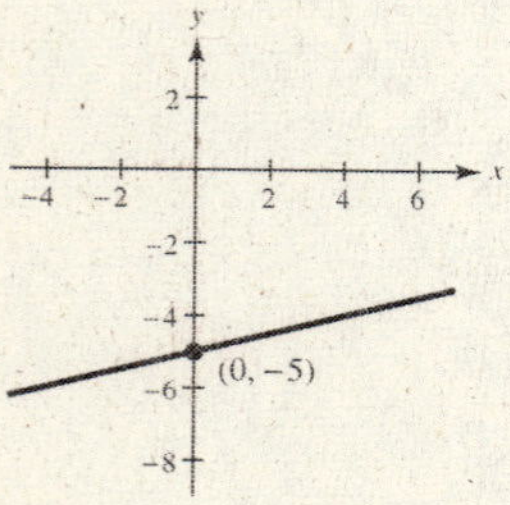

67.

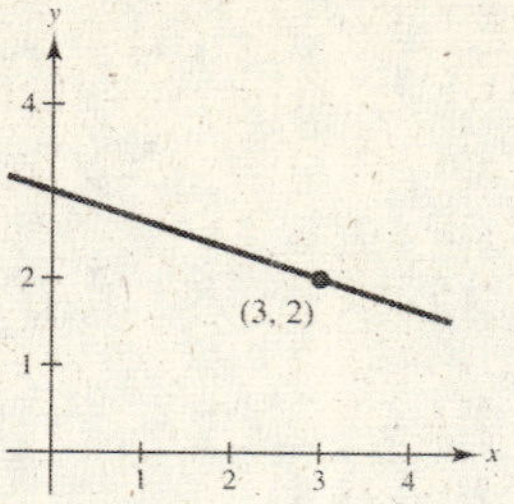

Locate a second point with the slope of $-\frac{1}{3}$.

$$m = -\frac{1}{3} = \frac{\text{Change in } y}{\text{Change in } x}$$

69.

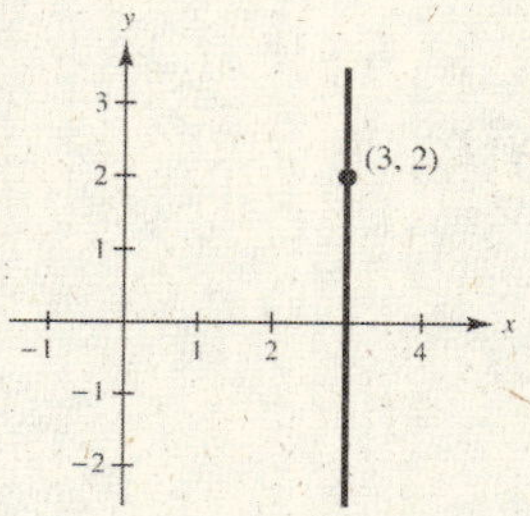

m is undefined so the line is vertical.

71. $3x - 5y - 15 = 0$

$$3(0) - 5y - 15 = 0$$

$$-5y = 15$$

$$y = -3 \qquad (0, -3)$$

$$3x - 5(0) - 15 = 0$$

$$3x = 15$$

$$x = 5 \qquad (5, 0)$$

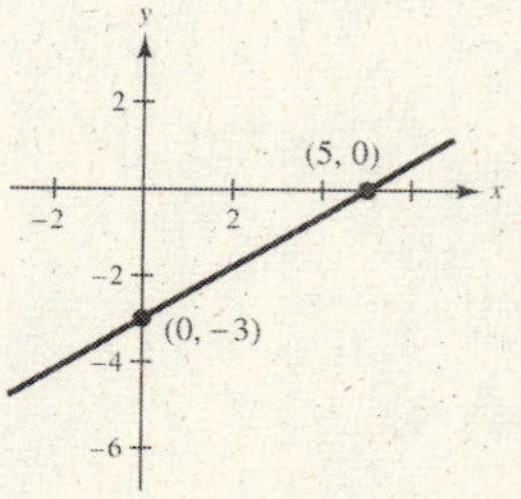

73. $-4x - 2y + 16 = 0$

$-4x - 2(0) + 16 = 0$

$-4x = -16$

$x = 4 \qquad (4, 0)$

$-4(0) - 2y + 16 = 0$

$-2y = -16$

$y = 8 \qquad (0, 8)$

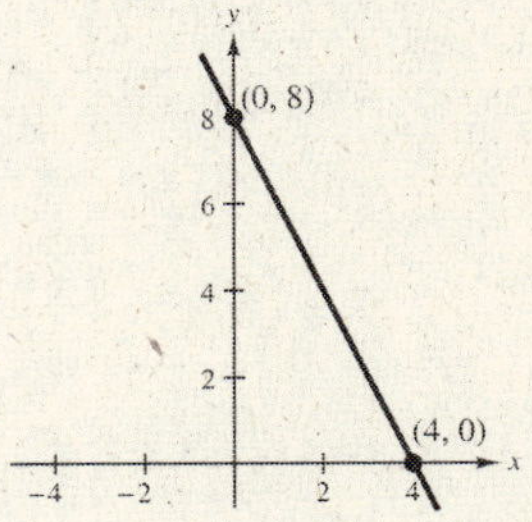

75. $L_1: y = \frac{1}{2}x - 2$

$L_2: y = \frac{1}{2}x + 3$

$m_1 = \frac{1}{2}$ and $m_2 = \frac{1}{2}$

$m_1 = m_2$ so the lines are parallel.

77. $L_1: y = \frac{3}{4}x - 3$

$L_2: y = -\frac{4}{3}x + 1$

$m_1 = \frac{3}{4}$ and $m_2 = -\frac{4}{3}$

$m_1 \cdot m_2 = -1$ so the lines are perpendicular.

79. $L_1: m_1 = \dfrac{8 - 4}{2 - 0} = \dfrac{4}{2} = 2$

$L_2: m_2 = \dfrac{5 - (-1)}{3 - 0} = \dfrac{6}{3} = 2$

$m_1 = m_2$ so the lines are parallel.

81. $L_1: m_1 = \dfrac{-2 - 2}{6 - 0} = \dfrac{-4}{6} = -\dfrac{2}{3}$

$L_2: m_2 = \dfrac{4 - 0}{8 - 2} = \dfrac{4}{6} = \dfrac{2}{3}$

$m_1 \neq m_2$ and $m_1 \cdot m_2 \neq -1$, so the lines are neither parallel nor perpendicular.

83. $-\dfrac{8}{100} = \dfrac{-2000}{x}$

$-8x = -200{,}000$

$x = 25{,}000$

The change in your horizontal position is 25,000 feet.

85. $\dfrac{3}{4} = \dfrac{h}{15}$

$45 = 4h$

$\dfrac{45}{4} = h$

The maximum height in the attic is

$\dfrac{45}{4}$ feet $= 11.25$ feet.

87. $y = 690.9t + 6505, \quad 1 \le t \le 6$

(a)
$y = 690.9(1) + 6505$
$= 690.9 + 6505$
$= 7195.9$

$y = 690.9(2) + 6505$
$= 1381.8 + 6505$
$= 7886.8$

$y = 690.9(3) + 6505$
$= 2072.7 + 6505$
$= 8577.7$

$y = 690.9(4) + 6505$
$= 2763.6 + 6505$
$= 9268.6$

$y = 690.9(5) + 6505$
$= 3454.5 + 6505$
$= 9959.5$

$y = 690.9(6) + 6505$
$= 4145.4 + 6505$
$= 10,650.4$

t	1	2	3	4	5	6
y	7195.9	7886.8	8577.7	9268.6	9959.5	10,650.4

(b)

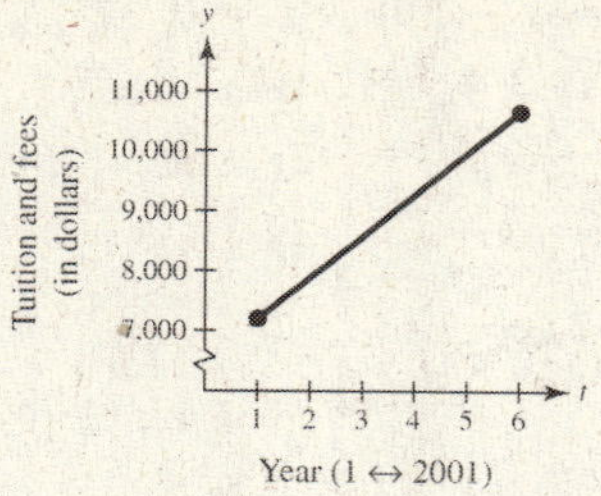

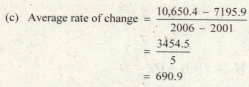

(c) Average rate of change $= \dfrac{10,650.4 - 7195.9}{2006 - 2001}$
$= \dfrac{3454.5}{5}$
$= 690.9$

The average rate of change of tuition and fees from 2001 to 2006 is $690.90.

(d) 2015: Let $t = 15$.
$y = 690.9t + 6505$
$= 690.9(15) + 6505$
$= 10,363.5 + 6505$
$= 16,868.5$

The amount of tuition and fees that would be paid in the year 2015 is $16,868.50.

89. Yes, any pair of points on a line can be used to calculate the slope of the line. When different pairs of points are selected, the change in y and the change in x are the lengths of the sides of similar triangles. Corresponding sides of similar triangles are proportional.

91. The x-coordinate of the x-intercept is the same as the solution of the equation when $y = 0$.

93. No. Two lines are perpendicular if their slopes, not y-intercepts, are negative reciprocals of each other.

95. $m \ge -3$

97. $16 \le n < 20$

99. $|g| = -4$

No solution

101. $\left|\frac{1}{5}m\right| = 2$

$\frac{1}{5}m = -2$ or $\frac{1}{5}m = 2$

$m = -10$ $\quad$ $m = 10$

103. $|2t - 3| = 11$

$2t - 3 = -11$ or $2t - 3 = 11$

$2t = -8$ $\quad$ $2t = 14$

$t = -4$ $\quad$ $t = 7$

105. $|n - 3| = |2n + 9|$

$n - 3 = -(2n + 9)$ or $n - 3 = 2n + 9$

$n - 3 = -2n - 9$ $\quad$ $-12 = n$

$3n = -6$

$n = -2$

Section 3.4 Equations of Lines

1. $y = \frac{2}{3}x + 2$ matches graph (b).

3. $y = -\frac{3}{2}x + 2$ matches graph (a).

5. $\dfrac{3}{1} = \dfrac{y - (-3)}{x - 2}$

$3(x - 2) = y + 3$

$3x - 6 = y + 3$

$3x - 9 = y$

7.
$$\begin{aligned} -\frac{1}{2} &= \frac{y-1}{x-(-3)} \\ x+3 &= -2(y-1) \\ x+3 &= -2y+2 \\ x+1 &= -2y \\ -\frac{1}{2}x-\frac{1}{2} &= y \end{aligned}$$

9.
$$\begin{aligned} \frac{4}{5} &= \frac{y-(-1)}{x-\frac{3}{4}} \\ 4\left(x-\frac{3}{4}\right) &= 5(y+1) \\ 4x-3 &= 5y+5 \\ 4x-8 &= 5y \\ \frac{4}{5}x-\frac{8}{5} &= y \end{aligned}$$

11. $y - 1 = 2(x - 3)$

$m = 2$. A point on the line is $(3, 1)$.

13. $y - (-1) = -5(x - 8)$

$m = -5$. A point on the line is $(8, -1)$.

15. $y - (-3) = \frac{1}{2}(x - (-6))$

$m = \frac{1}{2}$. A point on the line is $(-6, -3)$.

17. $y - 0 = -8(x - 0)$

$m = -8$. A point on the line is $(0, 0)$.

19.
$$\begin{aligned} y-0 &= -\tfrac{1}{2}(x-0) \\ y &= -\tfrac{1}{2}x \end{aligned}$$

21.
$$\begin{aligned} y+4 &= 3(x-0) \\ y+4 &= 3x \\ y &= 3x-4 \end{aligned}$$

23.
$$\begin{aligned} y-6 &= -\tfrac{3}{4}(x-0) \\ y-6 &= -\tfrac{3}{4}x \\ y &= -\tfrac{3}{4}x+6 \end{aligned}$$

25.
$$\begin{aligned} y-8 &= -2[x-(-2)] \\ y-8 &= -2(x+2) \\ y-8 &= -2x-4 \\ y &= -2x+4 \end{aligned}$$

27.
$$\begin{aligned} y-(-7) &= \tfrac{5}{4}[x-(-4)] \\ y+7 &= \tfrac{5}{4}(x+4) \\ y+7 &= \tfrac{5}{4}x+5 \\ y &= \tfrac{5}{4}x-2 \end{aligned}$$

29.
$$\begin{aligned} y-\tfrac{7}{2} &= -4[x-(-2)] \\ y-\tfrac{7}{2} &= -4(x+2) \\ y-\tfrac{7}{2} &= -4x-8 \\ y &= -4x-\tfrac{16}{2}+\tfrac{7}{2} \\ y &= -4x-\tfrac{9}{2} \end{aligned}$$

31.
$$\begin{aligned} y-\tfrac{5}{2} &= \tfrac{4}{3}\left(x-\tfrac{3}{4}\right) \\ y-\tfrac{5}{2} &= \tfrac{4}{3}x-1 \\ y &= \tfrac{4}{3}x-\tfrac{2}{2}+\tfrac{5}{2} \\ y &= \tfrac{4}{3}x+\tfrac{3}{2} \end{aligned}$$

33.
$$\begin{aligned} y-(-1) &= 0(x-2) \\ y+1 &= 0 \\ y &= -1 \end{aligned}$$

35.
$$\begin{aligned} m &= \frac{2-3}{5-0} = -\frac{1}{5} \\ y-2 &= -\frac{1}{5}(x-5) \\ y-2 &= -\frac{1}{5}x+1 \\ y &= -\frac{1}{5}x+3 \end{aligned}$$

37.
$$\begin{aligned} m &= \frac{6-\left(-\frac{2}{3}\right)}{-1-0} = \frac{-20}{3} \\ y-6 &= \frac{-20}{3}(x-(-1)) \\ y-6 &= \frac{-20}{3}x-\frac{20}{3} \\ y &= -\frac{20}{3}x-\frac{2}{3} \end{aligned}$$

39.
$$\begin{aligned} m &= \frac{-4-0}{3-0} = -\frac{4}{3} \\ y-(-4) &= -\frac{4}{3}(x-3) \\ y+4 &= -\frac{4}{3}x+4 \\ 3y+12 &= -4x+12 \\ 4x+3y &= 0 \end{aligned}$$

41. $m = \frac{0-4}{4-0} = \frac{-4}{4} = -1$

$$y - 4 = -1(x - 0)$$
$$y - 4 = -x$$
$$x + y - 4 = 0$$

43. $m = \frac{6-4}{5-1} = \frac{2}{4} = \frac{1}{2}$

$$y - 4 = \frac{1}{2}(x - 1)$$
$$2(y - 4) = x - 1$$
$$2y - 8 = x - 1$$
$$x - 2y + 7 = 0$$

45. $m = \frac{-2-2}{5+5} = -\frac{4}{10} = -\frac{2}{5}$

$$y - 2 = -\frac{2}{5}(x + 5)$$
$$5(y - 2) = -2(x + 5)$$
$$5y - 10 = -2x - 10$$
$$5y = -2x$$
$$2x + 5y = 0$$

47. $m = \frac{4-3}{\frac{9}{2}-\frac{3}{2}} = \frac{1}{\frac{6}{2}} = \frac{1}{3}$

$$y - 3 = \frac{1}{3}\left(x - \frac{3}{2}\right)$$
$$3(y - 3) = x - \frac{3}{2}$$
$$6y - 18 = 2x - 3$$
$$2x - 6y + 15 = 0$$

49. $m = \frac{\frac{7}{4}-\frac{1}{2}}{\left(\frac{3}{2}\right)-10} \cdot \frac{4}{4} = \frac{7-2}{6-40} = \frac{5}{-34}$

$$y - \frac{1}{2} = -\frac{5}{34}(x - 10)$$
$$34\left(y - \frac{1}{2}\right) = -5(x - 10)$$
$$34y - 17 = -5x + 50$$
$$5x + 34y - 67 = 0$$

51. $m = \frac{-1.4-9}{8-5} = \frac{-10.4}{3} = \frac{-104}{30} = -\frac{52}{15}$

$$y - 9 = -\frac{52}{15}(x - 5)$$
$$15(y - 9) = -52(x - 5)$$
$$15y - 135 = -52x + 260$$
$$52x + 15y - 395 = 0$$

53. $m = \frac{-4.2-0.6}{8-2} = -\frac{4.8}{6} = -0.8$

$$y - 0.6 = -0.8(x - 2)$$
$$y - 0.6 = -0.8x + 1.6$$
$$0.8x + y - 2.2 = 0$$
$$8x + 10y - 22 = 0$$
$$4x + 5y - 11 = 0$$

55. $m = \frac{7-1}{1-(-1)} = \frac{6}{2} = 3$

$$y - 1 = 3(x - (-1))$$
$$y - 1 = 3x + 3$$
$$y = 3x + 4$$

57. $m = \frac{3-3}{4+2} = \frac{0}{6} = 0$

$$y - 3 = 0(x - 4)$$
$$y - 3 = 0$$
$$y = 3$$

59. $x = -1$ because every x-coordinate is -1.

61. $y = -5$ because every y-coordinate is -5.

63. $x = -7$ because both points have an x-coordinate of -7.

65. $6x - 2y = 3$ slope $= 3$

$$-2y = -6x + 3$$
$$y = 3x - \tfrac{3}{2}$$

(a) $y - 1 = 3(x - 2)$

$$y - 1 = 3x - 6$$
$$y = 3x - 5$$

(b) $y - 1 = -\tfrac{1}{3}(x - 2)$

$$y - 1 = -\tfrac{1}{3}x + \tfrac{2}{3}$$
$$y = -\tfrac{1}{3}x + \tfrac{2}{3} + \tfrac{3}{3}$$
$$y = -\tfrac{1}{3}x + \tfrac{5}{3}$$

67. $5x + 4y = 24$

$4y = -5x + 24$

$y = -\frac{5}{4}x + 6 \qquad \text{slope} = -\frac{5}{4}$

(a) $y - 4 = -\frac{5}{4}[x - (-5)]$

$y - 4 = -\frac{5}{4}(x + 5)$

$y - 4 = -\frac{5}{4}x - \frac{25}{4}$

$y = -\frac{5}{4}x - \frac{25}{4} + \frac{16}{4}$

$y = -\frac{5}{4}x - \frac{9}{4}$

(b) $y - 4 = \frac{4}{5}[x - (-5)]$

$y - 4 = \frac{4}{5}(x + 5)$

$y - 4 = \frac{4}{5}x + 4$

$y = \frac{4}{5}x + 8$

69. $4x - y - 3 = 0$

$-y = -4x + 3$

$y = 4x + 3$

$\text{slope} = 4$

(a) $y - (-3) = 4(x - 5)$

$y + 3 = 4x - 20$

$y = 4x - 23$

(b) $y - (-3) = -\frac{1}{4}(x - 5)$

$y + 3 = -\frac{1}{4}x + \frac{5}{4}$

$y = -\frac{1}{4}x - \frac{7}{4}$

71. $x - 5 = 0$

$x = 5$ The slope is undefined.

(a) $x = \frac{2}{3}$

(b) $y = \frac{4}{3}$

73. $y + 5 = 0$

$y = -5$ The slope is zero.

(a) $y - 2 = 0(x + 1)$

$y - 2 = 0$

$y = 2$

(b) $x = -1$

75. $\frac{x}{3} + \frac{y}{2} = 1$

77. $\frac{x}{\frac{-5}{6}} + \frac{y}{\frac{-7}{3}} = 1$

$-\frac{6x}{5} - \frac{3y}{7} = 1$

79. $m = \frac{6000 - 5000}{50 - 0} = \frac{1000}{50} = 20$

$C - 5000 = 20(x - 0)$

$C = 20x + 5000$

$C = 20(400) + 5000$

$= 13{,}000$

The cost is $13,000.

81. $m = \frac{200{,}000 - 500{,}000}{2 - 5} = \frac{-300{,}000}{-3} = 100{,}000$

$S - 500{,}000 = 100{,}000(t - 5)$

$S - 500{,}000 = 100{,}000t - 500{,}000$

$S = 100{,}000t$

$S = 100{,}000(6) = 600{,}000$

The total sales for the sixth year are $600,000.

83. $m = \frac{1530 - 1500}{1000 - 0} = \frac{30}{1000} = \frac{3}{100}$

$S - 1500 = \frac{3}{100}(m - 0)$

$S = \frac{3}{100}m + 1500$ or

$S = 0.03m + 1500$

$0.03 = 3\%$

The commission rate is 3%.

85. (a) $S = L - 0.30L$

$S = 0.70L$

(b)

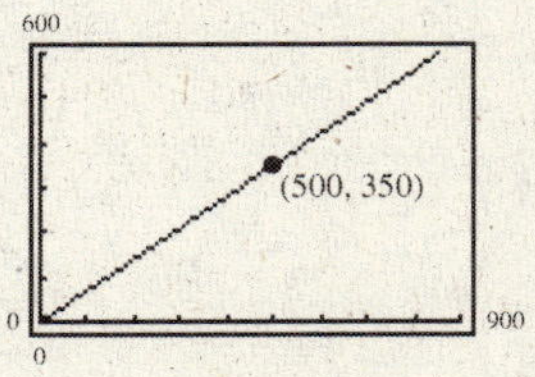

$S = 0.70(500)$

$S = 350$

The sale price is $350.

(c) $210 = 0.70L$

$300 = L$

The list price is $300.

87. (a) $(0, 7400), (4, 1500)$

$$m = \frac{7400 - 1500}{0 - 4} = \frac{5900}{-4} = -1475$$

$$V - 7400 = -1475(t - 0)$$
$$V - 7400 = -1475t$$
$$V = -1475t + 7400$$

(b) $V = -1475(2) + 7400$

$V = -2950 + 7400$

$V = 4450$

The photocopier has a value of \$4450 after 2 years.

89. (a) $(5, 1500), (6, 1560)$

$$m = \frac{1560 - 1500}{6 - 5} = \frac{60}{1} = 60$$

$$N - 1500 = 60(t - 5)$$
$$N - 1500 = 60t - 300$$
$$N = 60t + 1200$$

(b) $N = 60(25) + 1200$

$N = 2700$

Enrollment in 2015 will be 2700 students.

(c) $N = 60(10) + 1200$

$N = 1800$

Enrollment in 2000 was 1800 students.

91. (a) and (b)

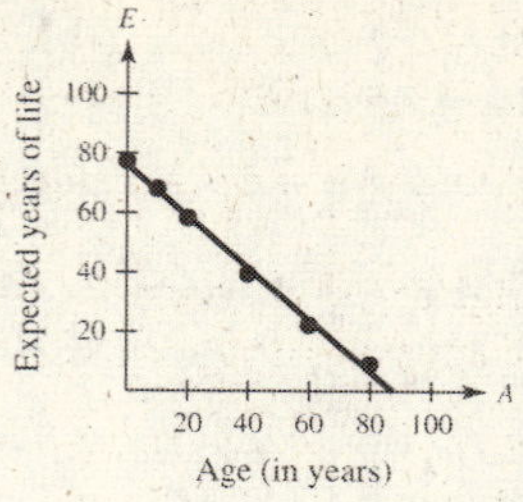

(c) $m = \dfrac{22.2 - 39.5}{60 - 40} = \dfrac{-17.3}{20} \approx -0.87$

$$E - 22.2 = -0.87(A - 60)$$
$$E - 22.2 = -0.87A + 52.2$$
$$E = -0.87A + 74.4$$

(d) $E = -0.87(30) + 74.4$

$E = 48.3$

The expected number of additional years of life is 48.3.

93. $(0, 0), (40, 5)$

$$m = \frac{5 - 0}{40 - 0} = \frac{5}{40} = \frac{1}{8}$$

$$y - 0 = \frac{1}{8}(x - 0)$$
$$y = \frac{1}{8}x$$
$$8y = x$$
$$x - 8y = 0$$

Depth of water $= 9 - y$

Distance from deep end	0	8	16	24	32	40
Depth of water	9	8	7	6	5	4

(a) $9 - y$

$9 = 9 - y$

$0 = y$

$x - 8(0) = 0$

$x = 0$

(b) $9 - y$

$8 = 9 - y$

$-1 = -y$

$1 = y$

$x - 8(1) = 0$

$x = 8$

(c) $9 - y$

$7 = 9 - y$

$-2 = -y$

$2 = y$

$x - 8(2) = 0$

$x = 16$

(d) $9 - y$

$6 = 9 - y$

$-3 = -y$

$3 = y$

$x - 8(3) = 0$

$x = 24$

(e) $9 - y$

$5 = 9 - y$

$-4 = -y$

$4 = y$

$x - 8(4) = 0$

$x = 32$

(f) $9 - y$

$4 = 9 - y$

$-5 = -y$

$5 = y$

$x - 8(5) = 0$

$x = 40$

95. Point-slope form: $y - y_1 = m(x - x_1)$

Slope-intercept form: $y = mx + b$

General form: $ax + by + c = 0$

97. The variable y is missing in the equation of a vertical line because any point on a vertical line is independent of y.

99.

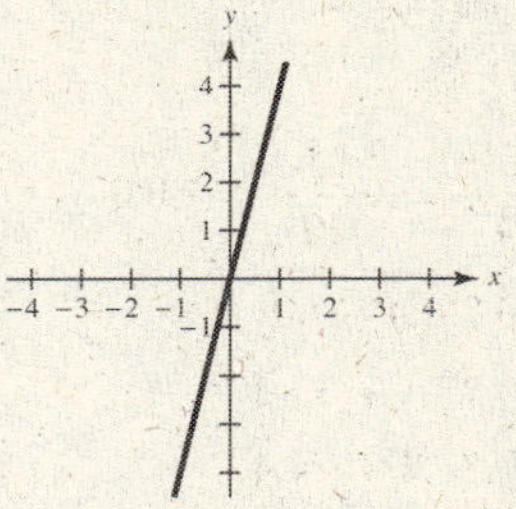

101.

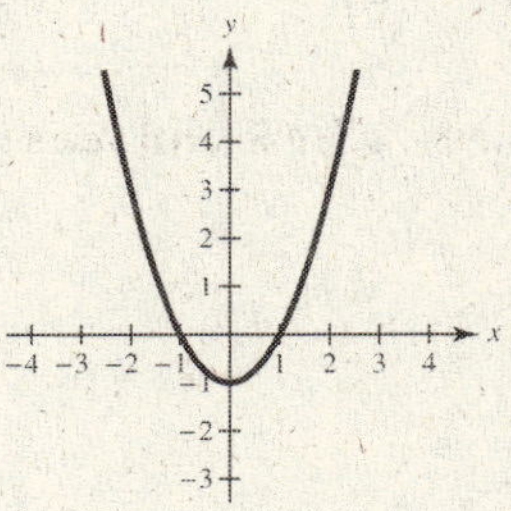

103. $\dfrac{4-2}{a-1} = 2$

$$\frac{2}{a-1} = 2$$
$$2 = 2(a-1)$$
$$2 = 2a - 2$$
$$4 = 2a$$
$$2 = a$$

104. $\dfrac{a-1}{2-0} = 3$

$$a - 1 = 6$$
$$a = 7$$

105. $\dfrac{3-a}{-2-(-4)} = \dfrac{1}{2}$

$$2(3-a) = 1(2)$$
$$6 - 2a = 2$$
$$-2a = -4$$
$$a = 2$$

109. $\dfrac{a-(-2)}{-1-(-7)} = -1$

$$a + 2 = -6$$
$$a = -8$$

Mid-Chapter Quiz for Chapter 3

1. Quadrant I or II. Because x can be any real number and y is 4, the point $(x, 4)$ can only be located in quadrants in which the y-coordinate is positive.

2. $4x - 3y = 10$

(a) $4(2) - 3(1) \stackrel{?}{=} 10$

$8 - 3 \stackrel{?}{=} 10$

$5 \neq 10$ Not a solution

(b) $4(1) - 3(-2) \stackrel{?}{=} 10$

$4 + 6 \stackrel{?}{=} 10$

$10 = 10$ Solution

(c) $4(2.5) - 3(0) \stackrel{?}{=} 10$

$10 - 0 \stackrel{?}{=} 10$

$10 = 10$ Solution

(d) $4(2) - 3\left(-\frac{2}{3}\right) \stackrel{?}{=} 10$

$8 + 2 \stackrel{?}{=} 10$

$10 = 10$ Solution

3.

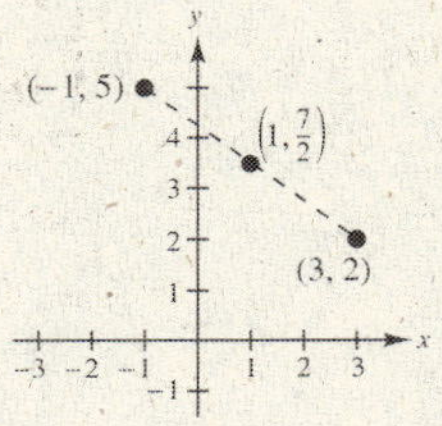

$$d = \sqrt{(-1-3)^2 + (5-2)^2}$$
$$= \sqrt{16+9}$$
$$= \sqrt{25}$$
$$= 5$$

$$M = \left(\frac{-1+3}{2}, \frac{5+2}{2}\right) = \left(1, \frac{7}{2}\right)$$

4.

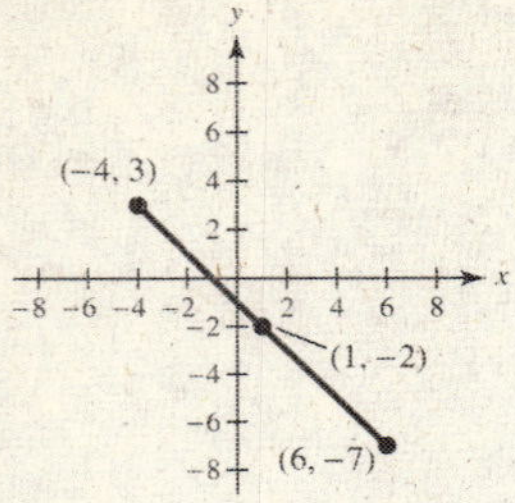

$$d = \sqrt{[6-(-4)]^2 + (-7-3)^2}$$
$$= \sqrt{10^2 + (-10)^2}$$
$$= \sqrt{100 + 100}$$
$$= \sqrt{200} = \sqrt{2 \cdot 100} = 10\sqrt{2}$$
$$M = \left(\frac{-4+6}{2}, \frac{3+(-7)}{2}\right) = \left(\frac{2}{2}, \frac{-4}{2}\right) = (1, -2)$$

5. $3x + y - 6 = 0$

$3(0) + y - 6 = 0$

$y = 6 \quad (0, 6)$

$3x + 0 - 6 = 0$

$3x = 6$

$x = 2 \quad (2, 0)$

$3(1) + y - 6 = 0$

$y = 3 \quad (1, 3)$

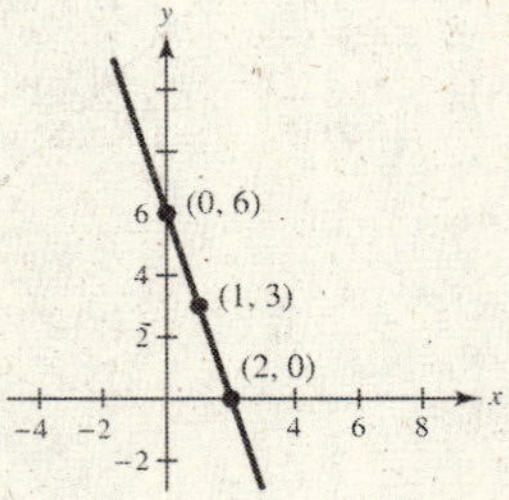

6. $y = 6x - x^2$

$y = 6(0) - 0^2$

$= 0 \quad (0, 0)$

$y = 6(6) - 6^2$

$= 0 \quad (6, 0)$

$y = 6(3) - 3^2$

$= 18 - 9$

$= 9 \quad (3, 9)$

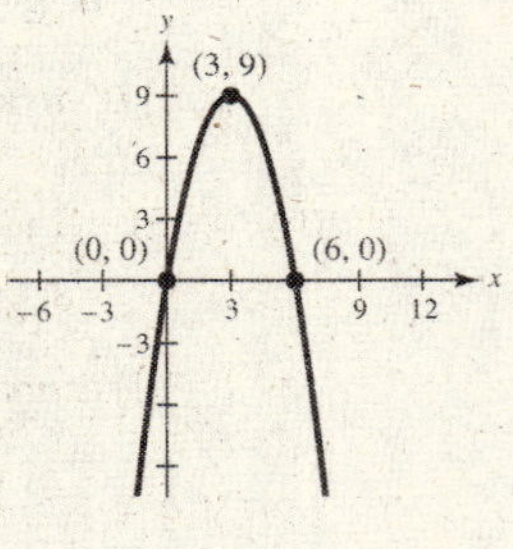

7. $y = |x - 2| - 3$

$y = |0 - 2| - 3$

$= -1 \quad (0, -1)$

$y = |5 - 2| - 3$

$= 0 \quad (5, 0)$

$y = |2 - 2| - 3$

$= -3 \quad (2, -3)$

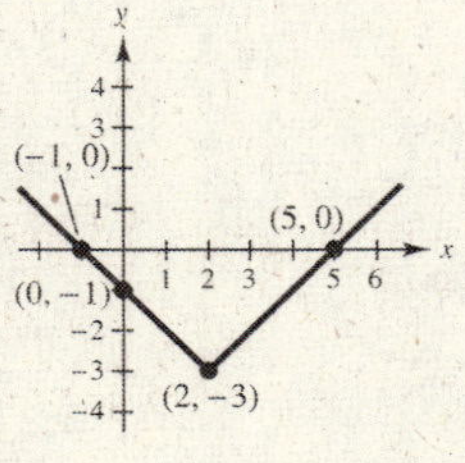

8. $m = \dfrac{8 - 8}{7 - (-3)} = \dfrac{0}{10} = 0$ Line is horizontal.

9. $m = \dfrac{5 - 0}{6 - 3} = \dfrac{5}{3}$ Line rises.

10. $m = \dfrac{-1 - 7}{4 - (-2)} = \dfrac{-8}{6} = -\dfrac{4}{3}$

Line falls.

11. $3x + 6y = 6$

$6y = -3x + 6$

$y = -\dfrac{1}{2}x + 1$

$m = -\dfrac{1}{2}$; y-intercept $= (0, 1)$

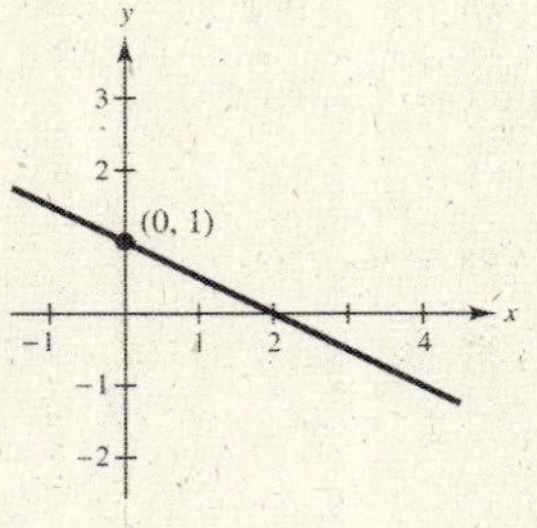

12. $6x - 4y = 12$

$-4y = -6x + 12$

$y = \dfrac{3}{2}x - 3$

$m = \dfrac{3}{2}$

y-intercept $= (0, -3)$

13. $y = 3x + 2$; $y = -\dfrac{1}{3}x - 4$

$m_1 = 3$

$m_2 = -\dfrac{1}{3}$

$m_1 \cdot m_2 = -1$

The lines are perpendicular.

14. $L_1: m_1 = \dfrac{(-9) - 3}{(-2) - 4} = \dfrac{-12}{-6} = 2 \Rightarrow y - 3 = 2(x - 4) \Rightarrow y = 2x - 5$

$L_2: m_2 = \dfrac{5 - (-5)}{5 - 0} = \dfrac{10}{5} = 2 \Rightarrow y = 2x - 5$

The lines are neither parallel nor perpendicular because they are the same line.

15.
$$\begin{aligned} y - (-1) &= \tfrac{1}{2}(x - 6) \\ y + 1 &= \tfrac{1}{2}(x - 6) \\ 2(y + 1) &= x - 6 \\ 2y + 2 &= x - 6 \\ x - 2y - 8 &= 0 \end{aligned}$$

16. The total depreciation over the 10-year period is $124,000 − $4000 = $120,000.

The annual depreciation is $\dfrac{\$120{,}000}{10} = \$12{,}000.$

$124,000 − (2)$12,000 = $100,000

$124,000 − (3)$12,000 = $88,000

$124,000 − (4)$12,000 = $76,000

So, $y = 124{,}000 - 12{,}000t,\ 0 \le t \le 10.$

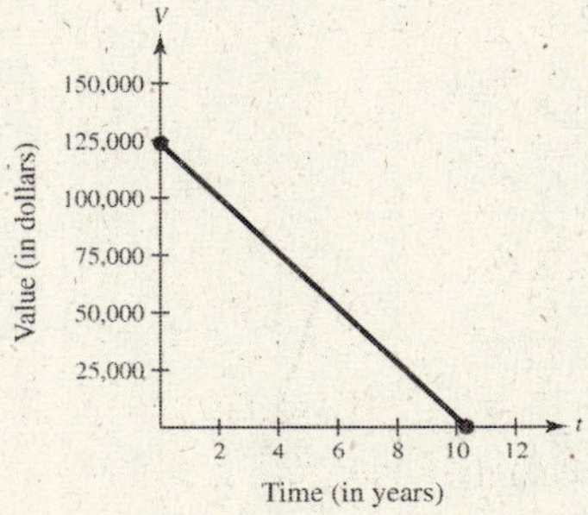

Section 3.5 Graphs of Linear Inequalities

1. $x - 2y < 4$

(a) $0 - 2(0) \overset{?}{<} 4$

$0 < 4$

(0, 0) *is* a solution.

(b) $2 - 2(-1) \overset{?}{<} 4$

$2 + 2 < 4$

$4 \not< 4$

(2, −1) *is not* a solution.

(c) $3 - 2(4) \overset{?}{<} 4$

$3 - 8 < 4$

$-5 < 4$

(3, 4) *is* a solution.

(d) $5 - 2(1) \overset{?}{<} 4$

$5 - 2 < 4$

$3 < 4$

(5, 1) *is* a solution.

3. $3x + y \ge 10$

(a) $3(1) + 3 \overset{?}{\ge} 10$

$9 \not\ge 10$

(1, 3) *is not* a solution.

(b) $3(-3) + 1 \overset{?}{\ge} 10$

$-8 \not\ge 10$

(−3, 1) *is not* a solution.

(c) $3(3) + 1 \overset{?}{\ge} 10$

$10 \ge 10$

(3, 1) *is* a solution.

(d) $3(2) + 15 \overset{?}{\ge} 10$

$21 \ge 10$

(2, 15) *is* a solution.

5. $y > 0.2x - 1$

(a) $2 \overset{?}{>} 0.2(0) - 1$

$2 > -1$

$(0, 2)$ *is* a solution.

(b) $0 \overset{?}{>} 0.2(6) - 1$

$0 > 0.2$

$(6, 0)$ *is not* a solution.

(c) $-1 \overset{?}{>} 0.2(4) - 1$

$-1 \not> -0.2$

$(4, -1)$ *is not* a solution.

(d) $7 \overset{?}{>} 0.2(-2) - 1$

$7 > -1.4$

$(-2, 7)$ *is* a solution.

7. $y \le 3 - |x|$

(a) $4 \overset{?}{\le} 3 - |-1|$

$4 \not\le 3 - 1$

$(-1, 4)$ *is not* a solution.

(b) $-2 \overset{?}{\le} 3 - |2|$

$-2 \le 3 - 2$

$(2, -2)$ *is* a solution.

(c) $0 \overset{?}{\le} 3 - |6|$

$0 \le 3 - 6$

$0 \not\le -3$

$(6, 0)$ *is not* a solution.

(d) $-2 \overset{?}{\le} 3 - |5|$

$-2 \le 3 - 5$

$-2 \le -2$

$(5, -2)$ *is* a solution.

9. $y \ge -2$; (b)

11. $3x - 2y < 0$; (d)

13. $x + y < 4$; (f)

15. Because the origin $(0, 0)$ does not satisfy the inequality, the graph consists of the half-plane lying below the line.

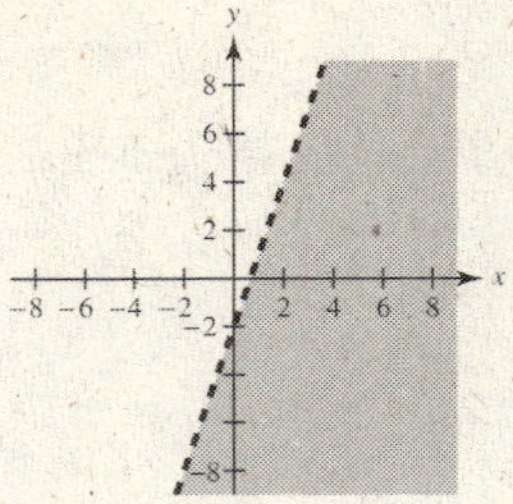

17. $x \ge 6$

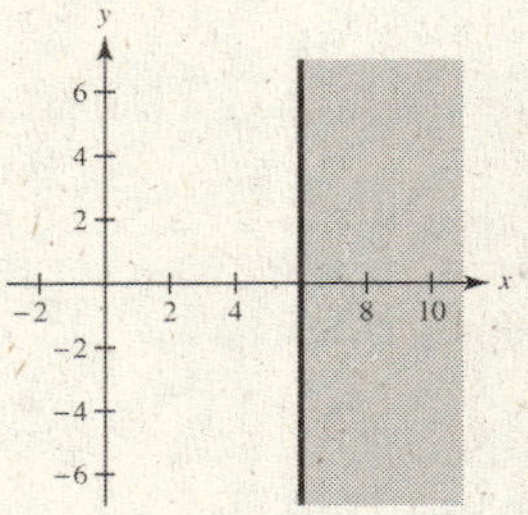

19. $y < 5$

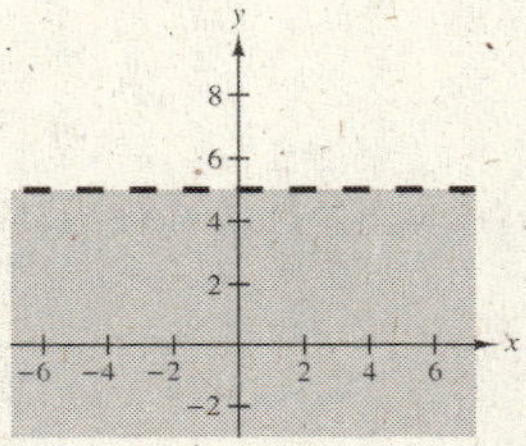

21. $y > \frac{1}{2}x$

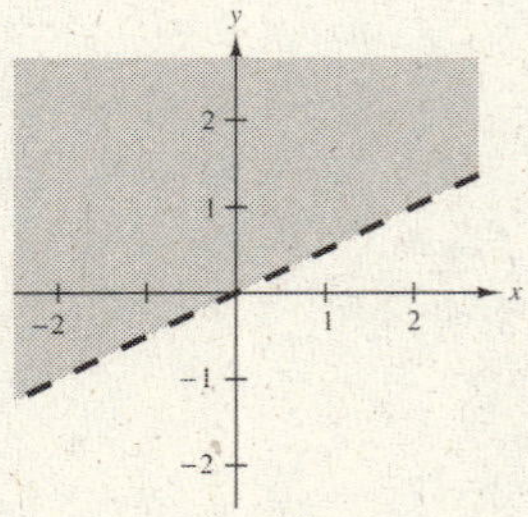

23. $y \ge 3 - x$

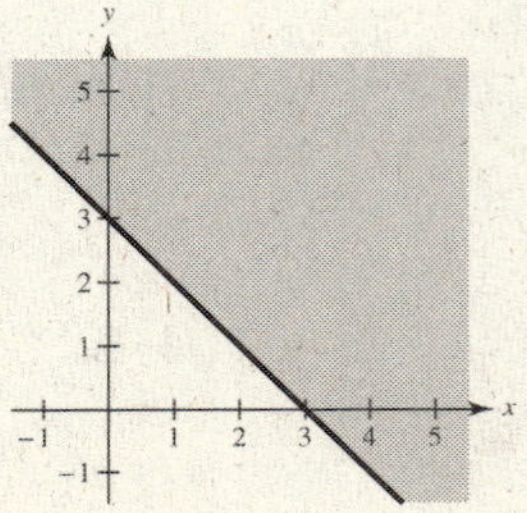

25. $y \le x + 2$

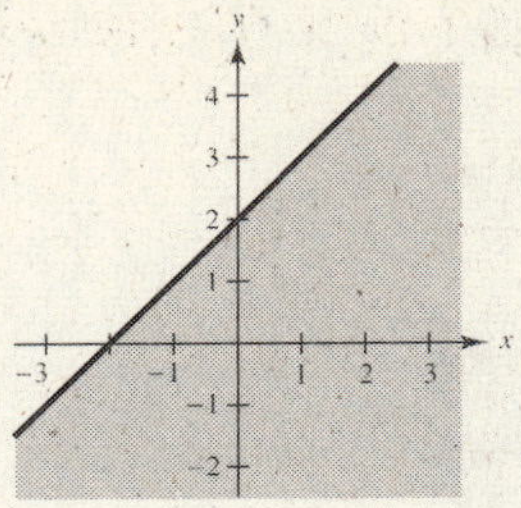

27. $x + y \ge 4$

$y \ge -x + 4$

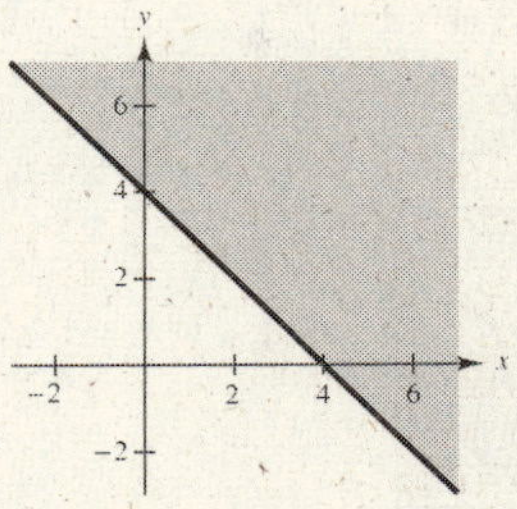

29. $x - 2y \ge 6$

$-2y \ge -x + 6$

$y \le \frac{1}{2}x - 3$

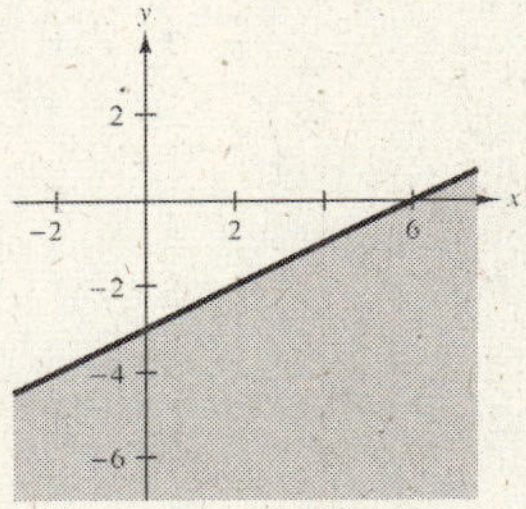

31. $3x + 2y \ge 2$

$2y \ge -3x + 2$

$y \ge -\frac{3}{2}x + 1$

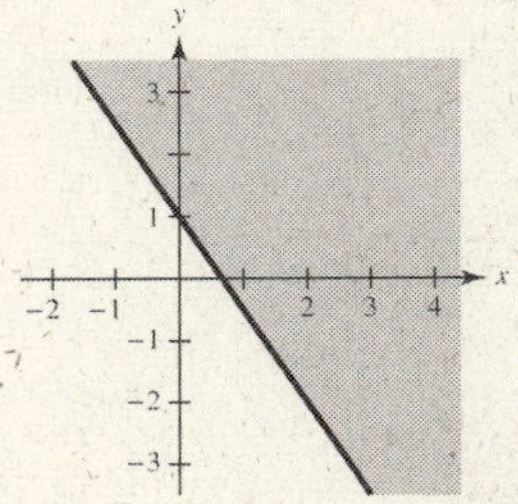

33. $5x + 4y < 20$

$4y < -5x + 20$

$y < -\frac{5}{4}x + 5$

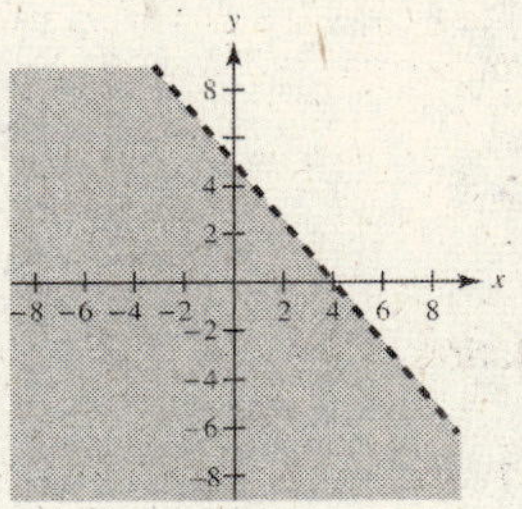

35. $x - 3y - 9 < 0$

$-3y < -x + 9$

$y > \frac{1}{3}x - 3$

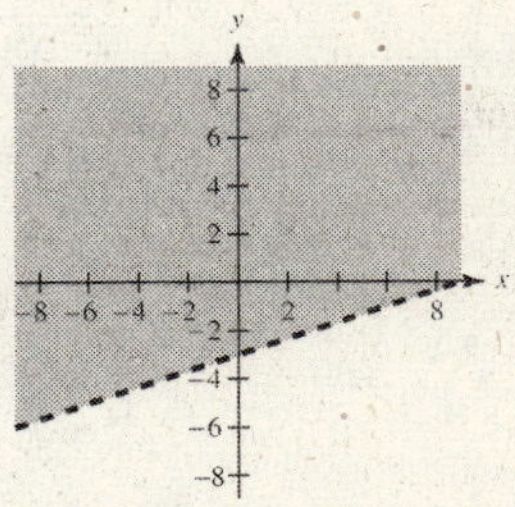

37. $3x - 2 \le 5x + y$

$-y \le 2x + 2$

$y \ge -2x - 2$

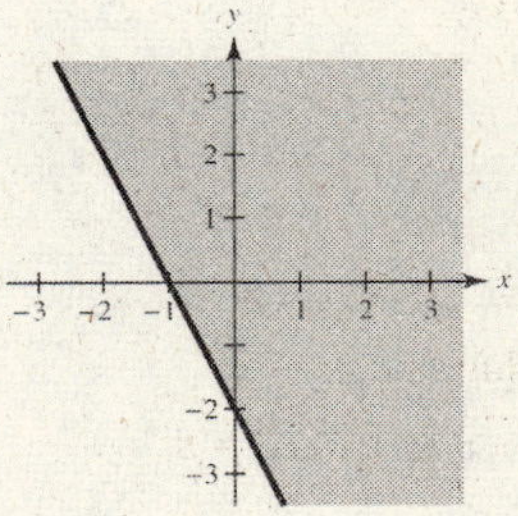

39. $0.2x + 0.3y < 2$

$2x + 3y < 20$

$3y < -2x + 20$

$y < -\frac{2}{3}x + \frac{20}{3}$

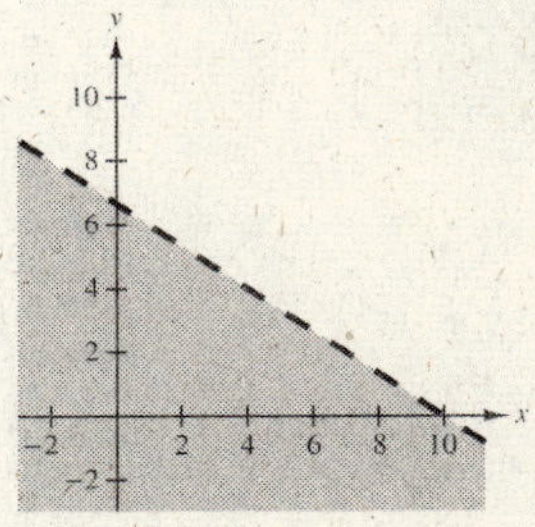

41. $y - 1 > -\frac{1}{2}(x - 2)$

$y - 1 = -\frac{1}{2}x + 1$

$y = -\frac{1}{2}x + 2$

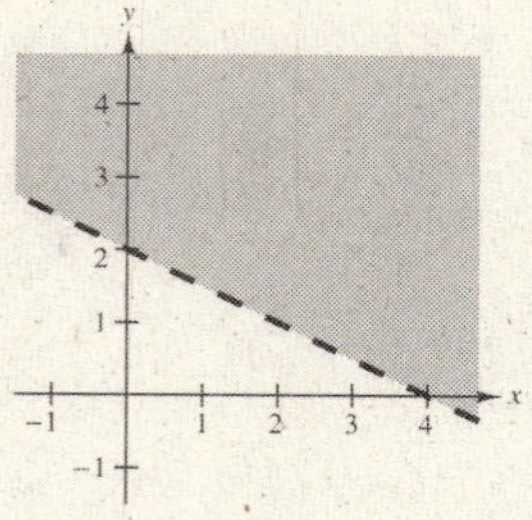

43. $\frac{x}{3} + \frac{y}{4} \le 1$

$4x + 3y \le 12$

$3y \le -4x + 12$

$y \le -\frac{4}{3}x + 4$

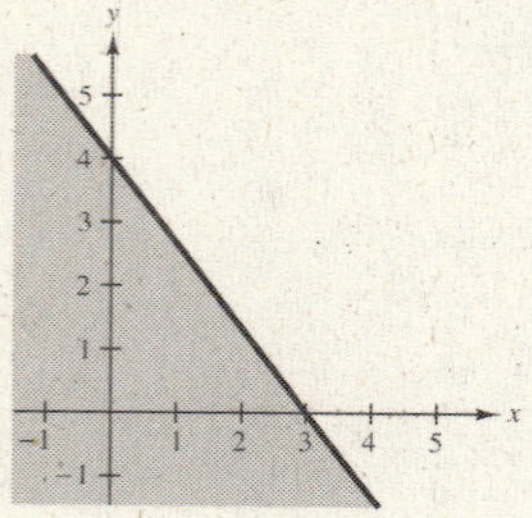

45. $y \ge \frac{3}{4}x - 1$

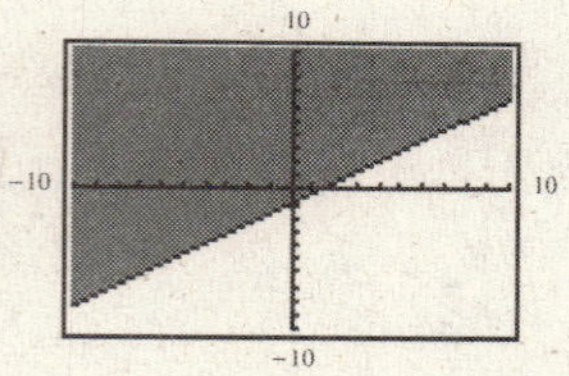

47. $y \le -\frac{2}{3}x + 6$

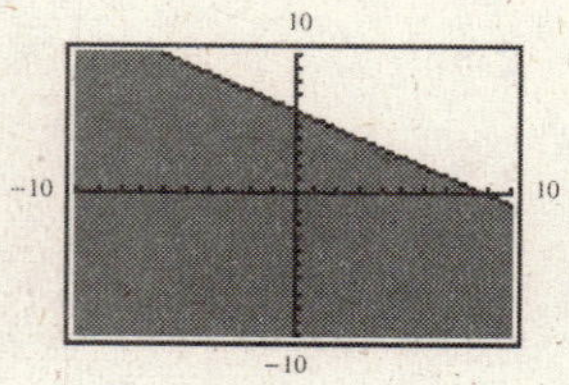

49. $x - 2y - 4 \ge 0$

$-2y \ge -x + 4$

$y \le \frac{1}{2}x - 2$

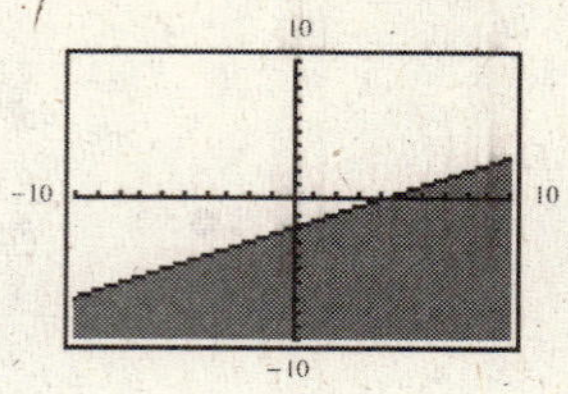

51. $2x + 3y - 12 \le 0$

$3y \le -2x + 12$

$y \le -\frac{2}{3}x + 4$

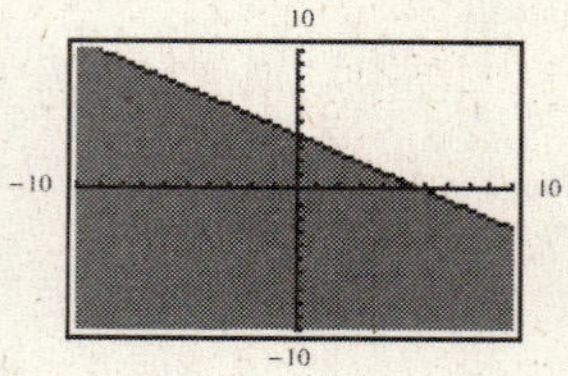

53. $m = \frac{2 - 5}{3 + 1} = -\frac{3}{4}$

$y - 2 > -\frac{3}{4}(x - 3)$

$4y - 8 > -3x + 9$

$3x + 4y > 17$

55. $y < 2$

57. $m = \frac{2 - 0}{1 - 0} = \frac{2}{1} = 2$

$y - 0 \ge 2(x - 0)$

$y \ge 2x$

59. (a) $P = 2x + 2y$

$2x + 2y \le 500, \quad x \ge 0, y \ge 0$

or

$y \le -x + 250, \quad x \ge 0, y \ge 0$

(b)

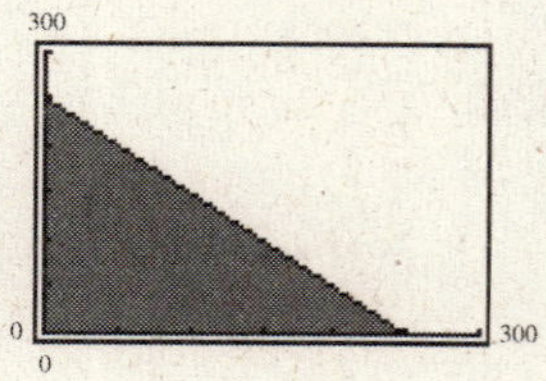

61. $10x + 15y \le 1000$

$15y \le -10x + 1000$

$y \le -\frac{2}{3}x + \frac{200}{3}, \quad x \ge 0, y \ge 0$

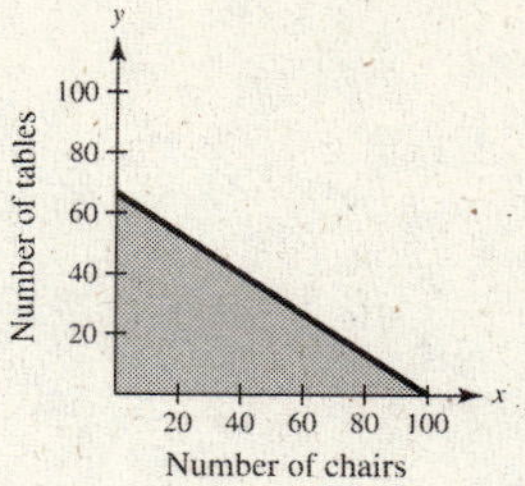

63. (a) *Verbal model:*

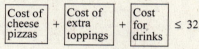

Cost of cheese pizzas	+	Cost of extra toppings	+	Cost for drinks	≤ 32

Labels: Cost of cheese pizzas = 2(10) = \$20

Cost for extra toppings = $0.60x$ (dollars)

Cost for drinks = y (dollars)

Inequality: $20 + 0.60x + y \le 32$

$0.60x + y \le 12, \quad x \ge 0, y \ge 0$

(b)

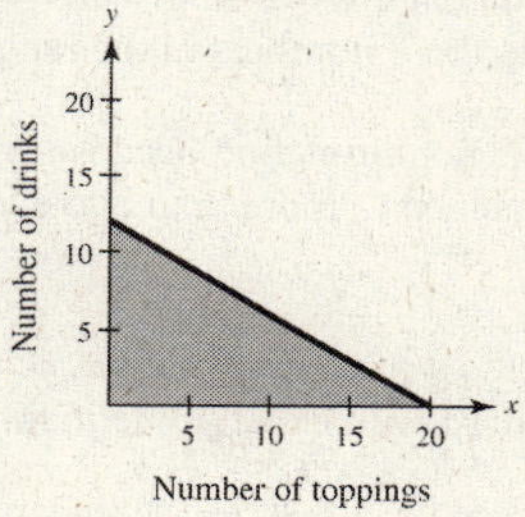

(c) $(6, 6)$

$0.60(6) + 6 \stackrel{?}{\le} 12$

$3.6 + 6 \stackrel{?}{\le} 12$

$9.6 \le 12$

$(6, 6)$ is a solution of the inequality.

65. (a) $12x + 16y \ge 250$

$16y \ge -12x + 250$

$y \ge -\frac{12}{16}x + \frac{250}{16}$

$y \ge -\frac{3}{4}x + \frac{125}{8}, \quad x \ge 0, y \ge 0$

(b)

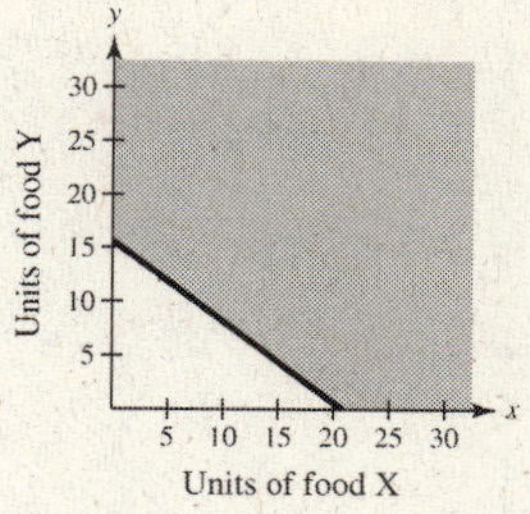

(x, y): $(1, 15), (10, 9), (15, 5)$

67. (a) $11x + 9y \ge 240$

$9y \ge -11x + 240$

$y \ge -\frac{11}{9}x + \frac{80}{3}; \quad x \ge 0, y \ge 0$

(b)

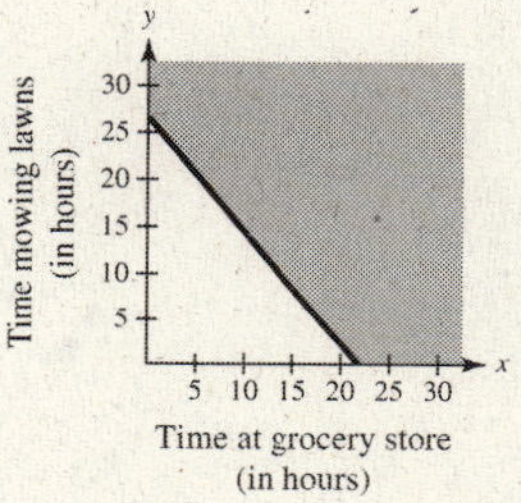

(x, y): $(2, 25), (4, 22), (10, 15)$

69. (a) $2w + t \ge 70$

$t \ge -2w + 70, \quad x \ge 0, y \ge 0$

(b)

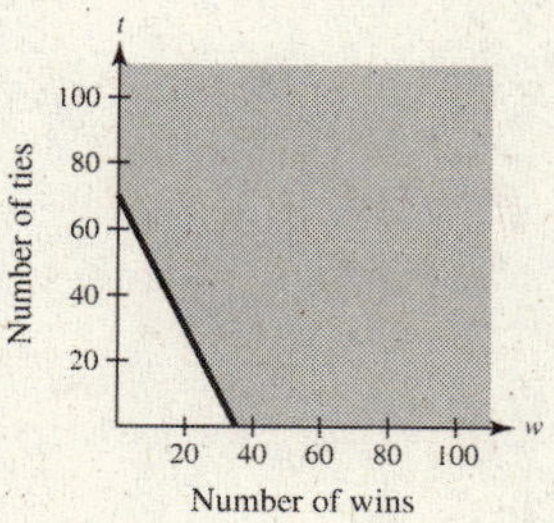

(w, t): $(10, 50), (20, 30), (30, 10)$

71. (x_1, y_1) is a solution of a linear inequality in x and y means the inequality is true when x_1 and y_1 are substituted for x and y, respectively.

73. To represent the other points in the plane, reverse the inequality symbol so that it is $\le$ rather than $>$.

75. If the point $(0, 0)$ cannot be used as a test point, the point $(0, 0)$ must lie on the boundary line. Therefore, the line passes through the origin, and the y-intercept is $(0, 0)$.

77. $|x| = 6$

$x = -6$ or $x = 6$

79. $|2x + 3| = 9$

$2x + 3 = -9$ or $2x + 3 = 9$

81.

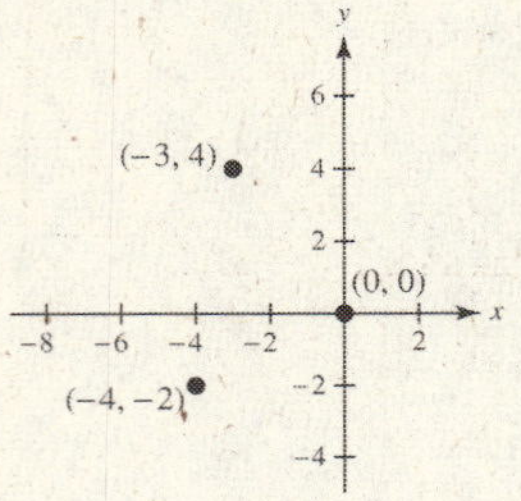

83.

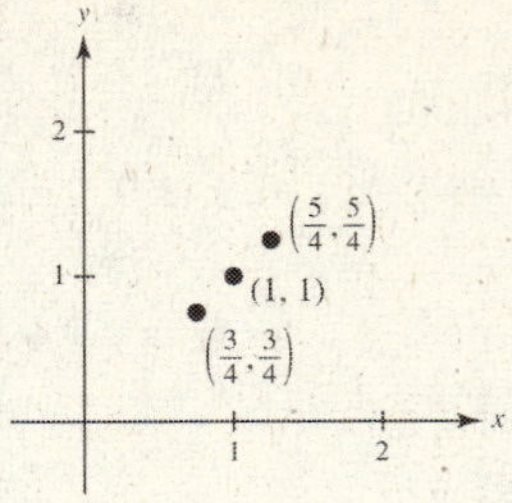

85. (a) $y - y_1 = m(x - x_1)$

$y - 0 = -2(x - 3)$

$y = -2x + 6$

(b) $y - y_1 = m(x - x_1)$

$y - 0 = \frac{1}{2}(x - 3)$

$y = \frac{1}{2}x - \frac{3}{2}$

Section 3.6 Relations and Functions

1. Domain $= \{-2, 0, 1\}$

Range $= \{-1, 0, 1, 4\}$

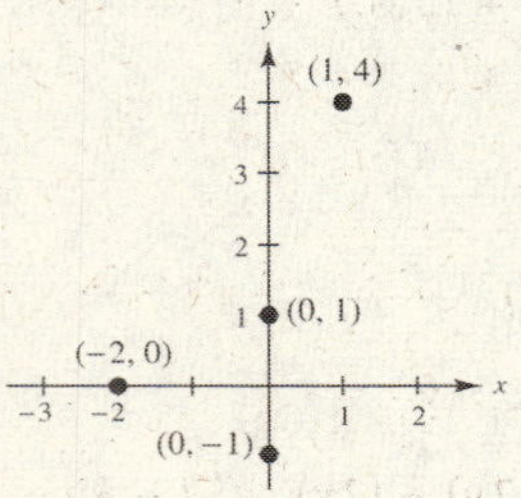

3. Domain $= \{0, 2, 4, 5, 6\}$

Range $= \{-3, 0, 5, 8\}$

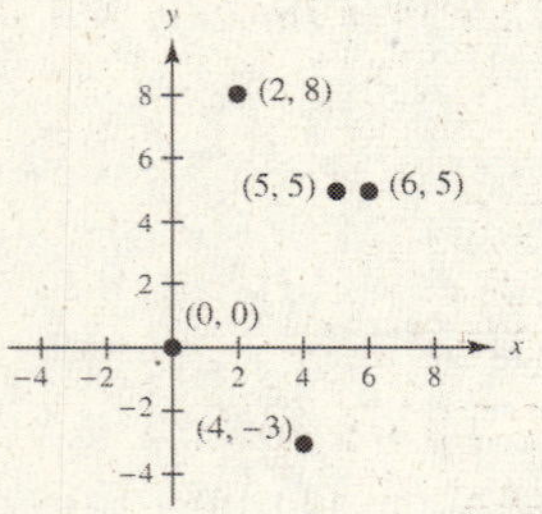

5. (1, 1), (2, 8), (3, 27), (4, 64), (5, 125), (6, 216), (7, 343)

7. {(2004, Boston Red Sox), (2005, Chicago White Sox), (2006, St. Louis Cardinals), (2007, Boston Red Sox)}

9. {(3, 9), (1, 3), (2, 6), (8, 24), (7, 21)}

11. No, this relation is not a function because −1 in the domain is matched to 2 numbers (6 and 7) in the range.

13. Yes, this relation is a function as each number in the domain is matched to exactly one number in the range.

15. No, this relation is not a function because 0 in the domain is matched to 2 numbers in the range (5 and 9).

17. No, the relation is not a function because an element in the domain is matched to more than one element in the range.

19. Yes, this relation is a function because each number in the domain is paired with exactly one number in the range.

21. No, this relation is not a function because the 4 and the 7 in the domain are each paired with 2 different numbers in the range.

23. (a) Yes, this relation is a function because each number in the domain is paired with exactly one number in the range.

(b) No, this relation is not a function because the 1 in the domain is paired with 2 different numbers in the range.

(c) Yes, this relation is a function because each number in the domain is paired with exactly one number in the range.

(d) No, this relation is not a function because each number in the domain is not paired with a number.

25. This set of ordered pairs does represent a function. Each input only has one output.

27. This set of ordered pairs does not represent a function. All integers from 1 to 20, except for 1, have more than 1 factor.

29. $x^2 + y^2 = 25$

$0^2 + 5^2 \stackrel{?}{=} 25 \qquad 0^2 + (-5)^2 \stackrel{?}{=} 25$

$25 = 25 \qquad 25 = 25$

There are two values of y associated with one value of x, which implies y is not a function of x.

31. $|y| = x + 2$

$|3| \stackrel{?}{=} 1 + 2 \qquad |-3| \stackrel{?}{=} 1 + 2$

$3 = 3 \qquad 3 = 3$

There are two values of y associated with one value of x, which implies y is not a function of x.

33. $y = 10x + 12$ represents y as a function of x because there is one value of y associated with each value of x.

35. $3x + 7y - 2 = 0$ represents y as a function of x because there is one value of y associated with each value of x.

37. $y = x(x - 10)$ represents y as a function of x because there is one value of y associated with each value of x.

39. $y^2 = x$

$y = \sqrt{x}$ or $-\sqrt{x}$

Because $(4, 2)$ and $(4, -2)$ are both solutions, y is not a function of x.

41. $y = |x|$

$y = |-3| = 3$

Because one value of x corresponds to one value of y, y is a function of x.

43. $f(x) = 3x + 5$

(a) $f(2) = 3(2) + 5 = 11$

(b) $f(-2) = 3(-2) + 5 = -1$

(c) $f(k) = 3(k) + 5 = 3k + 5$

(d) $f(k + 1) = 3(k + 1) + 5 = 3k + 3 + 5 = 3k + 8$

45. $f(x) = 3 - x^2$

(a) $f(0) = 3 - 0^2 = 3$

(b) $f(-3) = 3 - (-3)^2 = 3 - 9 = -6$

(c) $f(m) = 3 - m^2$

(d) $f(2t) = 3 - (2t)^2 = 3 - 4t^2$

47. $f(x) = \dfrac{x}{x + 2}$

(a) $f(3) = \dfrac{3}{3 + 2} = \dfrac{3}{5}$

(b) $f(-4) = \dfrac{-4}{-4 + 2} = \dfrac{-4}{-2} = 2$

(c) $f(s) = \dfrac{s}{s + 2}$

(d) $f(s - 2) = \dfrac{s - 2}{(s - 2) + 2} = \dfrac{s - 2}{s}$

49. $f(x) = 12x - 7$

(a) $f(3) = 12(3) - 7 = 29$

(b) $f\left(\frac{3}{2}\right) = 12\left(\frac{3}{2}\right) - 7 = 11$

(c) $f(a) + f(1) = [12(a) - 7] + [12(1) - 7]$

$= 12a - 7 + 12 - 7$

$= 12a - 2$

(d) $f(a + 1) = 12(a + 1) - 7$

$= 12a + 12 - 7$

$= 12a + 5$

51. $g(x) = 2 - 4x + x^2$

(a) $g(4) = 2 - 4(4) + 4^2 = 2 - 16 + 16 = 2$

(b) $g(0) = 2 - 4(0) + 0^2 = 2$

(c) $g(2y) = 2 - 4(2y) + (2y)^2 = 2 - 8y + 4y^2$

(d) $g(4) + g(6) = [2 - 4(4) + 4^2] + [2 - 4(6) + 6^2]$

$= (2 - 16 + 16) + (2 - 24 + 36)$

$= 2 + 14 = 16$

53. $f(x) = \sqrt{x + 5}$

(a) $f(-1) = \sqrt{-1 + 5} = 2$

(b) $f(4) = \sqrt{4 + 5} = 3$

(c) $f(z - 5) = \sqrt{z - 5 + 5} = \sqrt{z}$

(d) $f(5z) = \sqrt{5z + 5}$

55. $g(x) = 8 - |x - 4|$

(a) $g(0) = 8 - |0 - 4| = 8 - 4 = 4$

(b) $g(8) = 8 - |8 - 4| = 8 - 4 = 4$

(c) $g(16) - g(-1) = (8 - |16 - 4|) - (8 - |-1 - 4|)$

$= (8 - 12) - (8 - 5)$

$= -4 - 3 = -7$

(d) $g(x - 2) = 8 - |x - 2 - 4| = 8 - |x - 6|$

57. $f(x) = \dfrac{3x}{x - 5}$

(a) $f(0) = \dfrac{3(0)}{0 - 5} = 0$

(b) $f\left(\dfrac{5}{3}\right) = \dfrac{3\left(\dfrac{5}{3}\right)}{\dfrac{5}{3} - 5} \cdot \dfrac{3}{3} = \dfrac{15}{5 - 15} = \dfrac{15}{-10} = -\dfrac{3}{2}$

(c) $f(2) - f(-1) = \left[\dfrac{3(2)}{2 - 5}\right] - \left[\dfrac{3(-1)}{-1 - 5}\right]$

$= \dfrac{6}{-3} - \dfrac{-3}{-6} = -2 - \dfrac{1}{2} = -\dfrac{5}{2}$

(d) $f(x + 4) = \dfrac{3(x + 4)}{x + 4 - 5} = \dfrac{3x + 12}{x - 1}$

59. $f(x) = \begin{cases} x + 8, & \text{if } x < 0 \\ 10 - 2x, & \text{if } x \ge 0 \end{cases}$

(a) $f(4) = 10 - 2(4) = 10 - 8 = 2$

(b) $f(-10) = -10 + 8 = -2$

(c) $f(0) = 10 - 2(0) = 10$

(d) $f(6) - f(-2) = [10 - 2(6)] - [-2 + 8]$

$= 10 - 12 - 6 = -8$

61. $h(x) = \begin{cases} 4 - x^2, & \text{if } x \le 2 \\ x - 2, & \text{if } x > 2 \end{cases}$

(a) $h(2) = 4 - 2^2 = 0$

(b) $h\left(-\frac{3}{2}\right) = 4 - \left(-\frac{3}{2}\right)^2 = 4 - \frac{9}{4} = \frac{16}{4} - \frac{9}{4} = \frac{7}{4}$

(c) $h(5) = 5 - 2 = 3$

(d) $h(-3) + h(7) = \left[4 - (-3)^2\right] + [7 - 2]$

$= 4 - 9 + 5 = 0$

63. $f(x) = 2x + 5$

(a) $\dfrac{f(x + 2) - f(2)}{x} = \dfrac{[2(x + 2) + 5] - [2(2) + 5]}{x}$

$= \dfrac{2x + 4 + 5 - 4 - 5}{x}$

$= \dfrac{2x}{x} = 2$

(b) $\dfrac{f(x - 3) - f(3)}{x} = \dfrac{[2(x - 3) + 5] - [2(3) + 5]}{x}$

$= \dfrac{2x - 6 + 5 - 6 - 5}{x}$

$= \dfrac{2x - 12}{x}$

65. The domain of $f(x) = x^2 + x - 2$ is all real numbers x.

67. The domain of $f(t) = \dfrac{t + 3}{t(t + 2)}$ is all real numbers t such that $t(t + 2) \ne 0$. So, $t \ne 0$ and $t \ne -2$.

69. The domain of $g(x) = \sqrt{x + 4}$ is all real numbers x such that $x + 4 \ge 0$. So, $x \ge -4$.

71. The domain of $f(x) = \sqrt{2x - 1}$ is all real numbers x such that $2x - 1 \ge 0$. So, $x \ge \frac{1}{2}$.

73. The domain of $f(t) = |t - 4|$ is all real numbers t.

75. Domain: $\{0, 2, 4, 6\}$
Range: $\{0, 1, 8, 27\}$

76. Domain: $\{-3, -1, 2, 5\}$
Range: $\{0, 3, 4\}$

77. Domain: $\{-3, -1, 4, 10\}$
Range: $\left\{-\frac{17}{2}, -\frac{5}{2}, 2\right\}$

78. Domain: $\left\{\frac{1}{2}, \frac{3}{4}, 1, \frac{5}{4}\right\}$
Range: $\{4, 5, 6, 7\}$

79. Domain: All real numbers r such that $r > 0$
Range: All real numbers C such that $C > 0$

81. Domain: All real numbers r such that $r > 0$
Range: All real numbers A such that $A > 0$

83. *Verbal Model:* [Perimeter] = 4 · [Length of side]

Labels: Perimeter = P
Length of side = x

Function: $P = 4x$

85. *Verbal Model:* [Volume] = [Length of side]^3

Labels: Volume = V
Length of side = x

Function: $V = x^3$

87. *Verbal Model:*

[Total cost] = [Variable costs] + [Fixed costs]

Labels: Total cost = $C(x)$
Variable costs = $1.95x$
Fixed costs = 8000
Number of games = x

Function: $C(x) = 1.95x + 8000, x > 0$

89. *Verbal Model:* $\boxed{\text{Distance}} = \boxed{\text{Rate}} \cdot \boxed{\text{Time}}$

Labels: Distance $= d$

Rate $= 120$

Time $= t$

Function: $d = 120t$

91. *Verbal Model:* $\boxed{\text{Distance}} = \boxed{\text{Rate}} \cdot \boxed{\text{Time}}$

Labels: Distance $= d$

Rate $= 65$

Time $= t$

Function: $d = 65t$

$d = 65(4) = 260$

When $t = 4$, $d = 260$ miles.

93. *Verbal Model:* $\boxed{\text{Area}} = \boxed{\text{Side}} \cdot \boxed{\text{Side}}$

Label: Area $= A$

Function: $A = (32 - 2x)(32 - 2x)$

$= (32 - 2x)^2, x > 0$

95. *Verbal Model:*

$\boxed{\text{Volume}} = \boxed{\text{Length}} \cdot \boxed{\text{Width}} \cdot \boxed{\text{Height}}$

Labels: Volume $= V$

Length $= (24 - 2x)$

Width $= (24 - 2x)$

Height $= x$

Function: $V = x(24 - 2x)^2, x > 0$

97. (a) $P(1600) = 50\sqrt{1600} - 0.5(1600) - 500$

$= 50(40) - 800 - 500$

$= 2000 - 800 - 500$

$= \$700$

(b) $P(2500) = 50\sqrt{2500} - 0.5(2500) - 500$

$= 50(50) - 1250 - 500$

$= 2500 - 1250 - 500$

$= \$750$

99. (a) $w(30) = 12(30) = \$360$

$w(40) = 12(40) = \$480$

$w(45) = 18(45 - 40) + 480 = \$570$

$w(50) = 18(50 - 40) + 480 = \$660$

(b) Because $h < 0$ is not in the domain of w, you cannot use values of h for which $h < 0$.

101. For each year, there is only one value associated with the enrollment in public schools, so enrollment is a function of the year.

$f(2004) \approx 13$ million

103. Because each selling price has one associated sales tax, the use of the word function is mathematically correct.

105. If the subset of A only includes input values that are only mapped to one output in the range, then the subset of A is a function.

107. Statement: The number of ceramic tiles required to floor a kitchen is a function of the area of the floor.

Domain: The area of the floor

Range: The number of ceramic tiles

The relation is a function.

109. Multiplicative Identity Property

111. Multiplicative Inverse Property

113.

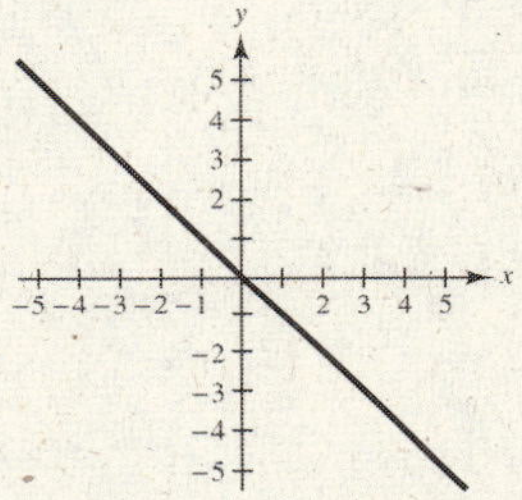

115. $x - 4y = 16$

$-4y = 16 - x$

$y = \frac{1}{4}x - 4$

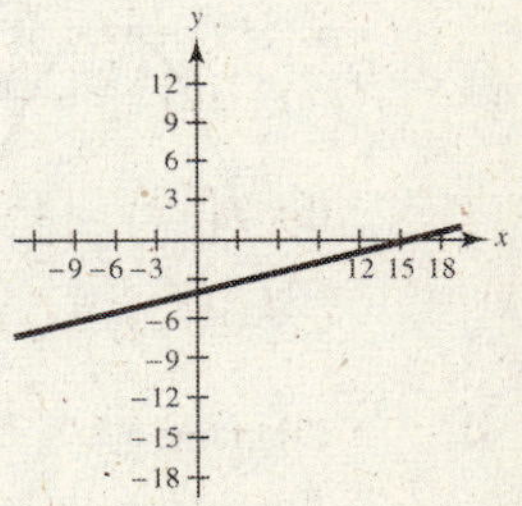

117.

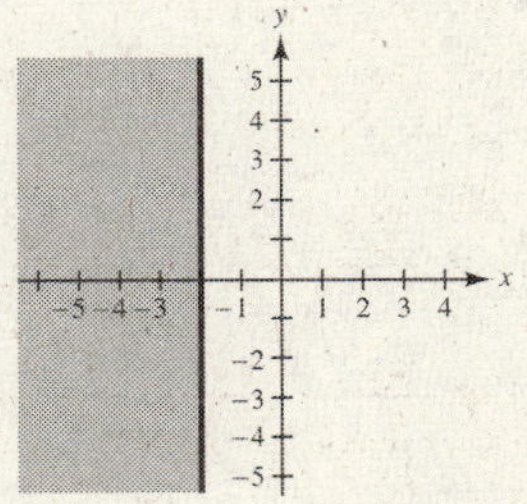

119.

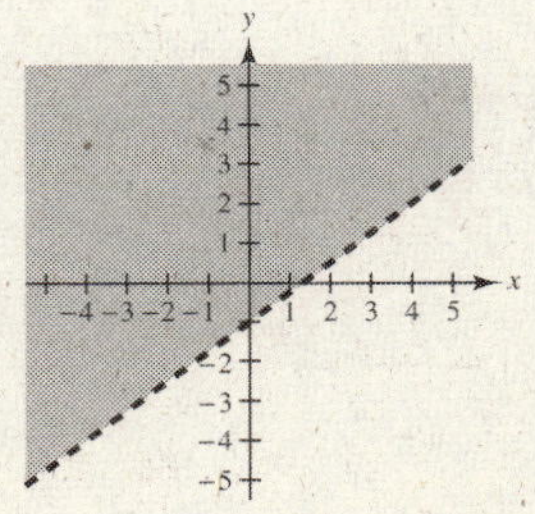

Section 3.7 Graphs of Functions

1.

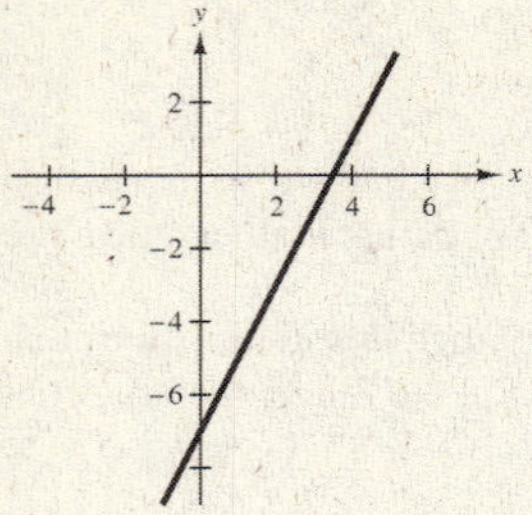

Domain: $-\infty < x < \infty$

Range: $-\infty < y < \infty$

3.

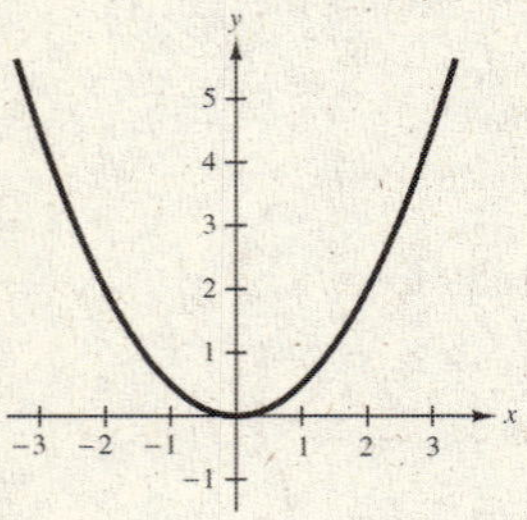

Domain: $-\infty < x < \infty$

Range: $0 \le y < \infty$

5.

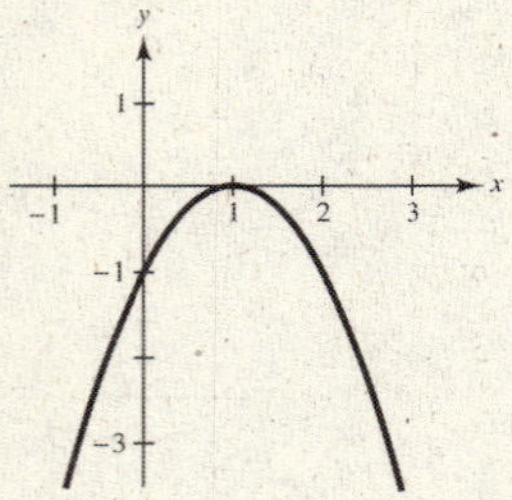

Domain: $-\infty < x < \infty$

Range: $-\infty < y \le 0$

7.

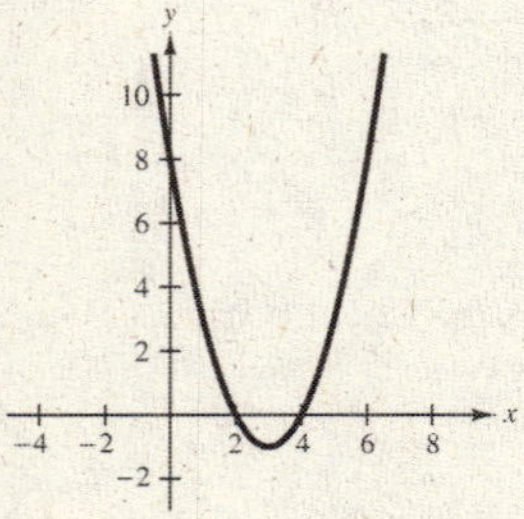

Domain: $-\infty < x < \infty$

Range: $-1 \le y < \infty$

9.

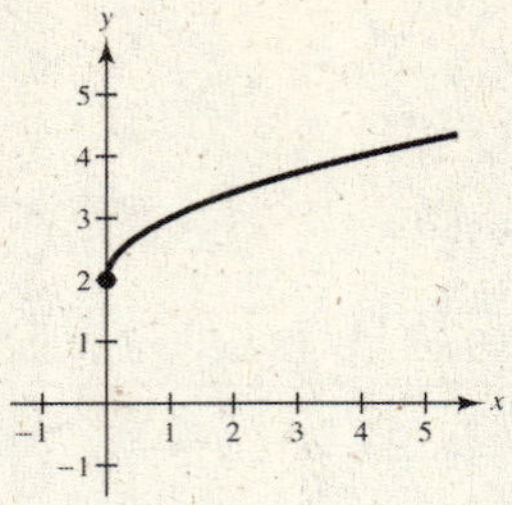

Domain: $0 \le x < \infty$

Range: $2 \le y < \infty$

11.

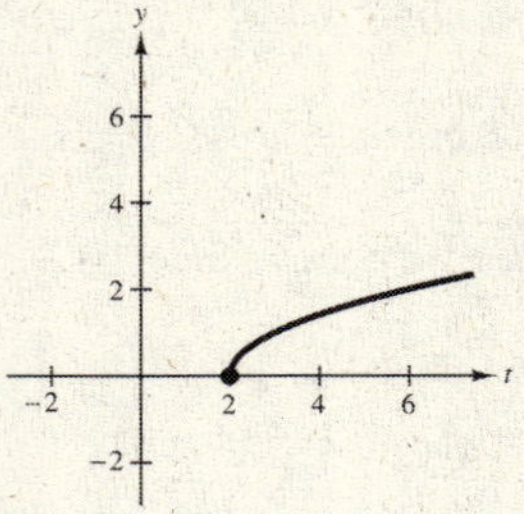

Domain: $2 \le t < \infty$

Range: $0 \le y < \infty$

13.

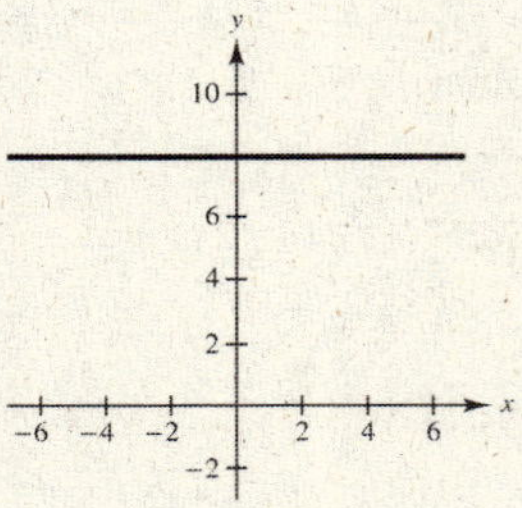

Domain: $-\infty < x < \infty$

Range: $y = 8$

15.

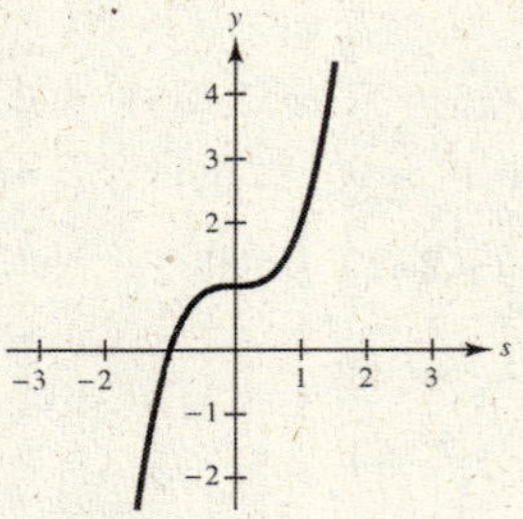

Domain: $-\infty < s < \infty$

Range: $-\infty < y < \infty$

17.

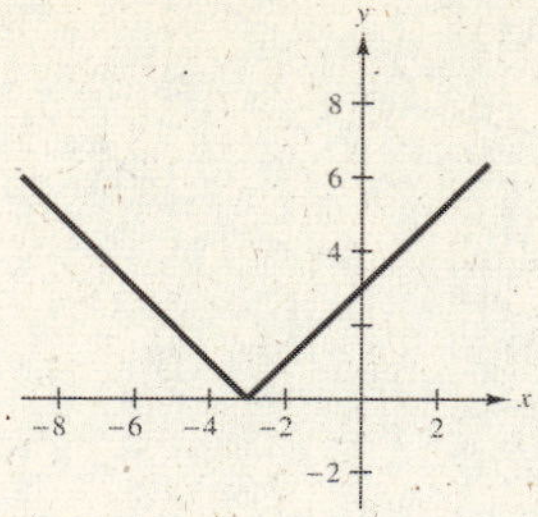

Domain: $-\infty < x < \infty$

Range: $0 \le y < \infty$

19.

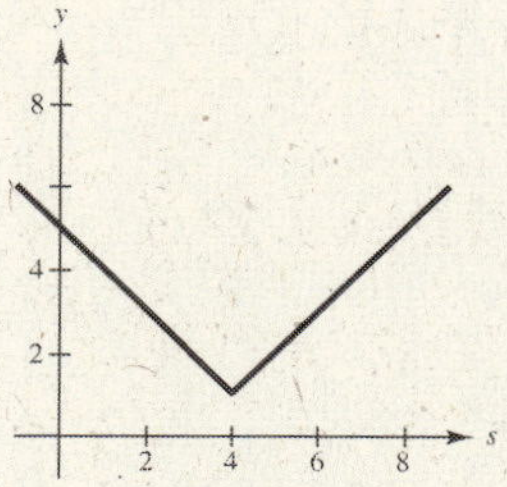

Domain: $-\infty < s < \infty$

Range: $1 \le y < \infty$

21.

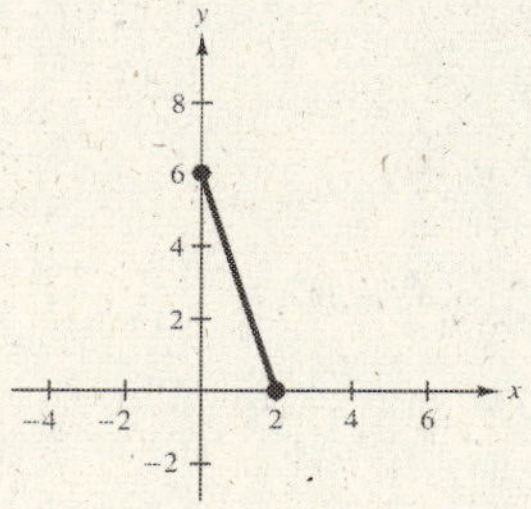

Domain: $0 \le x \le 2$

Range: $0 \le y \le 6$

23.

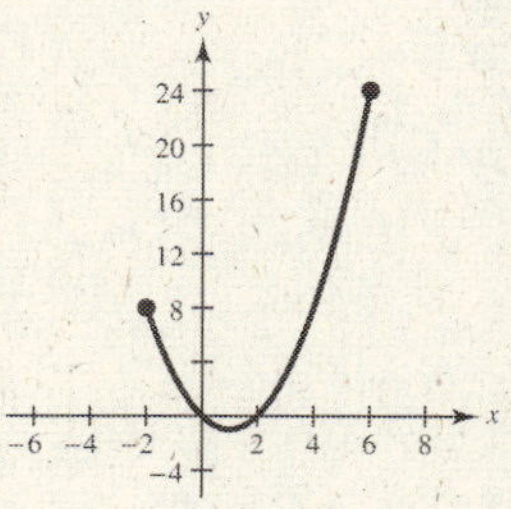

Domain: $-2 \le x \le 6$

Range: $-1 \le y \le 24$

25.

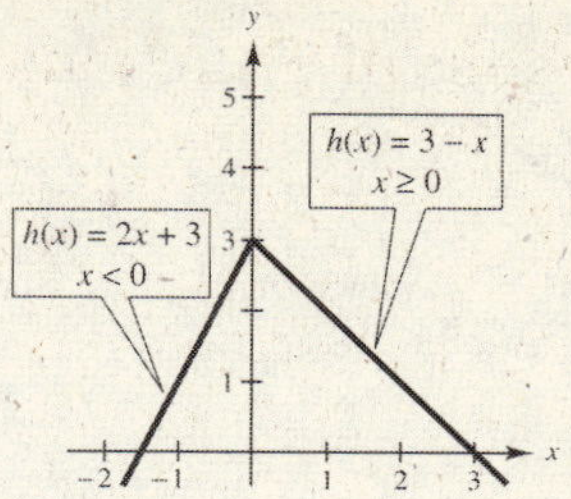

Domain: $-\infty < x < \infty$

Range: $\infty < y \le 3$

27.

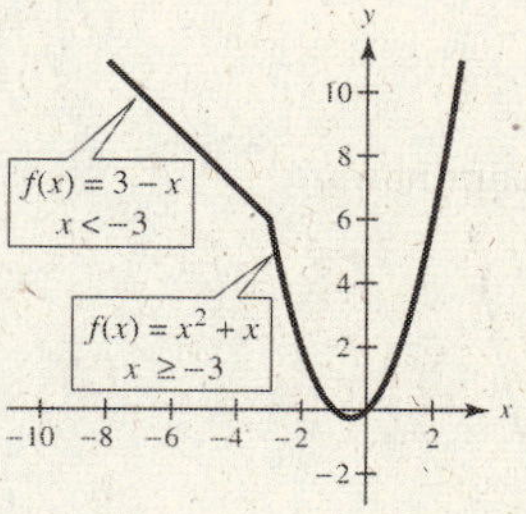

Domain: $-\infty < x < \infty$

Range: $-\frac{1}{4} \le y < \infty$

29.

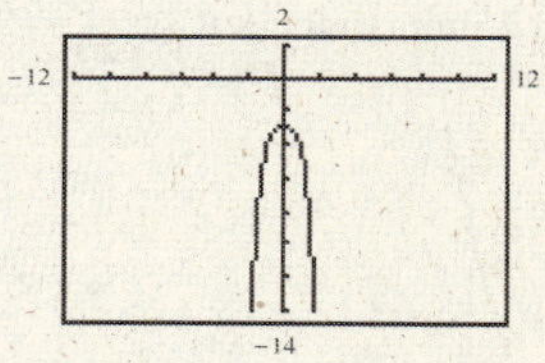

Domain: $-\infty < x < \infty$

Range: $-\infty < y \le -3$

31.

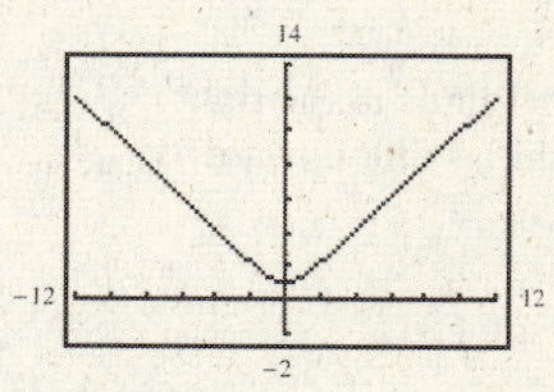

Domain: $-\infty < x < \infty$

Range: $1 \le y < \infty$

33. Yes, $y = \frac{1}{3}x^3$ passes the Vertical Line Test and is a function of x.

35. Yes, $y = -(x - 3)^2$ passes the Vertical Line Test and is a function of x.

37. No, y is not a function of x by the Vertical Line Test.

39.

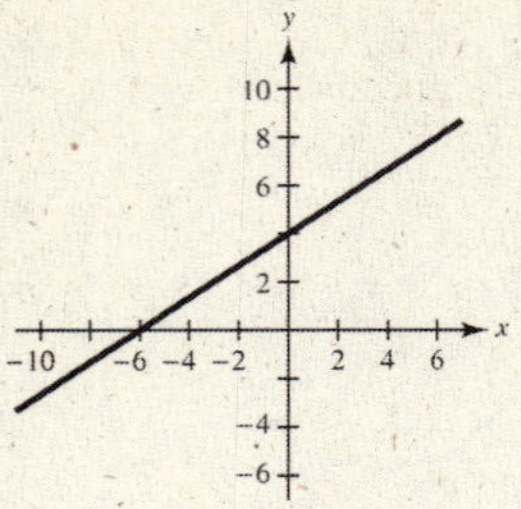

y is a function of x.

41.

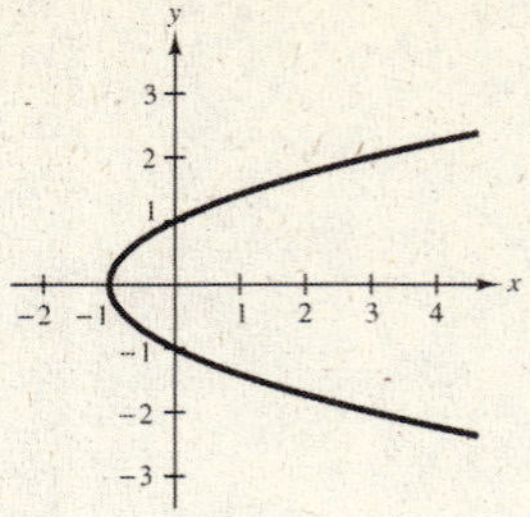

y is not a function of x.

43. Graph (b) matches $f(x) = x^2 - 1$.

45. Graph (a) matches $f(x) = 2 - |x|$.

47. (a) Vertical shift 2 units upward

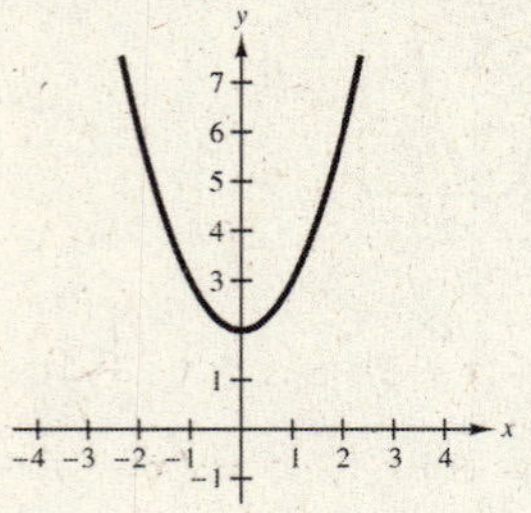

(b) Vertical shift 4 units downward

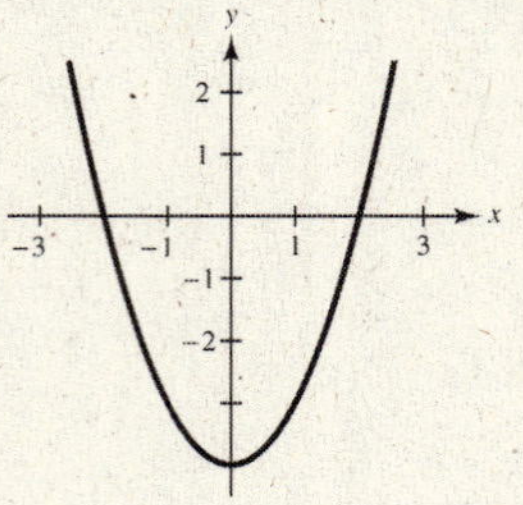

(c) Horizontal shift 2 units to the left

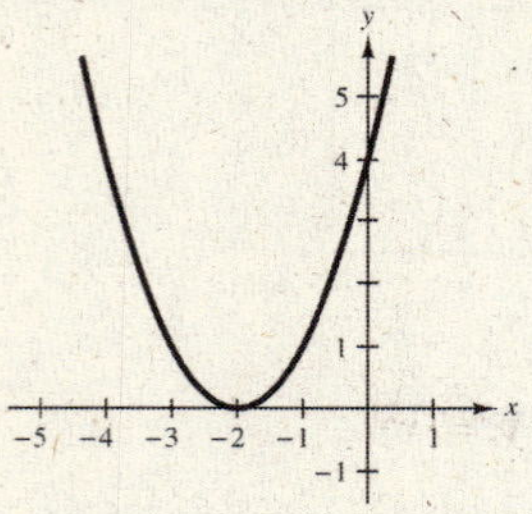

(d) Horizontal shift 4 units to the right

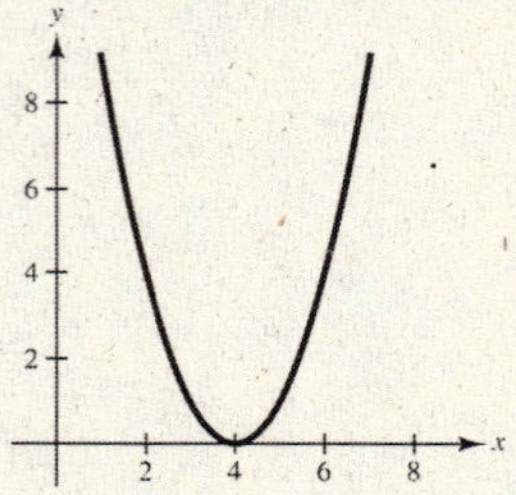

(e) Horizontal shift 3 units to the right and a vertical shift 1 unit upward

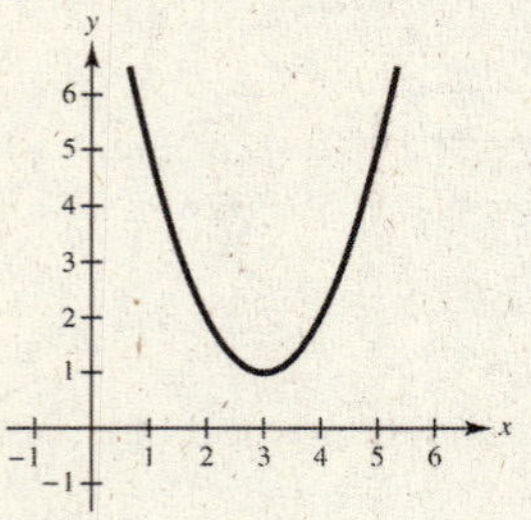

(f) Reflection in the x-axis and a vertical shift 4 units upward

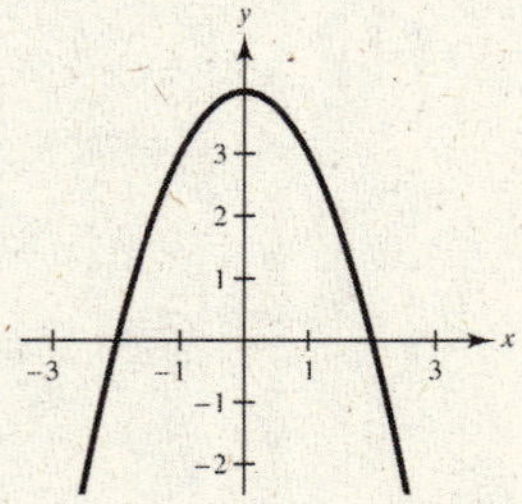

49.

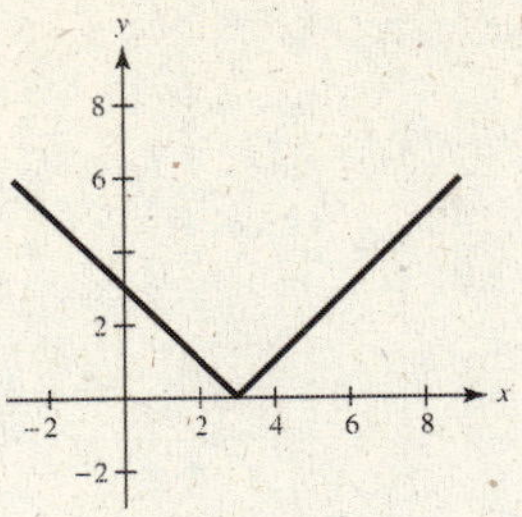

Horizontal shift 3 units to the right

51.

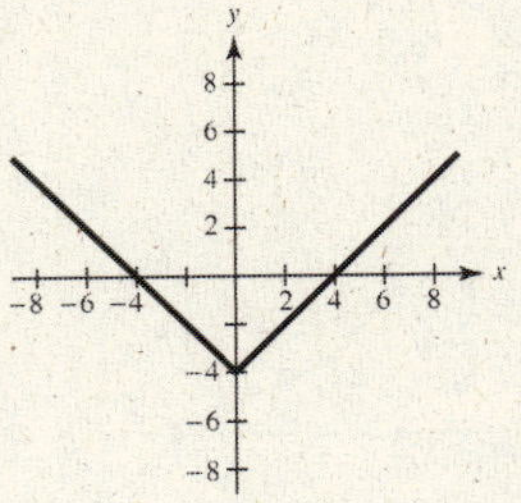

Vertical shift 4 units downward

53.

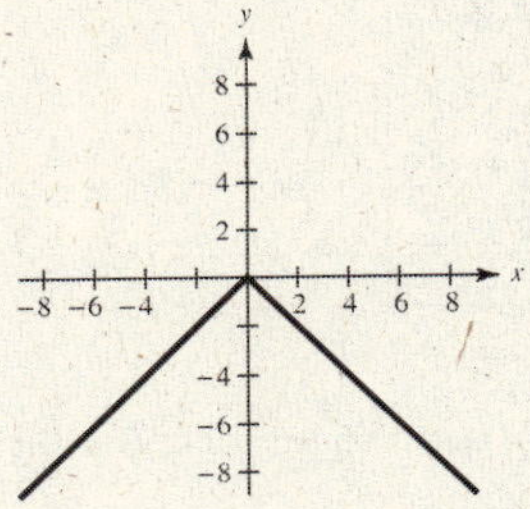

Reflection in the x-axis

55. The graph is shifted 3 units left.

$h(x) = (x + 3)^2$

57. The graph is reflected in the x-axis.

$h(x) = -x^2$

59. The graph is shifted 3 units upward.

$h(x) = -(x + 3)^2$

61. The graph is reflected across the x-axis.

$h(x) = -\sqrt{x}$

63. The graph is shifted 2 units left.

$h(x) = \sqrt{x + 2}$

65. The graph is reflected across the y-axis.

$h(x) = \sqrt{-x}$

67.

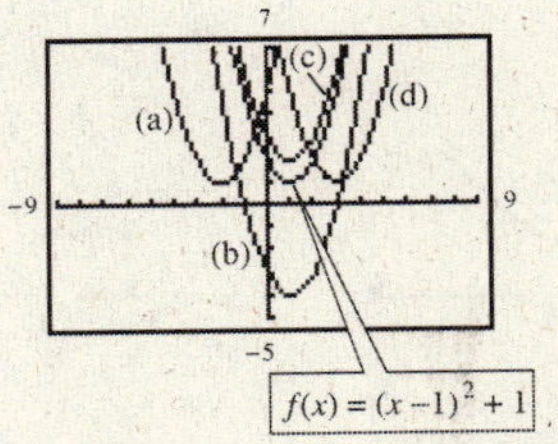

(a) $f(x) = (x - 1 + 3)^2 + 1 = (x + 2)^2 + 1$

(b) $f(x) = (x - 1)^2 + 1 - 5 = (x - 1)^2 - 4$

(c) $f(x) = (x - 1)^2 + 1 + 1 = (x - 1)^2 + 2$

(d) $f(x) = (x - 1 - 2)^2 + 1 = (x - 3)^2 + 1$

69. Basic function: $y = x^3$

Transformation: Horizontal shift 2 units right

Equation: $y = (x - 2)^3$

71. Basic function: $y = x^2$

Transformation: Reflection in the x-axis, horizontal shift 1 unit left, vertical shift 1 unit upward

Equation: $y = -(x + 1)^2 + 1$

73. Basic function: $y = x$

Transformation: Horizontal shift 3 units left or vertical shift 3 units upward

Equation: $y = x + 3$

75. (a) $y = f(x) + 2$

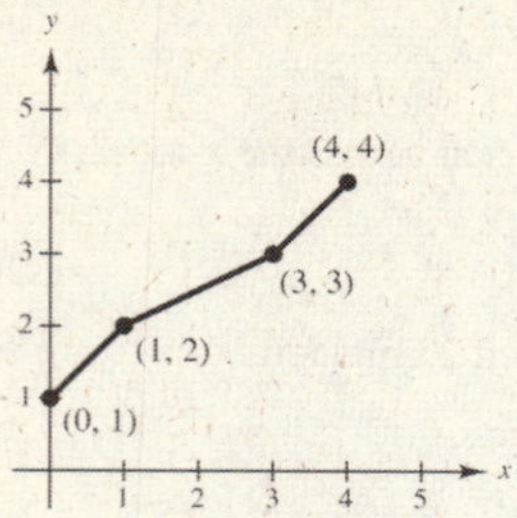

(b) $y = -f(x)$

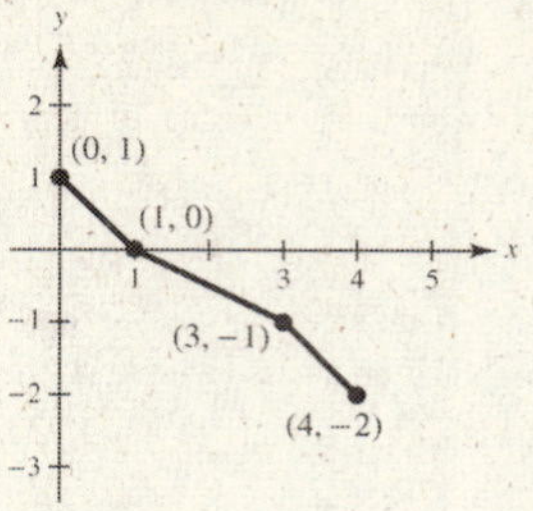

(c) $y = f(x - 2)$

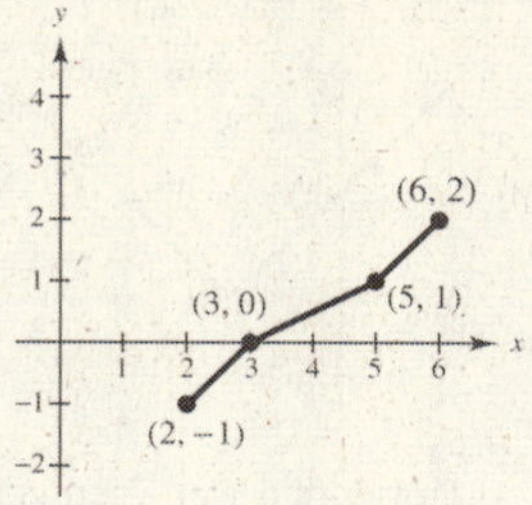

(d) $y = f(x + 2)$

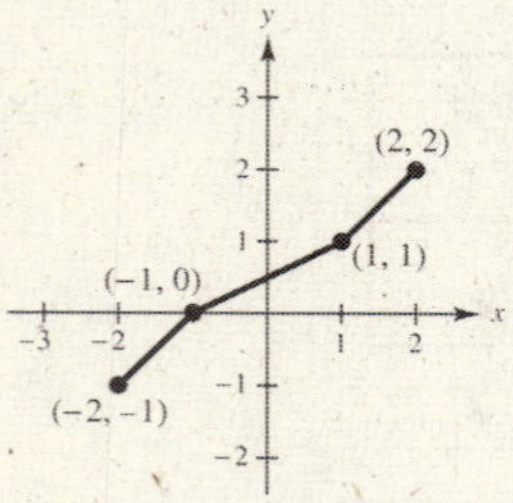

(e) $y = f(x) - 1$

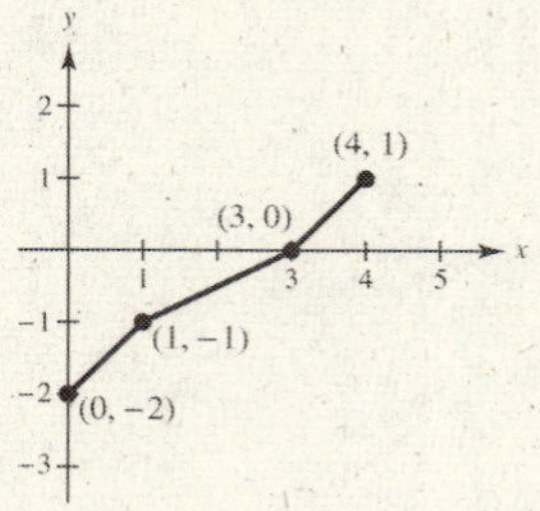

(f) $y = f(-x)$

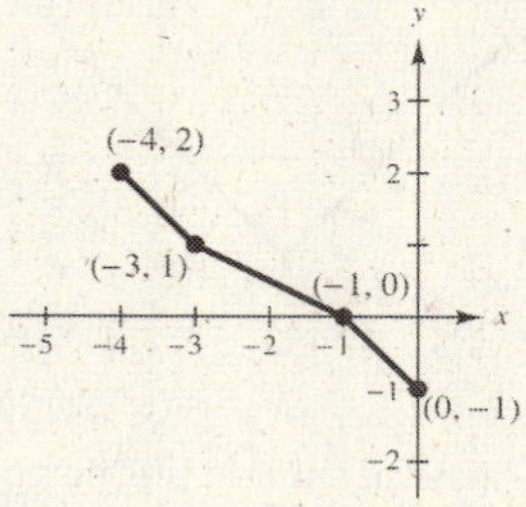

77. (a)

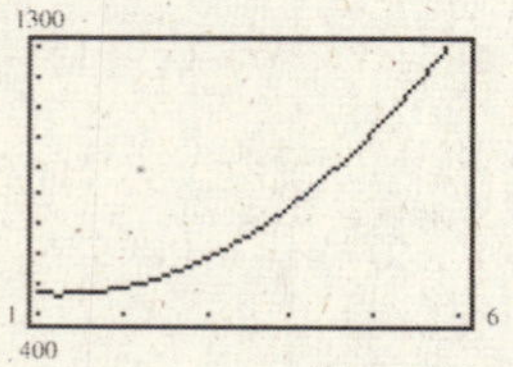

(b) 2003

79. (a)

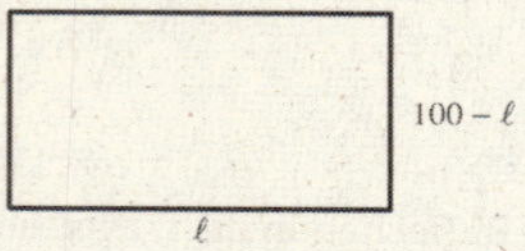

$A = l \cdot w$

$A = l(100 - l)$

Let $l =$ length.

$100 - l =$ width

$P = 2l + 2w$

$200 = 2l + 2w$

$100 = l + w$

$100 - l = w$

(b)

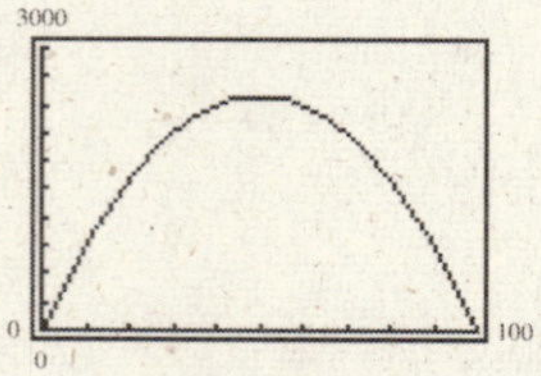

(c) When $l = 50$, the largest value of A is 2500. $(50, 2500)$ is the highest point on the graph of A giving the largest value of the function. The figure is a square.

81. (a)

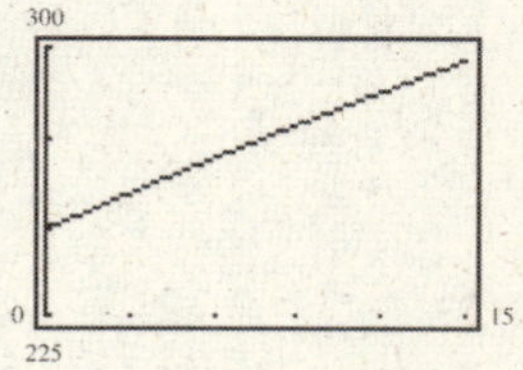

(b) P is a function of t because for each year there is one population.

(c) $P(5) = 265.9$ million people;

$P(10) = 281.7$ million people

(d) $P_1(t)$ corresponds to 2000, or a horizontal shift 10 years to the right.

(e)

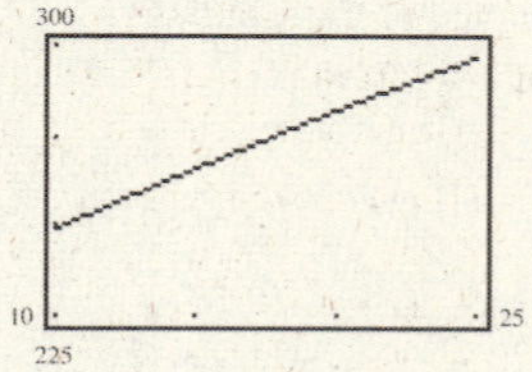

83. Use the Vertical Line Test to determine if an equation represents y as a function of x. If the graph of an equation has the property that no vertical line intersects the graph at two (or more) points, the equation represents y as a function of x.

85. If the domain of the function $f(x) = 2x$ changes from $0 \le x \le 2$ to $0 \le x \le 4$, then the range changes from $0 \le y \le 4$ to $0 \le y \le 8$.

87. The sum of four times a number and 1

89. The ratio of two times a number and 3

91. $2x + y = 4$

$y = 4 - 2x$

93. $-4x + 3y + 3 = 0$

$3y = 4x - 3$

$y = \frac{4}{3}x - 1$

95. $y < 2x + 1$

$1 \overset{?}{<} 2(0) + 1$

$1 \not< 1$; $(0, 1)$ is not a solution.

97. $2x - 3y > 2y$

$2(6) - 3(2) \overset{?}{>} 2(2)$

$12 - 6 \overset{?}{>} 4$

$6 > 4$; $(6, 2)$ is a solution.

Review Exercises for Chapter 3

1.

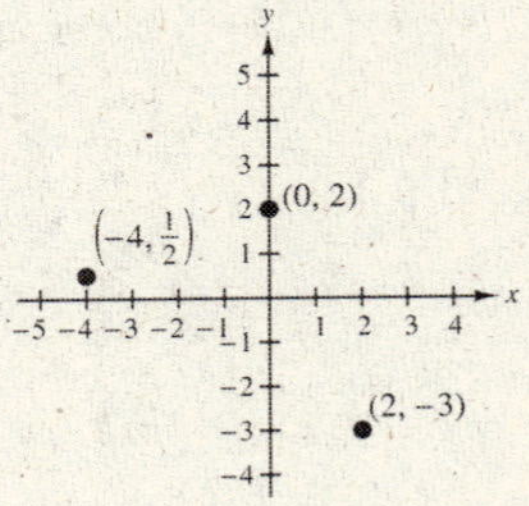

3. Quadrant I

5. Quadrant I or IV

7. (a) $(2, 3)$: $3 \overset{?}{=} 4 - \frac{1}{2}(2)$

$3 \overset{?}{=} 4 - 1$

$3 = 3$

$(2, 3)$ is a solution.

(b) $(-1, 5)$: $5 \overset{?}{=} 4 - \frac{1}{2}(-1)$

$5 \overset{?}{=} 4 + \frac{1}{2}$

$5 \neq 4\frac{1}{2}$

$(-1, 5)$ is not a solution.

(c) $(-6, 1)$: $1 \overset{?}{=} 4 - \frac{1}{2}(-6)$

$1 \overset{?}{=} 4 + 3$

$1 \neq 7$

$(-6, 1)$ is not a solution.

(d) $(8, 0)$: $0 \overset{?}{=} 4 - \frac{1}{2}(8)$

$0 \overset{?}{=} 4 - 4$

$0 = 0$

$(8, 0)$ is a solution.

9. $d = \sqrt{(4 - 4)^2 + (3 - 8)^2} = \sqrt{0 + 25} = \sqrt{25} = 5$

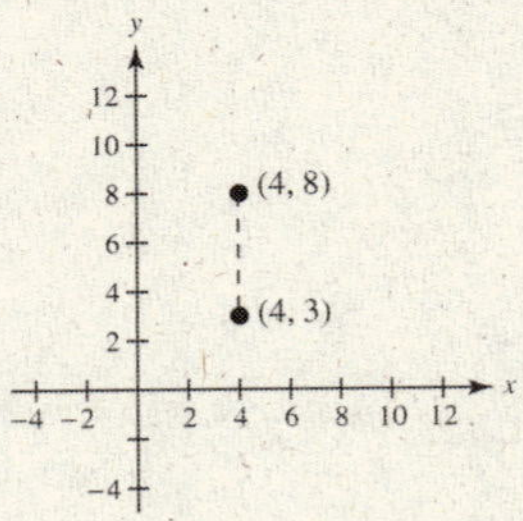

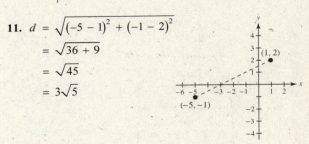

11. $d = \sqrt{(-5 - 1)^2 + (-1 - 2)^2}$

$= \sqrt{36 + 9}$

$= \sqrt{45}$

$= 3\sqrt{5}$

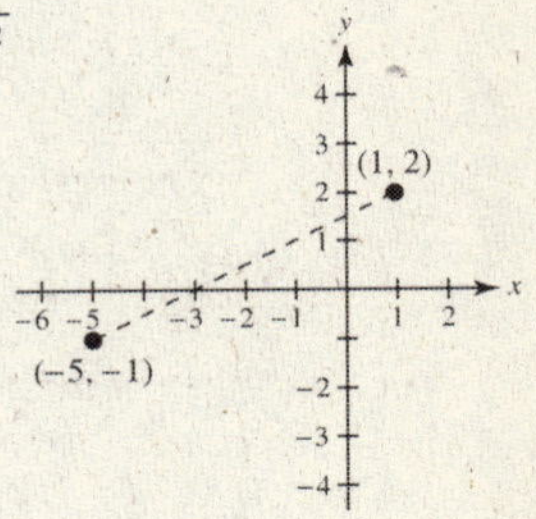

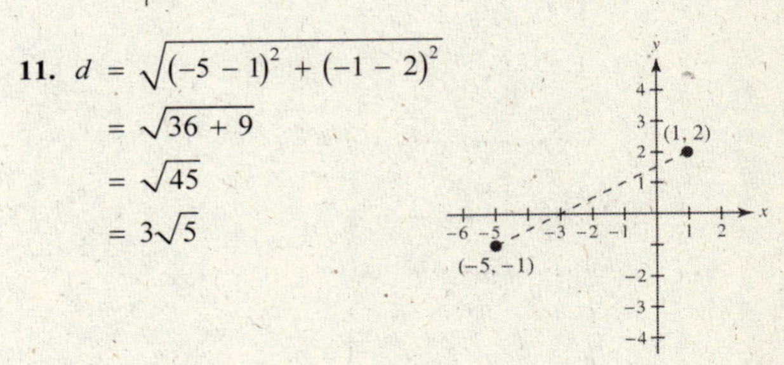

13. let $A = (-3, -2), B = (1, 0), C = (5, 2)$.

$$AB = \sqrt{(-3-1)^2 + (-2-0)^2} = \sqrt{(-4)^2 + (-2)^2} = \sqrt{16+4} = \sqrt{20}$$

$$BC = \sqrt{(1-5)^2 + (0-2)^2} = \sqrt{(-4)^2 + (-2)^2} = \sqrt{16+4} = \sqrt{20}$$

$$AC = \sqrt{(-3-5)^2 + (-2-2)^2} = \sqrt{(-8)^2 + (-4)^2} = \sqrt{64+16} = \sqrt{80} = 2\sqrt{20}$$

$$\sqrt{20} + \sqrt{20} = 2\sqrt{20}$$

The points are collinear.

15. $M = \left(\frac{1+7}{2}, \frac{4+2}{2}\right) = \left(\frac{8}{2}, \frac{6}{2}\right) = (4, 3)$

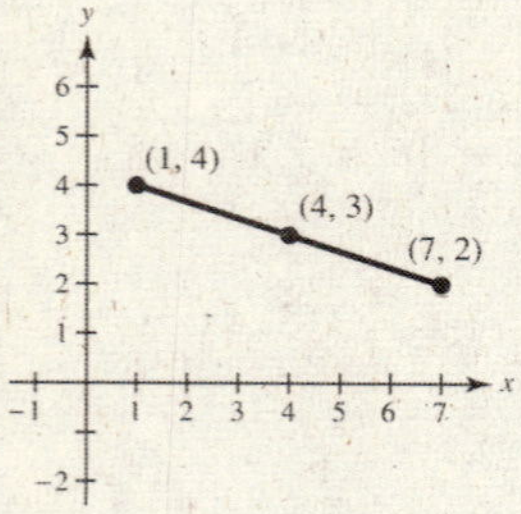

17. $M = \left(\frac{5+(-3)}{2}, \frac{-2+(5)}{2}\right) = \left(\frac{2}{2}, \frac{3}{2}\right) = \left(1, \frac{3}{2}\right)$

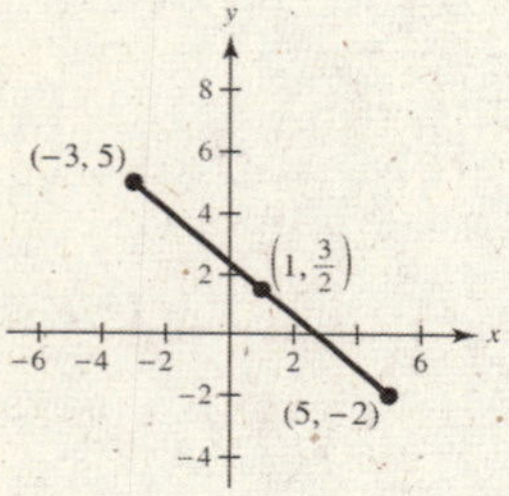

19.

$$3y - 2x - 3 = 0$$
$$3y - 2(0) - 3 = 0$$
$$3y = 3$$
$$y = 1 \quad (0, 1)$$
$$3(0) - 2x - 3 = 0$$
$$-2x = 3$$
$$x = -\tfrac{3}{2} \quad \left(-\tfrac{3}{2}, 0\right)$$

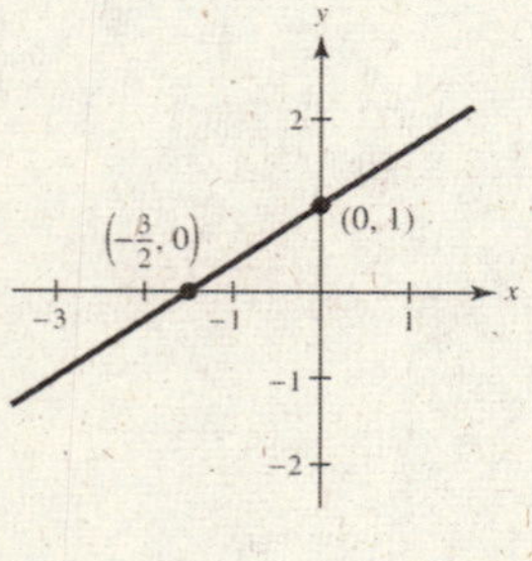

21.

$$y = x^2 - 1$$
$$y = 0^2 - 1$$
$$= -1 \qquad (0, -1)$$
$$0 = x^2 - 1$$
$$0 = (x-1)(x+1)$$
$$x = 1, x = -1 \qquad (1, 0), (-1, 0)$$

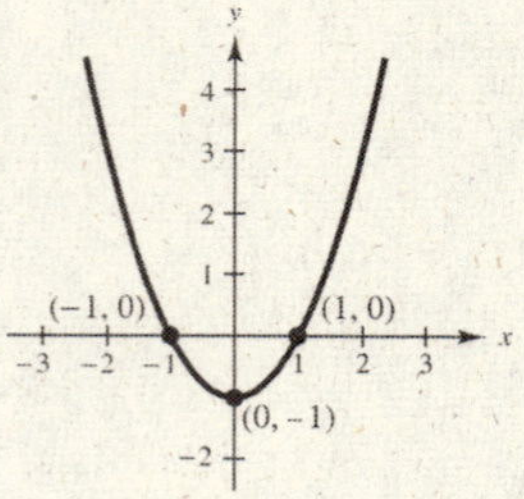

23.

$$y = |x| - 2$$
$$y = |0| - 2$$
$$= -2 \qquad (0, -2)$$
$$0 = |x| - 2$$
$$2 = |x|$$
$$\pm 2 = x \qquad (2, 0), (-2, 0)$$

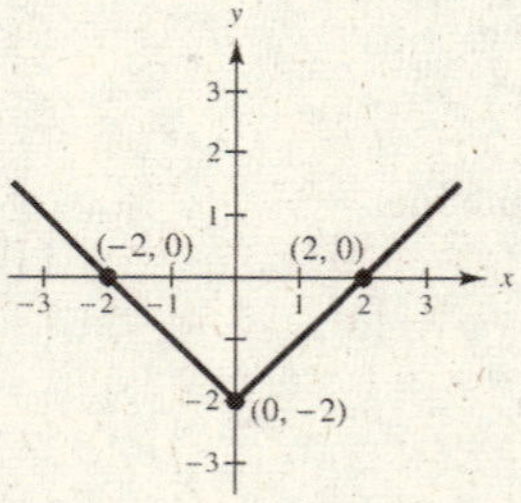

25. $8x - 2y = -4$

$$\begin{aligned} y\text{-intercept: } 8(0) - 2y &= -4 \\ -2y &= -4 \\ y &= 2 \quad (0, 2) \\ x\text{-intercept: } 8x - 2(0) &= -4 \\ 8x &= -4 \\ x &= -\tfrac{1}{2} \quad \left(-\tfrac{1}{2}, 0\right) \end{aligned}$$

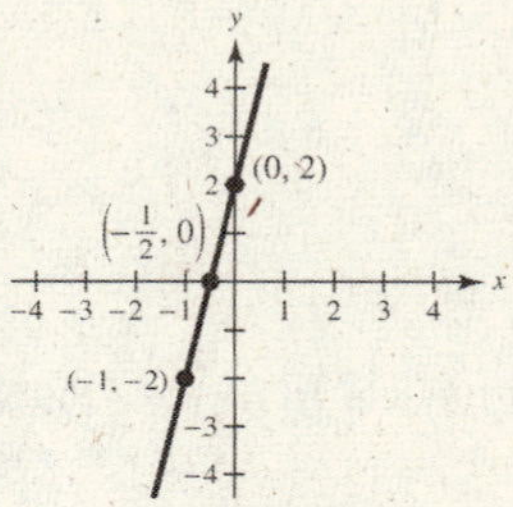

27. $y = 5 - |x|$

$$\begin{aligned} y\text{-intercept: } y &= 5 - |x| \\ y &= 5 \qquad (0, 5) \\ x\text{-intercepts: } 0 &= 5 - |x| \\ |x| &= 5 \\ x &= -5, x = 5 \quad (-5, 0), (5, 0) \end{aligned}$$

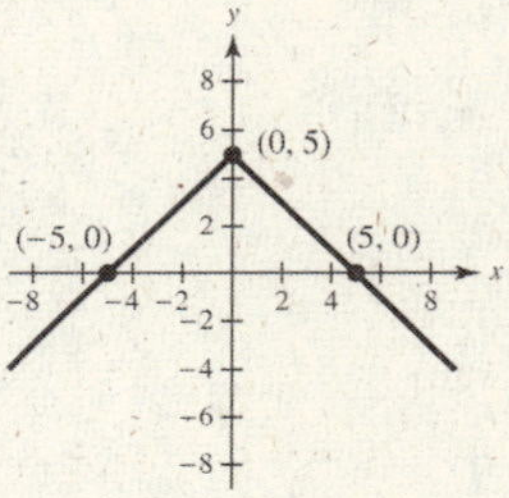

29. $y = |2x + 1| - 5$

$$\begin{aligned} y\text{-intercept: } y &= |2(0) + 1| - 5 \\ &= 1 - 5 \\ &= -4 \quad (0, -4) \\ x\text{-intercepts: } 0 &= |2x + 1| - 5 \\ 5 &= |2x + 1| \\ 5 &= 2x + 1 \quad \text{or} \quad -5 = 2x + 1 \\ 4 &= 2x \qquad\qquad -6 = 2x \\ 2 &= x \qquad\qquad -3 = x, \quad (2, 0), (-3, 0) \end{aligned}$$

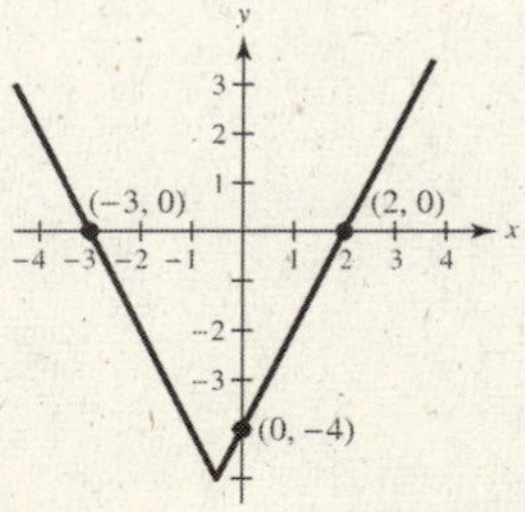

31.

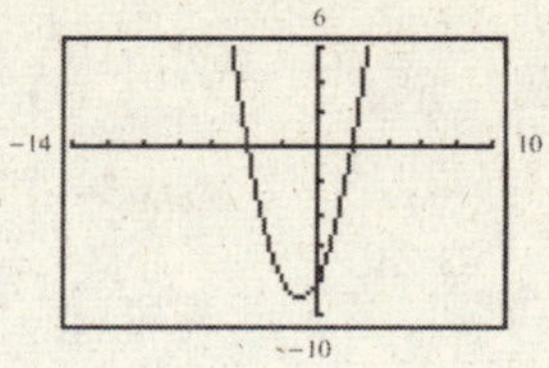

$(-4, 0), (0, -8), (2, 0)$

33.

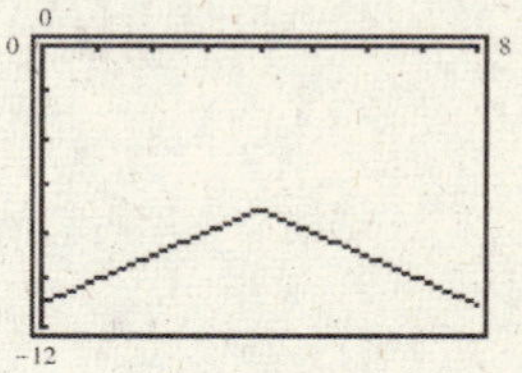

$(0, -11)$

no x-intercepts

35.

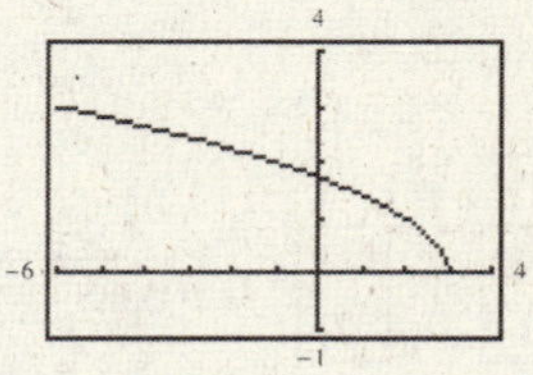

$(3, 0), (0, 1.73)$

37. (a) The annual depreciation is

$$\frac{35{,}000 - 15{,}000}{5} = 4000.$$

$$35{,}000 - 4000(1) = 31{,}000$$
$$35{,}000 - 4000(2) = 27{,}000$$
$$35{,}000 - 4000(3) = 23{,}000$$

So, $y = 35{,}000 - 4000t$.

(b)

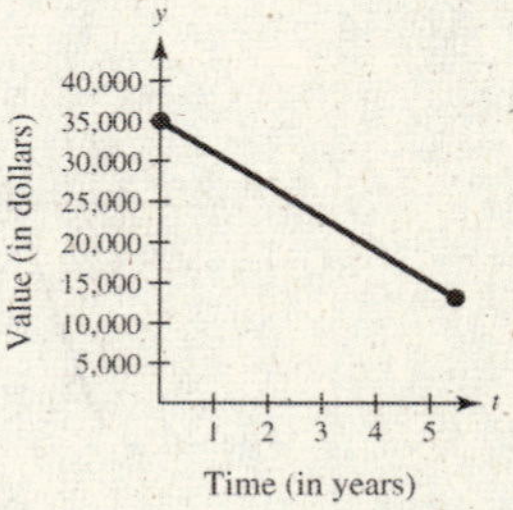

(c) $y = 35{,}000 - 4000(0)$

$y = 35{,}000$

$(0, 35{,}000)$

The y-intercept represents the value of the SUV when purchased.

39. $m = \dfrac{-1 - 2}{-3 - 4} = \dfrac{-3}{-7} = \dfrac{3}{7}$

41. $m = \dfrac{\frac{3}{4} - \frac{3}{4}}{4 - (-6)} = \dfrac{0}{10} = 0$

43. $m = \dfrac{0 - 6}{8 - 0} = \dfrac{-6}{8} = \dfrac{-3}{4} = -\dfrac{3}{4}$

45. $-3 = \dfrac{y + 4}{x - 2}$

$(0, 2), (1, -1)$

47. $\dfrac{1}{4} = \dfrac{y - \frac{1}{2}}{x - (-4)}$

$(-6, 0)$ and $(2, 2)$

49. *Verbal model:* $\boxed{\dfrac{\text{Rise}}{\text{Run}}} = \boxed{\dfrac{\text{Rise}}{\text{Run}}}$

Proportion: $\dfrac{1}{12} = \dfrac{4}{x}$

$x = 48$

Verbal model: $\boxed{\text{Leg } 1}^2 + \boxed{\text{Leg } 2}^2 = \boxed{\text{Hypotenuse}}^2$

Labels: Leg 1 = 4

Leg 2 = 48

Hypotenuse = x

Equation: $4^2 + 48^2 = x^2$

$16 + 2304 = x^2$

$\sqrt{2320} = x \approx 48.17$

The length of the ramp is about 48.17 feet.

51. $5x - 2y - 4 = 0$

$-2y = -5x + 4$

$y = \frac{5}{2}x - 2$

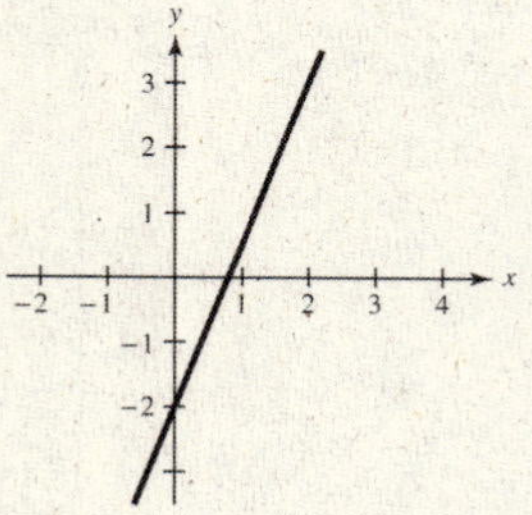

53. $6x + 2y + 5 = 0$

$2y = -6x - 5$

$y = -3x - \frac{5}{2}$

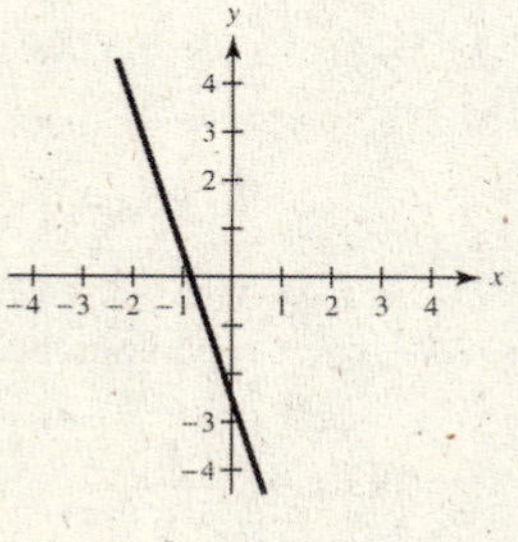

55. $L_1 : y = \frac{3}{2}x + 1$

$L_2 : y = \frac{2}{3}x - 1$

$m_1 = \frac{3}{2}, m_2 = \frac{2}{3}$

$m_1 \neq m_2, m_1 \cdot m_2 \neq -1$

The lines are neither parallel nor perpendicular.

57. $L_1 : y = \frac{3}{2}x - 2$

$L_2 : y = -\frac{2}{3}x + 1$

$m_1 = \frac{3}{2}, m_2 = -\frac{2}{3}$

$m_1 \cdot m_2 = -1$

The lines are perpendicular.

59. Let $(t_1, c_1) = (1998, 1.09)$ and $(t_2, c_2) = (2007, 3.13)$.

$$\begin{aligned}\text{Average rate of change} &= \frac{c_2 - c_1}{t_2 - t_1}\\ &= \frac{3.13 - 1.09}{2007 - 1998}\\ &= \frac{2.04}{9}\\ &\approx 0.227 \approx 0.23\end{aligned}$$

The average rate of change is about \$0.23 per gallon.

61. $y - (-4) = -4(x - 1)$

$y + 4 = -4x + 4$

$4x + y = 0$

63. $y - 5 = \frac{1}{4}[x - (-6)]$

$y - 5 = \frac{1}{4}(x + 6)$

$4(y - 5) = x + 6$

$4y - 20 = x + 6$

$4y - x - 26 = 0$

65. $m = \dfrac{0 + 3}{-6 - 0} = \dfrac{3}{-6} = -\dfrac{1}{2}$

$y - 0 = -\dfrac{1}{2}(x + 6)$

$y = -\dfrac{1}{2}x - 3$

67. $m = \dfrac{6 - (-3)}{4 - (-2)} = \dfrac{9}{6} = \dfrac{3}{2}$

$y - 6 = \dfrac{3}{2}(x - 4)$

$y - 6 = \dfrac{3}{2}x - 6$

$y = \dfrac{3}{2}x$

69. $y = -9$

71. $x = -5$

73. $3x + y = 2$

$y = -3x + 2$

(a) $y + \frac{4}{5} = -3\left(x - \frac{3}{5}\right)$

$y + \frac{4}{5} = -3x + \frac{9}{5}$

$y = -3x + 1$

(b) $y + \frac{4}{5} = \frac{1}{3}\left(x - \frac{3}{5}\right)$

$y + \frac{4}{5} = \frac{1}{3}x - \frac{3}{15}$

$y = \frac{1}{3}x - 1$

75. (a) $S = 32{,}000 + 1050(t - 2)$

$S = 32{,}000 + 1050t - 2100$

$S = 1050t + 29{,}900$

$t = 2$ corresponds to 2002.

(b) $t = 10$ corresponds to 2010.

$s = 32{,}000 + 1050(10 - 2)$

$= 32{,}000 + 8400$

$= 40{,}400$

Your annual salary in 2010 is predicted to be \$40,400.

(c) $t = 5$ corresponds to 2005.

$s = 32{,}000 + 1050(5 - 2)$

$= 32{,}000 + 3150$

$= 35{,}150$

Your annual salary in 2005 is estimated to be \$35,150.

77. $5x - 8y \geq 12$

(a) $5(-1) - 8(2) \overset{?}{\geq} 12$

$-5 - 16 \geq 12$

$-21 \ngeq 12$ Not a solution

(b) $5(3) - 8(-1) \overset{?}{\geq} 12$

$15 + 8 \geq 12$

$23 \geq 12$ Solution

(c) $5(4) - 8(0) \overset{?}{\geq} 12$

$20 - 0 \geq 12$

$20 \geq 12$ Solution

(d) $5(0) - 8(3) \overset{?}{\geq} 12$

$0 - 24 \geq 12$

$-24 \ngeq 12$ Not a solution

79. $y > -2$

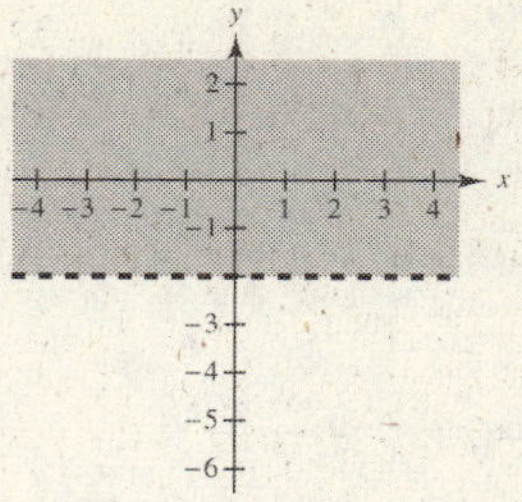

81. $x - 2 \geq 0$

$x \geq 2$

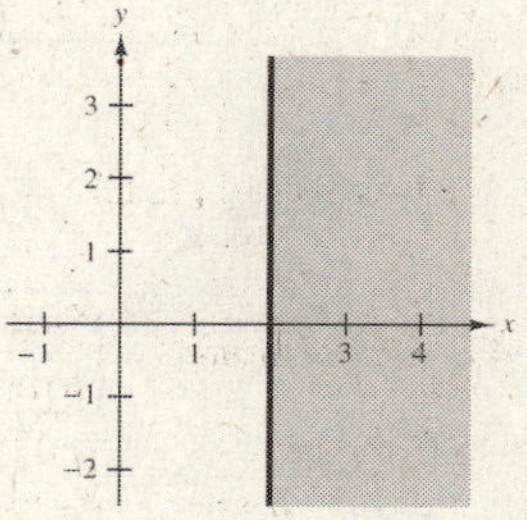

83. $2x + y < 1$

$y < -2x + 1$

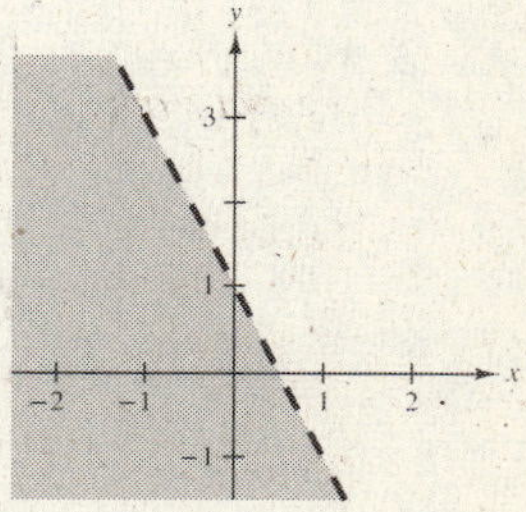

85. $-(x - 1) \leq 4y - 2$

$-x + 1 \leq 4y - 2$

$-x + 3 \leq 4y$

$-\frac{1}{4}x + \frac{3}{4} \leq y$

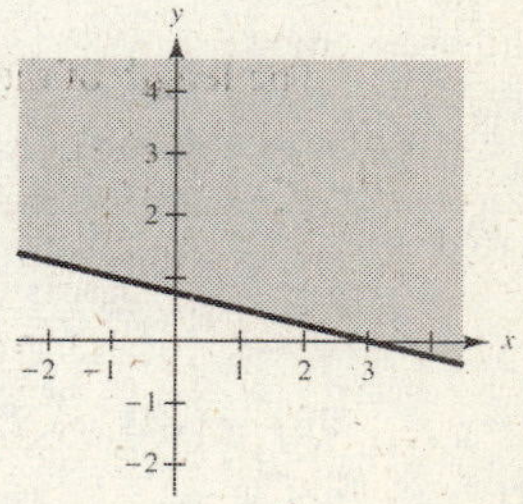

87. $y \leq 12 - \frac{3}{2}x$

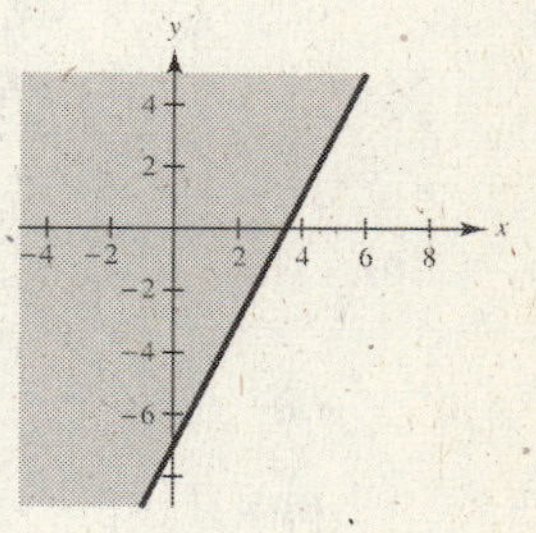

89. $x + y \geq 0$

$y \geq -x$

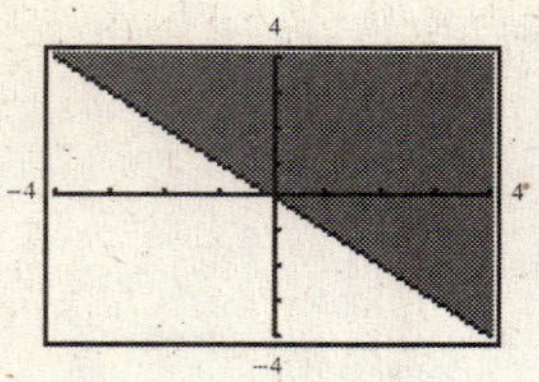

91. (a) $P = 2x + 2y$

$2x + 2y \leq 800$

$x + y \leq 400, \quad x \geq 0, y \geq 0$

(b)

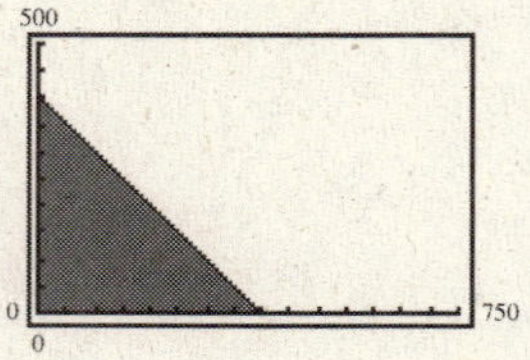

93. Domain: $\{-3, -1, 0, 1\}$

Range: $\{0, 1, 4, 5\}$

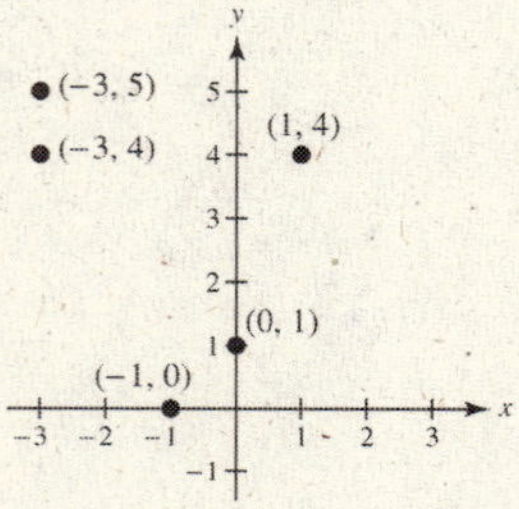

95. No, this relation is not a function because the 8 in the domain is matched with two numbers (1 and 2) in the range.

97. Yes, this relation is a function because each number in the domain is matched with only one number in the range.

99. $f(t) = \sqrt{5 - t}$

(a) $f(-4) = \sqrt{5 - (-4)} = \sqrt{9} = 3$

(b) $f(5) = \sqrt{5 - 5} = 0$

(c) $f(3) = \sqrt{5 - 3} = \sqrt{2}$

(d) $f(5z) = \sqrt{5 - 5z}$

101. $f(x) = \begin{cases} -3x, & x \leq 0 \\ 1 - x^2, & x > 0 \end{cases}$

(a) $f(3) = 1 - 3^2 = -8$

(b) $f\left(-\frac{2}{3}\right) = -3\left(-\frac{2}{3}\right) = 2$

(c) $f(0) = -3(0) = 0$

(d) $f(4) - f(3) = (1 - 4^2) - (1 - 3^2)$

$= -15 - (-8) = -15 + 8 = -7$

103. (a) $\dfrac{f(x + 2) - f(2)}{x} = \dfrac{[3 - 2(x + 2)] - [3 - 2(2)]}{x}$

$= \dfrac{3 - 2x - 4 - 3 + 4}{x}$

$= \dfrac{-2x}{x} = -2$

(b) $\dfrac{f(x - 3) - f(3)}{x} = \dfrac{[3 - 2(x - 3)] - [3 - 2(3)]}{x}$

$= \dfrac{3 - 2x + 6 - 3 + 6}{x}$

$= \dfrac{-2x + 12}{x}$

105. $h(x) = 4x^2 - 7$

Domain: $-\infty < x < \infty$

107. $f(x) = \sqrt{3x + 10}$

$3x + 10 \geq 0$

$3x \geq -10$

$x \geq -\frac{10}{3}$

Domain: $x \geq -\frac{10}{3}$

109. *Verbal model:* [Perimeter] = 2[Length] + 2[Width]

$150 = 2\text{Length} + 2x$

$\dfrac{150 - 2x}{2} = \text{Length}$

$75 - x = \text{Length}$

Verbal model: [Area] = [Length] · [Width]

Labels: Area $= A$

Length $= 75 - x$

Width $= x$

Function: $A = (75 - x)x$

Domain: $0 < x < \frac{75}{2}$

111.

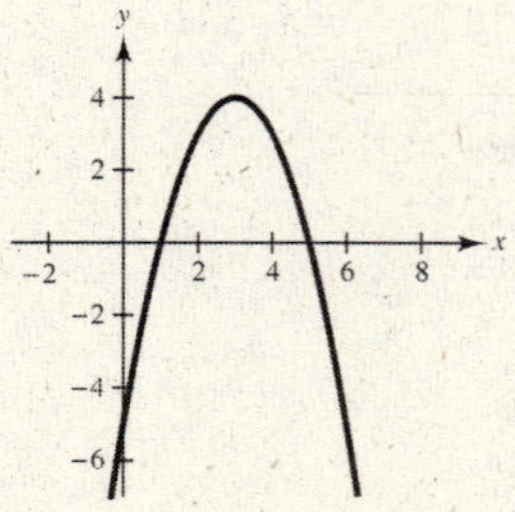

Domain: $-\infty < x < \infty$

Range: $-\infty < y \leq 4$

113.

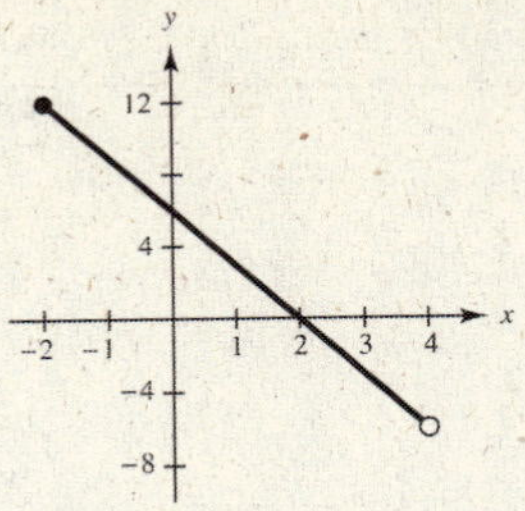

Domain: $-2 \le x < 4$

Range: $-6 < y \le 12$

115.

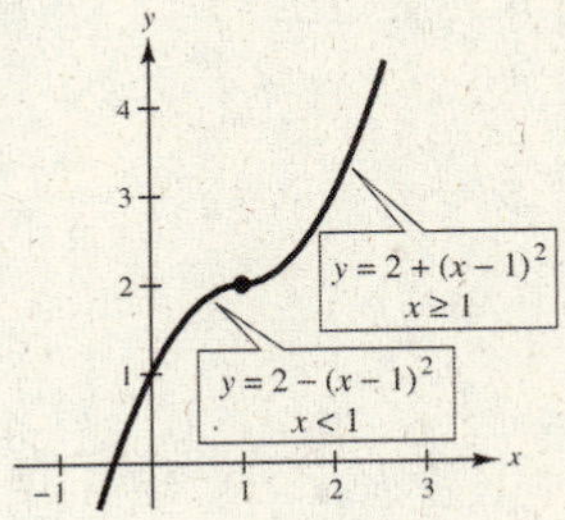

Domain: $-\infty < x < \infty$

Range: $-\infty < y < \infty$

117. $f(x) = -x^2 + 2$ (c)

119. $f(x) = -\sqrt{x}$ (b)

121. No, y is not a function of x.

123. $h(x) = -\sqrt{x}$ is a reflection in the x-axis of $f(x) = \sqrt{x}$.

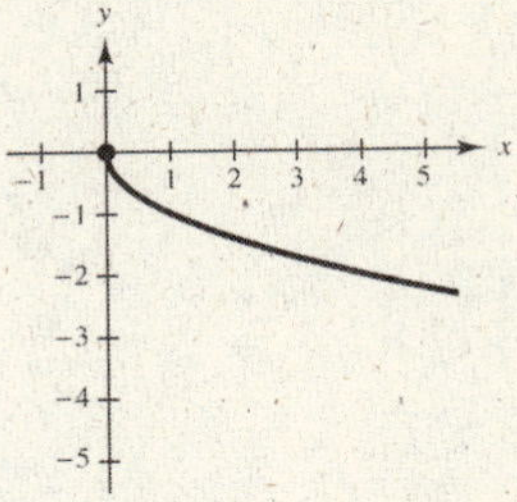

125. $h(x) = \sqrt{x - 2} - 1$

Horizontal shift 2 units to the right followed by a vertical shift 1 unit downward

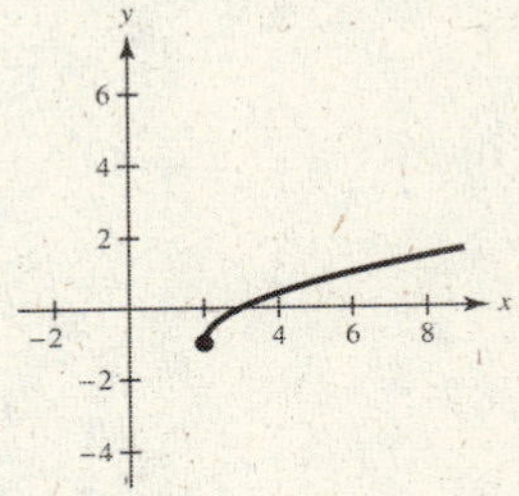

Chapter Test for Chapter 3

1. Quadrant IV

2.

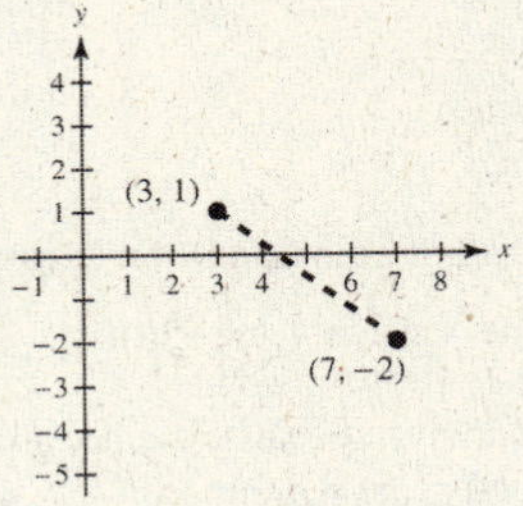

$$d = \sqrt{(7-3)^2 + (-2-1)^2} = \sqrt{16+9} = \sqrt{25} = 5$$

$$M = \left(\frac{7+3}{2}, \frac{-2+1}{2}\right) = \left(\frac{10}{2}, \frac{-1}{2}\right) = \left(5, -\frac{1}{2}\right)$$

3. $y = -3(x + 1)$

y-intercept: $y = -3(0 + 1)$

$y = -3,\ (0, -3)$

x-intercept: $0 = -3(x + 1)$

$0 = -3x - 3$

$3 = -3x$

$-1 = x,\ (-1, 0)$

4.

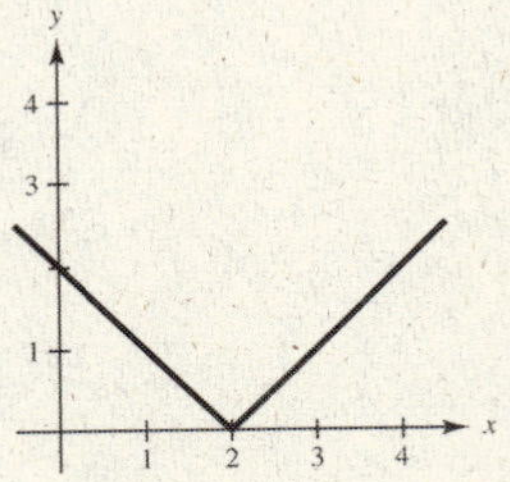

5. (a) $m = \dfrac{3-7}{2+4} = -\dfrac{4}{6} = -\dfrac{2}{3}$

(b) $m = \dfrac{6+2}{3-3} = \dfrac{8}{0} =$ undefined

6.

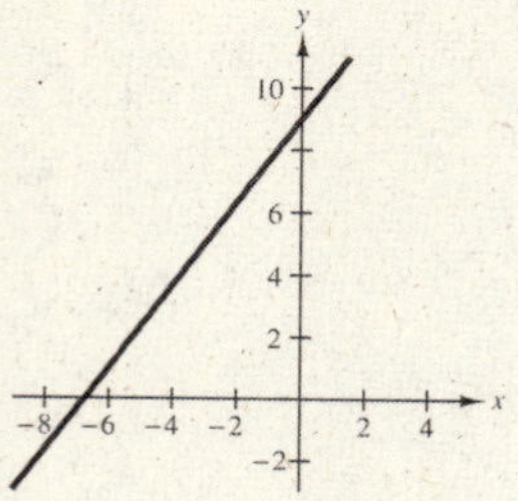

7.
$$2x + 5y = 10$$
$$2(0) + 5y = 10$$
$$5y = 10$$
$$y = 2 \quad (0, 2)$$
$$2x + 5(0) = 10$$
$$2x = 10$$
$$x = 5 \quad (5, 0)$$

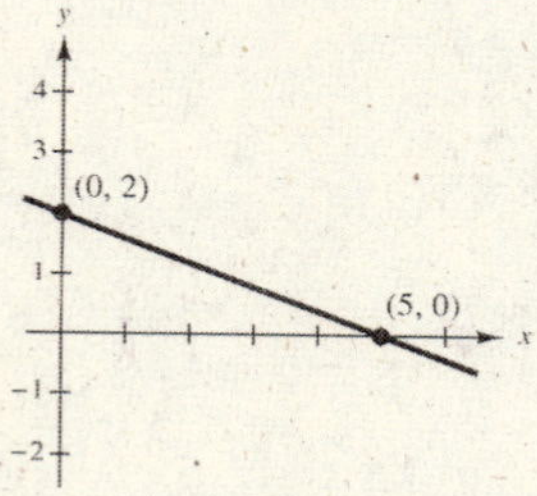

8. $m = \dfrac{3-(-6)}{8-2} = \dfrac{9}{6} = \dfrac{3}{2}$
$$y - 3 = \frac{3}{2}(x - 8)$$
$$2(y - 3) = 3(x - 8)$$
$$2y - 6 = 3x - 24$$
$$-3x + 2y + 18 = 0$$
$$3x - 2y - 18 = 0$$

9.
$$x = -2$$
$$x + 2 = 0$$

10.
$$3x - 5y = 4$$
$$-5y = -3x + 4$$
$$y = \frac{-3x}{-5} + \frac{4}{-5}$$
$$y = \frac{3}{5}x - \frac{4}{5}$$

slope $= \dfrac{3}{5}$

(a)
$$y - 3 = \frac{3}{5}(x + 2)$$
$$y - 3 = \frac{3}{5}x + \frac{6}{5}$$
$$y = \frac{3}{5}x + \frac{21}{5}$$

(b)
$$y - 3 = -\frac{5}{3}(x + 2)$$
$$y - 3 = -\frac{5}{3}x - \frac{10}{3}$$
$$y = -\frac{5}{3}x - \frac{1}{3}$$

11.
$$x + 4y \le 8$$
$$4y \le -x + 8$$
$$y \le -\frac{1}{4}x + 2$$

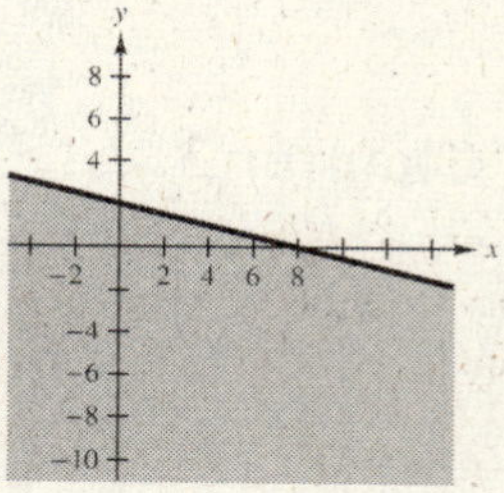

12. No, $y^2(4 - x) = x^3$ is not a function of x because the graph does not pass the Vertical Line Test.

13. (a) The relation is a function because each x number is matched with exactly one y number.

(b) The relation is not a function because 0 is matched with two numbers, 0 and -4.

14. (a) $g(2) = \dfrac{2}{2-3} = -2$

(b) $g\left(\dfrac{7}{2}\right) = \dfrac{\frac{7}{2}}{\frac{7}{2} - 3} = \dfrac{7}{7-6} = 7$

(c) $g(x + 2) = \dfrac{x+2}{(x+2)-3} = \dfrac{x+2}{x-1}$

15. (a) $h(t) = \sqrt{9 - t}$

$$9 - t \geq 0$$
$$-t \geq -9$$
$$t \leq 9$$

Domain: All real values of t such that $t \leq 9$

(b) $f(x) = \dfrac{x + 1}{x - 4}$

$$x - 4 \neq 0$$
$$x \neq 4$$

Domain: All real values of x such that $x \neq 4$

16.

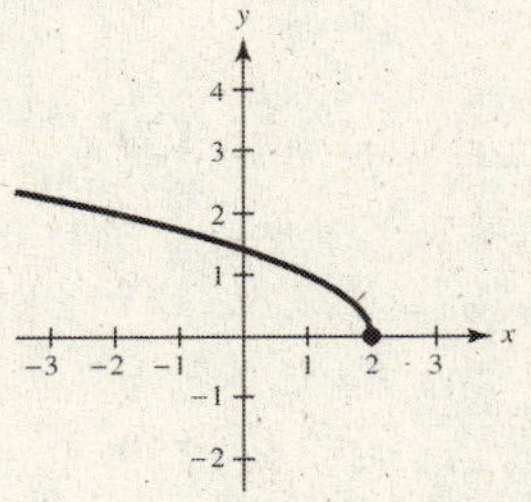

17. $g(x) = -(x - 2)^2 + 1$ is a reflection in the x-axis, horizontal shift 2 units to the right and a vertical shift 1 unit upward.

18. $(0, \$26{,}000), (4, \$10{,}000)$

$$m = \frac{10{,}000 - 26{,}000}{4 - 0} = \frac{-16{,}000}{4} = -4000$$

$$V - 26{,}000 = -4000(t - 0)$$
$$V = -4000t + 26{,}000$$
$$16{,}000 = -4000t + 26{,}000$$
$$-10{,}000 = -4000t$$
$$\frac{-10{,}000}{-4000} = t$$
$$2.5 = \frac{5}{2} = t$$

After 2.5 years, the car will be worth $16,000.

19. (a) $y = |x - 2|$

(b) $y = |x| - 2$

(c) $y = -|x| + 2$

CHAPTER 4
Systems of Equations and Inequalities

CHAPTER 4
Systems of Equations and Inequalities

Section 4.1 Systems of Equations

1. (a) $(1, 4)$

$$1 + 2(4) \stackrel{?}{=} 9$$
$$9 = 9$$
$$-2(1) + 3(4) \stackrel{?}{=} 10$$
$$-2 + 12 \stackrel{?}{=} 10$$
$$10 = 10$$

Solution

(b) $(3, -1)$

$$3 + 2(-1) \stackrel{?}{=} 9$$
$$3 - 2 \stackrel{?}{=} 9$$
$$1 \neq 9$$

Not a solution

3. (a) $(-3, 2)$

$$-2(-3) + 7(2) \stackrel{?}{=} 46$$
$$6 + 14 \stackrel{?}{=} 46$$
$$20 \neq 46$$

Not a solution

(b) $(-2, 6)$

$$-2(-2) + 7(6) \stackrel{?}{=} 46$$
$$4 + 42 \stackrel{?}{=} 46$$
$$46 = 46$$
$$3(-2) + 6 \stackrel{?}{=} 0$$
$$-6 + 6 \stackrel{?}{=} 0$$
$$0 = 0$$

Solution

5. (a) $(8, 4)$

$$4(8) - 5(4) \stackrel{?}{=} 12$$
$$12 = 12$$
$$3(8) + 2(4) \stackrel{?}{=} -2.5$$
$$32 \neq -2.5$$

Not a solution

(b) $\left(\frac{1}{2}, -2\right)$

$$4\left(\tfrac{1}{2}\right) - 5(2) \stackrel{?}{=} 12$$
$$12 = 12$$
$$3\left(\tfrac{1}{2}\right) + 2(-2) \stackrel{?}{=} -2.5$$
$$-2.5 = -2.5$$

Solution

7. (a) $(5, -2)$

$$-3(5) + 2(-2) \stackrel{?}{=} -19$$
$$-15 - 4 \stackrel{?}{=} -19$$
$$-19 = -19$$
$$5(5) - (-2) \stackrel{?}{=} 27$$
$$25 + 2 \stackrel{?}{=} 27$$
$$27 = 27$$

Solution

(b) $(3, 4)$

$$-3(3) + 2(4) \stackrel{?}{=} -19$$
$$-9 + 8 \stackrel{?}{=} -19$$
$$-1 \neq -19$$

Not a solution

9. The equations have the same slope, so the lines are parallel. The system of linear equations has no solution.

11. The equations have different slopes, so the lines intersect at one point. The system of linear equations has one solution.

13. The equations are the same, so the lines coincide. The system of linear equations has infinitely many solutions.

15. $x + 2y = 6$ $\quad$ $x + 2y = 3$

$2y = -x + 6$ $\quad$ $2y = -x + 3$

$y = -\frac{1}{2}x + 6$ $\quad$ $y = -\frac{1}{2}x + \frac{3}{2}$

The slopes are equal, so the system is inconsistent.

17. $2x - 3y = -12$ $\quad$ $-8x + 12y = -12$

$-3y = -2x - 12$ $\quad$ $12y = 8x - 12$

$y = \frac{2}{3}x + 4$ $\quad$ $y = \frac{2}{3}x - 1$

The slopes are equal, so the system is inconsistent.

19. $-x + 4y = 7$ $\quad$ $3x - 12y = -21$

$4y = x + 7$ $\quad$ $-12y = -3x - 21$

$y = \frac{1}{4}x + \frac{7}{4}$ $\quad$ $y = \frac{1}{4}x + \frac{7}{4}$

The lines are the same, so the system is consistent and dependent.

21. $5x - 3y = 1$ $\quad$ $6x - 4y = -3$

$-3y = -5x + 1$ $\quad$ $-4y = -6x - 3$

$y = \frac{5}{3}x - \frac{1}{3}$ $\quad$ $y = \frac{3}{2}x + \frac{3}{4}$

The slopes are not equal, so the system is consistent.

23. $\frac{1}{3}x - \frac{1}{2}y = 1$ $\quad$ $-2x + 3y = 6$

$-\frac{1}{2}y = -\frac{1}{3}x + 1$ $\quad$ $3y = 2x + 6$

$y = \frac{2}{3}x - 2$ $\quad$ $y = \frac{2}{3}x + 2$

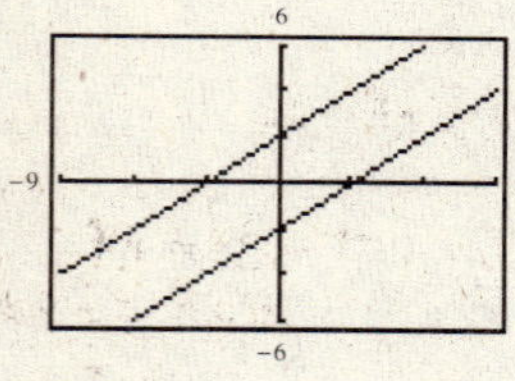

Inconsistent

25. $-2x + 3y = 6$ $\quad$ $x - y = -1$

$3y = 2x + 6$ $\quad$ $-y = -x - 1$

$y = \frac{2}{3}x + 2$ $\quad$ $y = x + 1$

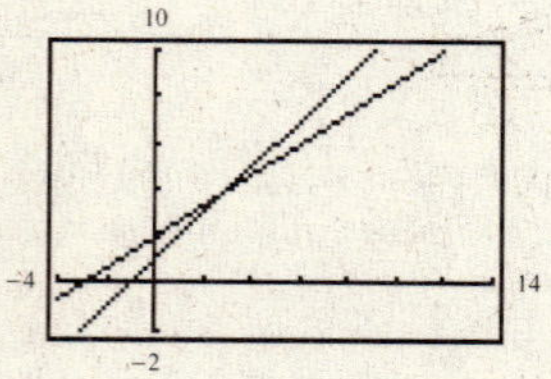

Consistent; One solution

27. The slopes are the same, so there is no solution.

29. There is one solution at $\left(1, \frac{1}{3}\right)$.

Check:

$5(1) - 3\left(\frac{1}{3}\right) \stackrel{?}{=} 4$ $\quad$ $2(1) + 3\left(\frac{1}{3}\right) \stackrel{?}{=} 3$

$4 = 4$ $\quad$ $3 = 3$

31. The lines are the same, so there are infinitely many solutions.

33. There is one solution at $(1, 2)$.

Check:

$0.5(1) + 0.5(2) \stackrel{?}{=} 1.5$ $\quad$ $-1 + 2 = 1$

$1.5 = 1.5$

35.

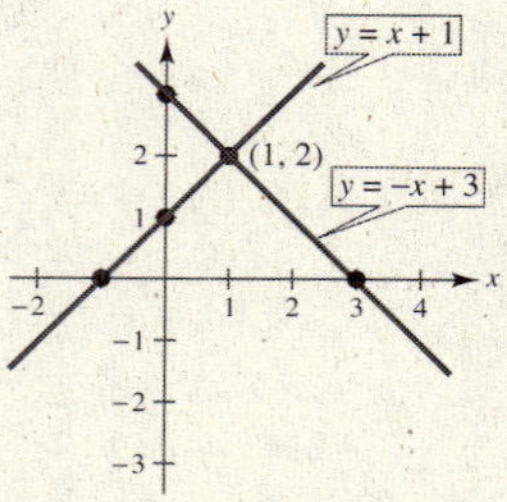

The point of intersection is $(1, 2)$.

37. $x - y = 2$ $\quad$ $x + y = 2$

$-y = -x + 2$ $\quad$ $y = -x + 2$

$y = x - 2$

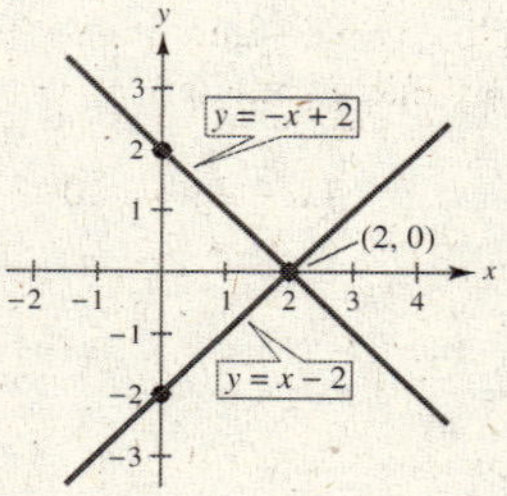

The point of intersection is $(2, 0)$.

39. Solve the first equation for y.

$$3x - 4y = 5$$
$$-4y = -3x + 5$$
$$y = \tfrac{3}{4}x - \tfrac{5}{4}$$

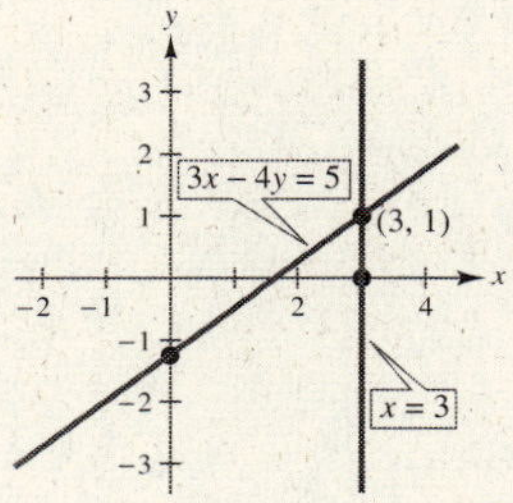

The point of intersection is $(3, 1)$.

41. $4x + 5y = 20$ $\qquad$ $\tfrac{4}{5}x + y = 4$

$$5y = -4x + 20 \qquad y = -\tfrac{4}{5}x + 4$$
$$y = -\tfrac{4}{5}x + 4$$

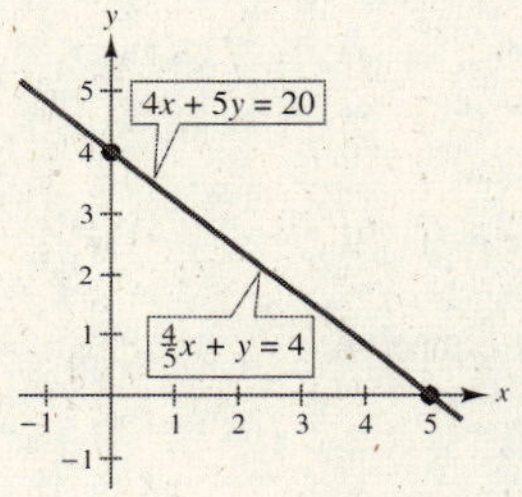

The lines representing the two equations are the same, so the system has infinitely many solutions.

43. $2x - 5y = 20$ $\qquad$ $4x - 5y = 40$

$$-5y = -2x + 20 \qquad -5y = -4x + 40$$
$$y = \tfrac{2}{5}x - 4 \qquad y = \tfrac{4}{5}x - 8$$

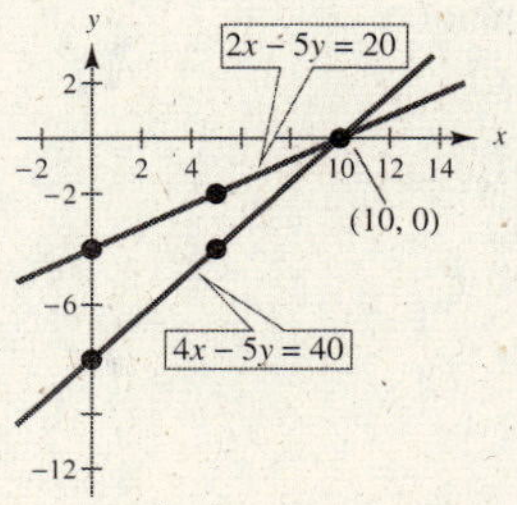

The point of intersection is $(10, 0)$.

45. $x + y = 3$ $\qquad$ $3x + 3y = 6$

$$y = -x + 3 \qquad 3y = -3x + 6$$
$$y = -x + 2$$

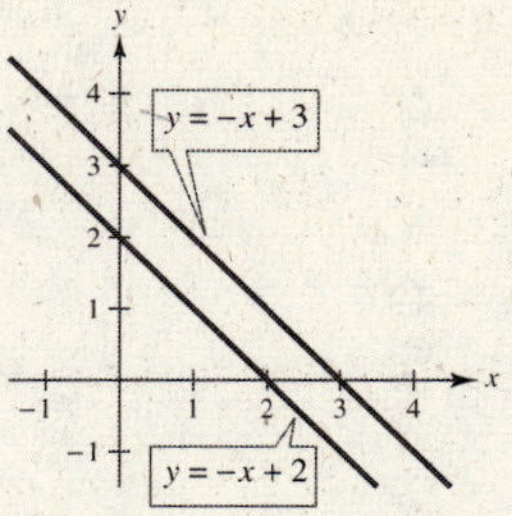

The lines have the same slope but are not the same line. Therefore, they are not parallel lines. The system of equations has no solution.

47. $4x + 5y = 7$ $\qquad$ $2x - 3y = 9$

$$5y = -4x + 7 \qquad -3y = -2x + 9$$
$$y = -\tfrac{4}{5}x + \tfrac{7}{5} \qquad y = \tfrac{2}{3}x - 3$$

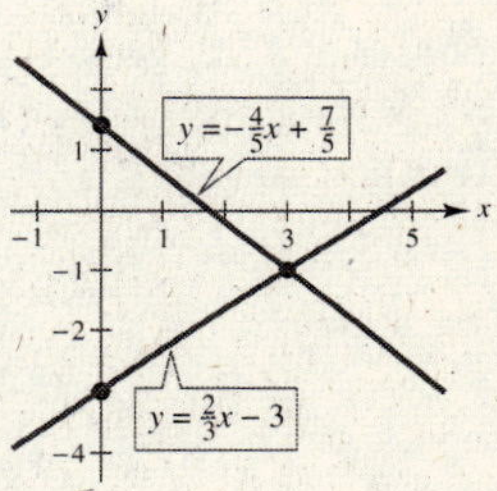

The point of intersection is $(3, -1)$.

49. $2x - y = 7.7$ $\qquad$ $-x - 3y = 1.4$

$$-y = -2x + 7.7 \qquad -3y = x + 1.4$$
$$y = 2x - 7.7 \qquad y = -\frac{1}{3}x - \frac{1.4}{3}$$

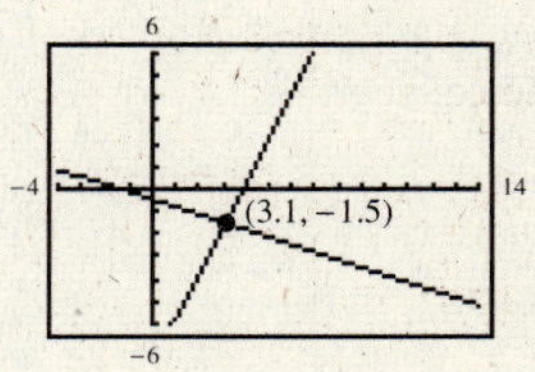

Solution: $(3.1, -1.5)$

51. $3.4x - 5.6y = 10.2$

$$-5.6y = -3.4x + 10.2$$

$$y = \frac{-3.4}{-5.6}x + \frac{10.2}{-5.6}$$

$$y = \frac{17}{28}x - \frac{51}{28}$$

$$5.8x + 1.4y = -33.6$$

$$1.4y = -5.8x - 33.6$$

$$y = \frac{-5.8}{1.4}x - \frac{33.6}{1.4}$$

$$y = -\frac{29}{7}x - 24$$

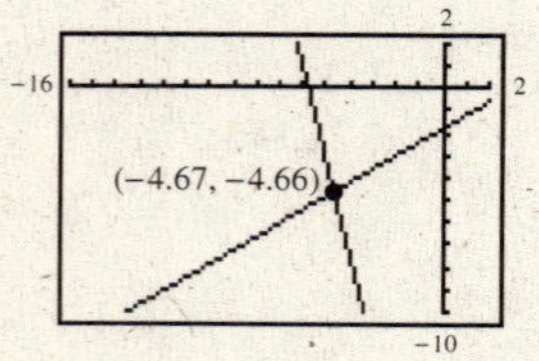

Solution: $(-4.67, -4.66)$

53. Solve for x in the first equation.

$x = 2y$

Substitute into second equation.

$$3(2y) + 2y = 8$$
$$6y + 2y = 8$$
$$8y = 8$$
$$y = 1$$

$$x = 2(1)$$
$$= 2$$

The solution is $(2, 1)$.

55. $x = 4$

Substitute into second equation.

$$4 - 2y = -2$$
$$-2y = -6$$
$$y = 3$$

The solution is $(4, 3)$.

57. Solve for y in the first equation.

$y = 3 - x$

Substitute into second equation.

$$2x - (3 - x) = 0$$
$$2x - 3 + x = 0$$
$$3x = 3$$
$$x = 1$$

$$y = 3 - 1$$
$$y = 2$$

The solution is $(1, 2)$.

59. Solve for x in the first equation.

$x = 2 - y$

Substitute into second equation.

$$2 - y - 4y = 12$$
$$-5y = 10$$
$$y = -2$$

$$x = 2 - (-2)$$
$$x = 4$$

The solution is $(4, -2)$.

61. Solve for y in the first equation.

$y = -3x + 8$

Substitute into second equation.

$$3x + (-3x + 8) = 6$$
$$8 \neq 6$$

This system of equations has no solution.

63. Solve for x in the second equation.

$x = -7 + 7y$

Substitute into first equation.

$$-7 + 7y + 6y = 19$$
$$13y = 26$$
$$y = 2$$

$$x = -7 + 7(2) = 7$$

The solution is $(7, 2)$.

65. Solve for y in the first equation.

$$5y = -8x + 100$$
$$y = -\tfrac{8}{5}x + 20$$

Substitute into second equation.

$$9x - 10\left(-\tfrac{8}{5}x + 20\right) = 50$$
$$9x + 16x - 200 = 50$$
$$25x = 250$$
$$x = 10$$

$$y = -\tfrac{8}{5}(10) + 20$$
$$y = 4$$

The solution is $(10, 4)$.

67. Solve for y in the first equation.

$$16y = 13x + 10$$
$$y = \tfrac{13}{16}x + \tfrac{10}{16}$$

Substitute into second equation.

$$5x + 16\left(\tfrac{13}{16}x + \tfrac{10}{16}\right) = -26$$
$$5x + 13x + 10 = -26$$
$$18x = -36$$
$$x = -2$$

$$y = \tfrac{13}{16}(-2) + \tfrac{10}{16}$$
$$y = -\tfrac{13}{8} + \tfrac{5}{8}$$
$$y = -1$$

The solution is $(-2, -1)$.

69. Solve for x in the first equation.

$$4x = -15 + 14y$$
$$x = \frac{-15 + 14y}{4}$$

Substitute into second equation.

$$18\left(\frac{-15 + 14y}{4}\right) - 12y = 9$$
$$18(-15 + 14y) - 48y = 36$$
$$-270 + 252y - 48y = 306$$
$$204y = 306$$
$$y = \frac{3}{2}$$

$$x = \frac{-15 + 14\left(\frac{3}{2}\right)}{4} = \frac{-15 + 21}{4} = \frac{3}{2}$$

The solution is $\left(\frac{3}{2}, \frac{3}{2}\right)$.

71. Solve for y in the second equation.

$$y = -x + 20$$

Substitute into first equation.

$$\tfrac{1}{5}x + \tfrac{1}{2}(-x + 20) = 8$$
$$\tfrac{1}{5}x - \tfrac{1}{2}x + 10 = 8$$
$$\tfrac{1}{5}x - \tfrac{1}{2}x = -2$$
$$2x - 5x = -20$$
$$-3x = -20$$
$$x = \tfrac{20}{3}$$

$$y = -\tfrac{20}{3} + 20$$
$$y = \tfrac{40}{3}$$

The solution is $\left(\frac{20}{3}, \frac{40}{3}\right)$.

73. Solve for x in the first equation.

$$8\left(\tfrac{1}{8}x + \tfrac{1}{2}y\right) = (1)8$$
$$x + 4y = 8$$
$$x = 8 - 4y$$

Substitute into second equation.

$$5\left(\tfrac{3}{5}x + y\right) = \left(\tfrac{3}{5}\right)5$$
$$3x + 5y = 3$$
$$3(8 - 4y) + 5y = 3$$
$$24 - 12y + 5y = 3$$
$$-7y = -21$$
$$y = 3$$

$$x = 8 - 4(3)$$
$$x = 8 - 12$$
$$x = -4$$

The solution is $(-4, 3)$.

75. Solve for y in the first equation.

$$y = -\tfrac{4}{3}x + 1$$

Substitute into second equation.

$$12x + 9\left(-\tfrac{4}{3}x + 1\right) = 6$$
$$12x - 12x + 9 = 6$$
$$9 \neq 6$$

This system of equations has no solution.

77. Solve for x in the first equation.

$$9\left(\tfrac{1}{9}x - \tfrac{2}{3}y\right) = 2(9)$$
$$x - 6y = 18$$
$$x = 6y + 18$$

Substitute into second equation.

$$3\left(\tfrac{2}{3}x - 4y\right) = 6(3)$$
$$2x - 12y = 18$$
$$2(6y + 18) - 12y = 18$$
$$12y + 36 - 12y = 18$$
$$36 \neq 18$$

This system of equations has no solution.

79.
$$x = -200 + 100y$$
$$3(-200 + 100y) - 275y = 198$$
$$-600 + 300y - 275y = 198$$
$$25y = 798$$
$$y = \frac{798}{25}$$
$$y = 31.92$$

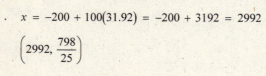

$$x = -200 + 100(31.92) = -200 + 3192 = 2992$$

$$\left(2992, \frac{798}{25}\right)$$

Solve each equation for y.

$$3x - 25y = 50$$
$$-25y = 50 - 3x$$
$$y = -2 + \tfrac{3}{25}x$$

or

$$y = 0.12x - 2$$

$$9x - 100y = 50$$
$$-100y = -9x + 50$$
$$y = \tfrac{9}{100}x - \tfrac{1}{2}$$

or

$$y = 0.09x - 0.5$$

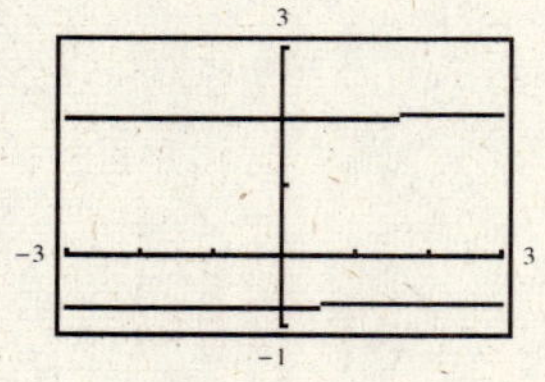

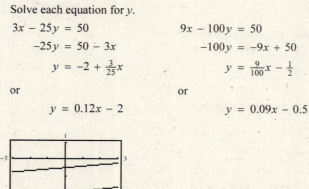

81.
$$3x - 25y = 50$$
$$3x = 50 + 25y$$
$$x = \tfrac{50}{3} + \tfrac{25}{3}y$$

$$9\left(\tfrac{50}{3} + \tfrac{25}{3}y\right) - 100y = 50$$
$$150 + 75y - 100y = 50$$
$$-25y = -100$$
$$y = 4$$

$$x = \tfrac{50}{3} + \tfrac{25}{3}(4)$$
$$x = \tfrac{50}{3} + \tfrac{100}{3}$$
$$x = \tfrac{150}{3}$$
$$x = 50$$

$(50, 4)$

Solve each equation for y.

$$3x - 25y = 50$$
$$-25y = 50 - 3x$$
$$y = -2 + \tfrac{3}{25}x$$

or

$$y = 0.12x - 2$$

$$9x - 100y = 50$$
$$-100y = -9x + 50$$
$$y = \tfrac{9}{100}x - \tfrac{1}{2}$$

or

$$y = 0.09x - 0.5$$

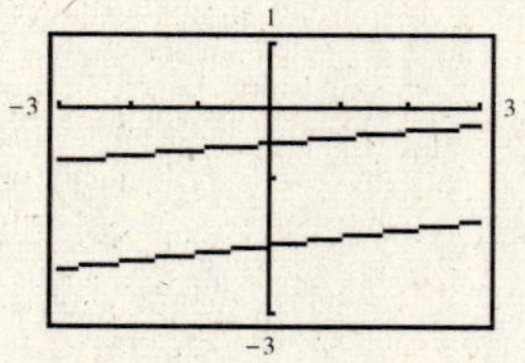

83. $3y = ax + b;$ $\qquad y = \frac{2}{3}x + 1$

$y = \frac{a}{3}x + \frac{b}{3}$

$\frac{a}{3} = \frac{2}{3}$ $\qquad \frac{b}{3} = 1$

$3a = 6$ $\qquad b = 3$

$a = 2$

85. $\frac{3}{4}x - ay + b = 0$

$-ay = -\frac{3}{4}x - b$

$y = \frac{-3x}{-4a} - \frac{b}{-a}$

$y = \frac{3}{4a}x + \frac{b}{a}$

$6x - 8y - 48 = 0$

$-8y = -6x + 48$

$y = \frac{3}{4}x - 6$

$\frac{3}{4a} = \frac{3}{4}$ $\qquad \frac{b}{a} = -6$

$12 = 12a$ $\qquad b = -6a$

$1 = a$ $\qquad = -6(1) = -6$

87. Answers will vary.

$$\begin{cases} 2x - 3y = -7 \\ x + y = 9 \end{cases}$$

89. Answers will vary.

$$\begin{cases} 7x + y = -9 \\ -x + 3y = -5 \end{cases}$$

91. Answers will vary.

$$\begin{cases} 3x + 10y = 7 \\ 6x + 20y = 14 \end{cases}$$

93. Answers will vary.

$$\begin{cases} y + x = 4 \\ y + x = 3 \end{cases}$$

95. *Verbal Model:* [Amount of hay 1] + [Amount of hay 2] = 100

$125 · [Amount of hay 1] + $75 · [Amount of hay 2] = $90 · 100

Labels: Amount of hay 1 = x

Amount of hay 2 = y

System: $x + y = 100$

$125x + 75y = 90(100)$

Solve for x in the first equation.

$x = 100 - y$

Substitute into second equation.

$125(100 - y) + 75y = 9000$

$12{,}500 - 125y + 75y = 9000$

$-50y = -3500$

$y = 70$

$x = 100 - 70$

$x = 30$

30 tons at \$125 per ton and 70 tons at \$75 per ton should be used.

97. *Verbal Model:* $\boxed{\text{Total cost}} = \boxed{\text{Cost per unit}} \cdot \boxed{\text{Number of units}} + \boxed{\text{Initial cost}}$

$\boxed{\text{Total revenue}} = \boxed{\text{Price per unit}} \cdot \boxed{\text{Number of units}}$

Labels: Total cost $= C$

Cost per unit $= 1.20$

Number of units $= x$

Initial cost $= 8000$

Total revenue $= R$

Price per unit $= 2.00$

System: $C = 1.20x + 8000$

$R = 2.00x$

Break-even point occurs when $R = C$ so

$$1.20x + 8000 = 2.00x$$
$$8000 = 0.80x$$
$$10{,}000 = x.$$

10,000 candy bars must be sold.

99. *Verbal Model:* $\boxed{\text{Total cost}} = \boxed{\text{Cost per unit}} \cdot \boxed{\text{Number of units}} + \boxed{\text{Initial cost}}$

$\boxed{\text{Total revenue}} = \boxed{\text{Price per unit}} \cdot \boxed{\text{Number of units}}$

Labels: Total cost $= C$

Cost per unit $= 1.65$

Number of units $= x$

Initial cost $= 10{,}000$

Total revenue $= R$

Price per unit $= 3.25$

System: $C = 1.65x + 10{,}000$

$R = 3.25x$

Break-even point occurs when $R = C$ so

$$1.65x + 10{,}000 = 3.25x$$
$$10{,}000 = 1.60x$$
$$6250 = x.$$

6250 bottles must be sold.

101. *Verbal Model:* [Amount in 8.5% bond] + [Amount in 10% bond] = 12,000

8.5% · [Amount in 8.5% bond] + 10% · [Amount in 10% bond] = 1140

Labels: Amount in 8.5% bond $= x$

Amount in 10% bond $= y$

System:

$$x + y = 12{,}000$$
$$0.085x + 0.10y = 1140$$

Solve for x in the first equation.

$$x = 12{,}000 - y$$

Substitute into second equation.

$$0.085(12{,}000 - y) + 0.10y = 1140$$
$$1020 - 0.085y + 0.10y = 1140$$
$$0.015y = 120$$
$$y = 8000$$

$$x = 12{,}000 - 8000$$
$$x = 4000$$

$4000 is at 8.5% and $8000 is at 10%.

103. *Verbal Model:* [Larger number] + [Smaller number] = 80

[Larger number] − [Smaller number] = 18

Labels: Larger number $= x$

Smaller number $= y$

System:

$$x + y = 80$$
$$x - y = 18$$

Solve for x in the second equation.

$$x = 18 + y$$

Substitute into first equation.

$$18 + y + y = 80$$
$$2y = 62$$
$$y = 31$$
$$x = 18 + 31 = 49$$

105. *Verbal Model:* [Larger number] + 3 · [Smaller number] = 51

[Larger number] − [Smaller number] = 3

Labels: Larger number $= x$

Smaller number $= y$

System:

$$x + 3y = 51$$
$$x - y = 3$$

Solve for x in the second equation.

$$x = y + 3$$

Substitute into first equation.

$$y + 3 + 3y = 51$$
$$4y = 48$$
$$y = 12$$
$$x = 12 + 3 = 15$$

107. *Verbal Model:* [Larger number] + [Smaller number] = 52

[Larger number] = 2 · [Smaller number] − 8

Labels: Larger number $= x$

Smaller number $= y$

System:

$$x + y = 52$$
$$x = 2y - 8$$

Substitute into first equation.

$$2y - 8 + y = 52$$
$$3y = 60$$
$$y = 20$$
$$x = 2(20) - 8 = 32$$

109. *Verbal Model:* $2 \cdot \boxed{\text{Smaller number}} - \boxed{\text{Larger number}} = 13$

$\boxed{\text{Smaller number}} + 2 \cdot \boxed{\text{Larger number}} = 114$

Labels: Larger number $= x$

Smaller number $= y$

System:
$$2y - x = 13$$
$$y + 2x = 114$$

Solve for x in the first equation.

$x = 2y - 13$

Substitute into second equation.

$$y + 2(2y - 13) = 114$$
$$y + 4y - 26 = 114$$
$$5y = 140$$
$$y = 28$$

$$x = 2(28) - 13$$
$$= 56 - 13$$
$$x = 43$$

111. *Verbal Model:* $2 \cdot \boxed{\text{Length}} + 2 \cdot \boxed{\text{Width}} = 50$

$\boxed{\text{Length}} = 5 + \boxed{\text{Width}}$

Labels: Length $= x$

Width $= y$

System:
$$2x + 2y = 50$$
$$x = 5 + y$$

Substitute into first equation.

$$2(5 + y) + 2y = 50$$
$$10 + 2y + 2y = 50$$
$$4y = 40$$
$$y = 10$$

$x = 5 + 10 = 15$

10 feet × 15 feet

113. *Verbal Model:* $2 \cdot \boxed{\text{Length}} + 2 \cdot \boxed{\text{Width}} = 68$

$\boxed{\text{Length}} = \frac{7}{10} \cdot \boxed{\text{Width}}$

Labels: Length $= x$

Width $= y$

System:
$$2x + 2y = 68$$
$$x = \tfrac{7}{10}y$$

Substitute into first equation.

$$2\left(\tfrac{7}{10}y\right) + 2y = 68$$
$$\tfrac{7}{5}y + 2y = 68$$
$$7y + 10y = 340$$
$$17y = 340$$
$$y = 20$$

$x = \frac{7}{10}(20) = 7(2) = 14$

14 yards × 20 yards

115. Using a graphing calculator, the population of South Carolina first exceeded the population of Kentucky in 2001.

117. A system of equations has no solution when a false statement, such as $2 = 0$, is produced by the substitution process.

119. The graphical method usually yields approximate solutions.

121. The system will have exactly one solution because the system is consistent.

123. The system has no solution. Because the lines have different slopes and only two of the lines have the same y-intercept, there is one point of intersection for each pair of lines. There is no point of intersection for all three lines.

125.
$$x - 4 = 1$$
$$x = 5$$

Check: $5 - 4 \stackrel{?}{=} 1$

$1 = 1$

127.
$$3x - 12 = x + 2$$
$$2x - 12 = 2$$
$$2x = 14$$
$$x = 7$$

Check: $3(7) - 12 \stackrel{?}{=} 7 + 2$

$21 - 12 \stackrel{?}{=} 9$

$9 = 9$

129.

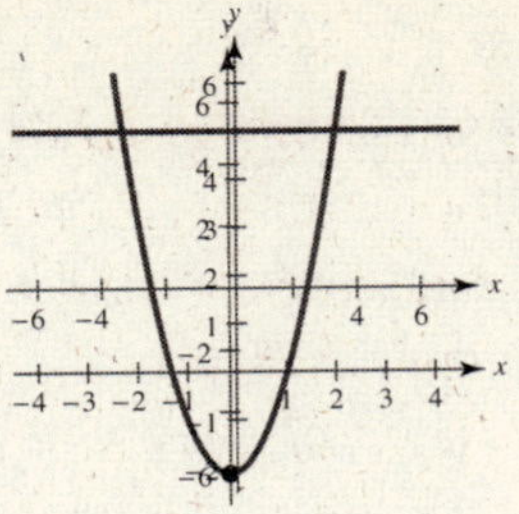

131.

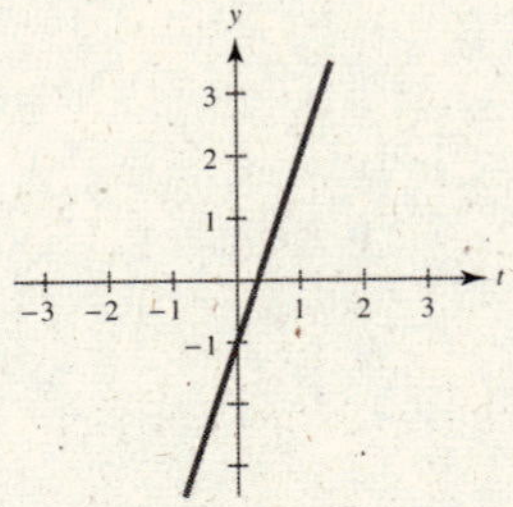

133.

$$\begin{aligned} -x + y^2 &= 0 \\ -4 + 2^2 &= 0 \\ -4 + 4 &= 0 \\ 0 &= 0 \end{aligned} \qquad \begin{aligned} -4 + (-2)^2 &= 0 \\ -4 + 4 &= 0 \\ 0 &= 0 \end{aligned}$$

There are two values of y associated with one value of x. So, y is not a function of x.

Section 4.2 Linear Systems in Two Variables

1.

$$\begin{aligned} 2x + y &= 4 \\ x - y &= 2 \\ \hline 3x \quad &= 6 \\ x \quad &= 2 \end{aligned}$$

$$\begin{aligned} 2 - y &= 2 \\ -y &= 0 \\ y &= 0 \end{aligned}$$

The solution is $(2, 0)$.

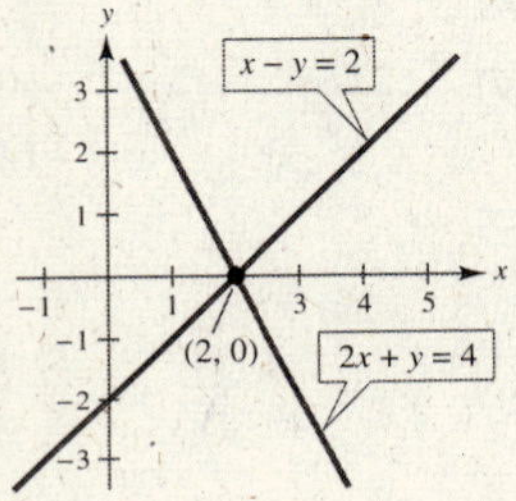

3.

$$\begin{aligned} -x + 2y &= 1 \\ x - y &= 2 \\ \hline y &= 3 \end{aligned}$$

$$\begin{aligned} x - 3 &= 2 \\ x &= 5 \end{aligned}$$

The solution is $(5, 3)$.

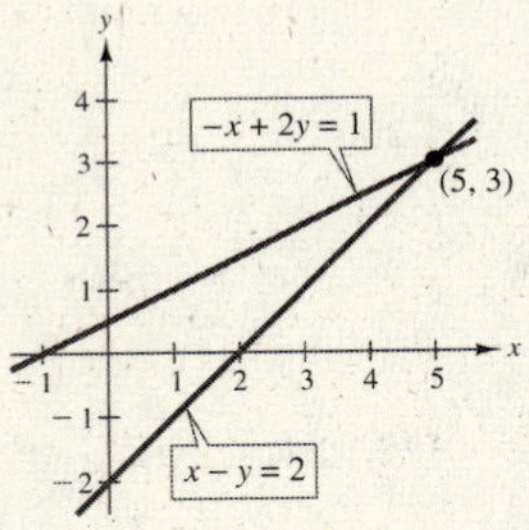

5.

$$\begin{aligned} 3x + y &= 3 \\ 2x - y &= 7 \\ \hline 5x \quad &= 10 \\ x \quad &= 2 \end{aligned}$$

$$\begin{aligned} 3(2) + y &= 3 \\ 6 + y &= 3 \\ y &= -3 \end{aligned}$$

The solution is $(2, -3)$.

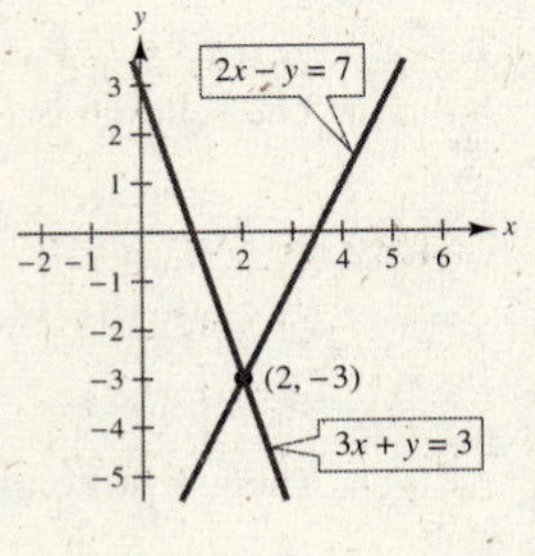

7.

$$\begin{aligned} -2x + 4y &= 8 \Rightarrow -4x + 8y = 16 \\ 0.4x - 0.8y &= 2.4 \Rightarrow \quad 4x - 8y = 24 \\ & \qquad\qquad\quad \overline{\quad 0 \neq 40 \quad} \end{aligned}$$

There is no solution.

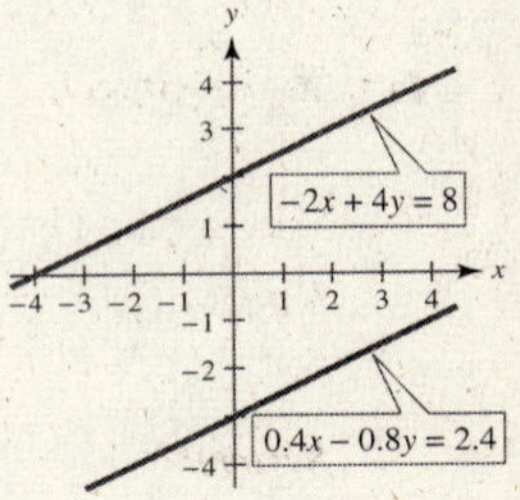

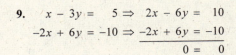

9.

$$\begin{aligned} x - 3y &= 5 \Rightarrow \quad 2x - 6y = 10 \\ -2x + 6y &= -10 \Rightarrow -2x + 6y = -10 \\ & \qquad\qquad\quad \overline{\quad 0 = \quad 0 \quad} \end{aligned}$$

There are infinitely many solutions.

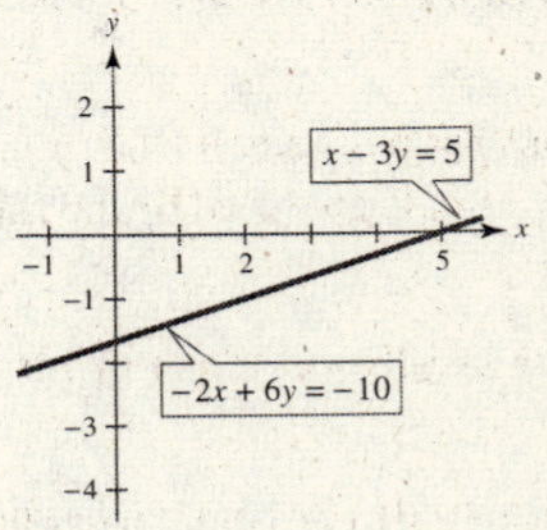

11.
$$\begin{aligned} 2x - 8y &= -11 \Rightarrow & 6x - 24y &= -33 \\ 5x + 3y &= 7 \Rightarrow & 40x + 24y &= 56 \\ & & 46x &= 23 \\ & & x &= \frac{23}{46} \\ & & x &= \frac{1}{2} \end{aligned}$$

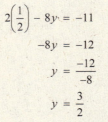

$$\begin{aligned} 2\left(\frac{1}{2}\right) - 8y &= -11 \\ -8y &= -12 \\ y &= \frac{-12}{-8} \\ y &= \frac{3}{2} \end{aligned}$$

The solution is $\left(\frac{1}{2}, \frac{3}{2}\right)$.

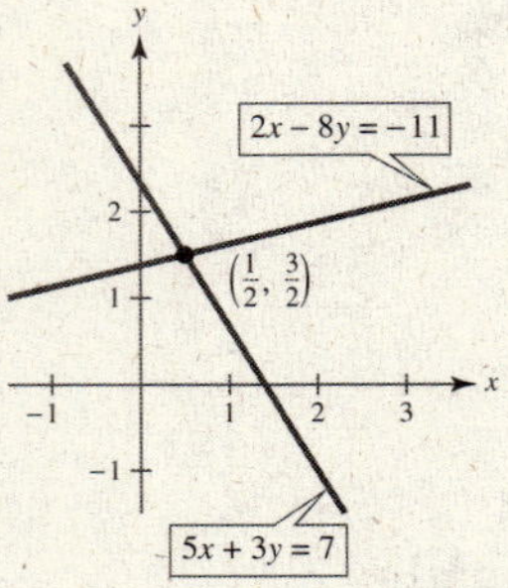

13.
$$\begin{aligned} 6x - 6y &= 25 \Rightarrow & 6x - 6y &= 25 \\ 3y &= 11 \Rightarrow & 6y &= 22 \\ & & 6x &= 47 \\ & & x &= \tfrac{47}{6} \end{aligned}$$

$$\begin{aligned} 3y &= 11 \\ y &= \tfrac{11}{3} \end{aligned}$$

The solution is $\left(\frac{47}{6}, \frac{11}{3}\right)$.

15.
$$\begin{aligned} x + y &= 0 \\ x - y &= 4 \\ 2x &= 4 \\ x &= 2 \end{aligned}$$

$$\begin{aligned} 2 + y &= 0 \\ y &= -2 \end{aligned}$$

The solution is $(2, -2)$.

17.
$$\begin{aligned} 3x - 2y &= 8 \\ 2x + 2y &= 7 \\ 5x &= 15 \\ x &= 3 \end{aligned}$$

$$\begin{aligned} 3(3) - 2y &= 8 \\ -2y &= -1 \\ y &= \tfrac{1}{2} \end{aligned}$$

The solution is $\left(3, \frac{1}{2}\right)$.

19.
$$\begin{aligned} 5x + 2y &= 7 \Rightarrow & 5x + 2y &= 7 \\ 3x - y &= 13 \Rightarrow & 6x - 2y &= 26 \\ & & 11x &= 33 \\ & & x &= 3 \end{aligned}$$

$$\begin{aligned} 3(3) - y &= 13 \\ -y &= 4 \\ y &= -4 \end{aligned}$$

The solution is $(3, -4)$.

21.
$$\begin{aligned} x - 3y &= 2 \Rightarrow & -3x + 9y &= -6 \\ 3x - 7y &= 4 \Rightarrow & 3x - 7y &= 4 \\ & & 2y &= -2 \\ & & y &= -1 \end{aligned}$$

$$\begin{aligned} x - 3(-1) &= 2 \\ x &= -1 \end{aligned}$$

The solution is $(-1, -1)$.

23.
$$\begin{aligned} 4x + 3y &= -10 \Rightarrow & 4x + 3y &= -10 \\ 3x - y &= -14 \Rightarrow & 9x - 3y &= -42 \\ & & 13x &= -52 \\ & & x &= -4 \end{aligned}$$

$$\begin{aligned} 3(-4) - y &= -14 \\ -y &= -2 \\ y &= 2 \end{aligned}$$

The solution is $(-4, 2)$.

25.
$$\begin{aligned} 2u + 3v &= 8 \Rightarrow & -6u - 9v &= -24 \\ 3u + 4v &= 13 \Rightarrow & 6u + 8v &= 26 \\ & & -v &= 2 \\ & & v &= -2 \end{aligned}$$

$$\begin{aligned} 2u + 3(-2) &= 8 \\ 2u &= 14 \\ u &= 7 \end{aligned}$$

The solution is $(7, -2)$.

27.
$$\begin{aligned} 12x - 5y &= 2 \Rightarrow & 24x - 10y &= 4 \\ -24x + 10y &= 6 \Rightarrow & -24x + 10y &= 6 \\ & & 0 &\neq 10 \end{aligned}$$

There is no solution.

29. $\frac{2}{3}r - s = 0 \Rightarrow 2r - 3s = 0 \Rightarrow 8r - 12s = 0$

$10r + 4s = 19 \Rightarrow 10r + 4s = 19 \Rightarrow 30r + 12s = 57$

$38r = 57$

$r = \frac{57}{38}$

$r = \frac{3}{2}$

$\frac{2}{3}\left(\frac{3}{2}\right) - s = 0$

$-s = -1$

$s = 1$

The solution is $\left(\frac{3}{2}, 1\right)$.

31. $0.05x - 0.03y = 0.21 \Rightarrow 5x - 3y = 21$

$x + y = 9 \Rightarrow 3x + 3y = 27$

$8x = 48$

$x = 6$

$x + y = 9$

$6 + y = 9$

$y = 3$

The solution is $(6, 3)$.

33. $0.7u - v = -0.4 \Rightarrow 7u - 10v = -4 \Rightarrow 21u - 30v = -12$

$0.3u - 0.8v = 0.2 \Rightarrow 3u - 8v = 2 \Rightarrow -21u + 56v = -14$

$26v = -26$

$v = -1$

$7u - 10(-1) = -4$

$7u = -14$

$u = -2$

The solution is $(-2, -1)$.

35. $5x + 7y = 25 \Rightarrow 5x + 7y = 25$

$x + 1.4y = 5 \Rightarrow -5x - 7y = -25$

$0 = 0$

There are infinitely many solutions.

37. $\frac{1}{2}x - \frac{1}{3}y = 1 \Rightarrow 3x - 2y = 6 \Rightarrow -6x + 4y = -12$

$\frac{1}{4}x - \frac{1}{9}y = \frac{2}{3} \Rightarrow 9x - 4y = 24 \Rightarrow 9x - 4y = 24$

$3x = 12$

$x = 4$

$3(4) - 2y = 6$

$-2y = -6$

$y = 3$

The solution is $(4, 3)$.

39. $\frac{1}{5}x + \frac{1}{5}y = 4 \Rightarrow x + y = 20 \Rightarrow 3x + 3y = 60$

$\frac{2}{3}x - y = \frac{8}{3} \Rightarrow 2x - 3y = 8 \Rightarrow 2x - 3y = 8$

$5x = 68$

$x = \frac{68}{5}$

41. $x + 7y = -6$

$x = -7y - 6$

$(-7y - 6) - 5y = 18$

$-7y - 6 - 5y = 18$

$-12y = 24$

$y = -2$

$x + 7(-2) = -6$

$x - 14 = -6$

$x = 8$

The solution is $(8, -2)$.

43. $y = 5x - 3 \Rightarrow y = 5x - 3$

$y = -2x + 11 \Rightarrow -y = 2x - 11$

$0 = 7x - 14$

$14 = 7x$

$2 = x$

$y = 5(2) - 3$

$y = 10 - 3$

$y = 7$

The solution is $(2, 7)$.

45. $2x - y = 20$

$-x + y = -5$

$x = 15$

$-15 + y = -5$

$y = 10$

The solution is $(15, 10)$.

47. $\frac{3}{2}x + 2y = 12$

$\frac{1}{4}x + \quad y = 4$

$y = 4 - \frac{1}{4}x$

$\frac{3}{2}x + 2\left(4 - \frac{1}{4}x\right) = 12$

$\frac{3}{2}x + 8 - \frac{1}{2}x = 12$

$x = 4$

$y = 4 - \frac{1}{4}(4)$

$= 4 - 1 = 3$

The solution is $(4, 3)$.

49. $4x - 5y = 3 \Rightarrow -5y = -4x + 3 \Rightarrow y = \frac{4}{5}x - \frac{3}{5}$

$-8x + 10y = -6 \Rightarrow 10y = 8x - 6 \Rightarrow y = \frac{4}{5}x - \frac{3}{5}$

Many solutions $\Rightarrow$ consistent

51. $-2x + 5y = 3 \Rightarrow 5y = 2x + 3 \Rightarrow y = \frac{2}{5}x + \frac{3}{5}$

$5x + 2y = 8 \Rightarrow 2y = -5x + 8 \Rightarrow y = -\frac{5}{2}x + 4$

One solution $\Rightarrow$ consistent

53.

$-10x + 15y = 25 \Rightarrow 15y = 10x + 25 \Rightarrow$

$2x - 3y = -24 \Rightarrow -3y = -2x - 24 \Rightarrow$

$y = \frac{2}{3}x + \frac{5}{3}$

$y = \frac{2}{3}x + 8$

No solution $\Rightarrow$ inconsistent

55. $5x - 10y = 40 \Rightarrow y = \frac{1}{2}x - 4$

$-2x + ky = 30 \Rightarrow y = \frac{2}{k}x + \frac{30}{k}$

So, $\frac{2}{k} = \frac{1}{2} \Rightarrow k = 4.$

$5x - 10y = 40 \Rightarrow 10x - 20y = 80$

$-2x + 4y = 30 \Rightarrow -10x + 20y = 150$

$0 \neq 230$

57. Answers will vary. Write equations so that $(3, -6)$ satisfies each equation.

$$\begin{cases} -2x + 3y = -24 \\ 2x + y = 0 \end{cases}$$

$-2(3) + 3(-6) \stackrel{?}{=} -24$

$-6 - 18 \stackrel{?}{=} -24$

$-24 = -24$

$2(3) + (-6) \stackrel{?}{=} 0$

$6 - 6 \stackrel{?}{=} 0$

$0 = 0$

59. Answers will vary. Write equations so that $\left(-\frac{1}{2}, 4\right)$ satisfies each equation.

$$\begin{cases} 4x - y = -6 \\ 2x + y = 3 \end{cases}$$

$4\left(-\frac{1}{2}\right) - 4 \stackrel{?}{=} -6$

$-2 - 4 \stackrel{?}{=} -6$

$-6 = -6$

$2\left(-\frac{1}{2}\right) + 4 \stackrel{?}{=} 3$

$-1 + 4 \stackrel{?}{=} 3$

$3 = 3$

61. *Verbal Model:* [Total cost] = [Cost per week] · [Number of weeks] + [Initial cost]

[Total revenue] = [Price per week] · [Number of weeks]

Labels: Total cost $= C$

Cost per week $= 7400$

Number of weeks $= x$

Initial cost $= 85{,}000$

Total revenue $= R$

Price per week $= 8300$

System: $C = 7400x + 85{,}000$

$R = 8300x$

Break-even point occurs when $R = C$.

$7400x + 85{,}000 = 8300x$

$85{,}000 = 900x$

$94.44 \approx x$

It will take 95 weeks to break even.

63. *Verbal Model:* [Band 4-hour rate] + [Band additional hour rate] · [Number of additional hours] = [Total cost]

[DJ 4-hour rate] + [DJ additional hour rate] · [Number of additional hours] = [Total cost]

Labels: Band 4-hour rate = 500

Band additional hour rate = 50

DJ 4-hour rate = 300

DJ additional hour rate = 75

Number of additional hours = x

Total cost = y

System:

$$\begin{aligned} 500 + 50x &= y \\ 300 + 75x &= y \\ \hline 200 - 25x &= 0 \\ -25x &= -200 \\ x &= 8 \end{aligned}$$

After 8 additional hours, the cost for the DJ will exceed the cost of the band. So, the total time will be 12 hours.

65. *Verbal Model:* [Amount in 8% bond] + [Amount in 9.5% bond] = [Total investment]

[Interest in 8% bond] + [Interest in 9.5% bond] = [Total interest]

Labels: Amount in 8% bond = x

Amount in 9.5% bond = y

System:

$$\begin{aligned} x + y &= 20{,}000 \Rightarrow & -0.08x - 0.08y &= -1600 \\ 0.08x + 0.095y &= 1780 \Rightarrow & 0.08x + 0.095y &= 1780 \\ \hline & & 0.015y &= 180 \\ & & y &= 12{,}000 \end{aligned}$$

$$\begin{aligned} x + 12{,}000 &= 20{,}000 \\ x &= 8000 \end{aligned}$$

\$8000 is invested at 8% and \$12,000 is invested at 9.5%.

67. *Verbal Model:* [Distance] = [Rate] · [Time]

Time at 55 mph = x

Labels: D_1 = distance at 40 mph for 2 hours + at 55 mph for x hours

D_2 = distance at 50 mph for $(2 + x)$ hours

System: $D_1 = D_2$

$$\begin{aligned} 40(2) + 55(x) &= 50(2 + x) \\ 80 + 55x &= 100 + 50x \\ 5x &= 20 \\ x &= 4 \end{aligned}$$

The van must travel 4 hours longer.

69. *Verbal Model:* [Plane speed (still air)] − [Speed of air] = [Speed into head wind]

[Plane speed (still air)] + [Speed of air] = [Speed into head wind]

Labels: Plane speed $= x$

Speed of air $= y$

System:

$$x - y = \frac{1800}{3.6} \Rightarrow x - y = 500$$

$$x + y = \frac{1800}{3} \Rightarrow \underline{x + y = 600}$$

$$2x = 1100$$

$$x = 550$$

$$550 - y = 500$$

$$-y = -50$$

$$y = 50$$

The speed of the plane is 550 miles per hour. The wind speed is 50 miles per hour.

71. *Verbal Model:* [Number of adult tickets] + [Number of children tickets] = [500]

[Value of adult tickets] + [Value of children tickets] = [3400]

Labels: Number of adult tickets $= x$

Number of children tickets $= y$

System:

$$x + y = 500$$

$$7.50x + 4.00y = 3400$$

$$y = 500 - x$$

$$7.50x + 4.00(500 - x) = 3400$$

$$7.50x + 2000 - 4.00x = 3400$$

$$3.5x = 1400$$

$$x = 400$$

$$y = 500 - 400 = 100$$

There were 400 adult tickets and 100 children tickets sold.

73. *Verbal Model:* 12([Cost of regular gasoline]) + 8([Cost of premium gasoline]) = [76.48]

[Cost of premium gasoline] = [0.11] + [Cost of regular gasoline]

Labels: Cost of regular gasoline $= x$

Cost of premium gasoline $= y$

System:

$$12x + 8y = 76.48$$

$$y = 0.11 + x$$

$$12x + 8(0.11 + x) = 76.48$$

$$12x + 0.88 + 8x = 76.48$$

$$20x = 75.60$$

$$x = 3.78$$

$$y = 0.11 + 3.78 = 3.89$$

Regular unleaded gasoline costs \$3.78 per gallon and premium unleaded gasoline costs \$3.89 per gallon.

75. *Verbal Model:* $\boxed{\text{Number of liters Solution 1}} + \boxed{\text{Number of liters Solution 2}} = \boxed{20}$

$\boxed{\text{Value of Solution 1}} + \boxed{\text{Value of Solution 2}} = \boxed{20(0.50)}$

Labels: Number of liters Solution 1 $= x$

Number of liters Solution 2 $= y$

System:

$$\begin{aligned} x + \quad y &= 20 \\ 0.40x + 0.65y &= 20(0.50) \\ x \quad &= 20 - y \end{aligned}$$

$$\begin{aligned} 40(20 - y) + 65y &= 20(50) \\ 800 - 40y + 65y &= 1000 \\ 25y &= 200 \\ y &= 8 \end{aligned}$$

$x = 20 - 8 = 12$

There are 12 liters of 40% solution and 8 liters of 65% solution.

77. *Verbal Model:* $\boxed{\text{Number of ounces Solution 1}} + \boxed{\text{Number of ounces Solution 2}} = 32$

$\boxed{\text{Value of Solution 1}} + \boxed{\text{Value of Solution 2}} = 32(0.75)$

Labels: Number of ounces Solution 1 $= x$

Number of ounces Solution 2 $= y$

System:

$$\begin{aligned} x + \quad y &= 32 \\ 0.50x + 0.90y &= 32(0.75) \\ x \quad &= 32 - y \end{aligned}$$

$$\begin{aligned} 0.50(32 - y) + 0.90y &= 32(0.75) \\ 50(32 - y) + 90y &= 32(75) \\ 1600 - 50y + 90y &= 2400 \\ 40y &= 800 \\ y &= 20 \end{aligned}$$

$x = 32 - 20 = 12$

There are 12 ounces of 50% solution and 20 ounces of 90% solution.

79. *Verbal Model:* $\boxed{\text{Number of liters Solution 1}} + \boxed{\text{Number of liters Solution 2}} = 30$

$\boxed{\text{Value of Solution 1}} + \boxed{\text{Value of Solution 2}} = 30(0.46)$

Labels: Number of liters Solution 1 $= x$

Number of liters Solution 2 $= y$

System:

$$\begin{aligned} x + \quad y &= 30 \\ 0.40x + 0.70y &= 30(0.46) \\ x \quad &= 30 - y \end{aligned}$$

$x = 30 - y$

$$\begin{aligned} 0.40(30 - y) + 0.70y &= 30(0.46) \\ 40(30 - y) + 70y &= 30(46) \\ 1200 - 40y + 70y &= 1380 \\ 30y &= 180 \\ y &= 6 \end{aligned}$$

$x = 30 - 6 = 24$

There are 24 liters of 40% solution and 6 liters of 70% solution.

81. *Verbal Model:* [Amount of peanuts] + [Amount of cashews] = 10

[Cost for peanuts] + [Cost for cashews] = [Total cost]

Labels: Amount of peanuts $= x$

Amount of cashews $= y$

System:

$$\begin{aligned} x + \quad y &= 10 \\ 5.70x + 8.70y &= 6.87(10) \\ y &= 10 - x \end{aligned}$$

$$\begin{aligned} 5.70x + 8.70(10 - x) &= 68.7 \\ 5.70x + 87 - 8.70x &= 68.7 \\ -3x &= -18.3 \\ x &\approx 6.1 \end{aligned}$$

$y = 10 - x = 10 - 6.1 \approx 3.9$

There are about 6.1 pounds of peanuts and 3.9 pounds of cashews.

83. (a)

$$\begin{aligned} 5m + 3b &= 7 \\ -3m - 3b &= -4 \\ \hline 2m &= 3 \\ m &= \tfrac{3}{2} \end{aligned}$$

$$\begin{aligned} 5\left(\tfrac{3}{2}\right) + 3b &= 7 \\ 15 + 6b &= 14 \\ 6b &= -1 \\ b &= -\tfrac{1}{6} \end{aligned}$$

$y = \frac{3}{2}x - \frac{1}{6}$

(b)

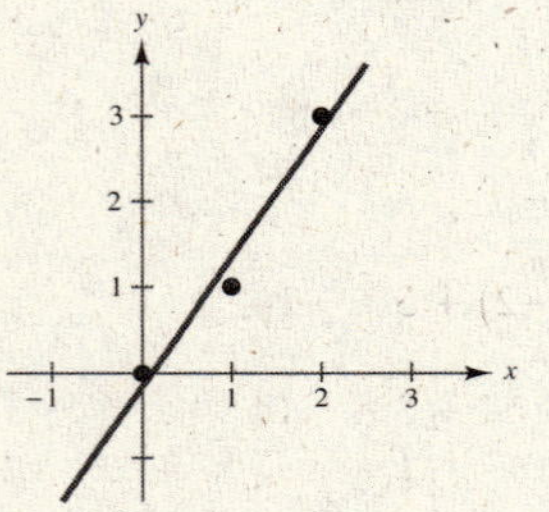

85. (a) and (b)

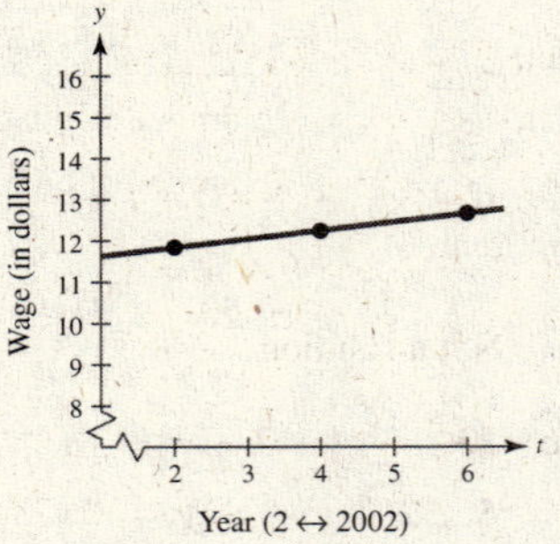

(b)

$$\begin{aligned} 3b + 15m &= 37.38 \Rightarrow & -15b - 75m &= -186.9 \\ 15b + 125m &= 197.3 \Rightarrow & 15b + 125m &= 197.3 \\ \hline & & 50m &= 10.4 \\ & & m &= 0.208 \end{aligned}$$

$$\begin{aligned} 3b + 15(0.208) &= 37.38 \\ 3b + 3.12 &= 37.38 \\ 3b &= 34.26 \\ b &= 11.42 \end{aligned}$$

$y = 0.208x + 11.42$

(c) The slope of the line in the context of this problem is the average annual increase in hourly wage, \$0.208.

87. When solving a system by elimination, you can recognize that it has infinitely many solutions when adding a nonzero multiple of one equation to another equation to eliminate a variable, and you get $0 = 0$ for the result.

89. When you add the equations to eliminate one variable, both are eliminated, yielding a contradiction. For example, when you add the equations in the system $x - y = 3$ and $x - y = 8$ after multiplying the first equation by -1, you obtain $0 = 5$.

91. $m = \dfrac{2 - 0}{4 - 0} = \dfrac{2}{4} = \dfrac{1}{2}$

$$y - 0 = \frac{1}{2}(x - 0)$$
$$y = \frac{1}{2}x$$
$$2y = x$$
$$x - 2y = 0$$

93. $m = \dfrac{2 - 2}{5 - (-1)} = \dfrac{0}{6} = 0$

$$y - 2 = 0[x - (-1)]$$
$$y - 2 = 0$$

95. This set of ordered pairs does represent a function. No input value is matched with two output values.

97. This set of ordered pairs does not represent a function. The input value 3 is matched with two different output values, −2 and −4.

99. (a)

$$3x - 4y = 10 \qquad 2x + 6y = -2$$
$$3(2) - 4(-1) \stackrel{?}{=} 10 \qquad 2(2) + 6(-1) \stackrel{?}{=} -2$$
$$6 + 4 \stackrel{?}{=} 10 \qquad 4 - 6 \stackrel{?}{=} -2$$
$$10 = 10 \qquad -2 = -2$$

$(2, -1)$ is a solution.

(b)

$$3(-1) - 4(0) \stackrel{?}{=} 10 \qquad 2(-1) + 6(0) \stackrel{?}{=} -2$$
$$-3 - 0 \stackrel{?}{=} 10 \qquad -2 + 0 \stackrel{?}{=} -2$$
$$-3 \neq 10 \qquad -2 = -2$$

$(-1, 0)$ is not a solution.

Section 4.3 Linear Systems in Three Variables

1. (a)

$$0 + 3(3) + 2(-2) \stackrel{?}{=} 1$$
$$9 - 4 \neq 1$$

Not a solution

(b)

$$12 + 3(5) + 2(-13) \stackrel{?}{=} 1$$
$$12 + 15 - 26 = 1$$
$$1 = 1$$
$$5(12) - 5 + 3(-13) \stackrel{?}{=} 16$$
$$60 - 5 - 39 \stackrel{?}{=} 16$$
$$16 = 16$$
$$-3(12) + 7(5) - 13 \stackrel{?}{=} -14$$
$$-36 + 35 - 13 \stackrel{?}{=} -14$$
$$-14 = -14$$

Solution

(c)

$$1 + 3(-2) + 2(3) \stackrel{?}{=} 1$$
$$1 - 6 + 6 = 1$$
$$1 = 1$$
$$5(1) - (-2) + 3(3) \stackrel{?}{=} 16$$
$$5 + 2 + 9 \stackrel{?}{=} 16$$
$$16 = 16$$
$$-3(1) + 7(-2) + 3 \stackrel{?}{=} -14$$
$$-3 - 14 + 3 \stackrel{?}{=} -14$$
$$-14 = -14$$

Solution

(d)

$$-2 + 3(5) + 2(-3) \stackrel{?}{=} 1$$
$$-2 + 15 - 6 = 1$$
$$7 \neq 1$$

Not a solution

3. $3y - (-5) = 2$

$3y = -3$

$y = -1$

$x - 2(-1) + 4(-5) = 4$

$x + 2 - 20 = 4$

$x - 18 = 4$

$x = 22$

The solution is $(22, -1, -5)$.

5. $3 + z = 2$

$z = -1$

$x - 2(3) + 4(-1) = 4$

$x - 6 - 4 = 4$

$x - 10 = 4$

$x = 14$

The solution is $(14, 3, -1)$.

7. $x - 2y = 8$

$\underline{-x + 3y = 6}$

$y = 14$

$$\begin{cases} x - 2y = 8 \\ y = 14 \end{cases}$$

This operation eliminated the x-term in Equation 2.

9. Yes. The first equation was multiplied by –2 and added to the second equation. Then the first equation was multiplied by –3 and added to the third equation.

11. $$\begin{cases} x + z = 4 \\ y = 2 \\ 4x + z = 7 \end{cases}$$

$$\begin{cases} x + z = 4 \\ y = 2 \\ -3z = -9 \end{cases}$$

$$\begin{cases} x + z = 4 \\ y = 2 \\ z = 3 \end{cases}$$

$$\begin{cases} x = 1 \\ y = 2 \\ z = 3 \end{cases}$$

The solution is $(1, 2, 3)$.

13. $$\begin{cases} x + y + z = 6 \\ 2x - y + z = 3 \\ 3x - z = 0 \end{cases}$$

$$\begin{cases} x + y + z = 6 \\ -3y - z = -9 \\ -3y - 4z = -18 \end{cases}$$

$$\begin{cases} x + y + z = 6 \\ y + \frac{1}{3}z = 3 \\ -3y - 4z = -18 \end{cases}$$

$$\begin{cases} x + y + z = 6 \\ y + \frac{1}{3}z = 3 \\ -3z = -9 \end{cases}$$

$$\begin{cases} x + y + z = 6 \\ y + \frac{1}{3}z = 3 \\ z = 3 \end{cases}$$

$y + \frac{1}{3}(3) = 3$

$y = 2$

$x + 2 + 3 = 6$

$x = 1$

The solution is $(1, 2, 3)$.

15. $$\begin{cases} x + y + z = -3 \\ 4x + y - 3z = 11 \\ 2x - 3y + 2z = 9 \end{cases}$$

$$\begin{cases} x + y + z = -3 \\ -3y - 7z = 23 \\ -5y = 15 \end{cases}$$

$$\begin{cases} x + y + z = -3 \\ y + \frac{7}{3}z = -\frac{23}{3} \\ y = -3 \end{cases}$$

$$\begin{cases} x + y + z = -3 \\ y + \frac{7}{3}z = -\frac{23}{3} \\ -\frac{7}{3}z = \frac{14}{3} \end{cases}$$

$$\begin{cases} x + y + z = -3 \\ y + \frac{7}{3}z = -\frac{23}{3} \\ z = -2 \end{cases}$$

$y = -3$

$x + (-3) + (-2) = -3$

$x - 5 = -3$

$x = 2$

The solution is $(2, -3, -2)$.

17. $\begin{cases} x + 2y + 6z = 5 \\ -x + y - 2z = 3 \\ x - 4y - 2z = 1 \end{cases}$

$\begin{cases} x + 2y + 6z = 5 \\ 3y + 4z = 8 \\ x - 4y - 2z = 1 \end{cases}$

$\begin{cases} x + 2y + 6z = 5 \\ 3y + 4z = 8 \\ -6y - 8z = -4 \end{cases}$

$\begin{cases} x + 2y + 6z = 5 \\ 3y + 4z = 8 \\ 0 = 12 \end{cases}$

There is no solution.

19. $\begin{cases} 2x + 2z = 2 \\ 5x + 3y = 4 \\ 3y - 4z = 4 \end{cases}$

$\begin{cases} x + z = 1 \\ 5x + 3y = 4 \\ 3y - 4z = 4 \end{cases}$

$\begin{cases} x + z = 1 \\ 3y - 5z = -1 \\ 3y - 4z = 4 \end{cases}$

$\begin{cases} x + z = 1 \\ y - \frac{5}{3}z = -\frac{1}{3} \\ 3y - 4z = 4 \end{cases}$

$\begin{cases} x + z = 1 \\ y - \frac{5}{3}z = -\frac{1}{3} \\ z = 5 \end{cases}$

$y - \frac{5}{3}(5) = -\frac{1}{3}$

$y = 8$

$x + 5 = 1$

$x = -4$

The solution is $(-4, 8, 5)$.

21. $\begin{cases} 6y + 4z = -12 \\ 3x + 3y = 9 \\ 2x - 3z = 10 \end{cases}$

$\begin{cases} x + y = 3 \\ 6y + 4z = -12 \\ 2x - 3z = 10 \end{cases}$

$\begin{cases} x + y = 3 \\ 6y + 4z = -12 \\ -2y - 3z = 4 \end{cases}$

$\begin{cases} x + y = 3 \\ y + \frac{2}{3}z = -2 \\ -2y - 3z = 4 \end{cases}$

$\begin{cases} x + y = 3 \\ y + \frac{2}{3}z = -2 \\ -\frac{5}{3}z = 0 \end{cases}$

$\begin{cases} x + y = 3 \\ y + \frac{2}{3}z = -2 \\ z = 0 \end{cases}$

$y + \frac{2}{3}(0) = -2$

$y = -2$

$x + (-2) = 3$

$x = 5$

The solution is $(5, -2, 0)$.

23. $\begin{cases} 2x + y + 3z = 1 \\ 2x + 6y + 8z = 3 \\ 6x + 8y + 18z = 5 \end{cases}$

$\begin{cases} 2x + y + 3z = 1 \\ 5y + 5z = 2 \\ 5y + 9z = 2 \end{cases}$

$\begin{cases} 2x + y + 3z = 1 \\ 5y + 5z = 2 \\ 4z = 0 \end{cases}$

$\begin{cases} x + \frac{1}{2}y + \frac{3}{2}z = \frac{1}{2} \\ y + z = \frac{2}{5} \\ z = 0 \end{cases}$

$y + 0 = \frac{2}{5}$

$y = \frac{2}{5}$

$x + \frac{1}{2}\left(\frac{2}{5}\right) + \frac{3}{2}(0) = \frac{1}{2}$

$x + \frac{1}{5} = \frac{1}{2}$

$x = \frac{5}{10} - \frac{2}{10} = \frac{3}{10}$

The solution is $\left(\frac{3}{10}, \frac{2}{5}, 0\right)$.

25. $\begin{cases} y + z = 5 \\ 2x + 4z = 4 \\ 2x - 3y = -14 \end{cases}$

$\begin{cases} y + z = 5 \\ 2x + 4z = 4 \\ -3y - 4z = -18 \end{cases}$

$\begin{cases} y + z = 5 \\ 2x + 4z = 4 \\ -z = -3 \end{cases}$

$\begin{cases} y + z = 5 \\ 2x + 4z = 4 \\ z = 3 \end{cases}$

$y + 3 = 5$

$y = 2$

$2x + 4(3) = 4$

$2x + 12 = 4$

$2x = -8$

$x = -4$

The solution is $(-4, 2, 3)$.

27. $\begin{cases} 2x + z = 1 \\ 5y - 3z = 2 \\ 6x + 20y - 9z = 11 \end{cases}$

$\begin{cases} x + \frac{1}{2}z = \frac{1}{2} \\ 5y - 3z = 2 \\ 6x + 20y - 9z = 11 \end{cases}$

$\begin{cases} x + \frac{1}{2}z = \frac{1}{2} \\ 5y - 3z = 2 \\ 20y - 12z = 8 \end{cases}$

$\begin{cases} x + \frac{1}{2}z = \frac{1}{2} \\ y - \frac{3}{5}z = \frac{2}{5} \\ 20y - 12z = 8 \end{cases}$

$\begin{cases} x + \frac{1}{2}z = \frac{1}{2} \\ y - \frac{3}{5}z = \frac{2}{5} \\ 0 = 0 \end{cases}$

$y = \frac{3}{5}z + \frac{2}{5}$

$x + \frac{1}{2}z = \frac{1}{2}$

$x = \frac{1}{2} - \frac{1}{2}z$

Let $z = a$; $\left(\frac{1}{2} - \frac{1}{2}a, \frac{3}{5}a + \frac{2}{5}, a\right)$.

29. $\begin{cases} 2x + 3z = 4 \\ 5x + y + z = 2 \\ 11x + 3y - 3z = 0 \end{cases}$

$\begin{cases} x + \frac{3}{2}z = 2 \\ 5x + y + z = 2 \\ 11x + 3y - 3z = 0 \end{cases}$

$\begin{cases} x + \frac{3}{2}z = 2 \\ y - \frac{13}{2}z = -8 \\ 3y - \frac{39}{2}z = -22 \end{cases}$

$\begin{cases} x + \frac{3}{2}z = 2 \\ y - \frac{13}{2}z = -8 \\ 0 = 2 \end{cases}$

There is no solution.

31. $\begin{cases} 0.2x + 1.3y + 0.6z = 0.1 \\ 0.1x + 0.3z = 0.7 \\ 2x + 10y + 8z = 8 \end{cases}$

$\begin{cases} 2x + 13y + 6z = 1 \\ 1x + 3z = 7 \\ 2x + 10y + 8z = 8 \end{cases}$

$\begin{cases} 1x + 3z = 7 \\ 13y = -13 \\ 10y + 2z = -6 \end{cases}$

$\begin{cases} 1x + 3z = 7 \\ y = -1 \\ 10y + 2z = -6 \end{cases}$

$\begin{cases} x + 3z = 7 \\ y = -1 \\ 2z = 4 \end{cases}$

$\begin{cases} x + 3z = 7 \\ y = -1 \\ z = 2 \end{cases}$

$x + 3(2) = 7$

$x = 1$

The solution is $(1, -1, 2)$.

33. $\begin{cases} x + 4y - 2z = 2 \\ -3x + y + z = -2 \\ 5x + 7y - 5z = 6 \end{cases}$

$\begin{cases} x + 4y - 2z = 2 \\ 13y - 5z = 4 \\ -13y + 5z = -4 \end{cases}$

$\begin{cases} x + 4y - 2z = 2 \\ y - \frac{5}{13}z = \frac{4}{13} \\ -13y + 5z = -4 \end{cases}$

$\begin{cases} x + 4y - 2z = 2 \\ y - \frac{5}{13}z = \frac{4}{13} \\ 0 = 0 \end{cases}$

$y = \frac{5}{13}z + \frac{4}{13}$

$x + 4\left(\frac{5}{13}z + \frac{4}{13}\right) - 2z = 2$

$x + \frac{20}{13}z + \frac{16}{13} - \frac{26}{13}z = \frac{26}{13}$

$x - \frac{6}{13}z = \frac{10}{13}$

$x = \frac{6}{13}z + \frac{10}{13}$

Let $z = a$; $\left(\frac{6}{13}a + \frac{10}{13}, \frac{5}{13}a + \frac{4}{13}, a\right)$.

35. $x + y + z = 3$, $2x + y + 2z = 9$, $x - 2z = 0$ or $x + 2y - z = -4$, $y + 2z = 1$, $3x + y + 3z = 15$

Many correct answers. Write equations so that $(4, -3, 2)$ satisfies each equation.

37. $\begin{cases} 128 = \frac{1}{2}a(1)^2 + v_0(1) + s_0 \\ 80 = \frac{1}{2}a(2)^2 + v_0(2) + s_0 \\ 0 = \frac{1}{2}a(3)^2 + v_0(3) + s_0 \end{cases}$

$\begin{cases} 128 = \frac{1}{2}a + v_0 + s_0 \\ 80 = 2a + 2v_0 + s_0 \\ 0 = \frac{9}{2}a + 3v_0 + s_0 \end{cases}$

$\begin{cases} 256 = a + 2v_0 + 2s_0 \\ 80 = 2a + 2v_0 + s_0 \\ 0 = \frac{9}{2}a + 3v_0 + s_0 \end{cases}$

$\begin{cases} 256 = a + 2v_0 + 2s_0 \\ -432 = -2v_0 - 3s_0 \\ -1152 = -6v_0 - 8s_0 \end{cases}$

$\begin{cases} 256 = a + 2v_0 + 2s_0 \\ 216 = v_0 + \frac{3}{2}s_0 \\ -1152 = -6v_0 - 8s_0 \end{cases}$

$\begin{cases} 256 = a + 2v_0 + 2s_0 \\ 216 = v_0 + \frac{3}{2}s_0 \\ 144 = + s_0 \end{cases}$

$216 = v_0 + \frac{3}{2}(144)$

$0 = v_0$

$256 = a + 0 + 288$

$-32 = a$

$s = -16t^2 + 144$

39. $\begin{cases} 32 = \frac{1}{2}a(1)^2 + v_0(1) + s_0 \\ 32 = \frac{1}{2}a(2)^2 + v_0(2) + s_0 \\ 0 = \frac{1}{2}a(3)^2 + v_0(3) + s_0 \end{cases}$

$\begin{cases} 64 = a + 2v_0 + 2s_0 \\ 32 = 2a + 2v_0 + s_0 \\ 0 = 9a + 6v_0 + s_0 \end{cases}$

$\begin{cases} 64 = a + 2v_0 + 2s_0 \\ -96 = -2v_0 - 3s_0 \\ -576 = -12v_0 - 16s_0 \end{cases}$

$\begin{cases} 64 = a + 2v_0 + 2s_0 \\ 48 = v_0 + \frac{3}{2}s_0 \\ -576 = -12v_0 - 16s_0 \end{cases}$

$\begin{cases} 64 = a + 2v_0 + 2s_0 \\ 48 = v_0 + \frac{3}{2}s_0 \\ 0 = + 2s_0 \end{cases}$

$0 = s_0$

$48 = v_0 + 0$

$64 = a + 2(48) + 0$

$-32 = a$

$s = -16t^2 + 48t$

41. Let x = measure of first angle, y = measure of second angle, and z = measure of third angle.

$\begin{cases} x + y + z = 180 \\ x + y = 2z \\ y = z - 28 \end{cases}$

$\begin{cases} x + y + z = 180 \\ x + y - 2z = 0 \\ y - z = -28 \end{cases}$

$x + y + z = 180$

$\underline{-x - y + 2z = 0}$

$3z = 180$

$z = 60$

$y - 60 = -28$

$y = 32$

$x + 32 - 2(60) = 0$

$x + 32 - 120 = 0$

$x = 88$

The measures of the three angles are 32°, 88°, and 60°.

43. Let x = amount at 6%, y = amount at 10%, and z = amount at 15%.

$\begin{cases} x + y + z = 80{,}000 \\ 0.06x + 0.10y + 0.15z = 8850 \\ y = z + 750 \end{cases}$

$\begin{cases} x + y + z = 80{,}000 \\ 6x + 10y + 15z = 885{,}000 \\ y - z = 750 \end{cases}$

$\begin{cases} -6x - 6y - 6z = -480{,}000 \\ 6x + 10y + 15z = 885{,}000 \\ y - z = 750 \end{cases}$

$\begin{cases} 4y + 9z = 405{,}000 \\ y - z = 750 \end{cases}$

$\begin{cases} 4y + 9z = 405{,}000 \\ 9y - 9z = 6750 \end{cases}$

$13y = 411{,}750$

$y \approx 31{,}673.08$

$31{,}673.08 - z = 750$

$z \approx 30{,}923.08$

$x + 31{,}673.08 + 30{,}923.08 = 80{,}000$

$x \approx 17{,}403.84$

The amounts invested are \$17,404 at 6%, \$31,673 at 10%, and \$30,923 at 15%.

45. Let x = amount at 6%, y = amount at 8%, and z = amount at 9%.

$$\begin{cases} 0.06x + 0.08y + 0.09z = 1150 \\ y = 2x \\ z = x - 1000 \end{cases}$$

$$\begin{cases} 6x + 8y + 9z = 115{,}000 \\ -2x + y = 0 \\ -x + z = -1000 \end{cases}$$

$$\begin{cases} 6x + 8y + 9z = 115{,}000 \\ -2x + y = 0 \\ 9x - 9z = 9000 \end{cases}$$

$$\begin{cases} 15x + 8y = 124{,}000 \\ -2x + y = 0 \\ 9x - 9z = 9000 \end{cases}$$

$$\begin{cases} 15x + 8y = 124{,}000 \\ 16x - 8y = 0 \\ 9x - 9z = 9000 \end{cases}$$

$$\begin{cases} 31x = 124{,}000 \\ 16x - 8y = 0 \\ 9x - 9z = 9000 \end{cases}$$

$x = 4000$

$-2(4000) + y = 0$ $\quad$ $-4000 + z = -1000$

$y = 8000$ $\quad$ $z = 3000$

You invested $4000 at 6%, $8000 at 8%, and $3000 at 9%.

47. Let x = gallons of spray X, y = gallons of spray Y, and z = gallons of spray Z.

	A	B	C	Total parts
X	1	2	2	5
Y			1	1
Z	1	1		2
	12	16	26	Total gallons

$$\begin{cases} 0.20x + 0.50z = 12 \\ 0.40x + 0.50z = 16 \\ 0.40x + 1y = 26 \end{cases}$$

$$\begin{cases} x + 2.5z = 60 \\ 0.4x + 0.5z = 16 \\ 0.4x + 1y = 26 \end{cases}$$

$$\begin{cases} x + 2.5z = 60 \\ -0.5z = -8 \\ 1y - 1z = 2 \end{cases}$$

$$\begin{cases} x + 2.5z = 60 \\ y - z = 2 \\ z = 16 \end{cases}$$

$$\begin{cases} x = 20 \\ y = 18 \\ z = 16 \end{cases}$$

The amounts needed are 20 gallons of Spray X, 18 gallons of Spray Y, and 16 gallons of Spray Z.

49. Let x = \$1.50 hot dogs, y = \$2.50 hot dogs, and z = \$3.25 hot dogs.

$$\begin{cases} x + y + z = 143 \\ 1.50x + 2.50y + 3.25z = 289.25 \\ x = 4z \end{cases}$$

$$\begin{cases} x + y + z = 143 \\ 1.50x + 2.50y + 3.25z = 289.25 \\ x - 4z = 0 \end{cases}$$

$$\begin{cases} -2.50x - 2.50y - 2.50z = -357.50 \\ 1.50x + 2.50y + 3.25z = 289.25 \\ x - 4z = 0 \end{cases}$$

$$\begin{array}{r} -x + 0.75z = -68.25 \\ x - 4z = 0 \\ \hline -3.25z = -68.25 \\ z = 21 \end{array}$$

$x - 4(21) = 0$

$x = 84$

$84 + y + 21 = 143$

$y = 38$

There were 84 hot dogs sold at $1.50, 38 hot dogs sold at $2.50, and 21 hot dogs sold at $3.25.

51. Let x = amount of first solution (10%), y = amount of second solution (15%), and z = amount of third solution (25%).

(a) $\begin{cases} x + y + z = 12 \\ 10x + 15y + 25z = 20(12) \\ z = 4 \end{cases}$

$\begin{cases} x + y + 4 = 12 \\ 10x + 15y + 100 = 240 \end{cases}$

$\begin{cases} x + y = 8 \\ 10x + 15y = 140 \end{cases}$

$$\begin{aligned} -10x - 10y &= -80 \\ 10x + 15y &= 140 \\ \hline 5y &= 60 \\ y &= 12 \end{aligned}$$

Not possible

(b) Cannot eliminate the 25% solution entirely because some of it is needed to raise the concentration of the weaker solutions. Because the 10% solution is the weakest concentration, let $x = 0$ to minimize the amount of 25% solution needed to mix with the 15% solution.

$\begin{cases} x + y = 8 \\ 10x + 15y = 140 \\ x = 0 \end{cases}$

$\begin{cases} y + z = 12 \\ 15y + 25z = 240 \end{cases}$

$$\begin{aligned} -15y - 15z &= -180 \\ 15y + 25z &= 240 \\ \hline 10z &= 60 \\ z &= 6 \end{aligned}$$

$y + 6 = 12$

$y = 6$

0 gallons of 10% solution, 6 gallons of 15% solution, 6 gallons of 25% solution

(c) Cannot use only the 25% solution because some of the weaker solutions are needed to dilute its concentration. Because the 15% solution is the stronger of the two remaining solutions, let $y = 0$ to eliminate it, maximizing the amount of 25% solution needed to mix with the 10% solution.

$\begin{cases} x + y + z = 12 \\ 10x + 15y + 25z = 240 \\ y = 0 \end{cases}$

$\begin{cases} x + z = 12 \\ 10x + 25z = 240 \end{cases}$

$$\begin{aligned} -10x - 10z &= -120 \\ 10x + 25z &= 240 \\ \hline 15z &= 120 \\ z &= 8 \end{aligned}$$

$x + 8 = 12$

$x = 4$

4 gallons of 10% solution, 0 gallons of 15% solution, 8 gallons of 25% solution

53. $\begin{cases} 0.40x + 0.30y + 0.50z = 30 \\ 0.20x + 0.25y + 0.25z = 17 \\ 0.10x + 0.15y + 0.25z = 10 \end{cases}$

$\begin{cases} x + 0.75y + 1.25z = 75 \\ 0.20x + 0.25y + 0.25z = 17 \\ 0.10x + 0.15y + 0.25z = 10 \end{cases}$

$\begin{cases} x + 0.75y + 1.25z = 75 \\ 0.1y = 2 \\ 0.075y + .125z = 2.5 \end{cases}$

$\begin{cases} x + 0.75y + 1.25z = 75 \\ y = 20 \\ 0.075y + 0.125z = 2.5 \end{cases}$

$\begin{cases} x + 1.25z = 60 \\ y = 20 \\ 0.125z = 1 \end{cases}$

$\begin{cases} x + 1.25z = 60 \\ y = 20 \\ z = 8 \end{cases}$

$\begin{cases} x = 50 \\ y = 20 \\ z = 8 \end{cases}$

String: 50, Wind: 20, Percussion: 8

55. The solution is apparent because the row-echelon form is

$\begin{cases} x = 1 \\ y = -3. \\ z = 4 \end{cases}$

57. Three planes have no point in common when two of the planes are parallel and the third plane intersects the other two planes.

59. The graphs are three planes with three possible situations. If all three planes intersect at one point, there is one solution. If all three planes intersect in one line, there is an infinite number of solutions. If each pair of planes intersects in a line, but the three lines of intersection are all parallel, there is no solution.

61. Terms: $3x$, 2

Coefficients: 3, 2

63. Terms: $14t^5$, $-t$, 25

Coefficients: 14, -1, 25

65. $2x + 3y = 17 \Rightarrow 8x + 12y = 68$

$4y = 12 \Rightarrow \quad -12y = -36$

$8x = 32$

$x = 4$

$2(4) + 3y = 17$

$3y = 9$

$y = 3$

The solution is $(4, 3)$.

67. $3x - 4y = -30$

$5x + 4y = 14$

$8x = -16$

$x = -2$

$3(-2) - 4y = -30$

$-4y = -24$

$y = 6$

The solution is $(-2, 6)$.

Mid-Chapter Quiz for Chapter 4

1. (a) $(1, -2)$

$5(1) - 12(-2) \stackrel{?}{=} 2$

$5 + 24 \neq 2$

This is not a solution.

(b) $(10, 4)$

$5(10) - 12(4) \stackrel{?}{=} 2$

$50 - 48 = 2$

$2 = 2$

$2(10) + 1.5(4) \stackrel{?}{=} 26$

$20 + 6 = 26$

$26 = 26$

This is a solution.

3.

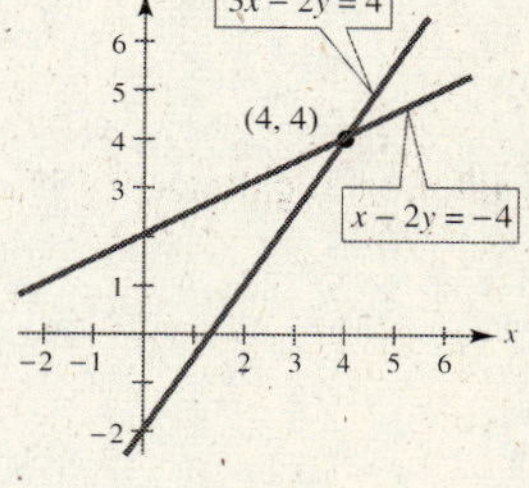

One solution

5. $x - y = 0 \qquad 2x = 8$

$y = x \qquad x = 4$

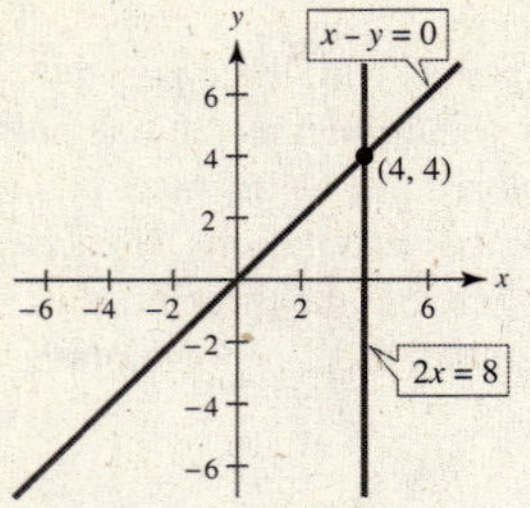

The solution is $(4, 4)$.

7.

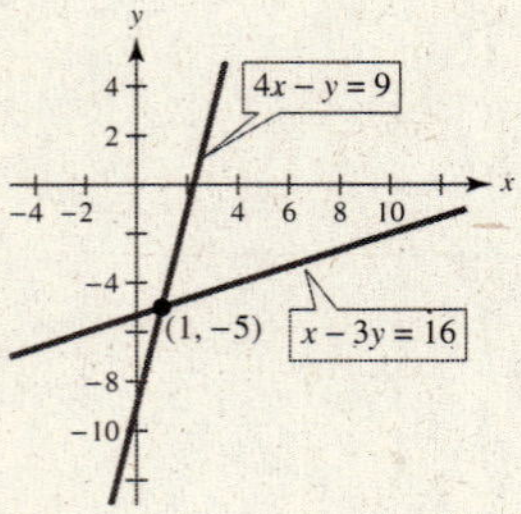

The solution is $(1, -5)$.

9. $5x - y = 32 \Rightarrow -y = -5x + 32 \Rightarrow y = 5x - 32$

$$6x - 9y = 18$$
$$6x - 9(5x - 32) = 18$$
$$6x - 45x + 288 = 18$$
$$-39x = -270$$
$$x = \frac{-270}{-39} = \frac{90}{13}$$
$$y = 5\left(\frac{90}{13}\right) - 32 = \frac{450}{13} - \frac{416}{13} = \frac{34}{13}$$

The solution is $\left(\frac{90}{13}, \frac{34}{13}\right)$.

11. $x + 10y = 18$

$$5x + 2y = 42$$
$$x + 10y = 18$$
$$-48y = -48$$
$$x + 10y = 18$$
$$y = 1$$
$$x = 8$$
$$y = 1$$

The solution is $(8, 1)$.

13. $\begin{cases} a + b + c = 1 \\ 4a + 2b + c = 2 \\ 9a + 3b + c = 4 \end{cases}$

$$\begin{cases} a + b + c = 1 \\ -2b - 3c = -2 \\ -6b - 8c = -5 \end{cases}$$

$$\begin{cases} a + b + c = 1 \\ b + \frac{3}{2}c = 1 \\ -6b - 8c = -5 \end{cases}$$

$$\begin{cases} a - \frac{1}{2}c = 0 \\ b + \frac{3}{2}c = 1 \\ c = 1 \end{cases}$$

$$a = \frac{1}{2}$$
$$b = -\frac{1}{2}$$
$$c = 1$$

The solution is $\left(\frac{1}{2}, -\frac{1}{2}, 1\right)$.

15. Answers will vary. Write any two equations that are satisfied by $(10, -12)$.

$$x + y = -2$$
$$2x - y = 32$$

17. *Verbal Model:* $\boxed{\text{Amount Solution 1}} + \boxed{\text{Amount Solution 2}} = \boxed{\text{Amount mixture}}$

$$0.20\boxed{\text{Amount Solution 1}} + 0.50\boxed{\text{Amount Solution 2}} = 0.30 \cdot 20$$

Labels: Amount Solution 1 $= x$

Amount Solution 2 $= y$

System:

$$x + y = 20$$
$$0.20x + 0.50y = 0.30(20)$$
$$x + y = 20$$
$$20x + 50y = 600$$

By substitution:

$$y = 20 - x$$
$$20x + 50(20 - x) = 600$$
$$20x + 1000 - 50x = 600$$
$$-30x = -400$$
$$x = 13\tfrac{1}{3} \text{ gallons at 20\% solution}$$
$$20 - x = 6\tfrac{2}{3} \text{ gallons at 50\% solution}$$

Section 4.4 Matrices and Linear Systems

1. 4×2

3. 2×2

5. 4×1

7. 1×1

9. 1×4

11. (a) $\begin{bmatrix} 4 & -5 \\ -1 & 8 \end{bmatrix}$ (b) $\left[\begin{array}{cc:c} 4 & -5 & -2 \\ -1 & 8 & 10 \end{array}\right]$

13. (a) $\begin{bmatrix} 1 & 1 & 0 \\ 5 & -2 & -2 \\ 2 & 4 & 1 \end{bmatrix}$ (b) $\left[\begin{array}{ccc:c} 1 & 1 & 0 & 0 \\ 5 & -2 & -2 & 12 \\ 2 & 4 & 1 & 5 \end{array}\right]$

15. (a) $\begin{bmatrix} 5 & 1 & -3 \\ 0 & 2 & 4 \end{bmatrix}$ (b) $\left[\begin{array}{ccc:c} 5 & 1 & -3 & 7 \\ 0 & 2 & 4 & 12 \end{array}\right]$

17. $\begin{cases} 4x + 3y = 8 \\ x - 2y = 3 \end{cases}$

19. $\begin{cases} x + 2z = -10 \\ 3y - z = 5 \\ 4x + 2y = 3 \end{cases}$

21. $\begin{cases} 5x + 8y + 2z = -1 \\ -2x + 15y + 5z + w = 9 \\ x + 6y - 7z = -3 \end{cases}$

23. $\begin{cases} 13x + y + 4z - 2w = -4 \\ 5x + 4y - w = 0 \\ x + 2y + 6z + 8w = 5 \\ -10x + 12y + 3z + w = -2 \end{cases}$

25. Interchange the first and second rows.

27. Multiply the first row by $-\frac{1}{3}$.

29. Add 5 times the first row to the third row.

31. $\begin{bmatrix} 1 & 1 & -4 & 2 \\ 0 & 0 & 8 & 3 \\ 0 & 4 & 5 & 5 \end{bmatrix}$ $\begin{matrix} \\ R_3 \\ R_2 \end{matrix}\begin{bmatrix} 1 & 1 & -4 & 2 \\ 0 & 4 & 5 & 5 \\ 0 & 0 & 8 & 3 \end{bmatrix}$

33. $\begin{bmatrix} 9 & -18 & 27 \\ 3 & 4 & 5 \end{bmatrix} \tfrac{1}{9}R_1 \rightarrow \begin{bmatrix} 1 & -2 & 3 \\ 3 & 4 & 5 \end{bmatrix}$

35. $\begin{bmatrix} 1 & 4 & 3 \\ 2 & 8 & 6 \end{bmatrix} -2R_1 + R_2 \rightarrow \begin{bmatrix} 1 & 4 & 3 \\ 0 & 0 & 0 \end{bmatrix}$

37.

$$\begin{bmatrix} 1 & 2 & 3 \\ 2 & -1 & -4 \end{bmatrix}$$

$$-2R_1 + R_2 \begin{bmatrix} 1 & 2 & 3 \\ 0 & -5 & -10 \end{bmatrix}$$

$$-\tfrac{1}{5}R_2 \begin{bmatrix} 1 & 2 & 3 \\ 0 & 1 & 2 \end{bmatrix}$$

39.

$$\begin{bmatrix} 4 & 6 & 1 \\ -2 & 2 & 5 \end{bmatrix}$$

$$\tfrac{1}{4}R_1 \quad \begin{bmatrix} 1 & \frac{3}{2} & \frac{1}{4} \\ -2 & 2 & 5 \end{bmatrix}$$

$$\begin{matrix} \\ 2R_1 + R_2 \end{matrix} \quad \begin{bmatrix} 1 & \frac{3}{2} & \frac{1}{4} \\ 0 & 5 & \frac{11}{2} \end{bmatrix}$$

$$\begin{matrix} \\ \tfrac{1}{5}R_2 \end{matrix} \quad \begin{bmatrix} 1 & \frac{3}{2} & \frac{1}{4} \\ 0 & 1 & \frac{11}{10} \end{bmatrix}$$

$$\begin{matrix} \\ -\tfrac{3}{2}R_2 + R_1 \end{matrix} \quad \begin{bmatrix} 1 & 0 & -\frac{7}{5} \\ 0 & 1 & \frac{11}{10} \end{bmatrix}$$

41.

$$\begin{bmatrix} 1 & 1 & 0 & 5 \\ -2 & -1 & 2 & -10 \\ 3 & 6 & 7 & 14 \end{bmatrix}$$

$$\begin{matrix} \\ 2R_1 + R_2 \\ -3R_1 + R_3 \end{matrix} \quad \begin{bmatrix} 1 & 1 & 0 & 5 \\ 0 & 1 & 2 & 0 \\ 0 & 3 & 7 & -1 \end{bmatrix}$$

$$\begin{matrix} \\ \\ -3R_2 + R_3 \end{matrix} \quad \begin{bmatrix} 1 & 1 & 0 & 5 \\ 0 & 1 & 2 & 0 \\ 0 & 0 & 1 & -1 \end{bmatrix}$$

43.

$$\begin{bmatrix} 1 & -1 & -1 & 1 \\ 4 & -4 & 1 & 8 \\ -6 & 8 & 18 & 0 \end{bmatrix}$$

$$\begin{matrix} \\ -4R_1 + R_2 \\ 6R_1 + R_3 \end{matrix} \quad \begin{bmatrix} 1 & -1 & -1 & 1 \\ 0 & 0 & 5 & 4 \\ 0 & 2 & 12 & 6 \end{bmatrix}$$

$$\begin{matrix} \\ R_3 \\ R_2 \end{matrix} \quad \begin{bmatrix} 1 & -1 & -1 & 1 \\ 0 & 2 & 12 & 6 \\ 0 & 0 & 5 & 4 \end{bmatrix}$$

$$\begin{matrix} \\ \tfrac{1}{2}R_2 \\ \tfrac{1}{5}R_3 \end{matrix} \quad \begin{bmatrix} 1 & -1 & -1 & 1 \\ 0 & 1 & 6 & 3 \\ 0 & 0 & 1 & \frac{4}{5} \end{bmatrix}$$

45.

$$\begin{bmatrix} 1 & 1 & -1 & 3 \\ 2 & 1 & 2 & 5 \\ 3 & 2 & 1 & 8 \end{bmatrix}$$

$$\begin{matrix} \\ -2R_1 + R_2 \\ -3R_1 + R_3 \end{matrix} \quad \begin{bmatrix} 1 & 1 & -1 & 3 \\ 0 & -1 & 4 & -1 \\ 0 & -1 & 4 & -1 \end{bmatrix}$$

$$\begin{matrix} \\ \\ -R_2 + R_3 \end{matrix} \quad \begin{bmatrix} 1 & 1 & -1 & 3 \\ 0 & -1 & 4 & -1 \\ 0 & 0 & 0 & 0 \end{bmatrix}$$

$$\begin{matrix} \\ -R_2 \\ \\ \end{matrix} \quad \begin{bmatrix} 1 & 1 & -1 & 3 \\ 0 & 1 & -4 & 1 \\ 0 & 0 & 0 & 0 \end{bmatrix}$$

47. $x - 2y = 4$

$y = -3$

$x - 2(-3) = 4$

$x + 6 = 4$

$x = -2$

The solution is $(-2, -3)$.

49. $x + 5y = 3$

$y = -2$

$x + 5(-2) = 3$

$x - 10 = 3$

$x = 13$

The solution is $(13, -2)$.

51. $x - y + 2z = 4$

$y - z = 2$

$z = -2$

$y - (-2) = 2$

$y + 2 = 2$

$y = 0$

$x - 0 + 2(-2) = 4$

$x - 4 = 4$

$x = 8$

The solution is $(8, 0, -2)$.

53. $x + 2y = 7$

$3x - 7y = 8$

$$\begin{bmatrix} 1 & 2 & \vdots & 7 \\ 3 & -7 & \vdots & 8 \end{bmatrix}$$

$$-3R_1 + R_2 \begin{bmatrix} 1 & 2 & \vdots & 7 \\ 0 & -13 & \vdots & -13 \end{bmatrix}$$

$$-\tfrac{1}{13}R_2 \begin{bmatrix} 1 & 2 & \vdots & 7 \\ 0 & 1 & \vdots & 1 \end{bmatrix}$$

$y = 1 \qquad x + 2(1) = 7$

$x = 5$

The solution is $(5, 1)$.

55. $6x - 4y = 2$

$5x + 2y = 7$

$$\begin{bmatrix} 6 & -4 & \vdots & 2 \\ 5 & 2 & \vdots & 7 \end{bmatrix}$$

$$\tfrac{1}{6}R_1 \begin{bmatrix} 1 & -\frac{2}{3} & \vdots & \frac{1}{3} \\ 5 & 2 & \vdots & 7 \end{bmatrix}$$

$$-5R_1 + R_2 \begin{bmatrix} 1 & -\frac{2}{3} & \vdots & \frac{1}{3} \\ 0 & \frac{16}{3} & \vdots & \frac{16}{3} \end{bmatrix}$$

$$-\tfrac{3}{16}R_2 \begin{bmatrix} 1 & -\frac{2}{3} & \vdots & \frac{1}{3} \\ 0 & 1 & \vdots & 1 \end{bmatrix}$$

$y = 1 \qquad x - \frac{2}{3}(1) = \frac{1}{3}$

$x = 1$

The solution is $(1, 1)$.

57. $12 + 10y = -14$

$4x - 3y = -11$

$$\begin{bmatrix} 12 & 10 & \vdots & -14 \\ 4 & -3 & \vdots & -11 \end{bmatrix}$$

$$\tfrac{1}{12}R_1 \begin{bmatrix} 1 & \frac{5}{6} & \vdots & -\frac{7}{6} \\ 4 & -3 & \vdots & -11 \end{bmatrix}$$

$$-4R_1 + R_2 \begin{bmatrix} 1 & \frac{5}{6} & \vdots & -\frac{7}{6} \\ 0 & -\frac{19}{3} & \vdots & -\frac{19}{3} \end{bmatrix}$$

$$-\tfrac{3}{19} \cdot R_2 \begin{bmatrix} 1 & \frac{5}{6} & \vdots & -\frac{7}{6} \\ 0 & 1 & \vdots & 1 \end{bmatrix}$$

$y = 1 \qquad x + \frac{5}{6}(1) = -\frac{7}{6}$

$x = -2$

The solution is $(-2, 1)$.

59. $-x + 2y = 1.5$

$2x - 4y = 3$

$$\begin{bmatrix} -1 & 2 & \vdots & 1.5 \\ 2 & -4 & \vdots & 3 \end{bmatrix}$$

$$-R_1 \begin{bmatrix} 1 & -2 & \vdots & -1.5 \\ 2 & -4 & \vdots & 3 \end{bmatrix}$$

$$-2R_1 + R_2 \begin{bmatrix} 1 & -2 & \vdots & -1.5 \\ 0 & 0 & \vdots & 6 \end{bmatrix}$$

Inconsistent; no solution

61. $x - 2y - z = 6$

$y + 4z = 5$

$4x + 2y + 3z = 8$

$$\begin{bmatrix} 1 & -2 & -1 & \vdots & 6 \\ 0 & 1 & 4 & \vdots & 5 \\ 4 & 2 & 3 & \vdots & 8 \end{bmatrix}$$

$$-4R_1 + R_3 \begin{bmatrix} 1 & -2 & -1 & \vdots & 6 \\ 0 & 1 & 4 & \vdots & 5 \\ 0 & 10 & 7 & \vdots & -16 \end{bmatrix}$$

$$-10R_2 + R_3 \begin{bmatrix} 1 & -2 & -1 & \vdots & 6 \\ 0 & 1 & 4 & \vdots & 5 \\ 0 & 0 & -33 & \vdots & -66 \end{bmatrix}$$

$$\frac{1}{-33}R_3 \begin{bmatrix} 1 & -2 & -1 & \vdots & 6 \\ 0 & 1 & 4 & \vdots & 5 \\ 0 & 0 & 1 & \vdots & 2 \end{bmatrix}$$

$z = 2 \qquad y + 4(2) = 5 \qquad x - 2(-3) - (2) = 6$

$y = -3 \qquad x + 6 - 2 = 6$

$x = 2$

The solution is $(2, -3, 2)$.

63. $x + y - 5z = 3$

$x \quad - 2z = 1$

$2x - y - z = 0$

$$\begin{bmatrix} 1 & 1 & -5 & \vdots & 3 \\ 1 & 0 & -2 & \vdots & 1 \\ 2 & -1 & -1 & \vdots & 0 \end{bmatrix}$$

$$\begin{matrix} \\ -R_1 + R_2 \\ -2R_1 + R_3 \end{matrix} \begin{bmatrix} 1 & 1 & -5 & \vdots & 3 \\ 0 & -1 & 3 & \vdots & -2 \\ 0 & -3 & 9 & \vdots & -6 \end{bmatrix}$$

$$-R_2 \begin{bmatrix} 1 & 1 & -5 & \vdots & 3 \\ 0 & 1 & -3 & \vdots & 2 \\ 0 & -3 & 9 & \vdots & -6 \end{bmatrix}$$

$$\begin{matrix} \\ \\ 3R_2 + R_3 \end{matrix} \begin{bmatrix} 1 & 1 & -5 & \vdots & 3 \\ 0 & 1 & -3 & \vdots & 2 \\ 0 & 0 & 0 & \vdots & 0 \end{bmatrix}$$

$y - 3z = 2$ $\qquad x + (2 + 3z) - 5z = 3$

$y = 2 + 3z$ $\qquad x = 1 + 2z$

Let $a = z$. (a is any real number).

$(1 + 2a, 2 + 3a, a)$

65. $2x + 4y \quad = 10$

$2x + 2y + 3z = 3$

$-3x + y + 2z = -3$

$$\begin{bmatrix} 2 & 4 & 0 & \vdots & 10 \\ 2 & 2 & 3 & \vdots & 3 \\ -3 & 1 & 2 & \vdots & -3 \end{bmatrix}$$

$$-R_1 + R_2 \begin{bmatrix} 2 & 4 & 0 & \vdots & 10 \\ 0 & -2 & 3 & \vdots & -7 \\ -3 & 1 & 2 & \vdots & -3 \end{bmatrix}$$

$$\tfrac{1}{2}R_1 \begin{bmatrix} 1 & 2 & 0 & \vdots & 5 \\ 0 & -2 & 3 & \vdots & -7 \\ -3 & 1 & 2 & \vdots & -3 \end{bmatrix}$$

$$\begin{matrix} \\ \\ 3R_1 + R_3 \end{matrix} \begin{bmatrix} 1 & 2 & 0 & \vdots & 5 \\ 0 & -2 & 3 & \vdots & -7 \\ 0 & 7 & 2 & \vdots & 12 \end{bmatrix}$$

$$-\tfrac{1}{2}R_2 \begin{bmatrix} 1 & 2 & 0 & \vdots & 5 \\ 0 & 1 & -\frac{3}{2} & \vdots & \frac{7}{2} \\ 0 & 7 & 2 & \vdots & 12 \end{bmatrix}$$

$$\begin{matrix} \\ \\ -7R_2 + R_3 \end{matrix} \begin{bmatrix} 1 & 2 & 0 & \vdots & 5 \\ 0 & 1 & -\frac{3}{2} & \vdots & \frac{7}{2} \\ 0 & 0 & \frac{25}{2} & \vdots & -\frac{25}{2} \end{bmatrix}$$

$$\begin{matrix} \\ \\ \frac{2}{25}R_3 \end{matrix} \begin{bmatrix} 1 & 2 & 0 & \vdots & 5 \\ 0 & 1 & -\frac{3}{2} & \vdots & \frac{7}{2} \\ 0 & 0 & 1 & \vdots & -1 \end{bmatrix}$$

$z = -1$ $\qquad y - \frac{3}{2}(-1) = \frac{7}{2}$ $\qquad x + 2(2) = 5$

$y = \frac{4}{2}$ $\qquad x + 4 = 5$

$y = 2$ $\qquad x = 1$

The solution is $(1, 2, -1)$.

67. $x - 3y + 2z = 8$
$2y - z = -4$
$x + z = 3$

$$\begin{bmatrix} 1 & -3 & 2 & \vdots & 8 \\ 0 & 2 & -1 & \vdots & -4 \\ 1 & 0 & 1 & \vdots & 3 \end{bmatrix}$$

$$\begin{array}{r} \\ \frac{1}{2}R_2 \\ -R_1 + R_3 \end{array} \begin{bmatrix} 1 & -3 & 2 & \vdots & 8 \\ 0 & 1 & -\frac{1}{2} & \vdots & -2 \\ 0 & 3 & -1 & \vdots & -5 \end{bmatrix}$$

$$\begin{array}{r} 3R_2 + R_1 \\ \\ -3R_2 + R_3 \end{array} \begin{bmatrix} 1 & 0 & \frac{1}{2} & \vdots & 2 \\ 0 & 1 & -\frac{1}{2} & \vdots & -2 \\ 0 & 0 & \frac{1}{2} & \vdots & 1 \end{bmatrix}$$

$$\begin{array}{r} \\ \\ 2R_3 \end{array} \begin{bmatrix} 1 & 0 & \frac{1}{2} & \vdots & 2 \\ 0 & 1 & -\frac{1}{2} & \vdots & -2 \\ 0 & 0 & 1 & \vdots & 2 \end{bmatrix}$$

$z = 2$ $\quad y - \frac{1}{2}(2) = -2$ $\quad x + \frac{1}{2}(2) = 2$

$y = -1$ $\quad x = 1$

The solution is $(1, -1, 2)$.

69. $-2x - 2y - 15z = 0$
$x + 2y + 2z = 18$
$3x + 3y + 22z = 2$

$$\begin{bmatrix} -2 & -2 & -15 & \vdots & 0 \\ 1 & 2 & 2 & \vdots & 18 \\ 3 & 3 & 22 & \vdots & 2 \end{bmatrix}$$

$$\begin{array}{r} R_2 \\ R_1 \\ \\ \end{array} \begin{bmatrix} 1 & 2 & 2 & \vdots & 18 \\ -2 & -2 & -15 & \vdots & 0 \\ 3 & 3 & 22 & \vdots & 2 \end{bmatrix}$$

$$\begin{array}{r} \\ 2R_1 + R_2 \\ -3R_1 + R_3 \end{array} \begin{bmatrix} 1 & 2 & 2 & \vdots & 18 \\ 0 & 2 & -11 & \vdots & 36 \\ 0 & -3 & 16 & \vdots & -52 \end{bmatrix}$$

$$\begin{array}{r} \\ \\ \frac{3}{2}R_2 + R_3 \end{array} \begin{bmatrix} 1 & 2 & 2 & \vdots & 18 \\ 0 & 2 & -11 & \vdots & 36 \\ 0 & 0 & -\frac{1}{2} & \vdots & 2 \end{bmatrix}$$

$$\begin{array}{r} \\ \frac{1}{2}R_2 \\ -2R_3 \end{array} \begin{bmatrix} 1 & 2 & 2 & \vdots & 18 \\ 0 & 1 & -\frac{11}{2} & \vdots & 18 \\ 0 & 0 & 1 & \vdots & -4 \end{bmatrix}$$

$z = -4$ $\quad y - \frac{11}{2}(-4) = 18$

$y + 22 = 18$

$y = -4$

$x + 2(-4) + 2(-4) = 18$

$x - 8 - 8 = 18$

$x - 16 = 18$

$x = 34$

$(34, -4, -4)$

71. $2x \qquad + 4z = 1$

$x + y + 3z = 0$

$x + 3y + 5z = 0$

$$\begin{bmatrix} 2 & 0 & 4 & \vdots & 1 \\ 1 & 1 & 3 & \vdots & 0 \\ 1 & 3 & 5 & \vdots & 0 \end{bmatrix}$$

$$\begin{matrix} R_2 \\ R_1 \\ \\ \end{matrix} \begin{bmatrix} 1 & 1 & 3 & \vdots & 0 \\ 2 & 0 & 4 & \vdots & 1 \\ 1 & 3 & 5 & \vdots & 0 \end{bmatrix}$$

$$\begin{matrix} \\ -2R_1 + R_2 \\ -R_1 + R_3 \end{matrix} \begin{bmatrix} 1 & 1 & 3 & \vdots & 0 \\ 0 & -2 & -2 & \vdots & 1 \\ 0 & 2 & 2 & \vdots & 0 \end{bmatrix}$$

$$\begin{matrix} \\ -\frac{1}{2}R_2 \\ \\ \end{matrix} \begin{bmatrix} 1 & 1 & 3 & \vdots & 0 \\ 0 & 1 & 1 & \vdots & -\frac{1}{2} \\ 0 & 2 & 2 & \vdots & 0 \end{bmatrix}$$

$$\begin{matrix} \\ \\ -2R_2 + R_3 \end{matrix} \begin{bmatrix} 1 & 1 & 3 & \vdots & 0 \\ 0 & 1 & 1 & \vdots & -\frac{1}{2} \\ 0 & 0 & 0 & \vdots & 1 \end{bmatrix}$$

Inconsistent; no solution

73. $x + 3y \qquad = 2$

$2x + 6y \qquad = 4$

$2x + 5y + 4z = 3$

$$\begin{bmatrix} 1 & 3 & 0 & \vdots & 2 \\ 2 & 6 & 0 & \vdots & 4 \\ 2 & 5 & 4 & \vdots & 3 \end{bmatrix}$$

$$\begin{matrix} \\ -2R_1 + R_2 \\ -2R_1 + R_3 \end{matrix} \begin{bmatrix} 1 & 3 & 0 & \vdots & 2 \\ 0 & 0 & 0 & \vdots & 0 \\ 0 & -1 & 4 & \vdots & -1 \end{bmatrix}$$

$$\begin{matrix} \\ -R_3 \\ R_2 \end{matrix} \begin{bmatrix} 1 & 3 & 0 & \vdots & 2 \\ 0 & 1 & -4 & \vdots & 1 \\ 0 & 0 & 0 & \vdots & 0 \end{bmatrix}$$

Let $z = a$, then $y = 1 + 4a$.

$x = 2 - 3(1 + 4a)$

$= 2 - 3 - 12a = -1 - 12a$

$(-12a - 1, 1 + 4a, a)$

75. $4x - y + z = 4$

$-6x + 3y - 2z = -5$

$2x + 5y - z = 7$

$$\begin{bmatrix} 4 & -1 & 1 & \vdots & 4 \\ -6 & 3 & -2 & \vdots & -5 \\ 2 & 5 & -1 & \vdots & 7 \end{bmatrix}$$

$$\begin{matrix} \frac{1}{4}R_1 \\ \\ \\ \end{matrix} \begin{bmatrix} 1 & -\frac{1}{4} & \frac{1}{4} & \vdots & 1 \\ -6 & 3 & -2 & \vdots & -5 \\ 2 & 5 & -1 & \vdots & 7 \end{bmatrix}$$

$$\begin{matrix} \\ 6R_1 + R_2 \\ -2R_1 + R_3 \end{matrix} \begin{bmatrix} 1 & -\frac{1}{4} & \frac{1}{4} & \vdots & 1 \\ 0 & \frac{3}{2} & -\frac{1}{2} & \vdots & 1 \\ 0 & \frac{11}{2} & -\frac{3}{2} & \vdots & 5 \end{bmatrix}$$

$$\begin{matrix} \\ \frac{2}{3}R_2 \\ 2R_3 \end{matrix} \begin{bmatrix} 1 & -\frac{1}{4} & \frac{1}{4} & \vdots & 1 \\ 0 & 1 & -\frac{1}{3} & \vdots & \frac{2}{3} \\ 0 & 11 & -3 & \vdots & 10 \end{bmatrix}$$

$$\begin{matrix} \frac{1}{4}R_2 + R_1 \\ -11R_2 + R_3 \\ \\ \end{matrix} \begin{bmatrix} 1 & 0 & \frac{1}{6} & \vdots & \frac{7}{6} \\ 0 & 1 & -\frac{1}{3} & \vdots & \frac{2}{3} \\ 0 & 0 & \frac{2}{3} & \vdots & \frac{8}{3} \end{bmatrix}$$

$$\begin{matrix} \frac{3}{2}R_3 \\ \\ \\ \end{matrix} \begin{bmatrix} 1 & 0 & \frac{1}{6} & \vdots & \frac{7}{6} \\ 0 & 1 & -\frac{1}{3} & \vdots & \frac{2}{3} \\ 0 & 0 & 1 & \vdots & 4 \end{bmatrix}$$

$z = 4 \qquad y - \frac{1}{3}(4) = \frac{2}{3} \qquad x + \frac{1}{6}(4) = \frac{7}{6}$

$y = 2 \qquad x = \frac{1}{2}$

The solution is $\left(\frac{1}{2}, 2, 4\right)$.

77. $2x + y - 2z = 4$

$3x - 2y + 4z = 6$

$-4x + y + 6z = 12$

$$\begin{bmatrix} 2 & 1 & -2 & \vdots & 4 \\ 3 & -2 & 4 & \vdots & 6 \\ -4 & 1 & 6 & \vdots & 12 \end{bmatrix}$$

$$\begin{matrix} \\ -\frac{3}{2}R_1 + R_2 \\ 2R_1 + R_3 \end{matrix} \begin{bmatrix} 2 & 1 & -2 & \vdots & 4 \\ 0 & -\frac{7}{2} & 7 & \vdots & 0 \\ 0 & 3 & 2 & \vdots & 20 \end{bmatrix}$$

$$\begin{matrix} \frac{1}{2}R_1 \\ -\frac{2}{7}R_2 \\ \\ \end{matrix} \begin{bmatrix} 1 & \frac{1}{2} & -1 & \vdots & 2 \\ 0 & 1 & -2 & \vdots & 0 \\ 0 & 3 & 2 & \vdots & 20 \end{bmatrix}$$

$$\begin{matrix} \\ \\ -3R_2 + R_3 \end{matrix} \begin{bmatrix} 1 & \frac{1}{2} & -1 & \vdots & 2 \\ 0 & 1 & -2 & \vdots & 0 \\ 0 & 0 & 8 & \vdots & 20 \end{bmatrix}$$

$$\begin{matrix} \\ \\ \frac{1}{8}R_3 \end{matrix} \begin{bmatrix} 1 & \frac{1}{2} & -1 & \vdots & 2 \\ 0 & 1 & -2 & \vdots & 0 \\ 0 & 0 & 1 & \vdots & \frac{5}{2} \end{bmatrix}$$

$z = \frac{5}{2}$ $\quad y - 2\left(\frac{5}{2}\right) = 0$ $\quad x + \frac{1}{2}(5) - \left(\frac{5}{2}\right) = 2$

$y - 5 = 0$ $\quad x + \frac{5}{2} - \frac{5}{2} = 2$

$y = 5$ $\quad x = 2$

$\left(2, 5, \frac{5}{2}\right)$

79. *Verbal Model:* $\boxed{\text{Money 1}} + \boxed{\text{Money 2}} + \boxed{\text{Money 3}} = \boxed{1{,}500{,}000}$

$0.08 \cdot \boxed{\text{Money 1}} + 0.09 \cdot \boxed{\text{Money 2}} + 0.12 \cdot \boxed{\text{Money 3}} = 113{,}000$

$\boxed{\text{Money 1}} = 4 \cdot \boxed{\text{Money 3}}$

Labels: $x = \text{Money 1}$

$y = \text{Money 2}$

$z = \text{Money 3}$

System:

$$\begin{aligned} x + y + z &= 1{,}500{,}000 \\ 0.08x + 0.09y + 0.12z &= 133{,}000 \\ x &= 4z \end{aligned}$$

$$\begin{bmatrix} 1 & 1 & 1 & \vdots & 1{,}500{,}000 \\ 8 & 9 & 12 & \vdots & 13{,}300{,}000 \\ 1 & 0 & -4 & \vdots & 0 \end{bmatrix}$$

$$\begin{matrix} \\ -8R_1 + R_2 \\ -R_1 + R_3 \end{matrix} \begin{bmatrix} 1 & 1 & 1 & \vdots & 1{,}500{,}000 \\ 0 & 1 & 4 & \vdots & 1{,}300{,}000 \\ 0 & -1 & -5 & \vdots & -1{,}500{,}000 \end{bmatrix}$$

$$\begin{matrix} \\ \\ R_2 + R_3 \end{matrix} \begin{bmatrix} 1 & 1 & 1 & \vdots & 1{,}500{,}000 \\ 0 & 1 & 4 & \vdots & 1{,}300{,}000 \\ 0 & 0 & -1 & \vdots & -200{,}000 \end{bmatrix}$$

$$\begin{matrix} \\ \\ -R_3 \end{matrix} \begin{bmatrix} 1 & 1 & 1 & \vdots & 1{,}500{,}000 \\ 0 & 1 & 4 & \vdots & 1{,}300{,}000 \\ 0 & 0 & 1 & \vdots & 200{,}000 \end{bmatrix}$$

$z = 200{,}000$

$y + 4(200{,}000) = 1{,}300{,}000$

$y = 500{,}000$

$x + 500{,}000 + 200{,}000 = 1{,}500{,}000$

$x = 800{,}000$

\$800,000 at 8%, \$500,000 at 90%, \$200,000 at 12%

81. *Verbal Model:* $\boxed{\text{Theater A tickets}} + \boxed{\text{Theater B tickets}} + \boxed{\text{Theater C tickets}} = 1500$

$1.50\boxed{\text{Theater A tickets}} + 7.50\boxed{\text{Theater B tickets}} + 8.50\boxed{\text{Theater C tickets}} = 10{,}050$

$\boxed{\text{Theater B tickets}} = 2\boxed{\text{Theater A tickets}}$

$\boxed{\text{Theater C tickets}} = 2\boxed{\text{Theater A tickets}}$

Labels: $x =$ Theater A tickets

$y =$ Theater B tickets

$z =$ Theater C tickets

System:

$$\begin{aligned} x + y + z &= 1500 \\ 1.5x + 7.5y + 8.5z &= 10{,}050 \\ -2x + y &= 0 \\ -2x + z &= 0 \end{aligned}$$

$$\left[\begin{array}{ccc:c} 1 & 1 & 1 & 1500 \\ 1.5 & 7.5 & 8.5 & 10{,}050 \\ -2 & 1 & 0 & 0 \\ -2 & 0 & 1 & 0 \end{array}\right]$$

$$\begin{array}{r} \\ -1.5R_1 + R_2 \\ -R_4 + R_3 \\ 2R_1 + R_4 \end{array} \left[\begin{array}{ccc:c} 1 & 1 & 1 & 1500 \\ 0 & 6 & 7 & 7800 \\ 0 & 1 & -1 & 0 \\ 0 & 2 & 3 & 3000 \end{array}\right]$$

$$\begin{array}{r} \\ \frac{1}{6}R_2 \\ \\ -2R_3 + R_4 \end{array} \left[\begin{array}{ccc:c} 1 & 1 & 1 & 1500 \\ 0 & 1 & \frac{7}{6} & 1300 \\ 0 & 1 & -1 & 0 \\ 0 & 0 & 5 & 3000 \end{array}\right]$$

$$\begin{array}{r} \\ \\ \\ \frac{1}{5}R_4 \end{array} \left[\begin{array}{ccc:c} 1 & 1 & 1 & 1500 \\ 0 & 1 & \frac{7}{6} & 1300 \\ 0 & 1 & -1 & 0 \\ 0 & 0 & 1 & 600 \end{array}\right]$$

$z = 600$ $\qquad$ $y - 1(600) = 0$ $\qquad$ $x + 600 + 600 = 1500$

$y = 600$ $\qquad$ $x = 300$

Theater A: 300 tickets
Theater B: 600 tickets
Theater C: 600 tickets

83. *Verbal Model:* [Number 1] + [Number 2] + [Number 3] = [33]

[Number 2] = 3 + [Number 1]

[Number 3] = 4 · [Number 1]

Labels: Number 1 = x

Number 2 = y

Number 3 = z

System:

$$\begin{aligned} x + y + z &= 33 \\ y &= 3 + x \\ z &= 4x \end{aligned}$$

$$\begin{bmatrix} 1 & 1 & 1 & \vdots & 33 \\ -1 & 1 & 0 & \vdots & 3 \\ -4 & 0 & 1 & \vdots & 0 \end{bmatrix}$$

$$\begin{matrix} \\ R_1 + R_2 \\ 4R_1 + R_3 \end{matrix} \begin{bmatrix} 1 & 1 & 1 & \vdots & 33 \\ 0 & 2 & 1 & \vdots & 36 \\ 0 & 4 & 5 & \vdots & 132 \end{bmatrix}$$

$$\begin{matrix} \\ \frac{1}{2}R_2 \\ \\ \end{matrix} \begin{bmatrix} 1 & 1 & 1 & \vdots & 33 \\ 0 & 1 & \frac{1}{2} & \vdots & 18 \\ 0 & 4 & 5 & \vdots & 132 \end{bmatrix}$$

$$\begin{matrix} \\ \\ -4R_2 + R_3 \end{matrix} \begin{bmatrix} 1 & 1 & 1 & \vdots & 33 \\ 0 & 1 & \frac{1}{2} & \vdots & 18 \\ 0 & 0 & 3 & \vdots & 60 \end{bmatrix}$$

$$\begin{matrix} \\ \\ \frac{1}{3}R_3 \end{matrix} \begin{bmatrix} 1 & 1 & 1 & \vdots & 33 \\ 0 & 1 & \frac{1}{2} & \vdots & 18 \\ 0 & 0 & 1 & \vdots & 20 \end{bmatrix}$$

$z = 20$ $\qquad y + \frac{1}{2}(20) = 18$ $\qquad x + 8 + 20 = 33$

$y = 8$ $\qquad x = 5$

The three numbers are 5, 8, and 20.

85. *Verbal Model:*

	Computer chips	Resistors	Transistors
Copper	2	1	3
Zinc	2	3	2
Glass	1	2	2

Labels: Computer chips $= x$

Resistors $= y$

Transistors $= z$

System:

$$\begin{aligned} 2x + y + 3z &= 70 \\ 2x + 3y + 2z &= 80 \\ x + 2y + 2z &= 55 \end{aligned}$$

$$\begin{bmatrix} 2 & 1 & 3 & \vdots & 70 \\ 2 & 3 & 2 & \vdots & 80 \\ 1 & 2 & 2 & \vdots & 55 \end{bmatrix}$$

$$\begin{matrix} R_1 \\ \\ R_3 \end{matrix} \begin{bmatrix} 1 & 2 & 2 & \vdots & 55 \\ 2 & 3 & 2 & \vdots & 80 \\ 2 & 1 & 3 & \vdots & 70 \end{bmatrix}$$

$$\begin{matrix} \\ -2R_1 + R_2 \\ -2R_1 + R_3 \end{matrix} \begin{bmatrix} 1 & 2 & 2 & \vdots & 55 \\ 0 & -1 & -2 & \vdots & -30 \\ 0 & -3 & -1 & \vdots & -40 \end{bmatrix}$$

$$\begin{matrix} -R_2 \\ \\ \\ \end{matrix} \begin{bmatrix} 1 & 2 & 2 & \vdots & 55 \\ 0 & 1 & 2 & \vdots & 30 \\ 0 & -3 & -1 & \vdots & -40 \end{bmatrix}$$

$$\begin{matrix} -2R_2 + R_1 \\ 3R_2 + R_3 \\ \\ \end{matrix} \begin{bmatrix} 1 & 0 & -2 & \vdots & -5 \\ 0 & 1 & 2 & \vdots & 30 \\ 0 & 0 & 5 & \vdots & 50 \end{bmatrix}$$

$$\begin{matrix} \\ \\ \frac{1}{5}R_3 \end{matrix} \begin{bmatrix} 1 & 0 & -2 & \vdots & -5 \\ 0 & 1 & 2 & \vdots & 30 \\ 0 & 0 & 1 & \vdots & 10 \end{bmatrix}$$

$$\begin{matrix} 2R_3 + R_1 \\ -2R_3 + R_2 \\ \\ \end{matrix} \begin{bmatrix} 1 & 0 & 0 & \vdots & 15 \\ 0 & 1 & 0 & \vdots & 10 \\ 0 & 0 & 1 & \vdots & 10 \end{bmatrix}$$

15 computer chips, 10 resistors, 10 transistors

87. *Verbal Model:* $0.10 \cdot \boxed{\text{CDs}} + 0.08\boxed{\text{Bonds}} + 0.12\boxed{\text{BC stocks}} + 0.13\boxed{\text{G stocks}} = 50{,}000$

$\boxed{\text{BC stocks}} + \boxed{\text{G stocks}} = 125{,}000$

$\boxed{\text{CDs}} + \boxed{\text{Bonds}} = 375{,}000$

Labels: x = certificates of deposit

y = municipal bonds

z = blue-chip stocks

w = growth stocks

System:

$$\begin{aligned} 0.10x + 0.08y + 0.12z + 0.13w &= 50{,}000 \\ z + w &= 125{,}000 \\ x + y &= 375{,}000 \end{aligned}$$

$$\begin{bmatrix} 10 & 8 & 12 & 13 & \vdots & 5{,}000{,}000 \\ 0 & 0 & 1 & 1 & \vdots & 125{,}000 \\ 1 & 1 & 0 & 0 & \vdots & 375{,}000 \end{bmatrix}$$

$$\begin{matrix} R_1 \\ \\ R_3 \end{matrix} \begin{bmatrix} 1 & 1 & 0 & 0 & \vdots & 375{,}000 \\ 0 & 0 & 1 & 1 & \vdots & 125{,}000 \\ 10 & 8 & 12 & 13 & \vdots & 5{,}000{,}000 \end{bmatrix}$$

$$\begin{matrix} \\ \\ -10R_1 + R_3 \end{matrix} \begin{bmatrix} 1 & 1 & 0 & 0 & \vdots & 375{,}000 \\ 0 & 0 & 1 & 1 & \vdots & 125{,}000 \\ 0 & -2 & 12 & 13 & \vdots & 1{,}250{,}000 \end{bmatrix}$$

$$\begin{matrix} \\ \\ -\frac{1}{2}R_3 \end{matrix} \begin{bmatrix} 1 & 1 & 0 & 0 & \vdots & 375{,}000 \\ 0 & 0 & 1 & 1 & \vdots & 125{,}000 \\ 0 & 1 & -6 & -\frac{13}{2} & \vdots & -625{,}000 \end{bmatrix}$$

$$\begin{matrix} -R_3 + R_1 \\ \\ \\ \end{matrix} \begin{bmatrix} 1 & 0 & 6 & \frac{13}{2} & \vdots & 1{,}000{,}000 \\ 0 & 0 & 1 & 1 & \vdots & 125{,}000 \\ 0 & 1 & -6 & -\frac{13}{2} & \vdots & -625{,}000 \end{bmatrix}$$

$$\begin{matrix} -6R_2 + R_1 \\ \\ 6R_2 + R_3 \end{matrix} \begin{bmatrix} 1 & 1 & 0 & 0.5 & \vdots & 250{,}000 \\ 0 & 0 & 1 & 1 & \vdots & 125{,}000 \\ 0 & 1 & 0 & -0.5 & \vdots & 125{,}000 \end{bmatrix}$$

So let $w = s$, then:

$x + 0.5w = 250{,}000$ $\quad z + w = 125{,}000$ $\quad y - 0.5w = 125{,}000$

$x = -0.5s + 250{,}000$ $\quad z = -s + 125{,}000$ $\quad y = 0.5s + 125{,}000$

Certificates of deposit: $250{,}000 - 0.5s$

Municipal bonds: $125{,}000 + 0.5s$

Blue-chip stocks: $125{,}000 - s$

Growth stocks: s

If $s = \$100{,}000$:

CD = \$200,000

M Bonds = \$175,000

BC Stocks = \$25,000

G Stocks = \$100,000

89. The order of the matrix is 3 × 5. There are 15 entries in the matrix, so the order is 3 × 5, 5 × 3, or 15 × 1. Because there are more columns than rows, the second number in the order must be larger than the first.

91. The row-echelon form of an augmented matrix that corresponds to a system of linear equations that is inconsistent occurs when there is a row in the matrix with all zero entries except in the last column.

93. The first entry in the first column is one and the other two entries are zero. In the second column, the first entry is a nonzero real number, the second entry is one, and the third entry is zero. In the third column, the first two entries are nonzero real numbers and the third entry is one.

95. $6(-7) = -42$

97. $5(4) - 3(-2) = 20 + 6 = 26$

99. $\begin{cases} x = 4 \\ 3y + 2z = -4 \\ x + y + z = 3 \end{cases}$

$\begin{cases} x = 4 \\ -3x - z = -13 \\ x + y + z = 3 \end{cases}$

$x = 4$

$-3(4) - z = -13$

$-z = -1$

$z = 1$

$4 + y + 1 = 3$

$y = -2$

The solution is $(4, -2, 1)$.

Section 4.5 Determinants and Linear Systems

1. $\det(A) = \begin{vmatrix} 2 & 1 \\ 3 & 4 \end{vmatrix} = 2(4) - 3(1) = 8 - 3 = 5$

3. $\det(A) = \begin{vmatrix} 5 & 2 \\ -6 & 3 \end{vmatrix} = 5(3) - (-6)(2) = 15 + 12 = 27$

5. $\det(A) = \begin{vmatrix} -4 & 0 \\ 9 & 0 \end{vmatrix} = (-4)(0) - 9(0) = 0$

7. $\det(A) = \begin{vmatrix} 3 & -3 \\ -6 & 6 \end{vmatrix} = 3(6) - (-6)(-3)$

$= 18 - 18 = 0$

9. $\det(A) = \begin{vmatrix} -7 & 6 \\ \frac{1}{2} & 3 \end{vmatrix} = (-7)(3) - \left(\frac{1}{2}\right)(6)$

$= -21 - 3 = -24$

11. $\det(A) = \begin{vmatrix} 0.4 & 0.7 \\ 0.7 & 0.4 \end{vmatrix} = 0.4(0.4) - 0.7(0.7)$

$= 0.16 - 0.49 = -0.33$

13. $\det(A) = \begin{vmatrix} 2 & 3 & -1 \\ 6 & 0 & 0 \\ 4 & 1 & 1 \end{vmatrix}$

$= -(6)\begin{vmatrix} 3 & -1 \\ 1 & 1 \end{vmatrix} + 0 + 0$ (second row)

$= (-6)(4)$

$= -24$

15. $\det(A) = \begin{vmatrix} 1 & 1 & 2 \\ 3 & 1 & 0 \\ -2 & 0 & 3 \end{vmatrix} = (2)\begin{vmatrix} 3 & 1 \\ -2 & 0 \end{vmatrix} - (0)\begin{vmatrix} 1 & 1 \\ -2 & 0 \end{vmatrix} + (3)\begin{vmatrix} 1 & 1 \\ 3 & 1 \end{vmatrix}$ (third column)

$= (2)(2) - 0 + (3)(-2) = 4 - 6 = -2$

17. $\det(A) = \begin{vmatrix} 2 & 4 & 6 \\ 0 & 3 & 1 \\ 0 & 0 & -5 \end{vmatrix}$

$= (2)\begin{vmatrix} 3 & 1 \\ 0 & -5 \end{vmatrix} - 0 + 0$ (first column)

$= (2)(-15) = -30$

19. $\det(A) = \begin{vmatrix} -2 & 2 & 3 \\ 1 & -1 & 0 \\ 0 & 1 & 4 \end{vmatrix}$

$= -(1)\begin{vmatrix} 2 & 3 \\ 1 & 4 \end{vmatrix} + (-1)\begin{vmatrix} -2 & 3 \\ 0 & 4 \end{vmatrix} - 0$ (second row)

$= (-1)(5) + (-1)(-8) = -5 + 8 = 3$

21. $\det(A) = \begin{vmatrix} 1 & 4 & -2 \\ 3 & 6 & -6 \\ -2 & 1 & 4 \end{vmatrix} = (1)\begin{vmatrix} 6 & -6 \\ 1 & 4 \end{vmatrix} - (4)\begin{vmatrix} 3 & -6 \\ -2 & 4 \end{vmatrix} + (-2)\begin{vmatrix} 3 & 6 \\ -2 & 1 \end{vmatrix}$ (first row)

$= (1)(30) - (4)(0) + (-2)(15) = 30 - 0 - 30 = 0$

23. $\det(A) = \begin{vmatrix} 2 & -2 & 7 \\ 1 & -3 & -2 \\ -2 & 6 & 4 \end{vmatrix} = 2\begin{vmatrix} -3 & -2 \\ 6 & 4 \end{vmatrix} - 1\begin{vmatrix} -2 & 7 \\ 6 & 4 \end{vmatrix} - 2\begin{vmatrix} -2 & 7 \\ -3 & -2 \end{vmatrix}$ (first column)

$= 2(0) - 1(-50) - 2(25) = 0$

25. $\det(A) = \begin{vmatrix} 2 & -5 & 0 \\ 4 & 7 & 0 \\ -7 & 25 & 3 \end{vmatrix} = 0 - 0 + 3\begin{vmatrix} 2 & -5 \\ 4 & 7 \end{vmatrix}$ (third column)

$= (3)(34) = 102$

27. $\det(A) = \begin{vmatrix} 0.1 & 0.2 & 0.3 \\ -0.3 & 0.2 & 0.2 \\ 5 & 4 & 4 \end{vmatrix} = (5)\begin{vmatrix} 0.2 & 0.3 \\ 0.2 & 0.2 \end{vmatrix} - (4)\begin{vmatrix} 0.1 & 0.3 \\ -0.3 & 0.2 \end{vmatrix} + (4)\begin{vmatrix} 0.1 & 0.2 \\ -0.3 & 0.2 \end{vmatrix}$ (third row)

$= (5)(-0.02) - (4)(0.11) + (4)(0.08) = -0.1 - 0.44 + 0.32 = -0.22$

29. $\det(A) = \begin{vmatrix} x & y & 1 \\ 3 & 1 & 1 \\ -2 & 0 & 1 \end{vmatrix} = (-2)\begin{vmatrix} y & 1 \\ 1 & 1 \end{vmatrix} - 0 + (1)\begin{vmatrix} x & y \\ 3 & 1 \end{vmatrix}$ (third row)

$= (-2)(y - 1) + (1)(x - 3y) = -2y + 2 + x - 3y = x - 5y + 2$

31. Solution is 248.

33. Solution is 105.625.

35. Solution is 4.32.

37. $\left[\begin{array}{rr:r} 1 & 2 & 5 \\ -1 & 1 & 1 \end{array}\right]$

$D = \begin{vmatrix} 1 & 2 \\ -1 & 1 \end{vmatrix} = 1 - (-2) = 3$

$x = \dfrac{D_x}{D} = \dfrac{\begin{vmatrix} 5 & 2 \\ 1 & 1 \end{vmatrix}}{3} = \dfrac{5 - 2}{3} = \dfrac{3}{3} = 1$

$y = \dfrac{D_y}{D} = \dfrac{\begin{vmatrix} 1 & 5 \\ -1 & 1 \end{vmatrix}}{3} = \dfrac{1 - (-5)}{3} = \dfrac{6}{3} = 2$

The solution is $(1, 2)$.

39. $\left[\begin{array}{rr:r} 3 & 4 & -2 \\ 5 & 3 & 4 \end{array}\right]$

$D = \begin{vmatrix} 3 & 4 \\ 5 & 3 \end{vmatrix} = 9 - 20 = -11$

$x = \dfrac{D_x}{D} = \dfrac{\begin{vmatrix} -2 & 4 \\ 4 & 3 \end{vmatrix}}{-11} = \dfrac{-6 - 16}{-11} = \dfrac{-22}{-11} = 2$

$y = \dfrac{D_y}{D} = \dfrac{\begin{vmatrix} 3 & -2 \\ 5 & 4 \end{vmatrix}}{-11} = \dfrac{12 - (-10)}{-11} = \dfrac{22}{-11} = -2$

The solution is $(2, -2)$.

41. $\begin{bmatrix} 20 & 8 & \vdots & 11 \\ 12 & -24 & \vdots & 21 \end{bmatrix}$

$$D = \begin{vmatrix} 20 & 8 \\ 12 & -24 \end{vmatrix} = -480 - 96 = -576$$

$$x = \frac{D_x}{D} = \frac{\begin{vmatrix} 11 & 8 \\ 21 & -24 \end{vmatrix}}{-576} = \frac{-264 - 168}{-576} = \frac{-432}{-576} = \frac{3}{4}$$

$$y = \frac{D_y}{D} = \frac{\begin{vmatrix} 20 & 11 \\ 12 & 21 \end{vmatrix}}{-576} = \frac{420 - 132}{-576} = \frac{288}{-576} = -\frac{1}{2}$$

The solution is $\left(\frac{3}{4}, -\frac{1}{2}\right)$.

43. $\begin{bmatrix} -0.4 & 0.8 & \vdots & 1.6 \\ 2 & -4 & \vdots & 5 \end{bmatrix}$

$$D = \begin{vmatrix} -0.4 & 0.8 \\ 2 & -4 \end{vmatrix} = 1.6 - 1.6 = 0$$

Cannot be solved by Cramer's Rule because $D = 0$.

45. $\begin{bmatrix} 3 & 6 & \vdots & 5 \\ 6 & 14 & \vdots & 11 \end{bmatrix}$

$$D = \begin{vmatrix} 3 & 6 \\ 6 & 14 \end{vmatrix} = 42 - 36 = 6$$

$$x = \frac{D_x}{D} = \frac{\begin{vmatrix} 5 & 6 \\ 11 & 14 \end{vmatrix}}{6} = \frac{70 - 66}{6} = \frac{4}{6} = \frac{2}{3}$$

$$y = \frac{D_y}{D} = \frac{\begin{vmatrix} 3 & 5 \\ 6 & 11 \end{vmatrix}}{6} = \frac{33 - 30}{6} = \frac{3}{6} = \frac{1}{2}$$

The solution is $\left(\frac{2}{3}, \frac{1}{2}\right)$.

47. $\begin{bmatrix} 4 & -1 & 1 & \vdots & -5 \\ 2 & 2 & 3 & \vdots & 10 \\ 5 & -2 & 6 & \vdots & 1 \end{bmatrix}$

$$D = \begin{vmatrix} 4 & -1 & 1 \\ 2 & 2 & 3 \\ 5 & -2 & 6 \end{vmatrix} = (1)\begin{vmatrix} 2 & 2 \\ 5 & -2 \end{vmatrix} - (3)\begin{vmatrix} 4 & -1 \\ 5 & -2 \end{vmatrix} + (6)\begin{vmatrix} 4 & -1 \\ 2 & 2 \end{vmatrix} = (1)(-14) + (-3)(-3) + (6)(10) = -14 + 9 + 60 = 55$$

$$x = \frac{\begin{vmatrix} -5 & -1 & 1 \\ 10 & 2 & 3 \\ 1 & -2 & 6 \end{vmatrix}}{55} = \frac{(1)\begin{vmatrix} 10 & 2 \\ 1 & -2 \end{vmatrix} - (3)\begin{vmatrix} -5 & -1 \\ 1 & -2 \end{vmatrix} + (6)\begin{vmatrix} -5 & -1 \\ 10 & 2 \end{vmatrix}}{55} = \frac{(1)(-22) + (-3)(11) + (6)(0)}{55} = \frac{-22 - 33}{55} = \frac{-55}{55} = -1$$

$$y = \frac{\begin{vmatrix} 4 & -5 & 1 \\ 2 & 10 & 3 \\ 5 & 1 & 6 \end{vmatrix}}{55} = \frac{(1)\begin{vmatrix} 2 & 10 \\ 5 & 1 \end{vmatrix} - (3)\begin{vmatrix} 4 & -5 \\ 5 & 1 \end{vmatrix} + (6)\begin{vmatrix} 4 & -5 \\ 2 & 10 \end{vmatrix}}{55} = \frac{(1)(-48) + (-3)(29) + (6)(50)}{55} = \frac{-48 - 87 + 300}{55} = \frac{165}{55} = 3$$

$$z = \frac{\begin{vmatrix} 4 & -1 & -5 \\ 2 & 2 & 10 \\ 5 & -2 & 1 \end{vmatrix}}{55} = \frac{(5)\begin{vmatrix} -1 & -5 \\ 2 & 10 \end{vmatrix} - (-2)\begin{vmatrix} 4 & -5 \\ 2 & 10 \end{vmatrix} + (1)\begin{vmatrix} 4 & -1 \\ 2 & 2 \end{vmatrix}}{55} = \frac{(5)(0) + (2)(50) + (1)(10)}{55} = \frac{0 + 100 + 10}{55} = \frac{110}{55} = 2$$

The solution is $(-1, 3, 2)$.

49. $\begin{bmatrix} 4 & 3 & 4 & \vdots & 1 \\ 4 & -6 & 8 & \vdots & 8 \\ -1 & 9 & -2 & \vdots & -7 \end{bmatrix}$

$$D = \begin{vmatrix} 4 & 3 & 4 \\ 4 & -6 & 8 \\ -1 & 9 & -2 \end{vmatrix} = 4\begin{vmatrix} -6 & 8 \\ 9 & -2 \end{vmatrix} - 4\begin{vmatrix} 3 & 4 \\ 9 & -2 \end{vmatrix} - 1\begin{vmatrix} 3 & 4 \\ -6 & 8 \end{vmatrix} = 4(-60) - 4(-42) - 1(48) = -240 + 168 - 48 = -120$$

$$a = \frac{\begin{vmatrix} 1 & 3 & 4 \\ 8 & -6 & 8 \\ -7 & 9 & -2 \end{vmatrix}}{-120} = \frac{1\begin{vmatrix} -6 & 8 \\ 9 & -2 \end{vmatrix} - 3\begin{vmatrix} 8 & 8 \\ -7 & -2 \end{vmatrix} + 4\begin{vmatrix} 8 & -6 \\ -7 & 9 \end{vmatrix}}{-120} = \frac{1(-60) - 3(40) + 4(30)}{-120} = \frac{-60 - 120 + 120}{-120} = \frac{-60}{-120} = \frac{1}{2}$$

$$b = \frac{\begin{vmatrix} 4 & 1 & 4 \\ 4 & 8 & 8 \\ -1 & -7 & -2 \end{vmatrix}}{-120} = \frac{4\begin{vmatrix} 8 & 8 \\ -7 & -2 \end{vmatrix} - 1\begin{vmatrix} 4 & 8 \\ -1 & -2 \end{vmatrix} + 4\begin{vmatrix} 4 & 8 \\ -1 & -7 \end{vmatrix}}{-120} = \frac{4(40) - 1(0) + 4(-20)}{-120} = \frac{160 - 0 - 80}{-120} = \frac{80}{-120} = -\frac{2}{3}$$

$$c = \frac{\begin{vmatrix} 4 & 3 & 1 \\ 4 & -6 & 8 \\ -1 & 9 & -7 \end{vmatrix}}{-120} = \frac{4\begin{vmatrix} -6 & 8 \\ 9 & -7 \end{vmatrix} - 4\begin{vmatrix} 3 & 1 \\ 9 & -7 \end{vmatrix} - 1\begin{vmatrix} 3 & 1 \\ -6 & 8 \end{vmatrix}}{-120} = \frac{4(-30) - 4(-30) - 1(30)}{-120} = \frac{-120 + 120 - 30}{-120} = \frac{-30}{-120} = \frac{1}{4}$$

The solution is $\left(\frac{1}{2}, -\frac{2}{3}, \frac{1}{4}\right)$.

51. $\begin{bmatrix} 5 & -3 & 2 & \vdots & 2 \\ 2 & 2 & -3 & \vdots & 3 \\ 1 & -7 & 8 & \vdots & -4 \end{bmatrix}$

$$D = \begin{vmatrix} 5 & -3 & 2 \\ 2 & 2 & -3 \\ 1 & 7 & 8 \end{vmatrix} = (5)\begin{vmatrix} 2 & -3 \\ -7 & 8 \end{vmatrix} - (2)\begin{vmatrix} -3 & 2 \\ -7 & 8 \end{vmatrix} + (1)\begin{vmatrix} -3 & 2 \\ 2 & -3 \end{vmatrix} = (5)(-5) - (2)(-10) + (1)(5) = -25 + 20 + 5 = 0$$

Cannot be solved by Cramer's Rule because $D = 0$.

53. $\begin{bmatrix} -3 & 10 & \vdots & 22 \\ 9 & -3 & \vdots & 0 \end{bmatrix}$

$$D = \begin{vmatrix} -3 & 10 \\ 9 & -3 \end{vmatrix} = -81$$

$$x = \frac{D_x}{D} = \frac{\begin{vmatrix} 22 & 10 \\ 0 & -3 \end{vmatrix}}{-81} = \frac{-66}{-81} = \frac{22}{27}$$

$$y = \frac{D_y}{D} = \frac{\begin{vmatrix} -3 & 22 \\ 9 & 0 \end{vmatrix}}{-81} = \frac{-198}{-81} = \frac{22}{9}$$

55. $D = \begin{vmatrix} 3 & -2 & 3 \\ 1 & 3 & 6 \\ 1 & 2 & 9 \end{vmatrix} = 48$

$$x = \frac{D_x}{D} = \frac{\begin{vmatrix} 8 & -2 & 3 \\ -3 & 3 & 6 \\ -5 & 2 & 9 \end{vmatrix}}{48} = \frac{153}{48} = \frac{51}{16}$$

$$y = \frac{D_y}{D} = \frac{\begin{vmatrix} 3 & 8 & 3 \\ 1 & -3 & 6 \\ 1 & -5 & 9 \end{vmatrix}}{48} = \frac{-21}{48} = \frac{-7}{16}$$

$$z = \frac{D_z}{D} = \frac{\begin{vmatrix} 3 & -2 & 8 \\ 1 & 3 & -3 \\ 1 & 2 & -5 \end{vmatrix}}{48} = \frac{-39}{48} = -\frac{13}{16}$$

$\left(\frac{51}{16}, -\frac{7}{16}, -\frac{13}{16}\right)$

57. $\begin{vmatrix} -3x & x \\ 4 & 3 \end{vmatrix} = 26$

$$\begin{aligned} -3x(3) - 4x &= 26 \\ -9x - 4x &= 26 \\ -13x &= 26 \\ x &= -2 \end{aligned}$$

59. $(x_1, y_1) = (0, 3), (x_2, y_2) = (4, 0), (x_3, y_3) = (8, 5)$

$$\begin{vmatrix} x_1 & y_1 & 1 \\ x_2 & y_2 & 1 \\ x_3 & y_3 & 1 \end{vmatrix} = \begin{vmatrix} 0 & 3 & 1 \\ 4 & 0 & 1 \\ 8 & 5 & 1 \end{vmatrix} = 32$$

$\text{Area} = +\frac{1}{2}(32) = 16$

61. $(x_1, y_1) = (-3, 4), (x_2, y_2) = (1, -2), (x_3, y_3) = (6, 1)$

$$\begin{aligned} \begin{vmatrix} x_1 & y_1 & 1 \\ x_2 & y_2 & 1 \\ x_3 & y_3 & 1 \end{vmatrix} &= \begin{vmatrix} -3 & 4 & 1 \\ 1 & -2 & 1 \\ 6 & 1 & 1 \end{vmatrix} \\ &= -3\begin{vmatrix} -2 & 1 \\ 1 & 1 \end{vmatrix} - 1\begin{vmatrix} 4 & 1 \\ 1 & 1 \end{vmatrix} + 6\begin{vmatrix} 4 & 1 \\ -2 & 1 \end{vmatrix} \\ &= -3(-3) - 1(3) + 6(6) \\ &= 9 - 3 + 36 = 42 \end{aligned}$$

$\text{Area} = +\frac{1}{2}(42) = 21$

63. $(x_1, y_1) = (-2, 1), (x_2, y_2) = (3, -1), (x_3, y_3) = (1, 6)$

$$\begin{vmatrix} x_1 & y_1 & 1 \\ x_2 & y_2 & 1 \\ x_3 & y_3 & 1 \end{vmatrix} = \begin{vmatrix} -2 & 1 & 1 \\ 3 & -1 & 1 \\ 1 & 6 & 1 \end{vmatrix} = (1)\begin{vmatrix} 3 & -1 \\ 1 & 6 \end{vmatrix} - (1)\begin{vmatrix} -2 & 1 \\ 1 & 6 \end{vmatrix} + (1)\begin{vmatrix} -2 & 1 \\ 3 & -1 \end{vmatrix} = (1)(19) - (1)(-13) + (1)(-1) = 19 + 13 - 1 = 31$$

$\text{Area} = +\frac{1}{2}(31) = \frac{31}{2} \text{ or } 15\frac{1}{2}$

65. $(x_1, y_1) = \left(0, \frac{1}{2}\right), (x_2, y_2) = \left(\frac{5}{2}, 0\right), (x_3, y_3) = (4, 3)$

$$\begin{aligned} \begin{vmatrix} x_1 & y_1 & 1 \\ x_2 & y_2 & 1 \\ x_3 & y_3 & 1 \end{vmatrix} = \begin{vmatrix} 0 & \frac{1}{2} & 1 \\ \frac{5}{2} & 0 & 1 \\ 4 & 3 & 1 \end{vmatrix} &= 0\begin{vmatrix} 0 & 1 \\ 3 & 1 \end{vmatrix} - \frac{1}{2}\begin{vmatrix} \frac{5}{2} & 1 \\ 4 & 1 \end{vmatrix} + 1\begin{vmatrix} \frac{5}{2} & 0 \\ 4 & 3 \end{vmatrix} \\ &= 0 - \frac{1}{2}\left(\frac{5}{2} - 4\right) + 1\left(\frac{15}{2} - 0\right) \\ &= -\frac{1}{2}\left(-\frac{3}{2}\right) + 1\left(\frac{15}{2}\right) \\ &= \frac{3}{4} + \frac{15}{2} \\ &= \frac{3}{4} + \frac{30}{4} \\ &= \frac{33}{4} \end{aligned}$$

$\text{Area} = \frac{1}{2}\left(\frac{33}{4}\right) = \frac{33}{8}$

67. Let $(x_1, y_1) = (-1, 2), (x_2, y_2) = (4, 0), (x_3, y_3) = (3, 5)$.

$$\begin{vmatrix} x_1 & y_1 & 1 \\ x_2 & y_2 & 1 \\ x_3 & y_3 & 1 \end{vmatrix} = \begin{vmatrix} -1 & 2 & 1 \\ 4 & 0 & 1 \\ 3 & 5 & 1 \end{vmatrix} = -4\begin{vmatrix} 2 & 1 \\ 5 & 1 \end{vmatrix} + 0 - 1\begin{vmatrix} -1 & 2 \\ 3 & 5 \end{vmatrix} = -4(-3) - 1(-11) = 12 + 11 = 23$$

$\text{Area} = \frac{1}{2}(23) = 11.5$

Let $(x_1, y_1) = (3, 5), (x_2, y_2) = (4, 0), (x_3, y_3) = (5, 4)$.

$$\begin{vmatrix} x_1 & y_1 & 1 \\ x_2 & y_2 & 1 \\ x_3 & y_3 & 1 \end{vmatrix} = \begin{vmatrix} 3 & 5 & 1 \\ 4 & 0 & 1 \\ 5 & 4 & 1 \end{vmatrix} = -4\begin{vmatrix} 5 & 1 \\ 4 & 1 \end{vmatrix} + 0 - 1\begin{vmatrix} 3 & 5 \\ 5 & 4 \end{vmatrix} = -4(1) - 1(-13) = -4 + 13 = 9$$

$\text{Area} = \frac{1}{2}(9) = 4.5$

Verbal Model: [Area of Shaded Region] = [Area of Triangle 1] + [Area of Triangle 2]

Equation: $A = 11.5 + 4.5 = 16$

69. Let $(x_1, y_1) = (-3, -1), (x_2, y_2) = (2, -2), (x_3, y_3) = (1, 2)$.

$$\begin{vmatrix} x_1 & y_1 & 1 \\ x_2 & y_2 & 1 \\ x_3 & y_3 & 1 \end{vmatrix} = \begin{vmatrix} -3 & -1 & 1 \\ 2 & -2 & 1 \\ 1 & 2 & 1 \end{vmatrix} = 19$$

$\text{Area} = \frac{1}{2}(19) = 9.5$

Verbal Model: [Area of Shaded Region] = [Area of Rectangle] − [Area of Triangle]

Equation: $A = (9)(4) - 9.5 = 36 - 9.5 = 26.5$

71. From the diagram the coordinates of A, B, and C are determined to be $A(0, 20)$, $B(10, -5)$ and $C(28, 0)$.

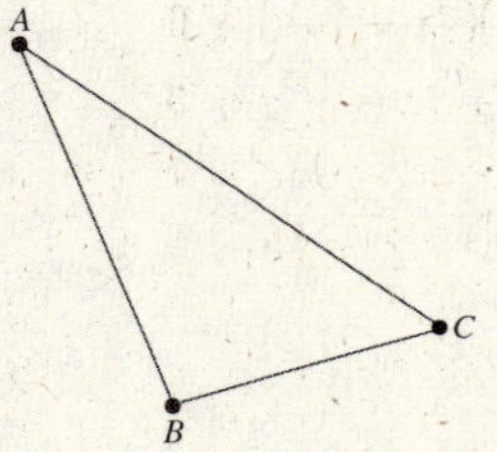

$$\begin{vmatrix} x_1 & y_1 & 1 \\ x_2 & y_2 & 1 \\ x_3 & y_3 & 1 \end{vmatrix} = \begin{vmatrix} 0 & 20 & 1 \\ 10 & -5 & 1 \\ 28 & 0 & 1 \end{vmatrix} = -500$$

$\text{Area} = -\frac{1}{2}(-500) = 250 \text{ mi}^2$

73. Let $(x_1, y_1) = (-1, 11), (x_2, y_2) = (0, 8), (x_3, y_3) = (2, 2)$.

$$\begin{vmatrix} x_1 & y_1 & 1 \\ x_2 & y_2 & 1 \\ x_3 & y_3 & 1 \end{vmatrix} = \begin{vmatrix} -1 & 11 & 1 \\ 0 & 8 & 1 \\ 2 & 2 & 1 \end{vmatrix} = (-1)\begin{vmatrix} 8 & 1 \\ 2 & 1 \end{vmatrix} + 0 + (2)\begin{vmatrix} 11 & 1 \\ 8 & 1 \end{vmatrix} = (-1)(6) + (2)(3) = -6 + 6 = 0$$

The three points are collinear.

75. Let $(x_1, y_1) = (2, -4), (x_2, y_2) = (5, 2)$, and $(x_3, y_3) = (10, 10)$.

$$\begin{vmatrix} x_1 & y_1 & 1 \\ x_2 & y_2 & 1 \\ x_3 & y_3 & 1 \end{vmatrix} = \begin{vmatrix} 2 & -4 & 1 \\ 5 & 2 & 1 \\ 10 & 10 & 1 \end{vmatrix} = 1\begin{vmatrix} 5 & 2 \\ 10 & 10 \end{vmatrix} - 1\begin{vmatrix} 2 & -4 \\ 10 & 10 \end{vmatrix} + 1\begin{vmatrix} 2 & -4 \\ 5 & 2 \end{vmatrix} = 1(30) - 1(60) + 1(24) = 30 - 60 + 24 = -6$$

The three points are not collinear.

77. Let $(x_1, y_1) = \left(-2, \frac{1}{3}\right), (x_2, y_2) = (2, 1), (x_3, y_3) = \left(3, \frac{1}{5}\right)$.

$$\begin{vmatrix} x_1 & y_1 & 1 \\ x_2 & y_2 & 1 \\ x_3 & y_3 & 1 \end{vmatrix} = \begin{vmatrix} -2 & \frac{1}{3} & 1 \\ 2 & 1 & 1 \\ 3 & \frac{1}{5} & 1 \end{vmatrix} = (1)\begin{vmatrix} 2 & 1 \\ 3 & \frac{1}{5} \end{vmatrix} - (1)\begin{vmatrix} -2 & \frac{1}{3} \\ 3 & \frac{1}{5} \end{vmatrix} + (1)\begin{vmatrix} -2 & \frac{1}{3} \\ 2 & 1 \end{vmatrix}$$

$$= (1)\left(-\frac{13}{5}\right) - (1)\left(-\frac{7}{5}\right) + (1)\left(-\frac{8}{3}\right) = -\frac{13}{5} + \frac{7}{5} - \frac{8}{3} = -\frac{18}{15} - \frac{40}{15} = -\frac{58}{15}$$

The three points are not collinear.

79. Let $(x_1, y_1) = (-2, -1)$ and $(x_2, y_2) = (4, 2)$.

$$\begin{vmatrix} x & y & 1 \\ -2 & -1 & 1 \\ 4 & 2 & 1 \end{vmatrix} = 0$$

$$1\begin{vmatrix} -2 & -1 \\ 4 & 2 \end{vmatrix} - 1\begin{vmatrix} x & y \\ 4 & 2 \end{vmatrix} + 1\begin{vmatrix} x & y \\ -2 & -1 \end{vmatrix} = 0$$

$$1(0) - (2x - 4y) + (-x + 2y) = 0$$

$$-2x + 4y - x + 2y = 0$$

$$-3x + 6y = 0$$

$$x - 2y = 0$$

81. $(x_1, y_1) = (10, 7), (x_2, y_2) = (-2, -7)$

$$\begin{vmatrix} x & y & 1 \\ 10 & 7 & 1 \\ -2 & -7 & 1 \end{vmatrix} = 0$$

$$(1)\begin{vmatrix} 10 & 7 \\ -2 & -7 \end{vmatrix} - (1)\begin{vmatrix} x & y \\ -2 & -7 \end{vmatrix} + (1)\begin{vmatrix} x & y \\ 10 & 7 \end{vmatrix} = 0$$

$$(1)(-56) - (-7x + 2y) + (1)(7x - 10y) = 0$$

$$-56 + 7x - 2y + 7x - 10y = 0$$

$$14x - 12y - 56 = 0$$

$$7x - 6y - 28 = 0$$

83. $(x_1, y_1) = \left(-2, \frac{3}{2}\right), (x_2, y_2) = (3, -3)$

$$\begin{vmatrix} x & y & 1 \\ -2 & \frac{3}{2} & 1 \\ 3 & -3 & 1 \end{vmatrix} = 0$$

$$x\begin{vmatrix} \frac{3}{2} & 1 \\ -3 & 1 \end{vmatrix} - y\begin{vmatrix} -2 & 1 \\ 3 & 1 \end{vmatrix} + 1\begin{vmatrix} -2 & \frac{3}{2} \\ 3 & -3 \end{vmatrix} = 0$$

$$\frac{9}{2}x + 5y + \frac{3}{2} = 0$$

$$9x + 10y + 3 = 0$$

85. $(x_1, y_1) = (2, 3.6), (x_2, y_2) = (8, 10)$

$$\begin{vmatrix} x & y & 1 \\ 2 & 3.6 & 1 \\ 8 & 10 & 1 \end{vmatrix} = 0$$

$$x\begin{vmatrix} 3.6 & 1 \\ 10 & 1 \end{vmatrix} - y\begin{vmatrix} 2 & 1 \\ 8 & 1 \end{vmatrix} + 1\begin{vmatrix} 2 & 3.6 \\ 8 & 10 \end{vmatrix} = 0$$

$$x(3.6 - 10) - y(2 - 8) + 1(20 - 28.8) = 0$$

$$-6.4x + 6y - 8.8 = 0$$

$$-3.2x + 3y - 4.4 = 0$$

$$32x - 30y + 44 = 0$$

$$16x - 15y + 22 = 0$$

87. $\begin{bmatrix} 1 & -1 & 1 & \vdots & 0 \\ 3 & 2 & 0 & \vdots & 7 \\ 0 & 2 & 4 & \vdots & 8 \end{bmatrix}$

$$D = \begin{vmatrix} 1 & -1 & 1 \\ 3 & 2 & 0 \\ 0 & 2 & 4 \end{vmatrix} = 1\begin{vmatrix} 2 & 0 \\ 2 & 4 \end{vmatrix} - 3\begin{vmatrix} -1 & 1 \\ 2 & 4 \end{vmatrix} + 0\begin{vmatrix} -1 & 1 \\ 2 & 0 \end{vmatrix} = (1)(8) - 3(-6) + 0 = 8 + 18 = 26$$

$$I_1 = \frac{\begin{vmatrix} 0 & -1 & 1 \\ 7 & 2 & 0 \\ 8 & 2 & 4 \end{vmatrix}}{26} = \frac{0\begin{vmatrix} 2 & 0 \\ 2 & 4 \end{vmatrix} - 7\begin{vmatrix} -1 & 1 \\ 2 & 4 \end{vmatrix} + 8\begin{vmatrix} -1 & 1 \\ 2 & 0 \end{vmatrix}}{26} = \frac{0 - (7)(-6) + (8)(-2)}{26} = \frac{42 - 16}{26} = \frac{26}{26} = 1$$

$$I_2 = \frac{\begin{vmatrix} 1 & 0 & 1 \\ 3 & 7 & 0 \\ 0 & 8 & 4 \end{vmatrix}}{26} = \frac{1\begin{vmatrix} 7 & 0 \\ 8 & 4 \end{vmatrix} - 3\begin{vmatrix} 0 & 1 \\ 8 & 4 \end{vmatrix} + 0\begin{vmatrix} 0 & 1 \\ 7 & 0 \end{vmatrix}}{26} = \frac{(1)(28) - (3)(-8) + 0}{26} = \frac{28 + 24}{26} = \frac{52}{26} = 2$$

$$I_3 = \frac{\begin{vmatrix} 1 & -1 & 0 \\ 3 & 2 & 7 \\ 0 & 2 & 8 \end{vmatrix}}{26} = \frac{1\begin{vmatrix} 2 & 7 \\ 2 & 8 \end{vmatrix} - 3\begin{vmatrix} -1 & 0 \\ 2 & 8 \end{vmatrix} + 0\begin{vmatrix} -1 & 0 \\ 2 & 7 \end{vmatrix}}{26} = \frac{(1)(2) - 3(-8) + 0}{26} = \frac{2 + 24}{26} = \frac{26}{26} = 1$$

89. $\begin{bmatrix} 1 & 1 & -1 & \vdots & 0 \\ 1 & 0 & 2 & \vdots & 12 \\ 1 & -2 & 0 & \vdots & -4 \end{bmatrix}$

$$D = \begin{vmatrix} 1 & 1 & -1 \\ 1 & 0 & 2 \\ 1 & -2 & 0 \end{vmatrix} = 1\begin{vmatrix} 0 & 2 \\ -2 & 0 \end{vmatrix} - 1\begin{vmatrix} 1 & -1 \\ -2 & 0 \end{vmatrix} + 1\begin{vmatrix} 1 & -1 \\ 0 & 2 \end{vmatrix} = (1)(4) - (1)(-2) + (1)(2) = 4 + 2 + 2 = 8$$

$$I_1 = \frac{\begin{vmatrix} 0 & 1 & -1 \\ 12 & 0 & 2 \\ -4 & -2 & 0 \end{vmatrix}}{8} = \frac{0\begin{vmatrix} 0 & 2 \\ -2 & 0 \end{vmatrix} - 12\begin{vmatrix} 1 & -1 \\ -2 & 0 \end{vmatrix} + (-4)\begin{vmatrix} 1 & -1 \\ 0 & 2 \end{vmatrix}}{8} = \frac{0 - (12)(-2) - (4)(2)}{8} = \frac{24 - 8}{8} = \frac{16}{8} = 2$$

$$I_2 = \frac{\begin{vmatrix} 1 & 0 & -1 \\ 1 & 12 & 2 \\ 1 & -4 & 0 \end{vmatrix}}{8} = \frac{1\begin{vmatrix} 12 & 2 \\ -4 & 0 \end{vmatrix} - 1\begin{vmatrix} 0 & -1 \\ -4 & 0 \end{vmatrix} + 1\begin{vmatrix} 0 & -1 \\ 12 & 2 \end{vmatrix}}{8} = \frac{(1)(8) - (1)(-4) + 1(12)}{8} = \frac{8 + 4 + 12}{8} = \frac{24}{8} = 3$$

$$I_3 = \frac{\begin{vmatrix} 1 & 1 & 0 \\ 1 & 0 & 12 \\ 1 & -2 & -4 \end{vmatrix}}{8} = \frac{1\begin{vmatrix} 0 & 12 \\ -2 & -4 \end{vmatrix} - 1\begin{vmatrix} 1 & 0 \\ -2 & -4 \end{vmatrix} + 1\begin{vmatrix} 1 & 0 \\ 0 & 12 \end{vmatrix}}{8} = \frac{(1)(24) - (1)(-4) + (1)(12)}{8} = \frac{24 + 4 + 12}{8} = \frac{40}{8} = 5$$

91. (a) $\begin{bmatrix} k & 3k & \vdots & 2 \\ 2+k & k & \vdots & 5 \end{bmatrix}$

$$D = \begin{vmatrix} k & 3k \\ 2+k & k \end{vmatrix} = k^2 - 3k(2+k)$$

$$= k^2 - 6k - 3k^2 = -2k^2 - 6k$$

$$x = \frac{D_x}{D} = \frac{\begin{vmatrix} 2 & 3k \\ 5 & k \end{vmatrix}}{-2k^2 - 6k}$$

$$= \frac{2k - 15k}{-2k^2 - 6k} = \frac{-13k}{-2k(k+3)}$$

$$= \frac{13}{2(k+3)} = \frac{13}{2k+6}$$

$$y = \frac{D_y}{D} = \frac{\begin{vmatrix} k & 2 \\ 2+k & 5 \end{vmatrix}}{-2k^2 - 6k}$$

$$= \frac{5k - 2(2+k)}{-2k^2 - 6k} = \frac{5k - 4 - 2k}{-2k^2 - 6k}$$

$$= \frac{3k - 4}{-2k^2 - 6k} = \frac{-1(4 - 3k)}{-1(2k^2 + 6k)}$$

$$= \frac{4 - 3k}{2k^2 + 6k}$$

(b) $-2k^2 - 6k = 0$

$-2k(k + 3) = 0$

$-2k = 0 \qquad k + 3 = 0$

$k = 0 \qquad k = -3$

93. A determinant is a real number associated with a square matrix.

95. The determinant is zero. Because two rows are identical, each term is zero when expanding by minors along the other row. Therefore, the sum is zero.

97. $4x - 2y < 0$

$-2y < -4x$

$y > 2x$

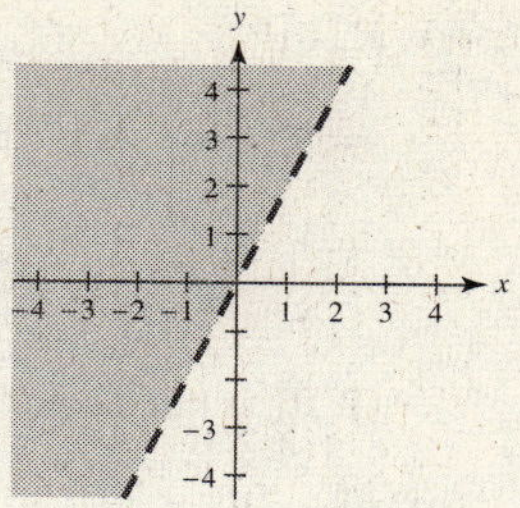

99. $-x + 3y > 12$

$3y > x + 12$

$y > \frac{1}{3}x + 4$

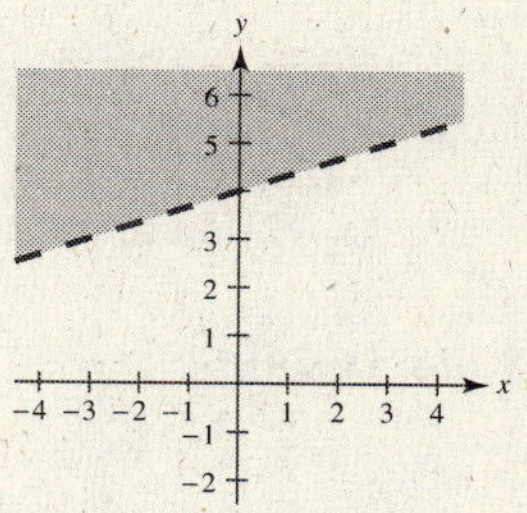

101. The graph of $h(x)$ has a vertical shift c units upward.

103. $f(x) - 2$

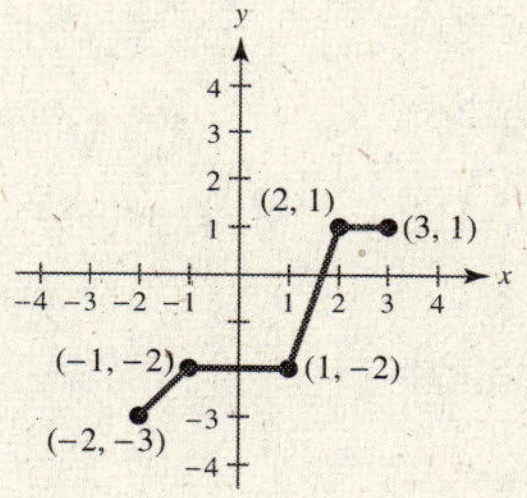

105. $f(-x)$

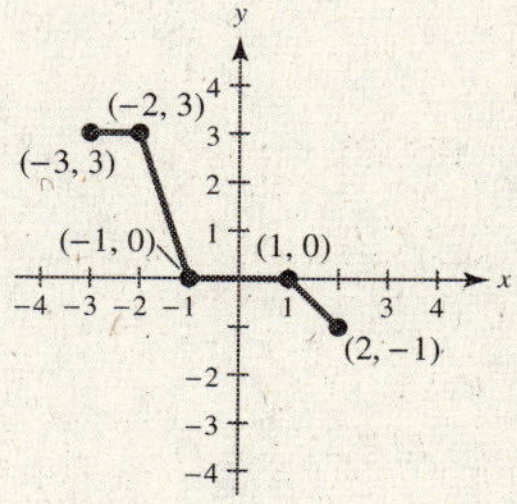

Section 4.6 Systems of Linear Inequalities

1. (d)

2. (b)

3. (f)

4. (c)

5. (a)

6. (e)

7. (a) $(2, 0)$:

$$\begin{array}{ll} 2x - y > 4 & x + 3y \le 6 \\ 2(2) - 0 \overset{?}{>} 4 & 2 + 3(0) \overset{?}{\le} 6 \\ 4 \ngtr 4 & 2 \le 6 \end{array}$$

$(2, 0)$ is not a solution.

(b) $(4, -2)$:

$$\begin{array}{ll} 2x - y > 4 & x + 3y \le 6 \\ 2(4) - (-2) \overset{?}{>} 4 & 4 + 3(-2) \overset{?}{\le} 6 \\ 10 > 4 & -2 \le 6 \end{array}$$

$(4, -2)$ is a solution.

9. (a) $(-3, 4)$:

$$\begin{array}{ll} -x + y < -2 & 4x + y < -3 \\ -(-3) + 4 \overset{?}{<} -2 & 4(-3) + 4 \overset{?}{<} -3 \\ 7 \nless -2 & -8 < -3 \end{array}$$

$(-3, 4)$ is not a solution.

(b) $(-1, -3)$:

$$\begin{array}{ll} -x + y < -2 & 4x + y < -3 \\ -(-1) + (-3) \overset{?}{<} -2 & 4(-1) + (-3) \overset{?}{<} -3 \\ -2 \nless -2 & -7 < -3 \end{array}$$

$(-1, -3)$ is not a solution.

11. $\begin{cases} x < 3 \\ x > -2 \end{cases}$

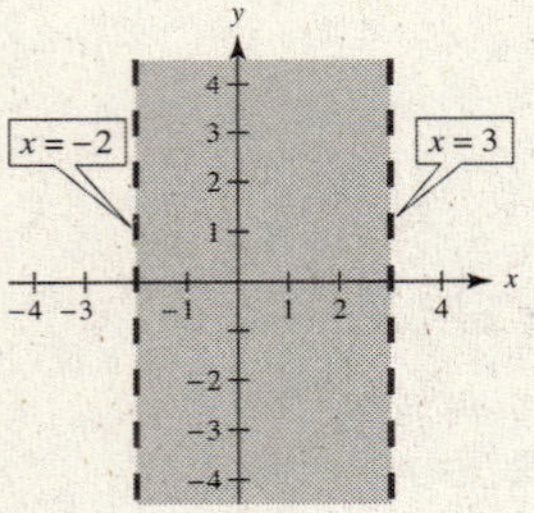

Dotted lines at $x = 3$ and $x = -2$. Shade to the right of $x = -2$ and to the left of $x = 3$.

13. $\begin{cases} x + y \le 3 \\ x - 1 \le 1 \end{cases} \Rightarrow \begin{array}{l} y \le -x + 3 \\ x \le 2 \end{array}$

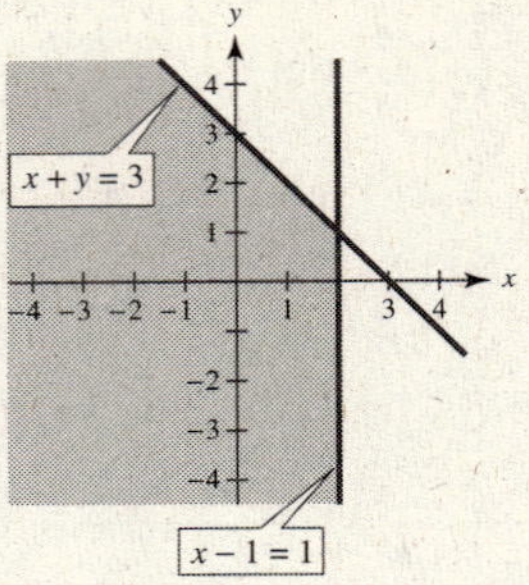

Solid lines at $x = 2$ and $y = -x + 3$. Shade to the left of $x = 2$ and below $y = -x + 3$.

15. $\begin{cases} 2x - 4y \le 6 \\ x + y \ge 2 \end{cases} \Rightarrow \begin{array}{l} y \ge \frac{1}{2}x - \frac{3}{2} \\ y \ge -x + 2 \end{array}$

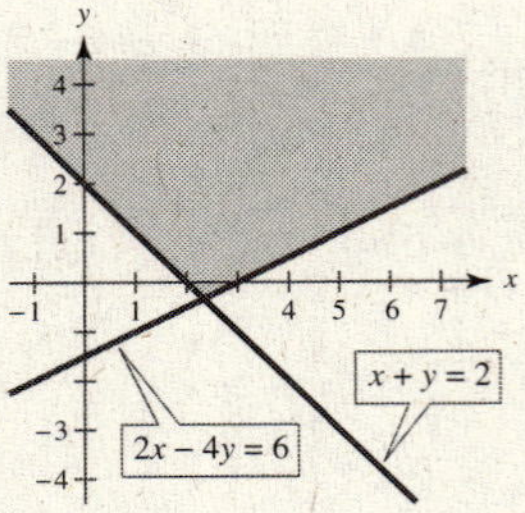

Solid lines at $y = \frac{1}{2}x - \frac{3}{2}$ and $y = -x + 2$. Shade above each line.

17. $\begin{cases} x + 2y \le 6 \\ x - 2y \le 0 \end{cases} \Rightarrow \begin{array}{l} y \le -\frac{1}{2}x + 3 \\ y \ge \frac{1}{2}x \end{array}$

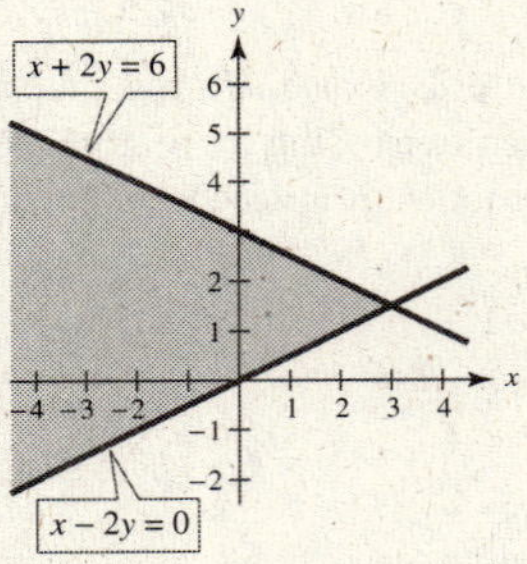

Solid lines at $y = -\frac{1}{2}x + 3$ and $y = \frac{1}{2}x$. Shade below $y = -\frac{1}{2}x + 3$ and above $y = \frac{1}{2}x$.

19. $\begin{cases} x - 2y > 4 \\ 2x + y > 6 \end{cases} \Rightarrow \begin{matrix} y < \frac{1}{2}x - 2 \\ y > -2x + 6 \end{matrix}$

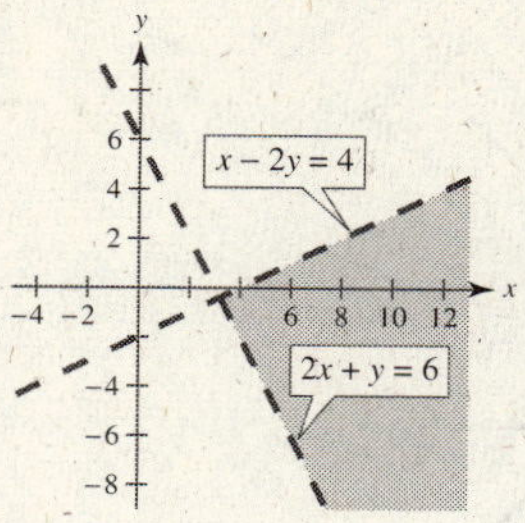

Dotted lines at $y = \frac{1}{2}x - 2$ and $y = -2x + 6$. Shade below $y = \frac{1}{2}x - 2$ and above $y = -2x + 6$.

21. $\begin{cases} x + y > -1 \\ x + y < 3 \end{cases} \Rightarrow \begin{matrix} y > -x - 1 \\ y < -x + 3 \end{matrix}$

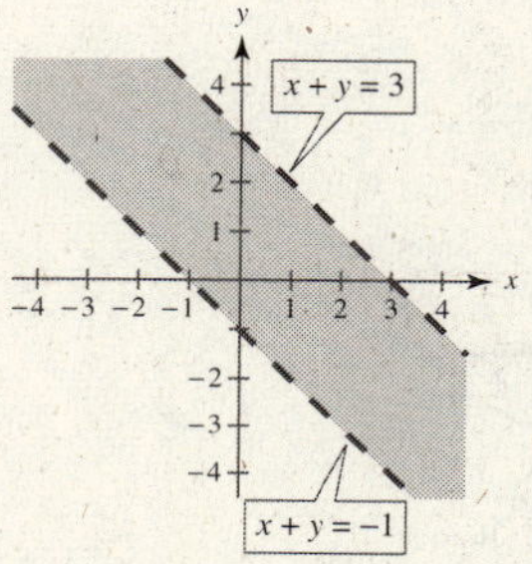

Dotted lines at $y = -x - 1$ and $y = -x + 3$. Shade above $y = -x - 1$ and below $y = -x + 3$.

23. $\begin{cases} y \ge \frac{4}{3}x + 1 \\ y \le 5x - 2 \end{cases}$

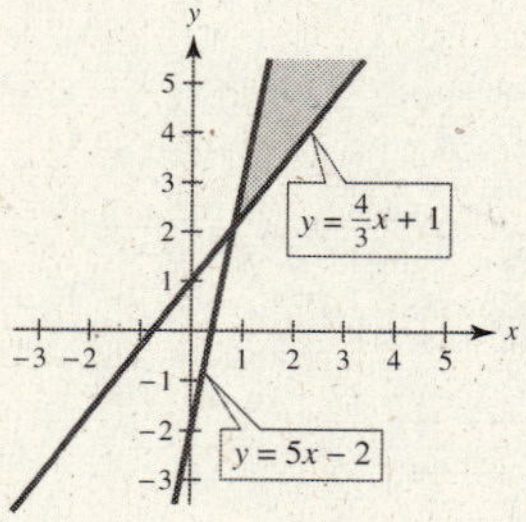

Solid lines at $y = \frac{4}{3}x + 1$ and $y = 5x - 2$. Shade above $y = \frac{4}{3}x + 1$ and below $5x - 2$.

25. $\begin{cases} y > x - 2 \\ y > -\frac{1}{3}x + 5 \end{cases}$

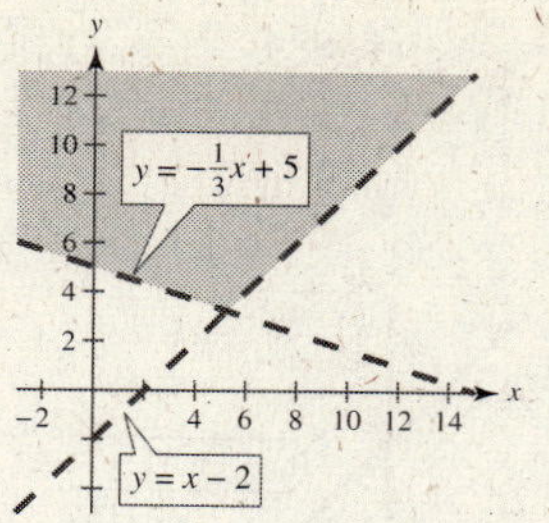

Dotted lines at $y = x - 2$ and $y = -\frac{1}{3}x + 5$. Shade above each line.

27. $\begin{cases} y \ge 3x - 3 \\ y \le -x + 1 \end{cases}$

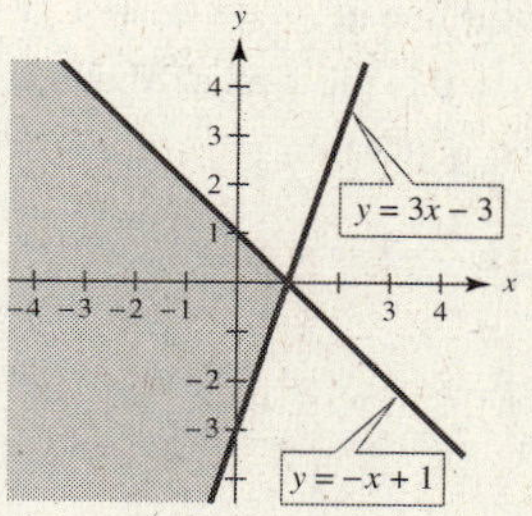

Solid lines at $y = 3x - 3$ and $y = -x + 1$. Shade above $y = 3x - 3$ and $y = -x + 1$.

29. $\begin{cases} x + 2y \le -4 \\ y \ge x + 5 \end{cases} \Rightarrow \begin{matrix} y \le -\frac{1}{2}x - 2 \\ y \ge x + 5 \end{matrix}$

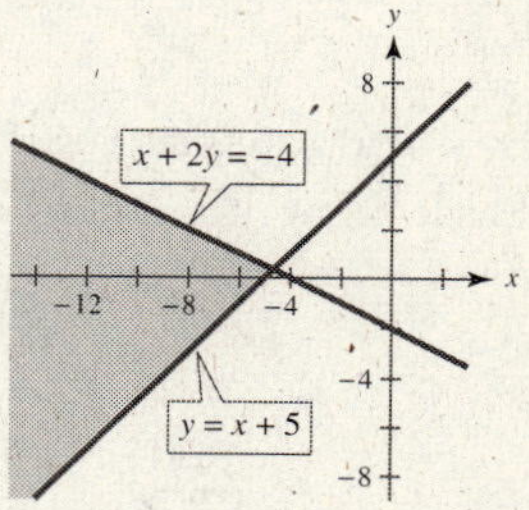

Solid lines at $y = -\frac{1}{2}x - 2$ and $y = x + 5$. Shade below $y = -\frac{1}{2}x - 2$ and above $y = x + 5$.

31. $\begin{cases} x + y \le 4 \\ x \quad\;\; \ge 0 \\ \quad\;\; y \ge 0 \end{cases} \Rightarrow \begin{array}{l} y \le -x + 4 \\ x \ge 0 \\ y \ge 0 \end{array}$

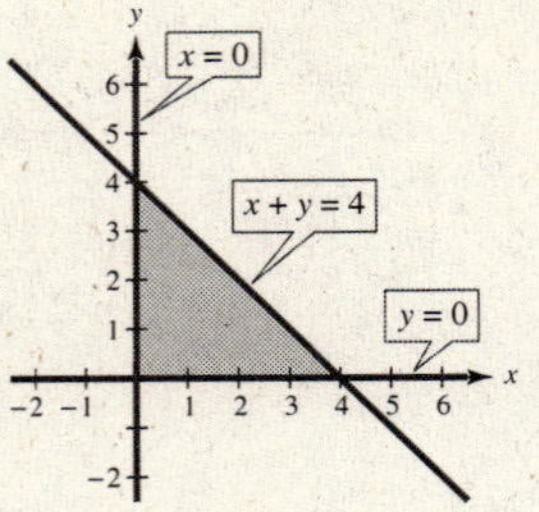

Solid line at $y = -x + 4$. $x = 0$ is the y-axis and $y = 0$ is the x-axis. Shade below $y = -x + 4$ in the first quadrant.

33. $\begin{cases} 4x - 2y > 8 \\ x \qquad\quad \ge 0 \\ \qquad\;\; y \le 0 \end{cases} \Rightarrow \begin{array}{l} y < 2x - 4 \\ x \ge 0 \\ y \le 0 \end{array}$

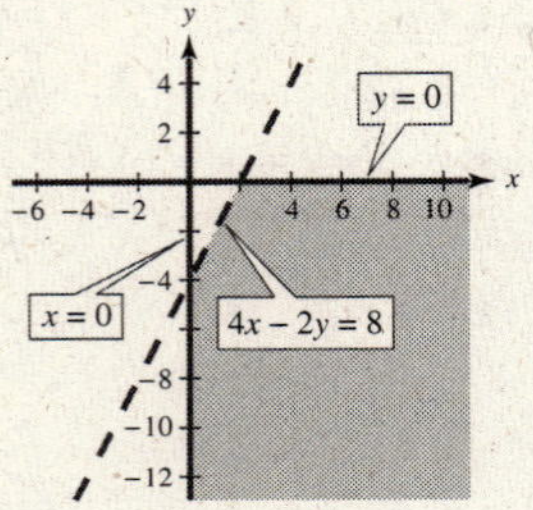

Dotted line at $y = 2x - 4$. $x = 0$ is the y-axis and $y = 0$ is the x-axis. Shade below $y = 2x - 4$ in the fourth quadrant.

35. $\begin{cases} y > -5 \\ x \le 2 \\ y \le x + 2 \end{cases}$

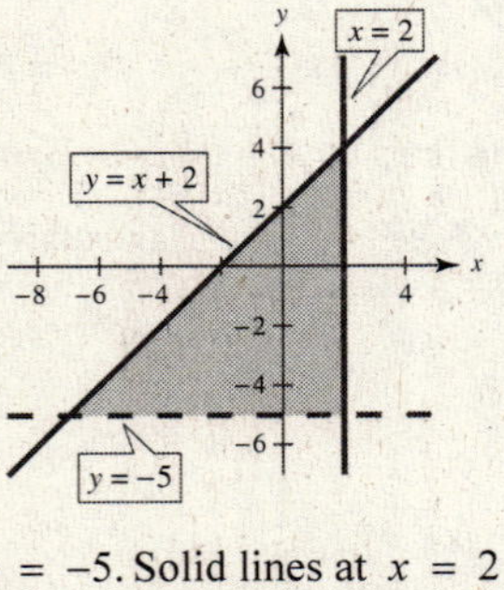

Dotted line at $y = -5$. Solid lines at $x = 2$ and $y = x + 2$. Shade above $y = -5$, to the left of $x = 2$ and below $y = x + 2$.

37. $\begin{cases} x + y \le 1 \\ -x + y \le 1 \\ \qquad\; y \ge 0 \end{cases} \Rightarrow \begin{array}{l} y \le -x + 1 \\ y \le x + 1 \\ y \ge 0 \end{array}$

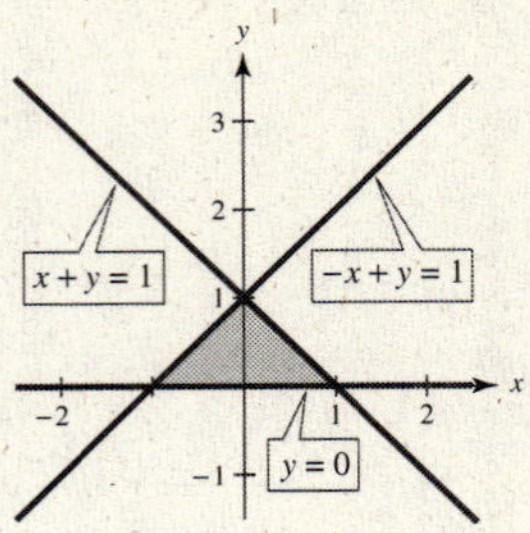

Solid lines at $y = -x + 1$ and $y = x + 1$. $y = 0$ is the x-axis. Shade below $y = -x + 1$ and $y = x + 1$ and above the x-axis.

39. $\begin{cases} x + \;\; y \le 5 \\ x - 2y \ge 2 \\ \qquad\; y \ge 3 \end{cases} \Rightarrow \begin{array}{l} y \le -x + 5 \\ y \le \frac{1}{2}x - 1 \\ y \ge 3 \end{array}$

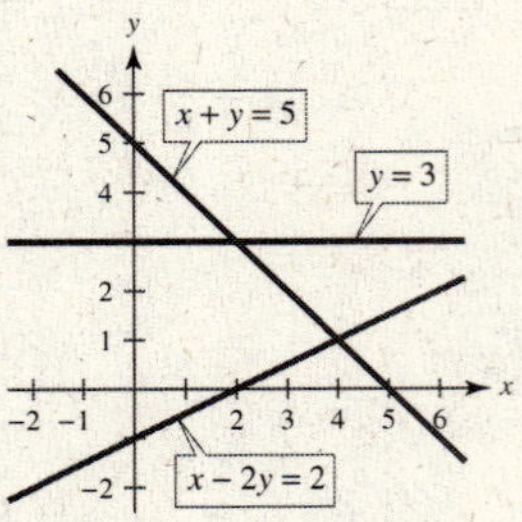

Solid lines at $y = -x + 5$, $y = \frac{1}{2}x - 1$, and $y = 3$.

Shade below $y = -x + 5$ and $y = \frac{1}{2}x - 1$ and above $y = 3$.

The half-planes do not all intersect, so there is no solution.

41. $\begin{cases} -3x + 2y < 6 \\ \;\;\; x - 4y > -2 \\ \;\; 2x + \;\; y < 3 \end{cases} \Rightarrow \begin{array}{l} y < -\frac{3}{2}x + 3 \\ y < \frac{1}{4}x + \frac{1}{2} \\ y < -2x + 3 \end{array}$

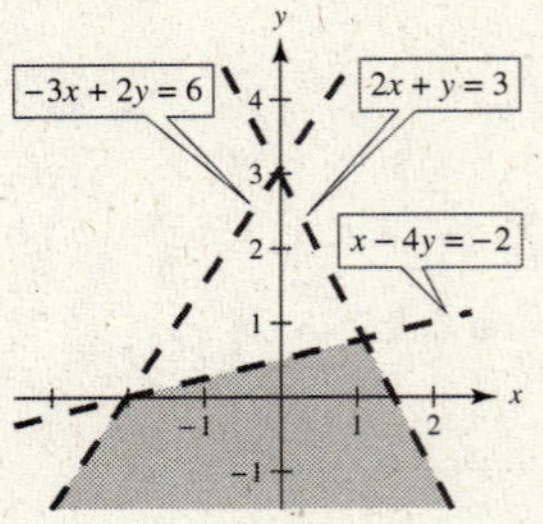

Dotted lines at $y = -\frac{3}{2}x + 3$, $y = \frac{1}{4}x + \frac{1}{2}$, and $y = -2x + 3$. Shade below all three lines.

43. $\begin{cases} x \geq 1 \\ x - 2y \leq 3 \\ 3x + 2y \geq 9 \\ x + y \leq 6 \end{cases} \Rightarrow \begin{array}{l} x \geq 1 \\ y \geq \frac{1}{2}x - \frac{3}{2} \\ y \geq -\frac{3}{2}x + \frac{9}{2} \\ y \leq -x + 6 \end{array}$

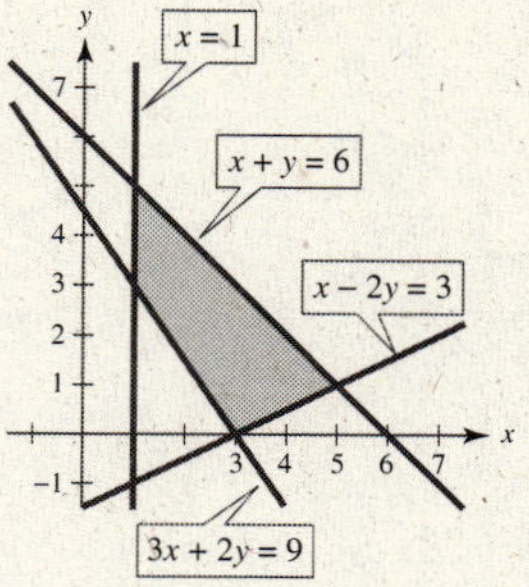

Solid lines at $x = 1$, $y = \frac{1}{2}x - \frac{3}{2}$, $y = -\frac{3}{2}x + \frac{9}{2}$, and $y = -x + 6$. Shade to the right of $x = 1$, above $y = \frac{1}{2}x - \frac{3}{2}$ and $y = -\frac{3}{2}x + \frac{9}{2}$ and below $y = -x + 6$.

45. $\begin{cases} 2x - 3y \leq 6 \\ y \leq 4 \end{cases} \Rightarrow \begin{array}{l} y \geq \frac{2}{3}x - 2 \\ y \leq 4 \end{array}$

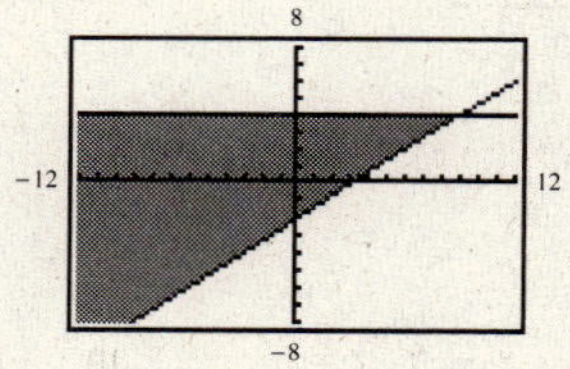

Shade above $y = \frac{2}{3}x - 2$ and below $y = 4$.

47. $\begin{cases} 2x - 2y \leq 5 \\ y \leq 6 \end{cases} \Rightarrow \begin{array}{l} y \geq x - \frac{5}{2} \\ y \leq 6 \end{array}$

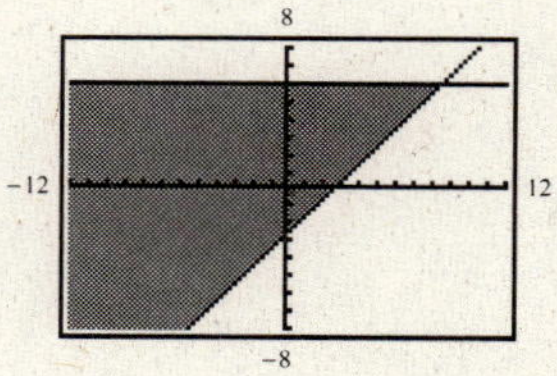

Shade above $y = x - \frac{5}{2}$ and below $y = 6$.

49. $\begin{cases} 2x + y \leq 2 \\ y \geq -4 \end{cases} \Rightarrow \begin{array}{l} y \leq -2x + 2 \\ y \geq -4 \end{array}$

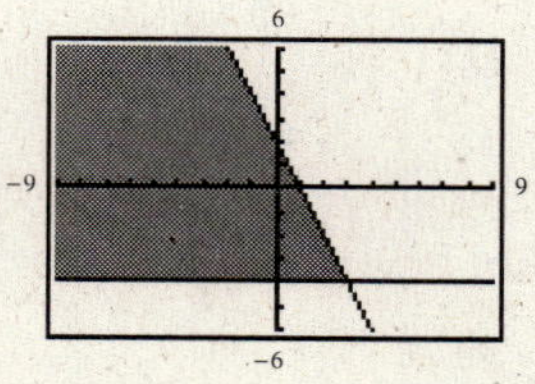

Shade below $y = -2x + 2$ and above $y = -4$.

51. Line 1: vertical $x = 1$

Line 2: points $(1, -2)$ and $(3, 0)$

$$m = \frac{0 + 2}{3 - 1} = \frac{2}{2} = 1$$

$$y - 0 = 1(x - 3)$$

$$y = x - 3$$

Line 3: points $(1, 4)$ and $(3, 0)$

$$m = \frac{0 - 4}{3 - 1} = \frac{-4}{2} = -2$$

$$y - 0 = -2(x - 3)$$

$$y = -2x + 6$$

System of linear inequalities:

$x \geq 1$	Region on and to the right of $x = 1$
$y \geq x - 3$	Region on and above line $y = x - 3$.
$y \leq -2x + 6$	Region on and below line $y = -2x + 6$.

53. Line 1: $x = -2$

Line 2: $x = 2$

Line 3: points $(-2, 3)$ and $(2, 1)$

$$m = \frac{1 - 3}{2 - (-2)} = \frac{-2}{4} = -\frac{1}{2}$$

$$y - 1 = -\frac{1}{2}(x - 2)$$

$$y - 1 = -\frac{1}{2}x + 1$$

$$y = -\frac{1}{2} + 2$$

Line 4: points $(-2, -3)$ and $(2, -5)$

$$m = \frac{-5 - (-3)}{2 - (-2)} = \frac{-2}{4} = -\frac{1}{2}$$

$$y - (-3) = -\frac{1}{2}(x - (-2))$$

$$y + 3 = -\frac{1}{2}x - 1$$

$$y = -\frac{1}{2}x - 4$$

System of linear inequalities:

$x \geq -2$:	Region on and to the right of $x = -2$
$x \leq 2$:	Region on and to the left of $x = 2$
$y \leq -\frac{1}{2}x + 2$:	Region on and below $y = -\frac{1}{2}x + 2$
$y \geq -\frac{1}{2}x - 4$:	Region on and above $y = -\frac{1}{2}x - 4$

55. Line 1: points $(-6, 3)$ and $(3, 9)$

$$m = \frac{9-3}{3+6} = \frac{6}{9} = \frac{2}{3}$$

$$y - 9 = \frac{2}{3}(x - 3)$$

$$y - 9 = \frac{2}{3}x - 2$$

$$y = \frac{2}{3}x + 7$$

Line 2: points $(-6, 3)$ and $(4, 12)$

$$m = \frac{12-3}{4+6} = \frac{9}{10}$$

$$y - 3 = \frac{9}{10}(x + 6)$$

$$y - 3 = \frac{9}{10}x + \frac{27}{5}$$

$$y = \frac{9}{10}x + \frac{42}{5}$$

Line 3: points $(3, 9)$ and $(4, 12)$

$$m = \frac{12-9}{4-3} = \frac{3}{1} = 3$$

$$y - 9 = 3(x - 3)$$

$$y - 9 = 3x - 9$$

$$y = 3x$$

System of linear inequalities:

$y \le \frac{9}{10}x + \frac{42}{5}$ Region on and below the line $y = \frac{9}{10}x + \frac{42}{5}$

$y \ge 3x$ Region on and above the line $y = 3x$.

$y \ge \frac{2}{3}x + 7$ Region on and above the line $y = \frac{2}{3}x + 7$

57. *Verbal Model:* [Number of hours in assembly] · [Number of tables] + [Number of hours in assembly] · [Number of chairs] ≤ 12

[Number of hours in finishing] · [Number of tables] + [Number of hours in finishing] · [Number of chairs] ≤ 16

Labels: Number of tables $= x$

Number of chairs $= y$

System:

$$1x + \tfrac{3}{2}y \le 12$$
$$\tfrac{4}{3}x + \tfrac{3}{4}y \le 16$$
$$x \ge 0$$
$$y \ge 0$$

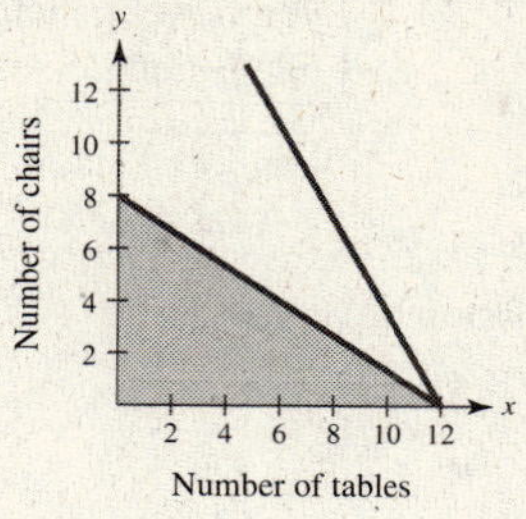

59. *Verbal Model:* [Amount in account X] + [Amount in account Y] ≤ 40,000

[Amount in account X] ≥ 10,000

[Amount in account Y] ≥ 2 · [Amount in account X]

Labels: Amount in account X $= x$

Amount in account Y $= y$

System:

$$x + y \le 40{,}000$$
$$x \ge 10{,}000$$
$$y \ge 2x$$

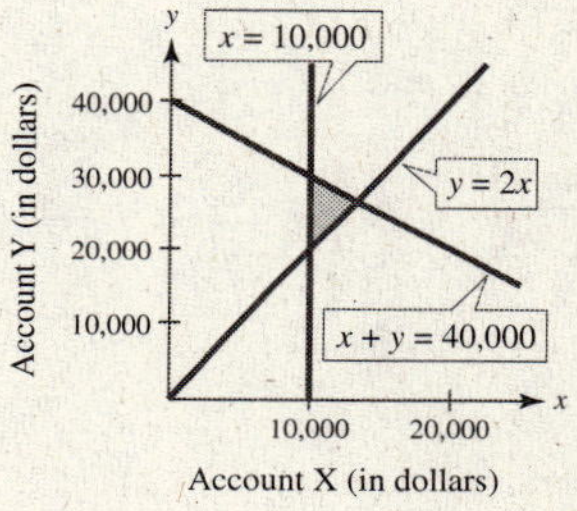

61. *Verbal Model:* 20 [Ounces of food X] + 10 [Ounces of food Y] ≥ 280

15 [Ounces of food X] + 10 [Ounces of food Y] ≥ 160

10 [Ounces of food X] + 20 [Ounces of food Y] ≥ 180

[Ounces of food X] ≥ 0

[Ounces of food Y] ≥ 0

Labels: Ounces of food X $= x$

Ounces of food Y $= y$

System:

$$20x + 10y \ge 280$$
$$15x + 10y \ge 160$$
$$10x + 20y \ge 180$$
$$x \ge 0$$
$$y \ge 0$$

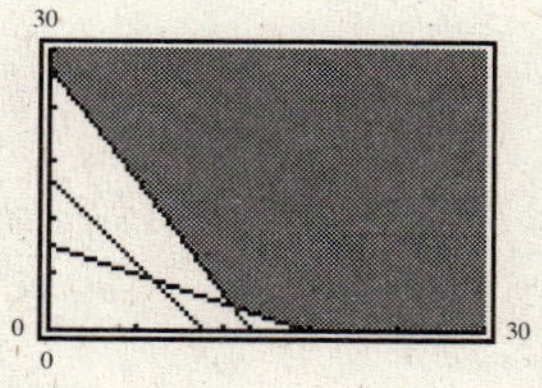

63. *Verbal Model:*

[Number of general admission tickets] + [Number of stadium seat tickets] ≥ 15,000

30 · [Number of general admission tickets] + 45 · [Number of stadium seat tickets] ≥ 525,000

[Number of general admission tickets] ≥ 8000

[Number of stadium seat tickets] ≥ 4000

Labels: Number of general admission tickets $= x$

Number of stadium seat tickets $= y$

System:

$$\begin{aligned} x + y &\geq 15{,}000 \\ 30x + 45y &\geq 525{,}000 \\ x &\geq 8000 \\ y &\geq 4000 \end{aligned}$$

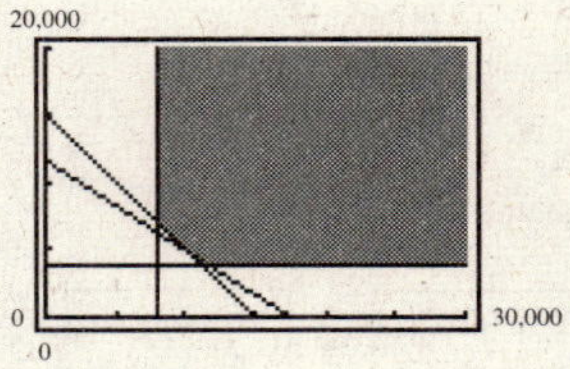

65. Line 1: vertical, $x = 90$

Line 2: horizontal, $y = 0$

Line 3: horizontal, $y = -10$

Line 4: points $(0, 0)$ and $(70, -10)$

$$m = \frac{-10 - 0}{70 - 0} = \frac{-10}{70} = \frac{-1}{7}$$

$$y - 0 = -\frac{1}{7}(x - 0)$$

$$y = -\frac{1}{7}x$$

System of linear inequalities:

$x \leq 90$: Region to the left of $x = 90$

$y \leq 0$: Region below $y = 0$

$y \geq -10$: Region on and above $y = -10$

$y \geq -\frac{1}{7}x$: Region on and above the line $y = -\frac{1}{7}x$

67. The graph of a linear equation splits the xy-plane into two parts, each of which is a half-plane. $y < 5$ is a half-plane.

69. Find all intersections between the lines corresponding to the inequalities.

71. Let $y = 0$:

$$\begin{aligned} y &= 4x + 2 \\ 0 &= 4x + 2 \\ -2 &= 4x \\ -\tfrac{1}{2} &= x \end{aligned}$$

The x-intercept is $\left(-\frac{1}{2}, 0\right)$.

Let $x = 0$:

$$\begin{aligned} y &= 4x + 2 \\ y &= 4(0) + 2 \\ y &= 2 \end{aligned}$$

The y-intercept is $(0, 2)$.

73. Let $y = 0$:

$$\begin{aligned} -x + 3y &= -3 \\ -x + 3(0) &= -3 \\ -x &= -3 \\ x &= 3 \end{aligned}$$

The x-intercept is $(3, 0)$.

Let $x = 0$:

$$\begin{aligned} -x + 3y &= -3 \\ -0 + 3y &= -3 \\ 3y &= -3 \\ y &= -1 \end{aligned}$$

The y-intercept is $(0, -1)$.

75. Let $y = 0$:

$$y = |x + 2|$$
$$0 = |x + 2|$$
$$0 = x + 2$$
$$-2 = x$$

The x-intercept is $(-2, 0)$.

Let $x = 0$:

$$y = |x + 2|$$
$$y = |0 + 2|$$
$$y = |2|$$
$$y = 2$$

The y-intercept is $(0, 2)$.

77. (a) $f(x) = 3x - 7$

$f(-1) = 3(-1) - 7 = -3 - 7 = -10$

(b) $f(x) = 3x - 7$

$f\left(\frac{2}{3}\right) = 3\left(\frac{2}{3}\right) - 7 = 2 - 7 = -5$

79. (a) $f(x) = 3x - x^2$

$f(0) = 3(0) - 0^2 = 0 - 0 = 0$

(b) $f(x) = 3x - x^2$

$f(2m) = 3(2m) - (2m)^2 = 6m - 4m^2$

Review Exercises for Chapter 4

1. (a) $(3, 4)$

$$3(3) + 7(4) \stackrel{?}{=} 2$$
$$37 \neq 2$$

Not a solution

(b) $(3, -1)$

$$3(3) + 7(-1) \stackrel{?}{=} 2 \qquad 5(3) + 6(-1) \stackrel{?}{=} 9$$
$$2 = 2 \qquad 9 = 9$$

Solution

3. (a) $(4, -5)$

$$26(4) + 13(-5) \stackrel{?}{=} 26$$
$$104 - 65 = 26$$
$$39 \neq 26$$

Not a solution

(b) $(7, 12)$

$$26(7) + 13(12) \stackrel{?}{=} 26$$
$$182 + 156 = 26$$
$$338 \neq 26$$

Not a solution

5. Solve each equation for y.

$$x + y = 2 \qquad x - y = 2$$
$$y = -x + 2 \qquad -y = -x + 2$$
$$y = x - 2$$

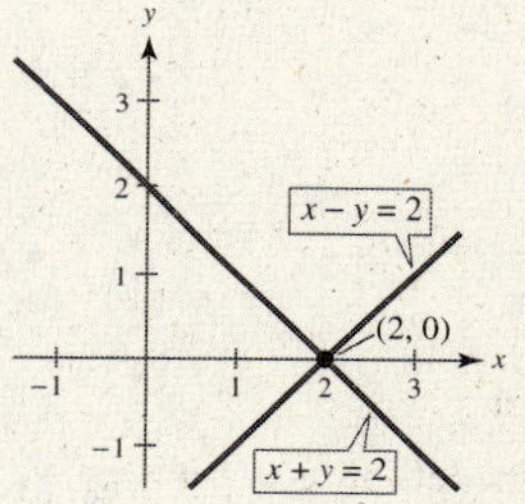

Point of intersection is $(2, 0)$.

7. Solve each equation for y.

$$x - y = 3 \qquad -x + y = 1$$
$$-y = -x + 3 \qquad y = x + 1$$
$$y = x - 3$$

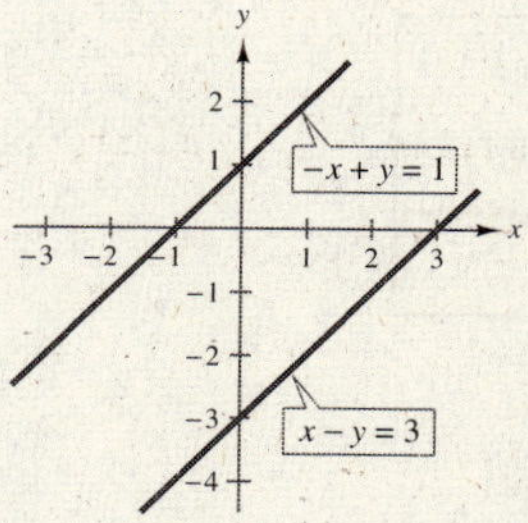

Inconsistent; no solution

9. Solve each equation for y.

$$2x - y = 0 \qquad -x + y = 4$$
$$-y = -2x \qquad y = x + 4$$
$$y = 2x$$

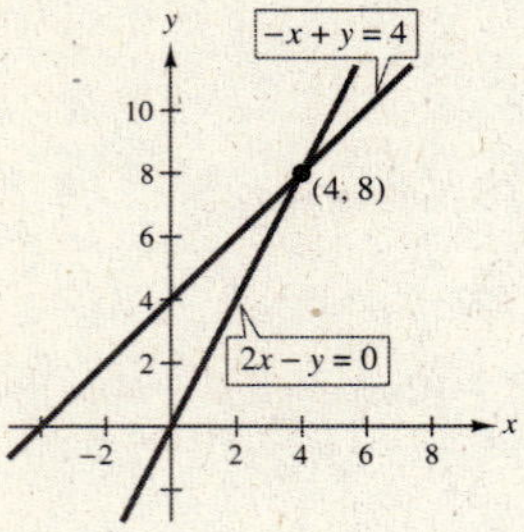

Point of intersection is $(4, 8)$.

11. Solve each equation for y.

$$2x + y = 4 \qquad -4x - 2y = -8$$
$$y = -2x + 4 \qquad -2y = 4x - 8$$
$$y = -2x + 4$$

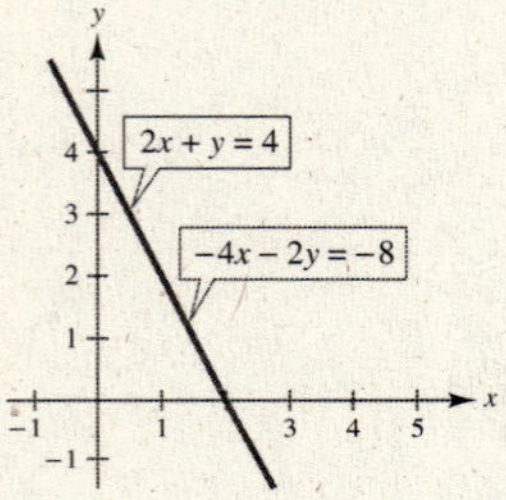

Infinitely many solutions

13. The equations have the same slope, so the lines are parallel. The system of linear equations has no solution.

15. Solve each equation for y.

$$5x - 3y = 3 \qquad 2x + 2y = 14$$
$$-3y = -5x + 3 \qquad 2y = -2x + 14$$
$$y = \tfrac{5}{3}x - 1 \qquad y = -1x + 7$$

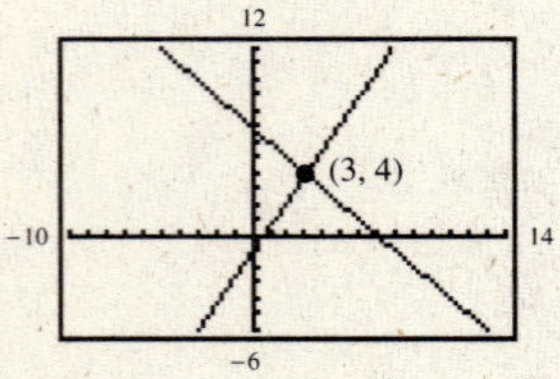

17. $2x - 3y = -1$

$x + 4y = 16$

$$x + 4y = 16$$
$$x = -4y + 16$$
$$2(-4y + 16) - 3y = -1$$
$$-8y + 32 - 3y = -1$$
$$-11y = -33$$
$$y = 3$$
$$x + 4(3) = 16$$
$$x = 4$$

The solution is $(4, 3)$.

19. $-5x + 2y = 4$

$10x - 4y = 7$

Solve for y and substitute into second equation.

$$2y = 5x + 4$$
$$y = \frac{5x + 4}{2}$$
$$10x - 4\left(\frac{5x + 4}{2}\right) = 7$$
$$10x - 2(5x + 4) = 7$$
$$10x - 10x - 8 = 7$$
$$-8 \neq 7$$

No solution

21. $3x - 7y = 5$

$5x - 9y = -5$

Solve for x and substitute into second equation.

$$3x = 7y + 5$$
$$x = \frac{7y + 5}{3}$$
$$5\left(\frac{7y + 5}{3}\right) - 9y = -5$$
$$5(7y + 5) - 27y = -15$$
$$35y + 25 - 27y = -15$$
$$8y = -40$$
$$y = -5$$
$$x = \frac{7(-5) + 5}{3}$$
$$x = -10$$

The solution is $(-10, -5)$.

23. $-x + y = 6$

$15x + y = -10$

Solve for y and substitute into second equation.

$$y = x + 6$$

$$15x + (x + 6) = -10$$

$$15x + x + 6 = -10$$

$$16x = -16$$

$$x = -1$$

$$y = (-1) + 6$$

$$y = 5$$

The solution is $(-1, 5)$.

25. $-3x - 3y = 3$

$x + y = -1$

Solve for y and substitute into first equation.

$$y = -x - 1$$

$$-3x - 3(-x - 1) = 3$$

$$-3x + 3x + 3 = 3$$

$$3 = 3$$

Infinitely many solutions

27. *Verbal Model:*

[Total cost] = [Cost per camera] · [Number of cameras] + [Initial cost]

[Total revenue] = [Price per camera] · [Number of cameras]

Labels: Total cost $= C$

Cost per camera $= 4.45$

Number of cameras $= x$

Initial cost $= 25{,}000$

Total revenue $= R$

Price per camera $= 8.95$

System: $C = 4.45x + 25{,}000$

$R = 8.95x$

Break-even point occurs when $R = C$.

$$8.95x = 4.45x + 25{,}000$$

$$4.5x = 25{,}000$$

$$x \approx 5555.56$$

5556 one-time-use cameras must be sold before the business breaks even.

29. $x + y = 0 \Rightarrow x + y = 0$

$2x + y = 0 \Rightarrow -2x - y = 0$

$$-x = 0$$

$$x = 0$$

$$0 + y = 0$$

$$y = 0$$

The solution is $(0, 0)$.

31. $2x - y = 2 \Rightarrow 16x - 8y = 16$

$6x + 8y = 39 \Rightarrow 6x + 8y = 39$

$$22x = 55$$

$$x = \frac{55}{22} = \frac{5}{2}$$

$$2\left(\frac{5}{2}\right) - y = 2$$

$$5 - y = 2$$

$$-y = -3$$

$$y = 3$$

The solution is $\left(\frac{5}{2}, 3\right)$.

33. $4x + y = -3$

$-4x + 3y = 23$

$$4y = 20$$

$$y = 5$$

$$4x + 5 = -3$$

$$4x = -8$$

$$x = -2$$

The solution is $(-2, 5)$.

35. $0.2x + 0.3y = 0.14 \Rightarrow 2x + 3y = 1.4 \Rightarrow -4x - 6y = -2.8$

$0.4x + 0.5y = 0.20 \Rightarrow 4x + 5y = 2 \Rightarrow 4x + 5y = 2$

$$-y = -0.8$$

$$y = 0.8$$

$$2x + 3(0.8) = 1.4$$

$$2x + 2.4 = 1.4$$

$$2x = -1$$

$$x = -\frac{1}{2} = -0.5$$

The solution is $(-0.5, 0.8)$.

37. *Verbal Model:* $\boxed{\text{Gallons of solution 1}} + \boxed{\text{Gallons of solution 2}} = 40$

$0.75 \cdot \boxed{\text{Gallons of solution 1}} + 0.50 \cdot \boxed{\text{Gallons of solution 2}} = 0.60(40)$

Labels: Gallons of solution 1 $= x$

Gallons of solution 2 $= y$

System:

$$x + y = 40$$
$$0.75x + 0.50y = 0.60(40)$$

$y = -x + 40$

$0.75x + 0.50(-x + 40) = 0.60(40)$

$0.75x - 0.5x + 20 = 24$

$0.25x = 4$

$x = 16$

$y = -16 + 40 = 24$

16 gallons of the 75% solution and 24 gallons of the 50% solution must be used.

39. *Verbal Model:* $\boxed{\text{Distance}} = \boxed{\text{Rate}} \cdot \boxed{\text{Time}}$

Labels: Time at 60 mi/h $= x$

D_1 = distance at 50 mi/h for 3 hours + at 60 mi/h for x hours

D_2 = distance at 52 mi/h for $3 + x$ hours

$D_1 = 50(3) + 60(x)$

$D_2 = 52(3 + x)$

System: Because $D_1 = D_2$,

$50(3) + 60(x) = 52(3 + x)$

$150 + 60x = 156 + 52x$

$8x = 6$

$x = \frac{6}{8} = \frac{3}{4}$ hour $= 45$ minutes.

41.
$$\begin{cases} x = 3 \\ x + 2y = 7 \\ -3x - y + 4z = 9 \end{cases}$$

$3 + 2y = 7$

$2y = 4$

$y = 2$

$-3(3) - 2 + 4z = 9$

$-9 - 2 + 4z = 9$

$4z = 20$

$z = 5$

The solution is $(3, 2, 5)$.

43.
$$\begin{cases} x + 2y = 6 \\ 3y = 9 \\ x + 2z = 12 \end{cases}$$

$3y = 9$

$y = 3$

$x + 2(3) = 6$

$x = 0$

$0 + 2z = 12$

$z = 6$

The solution is $(0, 3, 6)$.

45. $\begin{cases} x - y - 2z = -1 \\ 2x + 3y + z = -2 \\ 5x + 4y + 2z = 4 \end{cases}$

$\begin{cases} x - y - 2z = -1 \\ 5y + 5z = 0 \\ 9y + 12z = 9 \end{cases}$

$\begin{cases} x - y - 2z = -1 \\ y + z = 0 \\ 9y + 12z = 9 \end{cases}$

$\begin{cases} x - y - 2z = -1 \\ y + z = 0 \\ 3z = 9 \end{cases}$

$\begin{cases} x - y - 2z = -1 \\ y + z = 0 \\ z = 3 \end{cases}$

$y + 3 = 0$

$y = -3$

$x - (-3) - 2(3) = -1$

$x + 3 - 6 = -1$

$x = 2$

The solution is $(2, -3, 3)$.

47. $\begin{cases} x - y - 2z = 1 \\ -2x + y + 3z = -5 \\ 3x + 4y - z = 6 \end{cases}$

$\begin{cases} x - y - z = 1 \\ -y + z = -3 \\ 7y + 2z = 3 \end{cases}$

$x - y - z = 1$

$-y + z = -3$

$9z = -18$

$z = -2$

$-y + (-2) = -3$

$-y = -1$

$y = 1$

$x - 1 - (-2) = 1$

$x = 0$

The solution is $(0, 1, -2)$.

49. *Verbal Model:*

$$\boxed{\text{Amount in 7\% investment}} + \boxed{\text{Amount in 9\% investment}} + \boxed{\text{Amount in 11\% investment}} = \boxed{20{,}000}$$

$$\boxed{\text{Interest from 7\% investment}} + \boxed{\text{Interest from 9\% investment}} + \boxed{\text{Interest from 11\% investment}} = \boxed{1780}$$

$$\boxed{\text{Amount in 9\% investment}} - \boxed{\text{Amount in 11\% investment}} = \boxed{-2000}$$

System:

$$\begin{aligned} x + y + z &= 20{,}000 \\ 0.07x + 0.09y + 0.11z &= 1780 \\ y - z &= -2000 \end{aligned}$$

$$\begin{aligned} x + y + z &= 20{,}000 \\ 7x + 9y + 11z &= 178{,}000 \\ y - z &= -2000 \end{aligned}$$

$$\begin{aligned} x + y + z &= 20{,}000 \\ 2y + 4z &= 38{,}000 \\ y - z &= -2000 \end{aligned}$$

$$\begin{aligned} x + y + z &= 20{,}000 \\ y + 2z &= 19{,}000 \\ y - z &= -2000 \end{aligned}$$

$$\begin{aligned} x + y + z &= 20{,}000 \\ y + 2z &= 19{,}000 \\ -3z &= -21{,}000 \end{aligned}$$

$$\begin{aligned} x + y + z &= 20{,}000 \\ y + 2z &= 19{,}000 \\ z &= 7000 \end{aligned}$$

$$y + 2(7000) = 19{,}000 \qquad x + 5000 + 7000 = 20{,}000$$

$$y = 5000 \qquad x = 8000$$

7% investment: \$8000; 9% investment: \$5000; 11% investment: \$7000

51. 1×4

53. 2×3

55. (a) $\begin{bmatrix} 7 & -5 \\ 1 & -1 \end{bmatrix}$

(b) $\left[\begin{array}{cc:c} 7 & -5 & 11 \\ 1 & -1 & -5 \end{array}\right]$

57.

$$\begin{aligned} 4x - y &= 2 \\ 6x + 3y + 2z &= 1 \\ y + 4z &= 0 \end{aligned}$$

59.

$$\left[\begin{array}{cc:c} 5 & 4 & 2 \\ -1 & 1 & -22 \end{array}\right]$$

$$\begin{matrix} R_1 \\ R_2 \end{matrix} \left[\begin{array}{cc:c} -1 & 1 & -22 \\ 5 & 4 & 2 \end{array}\right]$$

$$-R_1 \left[\begin{array}{cc:c} 1 & -1 & 22 \\ 5 & 4 & 2 \end{array}\right]$$

$$-5R_1 + R_2 \left[\begin{array}{cc:c} 1 & -1 & 22 \\ 0 & 9 & -108 \end{array}\right]$$

$$\tfrac{1}{9}R_2 \left[\begin{array}{cc:c} 1 & -1 & 22 \\ 0 & 1 & -12 \end{array}\right]$$

$$y = -12 \qquad x - (-12) = 22$$

$$x = 10$$

The solution is $(10, -12)$.

61.

$$\begin{bmatrix} 0.2 & -0.1 & \vdots & 0.07 \\ 0.4 & -0.5 & \vdots & -0.01 \end{bmatrix}$$

$$\begin{matrix} 10R_1 \\ 10R_2 \end{matrix} \begin{bmatrix} 2 & -1 & \vdots & 0.7 \\ 4 & -5 & \vdots & -0.1 \end{bmatrix}$$

$$\tfrac{1}{2}R_1 \begin{bmatrix} 1 & -\frac{1}{2} & \vdots & 0.35 \\ 4 & -5 & \vdots & -0.1 \end{bmatrix}$$

$$-4R_1 + R_2 \begin{bmatrix} 1 & -\frac{1}{2} & \vdots & 0.35 \\ 0 & -3 & \vdots & -1.5 \end{bmatrix}$$

$$-\tfrac{1}{3}R_2 \begin{bmatrix} 1 & -\frac{1}{2} & \vdots & 0.35 \\ 0 & 1 & \vdots & 0.5 \end{bmatrix}$$

$y = 0.5 \qquad x - \frac{1}{2}(0.5) = 0.35$

$x = 0.6$

The solution is $(0.6, 0.5)$.

63.

$$\begin{aligned} x + 4y + 4z &= 7 \\ -3x + 2y + 3z &= 0 \\ 4x - 2z &= -2 \end{aligned}$$

$$\begin{bmatrix} 1 & 4 & 4 & \vdots & 7 \\ -3 & 2 & 3 & \vdots & 0 \\ 4 & 0 & -2 & \vdots & -2 \end{bmatrix}$$

$$\begin{matrix} \\ 3R_1 + R_2 \\ -4R_1 + R_3 \end{matrix} \begin{bmatrix} 1 & 4 & 4 & \vdots & 7 \\ 0 & 14 & 15 & \vdots & 21 \\ 0 & -16 & -18 & \vdots & -30 \end{bmatrix}$$

$$\tfrac{1}{14}R_2 \begin{bmatrix} 1 & 4 & 4 & \vdots & 7 \\ 0 & 1 & \frac{15}{14} & \vdots & \frac{3}{2} \\ 0 & -16 & -18 & \vdots & -30 \end{bmatrix}$$

$$16R_2 + R_3 \begin{bmatrix} 1 & 4 & 4 & \vdots & 7 \\ 0 & 1 & \frac{15}{14} & \vdots & \frac{3}{2} \\ 0 & 0 & -\frac{6}{7} & \vdots & -6 \end{bmatrix}$$

$$-\tfrac{7}{6}R_3 \begin{bmatrix} 1 & 4 & 4 & \vdots & 7 \\ 0 & 1 & \frac{15}{14} & \vdots & \frac{3}{2} \\ 0 & 0 & 1 & \vdots & 7 \end{bmatrix}$$

$z = 7$

$y + \frac{15}{14}(7) = \frac{3}{2}$

$y + \frac{15}{2} = \frac{3}{2}$

$y = -6$

$x + 4(-6) + 4(7) = 7$

$x + 4 = 7$

$x = 3$

The solution is $(3, -6, 7)$.

65.

$$\begin{bmatrix} 2 & 3 & 3 & \vdots & 3 \\ 6 & 6 & 12 & \vdots & 13 \\ 12 & 9 & -1 & \vdots & 2 \end{bmatrix}$$

$$\tfrac{1}{2}R_1 \begin{bmatrix} 1 & \frac{3}{2} & \frac{3}{2} & \vdots & \frac{3}{2} \\ 6 & 6 & 12 & \vdots & 13 \\ 12 & 9 & -1 & \vdots & 2 \end{bmatrix}$$

$$\begin{matrix} \\ -6R_1 + R_2 \\ -12R_1 + R_3 \end{matrix} \begin{bmatrix} 1 & \frac{3}{2} & \frac{3}{2} & \vdots & \frac{3}{2} \\ 0 & -3 & 3 & \vdots & 4 \\ 0 & -9 & -19 & \vdots & -16 \end{bmatrix}$$

$$-\tfrac{1}{3}R_2 \begin{bmatrix} 1 & \frac{3}{2} & \frac{3}{2} & \vdots & \frac{3}{2} \\ 0 & 1 & -1 & \vdots & -\frac{4}{3} \\ 0 & -9 & -19 & \vdots & -16 \end{bmatrix}$$

$$9R_2 + R_3 \begin{bmatrix} 1 & \frac{3}{2} & \frac{3}{2} & \vdots & \frac{3}{2} \\ 0 & 1 & -1 & \vdots & -\frac{4}{3} \\ 0 & 0 & -28 & \vdots & -28 \end{bmatrix}$$

$$-\tfrac{1}{28}R_3 \begin{bmatrix} 1 & \frac{3}{2} & \frac{3}{2} & \vdots & \frac{3}{2} \\ 0 & 1 & -1 & \vdots & -\frac{4}{3} \\ 0 & 0 & 1 & \vdots & 1 \end{bmatrix}$$

$x_3 = 1$

$x_2 - 1 = -\frac{4}{3}$

$x_2 = -\frac{1}{3}$

$x_1 + \frac{3}{2}\left(-\frac{1}{3}\right) + \frac{3}{2}(1) = \frac{3}{2}$

$x_1 - \frac{1}{2} + \frac{3}{2} = \frac{3}{2}$

$x_1 + 1 = \frac{3}{2}$

$x_1 = \frac{1}{2}$

The solution is $\left(\frac{1}{2}, -\frac{1}{3}, 1\right)$.

67.
$$\begin{aligned} x \qquad - 4z &= 17 \\ -2x + 4y + 3z &= -14 \\ 5x - y + 2z &= -3 \end{aligned}$$

$$\begin{bmatrix} 1 & 0 & -4 & \vdots & 17 \\ -2 & 4 & 3 & \vdots & -14 \\ 5 & -1 & 2 & \vdots & -3 \end{bmatrix}$$

$$\begin{matrix} \\ 2R_1 + R_2 \\ -5R_1 + R_3 \end{matrix} \begin{bmatrix} 1 & 0 & -4 & \vdots & 17 \\ 0 & 4 & -5 & \vdots & 20 \\ 0 & -1 & 22 & \vdots & -88 \end{bmatrix}$$

$$\begin{matrix} \\ R_2 \\ R_3 \end{matrix} \begin{bmatrix} 1 & 0 & -4 & \vdots & 17 \\ 0 & -1 & 22 & \vdots & -88 \\ 0 & 4 & -5 & \vdots & 20 \end{bmatrix}$$

$$\begin{matrix} \\ -R_2 \\ \\ \end{matrix} \begin{bmatrix} 1 & 0 & -4 & \vdots & 17 \\ 0 & 1 & -22 & \vdots & 88 \\ 0 & 4 & -5 & \vdots & 20 \end{bmatrix}$$

$$\begin{matrix} \\ \\ -4R_2 + R_3 \end{matrix} \begin{bmatrix} 1 & 0 & -4 & \vdots & 17 \\ 0 & 1 & -22 & \vdots & 88 \\ 0 & 0 & 83 & \vdots & -332 \end{bmatrix}$$

$$\begin{matrix} \\ \\ \frac{1}{83}R_3 \end{matrix} \begin{bmatrix} 1 & 0 & -4 & \vdots & 17 \\ 0 & 1 & -22 & \vdots & 88 \\ 0 & 0 & 1 & \vdots & -4 \end{bmatrix}$$

$$\begin{matrix} 4R_3 + R_1 \\ 22R_3 + R_2 \\ \\ \end{matrix} \begin{bmatrix} 1 & 0 & 0 & \vdots & 1 \\ 0 & 1 & 0 & \vdots & 0 \\ 0 & 0 & 1 & \vdots & -4 \end{bmatrix}$$

The solution is $(1, 0, -4)$.

69. $\begin{vmatrix} 9 & 8 \\ 10 & 10 \end{vmatrix} = 9(10) - 10(8) = 90 - 80 = 10$

71.
$$\begin{vmatrix} 8 & 6 & 3 \\ 6 & 3 & 0 \\ 3 & 0 & 2 \end{vmatrix} = (3)\begin{vmatrix} 6 & 3 \\ 3 & 0 \end{vmatrix} - 0\begin{vmatrix} 8 & 3 \\ 6 & 0 \end{vmatrix} + 2\begin{vmatrix} 8 & 6 \\ 6 & 3 \end{vmatrix} \text{ (third row)}$$
$$= (3)(-9) - 0 + (2)(-12)$$
$$= -27 - 24$$
$$= -51$$

73.
$$\begin{vmatrix} 8 & 3 & 2 \\ 1 & -2 & 4 \\ 6 & 0 & 5 \end{vmatrix} = 6\begin{vmatrix} 3 & 2 \\ -2 & 4 \end{vmatrix} - 0 + 5\begin{vmatrix} 8 & 3 \\ 1 & -2 \end{vmatrix} \text{ (third row)}$$
$$= (6)(16) + (5)(-19) = 1$$

75. $\begin{bmatrix} 7 & 12 & \vdots & 63 \\ 2 & 3 & \vdots & 15 \end{bmatrix}$

$$D = \begin{vmatrix} 7 & 12 \\ 2 & 3 \end{vmatrix} = 21 - 24 = -3$$

$$x = \frac{D_x}{D} = \frac{\begin{vmatrix} 63 & 12 \\ 15 & 3 \end{vmatrix}}{-3} = \frac{189 - 180}{-3} = \frac{9}{-3} = -3$$

$$y = \frac{D_y}{D} = \frac{\begin{vmatrix} 7 & 63 \\ 2 & 15 \end{vmatrix}}{-3} = \frac{105 - 126}{-3} = \frac{-21}{-3} = 7$$

The solution is $(-3, 7)$.

77. $\begin{bmatrix} -1 & 1 & 2 & \vdots & 1 \\ 2 & 3 & 1 & \vdots & -2 \\ 5 & 4 & 2 & \vdots & 4 \end{bmatrix}$

$$D = \begin{vmatrix} -1 & 1 & 2 \\ 2 & 3 & 1 \\ 5 & 4 & 2 \end{vmatrix} = (-1)\begin{vmatrix} 3 & 1 \\ 4 & 2 \end{vmatrix} - (1)\begin{vmatrix} 2 & 1 \\ 5 & 2 \end{vmatrix} + (2)\begin{vmatrix} 2 & 3 \\ 5 & 4 \end{vmatrix} = (-1)(2) - (1)(-1) + (2)(-7) = -2 + 1 - 14 = -15$$

$$x = \frac{\begin{vmatrix} 1 & 1 & 2 \\ -2 & 3 & 1 \\ 4 & 4 & 2 \end{vmatrix}}{-15} = \frac{(1)\begin{vmatrix} 3 & 1 \\ 4 & 2 \end{vmatrix} - (1)\begin{vmatrix} -2 & 1 \\ 4 & 2 \end{vmatrix} + (2)\begin{vmatrix} -2 & 3 \\ 4 & 4 \end{vmatrix}}{-15} = \frac{(1)(2) - (1)(-8) + (2)(-20)}{-15} = \frac{2 + 8 - 40}{-15} = \frac{-30}{-15} = 2$$

$$y = \frac{\begin{vmatrix} -1 & 1 & 2 \\ 2 & -2 & 1 \\ 5 & 4 & 2 \end{vmatrix}}{-15} = \frac{(-1)\begin{vmatrix} -2 & 1 \\ 4 & 2 \end{vmatrix} - (1)\begin{vmatrix} 2 & 1 \\ 5 & 2 \end{vmatrix} + (2)\begin{vmatrix} 2 & -2 \\ 5 & 4 \end{vmatrix}}{-15} = \frac{(-1)(-8) - (1)(-1) + (2)(18)}{-15} = \frac{8 + 1 + 36}{-15} = \frac{45}{-15} = -3$$

$$z = \frac{\begin{vmatrix} -1 & 1 & 1 \\ 2 & 3 & -2 \\ 5 & 4 & 4 \end{vmatrix}}{-15} = \frac{(-1)\begin{vmatrix} 3 & -2 \\ 4 & 4 \end{vmatrix} - (1)\begin{vmatrix} 2 & -2 \\ 5 & 4 \end{vmatrix} + (1)\begin{vmatrix} 2 & 3 \\ 5 & 4 \end{vmatrix}}{-15} = \frac{(-1)(20) - (1)(18) + (1)(-7)}{-15} = \frac{-20 - 18 - 7}{-15} = \frac{-45}{-15} = 3$$

The solution is $(2, -3, 3)$.

79. $(x_1, y_1) = (1, 0), (x_2, y_2) = (5, 0), (x_3, y_3) = (5, 8)$

$$\begin{vmatrix} x_1 & y_1 & 1 \\ x_2 & y_2 & 1 \\ x_3 & y_3 & 1 \end{vmatrix} = \begin{vmatrix} 1 & 0 & 1 \\ 5 & 0 & 1 \\ 5 & 8 & 1 \end{vmatrix} = -0 + 0 - (8)\begin{vmatrix} 1 & 1 \\ 5 & 1 \end{vmatrix} = (-8)(-4) = 32$$

Area $= +\frac{1}{2}(32) = 16$

81. $(x_1, y_1) = (1, 2), (x_2, y_2) = (4, -5), (x_3, y_3) = (3, 2)$

$$\begin{vmatrix} x_1 & y_1 & 1 \\ x_2 & y_2 & 1 \\ x_3 & y_3 & 1 \end{vmatrix} = \begin{vmatrix} 1 & 2 & 1 \\ 4 & -5 & 1 \\ 3 & 2 & 1 \end{vmatrix} = (1)\begin{vmatrix} 4 & -5 \\ 3 & 2 \end{vmatrix} - (1)\begin{vmatrix} 1 & 2 \\ 3 & 2 \end{vmatrix} + (1)\begin{vmatrix} 1 & 2 \\ 4 & -5 \end{vmatrix} = (1)(23) - (1)(-4) + (1)(-13) = 23 + 4 - 13 = 14$$

Area $= +\frac{1}{2}(14) = 7$

83. $(x_1, y_1) = (1, 2), (x_2, y_2) = (5, 0), (x_3, y_3) = (10, -2)$

$$\begin{vmatrix} x_1 & y_1 & 1 \\ x_2 & y_2 & 1 \\ x_3 & y_3 & 1 \end{vmatrix} = \begin{vmatrix} 1 & 2 & 1 \\ 5 & 0 & 1 \\ 10 & -2 & 1 \end{vmatrix} = 1\begin{vmatrix} 0 & 1 \\ -2 & 1 \end{vmatrix} - 2\begin{vmatrix} 5 & 1 \\ 10 & 1 \end{vmatrix} + 1\begin{vmatrix} 5 & 0 \\ 10 & -2 \end{vmatrix} = 1(2) - 2(-5) + 1(-10) = 2 + 10 - 10 = 2$$

The points are not collinear.

85. $$\begin{vmatrix} x & y & 1 \\ -4 & 0 & 1 \\ 4 & 4 & 1 \end{vmatrix} = 0$$

$$-(-4)\begin{vmatrix} y & 1 \\ 4 & 1 \end{vmatrix} + 0 - (1)\begin{vmatrix} x & y \\ 4 & 4 \end{vmatrix} = 0$$

$$(4)(y - 4) - (1)(4x - 4y) = 0$$

$$4y - 16 - 4x + 4y = 0$$

$$-4x + 8y - 16 = 0$$

$$x - 2y + 4 = 0$$

87. $\begin{cases} x + y < 5 \\ x > 2 \\ y \ge 0 \end{cases} \Rightarrow \begin{cases} y < -x + 5 \\ x > 2 \\ y \ge 0 \end{cases}$

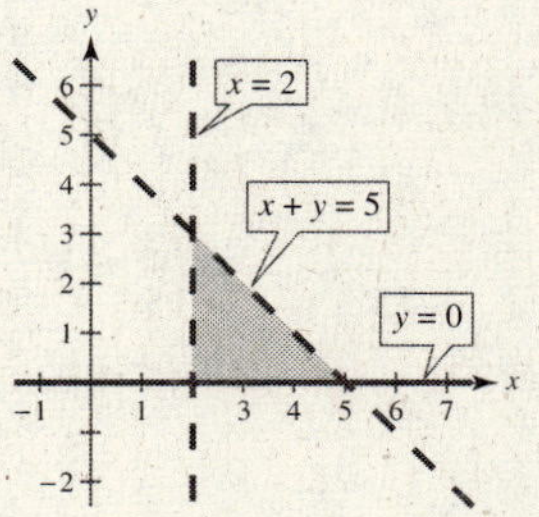

Dotted lines at $y = -x + 5$ and $x = 2$. $y = 0$ is the x-axis. Shade below $y = -x + 5$ and above $y = 0$. Shade to the right of $x = 2$.

89. $\begin{cases} x + 2y \le 160 \\ 3x + y \le 180 \\ x \ge 0 \\ y \ge 0 \end{cases} \Rightarrow \begin{cases} y \le -\frac{1}{2}x + 80 \\ y \le -3x + 180 \\ x \ge 0 \\ y \ge 0 \end{cases}$

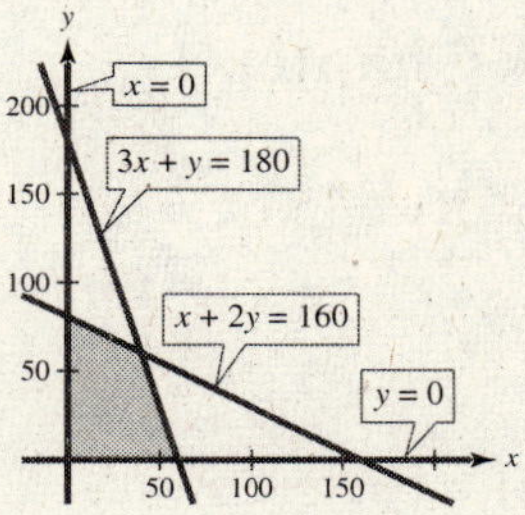

Solid lines at $y = -\frac{1}{2}x + 80$ and $y = -3x + 180$. $x = 0$ is the y-axis and $y = 0$ is the x-axis. Shade below $y = -\frac{1}{2}x + 80$ and $y = -3x + 180$. Shade to the right of $x = 0$ and above $y = 0$.

91. *Verbal Model:*

$\boxed{\text{Cartons of soup for soup kitchen}} + \boxed{\text{Cartons of soup for homeless shelter}} \le 500$

$\boxed{\text{Cartons of soup for soup kitchen}} \ge 150$

$\boxed{\text{Cartons of soup for homeless shelter}} \ge 220$

Labels:

Cartons of soup for soup kitchen $= x$

Cartons of soup for homeless shelter $= y$

System:

$$\begin{aligned} x + y &\le 500 \\ x &\ge 150 \\ y &\ge 220 \end{aligned}$$

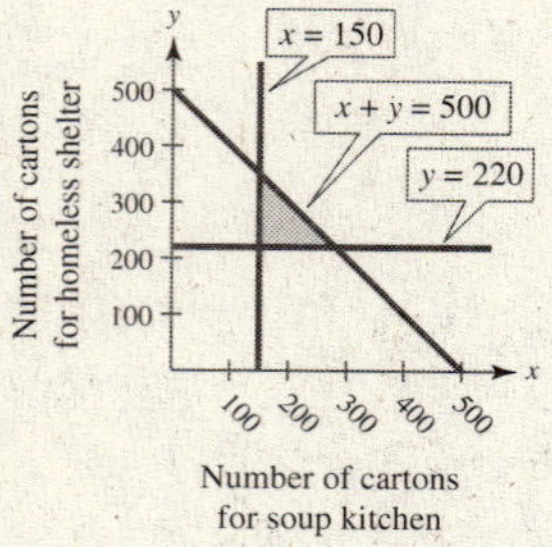

Chapter Test for Chapter 4

1. (a) $(2, 1)$

$$\begin{aligned} 2x - 2y &= 2 & -x + 2y &= 0 \\ 2(2) - 2(1) &\stackrel{?}{=} 2 & -2 + 2(1) &\stackrel{?}{=} 0 \\ 4 - 2 &\stackrel{?}{=} 2 & -2 + 2 &\stackrel{?}{=} 0 \\ 2 &= 2 & 0 &= 0 \end{aligned}$$

$(2, 1)$ is a solution.

(b) $(4, 3)$

$$\begin{aligned} 2x - 2y &= 2 & -x + 2y &= 0 \\ 2(4) - 2(3) &\stackrel{?}{=} 2 & -4 + 2(3) &\stackrel{?}{=} 0 \\ 8 - 6 &\stackrel{?}{=} 2 & -4 + 6 &\stackrel{?}{=} 0 \\ 2 &= 2 & 2 &\ne 0 \end{aligned}$$

$(4, 3)$ is not a solution.

2.

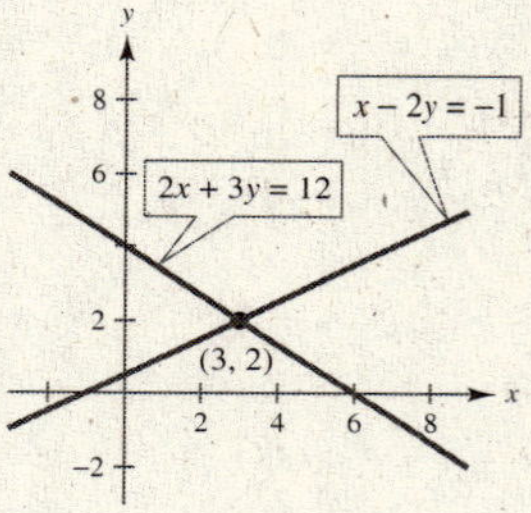

The solution is $(3, 2)$.

3. $4x - y = 1$
$4x - 3y = -5$

$4x - y = 1$
$-y = -4x + 1$
$y = 4x - 1$

$4x - 3(4x - 1) = -5$
$4x - 12x + 3 = -5$
$-8x + 3 = -5$
$-8x = -8$
$x = 1$

$4(1) - y = 1$
$-y = -3$
$y = 3$

The solution is $(1, 3)$.

4. $2x - 2y = -2$
$3x + y = 9$

Solve for y.
$y = -3x + 9$

Substitute into first equation.
$2x - 2(-3x + 9) = -2$
$2x + 6x - 18 = -2$
$8x = 16$
$x = 2$

$y = -3(2) + 9 = 3$

The solution is $(2, 3)$.

5.
$$\begin{array}{rcl} 3x - 4y &=& -14 \\ -3x + y &=& 8 \\ \hline -3y &=& -6 \\ y &=& 2 \end{array}$$

$3x - 4(2) = -14$
$3x = -6$
$x = -2$

The solution is $(-2, 2)$.

6.
$$\begin{cases} x + 2y - 4z = 0 \\ 3x + y - 2z = 5 \\ 3x - y + 2z = 7 \end{cases}$$

$$\begin{cases} x + 2y - 4z = 0 \\ -5y + 10z = 5 \\ -7y + 14z = 7 \end{cases}$$

$$\begin{cases} x + 2y - 4z = 0 \\ y - 2z = -1 \\ -7y + 14z = 7 \end{cases}$$

$$\begin{cases} x + 2y - 4z = 0 \\ y - 2z = -1 \\ 0 = 0 \end{cases}$$

Let $a = z$. (a is any real number.)

$y = 2z - 1$
$x = -2y + 4z$
$= -2(2z - 1) + 4z$
$= -4z + 2 + 4z$
$x = 2$

A solution is $(2, 2a - 1, a)$.

7.
$$\begin{bmatrix} 1 & 0 & -3 & \vdots & -10 \\ 0 & -2 & 2 & \vdots & 0 \\ 1 & -2 & 0 & \vdots & -7 \end{bmatrix}$$

$$\begin{matrix} \\ -\frac{1}{2}R_2 \\ -R_1 + R_3 \end{matrix} \begin{bmatrix} 1 & 0 & -3 & \vdots & -10 \\ 0 & 1 & -1 & \vdots & 0 \\ 0 & -2 & 3 & \vdots & 3 \end{bmatrix}$$

$$\begin{matrix} \\ \\ 2R_2 + R_3 \end{matrix} \begin{bmatrix} 1 & 0 & -3 & \vdots & -10 \\ 0 & 1 & -1 & \vdots & 0 \\ 0 & 0 & 1 & \vdots & 3 \end{bmatrix}$$

$z = 3$ $\quad$ $y - 3 = 0$ $\quad$ $x - 3(3) = -10$
$y = 3$ $\quad$ $x = -1$

The solution is $(-1, 3, 3)$.

8.

$$\begin{bmatrix} 1 & -3 & 1 & \vdots & -3 \\ 3 & 2 & -5 & \vdots & 18 \\ 0 & 1 & 1 & \vdots & -1 \end{bmatrix}$$

$$\begin{matrix} \\ -3R_1 + R_2 \\ \\ \end{matrix} \begin{bmatrix} 1 & -3 & 1 & \vdots & -3 \\ 0 & 11 & -8 & \vdots & 27 \\ 0 & 1 & 1 & \vdots & -1 \end{bmatrix}$$

$$\begin{matrix} \\ R_2 \\ R_3 \end{matrix} \begin{bmatrix} 1 & -3 & 1 & \vdots & -3 \\ 0 & 1 & 1 & \vdots & -1 \\ 0 & 11 & -8 & \vdots & 27 \end{bmatrix}$$

$$\begin{matrix} 3R_2 + R_1 \\ \\ -11R_2 + R_3 \end{matrix} \begin{bmatrix} 1 & 0 & 4 & \vdots & -6 \\ 0 & 1 & 1 & \vdots & -1 \\ 0 & 0 & -19 & \vdots & 38 \end{bmatrix}$$

$$\begin{matrix} \\ \\ -\frac{1}{19}R_3 \end{matrix} \begin{bmatrix} 1 & 0 & 4 & \vdots & -6 \\ 0 & 1 & 1 & \vdots & -1 \\ 0 & 0 & 1 & \vdots & -2 \end{bmatrix}$$

$z = -2 \qquad y + (-2) = -1 \qquad x + 4(-2) = -6$

$\qquad\qquad\quad y = 1 \qquad\qquad\qquad x = 2$

The solution is $(2, 1, -2)$.

9.

$$\begin{bmatrix} 2 & -7 & \vdots & 7 \\ 3 & 7 & \vdots & 13 \end{bmatrix}$$

$$D = \begin{vmatrix} 2 & -7 \\ 3 & 7 \end{vmatrix} = 14 + 21 = 35$$

$$x = \frac{D_x}{D} = \frac{\begin{vmatrix} 7 & -7 \\ 13 & 7 \end{vmatrix}}{35} = \frac{49 + 91}{35} = \frac{140}{35} = 4$$

$$y = \frac{D_y}{D} = \frac{\begin{vmatrix} 2 & 7 \\ 3 & 13 \end{vmatrix}}{35} = \frac{26 - 21}{35} = \frac{5}{35} = \frac{1}{7}$$

The solution is $\left(4, \frac{1}{7}\right)$.

10.

$$\begin{bmatrix} 3 & -2 & 1 & \vdots & 12 \\ 1 & -3 & 0 & \vdots & 2 \\ -3 & 0 & -9 & \vdots & -6 \end{bmatrix}$$

$$\begin{matrix} R_1 \\ \\ -\frac{1}{3}R_3 \end{matrix} \begin{bmatrix} 1 & 0 & 3 & \vdots & 2 \\ 1 & -3 & 0 & \vdots & 2 \\ 3 & -2 & 1 & \vdots & 12 \end{bmatrix}$$

$$\begin{matrix} \\ -1R_1 + R_2 \\ -3R_1 + R_3 \end{matrix} \begin{bmatrix} 1 & 0 & 3 & \vdots & 2 \\ 0 & -3 & -3 & \vdots & 0 \\ 0 & -2 & -8 & \vdots & 6 \end{bmatrix}$$

$$\begin{matrix} \\ -\frac{1}{3}R_2 \\ -\frac{1}{2}R_3 \end{matrix} \begin{bmatrix} 1 & 0 & 3 & \vdots & 2 \\ 0 & 1 & 1 & \vdots & 0 \\ 0 & 1 & 4 & \vdots & -3 \end{bmatrix}$$

$$\begin{matrix} \\ \\ -R_2 + R_3 \end{matrix} \begin{bmatrix} 1 & 0 & 3 & \vdots & 2 \\ 0 & 1 & 1 & \vdots & 0 \\ 0 & 0 & 3 & \vdots & -3 \end{bmatrix}$$

$$\begin{matrix} \\ \\ \frac{1}{3}R_3 \end{matrix} \begin{bmatrix} 1 & 0 & 3 & \vdots & 2 \\ 0 & 1 & 1 & \vdots & 0 \\ 0 & 0 & 1 & \vdots & -1 \end{bmatrix}$$

$$\begin{matrix} -3R_3 + R_1 \\ -R_3 + R_2 \\ \\ \end{matrix} \begin{bmatrix} 1 & 0 & 0 & \vdots & 5 \\ 0 & 1 & 0 & \vdots & 1 \\ 0 & 0 & 1 & \vdots & -1 \end{bmatrix}$$

The solution is $(5, 1, -1)$.

11.

$$\begin{bmatrix} 4 & 1 & 2 & \vdots & -4 \\ 0 & 3 & 1 & \vdots & 8 \\ -3 & 1 & -3 & \vdots & 5 \end{bmatrix}$$

$$\begin{matrix} \frac{1}{4}R_1 \\ \\ \\ \end{matrix} \begin{bmatrix} 1 & \frac{1}{4} & \frac{1}{2} & \vdots & -1 \\ 0 & 3 & 1 & \vdots & 8 \\ -3 & 1 & -3 & \vdots & 5 \end{bmatrix}$$

$$\begin{matrix} \\ \frac{1}{3}R_2 \\ 3R_1 + R_3 \end{matrix} \begin{bmatrix} 1 & \frac{1}{4} & \frac{1}{2} & \vdots & -1 \\ 0 & 1 & \frac{1}{3} & \vdots & \frac{8}{3} \\ 0 & \frac{7}{4} & -\frac{3}{2} & \vdots & 2 \end{bmatrix}$$

$$\begin{matrix} -\frac{1}{4}R_2 + R_1 \\ \\ -\frac{7}{4}R_2 + R_3 \end{matrix} \begin{bmatrix} 1 & 0 & \frac{5}{12} & \vdots & -\frac{5}{3} \\ 0 & 1 & \frac{1}{3} & \vdots & \frac{8}{3} \\ 0 & 0 & -\frac{25}{12} & \vdots & -\frac{8}{3} \end{bmatrix}$$

$$\begin{matrix} \\ \\ -\frac{12}{25}R_3 \end{matrix} \begin{bmatrix} 1 & 0 & \frac{5}{12} & \vdots & -\frac{5}{3} \\ 0 & 1 & \frac{1}{3} & \vdots & \frac{8}{3} \\ 0 & 0 & 1 & \vdots & \frac{32}{25} \end{bmatrix}$$

$$\begin{matrix} -\frac{5}{12}R_3 + R_1 \\ -\frac{1}{3}R_3 + R_2 \\ \\ \end{matrix} \begin{bmatrix} 1 & 0 & 0 & \vdots & -\frac{11}{5} \\ 0 & 1 & 0 & \vdots & \frac{56}{25} \\ 0 & 0 & 1 & \vdots & \frac{32}{25} \end{bmatrix}$$

The solution is $\left(-\frac{11}{5}, \frac{56}{25}, \frac{32}{25}\right)$.

12. $\begin{vmatrix} 2 & -2 & 0 \\ -1 & 3 & 1 \\ 2 & 8 & 1 \end{vmatrix} = 0\begin{vmatrix} -1 & 3 \\ 2 & 8 \end{vmatrix} - 1\begin{vmatrix} 2 & -2 \\ 2 & 8 \end{vmatrix} + 1\begin{vmatrix} 2 & -2 \\ -1 & 3 \end{vmatrix} = 0(-14) - 1(20) + 1(4) = 0 - 20 + 4 = -16$

13. $(x_1, y_1) = (0, 0), (x_2, y_2) = (5, 4), (x_3, y_3) = (6, 0)$

$$\begin{vmatrix} x_1 & y_1 & 1 \\ x_2 & y_2 & 1 \\ x_3 & y_3 & 1 \end{vmatrix} = \begin{vmatrix} 0 & 0 & 1 \\ 5 & 4 & 1 \\ 6 & 0 & 1 \end{vmatrix} = (1)\begin{vmatrix} 5 & 4 \\ 6 & 0 \end{vmatrix} = (1)(-24) = -24$$

Area $= -\frac{1}{2}(-24) = 12$

14. $\begin{cases} x - 2y > -3 \\ 2x + 3y \le 22 \\ y \ge 0 \end{cases} \Rightarrow \begin{array}{l} y < \frac{1}{2}x + \frac{3}{2} \\ y \le -\frac{2}{3}x + \frac{22}{3} \\ y \ge 0 \end{array}$

Dotted line at $y = \frac{1}{2}x + \frac{3}{2}$ and solid line at $y = -\frac{2}{3}x + \frac{22}{3}$. $y = 0$ is the x-axis.

Shade below $y = \frac{1}{2}x + \frac{3}{2}$ and $y = -\frac{2}{3}x + \frac{22}{3}$ and above $y = 0$.

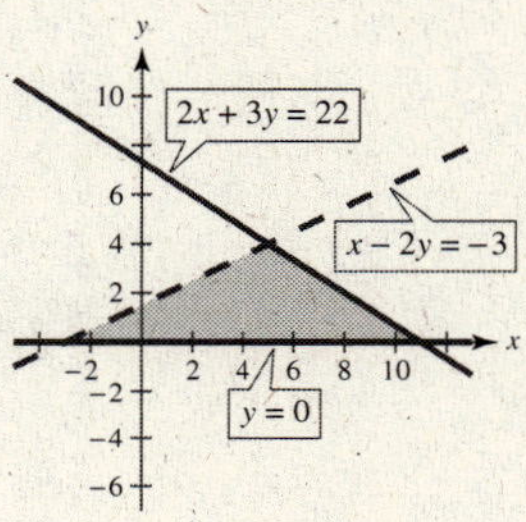

15. *Verbal Model:* $2 \cdot$ [Length] $+ 2 \cdot$ [Width] $= 68$

[Width] $= \frac{8}{9} \cdot$ [Length]

Labels: Length $= x$

Width $= y$

System: $2x + 2y = 68$

$y = \frac{8}{9}x$

Solve by substitution.

$$2x + 2\left(\frac{8}{9}x\right) = 68$$
$$18x + 16x = 612$$
$$34x = 612$$
$$x = 18$$
$$y = \frac{8}{9}(18)$$
$$y = 16$$

16 feet × 18 feet

16. *Verbal Model:* $\boxed{\text{Investment 1}} + \boxed{\text{Investment 2}} + \boxed{\text{Investment 3}} = \$25{,}000$

$4.5\% \cdot \boxed{\text{Investment 1}} + 5\% \cdot \boxed{\text{Investment 2}} + 8\% \cdot \boxed{\text{Investment 3}} = \$1275$

$\boxed{\text{Investment 1}} + 4000 = \boxed{\text{Investment 3}} + 10{,}000$

Labels: Investment 1 $= x$

Investment 2 $= y$

Investment 3 $= z$

System:

$$\begin{aligned} x + y + z &= 25{,}000 \\ 0.045x + 0.05y + 0.08z &= 1275 \end{aligned}$$

$y + 4000 = z + 10{,}000 \Rightarrow y - z = 6000$

Using determinants and Cramer's Rule:

$$\begin{bmatrix} 1 & 1 & 1 & 25{,}000 \\ 0.045 & 0.05 & 0.08 & 1275 \\ 0 & 1 & -1 & 6000 \end{bmatrix} \qquad D = \begin{vmatrix} 1 & 1 & 1 \\ 0.045 & 0.05 & 0.08 \\ 0 & 1 & -1 \end{vmatrix} = -0.04$$

$$x = \frac{D_x}{D} = \frac{\begin{vmatrix} 25{,}000 & 1 & 1 \\ 1275 & 0.05 & 0.08 \\ 6000 & 1 & -1 \end{vmatrix}}{-0.04} = \frac{-520}{-0.04} = \$13{,}000$$

$$y = \frac{D_y}{D} = \frac{\begin{vmatrix} 1 & 25{,}000 & 1 \\ 0.045 & 1275 & 0.08 \\ 0 & 6000 & -1 \end{vmatrix}}{-0.04} = \frac{-360}{-0.04} = \$9000$$

$$z = \frac{D_z}{D} = \frac{\begin{vmatrix} 1 & 1 & 25{,}000 \\ 0.045 & 0.05 & 1275 \\ 0 & 1 & 6000 \end{vmatrix}}{-0.04} = \frac{-120}{-0.04} = \$3000$$

Use your graphing calculator to find each determinant.

\$13,000 at 4.5%, \$9,000 at 5%, \$3,000 at 8%

17. *Verbal Model:* $30 \cdot \boxed{\text{Reserved seat tickets}} + 40 \cdot \boxed{\text{Floor seat tickets}} \geq 300{,}000$

$\boxed{\text{Reserved seat tickets}} \leq 9000$

$\boxed{\text{Floor seat tickets}} \leq 4000$

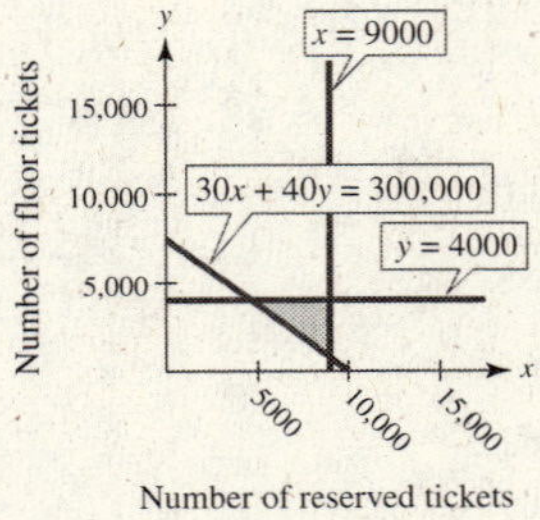

Labels: Reserved seat tickets $= x$

Floor seat tickets $= y$

System:

$$\begin{aligned} 30x + 40y &\geq 300{,}000 \\ x &\leq 9000 \\ y &\leq 4000 \end{aligned}$$

Cumulative Test for Chapters 1–4

1. (a) $-2 > -4$

(b) $\frac{2}{3} > \frac{1}{2}$

(c) $-4.5 = -|-4.5|$

2. "The number n is tripled and the product is decreased by 8," is expressed by $3n - 8$.

3. $t(3t - 1) - 2(t + 4) = 3t^2 - t - 2t - 8$
$= 3t^2 - 3t - 8$

4. $4x(x + x^2) - 6(x^2 + 4) = 4x^2 + 4x^3 - 6x^2 - 24$
$= 4x^3 + 4x^2 - 6x^2 - 24$
$= 4x^3 - 2x^2 - 24$

5.
$$\begin{aligned} 12 - 5(3 - x) &= x + 3 \\ 12 - 15 + 5x &= x + 3 \\ -3 + 5x &= x + 3 \\ -3 + 5x - x &= x + 3 - x \\ 3 - 3 + 4x &= 3 + 3 \\ 4x &= 6 \\ \frac{4x}{4} &= \frac{6}{4} \\ x &= \frac{3}{2} \end{aligned}$$

6.
$$\begin{aligned} 1 - \frac{x + 2}{4} &= \frac{7}{8} \\ 8\left[1 - \frac{x + 2}{4}\right] &= \left[\frac{7}{8}\right]8 \\ 8 - 2(x + 2) &= 7 \\ 8 - 2x - 4 &= 7 \\ 4 - 2x &= 7 \\ 4 - 4 - 2x &= 7 - 4 \\ -2x &= 3 \\ \frac{-2x}{-2} &= -\frac{3}{2} \\ x &= -\frac{3}{2} \end{aligned}$$

7. $|x - 2| \geq 3$

$x - 2 \leq -3$ or $x - 2 \geq 3$

$x \leq -1$ $\qquad$ $x \geq 5$

8.
$$\begin{aligned} -12 \leq 4x - 6 &< 10 \\ -6 \leq 4x &< 16 \\ -\tfrac{6}{4} \leq x &< 4 \\ -\tfrac{3}{2} \leq x &< 4 \end{aligned}$$

9. $1150 + 0.20(1150) = 1150 + 230 = 1380$

Your new premium is \$1380.

10.
$$\begin{aligned} \frac{9}{4.5} &= \frac{13}{x} \\ 9x &= 13(4.5) \\ x &= \frac{13(4.5)}{9} \\ x &= 6.5 \end{aligned}$$

11. *Verbal Model:* [Revenue] > [Cost]

Equation: $12.90x > 8.50x + 450$

$4.4x > 450$

$x > 102.27273$

Because a fractional part of a unit cannot be sold, $x \geq 103$.

12. No, $x - y^2 = 0$ does not represent y as a function of x.

13. $f(x) = \sqrt{x - 2}$

$D = x \geq 2$

$x - 2 \geq 0$

$2 \leq x < \infty$

14. (a) $f(x) = x^2 - 2x$

$f(3) = 3^2 - 2(3)$
$= 9 - 6$
$= 3$

(b) $f(x) = x^2 - 2x$

$f(-3c) = (-3c)^2 - 2(-3c)$
$= 9c^2 + 6c$

15. $m = \frac{6 - 0}{4 + 4} = \frac{6}{8} = \frac{3}{4}$

$$\begin{aligned} d &= \sqrt{(-4 - 4)^2 + (0 - 6)^2} \\ &= \sqrt{64 + 36} \\ &= \sqrt{100} \\ &= 10 \end{aligned}$$

Midpoint $= \left(\frac{-4 + 4}{2}, \frac{0 + 6}{2}\right) = (0, 3)$

16. (a) $2x - y = 1$

$-y = -2x + 1$

$y = 2x - 1$

$m = 2$

$y - 1 = 2(x + 2)$

$y - 1 = 2x + 4$

$2x - y + 5 = 0$

(b) $3x + 2y = 5$

$2y = -3x + 5$

$y = -\frac{3}{2}x + 5$

$m = \frac{2}{3}$

$y - 1 = \frac{2}{3}(x + 2)$

$y - 1 = \frac{2}{3}x + \frac{4}{3}$

$y = \frac{2}{3}x + \frac{7}{3}$

$3y = 2x + 7$

$2x - 3y + 7 = 0$

17. $4x + 3y - 12 = 0$

$4(0) + 3y - 12 = 0$

$3y = 12$

$y = 4$

$(0, 4)$

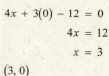

$4x + 3(0) - 12 = 0$

$4x = 12$

$x = 3$

$(3, 0)$

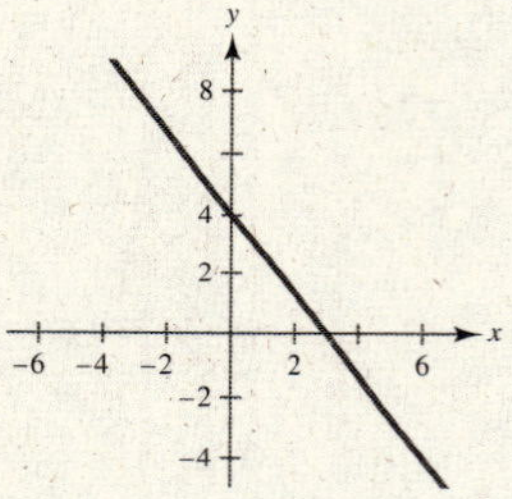

18. $y = 2 - (x - 3)^2$

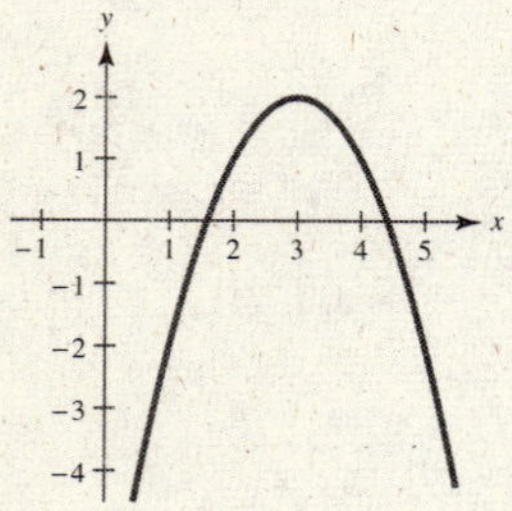

19. $\begin{cases} x + y = 6 \\ 2x - y = 3 \end{cases}$

Solve for y.

$y = 6 - x$

Substitute into second equation.

$2x - (6 - x) = 3$

$2x - 6 + x = 3$

$3x = 9$

$x = 3$

$y = 6 - 3$

$= 3$

The solution is $(3, 3)$.

20. $\begin{cases} 2x + y = 6 \\ 3x - 2y = 16 \end{cases}$

$4x + 2y = 12$

$3x - 2y = 16$

$7x = 28$

$x = 4$

$2(4) + y = 6$

$8 + y = 6$

$y = -2$

The solution is $(4, -2)$.

21. $\begin{cases} 2x + y - 2z = 1 \\ x - z = 1 \\ 3x + 3y + z = 12 \end{cases}$

$$\begin{bmatrix} 2 & 1 & -2 & \vdots & 1 \\ 1 & 0 & -1 & \vdots & 1 \\ 3 & 3 & 1 & \vdots & 12 \end{bmatrix}$$

$$\begin{matrix} R_1 \\ R_2 \\ \\ \end{matrix} \begin{bmatrix} 1 & 0 & -1 & \vdots & 1 \\ 2 & 1 & -2 & \vdots & 1 \\ 3 & 3 & 1 & \vdots & 12 \end{bmatrix}$$

$$\begin{matrix} \\ -2R_1 + R_2 \\ -3R_1 + R_3 \end{matrix} \begin{bmatrix} 1 & 0 & -1 & \vdots & 1 \\ 0 & 1 & 0 & \vdots & -1 \\ 0 & 3 & 4 & \vdots & 9 \end{bmatrix}$$

$$\begin{matrix} \\ \\ -3R_2 + R_3 \end{matrix} \begin{bmatrix} 1 & 0 & -1 & \vdots & 1 \\ 0 & 1 & 0 & \vdots & -1 \\ 0 & 0 & 4 & \vdots & 12 \end{bmatrix}$$

$$\begin{matrix} \\ \\ \frac{1}{4}R_3 \end{matrix} \begin{bmatrix} 1 & 0 & -1 & \vdots & 1 \\ 0 & 1 & 0 & \vdots & -1 \\ 0 & 0 & 1 & \vdots & 3 \end{bmatrix}$$

$$\begin{matrix} R_3 + R_1 \\ \\ \\ \end{matrix} \begin{bmatrix} 1 & 0 & 0 & \vdots & 4 \\ 0 & 1 & 0 & \vdots & -1 \\ 0 & 0 & 1 & \vdots & 3 \end{bmatrix}$$

The solution is $(4, -1, 3)$.

CHAPTER 5
Polynomials and Factoring

CHAPTER 5
Polynomials and Factoring

Section 5.1 Integer Exponents and Scientific Notation

1. (a) $-3x^3 \cdot x^5 = -3(x^3 \cdot x^5) = -3x^{3+5} = -3x^8$

(b) $(-3x)^2 \cdot x^5 = 9x^2 \cdot x^5 = 9x^{2+5} = 9x^7$

3. (a) $(-5z^2)^3 = (-5)^3 \cdot (z^2)^3 = -125z^{2\cdot 3} = -125z^6$

(b) $(-5z^4)^2 = (-5)^2(z^4)^2 = 25z^{4\cdot 2} = 25z^8$

5. (a) $(u^3v)(2v^2) = 2 \cdot u^3 \cdot v^{1+2} = 2u^3v^3$

(b) $(-4u^4)(u^5v) = -4 \cdot u^{4+5} \cdot v = -4u^9v$

7. (a) $5u^2 \cdot (-3u^6) = 5 \cdot -3 \cdot u^2 \cdot u^6$

$= -15u^{2+6} = -15u^8$

(b) $(2u)^4(4u) = 2^4u^4 \cdot 4u = 16 \cdot 4 \cdot u^{4+1} = 64u^5$

9. (a) $-(m^5n)^3(-m^2n^2)^2 = -m^{5\cdot 3}n^3 \cdot m^{2\cdot 2}n^{2\cdot 2}$

$= -m^{15}n^3 \cdot m^4n^4$

$= -m^{15+4} \cdot n^{3+4} = -m^{19}n^7$

(b) $(-m^5n)(m^2n^2) = -m^{5+2}n^{1+2} = -m^7n^3$

11. (a) $\dfrac{27m^5n^6}{9mn^3} = \dfrac{27}{9} \cdot \dfrac{m^5}{m} \cdot \dfrac{n^6}{n^3}$

$= 3 \cdot m^{5-1} \cdot n^{6-3} = 3m^4n^3$

(b) $\dfrac{-18m^3n^6}{-6mn^3} = \dfrac{-18}{-6} \cdot \dfrac{m^3}{m} \cdot \dfrac{n^6}{n^3}$

$= 3 \cdot m^{3-1} \cdot n^{6-3} = 3m^2n^3$

13. (a) $\left(\dfrac{3x}{4y}\right)^2 = \dfrac{3^2 \cdot x^2}{4^2 \cdot y^2} = \dfrac{9x^2}{16y^2}$

(b) $\left(\dfrac{5u}{3v}\right)^3 = \dfrac{5^3 \cdot u^3}{3^3 \cdot v^3} = \dfrac{125u^3}{27v^3}$

15. (a) $-\dfrac{(-2x^2y)^3}{9x^2y^2} = -\dfrac{(-2)^3(x^2)^3y^3}{9x^2y^2}$

$= -\dfrac{(-8)x^6y^3}{9x^2y^2} = \dfrac{8x^{6-2}y^{3-2}}{9} = \dfrac{8x^4y}{9}$

(b) $-\dfrac{(-2xy^3)^2}{6y^2} = -\dfrac{(-2)^2x^2(y^3)^2}{6y^2}$

$= -\dfrac{4x^2y^6}{6y^2} = -\dfrac{2x^2y^{6-2}}{3} = -\dfrac{2x^2y^4}{3}$

17. (a) $\left[\dfrac{(-5u^3v)^2}{10u^2v}\right]^2 = \left[\dfrac{(-5)^2 \cdot (u^3)^2 \cdot (v)^2}{10u^2v}\right]^2$

$= \left[\dfrac{25u^6v^2}{10u^2v}\right]^2$

$= \left[\dfrac{25}{10} \cdot \dfrac{u^6}{u^2} \cdot \dfrac{v^2}{v}\right]^2$

$= \left[\dfrac{5}{2} \cdot u^{6-2} \cdot v^{2-1}\right]^2$

$= \left[\dfrac{5}{2}u^4v\right]^2$

$= \dfrac{25}{4}u^8v^2$

(b) $\left[\dfrac{-5(u^3v)^2}{10u^2v}\right]^2 = \left[\dfrac{-5 \cdot (u^3)^2 \cdot (v)^2}{10u^2v}\right]^2$

$= \left[\dfrac{-5u^6v^2}{10u^2v}\right]^2$

$= \left[\dfrac{-5}{10} \cdot \dfrac{u^6}{u^2} \cdot \dfrac{v^2}{v}\right]^2$

$= \left[-\dfrac{1}{2} \cdot u^{6-2} \cdot v^{2-1}\right]^2$

$= \left[\dfrac{-1}{2}u^4v\right]^2$

$= \dfrac{1}{4}u^8v^2$

19. (a) $\dfrac{x^{2n+4}y^{4n}}{x^5y^{2n+1}} = x^{2n+4-5}y^{4n-(2n+1)}$

$= x^{2n-1}y^{4n-2n-1} = x^{2n-1}y^{2n-1}$

(b) $\dfrac{x^{6n}y^{n-7}}{x^{4n+2}y^5} = x^{6n-(4n+2)}y^{n-7-5}$

$= x^{6n-4n-2}y^{n-12} = x^{2n-2}y^{n-12}$

21. $5^{-2} = \dfrac{1}{5^2} = \dfrac{1}{25}$

23. $-10^{-3} = -\dfrac{1}{10^3} = -\dfrac{1}{1000}$

25. $(-3)^0 = 1$

27. $\dfrac{1}{4^{-3}} = \dfrac{1}{\left(\frac{1}{4}\right)^3} = 4^3 = 64$

29. $\dfrac{1}{(-2)^{-5}} = \dfrac{1}{\left(-\frac{1}{2}\right)^5} = \dfrac{1}{-\frac{1}{32}} = -32$

31. $\left(\frac{2}{3}\right)^{-1} = \frac{3}{2}$

33. $\left(\frac{3}{16}\right)^0 = 1$

35. $27 \cdot 3^{-3} = 3^3 \cdot 3^{-3} = 3^{3+(-3)} = 3^0 = 1$

37. $\dfrac{3^4}{3^{-2}} = 3^{4-(-2)} = 3^{4+2} = 3^6 = 729$

39. $\dfrac{10^3}{10^{-2}} = 10^{3-(-2)} = 10^{3+2} = 10^5 = 100{,}000$

41. $\left(4^2 \cdot 4^{-1}\right)^{-2} = \left(4^{2+(-1)}\right)^{-2}$
$= \left(4^1\right)^{-2}$
$= 4^{-2}$
$= \dfrac{1}{4^2}$
$= \dfrac{1}{16}$

43. $\left(2^{-3}\right)^2 = 2^{-6} = \dfrac{1}{2^6} = \dfrac{1}{64}$

45. $2^{-3} + 2^{-4} = \dfrac{1}{2^3} + \dfrac{1}{2^4}$
$= \dfrac{1}{8} + \dfrac{1}{16}$
$= \dfrac{2}{16} + \dfrac{1}{16}$
$= \dfrac{3}{16}$

47. $\left(\frac{3}{4} + \frac{5}{8}\right)^{-2} = \left(\frac{6}{8} + \frac{5}{8}\right)^{-2} = \left(\frac{11}{8}\right)^{-2} = \left(\frac{8}{11}\right)^2 = \frac{64}{121}$

49. $\left(5^0 - 4^{-2}\right)^{-1} = \left(1 - \dfrac{1}{4^2}\right)^{-1}$
$= \left(\dfrac{16}{16} - \dfrac{1}{16}\right)^{-1}$
$= \left(\dfrac{15}{16}\right)^{-1}$
$= \dfrac{16}{15}$

51. $y^4 \cdot y^{-2} = y^{4+(-2)} = y^2$

53. $z^5 \cdot z^{-3} = z^{5+(-3)} = z^2$

55. $7x^{-4} = \dfrac{7}{x^4}$

57. $(4x)^{-3} = \dfrac{1}{(4x)^3} = \dfrac{1}{64x^3}$

59. $\dfrac{1}{x^{-6}} = x^6$

61. $\dfrac{8a^{-6}}{6a^{-7}} = \dfrac{4}{3}a^{(-6)-(-7)} = \dfrac{4}{3}a^{-6+7} = \dfrac{4}{3}a$

63. $\dfrac{(4t)^0}{t^{-2}} = \dfrac{1}{t^{-2}} = t^2$

65. $\left(2x^2\right)^{-2} = \dfrac{1}{\left(2x^2\right)^2} = \dfrac{1}{4x^4}$

67. $\left(-3x^{-3}y^2\right)\left(4x^2y^{-5}\right) = -3 \cdot 4 \cdot x^{-3+2} \cdot y^{2+(-5)}$
$= -12x^{-1}y^{-3} = -\dfrac{12}{xy^3}$

69. $\left(3x^2y^{-2}\right)^{-2} = 3^{-2}x^{-4}y^4 = \dfrac{y^4}{9x^4}$

71. $\left(\dfrac{x}{10}\right)^{-1} = \dfrac{10}{x}$

73. $\dfrac{6x^3y^{-3}}{12x^{-2}y} = \dfrac{6x^{3-(-2)}y^{-3-1}}{6 \cdot 2} = \dfrac{x^5y^{-4}}{2} = \dfrac{x^5}{2y^4}$

75. $\left(\dfrac{3u^2v^{-1}}{3^3u^{-1}v^3}\right)^{-2} = \left(\dfrac{3u^{2-(-1)}v^{-1-3}}{3^3}\right)^{-2}$
$= \left(\dfrac{u^3v^{-4}}{3^2}\right)^{-2}$
$= \left(\dfrac{3^2}{u^3v^{-4}}\right)^2$
$= \dfrac{3^4}{u^6v^{-8}}$
$= \dfrac{81v^8}{u^6}$

77. $\left(\dfrac{a^{-2}}{b^{-2}}\right)\left(\dfrac{b}{a}\right)^3 = \left(\dfrac{b^2}{a^2}\right)\left(\dfrac{b^3}{a^3}\right) = \dfrac{b^5}{a^5}$

79. $(2x^3y^{-1})^{-3}(4xy^{-6}) = (2^{-3}x^{-9}y^3)(4xy^{-6})$
$= \frac{4x^{-9+1}y^{3+(-6)}}{2^3}$
$= \frac{4x^{-8}y^{-3}}{8}$
$= \frac{1}{2x^8y^3}$

81. $u^4(6u^{-3}v^0)(7v)^0 = u^4(6u^{-3})(1) = 6u^{4+(-3)} = 6u$

83. $\left[(x^{-4}y^{-6})^{-1}\right]^2 = (x^4y^6)^2 = x^8y^{12}$

85. $\frac{(2a^{-2}b^4)^3 b}{(10a^3b)^2} = \frac{2^3a^{-6}b^{12} \cdot b}{10^2a^6b^2}$
$= \frac{8a^{-6-6}b^{12+1-2}}{100}$
$= \frac{2a^{-12}b^{11}}{25}$
$= \frac{2b^{11}}{25a^{12}}$

87. $(u + v^{-2})^{-1} = \frac{1}{u + v^{-2}}$
$= \frac{1}{u + \left(\frac{1}{v^2}\right)} \cdot \frac{v^2}{v^2}$
$= \frac{v^2}{uv^2 + 1}$

89. $\frac{a + b}{b^{-1}a + 1} = \frac{a + b}{\frac{a}{b} + 1} \cdot \frac{b}{b} = \frac{b(a + b)}{a + b} = b$

91. $x^2 \cdot x^{-3} \cdot x \cdot y = (-3)^2(-3)^{-3}(-3)(4)$
$= (-3)^{2-3+1}(4)$
$= (-3)^0(4)$
$= 4$

93. $\frac{x^2}{y^{-2}} = \frac{(-3)^2}{(4)^{-2}} = \frac{9}{\left(\frac{1}{4}\right)^2} = \frac{9}{\frac{1}{16}} = 9 \cdot 16 = 144$

95. $(x + y)^{-4} = [(-3) + 4]^{-4} = 1^{-4} = 1$

97. $\left(\frac{5x}{3y}\right)^{-1} = \left[\frac{5(-3)}{3(4)}\right]^{-1} = \left[\frac{-15}{12}\right]^{-1} = \left[\frac{-5}{4}\right]^{-1} = -\frac{4}{5}$

99. $(xy)^{-2} = [(-3)(4)]^{-2} = (-12)^{-2} = \frac{1}{(-12)^2} = \frac{1}{144}$

101. $3{,}600{,}000 = 3.6 \times 10^6$

103. $47{,}620{,}000 = 4.762 \times 10^7$

105. $0.00031 = 3.1 \times 10^{-4}$

107. $0.0000000381 = 3.81 \times 10^{-8}$

109. $57{,}300{,}000 = 5.73 \times 10^7$

111. $9{,}460{,}800{,}000{,}000 = 9.4608 \times 10^{12}$

113. $0.0899 = 8.99 \times 10^{-2}$

115. $7.2 \times 10^8 = 7.2 \times 100{,}000{,}000 = 720{,}000{,}000$

117. $1.359 \times 10^{-7} = 0.0000001359$

119. $\$3.4659 \times 10^{10} = \$3.4659 \times 10{,}000{,}000{,}000$
$= \$34{,}659{,}000{,}000$

121. $1.5 \times 10^7 = 15{,}000{,}000$

123. $4.8 \times 10^{-10} = 0.00000000048$

125. $(2 \times 10^9)(3.4 \times 10^{-4}) = (2)(3.4)(10^5) = 6.8 \times 10^5$

127. $(5 \times 10^4)^2 = 5^2 \times 10^8 = 25 \times 10^8 = 2.5 \times 10^9$

129. $\frac{3.6 \times 10^{12}}{6 \times 10^5} = \frac{3.6}{6} \times 10^{12-5} = 0.6 \times 10^7 = 6.0 \times 10^6$

131. $(4{,}500{,}000)(2{,}000{,}000{,}000) = (4.5 \times 10^6)(2 \times 10^9)$
$= (4.5)(2) \times 10^{15}$
$= 9 \times 10^{15}$

133. $\frac{64{,}000{,}000}{0.00004} = \frac{6.4 \times 10^7}{4.0 \times 10^{-5}}$
$= 1.6 \times 10^{7-(-5)}$
$= 1.6 \times 10^{12}$

135. $\frac{(0.0000565)(2{,}850{,}000{,}000{,}000)}{0.00465} = \frac{(5.65 \times 10^{-5})(2.85 \times 10^{12})}{4.65 \times 10^{-3}} = \frac{(5.65)(2.85)}{4.65} \times 10^{10} \approx 3.4629032 \times 10^{10} \approx 3.46 \times 10^{10}$

137. $\dfrac{1.357 \times 10^{12}}{(4.2 \times 10^{2})(6.87 \times 10^{-3})} = \dfrac{1.357}{(4.2)(6.87)} \times 10^{13}$

$= 0.0470299 \times 10^{13}$

$= 4.70299 \times 10^{11}$

$\approx 4.70 \times 10^{11}$

139. $(2.58 \times 10^{6})^{4} = 2.58^{4} \times (10^{6})^{4}$

$= 44.3 \times 10^{24}$

$= 4.43 \times 10^{25}$

141. $\dfrac{(5{,}000{,}000)^{3}(0.000037)^{2}}{(0.005)^{4}} = \dfrac{(5.0 \times 10^{6})^{3}(3.7 \times 10^{-5})^{2}}{(5.0 \times 10^{-3})^{4}}$

$= \dfrac{(5^{3} \times 10^{18})(3.7^{2} \times 10^{-10})}{5^{4} \times 10^{-12}}$

$= \dfrac{(125)(13.69)}{625} \times 10^{18+(-10)-(-12)}$

$= 2.738 \times 10^{18-10+12}$

$= 2.738 \times 10^{20}$

$\approx 2.74 \times 10^{20}$

145. $8.483 \times 10^{22} = 84{,}830{,}000{,}000{,}000{,}000{,}000{,}000$

147. $\dfrac{1.50 \times 10^{11}}{9.46 \times 10^{15}} = \dfrac{1.50}{9.46} \times 10^{-4}$

$\approx 0.158562 \times 10^{-4}$

$\approx 1.59 \times 10^{-5}$ year

≈ 8.4 minutes

149. $(95)(9.46 \times 10^{15}) = 899 \times 10^{15}$

$= 8.99 \times 10^{17}$ meters

151. $(2.96 \times 10^{8})(2.66 \times 10^{4}) = (2.96 \times 2.66) \times (10^{8} \times 10^{4})$

$\approx 7.87 \times 10^{12}$

$\approx 7{,}870{,}000{,}000{,}000$

153. Scientific notation makes it easier to multiply or divide very large or very small numbers because the properties of exponents make it more efficient.

155. False. $\dfrac{1}{3^{-3}} = \dfrac{1}{\left(\frac{1}{3}\right)^{3}} = \dfrac{1}{\frac{1}{27}} = 27$, which is greater than 1.

157. $a^{m} \cdot b^{n} = ab^{m+n}$ is false because the product rule can be applied only to exponential expressions with the same base.

159. $(a^{m})^{n} = a^{m+n}$ is false because the power-to-power rule applied to this expression raises the base to the product of the exponents.

161. $3x + 4x - x = (3 + 4 - 1)x = 6x$

163. $a^{2} + 2ab - b^{2} + ab + 4b^{2} = a^{2} + (2 + 1)ab + (-1 + 4)b^{2} = a^{2} + 3ab + 3b^{2}$

165.

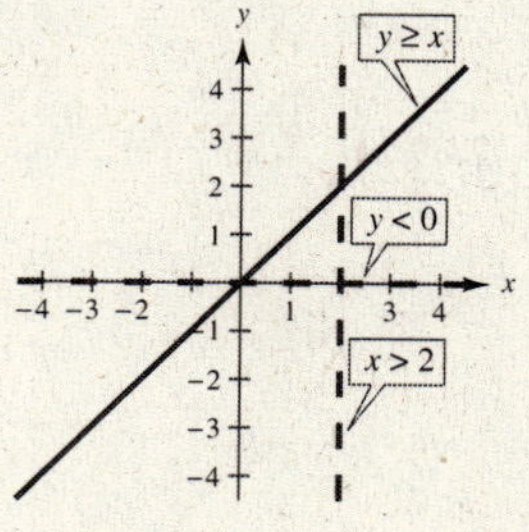

$$\begin{cases} x > 2 \\ x - y \le 0 \Rightarrow \\ y < 0 \end{cases} \quad \begin{matrix} x > 2 \\ y \ge x \\ y < 0 \end{matrix}$$

Section 5.2 Adding and Subtracting Polynomials

1. Standard form: $4y + 16$

Degree: 1

Leading coefficient: 4

3. Standard form: $x^2 + 2x - 6$

Degree: 2

Leading coefficient: 1

5. Standard form: $-42x^3 - 10x^2 + 3x + 5$

Degree: 3

Leading coefficient: -42

7. Standard form: $t^5 - 14t^4 - 20t + 4$

Degree: 5

Leading coefficient: 1

9. Standard form: -4

Degree: 0

Leading coefficient: -4

11. Standard form: $-16t^2 + v_0t$

Degree: 2

Leading coefficient: -16

13. $12 - 5y^2$ is a binomial.

15. $x^3 + 2x^2 - 4$ is a trinomial.

17. 5 is a monomial.

19. A monomial of degree 2 is any monomial in x with x^2 and any leading coefficient such as $3x^2$.

21. A binomial of degree 3 and leading coefficient 8 is any binomial in x with leading term $8x^3$ and one other term of degree less than three such as $8x^3 + 5$.

23. $y^{-3} - 2$ is not a polynomial because the first term is not of the form ax^k (k must be nonnegative)

25. $6 - \sqrt{n}$ is not a polynomial because the second term is not of the form ax^k (k must be an integer).

27. $5 + (2 + 3x) = (5 + 2) + 3x = 3x + 7$

29. $(2x^2 - 3) + (5x^2 + 6) = (2x^2 + 5x^2) + (-3 + 6)$
$= 7x^2 + 3$

31. $(5y + 6) + (4y^2 - 6y - 3) = 4y^2 + (5y - 6y) + (6 - 3) = 4y^2 - y + 3$

33. $(2 - 8y) + (-2y^4 + 3y + 2) = (-2y^4) + (-8y + 3y) + (2 + 2) = -2y^4 - 5y + 4$

35. $(x^3 + 9) + (2x^2 + 5) + (x^3 - 14) = (x^3 + x^3) + 2x^2 + (9 + 5 - 14) = 2x^3 + 2x^2$

37. $(x^2 - 3x + 8) + (2x^2 - 4x) + 3x^2 = (x^2 + 2x^2 + 3x^2) + (-3x - 4x) + (8) = 6x^2 - 7x + 8$

39. $\left(\frac{2}{3}x^3 - 4x + 1\right) + \left(-\frac{3}{5} + 7x - \frac{1}{2}x^3\right) = \left(\frac{2}{3}x^3 - \frac{1}{2}x^3\right) + (-4x + 7x) + \left(1 - \frac{3}{5}\right)$
$= \left(\frac{4}{6}x^3 - \frac{3}{6}x^3\right) + 3x + \left(\frac{5}{5} - \frac{3}{5}\right) = \frac{1}{6}x^3 + 3x + \frac{2}{5}$

41. $(6.32t - 4.51t^2) + (7.2t^2 + 1.03t - 4.2) = (-4.51t^2 + 7.2t^2) + (6.32t + 1.03t) - 4.2 = 2.69t^2 + 7.35t - 4.2$

43.
$$\begin{array}{r} 5x^2 - 3x + 4 \\ -3x^2 - 4 \\ \hline 2x^2 - 3x \end{array}$$

45.
$$\begin{array}{r} 4x^3 - 2x^2 + 8x \\ 4x^2 + x - 6 \\ \hline 4x^3 + 2x^2 + 9x - 6 \end{array}$$

47.
$$\begin{array}{r} 5p^2 - 4p + 2 \\ -3p^2 + 2p - 7 \\ \hline 2p^2 - 2p - 5 \end{array}$$

49.
$$\begin{array}{r} -3.6b^2 + 2.5b \\ -2.4b^2 - 3.1b + 7.1 \\ 6.6b^2 \\ \hline 0.6b^2 - 0.6b + 7.1 \end{array}$$

51. $(4 - y^3) - (4 + y^3) = (4 - y^3) + (-4 - y^3)$
$= (4 - 4) + (-y^3 - y^3)$
$= -2y^3$

53. $(3x^2 - 2x + 1) - (2x^2 + x - 1) = (3x^2 - 2x + 1) + (-2x^2 - x + 1)$
$= (3x^2 - 2x^2) + (-2x - x) + (1 + 1)$
$= x^2 - 3x + 2$

55. $(6t^3 - 12) - (-t^3 + t - 2) = (6t^3 - 12) + (t^3 - t + 2) = (6t^3 + t^3) - t + (-12 + 2) = 7t^3 - t - 10$

57. $\left(\frac{1}{4}y^2 - 5y\right) - \left(12 + 4y - \frac{3}{2}y^2\right) = \left(\frac{1}{4}y^2 - 5y\right) + \left(-12 - 4y + \frac{3}{2}y^2\right)$
$= \left(\frac{1}{4}y^2 + \frac{3}{2}y^2\right) + (-5y - 4y) - 12$
$= \frac{7}{4}y^2 - 9y - 12$

59. $(10.4t^4 - 0.23t^5 + 1.3t^2) - (2.6 - 7.35t + 6.7t^2 - 9.6t^5) = (10.4t^4 - 0.23t^5 + 1.3t^2) + (-2.6 + 7.35t - 6.7t^2 + 9.6t^5)$
$= (-0.23t^5 + 9.6t^5) + 10.4t^4 + (1.3t^2 - 6.7t^2) + 7.35t - 2.6$
$= 9.37t^5 + 10.4t^4 - 5.4t^2 + 7.35t - 2.6$

61. $(x^3 - 3x) - [3x^3 - (x^2 + 5x)] = (x^3 - 3x) - [3x^3 - x^2 - 5x]$
$= x^3 - 3x - 3x^3 + x^2 + 5x$
$= -2x^3 + x^2 + 2x$

63.
$$\begin{array}{rcr} x^2 - x + 3 & \Rightarrow & x^2 - x + 3 \\ -\quad (x - 2) & \Rightarrow & - x + 2 \\ \hline & & x^2 - 2x + 5 \end{array}$$

65.
$$\begin{array}{rcr} 2x^2 - 4x + 5 & \Rightarrow & 2x^2 - 4x + 5 \\ -(-4x^2 + 5x - 6) & \Rightarrow & -4x^2 - 5x + 6 \\ \hline & & -2x^2 - 9x + 11 \end{array}$$

67.
$$\begin{array}{rcr} -3x^7 \qquad + 6x^4 + 4 & \Rightarrow & -3x^7 \qquad + 6x^4 + 4 \\ -(8x^7 + 10x^5 - 2x^4 - 12) & \Rightarrow & -8x^7 - 10x^5 + 2x^4 + 12 \\ \hline & & -11x^7 - 10x^5 + 8x^4 + 16 \end{array}$$

69. $-(2x^3 - 3) + (4x^3 - 2x) = -2x^3 + 3 + 4x^3 - 2x = (-2x^3 + 4x^3) + (-2x) + (3) = 2x^3 - 2x + 3$

71. $(4x^5 - 10x^3 + 6x) - (8x^5 - 3x^3 + 11) + (4x^5 + 5x^3 - x^2) = (4x^5 - 10x^3 + 6x)$
$+ (-8x^5 + 3x^3 - 11) + (4x^5 + 5x^3 - x^2)$
$= (4x^5 - 8x^5 + 4x^5) + (-10x^3 + 3x^3 + 5x^3)$
$- x^2 + 6x - 11$
$= -2x^3 - x^2 + 6x - 11$

73. $(5n^2 + 6) + [(2n - 3n^2) - (2n^2 + 2n + 6)] = (5n^2 + 6) + [2n - 3n^2 - 2n^2 - 2n - 6]$
$= (5n^2 + 6) + [(-3n^2 - 2n^2) + (2n - 2n) - 6]$
$= (5n^2 + 6) + (-5n^2 + 0 - 6)$
$= (5n^2 - 5n^2) + (6 - 6) = 0$

75. $(8x^3 - 4x^2 + 3x) - [(x^3 - 4x^2 + 5) + (x - 5)] = (8x^3 - 4x^2 + 3x) - [x^3 - 4x^2 + x]$
$= (8x^3 - 4x^2 + 3x) + (-x^3 + 4x^2 - x)$
$= (8x^3 - x^3) + (-4x^2 + 4x^2) + (3x - x)$
$= 7x^3 + 2x$

77. $3(4x^2 - 1) + (3x^3 - 7x^2 + 5) = 12x^2 - 3 + 3x^3 - 7x^2 + 5 = 3x^3 + 5x^2 + 2$

79. $2(t^2 + 12) - 5(t^2 + 5) + 6(t^2 + 5) = 2t^2 + 24 - 5t^2 - 25 + 6t^2 + 30 = (2t^2 - 5t^2 + 6t^2) + (24 - 25 + 30) = 3t^2 + 29$

81. $15v - 3(3v - v^2) + 9(8v + 3) = 15v - 9v + 3v^2 + 72v + 27 = (3v^2) + (15v - 9v + 72v) + 27 = 3v^2 + 78v + 27$

83. $5s - [6s - (30s + 8)] = 5s - [6s - 30s - 8] = (5s - 6s + 30s) + (8) = 29s + 8$

85. $(2x^{2r} - 6x^r - 3) + (3x^{2r} - 2x^r + 6) = (2x^{2r} + 3x^{2r}) + (-6x^r - 2x^r) + (-3 + 6) = 5x^{2r} - 8x^r + 3$

87. $(3x^{2m} + 2x^m - 8) - (x^{2m} - 4x^m + 3) = (3x^{2m} + 2x^m - 8) + (-x^{2m} + 4x^m - 3)$
$= (3x^{2m} - x^{2m}) + (2x^m + 4x^m) + (-8 - 3)$
$= 2x^{2m} + 6x^m - 11$

89. $(7x^{4n} - 3x^{2n} - 1) - (4x^{4n} + x^{3n} - 6x^{2n}) = (7x^{4n} - 3x^{2n} - 1) + (-4x^{4n} - x^{3n} + 6x^{2n})$
$= (7x^{4n} - 4x^{4n}) + (-x^{3n}) + (-3x^{2n} + 6x^{2n}) - 1$
$= 3x^{4n} - x^{3n} + 3x^{2n} - 1$

91. *Keystrokes:*

y_1 [Y=] [(] [X, T, θ] [^] 3 [−] 3 [X,T,θ] [x^2] [−] 2 [)] [−] [(] [X,T,θ] [x^2] [+] 1 [)] [ENTER]

y_2 [X, T, θ] [^] 3 [−] 4 [X,T,θ] [x^2] [−] 3 [GRAPH]

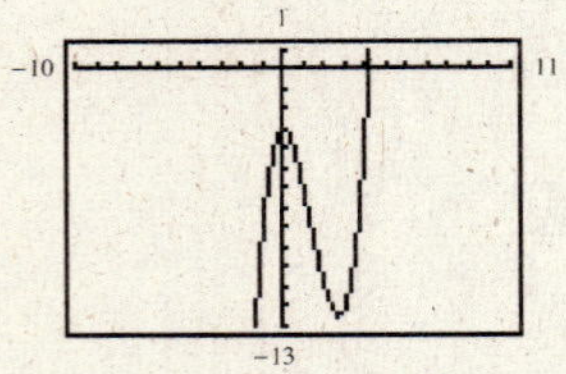

y_1 and y_2 represent equivalent expressions since the graphs of y_1 and y_2 are identical.

93. $h(x) = f(x) + g(x)$
$= (4x^3 - 3x^2 + 7) + (9 - x - x^2 - 5x^3)$
$= (4x^3 - 5x^3) + (-3x^2 - x^2) - x + (7 + 9)$
$= -x^3 - 4x^2 - x + 16$

95. $h(x) = -f(x) - g(x)$
$= -(4x^3 - 3x^2 + 7) - (9 - x - x^2 - 5x^3)$
$= -4x^3 + 3x^2 - 7 - 9 + x + x^2 + 5x^3$
$= (-4x^3 + 5x^3) + (3x^2 + x^2) + x + (-7 - 9)$
$= x^3 + 4x^2 + x - 16$

97.

Polynomial	*Value*	*Substitute*	*Simplify*
$h(t) = -16t^2 + 64$	(a) $t = 0$	$-16(0)^2 + 64$	64 feet
	(b) $t = \frac{1}{2}$	$-16\left(\frac{1}{2}\right)^2 + 64$	60 feet
	(c) $t = 1$	$-16(1)^2 + 64$	48 feet
	(d) $t = 2$	$-16(2)^2 + 64$	0 feet

At time $t = 0$, the object is dropped from a height of 64 feet and continues to fall, reaching the ground at time $t = 2$.

99. The free-falling object was dropped.

$-16(0)^2 + 100 = 100$ feet

101. The free-falling object was thrown upward.

$h(0) = -16(0)^2 + 40(0) + 12 = 12$ feet

103. $h = -16(1)^2 + 40(1) + 984 = 1008$ feet

$h = -16(5)^2 + 40(5) + 984 = 784$ feet

$h = -16(9)^2 + 40(9) + 984 = 48$ feet

105. *Verbal Model:* $\boxed{\text{Profit}} = \boxed{\text{Revenue}} - \boxed{\text{Cost}}$

Equation: $P = R - C$

$P = 14x - (8x + 15{,}000)$

$P = 6x - 15{,}000$

$P = 6(5000) - 15{,}000$

$P = \$15{,}000$

107. $P = (2x + 4) + (2x + 4) + x + x + (3x + 1) + (3x + 1) + x + x$

$= (2x + 2x + x + x + 3x + 3x + x + x) + (4 + 4 + 1 + 1)$

$= 14x + 10$

109. Area $= 12(x + 6) - 7x = 12x + 72 - 7x = 5x + 72$

111. Area of region $= \left(6 \cdot \frac{3}{2}x\right) + \left(6 \cdot \frac{9}{2}x\right)$ or $6 \cdot \left[\frac{3}{2}x + \frac{9}{2}x\right]$

$= 9x + 27x$ or $6\left[\frac{12}{2}x\right]$

$= 36x$

113. (a) *Verbal Model:* $\boxed{\text{Total stopping distance}} = \boxed{\text{Distance traveled during reaction time}} + \boxed{\text{Distance traveled braking}}$

Equation: $T = R + B$

$T = (1.1x) + (0.0475x^2 - 0.001x) = (0.0475x^2) + (1.1x - 0.001x) = 0.0475x^2 + 1.099x$

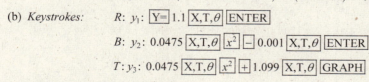

(b) *Keystrokes:* R: y_1: [Y=] 1.1 [X,T,θ] [ENTER]

B: y_2: 0.0475 [X,T,θ] [x^2] [−] 0.001 [X,T,θ] [ENTER]

T: y_3: 0.0475 [X,T,θ] [x^2] [+] 1.099 [X,T,θ] [GRAPH]

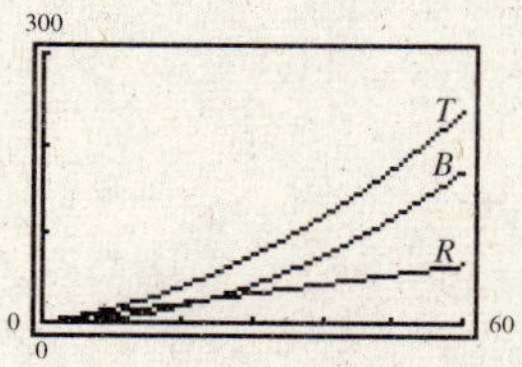

(c) Using the graph when $x = 25$, the approximate total stopping distance is 57.2 feet and when $x = 50$, the approximate total stopping distance is 173.7 feet.

115. The degree of the term ax^k is k. The degree of a polynomial is the degree of its highest-degree term.

117. (a) A polynomial is a trinomial is not always false. A polynomial is only a trinomial when it has three terms.

(b) A trinomial is a polynomial is always true.

119. No, not every trinomial is a second-degree polynomial. For example, $x^3 + 2x + 3$ is a trinomial of third-degree.

121. B has 2 rows and 3 columns so the order is 2×3.

123. A is a square matrix because it has an equal number of rows and columns.

125.

$$\begin{bmatrix} 2 & 0 & -2 \\ 4 & -1 & 1 \\ 0 & 4 & -6 \end{bmatrix}$$

$$\tfrac{1}{2}R_1 \begin{bmatrix} 1 & 0 & -1 \\ 4 & -1 & 1 \\ 0 & 4 & -6 \end{bmatrix}$$

$$-4R_1 + R_2 \begin{bmatrix} 1 & 0 & -1 \\ 0 & -1 & 5 \\ 0 & 4 & -6 \end{bmatrix}$$

$$-R_2 \begin{bmatrix} 1 & 0 & -1 \\ 0 & 1 & -5 \\ 0 & 4 & -6 \end{bmatrix}$$

$$-4R_2 + R_3 \begin{bmatrix} 1 & 0 & -1 \\ 0 & 1 & -5 \\ 0 & 0 & 14 \end{bmatrix} \Rightarrow \tfrac{1}{14}R_1 \begin{bmatrix} 1 & 0 & -1 \\ 0 & 1 & -5 \\ 0 & 0 & 1 \end{bmatrix}$$

127.

$$\begin{aligned} 6x - 3y &= 0 \Rightarrow & 6x - 3y &= 0 \\ 2x + y &= 4 \Rightarrow & 6x + 3y &= 12 \\ & & 12x &= 12 \\ & & x &= 1 \end{aligned}$$

$$2(1) + y = 4$$
$$y = 2$$

$(1, 2)$

129.

$$|C| = \begin{vmatrix} 2 & 0 & -2 \\ 4 & -1 & 1 \\ 0 & 4 & -6 \end{vmatrix}$$

$$\begin{aligned} &= 2\begin{vmatrix} -1 & 1 \\ 4 & -6 \end{vmatrix} - 0 - 2\begin{vmatrix} 4 & -1 \\ 0 & 4 \end{vmatrix} \\ &= 2(6 - 4) - 2(16 - 0) \\ &= 2(2) - 2(16) \\ &= 4 - 32 \\ &= -28 \end{aligned}$$

Section 5.3 Multiplying Polynomials

1. $(-2a^2)(-8a) = (-2)(-8)a^2 \cdot a = 16a^{2+1} = 16a^3$

3. $2y(5 - y) = (2y)(5) - (2y)(y) = -2y^2 + 10y$

5. $4x(2x^2 - 3x + 5) = (4x)(2x^2) - (4x)(3x) + (4x)(5)$
$= 8x^3 - 12x^2 + 20x$

7. $-2m^2(7 - 4m + 2m^2) = -2m^2(7) - 2m^2(-4m) - 2m^2(2m^2) = -14m^2 + 8m^3 - 4m^4 = -4m^4 + 8m^3 - 14m^2$

9. $-x^3(x^4 - 2x^3 + 5x - 6) = -x^3(x^4) - x^3(-2x^3) - x^3(5x) - x^3(-6) = -x^7 + 2x^6 - 5x^4 + 6x^3$

11. $-3x(-5x)(5x + 2) = (-3x)(-5x)(5x + 2) = 15x^2(5x + 2) = 15x^2(5x) + 15x^2(2) = 75x^3 + 30x^2$

13. $u^2v(3u^4 - 5u^2 + 6uv^3) = u^2v(3u^4) + u^2v(-5u^2) + u^2v(6uv^3) = 3u^6v - 5u^4v + 6u^3v^4$

15. $(x + 2)(x + 4) = x^2 + 4x + 2x + 8 = x^2 + 6x + 8$

17. $(x - 4)(x + 4) = x^2 + 4x - 4x - 16 = x^2 - 16$

19. $(x - 3)(x - 3) = x^2 - 3x - 3x + 9 = x^2 - 6x + 9$

21. $(2x - 3)(x + 5) = 2x^2 + 10x - 3x - 15$
$= 2x^2 + 7x - 15$

23. $(5x - 2)(2x - 6) = 10x^2 - 30x - 4x + 12$
$= 10x^2 - 34x + 12$

25. $(2x^2 - 1)(x + 2) = 2x^3 + 4x^2 - x - 2$

27. $\left(4y - \frac{1}{3}\right)(12y + 9) = 48y^2 + 36y - 4y - 3$
$= 48y^2 + 32y - 3$

29. $(2x + y)(3x + 2y) = 6x^2 + 4xy + 3xy + 2y^2$
$= 6x^2 + 7xy + 2y^2$

31. $(2t - 1)(t + 1) + 1(2t - 5)(t - 1) = 2t^2 + 2t - t - 1 + 2t^2 - 2t - 5t + 5 = 4t^2 - 6t + 4$

33. $(x - 1)(x^2 - 4x + 6) = (x - 1)(x^2) + (x - 1)(-4x) + (x - 1)(6) = x^3 - x^2 - 4x^2 + 4x + 6x - 6 = x^3 - 5x^2 + 10x - 6$

35. $(3a + 2)(a^2 + 3a + 1) = (3a + 2)(a^2) + (3a + 2)(3a) + (3a + 2)(1)$
$= 3a^3 + 2a^2 + 9a^2 + 6a + 3a + 2$
$= 3a^3 + 11a^2 + 9a + 2$

37. $(2u^2 + 3u - 4)(4u + 5) = (4u + 5)(2u^2) + (4u + 5)(3u) + (4u + 5)(-4)$
$= 8u^3 + 10u^2 + 12u^2 + 15u - 16u - 20$
$= 8u^3 + 22u^2 - u - 20$

39. $(x^3 - 3x + 2)(x - 2) = x^3(x - 2) + (-3x)(x - 2) + 2(x - 2)$
$= x^4 - 2x^3 - 3x^2 + 6x + 2x - 4$
$= x^4 - 2x^3 - 3x^2 + 8x - 4$

41. $(5x^2 + 2)(x^2 + 4x - 1) = (5x^2 + 2)(x^2) + (5x^2 + 2)(4x) + (5x^2 + 2)(-1)$
$= 5x^4 + 2x^2 + 20x^3 + 8x - 5x^2 - 2$
$= 5x^4 + 20x^3 - 3x^2 + 8x - 2$

43. $(t^2 + t - 2)(t^2 - t + 2) = t^2(t^2 - t + 2) + t(t^2 - t + 2) - 2(t^2 - t + 2)$
$= t^4 - t^3 + 2t^2 + t^3 - t^2 + 2t - 2t^2 + 2t - 4$
$= t^4 - t^2 + 4t - 4$

45.
$$\begin{array}{rrrr}
 & 7x^2 & -\ 14x & +\ 9 \\
 & \times & 2x & +\ 1 \\ \hline
 & 7x^2 & -\ 14x & +\ 9 \\
14x^3 & -\ 28x^2 & +\ 18x & \\ \hline
14x^3 & -\ 21x^2 & +\ 4x & +\ 9
\end{array}$$

47.
$$\begin{array}{rrrr}
 & 2u^2 & +\ 5u & +\ 3 \\
 & & u & -\ 2 \\ \hline
 & -4u^2 & -\ 10u & -\ 6 \\
2u^3 & +\ 5u^2 & +\ 3u & \\ \hline
2u^3 & +\ u^2 & -\ 7u & -\ 6
\end{array}$$

49.
$$\begin{array}{rrrr}
 & -x^2 & +\ 2x & -\ 1 \\
 & & 2x & +\ 1 \\ \hline
 & -x^2 & +\ 2x & -\ 1 \\
-2x^3 & +\ 4x^2 & -\ 2x & \\ \hline
-2x^3 & +\ 3x^2 & & -\ 1
\end{array}$$

51.
$$\begin{array}{rrrrr}
 & & t^2 & +\ t & -\ 2 \\
 & & t^2 & -\ t & +\ 2 \\ \hline
 & & +\ 2t^2 & +\ 2t & -\ 4 \\
 & -\ t^3 & -\ t^2 & +\ 2t & \\
t^4 & +\ t^3 & -\ 2t^2 & & \\ \hline
t^4 & & -\ t^2 & +\ 4t & -\ 4
\end{array}$$

53. $(x + 2)(x - 2) = (x)^2 - (2)^2 = x^2 - 4$

55. $(x - 8)(x + 8) = (x)^2 - (8)^2 = x^2 - 64$

57. $(2 + 7y)(2 - 7y) = (2)^2 - (7y)^2 = 4 - 49y^2$

59. $(3 - 2x^2)(3 + 2x^2) = (3)^2 - (2x^2)^2 = 9 - 4x^4$

61. $(2a + 5b)(2a - 5b) = (2a)^2 - (5b)^2 = 4a^2 - 25b^2$

63. $(6x - 9y)(6x + 9y) = (6x)^2 - (9y)^2 = 36x^2 - 81y^2$

65. $\left(2x - \frac{1}{4}\right)\left(2x + \frac{1}{4}\right) = (2x)^2 - \left(\frac{1}{4}\right)^2 = 4x^2 - \frac{1}{16}$

67. $(0.2t + 0.5)(0.2t - 0.5) = (0.2t)^2 - (0.5)^2$
$= 0.04t^2 - 0.25$

69. $(x + 5)^2 = (x^2) + 2(x)(5) + (5)^2 = x^2 + 10x + 25$

71. $(x - 10)^2 = (x)^2 - 2(x)(10) + 10^2 = x^2 - 20x + 100$

73. $(2x + 5)^2 = (2x)^2 + 2(2x)(5) + (5)^2$
$= 4x^2 + 20x + 25$

75. $(6x - 1)^2 = (6x)^2 - 2(6x)(1) + (1)^2 = 36x^2 - 12x + 1$

77. $(2x - 7y)^2 = (2x)^2 - 2(2x)(7y) + (7y)^2$
$= 4x^2 - 28xy + 49y^2$

79. $[(x + 2) + y]^2 = (x + 2)^2 + 2(x + 2)y + y^2$
$= (x)^2 + 2(x)(2) + (2)^2 + 2xy + 4y + y^2$
$= x^2 + 4x + 4 + 2xy + 4y + y^2$
or $x^2 + 2xy + y^2 + 4x + 4y + 4$

81. $[u - (v - 3)][u + (v - 3)] = (u)^2 - (v - 3)^2 = u^2 - \left[v^2 - 2(v)(3) + (3)^2\right] = u^2 - (v^2 - 6v + 9) = u^2 - v^2 + 6v - 9$

83. $(k + 5)^3 = (k + 5)^2(k + 5)$
$= (k^2 + 10k + 25)(k + 5)$
$= k^2(k + 5) + 10k(k + 5) + 25(k + 5)$
$= k^3 + 5k^2 + 10k^2 + 50k + 25k + 125$
$= k^3 + 15k^2 + 75k + 125$

85. $(u + v)^3 = (u + v)(u + v)(u + v)$
$= (u^2 + uv + uv + v^2)(u + v)$
$= (u^2 + 2uv + v^2)(u + v)$

$$\begin{array}{rrrr} & u^2 + & 2uv + & v^2 \\ & & u + & v \\ \hline & u^2v + & 2uv^2 + & v^3 \\ u^3 + & 2u^2v + & uv^2 & \\ \hline u^3 + & 3u^2v + & 3uv^2 + & v^3 \end{array}$$

87. $3x^r(5x^{2r} + 4x^{3r-1}) = 3x^r(5x^{2r}) + 3x^r(4x^{3r-1}) = 15x^{r+2r} + 12x^{r+3r-1} = 15x^{3r} + 12x^{4r-1}$

89. $(6x^m - 5)(2x^{2m} - 3) = 6x^m(2x^{2m} - 3) + (-5)(2x^{2m} - 3) = 12x^{m+2m} - 18x^m - 10x^{2m} + 15 = 12x^{3m} - 10x^{2m} - 18x^m + 15$

91. $(x^{m-n})^{m+n} = x^{(m-n)(m+n)} = x^{m^2+mn-mn-n^2} = x^{m^2-n^2}$

93. *Keystrokes:*

y_1 [Y=] [(] [X,T,θ] [+] 1 [)] [(] [X,T,θ] [x^2] [−] [X,T,θ] 2 [+] [)] [ENTER]

y_2 [X,T,θ] [^] 3 [+] [X,T,θ] [+] 2 [GRAPH]

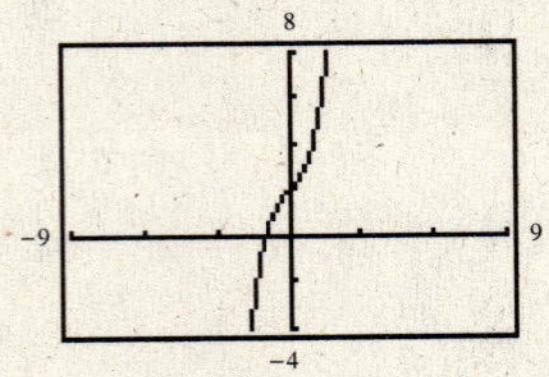

$y_1 = y_2$ because $(x + 1)(x^2 - x + 2) = x^3 - x^2 + 2x + x^2 - x + 2 = x^3 + x + 2$.

95. *Keystrokes:*

y_1 [Y=] [(] [X,T,θ] [+] [(] 1 [÷] 2 [)] [)] [(] [X,T,θ] [−] [(] 1 [÷] 2 [)] [)] [ENTER]

y_2 [X,T,θ] [x^2] [−] [(] 1 [÷] 4 [)] [GRAPH]

$y_1 = y_2$ because $\left(x - \frac{1}{2}\right)\left(x + \frac{1}{2}\right) = x^2 - \frac{1}{4}$

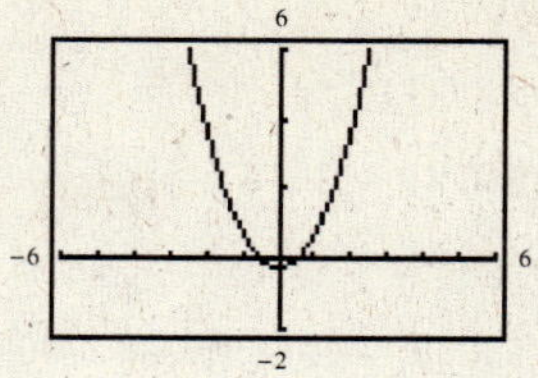

97. (a) $f(w + 2) = (w + 2)^2 - 2(w + 2) = w^2 + 2w + 2w + 4 - 2w - 4 = w^2 + 2w$

(b) $f(a - 4) - f(3) = (a - 4)^2 - 2(a - 4) + (3)^2 - 2(3) = a^2 - 4a - 4a + 16 - 2a + 8 + 9 - 6 = a^2 - 10a + 27$

99. (a) *Verbal Model:* [Volume] = [Length] · [Width] · [Height]

Function: $V(n) = n \cdot (n + 2) \cdot (n + 4) = n\left(n^2 + 6n + 8\right) = n^3 + 6n^2 + 8n$

(b) $V(3) = 3^3 + 6(3)^2 + 8(3) = 27 + 54 + 24 = 105$ cubic inches

(c) *Verbal Model:* [Area] = [Length] · [Width]

Function: $A(n) = n \cdot (n + 2) = n^2 + 2n$

(d) *Function:*

$$\begin{aligned} A(n + 5) &= (n + 5)(n + 5 + 2) \\ &= (n + 5)(n + 7) \\ &= n^2 + 7n + 5n + 35 \\ &= n^2 + 12n + 35 \\ A(n + 5) &= (n + 5)^2 + 2(n + 5) \\ &= n^2 + 10n + 25 + 2n + 10 \\ &= n^2 + 12n + 35 \end{aligned}$$

101. *Verbal Model:* [Area of shaded region] = [Area of outside rectangle] − [Area of inside rectangle]

Function: $A(x) = 3x(3x + 10) - x(x + 4) = 9x^2 + 30x - x^2 - 4x = 8x^2 + 26x$

103. *Verbal Model:* [Area of shaded region] = [Area of larger triangle] − [Area of smaller triangle]

Function: $A(x) = \frac{1}{2}(2x)(1.6x) = \frac{1}{2}x(0.8x) = 1.6x^2 - 0.4x^2 = 1.2x^2$

105. (a) *Verbal Model:* [Perimeter] = 2 [Length] + 2 [Width]

$P = 2\left[\frac{5}{2}(2w)\right] + 2(2w) = 10w + 4w = 14w$

(b) *Verbal Model:* [Area] = [Length] · [Width]

$A = \frac{5}{2}(2w) \cdot 2w = 10w^2$

107. *Verbal Model:* [Revenue] = [Number of units sold] · [Price per unit]

Equation: $R = x(175 - 0.02x) = 175x - 0.02x^2 = -0.02x^2 + 175x$

$R(3000) = -0.02(3000)^2 + 175(3000) = -180{,}000 + 525{,}000 = 345{,}000$ cents $= \$3450$

109. $\text{Interest} = 5000(1 + r)^2$

$= 5000(1 + r)(1 + r)$

$= 5000(1 + 2r + r^2)$

$= 5000 + 10{,}000r + 5{,}000r^2$

$= 5000r^2 + 10{,}000r + 5000$

111. $\text{Area} = l \cdot w = (x + a)(x + b) = x^2 + ax + bx + ab$

$\text{Area} = (x \cdot x) + (x \cdot a) + (x \cdot b) + (a \cdot b)$

$= x^2 + ax + bx + ab$

Formula: $(x + a)(x + b) = x^2 + bx + ax + ab$

FOIL Method

113. (a) $(x - 1)(x + 1) = x^2 - 1$

(b) $(x - 1)(x^2 + x + 1) = x^3 + x^2 + x - x^2 - x - 1 = x^3 - 1$

(c) $(x - 1)(x^3 + x^2 + x + 1) = x^4 + x^3 + x^2 + x - x^3 - x^2 - x - 1 = x^4 - 1$

$(x - 1)(x^4 + x^3 + x^2 + x + 1) = x^5 - 1$

115. When two polynomials are multiplied together, an understanding of the Distributive Property is essential because the Distributive Property is used to multiply each term of the first polynomial by each term of the second polynomial.

117. The degree of the product of two polynomials of degrees m and n is $m + n$.

119. $y = 5 - \frac{1}{2}x$

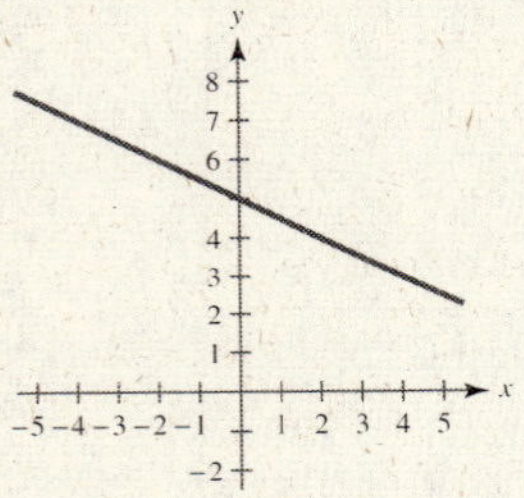

Function

121. $y - 4x + 1 = 0$

$y = 4x - 1$

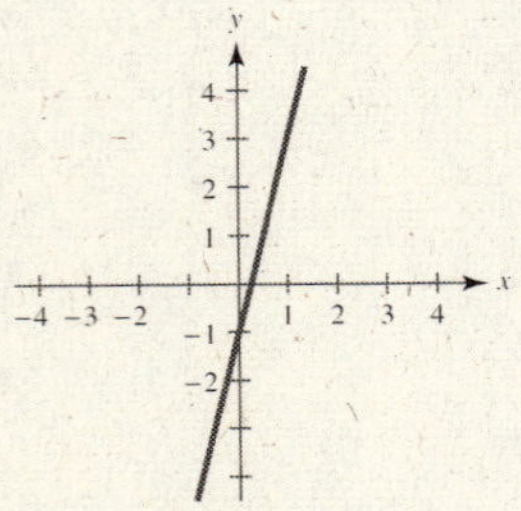

Function

123. To graph the equation $|y| + 2x = 0$, plot a few points and draw the graph. Use the Vertical Line Test to then determine that this equation is not a function.

x	$\lvert y\rvert + 2x$	point
0	0	(0, 0)
−1	2	(−1, 2)
−1	−2	(−1, −2)
−2	4	(−2, 4)
−2	−4	(−2, −4)

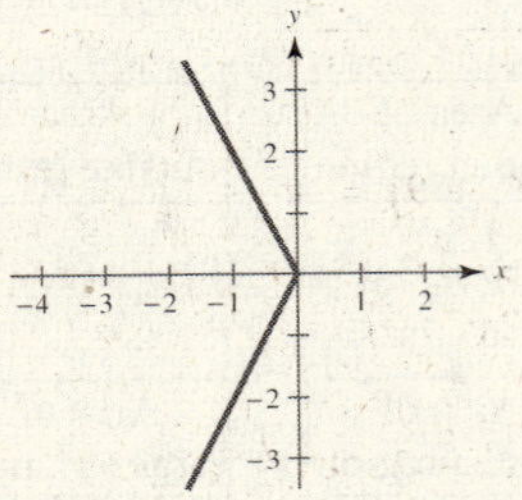

125. $2^{-5} = \dfrac{1}{2^5} = \dfrac{1}{32}$

127. $\dfrac{4^2}{4^{-1}} = 4^{2-(-1)} = 4^3 = 64$

129. $(6^3 + 3^{-6})^0 = 1$

Mid-Chapter Quiz for Chapter 5

1. Degree $= 4$

Leading coefficient $= -2$

2. $x^{-3} + 2x^2 - 6$ is not a polynomial because the first term is not of the form ax^k. (k must be non negative.)

3. $(5y^2)(-y^4)(2y^3) = (5)(-1)(2)y^{2+4+3} = -10y^9$

4. $(-6x)(-3x^2)^2 = (-6x)(-3)^2(x^2)^2$
$= (-6x)(9)x^4$
$= (-6)(9)x^{1+4}$
$= -54x^5$

5. $(-5n^2)(-2n^3) = 10n^5$

6. $(3m^3)^2(-2m^4) = (9m^6)(-2m^4) = -18m^{10}$

7. $\dfrac{6x^{-7}}{(-2x^2)^{-3}} = \dfrac{6(-2x^2)^3}{x^7}$
$= \dfrac{6(-2)^3(x^2)^3}{x^7}$
$= \dfrac{6(-8)x^6}{x^7}$
$= -\dfrac{48}{x}$

8. $\left(\dfrac{4y^2}{5x}\right)^{-2} = \left(\dfrac{5x}{4y^2}\right)^2$
$= \left(\dfrac{5x}{4y^2}\right)\left(\dfrac{5x}{4y^2}\right)$
$= \dfrac{25x^{1+1}}{16y^{2+2}}$
$= \dfrac{25x^2}{16y^4}$

9. $\left(\dfrac{3a^{-2}b^5}{9a^{-4}b^0}\right)^{-2} = \left(\dfrac{9a^{-4}b^0}{3a^{-2}b^5}\right)^2$
$= \left(\dfrac{9a^{-4}b^0}{3a^{-2}b^5}\right)\left(\dfrac{9a^{-4}b^0}{3a^{-2}b^5}\right)$
$= \dfrac{81a^{-4+(-4)}b^{0+0}}{9a^{-2+(-2)}b^{5+5}}$
$= \dfrac{9a^{-8}b^0}{a^{-4}b^{10}}$
$= 9a^{-8-(-4)}b^{0-10}$
$= 9a^{-4}b^{-10}$
$= \dfrac{9}{a^4b^{10}}$

10. $\left(\dfrac{5x^0y^{-7}}{2x^{-2}y^4}\right)^{-3} = \left(\dfrac{2x^{-2}y^4}{5x^0y^{-7}}\right)^3$
$= \left(\dfrac{2x^{-2}y^4}{5x^0y^{-7}}\right)\left(\dfrac{2x^{-2}y^4}{5x^0y^{-7}}\right)\left(\dfrac{2x^{-2}y^4}{5x^0y^{-7}}\right) = \dfrac{8x^{-2+(-2)+(-2)}y^{4+4+4}}{125x^{0+0+0}y^{-7+(-7)+(-7)}} = \dfrac{8x^{-6}y^{12}}{125x^0y^{-21}} = \dfrac{8x^{-6-0}y^{12-(-21)}}{125} = \dfrac{8x^{-6}y^{33}}{125} = \dfrac{8y^{33}}{125x^6}$

11. $(2t^3 + 3t^2 - 2) + (t^3 + 9) = 3t^3 + 3t^2 + 7$

12. $(3 - 7y) + (7y^2 + 2y - 3) = 7y^2 - 5y$

13. $(7x^3 - 3x^2 + 1) - (x^2 - 2x^3) = 7x^3 - 3x^2 + 1 - x^2 + 2x^3 = 9x^3 - 4x^2 + 1$

14. $(5 - u) - 2[3 - (u^2 + 1)] = (5 - u) - 2[3 - u^2 - 1] = (5 - u) - 2[2 - u^2] = 5 - u - 4 + 2u^2 = 2u^2 - u + 1$

15. $7y(4 - 3y) = 28y - 21y^2$

16. $(k + 8)(k + 5) = k^2 + 5k + 8k + 40 = k^2 + 13k + 40$

17. $(4x - y)(6x - 5y) = 24x^2 - 20xy - 6xy + 5y^2 = 24x^2 - 26xy + 5y^2$

18. $2z(z + 5) - 7(z + 5) = 2z^2 + 10z - 7z - 35 = 2z^2 + 3z - 35$

19. $(6r + 5)(6r - 5) = 36r^2 - 25$

20. $(2x - 3)^2 = (2x - 3)(2x - 3) = 4x^2 - 12x + 9$

21. $(x+1)(x^2-x+1) = x^3 - x^2 + x + x^2 - x + 1 = x^3 + 1$

22. $(x^2-3x+2)(x^2+5x-10) = x^2(x^2+5x-10) - 3x(x^2+5x-10) + 2(x^2+5x-10)$

$= x^4 + 5x^3 - 10x^2 - 3x^3 - 15x^2 + 30x + 2x^2 + 10x - 20$

$= x^4 + 2x^3 - 23x^2 + 40x - 20$

23. *Verbal Model:* [Area of shaded region] = [Area of large triangle] − [Area of small triangle]

Equation: $A = \frac{1}{2}(x+2)^2 - \frac{1}{2}x^2$

$= \frac{1}{2}(x^2 + 4x + 4) - \frac{1}{2}x^2$

$= \frac{1}{2}x^2 + 2x + 2 - \frac{1}{2}x^2$

$= 2x + 2$

24. $h\left(\frac{3}{2}\right) = -16\left(\frac{3}{2}\right)^2 + 32\left(\frac{3}{2}\right) + 100 = -16\left(\frac{9}{4}\right) + 16(3) + 100 = -4(9) + 48 + 100 = -36 + 148 = 112$ feet

$h(3) = -16(3)^2 + 32(3) + 100 = -16(9) + 96 + 100 = -144 + 196 = 52$ feet

25. $P = R - C = 24x - (5x + 2250) = 24x - 5x - 2250 = 19x - 2250$

$P(1500) = 19(1500) - 2250 = 28500 - 2250 = \$26,250$

Section 5.4 Factoring by Grouping and Special Forms

1. $6 = 2 \cdot 3$

3. $8 = 2 \cdot 2 \cdot 2$

5. $30 = 6 \cdot 5 = 2 \cdot 3 \cdot 5$

7. $27 = 3 \cdot 3 \cdot 3$

9. $16 = 2 \cdot 2 \cdot 2 \cdot 2$

$24 = 2 \cdot 2 \cdot 2 \cdot 3$

$\text{GCF} = 2 \cdot 2 \cdot 2 = 8$

11. $6 = 2 \cdot 3$

$12 = 2 \cdot 2 \cdot 3$

$16 = 2 \cdot 2 \cdot 2 \cdot 2$

$\text{GCF} = 2$

13. $x^3 = x \cdot x \cdot x$

$x^4 = x \cdot x \cdot x \cdot x$

$\text{GCF} = x \cdot x \cdot x = x^3$

15. $3x^2 = 3 \cdot x \cdot x$

$12x = 2^2 \cdot 3 \cdot x$

$\text{GCF} = 3x$

17. $16ab^2 = 2 \cdot 2 \cdot 2 \cdot 2 \cdot a \cdot b \cdot b$

$40a^2b^3 = 2 \cdot 2 \cdot 2 \cdot 5 \cdot a \cdot a \cdot b \cdot b \cdot b$

$\text{GCF} = 2 \cdot 2 \cdot 2 \cdot a \cdot b \cdot b = 8ab^2$

19. $9(x-2)^2 = 3 \cdot 3 \cdot (x-2) \cdot (x-2)$

$6(x-2)^3 = 3 \cdot 2 \cdot (x-2) \cdot (x-2) \cdot (x-2)$

$\text{GCF} = 3 \cdot (x-2) \cdot (x-2) = 3(x-2)^2$

21. $4x + 4 = 4(x+1)$

23. $6y - 20 = 2(3y - 10)$

25. $24t^2 - 36 = 12(2t^2 - 3)$

27. $x^2 + 9x = x(x+9)$

29. $8t^2 + 8t = 8t(t+1)$

31. $11u^2 + 9$ is prime. (No common factor other than 1.)

33. $3x^2y^2 - 15y = 3y(x^2y - 5)$

35. $28x^2 + 16x - 8 = 4(7x^2 + 4x - 2)$

37. $45x^2 - 15x + 30 = 15(3x^2 - x + 2)$

39. $14x^4y^3 + 21x^3y^2 + 9x^2 = x^2(14x^2y^3 + 21xy^2 + 9)$

41. $7 - 14x = -7(-1 + 2x) = -7(2x - 1)$

43. $6 - x = -1(-6 + x) = -1(x - 6) = -(x - 6)$

45. $7 - y^2 = -(-7 + y^2) = -(y^2 - 7)$

47. $4 + x - x^2 = -1(-4 - x + x^2) = -1(x^2 - x - 4)$
$= -(x^2 - x - 4)$

49. $2y - 2 - 6y^2 = -2(-y + 1 + 3y^2) = -2(3y^2 - y + 1)$

51. $40y - 12 = 4(10y - 3)$

53. $30x^2 + 25x = 5x(6x + 5)$

55. $2y(y - 4) + 5(y - 4) = (y - 4)(2y + 5)$

57. $5x(3x + 2) - 3(3x + 2) = (3x + 2)(5x - 3)$

59. $2(7a + 6) - 3a^2(7a + 6) = (7a + 6)(2 - 3a^2)$

61. $8t^3(4t - 1)^2 + 3(4t - 1)^2 = (4t - 1)^2(8t^3 + 3)$

63. $(x - 5)(4x + 9) - (3x + 4)(4x + 9) = (4x + 9)(x - 5 - 3x - 4) = (4x + 9)(-2x - 9)$

65. $x^2 + 25x + x + 25 = (x^2 + 25x) + (x + 25) = x(x + 25) + 1(x + 25) = (x + 25)(x + 1)$

67. $y^2 - 6y + 2y - 12 = (y^2 - 6y) + (2y - 12)$
$= y(y - 6) + 2(y - 6)$
$= (y - 6)(y + 2)$

69. $x^3 + 2x^2 + x + 2 = (x^3 + 2x^2) + (x + 2)$
$= x^2(x + 2) + 1(x + 2)$
$= (x + 2)(x^2 + 1)$

71. $3a^3 - 12a^2 - 2a + 8 = (3a^3 - 12a^2) + (-2a + 8)$
$= 3a^2(a - 4) - 2(a - 4)$
$= (a - 4)(3a^2 - 2)$

73. $z^4 - 2z + 3z^3 - 6 = (z^4 - 2z) + (3z^3 - 6)$
$= z(z^3 - 2) + 3(z^3 - 2)$
$= (z^3 - 2)(z + 3)$

75. $5x^3 - 10x^2y + 7xy^2 - 14y^3 = (5x^3 - 10x^2y) + (7xy^2 - 14y^3) = 5x^2(x - 2y) + 7y^2(x - 2y) = (x - 2y)(5x^2 + 7y^2)$

77. $x^2 - 9 = x^2 - 3^2 = (x + 3)(x - 3)$

79. $1 - a^2 = 1^2 - a^2 = (1 - a)(1 + a)$

81. $16y^2 - 9 = (4y)^2 - 3^2 = (4y - 3)(4y + 3)$

83. $81 - 4x^2 = 9^2 - (2x)^2 = (9 - 2x)(9 + 2x)$

85. $4z^2 - y^2 = (2z - y)(2z + y)$

87. $36x^2 - 25y^2 = (6x)^2 - (5y)^2$
$= (6x - 5y)(6x + 5y)$

89. $u^2 - \frac{1}{16} = u^2 - \left(\frac{1}{4}\right)^2 = \left(u - \frac{1}{4}\right)\left(u + \frac{1}{4}\right)$

91. $\frac{4}{9}x^2 - \frac{16}{25}y^2 = \left(\frac{2}{3}x\right)^2 - \left(\frac{4}{5}y\right)^2$
$= \left(\frac{2}{3}x - \frac{4}{5}y\right)\left(\frac{2}{3}x + \frac{4}{5}y\right)$

93. $(x - 1)^2 - 16 = [(x - 1) - 4][(x - 1) + 4] = (x - 5)(x + 3)$

95. $81 - (z + 5)^2 = 9^2 - (z + 5)^2 = [9 - (z + 5)][9 + (z + 5)] = [9 - z - 5][9 + z + 5] = (4 - z)(14 + z)$

97. $(2x + 5)^2 - (x - 4)^2 = [(2x + 5) - (x - 4)][(2x + 5) + (x - 4)]$
$= [2x + 5 - x + 4][2x + 5 + x - 4] = (x + 9)(3x + 1)$

99. $x^3 - 8 = x^3 - 2^3$
$= (x - 2)(x^2 + 2x + 4)$

101. $y^3 + 64 = y^3 + 4^3$
$= (y + 4)(y^2 - 4y + 16)$

103. $8t^3 - 27 = (2t)^3 - 3^3$
$= (2t - 3)(4t^2 + 6t + 9)$

105. $27u^3 + 1 = (3u)^3 + 1^3$
$= (3u + 1)(9u^2 - 3u + 1)$

107. $64a^3 + b^3 = (4a)^3 + b^3$
$= (4a + b)(16a^2 - 4ab + b^2)$

109. $x^3 + 27y^3 = x^3 + (3y)^3$
$= (x + 3y)(x^2 - 3xy + 9y^2)$

113. $8x^3 + 64 = 8(x^3 + 8)$
$= 8(x^3 + 2^3)$
$= 8(x + 2)(x^2 - 2x + 4)$

115. $y^4 - 81 = (y^2)^2 - 9^2$
$= (y^2 - 9)(y^2 + 9)$
$= (y - 3)(y + 3)(y^2 + 9)$

117. $3x^4 - 300x^2 = 3x^2(x^2 - 100)$
$= 3x^2(x - 10)(x + 10)$

119. $6x^6 - 48y^6 = 6(x^6 - 8y^6)$
$= 6\left[(x^2)^3 - (2y^2)^3\right]$
$= 6(x^2 - 2y^2)(x^4 + 2x^2y^2 + 4y^4)$

121. $4x^{2n} - 25 = (2x^n)^2 - 5^2$
$= (2x^n - 5)(2x^n + 5)$

123. $2x^{3r} + 8x^r + 4x^{2r} = 2x^r(x^{2r} + 4 + 2x^r)$
$= 2x^r(x^{2r} + 2x^r + 4)$

125. $4y^{m+n} + 7y^{2m+n} - y^{m+2n} = y^{m+n}(4 + 7y^m - y^n)$
$= y^{m+n}(7y^m - y^n + 4)$

127. *Keystrokes:*

y_1: [Y=] 3 [X,T,θ] [−] 6 [ENTER]

y_2: 3 [(] [X,T,θ] [−] 2 [)] [GRAPH]

$y_1 = y_2$

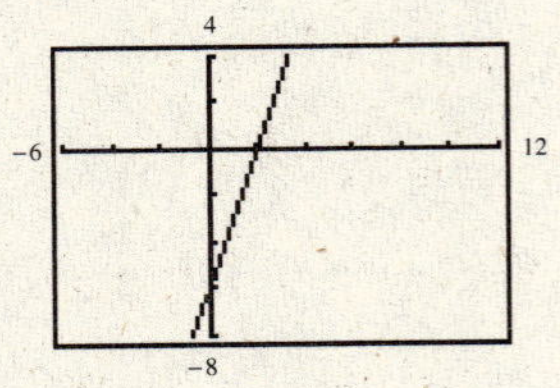

129. *Keystrokes:*

y_1: [Y=] [X,T,θ] [x^2] [−] 4 [ENTER]

y_2: [(] [X,T,θ] [+] 2 [)] [(] [X,T,θ] [−] 2 [)] [GRAPH]

$y_1 = y_2$

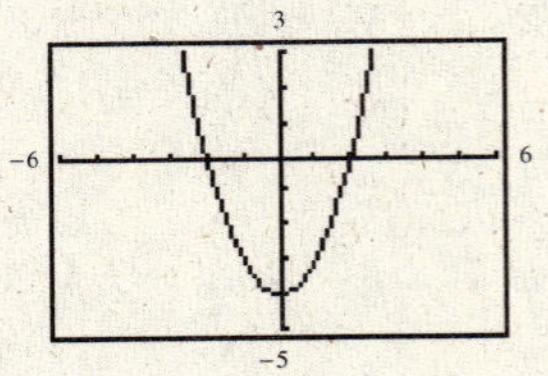

131. $3x^3 + 4x^2 - 3x - 4 = (3x^3 + 4x^2) + (-3x - 4)$
$= x^2(3x + 4) - 1(3x + 4)$
$= (x^2 - 1)(3x + 4)$
$= (x - 1)(x + 1)(3x + 4)$

or

$= (3x^3 - 3x) + (4x^2 - 4)$
$= 3x(x^2 - 1) + 4(x^2 - 1)$
$= (x^2 - 1)(3x + 4)$
$= (x - 1)(x + 1)(3x + 4)$

133. $R = 800x - 0.25x^2 = x(800 - 0.25x)$
$R = xp$
$p = 800 - 0.25x$

135. $P + Prt = P(1 + rt)$

137. $A = 45l - l^2 = l(45 - l)$ so
$w = 45 - l$

139. $S = 2x^2 + 4xh$
$S = 2x(x + 2h)$

141. $A = \pi R^2 - \pi r^2$
$= \pi(R^2 - r^2)$
$= \pi(R - r)(R + r)$

143. A polynomial is in factored form when the polynomial is written as a product of polynomials.

145. Check a result after factoring by multiplying the factors to see if the product is the original polynomial.

147. The Distributive Property is used to factor a polynomial in this example $x^2 + 2x = x(x + 2)$.

149. The three binomial factors of the expression are the sum of the squares of the monomials, the sum of the monomials, and the difference of the monomials.

$$a^4 - b^4 = (a^2 + b^2)(a^2 - b^2)$$
$$= (a^2 + b^2)(a + b)(a - b)$$

151. $\begin{vmatrix} 3 & 4 \\ 2 & 1 \end{vmatrix} = (3)(1) - (2)(4) = 3 - 8 = -5$

153. $\begin{vmatrix} -1 & 3 & 0 \\ -2 & 0 & 6 \\ 0 & 4 & 2 \end{vmatrix} = -1\begin{vmatrix} 0 & 6 \\ 4 & 2 \end{vmatrix} - 3\begin{vmatrix} -2 & 6 \\ 0 & 2 \end{vmatrix} + 0$

$$= (-1)[0 - 24] - (3)[-4 - 0]$$
$$= 24 + 12 = 36$$

155. $(x + 7)(x - 7) = x^2 - 7^2 = x^2 - 49$

157. $(2x - 3)^2 = (2x)^2 - 2(2x)(3) + 3^2$

$$= 4x^2 - 12x + 9$$

Section 5.5 Factoring Trinomials

1. $x^2 + 4x + 4 = x^2 + 2(2x) + 2^2 = (x + 2)^2$

3. $a^2 - 10a + 25 = a^2 - 2(5a) + 5^2 = (a - 5)^2$

5. $25y^2 - 10y + 1 = (5y)^2 - 2(5y) + 1 = (5y - 1)^2$

7. $9b^2 + 12b + 4 = (3b)^2 + 2(3b)(2) + 2^2 = (3b + 2)^2$

9. $u^2 + 8uv + 16v^2 = u^2 + 2(4uv) + (4v)^2 = (u + 4v)^2$

11. $36x^2 - 60xy + 25y^2 = (6x)^2 - 2(6x)(5y) + (5y)^2$

$$= (6x - 5y)^2$$

13. $5x^2 + 30x + 45 = 5(x^2 + 6x + 9)$

$$= 5[x^2 + 2(3)(x) + 3^2]$$
$$= 5(x + 3)^2$$

15. $3m^3 - 18m^2 + 27m = 3m(m^2 - 6m + 9)$

$$= 3m(m^2 - 2(3m) + 3^2)$$
$$= 3m(m - 3)^2$$

17. $20v^4 - 60v^3 + 45v^2 = 5v^2(4v^2 - 12v + 9)$

$$= 5v^2[(2v)^2 - 2(2v)(3) + 3^2]$$
$$= 5v^2(2v - 3)^2$$

19. $\frac{1}{4}x^2 - \frac{2}{3}x + \frac{4}{9} = \left(\frac{1}{2}x\right)^2 - 2\left(\frac{1}{2}x\right)\left(\frac{2}{3}\right) + \left(\frac{2}{3}\right)^2$

$$= \left(\frac{1}{2}x - \frac{2}{3}\right)^2$$

or

$$= \frac{9}{36}x^2 - \frac{24}{36}x + \frac{16}{36}$$
$$= \frac{1}{36}(9x^2 - 24x + 16)$$
$$= \frac{1}{36}[(3x)^2 - 2(3x)(4) + 4^2]$$
$$= \frac{1}{36}(3x - 4)^2$$

21. $x^2 + bx + 81 = x^2 + bx + 9^2$

(a) $b = 18$

$$x^2 + 18x + 9^2 = x^2 + 2(9x) + 9^2 = (x + 9)^2$$

or

(b) $b = -18$

$$x^2 - 18x + 9^2 = x^2 - 2(9x) + 9^2 = (x - 9)^2$$

23. $4x^2 + bx + 9 = (2x)^2 + bx + 3^2$

(a) $b = 12$

$$(2x)^2 + 12x + 3^2 = (2x)^2 + 2(2x)(3) + 3^2$$
$$= (2x + 3)^2$$

or

(b) $b = -12$

$$(2x)^2 - 12x + 3^2 = (2x)^2 - 2(2x)(3) + 3^2$$
$$= (2x - 3)^2$$

25. $c = 16$

$x^2 + 8x + c = x^2 + 2(4x) + c$

$= x^2 + 2(4x) + 4^2$

$= (x + 4)^2$

27. $c = 9$

$y^2 - 6y + c = y^2 - 2(3y) + c$

$= y^2 - 2(3y) + 3^2 = (y - 3)^2$

29. $a^2 + 6a + 8 = (a + 4)(a + 2)$

31. $y^2 - y - 20 = (y + 4)(y - 5)$

33. $x^2 + 10x + 24 = (x + 4)(x + 6)$

35. $z^2 - 6z + 8 = (z - 4)(z - 2)$

37. $x^2 + 6x + 5 = (x + 1)(x + 5)$

39. $x^2 - 5x + 6 = (x - 3)(x - 2)$

41. $y^2 + 7y - 30 = (y + 10)(y - 3)$

43. $t^2 - 6t - 16 = (t - 8)(t + 2)$

45. $x^2 - 20x + 96 = (x - 12)(x - 8)$

47. $x^2 - 2xy - 35y^2 = (x - 7y)(x + 5y)$

49. $x^2 + 30xy + 216y^2 = (x + 12y)(x + 18y)$

51. $x^2 + bx + 8$

$b = 6$ $\quad x^2 + 6x + 8 = (x + 4)(x + 2)$

$b = -6$ $\quad x^2 - 6x + 8 = (x - 4)(x - 2)$

$b = 9$ $\quad x^2 + 9x + 8 = (x + 8)(x + 1)$

$b = -9$ $\quad x^2 - 9x + 8 = (x - 8)(x - 1)$

53. $b = 20$: $\quad x^2 + 20x - 21 = (x + 21)(x - 1)$

$b = -20$: $\quad x^2 - 20x - 21 = (x - 21)(x + 1)$

$b = 4$: $\quad x^2 + 4x - 21 = (x + 7)(x - 3)$

$b = -4$: $\quad x^2 - 4x - 21 = (x - 7)(x + 3)$

55. $b = 36$: $\quad x^2 + 36x + 35 = (x + 35)(x + 1)$

$b = -36$: $\quad x^2 - 36x + 35 = (x - 35)(x - 1)$

$b = 12$: $\quad x^2 + 12x + 35 = (x + 7)(x + 5)$

$b = -12$: $\quad x^2 - 12x + 35 = (x - 7)(x - 5)$

57. There are many possibilities, such as:

$c = 5$ $\quad x^2 + 6x + 5 = (x + 5)(x + 1)$

$c = 8$ $\quad x^2 + 6x + 8 = (x + 4)(x + 2)$

$c = 9$ $\quad x^2 + 6x + 9 = (x + 3)(x + 3)$

Also note that if $c =$ a negative number, there are many possibilities for c such as the following.

$c = -7$ $\quad x^2 + 6x - 7 = (x + 7)(x - 1)$

$c = -16$ $\quad x^2 + 6x - 16 = (x + 8)(x - 2)$

$c = -27$ $\quad x^2 + 6x - 27 = (x + 9)(x - 3)$

59. There are many possibilities, such as:

$c = 2$ $\quad x^2 - 3x + 2 = (x - 2)(x - 1)$

$c = -4$ $\quad x^2 - 3x - 4 = (x - 4)(x + 1)$

$c = -10$ $\quad x^2 - 3x - 10 = (x - 5)(x + 2)$

$c = -18$ $\quad x^2 - 3x - 18 = (x - 6)(x + 3)$

There are more possibilities.

61. $5x^2 + 18x + 9 = (x + 3)(5x + 3)$

63. $5a^2 + 12a - 9 = (a + 3)(5a - 3)$

65. $2y^2 - 3y - 27 = (y + 3)(2y - 9)$

67. $6x^2 - 5x - 25 = (3x + 5)(2x - 5)$

69. $10y^2 - 7y - 12 = (5y + 4)(2y - 3)$

71. $12x^2 - 7x + 1 = (4x - 1)(3x - 1)$

73. $5z^2 + 2z - 3 = (5z - 3)(z + 1)$

75. $2t^2 - 7t - 4 = (2t + 1)(t - 4)$

77. $6b^2 + 19b - 7 = (3b - 1)(2b + 7)$

79. $-2x^2 - x + 6 = -1(2x^2 + x - 6)$

$= -1(2x - 3)(x + 2)$

or $(3 - 2x)(2 + x)$

81. $-15d^2 + 19d - 6 = -1(15d^2 - 19d + 6)$

$= -1(3d - 2)(5d - 3)$

83. $2 + 5x - 12x^2 = (2 - 3x)(1 + 4x)$

85. $4w^2 - 3w + 8$ is prime.

87. $60y^3 + 35y^2 - 50y = 5y(12y^2 + 7y - 10)$
$= 5y(3y - 2)(4y + 5)$

89. $10a^2 + 23ab + 6b^2 = (a + 2b)(10a + 3b)$

91. $24x^2 - 14xy - 3y^2 = (6x + y)(4x - 3y)$

93. $3x^2 + 10x + 8 = 3x^2 + 6x + 4x + 8$
$= (3x^2 + 6x) + (4x + 8)$
$= 3x(x + 2) + 4(x + 2)$
$= (3x + 4)(x + 2)$

95. $5x^2 - 12x - 9 = 5x^2 - 15x + 3x - 9$
$= 5x(x - 3) + 3(x - 3)$
$= (5x + 3)(x - 3)$

97. $15x^2 - 11x + 2 = 15x^2 - 6x - 5x + 2$
$= (15x^2 - 6x) + (-5x + 2)$
$= 3x(5x - 2) - 1(5x - 2)$
$= (3x - 1)(5x - 2)$

99. $3x^3 - 3x = 3x(x^2 - 1) = 3x(x^2 - 1^2) = 3x(x + 1)(x - 1)$

101. $10t^3 + 2t^2 - 36t = 2t(5t^2 + t - 18)$
$= 2t(5t - 9)(t + 2)$

103. $54x^3 - 2 = 2(27x^3 - 1)$
$= 2(3x - 1)(9x^2 + 3x + 1)$

104. $3t^3 - 24 = 3(t^3 - 8)$
$= 3[t^3 - (2)^3]$
$= 3(t - 2)(t^2 + 2t + 4)$

105. $27a^3b^4 - 9a^2b^3 - 18ab^2 = 9ab^2(3a^2b^2 - ab - 2)$
$= 9ab^2(3ab + 2)(ab - 1)$

107. $x^3 + 2x^2 - 16x - 32 = (x^3 + 2x^2) + (-16x - 32)$
$= x^2(x + 2) - 16(x + 2)$
$= (x + 2)(x^2 - 16)$
$= (x + 2)(x - 4)(x + 4)$

109. $49 - (r - 2)^2 = -1[-49 + (r - 2)^2]$
$= -1[(r - 2)^2 - 49]$
$= -1[(r - 2)^2 - 7^2]$
$= -[(r - 2) + 7][(r - 2) - 7]$
$= -(r + 5)(r - 9)$

111. $x^2 - 10x + 25 - y^2 = (x - 5)^2 - y^2$
$= [(x - 5) + y][(x - 5) - y]$
$= (x - 5 + y)(x - 5 - y)$

113. $x^8 - 1 = (x^4)^2 - 1^2 = (x^4 - 1)(x^4 + 1)$
$= [(x^2)^2 - 1^2](x^4 + 1)$
$= (x^2 - 1)(x^2 + 1)(x^4 + 1)$
$= (x - 1)(x + 1)(x^2 + 1)(x^4 + 1)$

115. $x^{2n} - 5x^n - 24 = (x^n - 8)(x^n + 3)$

117. $x^{2n} + 3x^n - 10 = (x^n + 5)(x^n - 2)$

119. $6y^{2n} + 13y^n + 6 = (2y^n + 3)(3y^n + 2)$

121. *Keystrokes:*

y_1: [Y=] [X,T,θ] [x^2] [+] 6 [X,T,θ] [+] 9 [ENTER]

y_2: [(] [X,T,θ] [+] 3 [)] [x^2] [GRAPH]

$y_1 = y_2$

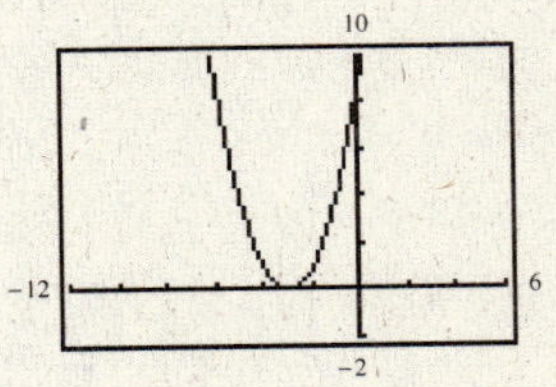

123. *Keystrokes:*

y_1: [Y=] 4 [X,T,θ] [x^2] [−] 13 [X,T,θ] [−] 12 [ENTER]

y_2: [(] 4 [X,T,θ] [+] 3 [)] [(] [X,T,θ] [−] 4 [)] [GRAPH]

$y_1 = y_2$

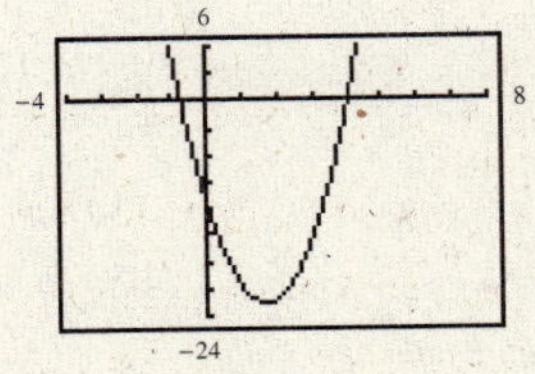

125. $a^2 - b^2 = (a + b)(a - b)$ matches graph (c).

127. $a^2 + 2ab + b^2 = (a + b)^2$ matches graph (b).

129. *Verbal Model:*

[Area of shaded region] = [Area of rectangle] − [Area of squares]

Equation:

$$\text{Area} = (8 \cdot 18) - 4 \cdot x^2$$
$$= 144 - 4x^2$$
$$= 4(36 - x^2)$$
$$= 4(6 + x)(6 - x)$$

131. (a) $8n^3 + 24n^2 + 16n = 2n(4n^2 + 12n + 8)$
$$= 2n(2n + 2)(2n + 4)$$

(b) If $n = 10$,
$$2n = 2(10) = 20$$
$$2n + 2 = 2(10) + 2 = 22$$
$$2n + 4 = 2(10) + 4 = 24$$

133. To factor $x^2 - 5x + 6$ begin by finding the factors of 6 whose sum is −5. They are −2 and −3. The factorization is $(x - 2)(x - 3)$.

135. An example of a prime trinomial is $x^2 + x + 1$.

137. No, $x(x + 2) - 2(x + 2)$ is not in factored form. It is not yet a product.

$$x(x + 2) - 2(x + 2) = (x + 2)(x - 2)$$

139. Problems will vary. It is possible to create factorable polynomials by working backward: first list several factors, and then multiply them to form a single polynomial.

141. *Verbal Model:* [Compared number] = [Percent] · [Base number]

Labels: Compared number $= a$

Percent $= p$

Base number $= b$

Equation:

$$a = p \cdot b$$
$$a = 1.25 \cdot 340$$
$$a = 425$$

143. *Verbal Model:* [Compared number] = [Percent] · [Base number]

Labels: Compared number $= a$

Percent $= p$

Base number $= b$

Equation:

$$a = p \cdot b$$
$$725 = p \cdot 2000$$
$$\frac{725}{2000} = p$$
$$36.25\% = p$$

145. *Verbal Model:* [Amount of solution 1] + [Amount of solution 2] = [Amount of final solution]

Labels: Percent of solution 1 = 20%

Liters of solution 1 $= x$

Percent of solution 2 = 60%

Liters of solution 2 $= 10 - x$

Percent of final solution = 30%

Liters of final solution = 10 L

Equation:

$$0.20x + 0.60(10 - x) = 0.30(10)$$
$$0.20x + 6 - 0.60x = 3$$
$$-0.40x = -3$$
$$x = 7.5 \text{ L}$$
$$10 - x = 2.5 \text{ L}$$

147. *Verbal Model:* [Amount of solution 1] + [Amount of solution 2] = [Amount of final solution]

Labels: Percent of solution 1 = 60%

Gallons of solution 1 = x

Percent of solution 2 = 90%

Gallons of solution 2 = $120 - x$

Percent of final solution = 85%

Gallons of final solution = 120 gal

Equation:

$$0.60x + 0.90(120 - x) = 0.85(120)$$
$$0.60x + 108 - 0.90x = 102$$
$$-0.30x = -6$$
$$x = 20 \text{ gal}$$
$$120 - x = 100 \text{ gal}$$

Section 5.6 Solving Polynomial Equations by Factoring

1. $x(x - 4) = 0$

$x = 0$

$x - 4 = 0$

$x = 4$

3. $(y - 3)(y + 10) = 0$

$y - 3 = 0 \qquad y + 10 = 0$

$y = 3 \qquad y = -10$

5. $25(a + 4)(a - 2) = 0$

$a + 4 = 0 \qquad a - 2 = 0$

$a = -4 \qquad a = 2$

7. $(2t + 5)(3t + 1) = 0$

$2t + 5 = 0 \qquad 3t + 1 = 0$

$t = -\frac{5}{2} \qquad t = -\frac{1}{3}$

9. $4x(2x - 3)(2x + 25) = 0$

$4x = 0 \qquad 2x - 3 = 0 \qquad 2x + 25 = 0$

$x = 0 \qquad x = \frac{3}{2} \qquad x = -\frac{25}{2}$

11. $(x - 3)(2x + 1)(x + 4) = 0$

$x - 3 = 0 \qquad 2x + 1 = 0 \qquad x + 4 = 0$

$x = 3 \qquad x = -\frac{1}{2} \qquad x = -4$

13. $5y - y^2 = 0$

$y(5 - y) = 0$

$y = 0 \qquad 5 - y = 0$

$5 = y$

15. $9x^2 + 15x = 0$

$3x(3x + 5) = 0$

$3x = 0 \qquad 3x + 5 = 0$

$x = 0 \qquad x = -\frac{5}{3}$

17. $2x^2 = 32x$

$2x^2 - 32x = 0$

$2x(x - 16) = 0$

$2x = 0 \qquad x - 16 = 0$

$x = 0 \qquad x = 16$

19. $5y^2 = 15y$

$5y^2 - 15y = 0$

$5y(y - 3) = 0$

$5y = 0 \qquad y - 3 = 0$

$y = 0 \qquad y = 3$

21. $x^2 - 25 = 0$

$(x + 5)(x - 5) = 0$

$x + 5 = 0 \qquad x - 5 = 0$

$x = -5 \qquad x = 5$

23. $3y^2 - 48 = 0$

$3(y^2 - 16) = 0$

$3(y + 4)(y - 4) = 0$

$y + 4 = 0 \qquad y - 4 = 0$

$y = -4 \qquad y = 4$

25. $x^2 - 3x - 10 = 0$

$(x - 5)(x + 2) = 0$

$x - 5 = 0 \qquad x + 2 = 0$

$x = 5 \qquad x = -2$

27. $x^2 - 10x + 24 = 0$

$(x - 6)(x - 4) = 0$

$x - 6 = 0 \qquad x - 4 = 0$

$x = 6 \qquad x = 4$

29. $4x^2 + 15x - 25 = 0$

$(4x - 5)(x + 5) = 0$

$4x - 5 = 0 \qquad x + 5 = 0$

$4x = 5 \qquad x = -5$

$x = \frac{5}{4}$

31. $7 + 13x - 2x^2 = 0$

$(7 - x)(1 + 2x) = 0$

$7 - x = 0 \qquad 1 + 2x = 0$

$7 = x \qquad -\frac{1}{2} = x$

33. $3y^2 - 2 = -y$

$3y^2 + y - 2 = 0$

$(3y - 2)(y + 1) = 0$

$3y - 2 = 0 \qquad y + 1 = 0$

$y = \frac{2}{3} \qquad y = -1$

35. $-13x + 36 = -x^2$

$x^2 - 13x + 36 = 0$

$(x - 9)(x - 4) = 0$

$x - 9 = 0 \qquad x - 4 = 0$

$x = 9 \qquad x = 4$

37. $m^2 - 8m + 18 = 2$

$m^2 - 8m + 16 = 0$

$(m - 4)^2 = 0$

$m - 4 = 0$

$m = 4$

39. $x^2 + 16x + 57 = -7$

$x^2 + 16x + 64 = 0$

$(x + 8)^2 = 0$

$x + 8 = 0$

$x = -8$

41. $4z^2 - 12z + 15 = 6$

$4z^2 - 12z + 9 = 0$

$(2z - 3)^2 = 0$

$2z - 3 = 0$

$z = \frac{3}{2}$

43. $x(x + 2) - 10(x + 2) = 0$

$(x + 2)(x - 10) = 0$

$x + 2 = 0 \qquad x - 10 = 0$

$x = -2 \qquad x = 10$

45. $u(u - 3) + 3(u - 3) = 0$

$(u - 3)(u + 3) = 0$

$u - 3 = 0 \qquad u + 3 = 0$

$u = 3 \qquad u = -3$

47. $x(x - 5) = 36$

$x^2 - 5x = 36$

$x^2 - 5x - 36 = 0$

$(x - 9)(x + 4) = 0$

$x - 9 = 0 \qquad x + 4 = 0$

$x = 9 \qquad x = -4$

49. $y(y + 6) = 72$

$y^2 + 6y - 72 = 0$

$(y + 12)(y - 6) = 0$

$y + 12 = 0 \qquad y - 6 = 0$

$y = -12 \qquad y = 6$

51. $3t(2t - 3) = 15$

$6t^2 - 9t - 15 = 0$

$3(2t^2 - 3t - 5) = 0$

$3(2t - 5)(t + 1) = 0$

$2t - 5 = 0 \qquad t + 1 = 0$

$2t = 5 \qquad t = -1$

$t = \frac{5}{2}$

53. $(a + 2)(a + 5) = 10$

$a^2 + 7a + 10 - 10 = 0$

$a^2 + 7a = 0$

$a(a + 7) = 0$

$a = 0 \qquad a + 7 = 0$

$a = 0 \qquad a = -7$

55. $(x - 4)(x + 5) = 10$

$x^2 + x - 20 - 10 = 0$

$x^2 + x - 30 = 0$

$(x + 6)(x - 5) = 0$

$x + 6 = 0 \qquad x - 5 = 0$

$x = -6 \qquad x = 5$

57. $(t - 2)^2 = 16$

$(t - 2)^2 - 16 = 0$

$(t - 2 + 4)(t - 2 - 4) = 0$

$(t + 2)(t - 6) = 0$

$t + 2 = 0 \qquad t - 6 = 0$

$t = -2 \qquad t = 6$

59. $9 = (x + 2)^2$

$0 = [(x + 2) - 3][(x + 2) + 3]$

$0 = (x - 1)(x + 5)$

$0 = (x - 1) \qquad 0 = (x + 5)$

$x = 1 \qquad x = -5$

61. $(x - 3)^2 - 25 = 0$

$[(x - 3) - 5][(x - 3) + 5] = 0$

$(x - 8)(x + 2) = 0$

$x - 8 = 0 \qquad x + 2 = 0$

$x = 8 \qquad x = -2$

63. $81 - (x + 4)^2 = 0$

$[9 - (x + 4)][9 + (x + 4)] = 0$

$(5 - x)(13 + x) = 0$

$5 - x = 0 \qquad 13 + x = 0$

$-x = -5 \qquad x = -13$

$x = 5$

65. $x^3 - 19x^2 + 84x = 0$

$x(x^2 - 19x + 84) = 0$

$x(x - 12)(x - 7) = 0$

$x = 0 \qquad x - 12 = 0 \qquad x - 7 = 0$

$x = 0 \qquad x = 12 \qquad x = 7$

67. $6t^3 = t^2 + t$

$6t^3 - t^2 - t = 0$

$t(6t^2 - t - 1) = 0$

$t(3t + 1)(2t - 1) = 0$

$t = 0 \qquad 3t + 1 = 0 \qquad 2t - 1 = 0$

$t = 0 \qquad t = -\frac{1}{3} \qquad t = \frac{1}{2}$

69. $z^2(z + 2) - 4(z + 2) = 0$

$(z + 2)(z^2 - 4) = 0$

$(z + 2)(z - 2)(z + 2) = 0$

$z + 2 = 0 \qquad z - 2 = 0 \qquad z + 2 = 0$

$z = -2 \qquad z = 2 \qquad z = -2$

71. $a^3 + 2a^2 - 9a - 18 = 0$

$(a^3 + 2a^2) + (-9a - 18) = 0$

$a^2(a + 2) - 9(a + 2) = 0$

$(a + 2)(a^2 - 9) = 0$

$(a + 2)(a - 3)(a + 3) = 0$

$a + 2 = 0 \qquad a - 3 = 0 \qquad a + 3 = 0$

$a = -2 \qquad a = 3 \qquad a = -3$

73. $c^3 - 3c^2 - 9c + 27 = 0$

$c^2(c - 3) - 9(c - 3) = 0$

$(c - 3)(c^2 - 9) = 0$

$(c - 3)(c - 3)(c + 3) = 0$

$c - 3 = 0 \qquad c - 3 = 0 \qquad c + 3 = 0$

$c = 3 \qquad c = 3 \qquad c = -3$

75. $x^4 - 3x^3 - x^2 + 3x = 0$

$x^3(x - 3) - x(x - 3) = 0$

$(x - 3)(x^3 - x) = 0$

$(x - 3)x(x^2 - 1) = 0$

$(x - 3)x(x - 1)(x + 1) = 0$

$x - 3 = 0 \qquad x = 0 \qquad x - 1 = 0 \qquad x + 1 = 0$

$x = 3 \qquad x = 1 \qquad x = -1$

77. $8x^4 + 12x^3 - 32x^2 - 48x = 0$

$4x^3(2x + 3) - 16x(2x + 3) = 0$

$(2x + 3)(4x^3 - 16x) = 0$

$(2x + 3)\,4x(x^2 - 4) = 0$

$(2x + 3)(4x)(x - 2)(x + 2) = 0$

$2x + 3 = 0 \quad 4x = 0 \quad x - 2 = 0 \quad x + 2 = 0$

$x = -\frac{3}{2} \quad x = 0 \quad x = 2 \quad x = -2$

79. From the graph, the x-intercepts are $(-3, 0)$ and $(3, 0)$. The solutions of the equation $0 = x^2 - 9$ are 3 and -3.

$0 = (x - 3)(x + 3)$

$0 = x - 3 \quad 0 = x + 3$

$3 = x \quad -3 = x$

81. From the graph, the x-intercepts are $(0, 0)$ and $(3, 0)$. The solutions of the equation $0 = x^3 - 6x^2 + 9x$ are 0 and 3.

$x^3 - 6x^2 + 9x = 0$

$x(x^2 - 6x + 9) = 0$

$x(x - 3)(x - 3) = 0$

$x = 0 \quad x - 3 = 0 \quad x - 3 = 0$

$x = 3 \quad x = 3$

83. *Keystrokes:*

[Y=] [X,T,θ] [x^2] [+] 5 [X,T,θ] [GRAPH]

The x-intercepts are $(-5, 0)$ and $(0, 0)$.

$y = x^2 + 5x$

$0 = x(x + 5)$

$x = 0 \quad x + 5 = 0$

$x = -5$

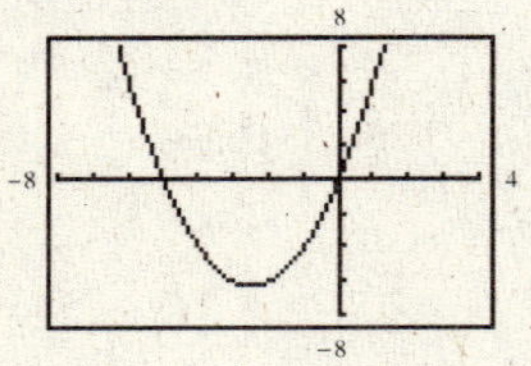

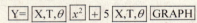

85. *Keystrokes:*

[Y=] [X,T,θ] [x^2] [−] 8 [X,T,θ] [+] 12 [GRAPH]

The x-intercepts are 2 and 6, so the solutions are 2 and 6.

$y = x^2 - 8x + 12$

$0 = (x - 2)(x - 6)$

$x - 2 = 0 \quad x - 6 = 0$

$x = 2 \quad x = 6$

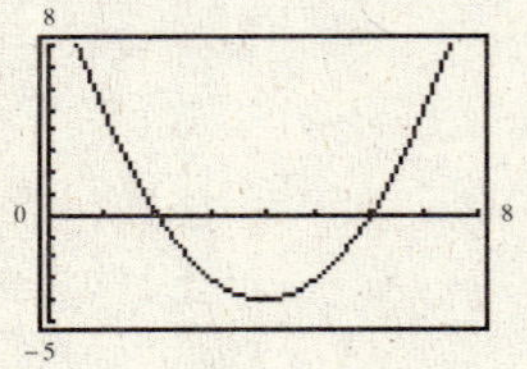

87. *Keystrokes:*

[Y=] 2 [X,T,θ] [x^2] [+] 5 [X,T,θ] [−] 12 [GRAPH]

The x-intercepts are -4 and $\frac{3}{2}$, so the solutions are -4 and $\frac{3}{2}$.

$y = 2x^2 + 5x - 12$

$0 = (2x - 3)(x + 4)$

$2x - 3 = 0 \quad x + 4 = 0$

$x = \frac{3}{2} \quad x = -4$

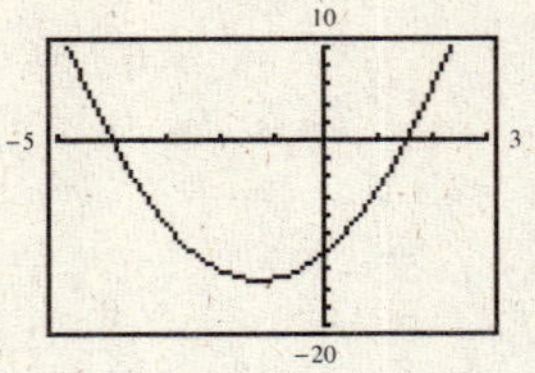

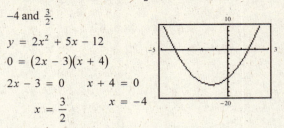

89. *Keystrokes:*

[Y=] 2 [X,T,θ] [^] [3] [−] 5 [X, T, θ]

[x^2] [−] 12 [X,T,θ] [GRAPH]

The x-intercepts are $-\frac{3}{2}$, 0, and 4, so the solutions are $-\frac{3}{2}$, 0, and 4.

$y = 2x^3 - 5x^2 - 12x$

$0 = x(2x^2 - 5x - 12)$

$0 = x(2x + 3)(x - 4)$

$x = 0 \quad 2x + 3 = 0 \quad x - 4 = 0$

$x = 0 \quad x = -\frac{3}{2} \quad x = 4$

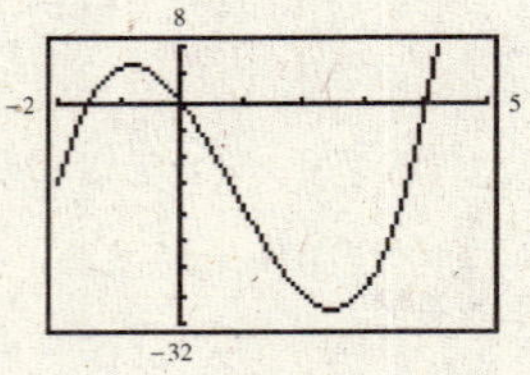

91. $ax^2 + bx = 0$

$x(ax + b) = 0$

$x = 0 \quad ax + b = 0$

$ax = -b$

$x = -\frac{b}{a}$

93. $x = -2, \quad x = 6$

$[x - (-2)][x - 6] = 0$

$(x + 2)(x - 6) = 0$

$x^2 - 4x - 12 = 0$

95. *Verbal Model:* [Number] + [Its square] = [240]

Labels: Number $= x$

Its square $= x^2$

Equation: $x + x^2 = 240$

$x^2 + x - 240 = 0$

$(x + 16)(x - 15) = 0$

$x + 16 = 0 \quad x - 15 = 0$

$x = -16 \quad x = 15$

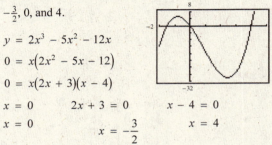

Reject

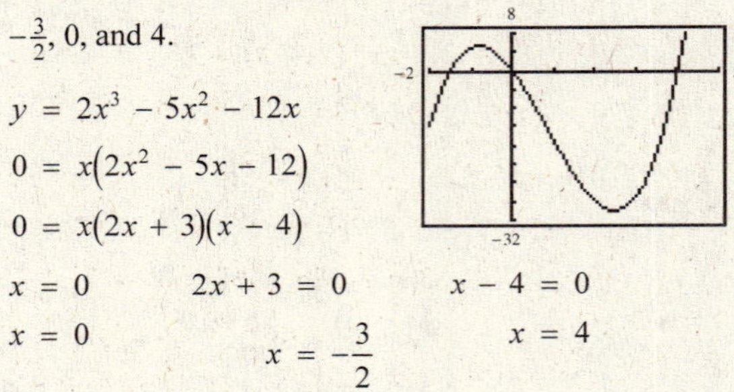

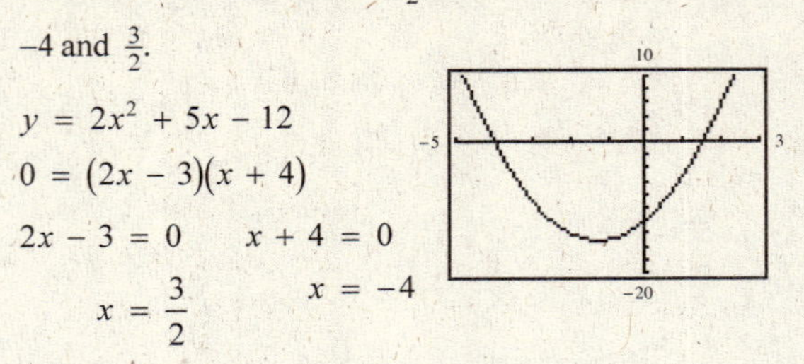

97. *Verbal Model:* $\boxed{\text{Length}} \cdot \boxed{\text{Width}} = \boxed{\text{Area}}$

Labels: Length $= w + 7$

Width $= w$

Equation: $(w + 7) \cdot w = 540$

$w^2 + 7w - 540 = 0$

$(w + 27)(w - 20) = 0$

$w + 27 = 0 \qquad w - 20 = 0$

$w = -27 \qquad w = 20$ feet

Reject $\qquad w + 7 = 27$ feet

99. *Verbal Model:* $\boxed{\frac{1}{2}} \cdot \boxed{\text{Base}} \cdot \boxed{\text{Height}} = \boxed{\text{Area}}$

Labels: Base $= x$

Height $= \frac{3}{2}x$

Equation: $\frac{1}{2} \cdot x \cdot \frac{3}{2}x = 27$

$\frac{3}{4}x^2 - 27 = 0$

$3x^2 - 108 = 0$

$3(x^2 - 36) = 0$

$3(x + 6)(x - 6) = 0$

$x + 6 = 0 \qquad x - 6 = 0$

$x = -6 \qquad x = 6$ inches

Reject $\qquad \frac{3}{2}x = 9$ inches

101. $h = -16t^2 + 400$

$0 = -16t^2 + 400$

$0 = -16(t^2 - 25)$

$0 = -16(t + 5)(t - 5)$

$t + 5 = 0 \qquad t - 5 = 0$

$t = -5 \qquad t = 5$ seconds

Reject

103. $h = -16t^2 + 80$

$16 = -16t^2 + 80$

$0 = -16t^2 + 64$

$0 = -16(t^2 - 4)$

$0 = -16(t + 2)(t - 2)$

$t + 2 = 0 \qquad t - 2 = 0$

$t = -2 \qquad t = 2$

Reject $\qquad$ Seconds

105. $-16t^2 + 48t + 1053 = 0$

$16t^2 - 48t - 1053 = 0$

$(4t - 39)(4t + 27) = 0$

$4t - 39 = 0 \qquad 4t + 27 = 0$

$t = \frac{39}{4} \qquad t = -\frac{27}{4}$

$t = 9.75 \qquad$ Reject

107. *Verbal Model:* $\boxed{\text{Revenue}} = \boxed{\text{Cost}}$

Equation: $R = C$

$140x - x^2 = 2000 + 50x$

$0 = x^2 - 90x + 2000$

$0 = (x - 40)(x - 50)$

$x - 40 = 0 \qquad x - 50 = 0$

$x = 40 \qquad x = 50$

Units $\qquad$ Units

111. (a) Volume V = Length · Width · Height

$$V = (5 - 2x)(4 - 2x)x$$

(b) $0 = (5 - 2x)(4 - 2x)x$

$5 - 2x = 0$ $\quad$ $4 - 2x = 0$ $\quad$ $x = 0$

$x = \frac{5}{2}$ $\quad$ $x = 2$ $\quad$ $x = 0$

Domain: Each side must be positive.

$x > 0$ $\quad$ $5 - 2x > 0$ $\quad$ $4 - 2x > 0$ so $0 < x < 2$

$x < \frac{5}{2}$ $\quad$ $x < 2$

(c)

x	0.25	0.50	0.75	1.00	1.25	1.50	1.75
V	3.94	6	6.56	6	4.69	3	1.31

(d) If $V = 3$, then $x = 1.5$.

$3 = [5 - 2(1.5)][4 - 2(1.5)](1.5)$

$3 = (5 - 3)(4 - 3)(1.5)$

$3 = (2)(1)(1.5)$

$3 = 3$

(e) *Keystrokes:*

Y= (5 − 2 X,T,θ) (4 − 2 X,T,θ) X,T,θ GRAPH

$x = 0.74$ yields the box of greatest volume.

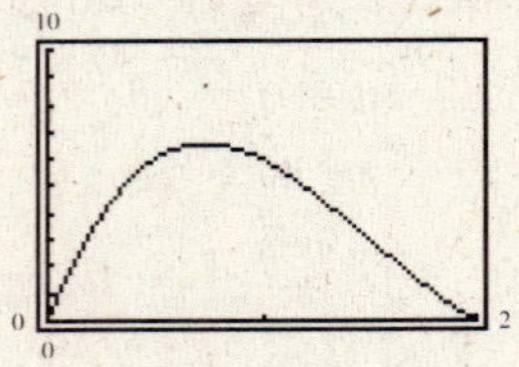

113. The maximum number of solutions of an nth degree polynomial equation is n. The third-degree equation $(x + 1)^3 = 0$ has only one solution, $x = -1$.

115. When a quadratic equation has a repeated solution, the graph of the equation has one x-intercept, which is the vertex of the graph.

117. A solution to a polynomial expression is the value of x when y is zero. If a polynomial is not factorable, the equation can still have real number solutions for x when y is zero

119. Unit price $= \dfrac{\$0.75}{12} = \$0.0625$ per ounce

121. Unit price $= \dfrac{\$2.13}{30} = \$0.071$ per ounce

123. The domain of $f(x) = \dfrac{x+3}{x+1}$ is all real numbers such that $x \neq -1$ because $x + 1 \neq 0$ means $x \neq -1$.

125. The domain of $g(x) = \sqrt{3 - x}$ is all real numbers such that $x \leq 3$ because $3 - x \geq 0$ means $-x \geq -3$ and $x \leq 3$.

Review Exercises for Chapter 5

1. $x^4 \cdot x^5 = x^{4+5} = x^9$

3. $(u^2)^3 = u^{2\cdot 3} = u^6$

5. $(-2z)^3 = (-2)^3 z^3 = -8z^3$

7. $-(u^2v)^2(-4u^3v) = -(u^4v^2)(-4u^3v)$

$= 4u^{4+3}v^{2+1}$

$= 4u^7v^3$

9. $\dfrac{12z^5}{6z^2} = \left(\dfrac{12}{6}\right) \cdot z^{5-2} = 2z^3$

11. $\dfrac{25g^4d^2}{80g^2d^2} = \dfrac{25}{80} \cdot \dfrac{g^4}{g^2} \cdot \dfrac{d^2}{d^2} = \dfrac{5}{16} \cdot g^{4-2}d^{2-2} = \dfrac{5}{16}g^2$

13. $\left(\dfrac{72x^4}{6x^2}\right)^2 = \left(12x^{4-2}\right)^2 = \left(12x^2\right)^2 = 144x^4$

15. $\left(2^3 \cdot 3^2\right)^{-1} = \left(8 \cdot 9\right)^{-1} = 72^{-1} = \frac{1}{72}$

17. $\left(\dfrac{3}{4}\right)^{-3} = \left(\dfrac{4}{3}\right)^3 = \dfrac{4^3}{3^3} = \dfrac{64}{27}$

19. $\left(6y^4\right)\left(2y^{-3}\right) = 12y^{4+(-3)} = 12y^1 = 12y$

21. $\dfrac{4x^{-2}}{2x} = 2x^{-2-1} = 2x^{-3} = \dfrac{2}{x^3}$

23. $\left(x^3y^{-4}\right)^0 = 1$

25. $\dfrac{7a^6b^{-2}}{14a^{-1}b^4} = \dfrac{7}{14} \cdot \dfrac{a^6}{a^{-1}} \cdot \dfrac{b^{-2}}{b^4} = \dfrac{1}{2} \cdot a^{6-(-1)} \cdot b^{-2-4}$

$= \dfrac{1}{2}a^7b^{-6} = \dfrac{a^7}{2b^6}$

27. $\left(\dfrac{3x^{-1}y^2}{12x^5y^{-3}}\right)^{-1} = \dfrac{12x^5y^{-3}}{3x^{-1}y^2}$

$= 4x^{5-(-1)}y^{(-3)-2}$

$= 4x^6y^{-5}$

$= \dfrac{4x^6}{y^5}$

29. $u^3\left(5u^0v^{-1}\right)\left(9u\right)^2 = u^3\left(5v^{-1}\right)\left(81u^2\right)$

$= (5)(81)u^{3+2}v^{-1}$

$= \dfrac{405u^5}{v}$

31. $0.0000319 = 3.19 \times 10^{-5}$

33. $17{,}350{,}000 = 1.735 \times 10^7$

35. $1.95 \times 10^6 = 1{,}950{,}000$

37. $2.05 \times 10^{-5} = 0.0000205$

39. $\left(6 \times 10^3\right)^2 = 6^2 \times 10^6 = 36 \times 10^6 = 36{,}000{,}000$

41. $\dfrac{3.5 \times 10^7}{7 \times 10^4} = \dfrac{3.5}{7} \times 10^{7-4}$

$= 0.5 \times 10^3$

$= 5 \times 10^2$

$= 500$

43. Standard form:

$-x^4 + 6x^3 + 5x^2 - 4x$

Leading coefficient: –1

Degree: 4

45. $x^4 + 3x^5 - 4 - 6x$

Standard form: $3x^5 + x^4 - 6x - 4$

Degree: 5

Leading coefficient: 3

47. Trinomial of degree 5 and leading coefficient –6:

$-6x^5 + 2x - 4$

49. $(10x + 8) + \left(x^2 + 3x\right) = x^2 + (10x + 3x) + 8$

$= x^2 + 13x + 8$

51. $\left(5x^3 - 6x + 11\right) + \left(5 + 6x - x^2 - 8x^3\right) = \left(5x^3 - 8x^3\right) - x^2 + \left(-6x + 6x\right) + \left(11 + 5\right) = -3x^3 - x^2 + 16$

53. $(3y - 4) - \left(2y^2 + 1\right) = (3y - 4) + \left(-2y^2 - 1\right) = -2y^2 + 3y + (-4 - 1) = -2y^2 + 3y - 5$

55. $\left(-x^3 - 3x\right) - 4\left(2x^3 - 3x + 1\right) = -x^3 - 3x - 8x^3 + 12x - 4 = \left(-x^3 - 8x^3\right) + \left(-3x + 12x\right) + (-4) = -9x^3 + 9x - 4$

57. $3y^2 - \left[2y + 3\left(y^2 + 5\right)\right] = 3y^2 - \left[2y + 3y^2 + 15\right] = 3y^2 - 2y - 3y^2 - 15 = \left(3y^2 - 3y^2\right) - 2y - 15 = -2y - 15$

59. $\left(3x^5 + 4x^2 - 8x + 12\right) - \left(2x^5 + x\right) + \left(3x^2 - 4x^3 - 9\right) = \left(3x^5 - 2x^5\right) - 4x^3 + \left(4x^2 + 3x^2\right) + \left(-8x - x\right) + (12 - 9)$

$= x^5 - 4x^3 + 7x^2 - 9x + 3$

61.
$$\begin{array}{r} 3x^2 + 5x \\ \underline{-4x^2 - x + 6} \\ -x^2 + 4x + 6 \end{array}$$

63.
$$\begin{array}{r} 3t - 5 \\ \underline{-(t^2 - t - 5)} \\ \end{array} \Rightarrow \begin{array}{r} 3t - 5 \\ \underline{-t^2 + t + 5} \\ -t^2 + 4t \end{array}$$

65. Perimeter $= 3x + x + 3x + (3x + 5) + 6x + (2x + 5) = 18x + 10$

67. *Verbal Model:* [Profit] = [Revenue] − [Cost]

Equation:
$$\begin{aligned} P(x) &= 35x - (16x + 3000) \\ &= 35x - 16x - 3000 \\ &= 19x - 3000 \\ P(1200) &= 19(1200) - 3000 \\ &= 22{,}800 - 3000 \\ &= \$19{,}800 \end{aligned}$$

69. $(-2x)^3(x + 4) = -8x^3(x + 4) = -8x^4 - 32x^3$

71. $3x(2x^2 - 5x + 3) = 6x^3 - 15x^2 + 9x$

73.
$$\begin{aligned} (x - 2)(x + 7) &= x^2 + 7x - 2x - 14 \\ &= x^2 + 5x - 14 \end{aligned}$$

75.
$$\begin{aligned} (5x + 3)(3x - 4) &= 15x^2 - 20x + 9x - 12 \\ &= 15x^2 - 11x - 12 \end{aligned}$$

77. $(4x^2 + 3)(6x^2 + 1) = 24x^4 + 4x^2 + 18x^2 + 3 = 24x^4 + 22x^2 + 3$

79.
$$\begin{aligned} (2x^2 - 3x + 2)(2x + 3) &= 2x^2(2x + 3) - 3x(2x + 3) + 2(2x + 3) \\ &= 4x^3 + 6x^2 - 6x^2 - 9x + 4x + 6 \\ &= 4x^3 + (6x^2 - 6x^2) + (-9x + 4x) + 6 \\ &= 4x^3 - 5x + 6 \end{aligned}$$

81.
$$\begin{aligned} 2u(u - 7) - (u + 1)(u - 7) &= 2u(u - 7) - u(u - 7) - 1(u - 7) \\ &= 2u^2 - 14u - u^2 + 7u - u + 7 \\ &= (2u^2 - u^2) + (-14u + 7u - u) + 7 \\ &= u^2 - 8u + 7 \end{aligned}$$

83. $(4x - 7)^2 = (4x)^2 - 2(4x)(7) + (-7)^2 = 16x^2 - 56x + 49$

85. $(6v + 9)(6v - 9) = (6v)^2 - 9^2 = 36v^2 - 81$

87.
$$\begin{aligned} [(u - 3) + v][(u - 3) - v] &= (u - 3)^2 - v^2 \\ &= u^2 - 2(u)(3) + (-3)^2 - v^2 \\ &= u^2 - 6u + 9 - v^2 \end{aligned}$$

89. *Verbal Model:* [Area of shaded region] = [Area of larger rectangle] − [Area of smaller rectangle]

Labels: Width of larger rectangle $= 2x$

Length of larger rectangle $= 2x + 5$

Width of smaller rectangle $= 2x - 3$

Length of smaller rectangle $= 2x + 1$

Equation:
$$\begin{aligned} \text{Area} &= 2x(2x + 5) - (2x + 1)(2x - 3) \\ &= 4x^2 + 10x - (4x^2 - 6x + 2x - 3) \\ &= 4x^2 + 10x - 4x^2 + 4x + 3 \\ &= 14x + 3 \end{aligned}$$

91. Interest $= 1000(1 + 0.06)^2 = \$1123.60$

93. $24x^2 - 18 = 6(4x^2 - 3)$

95. $-3b^2 + b = -b(3b - 1)$

97. $6x^2 + 15x^3 - 3x = 3x(2x + 5x^2 - 1)$

99. $28(x + 5) - 70(x + 5) = (28 - 70)(x + 5)$
$= -42(x + 5)$

101. $v^3 - 2v^2 - v + 2 = v^2(v - 2) - 1(v - 2)$
$= (v - 2)(v^2 - 1)$
$= (v - 2)(v - 1)(v + 1)$

103. $t^3 + 3t^2 + 3t + 9 = t^2(t + 3) + 3(t + 3)$
$= (t^2 + 3)(t + 3)$

105. $x^2 - 36 = x^2 - 6^2 = (x - 6)(x + 6)$

107. $(u + 6)^2 - 81 = (u + 6 - 9)(u + 6 + 9)$
$= (u - 3)(u + 15)$

109. $u^3 - 1 = (u - 1)(u^2 + u + 1)$

111. $8x^3 + 27 = (2x)^3 + (3)^3$
$= (2x + 3)(4x^2 - 6x + 9)$

113. $x^3 - x = x(x^2 - 1)$
$= x(x - 1)(x + 1)$

115. $24 + 3u^3 = 3u^3 + 24$
$= 3(u^3 + 8)$
$= 3(u^3 + 2^3)$
$= 3(u + 2)(u^2 - 2u + 4)$

117. $x^2 - 18x + 81 = x^2 - 2(9)x + 9^2$
$= (x - 9)^2$

119. $4s^2 + 40st + 100t^2 = (2s)^2 + 2(2s)(10) + (10t)^2$
$= (2s + 10t)^2$

121. $x^2 + 2x - 35 = (x + 7)(x - 5)$

123. $2x^2 - 7x + 6 = (2x - 3)(x - 2)$

125. $18x^2 + 27x + 10 = (3x + 2)(6x + 5)$

127. $4x^2 - 3x - 1 = 4x^2 - 4x + x - 1$
$= 4x(x - 1) + 1(x - 1)$
$= (4x + 1)(x - 1)$

129. $5x^2 - 12x + 7 = 5x^2 - 5x - 7x + 7$
$= 5x(x - 1) - 7(x - 1)$
$= (5x - 7)(x - 1)$

131. $7s^2 + 10s - 8 = 7s^2 + 14s - 4s - 8$
$= 7s(s + 2) - 4(s + 2)$
$= (7s - 4)(s + 2)$

133. $4a - 64a^3 = 4a(1 - 16a^2)$
$= 4a(1 - 4a)(1 + 4a)$

135. $z^3 + z^2 + 3z + 3 = z^2(z + 1) + 3(z + 1)$
$= (z^2 + 3)(z + 1)$

137. $\frac{1}{4}x^2 + xy + y^2 = \left(\frac{1}{2}x\right)^2 + 2\left(\frac{1}{2}\right)xy + y^2$
$= \left(\frac{1}{2}x + y\right)^2$

139. $x^2 - 10x + 25 - y^2 = (x - 5)^2 - y^2$
$= [(x - 5) - y][(x - 5) + y]$
$= (x - 5 - y)(x - 5 + y)$
$= (x - y - 5)(x + y - 5)$

141. $4x(x - 2) = 0$

$4x = 0 \qquad x - 2 = 0$
$x = 0 \qquad x = 2$

143. $(2x + 1)(x - 3) = 0$

$2x + 1 = 0 \qquad x - 3 = 0$
$x = -\frac{1}{2} \qquad x = 3$

145. $(x + 10)(4x - 1)(5x + 9) = 0$

$x + 10 = 0 \qquad 4x - 1 = 0 \qquad 5x + 9 = 0$
$x = -10 \qquad x = \frac{1}{4} \qquad x = -\frac{9}{5}$

147. $3s^2 - 2s - 8 = 0$
$(3s + 4)(s - 2) = 0$

$3s + 4 = 0 \qquad s - 2 = 0$
$s = -\frac{4}{3} \qquad s = 2$

149. $m(2m - 1) + 3(2m - 1) = 0$

$(m + 3)(2m - 1) = 0$

$m + 3 = 0 \qquad 2m - 1 = 0$

$m = -3 \qquad 2m = 1$

$m = \frac{1}{2}$

151. $z(5 - z) + 36 = 0$

$5z - z^2 + 36 = 0$

$z^2 - 5z - 36 = 0$

$(z - 9)(z + 4) = 0$

$z - 9 = 0 \qquad z + 4 = 0$

$z = 9 \qquad z = -4$

153. $v^2 - 100 = 0$

$(v - 10)(v + 10) = 0$

$v - 10 = 0 \qquad v + 10 = 0$

$v = 10 \qquad v = -10$

155. $2y^4 + 2y^3 - 24y^2 = 0$

$2y^2(y^2 + y - 12) = 0$

$2y^2(y + 4)(y - 3) = 0$

$2y^2 = 0 \qquad y + 4 = 0 \qquad y - 3 = 0$

$y = 0 \qquad y = -4 \qquad y = 3$

157. $x^3 - 11x^2 + 18x = 0$

$x(x^2 - 11x + 18) = 0$

$x(x - 9)(x - 2) = 0$

$x = 0 \qquad x - 9 = 0 \qquad x - 2 = 0$

$x = 9 \qquad x = 2$

159. $b^3 - 6b^2 - b + 6 = 0$

$b^2(b - 6) - (b - 6) = 0$

$(b - 6)(b^2 - 1) = 0$

$(b - 6)(b - 1)(b + 1) = 0$

$b - 6 = 0 \qquad b - 1 = 0 \qquad b + 1 = 0$

$b = 6 \qquad b = 1 \qquad b = -1$

161. $x^4 - 5x^3 - 9x^2 + 45x = 0$

$x^3(x - 5) - 9x(x - 5) = 0$

$(x^3 - 9x)(x - 5) = 0$

$x(x^2 - 9)(x - 5) = 0$

$x(x - 3)(x + 3)(x - 5) = 0$

$x = 0 \qquad x - 3 = 0 \qquad x + 3 = 0 \qquad x - 5 = 0$

$x = 3 \qquad x = -3 \qquad x = 5$

163. *Verbal Model:* [First odd integer] · [Second odd integer] = 99

Labels: First odd integer $= 2n + 1$

Second odd integer $= 2n + 3$

Equation: $(2n + 1)(2n + 3) = 99$

$4n^2 + 8n + 3 = 99$

$4n^2 + 8n - 96 = 0$

$4(n^2 + 2n - 24) = 0$

$4(n + 6)(n - 4) = 0$

$n + 6 = 0 \qquad n - 4 = 0 \qquad 2n + 1 = 9$

$n = -6 \qquad n = 4 \qquad 2n + 3 = 11$

Reject

165. *Verbal Model:* Area = [Length] · [Width]

Labels: Length $= 2\frac{1}{4}x$

Width $= x$

Equation: $900 = 2\frac{1}{4}x \cdot x$

$900 = \frac{9}{4}x^2$

$3600 = 9x^2$

$400 = x^2$

$20 = x$

$45 = 2\frac{1}{4}x$

45 inches × 20 inches

167. *Verbal Model:* $\boxed{\text{Surface Area}} = 2 \cdot \boxed{\text{Area of base}} + 4 \cdot \boxed{\text{Area of sides}}$

Labels: Length of base $= x$

Height of box $= h$

Equation:

$$s = 2x^2 + 4xh$$
$$512 = 2x^2 + 48x$$
$$0 = 2x^2 + 48x - 512$$
$$0 = 2(x^2 + 24x - 256)$$
$$0 = 2(x + 32)(x - 8)$$

$x + 32 = 0$ $\quad$ $x - 8 = 0$

$x = -32$ $\quad$ $x = 8$

Reject $\quad$ 8 inches × 8 inches

169. $h = -16t^2 + 3600$

$$0 = -16t^2 + 3600$$
$$0 = -16(t^2 - 225)$$
$$0 = -16(t + 15)(t - 15)$$

$t + 15 = 0$ $\quad$ $t - 15 = 0$

$t = -15$ $\quad$ $t = 15$

Reject $\quad$ 15 seconds

Chapter Test for Chapter 5

1. $3 - 4.5x + 8.2x^3 = 8.2x^3 - 4.5x + 3$

Degree: 3

Leading coefficient: 8.2

2. $\dfrac{4}{x^2 + 2}$ is not a polynomial because the variable appears in the denominator.

3. (a) $\dfrac{2^{-1}x^5y^{-3}}{4x^{-2}y^2} = \dfrac{1}{2 \cdot 4} \cdot x^{5-(-2)} \cdot y^{-3-2}$

$$= \frac{1}{8} \cdot x^7y^{-5} = \frac{x^7}{8y^5}$$

(b) $\left(\dfrac{-2x^2y}{z^{-3}}\right)^{-2} = \left(\dfrac{z^{-3}}{-2x^2y}\right)^2 = \dfrac{z^{-6}}{4x^4y^2} = \dfrac{1}{4x^4y^2z^6}$

4. (a) $\left(\dfrac{-2u^2}{v^{-1}}\right)^3\left(\dfrac{3v^2}{u^{-3}}\right) = \left(-\dfrac{8u^6}{v^{-3}}\right)\left(\dfrac{3v^2}{u^{-3}}\right)$

$$= -24u^{6-(-3)}v^{2-(-3)} = -24u^9v^5$$

(b) $\dfrac{(-3x^2y^{-1})^4}{6x^2y^0} = \dfrac{81x^8y^{-4}}{6x^2} = \dfrac{27x^{8-2}}{2y^4} = \dfrac{27x^6}{2y^4}$

5. (a) $(5a^2 - 3a + 4) + (a^2 - 4) = 6a^2 - 3a$

(b) $(16 - y^2) - (16 + 2y + y^2) = 16 - y^2 - 16 - 2y - y^2 = -2y^2 - 2y$

6. (a) $-2(2x^4 - 5) + 4x(x^3 + 2x - 1) = -4x^4 + 10 + 4x^4 + 8x^2 - 4x = 8x^2 - 4x + 10$

(b) $4t - [3t - (10t + 7)] = 4t - [3t - 10t - 7] = 4t - 3t + 10t + 7 = 11t + 7$

7. (a) $-3x(x - 4) = -3x^2 + 12x$

(b) $(2x - 3y)(x + 5y) = 2x^2 + 7xy - 15y^2$

8. (a) $(x-1)[2x+(x-3)] = (x-1)(3x-3) = 3x^2 - 6x + 3$

(b) $(2s-3)(3s^2-4s+7) = 6s^3 - 8s^2 + 14s - 9s^2 + 12s - 21 = 6s^3 - 17s^2 + 26s - 21$

9. (a) $(2w-7)^2 = (2w)^2 - 2(2w)(7) + 7^2 = 4w^2 - 28w + 49$

(b) $(4-(a+b))(4+(a+b)) = 4^2 - (a+b)^2 = 16 - (a^2 + 2ab + b^2) = 16 - a^2 - 2ab$

10. $18y^2 - 12y = 6y(3y-2)$

11. $v^2 - \frac{16}{9} = \left(v - \frac{4}{3}\right)\left(v + \frac{4}{3}\right)$

12. $x^3 - 3x^2 - 4x + 12 = x^2(x-3) - 4(x-3)$
$= (x-3)(x^2-4)$
$= (x-3)(x-2)(x+2)$

13. $9u^2 - 6u + 1 = (3u-1)(3u-1)$ or $(3u-1)^2$

14. $6x^2 - 26x - 20 = 2(3x^2 - 13x - 10)$
$= 2(3x+2)(x-5)$

15. $x^3 + 27 = (x+3)(x^2 - 3x + 9)$

16. $(x-3)(x+2) = 14$
$x^2 + 2x - 3x - 6 = 14$
$x^2 - x - 20 = 0$
$(x+4)(x-5) = 0$
$x + 4 = 0 \qquad x - 5 = 0$
$x = -4 \qquad x = 5$

17. $(y+2)^2 - 9 = 0$
$[(y+2)-3][(y+2)+3] = 0$
$y - 1 = 0 \qquad y + 5 = 0$
$y = 1 \qquad y = -5$

18. $12 + 5y - 3y^2 = 0$
$(3-y)(4+3y) = 0$
$3 - y = 0 \qquad 4 + 3y = 0$
$3 = y \qquad -\frac{4}{3} = y$

19. $2x^3 + 10x^2 + 8x = 0$
$2x(x^2 + 5x + 4) = 0$
$2x(x+4)(x+1) = 0$
$2x = 0 \qquad x + 4 = 0 \qquad x + 1 = 0$
$x = 0 \qquad x = -4 \qquad x = -1$

20. Area $= 2x(x+15) - x(x+4)$

Shaded region $= 2x^2 + 30x - x^2 - 4x$
$= x^2 + 26x$

21. *Verbal Model:* [Area rectangle] = [Length] · [Width]

Labels: Length $= \frac{3}{2}w$
Width $= w$

Equation: $54 = \frac{3}{2}w \cdot w$
$108 = 3w^2$
$36 = w^2$
6 centimeters = width
9 centimeters = length

22. *Verbal Model:* [Area of a triangle] $= \frac{1}{2} \cdot$ [Base] · [Height]

Labels: Base $= x$
Height $= 2x + 2$

Equation: $20 = \frac{1}{2} \cdot x \cdot (2x+2)$
$20 = x^2 + x$
$0 = x^2 + x - 20$
$0 = (x+5)(x-4)$
$x + 5 = 0 \qquad x - 4 = 0$
$x = -5 \qquad x = 4$ feet
Reject $\qquad 2x + 2 = 10$ feet

23. *Verbal Model:* [Revenue] = [Cost]

Labels: Revenue $= R$
Cost $= C$

Equation: $R = C$
$x^2 - 35x = 150 + 12x$
$x^2 - 47x - 150 = 0$
$(x-50)(x+3) = 0$
$x - 50 = 0 \qquad x + 3 = 0$
$x = 50 \qquad x = -3$

50 computer desks

CHAPTER 6
Rational Expressions, Equations, and Functions

CHAPTER 6
Rational Expressions, Equations, and Functions

Section 6.1 Rational Expressions and Functions

1. $4 \neq 0$

$D = (-\infty, \infty)$

3. $x - 3 \neq 0$

$x \neq 3$

$D = (-\infty, 3) \cup (3, \infty)$

5. $9 - x \neq 0$

$-x \neq -9$

$x \neq 9$

$D = (-\infty, 9) \cup (9, \infty)$

7. $x + 10 \neq 0$

$x \neq -10$

$D = (-\infty, -10) \cup (-10, \infty)$

9. $x^2 + 4 \neq 0$

$D = (-\infty, \infty)$

11. $y(y + 3) \neq 0$

$y \neq 0 \quad y \neq -3$

$D = (-\infty, -3) \cup (-3, 0) \cup (0, \infty)$

13. $x(x - 1) \neq 0$

$x \neq 0 \quad x - 1 \neq 0$

$x \neq 1$

$D = (-\infty, 0) \cup (0, 1) \cup (1, \infty)$

15. $t^2 - 16 \neq 0$

$(t - 4)(t + 4) \neq 0$

$t \neq 4 \quad t \neq -4$

$D = (-\infty, -4) \cup (-4, 4) \cup (4, \infty)$

17. $y^2 - 3y \neq 0$

$y(y - 3) \neq 0$

$y \neq 0 \ y \neq 3$

$D = (-\infty, 0) \cup (0, 3) \cup (3, \infty)$

19. $x^2 - 5x + 6 \neq 0$

$(x - 3)(x - 2) \neq 0$

$x \neq 3 \quad x \neq 2$

$D = (-\infty, 2) \cup (2, 3) \cup (3, \infty)$

21. $3u^2 - 2u - 5 \neq 0$

$(3u - 5)(u + 1) \neq 0$

$u \neq \frac{5}{3} \quad u \neq -1$

$D = (-\infty, -1) \cup \left(-1, \frac{5}{3}\right) \cup \left(\frac{5}{3}, \infty\right)$

23. (a) $f(1) = \dfrac{4(1)}{1 + 3} = \dfrac{4}{4} = 1$

(b) $f(-2) = \dfrac{4(-2)}{-2 + 3} = \dfrac{-8}{1} = -8$

(c) $f(-3) = \dfrac{4(-3)}{-3 + 3} = \dfrac{-12}{0}$

= not possible; undefined

(d) $f(0) = \dfrac{4(0)}{0 + 3} = \dfrac{0}{3} = 0$

25. (a) $g(0) = \dfrac{0^2 - 4(0)}{0^2 - 9} = 0$

(b) $g(4) = \dfrac{4^2 - 4(4)}{4^2 - 9} = \dfrac{16 - 16}{16 - 9} = \dfrac{0}{7} = 0$

(c) $g(3) = \dfrac{3^2 - 4(3)}{3^2 - 9} = \dfrac{9 - 12}{9 - 9} = \dfrac{-3}{0}$

= not possible; undefined

(d) $g(-3) = \dfrac{(-3)^2 - 4(-3)}{(-3)^2 - 9} = \dfrac{9 + 12}{9 - 9} = \dfrac{21}{0}$

= not possible; undefined

27. (a) $h(10) = \dfrac{10^2}{10^2 - 10 - 2} = \dfrac{100}{88} = \dfrac{25}{22}$

(b) $h(0) = \dfrac{0^2}{0^2 - 0 - 2} = \dfrac{0}{-2} = 0$

(c) $h(-1) = \dfrac{(-1)^2}{(-1)^2 - (-1) - 2} = \dfrac{1}{1 + 1 - 2} = \dfrac{1}{0}$

= not possible; undefined

(d) $h(2) = \dfrac{2^2}{2^2 - 2 - 2} = \dfrac{4}{4 - 2 - 2} = \dfrac{4}{0}$

= not possible; undefined

29. Since length must be positive, $x \ge 0$. Since $\frac{500}{x}$ must be defined, $x \ne 0$. Therefore, the domain is $x > 0$ or $(0, \infty)$.

31. $x =$ units of a product
$D = \{1, 2, 3, 4, \ldots\}$

33. Since p is the percent of air pollutants in the stack emission of a utility, $0 \le p \le 100$. Since $\frac{60{,}000p}{100 - p}$ must be defined, $p \ne 100$. Therefore, the domain is $[0, 100)$.

35. $\frac{5}{6} = \frac{5(x + 3)}{6(x + 3)}, \quad x \ne -3$

37. $\frac{x}{2} = \frac{3x(x + 16)^2}{2\left(3(x + 16)^2\right)}, \quad x \ne -16$

39. $\frac{x + 5}{3x} = \frac{(x + 5)\left(x(x - 2)\right)}{3x^2(x - 2)}, \quad x \ne 2$

41. $\frac{8x(\ )}{x^2 - 2x - 15} = \frac{8x(\ )}{(x - 5)(x + 3)}$
$= \frac{8x(x + 3)}{(x - 5)(x + 3)}$
$= \frac{8x}{x - 5}, \quad x \ne -3$

43. $\frac{5x}{25} = \frac{5x}{5 \cdot 5} = \frac{x}{5}$

45. $\frac{12x^2}{12x} = \frac{12}{12} \cdot \frac{x \cdot x}{x} = x, \quad x \ne 0$

47. $\frac{18x^2y}{15xy^4} = \frac{3 \cdot 6 \cdot x \cdot x \cdot y}{3 \cdot 5 \cdot x \cdot y \cdot y^3}$
$= \frac{6x}{5y^3}, \quad x \ne 0$

49. $\frac{3x^2 - 9x}{12x^2} = \frac{3x(x - 3)}{12x^2} = \frac{(x - 3)}{4x}$

51. $\frac{x^2(x - 8)}{x(x - 8)} = \frac{x \cdot x(x - 8)}{x(x - 8)}$
$= x, \quad x \ne 0, x \ne 8$

53. $\frac{2x - 3}{4x - 6} = \frac{2x - 3}{2(2x - 3)} = \frac{1}{2}, \quad x \ne \frac{3}{2}$

55. $\frac{y^2 - 49}{2y - 14} = \frac{(y - 7)(y + 7)}{2(y - 7)} = \frac{y + 7}{2}, \quad y \ne 7$

57. $\frac{a + 3}{a^2 + 6a + 9} = \frac{a + 3}{(a + 3)(a + 3)}$
$= \frac{1}{a + 3}$

59. $\frac{x^2 - 7x}{x^2 - 14x + 49} = \frac{x(x - 7)}{(x - 7)(x - 7)}$
$= \frac{x}{x - 7}$

61. $\frac{y^3 - 4y}{y^2 + 4y - 12} = \frac{y(y^2 - 4)}{(y + 6)(y - 2)}$
$= \frac{y(y - 2)(y + 2)}{(y + 6)(y - 2)}$
$= \frac{y(y + 2)}{y + 6}, \quad y \ne 2$

63. $\frac{y^4 - 16y^2}{y^2 + y - 12} = \frac{y^2(y^2 - 16)}{(y + 4)(y - 3)}$
$= \frac{y^2(y - 4)(y + 4)}{(y + 4)(y - 3)}$
$= \frac{y^2(y - 4)}{y - 3}, \quad y \ne -4$

65. $\frac{3x^2 - 7x - 20}{12 + x - x^2} = \frac{(3x + 5)(x - 4)}{-1(x^2 - x - 12)}$
$= \frac{(3x + 5)(x - 4)}{-1(x - 4)(x + 3)}$
$= -\frac{3x + 5}{x + 3}, \quad x \ne 4$

67. $\frac{2x^2 + 19x + 24}{2x^2 - 3x - 9} = \frac{(2x + 3)(x + 8)}{(2x + 3)(x - 3)}$
$= \frac{x + 8}{x - 3}, \quad x \ne -\frac{3}{2}$

69. $\frac{15x^2 + 7x - 4}{25x^2 - 16} = \frac{(5x + 4)(3x - 1)}{(5x + 4)(5x - 4)}$
$= \frac{3x - 1}{5x - 4}, \quad x \ne -\frac{4}{5}$

71. $\frac{3xy^2}{xy^2 + x} = \frac{3xy^2}{x(y^2 + 1)} = \frac{3y^2}{y^2 + 1}, \quad x \ne 0$

73. $\dfrac{y^2 - 64x^2}{5(3y + 24x)} = \dfrac{(y - 8x)(y + 8x)}{15(y + 8x)}$

$= \dfrac{y - 8x}{15}, \quad y \neq -8x$

75. $\dfrac{5xy + 3x^2y^2}{xy^3} = \dfrac{xy(5 + 3xy)}{xy \cdot y^2} = \dfrac{5 + 3xy}{y^2}, \quad x \neq 0$

77. $\dfrac{u^2 - 4v^2}{u^2 + uv - 2v^2} = \dfrac{(u - 2v)(u + 2v)}{(u - v)(u + 2v)}$

$= \dfrac{u - 2v}{u - v}, \quad u \neq -2v$

79. $\dfrac{3m^2 - 12n^2}{m^2 + 4mn + 4n^2} = \dfrac{3(m^2 - 4n^2)}{(m + 2n)(m + 2n)}$

$= \dfrac{3(m - 2n)(m + 2n)}{(m + 2n)(m + 2n)}$

$= \dfrac{3(m - 2n)}{m + 2n}$

81.

x	-2	-1	0	1	2	3	4
$\dfrac{x^2 - x - 2}{x - 2}$	-1	0	1	2	Undefined	4	5
$x + 1$	-1	0	1	2	3	4	5

Domain of $\dfrac{x^2 - x - 2}{x - 2}$ is $(-\infty, 2) \cup (2, \infty)$.

Domain of $x + 1$ is $(-\infty, \infty)$.

The two expressions are equal for all replacements of the variable x except 2.

83. $\dfrac{\text{Area of shaded portion}}{\text{Area of total figure}} = \dfrac{x(x + 1)}{(x + 1)(x + 3)}$

$= \dfrac{x}{x + 3}, \quad x > 0$

85. $\dfrac{\text{Area of shaded portion}}{\text{Area of total figure}} = \dfrac{\frac{1}{2}x(x + 1)}{\frac{1}{2}(2x)(2x + 2)}$

$= \dfrac{1}{4}, \quad x > 0$

87. (a) *Verbal Model:*

[Total cost] = [Number of units] · [Cost per unit] + [Initial cost]

Labels: Total cost $= C$

Number of units $= x$

Equation: $2500 + 9.25x = C$

(b) *Verbal Model:* [Average cost] = [Total cost] ÷ [Number of units]

Label: Average cost $= \overline{C}$

Equation: $\overline{C} = \dfrac{2500 + 9.25x}{x}$

(c) Domain $= \{1, 2, 3, 4, \ldots\}$

(d) $\dfrac{2500 + 9.25(100)}{100} = \$34.25$

89. (a) *Verbal Model:* [Distance] = [Rate] · [Time]

Van: $45(t + 3)$

Car: $60t$

(b) Distance between van and car $= d$

$= |45(t + 3) - 60t|$

$= |45t + 135 - 60t|$

$= |135 - 15t|$

$= |15(9 - t)|$

(c) $\dfrac{\text{Distance of car}}{\text{Distance of van}} = \dfrac{60t}{45(t + 3)} = \dfrac{4t}{3(t + 3)}$

91. $\dfrac{\text{Circular pool volume}}{\text{Rectangular pool volume}} = \dfrac{\pi(3d)^2(d + 2)}{d(3d)(3d + 6)}$

$= \dfrac{\pi(3d)^2(d + 2)}{3d^2 \cdot 3(d + 2)}$

$= \dfrac{\pi(3d)^2(d + 2)}{(3d)^2(d + 2)}$

$= \pi, \quad d > 0$

93. Average cable revenue per subscriber $= \dfrac{R}{S}$

$$= \frac{1189.2t + 25{,}266}{-0.35t + 67.1}, \quad 1 \le t \le 5$$

95. The rational expression is in simplified form if the numerator and denominator have no factors in common (other than ± 1).

97. $\dfrac{1}{x^2 + 1}$

There are many correct answers.

99. The student incorrectly divided out; the denominator may not be split up.

Correct solution:

$$\frac{x^2 + 7x}{x + 7} = \frac{x(x + 7)}{x + 7} = x, \quad x \ne -7$$

101. To write the polynomial $g(x)$, multiply $f(x)$ by $(x - 2)$ and divide by $(x - 2)$.

$$g(x) = \frac{f(x)(x - 2)}{(x - 2)}, \quad x \ne 2$$

103. $\left(\dfrac{1}{4}\right)\left(\dfrac{3}{4}\right) = \dfrac{1 \cdot 3}{4 \cdot 4} = \dfrac{3}{16}$

105. $\dfrac{1}{3}\left(\dfrac{3}{5}\right)(5) = \dfrac{1 \cdot 3 \cdot 5}{3 \cdot 5 \cdot 1} = 1$

107. $(-2a^3)(-2a) = -2 \cdot -2 \cdot a^3 \cdot a = 4a^4$

109. $(-3b)(b^2 - 3b + 5) = -3b(b^2) - 3b(-3b) - 3b(5)$

$$= -3b^3 + 9b^2 - 15b$$

Section 6.2 Multiplying and Dividing Rational Expressions

1. $\dfrac{7}{3y} = \dfrac{7x^2}{3y(x^2)}, \quad x \ne 0$

3. $\dfrac{3x}{x - 4} = \dfrac{3x(x + 2)^2}{(x - 4)(x + 2)^2}, \quad x \ne -2$

5. $\dfrac{3u}{7v} = \dfrac{3u(u + 1)}{7v(u + 1)}, \quad u \ne -1$

7. $\dfrac{13x}{x - 2} = \dfrac{13x((-1)(2 + x))}{4 - x^2}, \quad x \ne -2$

9. $4x \cdot \dfrac{7}{12x} = \dfrac{4x(7)}{12x} = \dfrac{7}{3}, \quad x \ne 0$

11. $\dfrac{8s^3}{9s} \cdot \dfrac{6s^2}{32s} = \dfrac{8s^3 \cdot 3 \cdot 2s \cdot s}{3 \cdot 3 \cdot s \cdot 8 \cdot 2 \cdot 2 \cdot s} = \dfrac{s^3}{6}, \quad s \ne 0$

13. $16u^4 \cdot \dfrac{12}{8u^2} = \dfrac{8 \cdot 2 \cdot u^2 \cdot u^2 \cdot 12}{8 \cdot u^2} = 24u^2, \quad u \ne 0$

15. $\dfrac{8}{3 + 4x} \cdot (9 + 12x) = \dfrac{8 \cdot 3(3 + 4x)}{3 + 4x} = 24, \quad x \ne -\dfrac{3}{4}$

17. $\dfrac{8u^2v}{3u + v} \cdot \dfrac{u + v}{12u} = \dfrac{4 \cdot 2 \cdot u \cdot u \cdot v(u + v)}{(3u + v) \cdot 4 \cdot 3 \cdot u}$

$$= \frac{2uv(u + v)}{3(3u + v)}, \quad u \ne 0$$

19. $\dfrac{12 - r}{3} \cdot \dfrac{3}{r - 12} = \dfrac{-1(r - 12) \cdot 3}{3(r - 12)} = -1, \quad r \ne 12$

21. $\dfrac{(2x - 3)(x + 8)}{x^3} \cdot \dfrac{x}{3 - 2x} = \dfrac{(2x - 3)(x + 8)x}{x \cdot x^2 \cdot -1(2x - 3)}$

$$= \frac{x + 8}{-x^2}, \quad x \ne \frac{3}{2}$$

23. $\dfrac{4r - 12}{r - 2} \cdot \dfrac{r^2 - 4}{r - 3} = \dfrac{4(r - 3)(r - 2)(r + 2)}{(r - 2) \cdot (r - 3)}$

$$= 4(r + 2), \quad r \ne 3, r \ne 2$$

25. $\dfrac{2t^2 - t - 15}{t + 2} \cdot \dfrac{t^2 - t - 6}{t^2 - 6t + 9} = \dfrac{(2t + 5)(t - 3)(t - 3)(t + 2)}{(t + 2)(t - 3)(t - 3)} = 2t + 5, \quad t \ne 3, t \ne -2$

27. $(4y^2 - x^2) \cdot \dfrac{xy}{(x - 2y)^2} = \dfrac{-1(x^2 - 4y^2)(xy)}{(x - 2y)^2} = -\dfrac{(x - 2y)(x + 2y)(xy)}{(x - 2y)(x - 2y)} = -\dfrac{xy(x + 2y)}{(x - 2y)}$

29. $\dfrac{x^2 + 2xy - 3y^2}{(x + y)^2} \cdot \dfrac{x^2 - y^2}{x + 3y} = \dfrac{(x + 3y)(x - y)}{(x + y)^2} \cdot \dfrac{(x - y)(x + y)}{x + 3y} = \dfrac{(x - y)^2}{x + y}, \quad x \ne -3y$

31. $\dfrac{x+5}{x-5}\cdot\dfrac{2x^2-9x-5}{3x^2+x-2}\cdot\dfrac{x^2-1}{x^2+7x+10}=\dfrac{x+5}{x-5}\cdot\dfrac{(2x+1)(x-5)}{(3x-2)(x+1)}\cdot\dfrac{(x-1)(x+1)}{(x+5)(x+2)}$

$$=\frac{(x+5)(2x+1)(x-5)(x-1)(x+1)}{(x-5)(3x-2)(x+1)(x+5)(x+2)}$$

$$=\frac{(2x+1)(x-1)}{(3x-2)(x+2)},\quad x\neq\pm5,-1$$

33. $\dfrac{9-x^2}{2x+3}\cdot\dfrac{4x^2+8x-5}{4x^2-8x+3}\cdot\dfrac{6x^4-2x^3}{8x^2+4x}=\dfrac{(3-x)(3+x)}{2x+3}\cdot\dfrac{(2x+5)(2x-1)}{(2x-3)(2x-1)}\cdot\dfrac{2x^3(3x-1)}{4x(2x+1)}$

$$=\frac{-1(x-3)(x+3)(2x+5)x^2(3x-1)}{(2x+3)(2x-3)2(2x+1)}$$

$$=\frac{x^2(x^2-9)(2x+5)(3x-1)}{2(2x+3)(3-2x)(2x+1)},\quad x\neq 0,\frac{1}{2}$$

35. $\dfrac{x^3+3x^2-4x-12}{x^3-3x^2-4x+12}\cdot\dfrac{x^2-9}{x}=\dfrac{x^2(x+3)-4(x+3)}{x^2(x-3)-4(x-3)}\cdot\dfrac{(x+3)(x-3)}{x}$

$$=\frac{(x+3)(x^2-4)(x+3)(x-3)}{(x-3)(x^2-4)\cdot x}$$

$$=\frac{(x+3)^2}{x},\quad x\neq-2,2,3$$

37. $\dfrac{x}{x+2}\div\dfrac{3}{x+1}=\dfrac{x}{x+2}\cdot\dfrac{x+1}{3}=\dfrac{x(x+1)}{3(x+2)},\quad x\neq-1$

39. $x^2\div\dfrac{3x}{4}=x^2\cdot\dfrac{4}{3x}=\dfrac{4x}{3},\quad x\neq 0$

41. $\dfrac{2x}{5}\div\dfrac{x^2}{15}=\dfrac{2x}{5}\cdot\dfrac{15}{x^2}=\dfrac{2(3)(5)x}{5x^2}=\dfrac{6}{x}$

43. $\dfrac{7xy^2}{10u^2v}\div\dfrac{21x^3}{45uv}=\dfrac{7xy^2}{10u^2v}\cdot\dfrac{45uv}{21x^3}=\dfrac{7xy^2\cdot3\cdot3\cdot5\cdot u\cdot v}{5\cdot2\cdot u\cdot u\cdot v\cdot7\cdot3x\cdot x^2}=\dfrac{3y^2}{2ux^2},\quad v\neq 0$

45. $\dfrac{3(a+b)}{4}\div\dfrac{(a+b)^2}{2}=\dfrac{3(a+b)}{4}\cdot\dfrac{2}{(a+b)^2}=\dfrac{3(a+b)\cdot2}{2\cdot2\cdot(a+b)(a+b)}=\dfrac{3}{2(a+b)}$

47. $\dfrac{4x}{3x-3}\div\dfrac{x^2+2x}{x^2+x-2}=\dfrac{4x}{3(x-1)}\cdot\dfrac{(x+2)(x-1)}{x(x+2)}=\dfrac{4x}{3x}=\dfrac{4}{3},\quad x\neq-2,0,1$

49. $\dfrac{(x^3y)^2}{(x+2y)^2}\div\dfrac{x^2y}{(x+2y)^3}=\dfrac{(x^3y)^2}{(x+2y)^2}\cdot\dfrac{(x+2y)^3}{x^2y}$

$$=\frac{(x^3y)(x^3y)(x+2y)^2(x+2y)}{(x+2y)^2x^2y}$$

$$=\frac{(x^3y)(x^2\cdot xy)(x+2y)}{x^2y}=x^4y(x+2y),\quad x\neq0,\ y\neq0,\ x\neq-2y$$

51. $\dfrac{x^2+2x-15}{x^2+11x+30} \div \dfrac{x^2-8x+15}{x^2+2x-24} = \dfrac{(x+5)(x-3)}{(x+5)(x+6)} \cdot \dfrac{(x+6)(x-4)}{(x-3)(x-5)}$

$= \dfrac{(x+5)(x-3)(x+6)(x-4)}{(x+5)(x+6)(x-3)(x-5)}$

$= \dfrac{x-4}{x-5}, \quad x \neq -6, x \neq -5, x \neq 3$

53. $\left[\dfrac{x^2}{9} \cdot \dfrac{3(x+4)}{x^2+2x}\right] \div \dfrac{x}{x+2} = \dfrac{x^2}{9} \cdot \dfrac{3(x+4)}{x(x+2)} \cdot \dfrac{x+2}{x}$

$= \dfrac{x+4}{3}, \quad x \neq -2, 0$

55. $\left[\dfrac{xy+y}{4x} \div (3x+3)\right] \div \dfrac{y}{3x} = \dfrac{y(x+1)}{4x} \cdot \dfrac{1}{3(x+1)} \cdot \dfrac{3x}{y}$

$= \dfrac{1}{4}, \quad x \neq -1, 0, y \neq 0$

57. $\dfrac{2x^2+5x-25}{3x^2+5x+2} \cdot \dfrac{3x^2+2x}{x+5} \div \left(\dfrac{x}{x+1}\right)^2 = \dfrac{(2x-5)(x+5)}{(3x+2)(x+1)} \cdot \dfrac{x(3x+2)}{x+5} \cdot \left(\dfrac{x+1}{x}\right)^2$

$= \dfrac{(2x-5)(x+5)x(3x+2)(x+1)(x+1)}{(3x+2)(x+1)(x+5)x \cdot x}$

$= \dfrac{(2x-5)(x+1)}{x}, \quad x \neq -1, -5, -\dfrac{2}{3}$

59. $x^3 \cdot \dfrac{x^{2n}-9}{x^{2n}+4x^n+3} \div \dfrac{x^{2n}-2x^n-3}{x} = x^3 \cdot \dfrac{(x^n-3)(x^n+3)}{(x^n+3)(x^n+1)} \cdot \dfrac{x}{(x^n-3)(x^n+1)}$

$= \dfrac{x^4}{(x^n+1)^2}, \quad x^n \neq -3, 3, 0$

61. *Keystrokes:*

y_1: [(] [(] [X,T,θ] [x²] [−] 10 [X,T,θ] [+] 25 [)] [÷] [(] [X,T,θ] [x²] [−] 25 [)] [)]
[×] [(] [(] [X,T,θ] [+] 5 [)] [÷] 2 [)] [ENTER]

y_2: [(] [X,T,θ] [−] 5 [)] [÷] 2 [GRAPH]

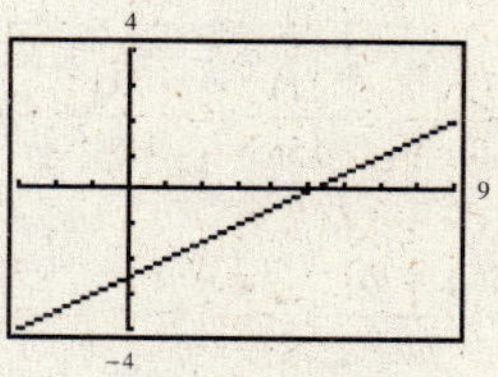

$\dfrac{x^2-10x+25}{x^2-25} \cdot \dfrac{x+5}{2} = \dfrac{(x-5)^2}{(x-5)(x+5)} \cdot \dfrac{x+5}{2} = \dfrac{x-5}{2}, \quad x \neq \pm 5$

63. $\text{Area} = \left(\dfrac{2w+3}{3}\right)\left(\dfrac{w}{2}\right) = \dfrac{(2w+3)w}{6}$

65. $\dfrac{\text{Unshaded Area}}{\text{Total Area}} = \dfrac{x \cdot x}{2x(4x+2)} = \dfrac{x}{2(2)(2x+1)} = \dfrac{x}{4(2x+1)}, \quad x > 0$

67. $\dfrac{\text{Unshaded Area}}{\text{Total Area}} = \dfrac{\pi x^2}{2x(4x+2)} = \dfrac{\pi x}{2(2)(2x+1)} = \dfrac{\pi x}{4(2x+1)}, \quad x > 0$

69. *Verbal Model:* [Number of jobs per person] = [Number of jobs] ÷ [Population]

Model:
$$Y = \frac{-0.696t + 8.94}{-0.092t + 1} \div 0.352t + 15.97$$
$$= \frac{-0.696t + 8.94}{-0.092t + 1} \cdot \frac{1}{0.352t + 15.97}$$
$$= \frac{-0.696t + 8.94}{(-0.092t + 1)(0.352t + 15.97)}, \quad 1 \le t \le 6$$

71. In simplifying a product of national expressions, you divide the common factors out of the numerator and denominator.

73. The domain needs to be restricted, $x \ne a, x \ne b$.

75. The first expression needs to be multiplied by the reciprocal of the second expression (not the second by the reciprocal of the first), and the domain needs to be restricted.

$$\frac{x^2 - 4}{5x} \div \frac{x + 2}{x - 2} = \frac{x^2 - 4}{5x} \cdot \frac{x - 2}{x + 2} = \frac{(x - 2)^2(x + 2)}{5x(x + 2)} = \frac{(x - 2)^2}{5x}, \; x \ne \pm 2$$

77. $\frac{1}{8} + \frac{3}{8} + \frac{5}{8} = \frac{1 + 3 + 5}{8} = \frac{9}{8}$

79. $\frac{3}{5} + \frac{4}{15} = \frac{9}{15} + \frac{4}{15} = \frac{9 + 4}{15} = \frac{13}{15}$

81. $x^2 + 3x = 0$

$x(x + 3) = 0$

$x = 0 \qquad x + 3 = 0$

$\qquad\qquad x = -3$

83. $4x^2 - 25 = 0$

$(2x + 5)(2x - 5) = 0$

$2x + 5 = 0 \qquad 2x - 5 = 0$

$2x = -5 \qquad 2x = 5$

$x = -\frac{5}{2} \qquad x = \frac{5}{2}$

Section 6.3 Adding and Subtracting Rational Expressions

1. $\frac{5x}{6} + \frac{4x}{6} = \frac{5x + 4x}{6} = \frac{9x}{6} = \frac{3x}{2}$

3. $\frac{2}{3a} - \frac{11}{3a} = \frac{2 - 11}{3a} = \frac{-9}{3a} = -\frac{3}{a}$

5. $\frac{x}{9} - \frac{x + 2}{9} = \frac{x - (x + 2)}{9} = \frac{x - x - 2}{9} = -\frac{2}{9}$

7. $\frac{z^2}{3} + \frac{z^2 - 2}{3} = \frac{z^2 + z^2 - 2}{3} = \frac{2z^2 - 2}{3}$

9. $\frac{2x + 5}{3x} + \frac{1 - x}{3x} = \frac{2x + 5 + 1 - x}{3x} = \frac{x + 6}{3x}$

11. $\frac{3y - 22}{y - 6} - \frac{2y - 16}{y - 6} = \frac{3y - 22 - (2y - 16)}{y - 6}$

$= \frac{3y - 22 - 2y + 16}{y - 6}$

$= \frac{y - 6}{y - 6}$

$= 1, y \ne 6$

13. $\frac{2x - 1}{x(x - 3)} + \frac{1 - x}{x(x - 3)} = \frac{2x - 1 + 1 - x}{x(x - 3)}$

$= \frac{x}{x(x - 3)}$

$= \frac{1}{x - 3}, \quad x \ne 0$

15. $\frac{w}{w^2 - 4} + \frac{2}{w^2 - 4} = \frac{w + 2}{(w + 2)(w - 2)}$

$= \frac{1}{w - 2}, \quad w \ne -2$

17. $\frac{c}{c^2 + 3c - 4} - \frac{1}{c^2 + 3c - 4} = \frac{c - 1}{(c + 4)(c - 1)}$

$= \frac{1}{c + 4}, \quad c \ne 1$

19. $\frac{3y}{3} - \frac{3y - 3}{3} - \frac{7}{3} = \frac{3y - (3y - 3) - 7}{3}$

$= \frac{3y - 3y + 3 - 7}{3} = -\frac{4}{3}$

21. $\dfrac{x^2 - 4x}{x - 3} + \dfrac{10 - 4x}{x - 3} - \dfrac{x - 8}{x - 3} = \dfrac{x^2 - 4x + 10 - 4x - x + 8}{x - 3}$

$= \dfrac{x^2 - 9x + 18}{x - 3}$

$= \dfrac{(x - 6)(x - 3)}{x - 3}, \quad x \neq 3$

23. $5x^2 = 5 \cdot x \cdot x$

$20x^3 = 5 \cdot 2 \cdot 2 \cdot x \cdot x \cdot x$

$\text{LCM} = 20x^3$

25. $9y^3 = 3 \cdot 3 \cdot y \cdot y \cdot y$

$12y = 2 \cdot 2 \cdot 3 \cdot y$

$\text{LCM} = 3 \cdot 3 \cdot 2 \cdot 2 \cdot y \cdot y \cdot y = 36y^3$

27. $15x^2 = 5 \cdot 3 \cdot x \cdot x$

$3(x + 5) = 3 \cdot (x + 5)$

$\text{LCM} = 15x^2(x + 5)$

29. $63z^2(z + 1) = 7 \cdot 9 \cdot z \cdot z(z + 1)$

$14(z + 1)^4 = 7 \cdot 2 \cdot (z + 1)^4$

$\text{LCM} = 126z^2(z + 1)^4$

31. $8t(t + 2) = 2 \cdot 2 \cdot 2 \cdot t \cdot (t + 2)$

$14(t^2 - 4) = 2 \cdot 7 \cdot (t + 2)(t - 2)$

$\text{LCM} = 2 \cdot 2 \cdot 2 \cdot 7 \cdot t \cdot (t + 2)(t - 2) = 56t(t^2 - 4)$

33. $2y^2 + y - 1 = (2y - 1)(y + 1)$

$4y^2 - 2y = 2y(2y - 1)$

$\text{LCM} = 2y(2y - 1)(y + 1)$

35. $\dfrac{7x^2}{4a(x^2)} = \dfrac{7}{4a}, \quad x \neq 0$

37. $\dfrac{5r(u + 1)}{3v(u + 1)} = \dfrac{5r}{3v}, \quad u \neq -1$

39. $\dfrac{7y(-1(x + 2))}{4 - x^2} = \dfrac{7y}{x - 2}, \quad x \neq -2$

$4 - x^2 = (2 - x)(2 + x)$

$= -1(x - 2)(2 + x)$

41. $\dfrac{n + 8}{3n - 12} = \dfrac{n + 8}{3(n - 4)} = \dfrac{n + 8(2n^2)}{3(n - 4)(2n^2)} = \dfrac{2n^2(n + 8)}{6n^2(n - 4)}$

$\dfrac{10}{6n^2} = \dfrac{10}{3 \cdot 2n^2} = \dfrac{10(n - 4)}{3 \cdot 2n^2(n - 4)} = \dfrac{10(n - 4)}{6n^2(n - 4)}$

$\text{LCD} = 6n^2(n - 4)$

43. $\dfrac{2}{x^2(x - 3)} = \dfrac{2(x + 3)}{x^2(x - 3)(x + 3)}$

$\dfrac{5}{x(x + 3)} = \dfrac{5x(x - 3)}{x^2(x + 3)(x - 3)}$

$\text{LCD} = x^2(x - 3)(x + 3) = x^2(x^2 - 9)$

45. $\dfrac{v}{2v^2 + 2v} = \dfrac{v}{2v(v + 1)} = \dfrac{v(3v)}{2v(v + 1)(3v)} = \dfrac{3v^2}{6v^2(v + 1)}$

$\dfrac{4}{3v^2} = \dfrac{4(2(v + 1))}{3v^2(2(v + 1))} = \dfrac{8v + 8}{6v^2(v + 1)}$

$\text{LCD} = 6v^2(v + 1)$

47. $\dfrac{x - 8}{x^2 - 25} = \dfrac{x - 8}{(x - 5)(x + 5)} = \dfrac{(x - 8)(x - 5)}{(x - 5)(x + 5)(x - 5)} = \dfrac{(x - 8)(x - 5)}{(x - 5)^2(x + 5)}$

$\dfrac{9x}{x^2 - 10x + 25} = \dfrac{9x}{(x - 5)^2} = \dfrac{9x(x + 5)}{(x - 5)^2(x + 5)} = \dfrac{9x(x + 5)}{(x - 5)^2(x + 5)}$

$\text{LCD} = (x - 5)^2(x + 5)$

49. $\dfrac{5}{4x} - \dfrac{3}{5} = \dfrac{5(5)}{4x(5)} - \dfrac{3(4x)}{5(4x)} = \dfrac{25}{20x} - \dfrac{12x}{20x} = \dfrac{25 - 12x}{20x}$

51. $\dfrac{7}{a} + \dfrac{14}{a^2} = \dfrac{7(a)}{a(a)} + \dfrac{14(1)}{a^2(1)} = \dfrac{7a}{a^2} + \dfrac{14}{a^2} = \dfrac{7a + 14}{a^2}$

53. $25 + \dfrac{10}{x + 4} = \dfrac{25(x + 4)}{1(x + 4)} + \dfrac{10(1)}{(x + 4)(1)}$

$= \dfrac{25(x + 4)}{x + 4} + \dfrac{10}{x + 4}$

$= \dfrac{25x + 100 + 10}{x + 4} = \dfrac{25x + 110}{x + 4}$

55. $\dfrac{20}{x-4}+\dfrac{20}{4-x}=\dfrac{20(1)}{(x-4)(1)}+\dfrac{20(-1)}{(4-x)(-1)}$
$=\dfrac{20}{(x-4)}-\dfrac{20}{x-4}$
$=\dfrac{20-20}{x-4}=0,\quad x\neq 4$

57. $\dfrac{3x}{x-8}-\dfrac{6}{8-x}=\dfrac{3x(1)}{(x-8)(1)}-\dfrac{6(-1)}{(8-x)(-1)}$
$=\dfrac{3x}{x-8}+\dfrac{6}{x-8}$
$=\dfrac{3x+6}{x-8}$

59. $\dfrac{3x}{3x-2}+\dfrac{2}{2-3x}=\dfrac{3x(1)}{3x-2(1)}+\dfrac{2(-1)}{(2-3x)(-1)}$
$=\dfrac{3x}{3x-2}+\dfrac{-2}{3x-2}$
$=\dfrac{3x-2}{3x-2}=1,\quad x\neq\dfrac{2}{3}$

61. $\dfrac{9}{5v}+\dfrac{3}{v-1}=\dfrac{9(v-1)}{5v(v-1)}+\dfrac{3(5v)}{5v(v-1)}$
$=\dfrac{9v-9+15v}{5v(v-1)}$
$=\dfrac{24v-9}{5v(v-1)}$
$=\dfrac{3(8v-3)}{5v(v-1)}$

63. $\dfrac{x}{x+3}-\dfrac{5}{x-2}=\dfrac{x(x-2)}{(x+3)(x-2)}-\dfrac{5(x+3)}{(x-2)(x+3)}$
$=\dfrac{x^2-2x-5x-15}{(x+3)(x-2)}$
$=\dfrac{x^2-7x-15}{(x+3)(x-2)}$

65. $\dfrac{12}{x^2-9}-\dfrac{2}{x-3}=\dfrac{12}{(x-3)(x+3)}-\dfrac{2}{x-3}$
$=\dfrac{12}{(x-3)(x+3)}-\dfrac{2(x+3)}{(x-3)(x+3)}$
$=\dfrac{12-2(x+3)}{(x-3)(x+3)}$
$=\dfrac{12-2x-6}{(x-3)(x+3)}$
$=\dfrac{6-2x}{(x-3)(x+3)}$
$=\dfrac{-2(-3+x)}{(x-3)(x+3)}$
$=\dfrac{-2}{x+3}$
$=-\dfrac{2}{x+3},\quad x\neq 3$

67. $\dfrac{3}{x-5}+\dfrac{2}{x+5}=\dfrac{3(x+5)}{(x-5)(x+5)}+\dfrac{2(x-5)}{(x+5)(x-5)}$
$=\dfrac{3x+15+2x-10}{(x-5)(x+5)}$
$=\dfrac{5x+5}{(x-5)(x+5)}$

69. $\dfrac{4}{x^2}-\dfrac{4}{x^2+1}=\dfrac{4(x^2+1)}{x^2(x^2+1)}-\dfrac{4x^2}{(x^2+1)x^2}$
$=\dfrac{4(x^2+1)}{x^2(x^2+1)}-\dfrac{4x^2}{x^2(x^2+1)}$
$=\dfrac{4x^2+4-4x^2}{x^2(x^2+1)}$
$=\dfrac{4}{x^2(x^2+1)}$

71. $\dfrac{x}{x^2-x-30}-\dfrac{1}{x+5}=\dfrac{x}{(x+5)(x-6)}-\dfrac{(x-6)}{(x+5)(x-6)}=\dfrac{x-(x-6)}{(x+5)(x-6)}=\dfrac{x-x+6}{(x+5)(x-6)}=\dfrac{6}{(x+5)(x-6)}$

73. $\dfrac{4}{x-4}+\dfrac{16}{(x-4)^2}=\dfrac{4(x-4)}{(x-4)(x-4)}+\dfrac{16(1)}{(x-4)^2(1)}$
$=\dfrac{4x-16}{(x-4)^2}+\dfrac{16}{(x-4)^2}$
$=\dfrac{4x-16+16}{(x-4)^2}$
$=\dfrac{4x}{(x-4)^2}$

75. $\dfrac{y}{x^2 + xy} - \dfrac{x}{xy + y^2} = \dfrac{y}{x(x + y)} - \dfrac{x}{y(x + y)}$

$= \dfrac{y(y)}{x(x + y)(y)} - \dfrac{x(x)}{y(x + y)(x)}$

$= \dfrac{y^2}{xy(x + y)} - \dfrac{x^2}{xy(x + y)}$

$= \dfrac{y^2 - x^2}{xy(x + y)}$

$= \dfrac{(y - x)(y + x)}{xy(x + y)}$

$= \dfrac{y - x}{xy}, \quad x \neq -y$

77. $\dfrac{4}{x} - \dfrac{2}{x^2} + \dfrac{4}{x + 3} = \dfrac{4x(x + 3)}{x(x)(x + 3)} - \dfrac{2(x + 3)}{x^2(x + 3)} + \dfrac{4(x^2)}{(x + 3)x^2}$

$= \dfrac{4x^2 + 12x}{x^2(x + 3)} - \dfrac{2x + 6}{x^2(x + 3)} + \dfrac{4x^2}{x^2(x + 3)}$

$= \dfrac{4x^2 + 12x - 2x - 6 + 4x^2}{x^2(x + 3)}$

$= \dfrac{8x^2 + 10x - 6}{x^2(x + 3)}$

$= \dfrac{2(4x^2 + 5x - 3)}{x^2(x + 3)}$

79. $\dfrac{3u}{u^2 - 2uv + v^2} + \dfrac{2}{u - v} - \dfrac{u}{u - v} = \dfrac{3u}{(u - v)^2} + \dfrac{2 - u}{u - v}$

$= \dfrac{3u(1)}{(u - v)^2(1)} + \dfrac{(2 - u)(u - v)}{(u - v)(u - v)}$

$= \dfrac{3u}{(u - v)^2} + \dfrac{2u - 2v - u^2 + uv}{(u - v)^2}$

$= \dfrac{3u + 2u - 2v - u^2 + uv}{(u - v)^2}$

$= \dfrac{5u - 2v - u^2 + uv}{(u - v)^2}$

$= -\dfrac{u^2 - uv - 5u + 2v}{(u - v)^2}$

81. $\dfrac{x + 2}{x - 1} - \dfrac{2}{x + 6} - \dfrac{14}{x^2 + 5x - 6} = \dfrac{(x + 2)(x + 6)}{(x - 1)(x + 6)} - \dfrac{2(x - 1)}{(x + 6)(x - 1)} - \dfrac{14(1)}{(x + 6)(x - 1)(1)}$

$= \dfrac{x^2 + 8x + 12}{(x - 1)(x + 6)} - \dfrac{2x - 2}{(x + 6)(x - 1)} - \dfrac{14}{(x + 6)(x - 1)}$

$= \dfrac{x^2 + 8x + 12 - 2x + 2 - 14}{(x - 1)(x + 6)}$

$= \dfrac{x^2 + 6x}{(x - 1)(x + 6)}$

$= \dfrac{x(x + 6)}{(x - 1)(x + 6)}$

$= \dfrac{x}{x - 1}, \quad x \neq -6$

83. *Keystrokes:*

y_1 [Y=] [(] 2 [÷] [X,T,θ] [)] [+] [(] 4 [÷] [(] [X,T,θ] [−] 2 [)] [)] [ENTER]

y_2 [(] 6 [X,T,θ] [−] 4 [)] [÷] [(] [X,T,θ] [(] [X,T,θ] [−] 2 [)] [)] [GRAPH]

$\dfrac{2}{x} + \dfrac{4}{(x - 2)} = \dfrac{2(x - 2)}{x(x - 2)} + \dfrac{4x}{x(x - 2)} = \dfrac{2x - 4 + 4x}{x(x - 2)} = \dfrac{6x - 4}{x(x - 2)}$

$y_1 = y_2$

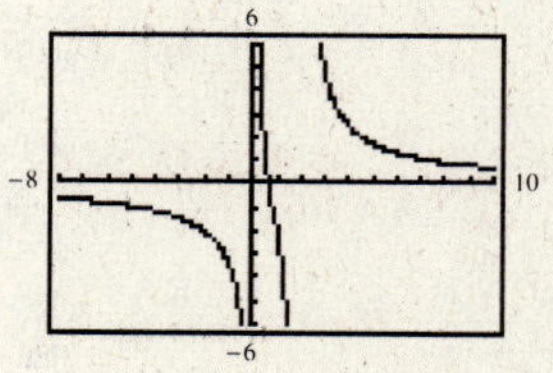

85. $\frac{t}{4} + \frac{t}{6} = \frac{t(3)}{4(3)} + \frac{t(2)}{6(2)} = \frac{3t}{12} + \frac{2t}{12} = \frac{5t}{12}$

87. $A + B + C = 0$

$-B + C = 0$

$-A = 4$

$A = -4$

Substitute into first equation.

$-4 + B + C = 0$

$B + C = 4$

Solve first and second equations.

$B + C = 4$

$-B + C = 0$

$2C = 4$

$C = 2$

$B + 2 = 4$

$B = 2$

$$\begin{aligned}\frac{4}{x^3 - x} &= \frac{-4}{x} + \frac{2}{x+1} + \frac{2}{x-1}\\ &= \frac{-4(x+1)(x-1)}{x(x+1)(x-1)} + \frac{2(x)(x-1)}{(x+1)(x)(x-1)} + \frac{2(x)(x+1)}{(x-1)(x)(x+1)}\\ &= \frac{-4(x^2-1) + 2x(x-1) + 2x(x+1)}{x(x^2-1)}\\ &= \frac{-4x^2 + 4 + 2x^2 - 2x + 2x^2 + 2x}{x^3 - x}\\ &= \frac{4}{x^3 - x}\end{aligned}$$

89. *Verbal Model:* [Total number of students] = [Number of males] + [Number of females]

Model:

$$\begin{aligned}T &= \frac{1434.4t + 5797.28}{0.205t + 1} + \frac{1809.8t + 7362.51}{0.183t + 1}\\ &= \frac{(1434.4t + 5797.28)(0.183t + 1) + (1809.8t + 7362.51)(0.205t + 1)}{(0.205t + 1)(0.183t + 1)}\\ &= \frac{262.4952t^2 + 2495.30224t + 5797.28 + 371.009t^2 + 3319.11455t + 7362.51}{(0.205t + 1)(0.183t + 1)}\\ &= \frac{633.5t^2 + 5814.4t + 13{,}159.79}{(0.205t + 1)(0.183t + 1)}, \quad 0 \le t \le 5\end{aligned}$$

91. $\frac{x-1}{x+4} - \frac{4x-11}{x+4} = \frac{(x-1)-(4x-11)}{x+4} = \frac{x-1-4x+11}{x+4} = \frac{-3x+10}{x+4}$

The subtraction must be distributed to both terms of the numerator of the second fraction.

93. Yes. $\frac{3}{2}(x+2) + \frac{x}{x+2}$

95. $5v + (4 - 3v) = (5v - 3v) + 4 = 2v + 4$

97. $(x^2 - 4x + 3) - (6 - 2x) = x^2 - 4x + 3 - 6 + 2x = x^2 - 2x - 3$

99. $x^2 - 7x + 12 = (x-3)(x-4)$

101. $2a^2 - 9a - 18 = (a-6)(2a+3)$

Section 6.4 Complex Fractions

1. $\dfrac{\left(\dfrac{3}{16}\right)}{\left(\dfrac{9}{12}\right)} = \dfrac{3}{16} \div \dfrac{9}{12} = \dfrac{3}{16} \cdot \dfrac{12}{9} = \dfrac{3 \cdot 2 \cdot 2 \cdot 3}{2 \cdot 2 \cdot 2 \cdot 2 \cdot 3 \cdot 3} = \dfrac{1}{4}$

3. $\dfrac{\left(\dfrac{8x^2y}{3z^2}\right)}{\left(\dfrac{4xy}{9z^5}\right)} = \dfrac{8x^2y}{3z^2} \div \dfrac{4xy}{9z^5}$

$= \dfrac{8x^2y}{3z^2} \cdot \dfrac{9z^5}{4xy}$

$= \dfrac{4 \cdot 2 \cdot 3 \cdot 3x^2 \cdot y \cdot z^5}{3 \cdot 4 \cdot x \cdot y \cdot z^2}$

$= 6xz^3, \; x, y, z \neq 0$

5. $\dfrac{\left(\dfrac{6x^3}{(5y)^2}\right)}{\left(\dfrac{(3x)^2}{15y^4}\right)} = \dfrac{6x^3}{25y^2} \div \dfrac{9x^2}{15y^4}$

$= \dfrac{6x^3}{25y^2} \cdot \dfrac{15y^4}{9x^2}$

$= \dfrac{3 \cdot 2 \cdot 5 \cdot 3 \cdot x^3 \cdot y^4}{5 \cdot 5 \cdot 3 \cdot 3 \cdot x^2 \cdot y^2}$

$= \dfrac{2xy^2}{5}, \quad x \neq 0, y \neq 0$

7. $\dfrac{\left(\dfrac{y}{3-y}\right)}{\left(\dfrac{y^2}{y-3}\right)} = \dfrac{y}{3-y} \div \dfrac{y^2}{y-3}$

$= \dfrac{y}{-1(y-3)} \cdot \dfrac{y-3}{y^2}$

$= \dfrac{y(y-3)}{-1y^2(y-3)}$

$= -\dfrac{1}{y}, \quad y \neq 3$

9. $\dfrac{\left(\dfrac{25x^2}{x-5}\right)}{\left(\dfrac{10x}{5+4x-x^2}\right)} = \dfrac{25x^2}{x-5} \div \dfrac{10x}{5+4x-x^2}$

$= \dfrac{25x^2}{x-5} \cdot \dfrac{5+4x-x^2}{10x}$

$= \dfrac{5 \cdot 5 \cdot x \cdot x \cdot (-1)(x^2-4x-5)}{(x-5) \cdot 5 \cdot 2 \cdot x}$

$= \dfrac{5 \cdot x \cdot (-1)(x-5)(x+1)}{(x-5)2}$

$= -\dfrac{5x(x+1)}{2}, \quad x \neq 0, 5, -1$

11. $\dfrac{\left(\dfrac{x^2+3x-10}{x+4}\right)}{3x-6} = \dfrac{x^2+3x-10}{x+4} \div \dfrac{3x-6}{1}$

$= \dfrac{(x+5)(x-2)}{x+4} \cdot \dfrac{1}{3(x-2)}$

$= \dfrac{(x+5)(x-2)}{(x+4)3(x-2)}$

$= \dfrac{x+5}{3(x+4)}, \quad x \neq 2$

13. $\dfrac{2x-14}{\left(\dfrac{x^2-9x+14}{x+3}\right)} = \dfrac{2x-14}{1} \div \dfrac{x^2-9x+14}{x+3}$

$= \dfrac{2(x-7)}{1} \cdot \dfrac{x+3}{(x-7)(x-2)}$

$= \dfrac{2(x-7)(x+3)}{(x-7)(x-2)}$

$= \dfrac{2(x+3)}{x-2}, \quad x \neq -3, 7$

15. $\dfrac{\left(\dfrac{6x^2-17x+5}{3x^2+3x}\right)}{\dfrac{3x-1}{3x+1}} = \dfrac{6x^2-17x+5}{3x^2+3x} \div \dfrac{3x-1}{3x+1}$

$= \dfrac{(3x-1)(2x-5)}{3x(x+1)} \cdot \dfrac{3x+1}{3x-1}$

$= \dfrac{(3x-1)(2x-5)(3x+1)}{3x(x+1)(3x-1)}$

$= \dfrac{(2x-5)(3x+1)}{3x(x+1)}, \quad x \neq \pm\dfrac{1}{3}$

17. $\dfrac{16x^2+8x+1}{3x^2+8x-3} \div \dfrac{4x^2-3x-1}{x^2+6x+9} = \dfrac{16x^2+8x+1}{3x^2+8x-3} \cdot \dfrac{x^2+6x+9}{4x^2-3x-1}$

$$= \frac{(4x+1)(4x+1)}{(3x-1)(x+3)} \cdot \frac{(x+3)(x+3)}{(4x+1)(x-1)}$$

$$= \frac{(4x+1)(4x+1)(x+3)(x+3)}{(3x-1)(x+3)(4x+1)(x-1)}$$

$$= \frac{(4x+1)(x+3)}{(3x-1)(x-1)}, \quad x \neq -3, -\frac{1}{4}$$

19. $\dfrac{x^2+3x-2x-6}{x^2-4} \div \dfrac{x+3}{x^2+4x+4} = \dfrac{x(x+3)-2(x+3)}{x^2-4} \cdot \dfrac{x^2+4x+4}{x+3}$

$$= \frac{(x+3)(x-2)}{(x-2)(x+2)} \cdot \frac{(x+2)(x+2)}{x+3}$$

$$= \frac{(x+3)(x-2)(x+2)(x+2)}{(x-2)(x+2)(x+3)}$$

$$= (x+2), \quad x \neq \pm 2, -3$$

21. $\dfrac{\left(\dfrac{x^2-3x-10}{x^2-4x+4}\right)}{\left(\dfrac{21+4x-x^2}{x^2-5x-14}\right)} = \dfrac{x^2-3x-10}{x^2-4x+4} \div \dfrac{21+4x-x^2}{x^2-5x-14}$

$$= \frac{x^2-3x-10}{x^2-4x+4} \cdot \frac{x^2-5x-14}{-1(x^2-4x-21)}$$

$$= \frac{(x-5)(x+2)}{(x-2)(x-2)} \cdot \frac{(x-7)(x+2)}{-1(x-7)(x+3)}$$

$$= -\frac{(x^2-3x-10)(x+2)}{(x^2-4x+4)(x+3)}$$

$$= -\frac{(x-5)(x+2)(x+2)}{(x-2)^2(x+3)}$$

$$= -\frac{(x+2)^2(x-5)}{(x-2)^2(x+3)}, \quad x \neq \pm 2, 7$$

23. $\dfrac{\left(1+\dfrac{4}{y}\right)}{y} = \dfrac{\left(1+\dfrac{4}{y}\right)}{y} \cdot \dfrac{y}{y} = \dfrac{y+4}{y^2}$

25. $\dfrac{\left(\dfrac{4}{x}+3\right)}{\left(\dfrac{4}{x}-3\right)} = \dfrac{\left(\dfrac{4}{x}+3\right)}{\left(\dfrac{4}{x}-3\right)} \cdot \dfrac{x}{x} = \dfrac{4+3x}{4-3x}, \quad x \neq 0$

27. $\dfrac{\left(\dfrac{x}{2}\right)}{\left(2+\dfrac{3}{x}\right)} = \dfrac{\dfrac{x}{2}}{2+\dfrac{3}{x}} \cdot \dfrac{2x}{2x}$

$$= \frac{x^2}{4x+6} = \frac{x^2}{2(2x+3)}, \quad x \neq 0$$

29. $\dfrac{\left(3+\dfrac{9}{x-3}\right)}{\left(4+\dfrac{12}{x-3}\right)} = \dfrac{\left(3+\dfrac{9}{x-3}\right)}{\left(4+\dfrac{12}{x-3}\right)} \cdot \dfrac{x-3}{x-3}$

$$= \frac{3(x-3)+\dfrac{9}{x-3}(x-3)}{4(x-3)+\dfrac{12}{x-3}(x-3)}$$

$$= \frac{3x-9+9}{4x-12+12}$$

$$= \frac{3x}{4x} = \frac{3}{4}, \quad x \neq 0, 3$$

31. $$\frac{\left(\frac{3}{x^2}+\frac{1}{x}\right)}{\left(2-\frac{4}{5x}\right)} = \frac{\left(\frac{3}{x^2}+\frac{1}{x}\right)}{\left(2-\frac{4}{5x}\right)}\cdot\frac{5x^2}{5x^2}$$
$$= \frac{15+5x}{10x^2-4x}$$
$$= \frac{5(3+x)}{2x(5x-2)}$$

33. $$\frac{\left(\frac{y}{x}-\frac{x}{y}\right)}{\left(\frac{x+y}{xy}\right)} = \frac{\left(\frac{y}{x}-\frac{x}{y}\right)}{\left(\frac{x+y}{xy}\right)}\cdot\frac{xy}{xy}$$
$$= \frac{\frac{y}{x}(xy)-\frac{x}{y}(xy)}{\left(\frac{x+y}{xy}\right)xy}$$
$$= \frac{y^2-x^2}{x+y}$$
$$= \frac{(y-x)(y+x)}{x+y}$$
$$= y-x, \quad x\neq 0, y\neq 0, x\neq -y$$

35. $$\frac{\left(x-\frac{2y^2}{x-y}\right)}{x-2y} = \frac{\left(x-\frac{2y^2}{x-y}\right)}{x-2y}\cdot\frac{(x-y)}{(x-y)}$$
$$= \frac{x(x-y)-2y^2}{(x-2y)(x-y)}$$
$$= \frac{x^2-xy-2y^2}{(x-2y)(x-y)}$$
$$= \frac{(x-2y)(x+y)}{(x-2y)(x-y)}$$
$$= \frac{x+y}{x-y}, \quad x\neq 2y$$

37. $$\frac{\left(1-\frac{1}{y}\right)}{\left(\frac{1-4y}{y-3}\right)} = \frac{\left(1-\frac{1}{y}\right)}{\left(\frac{1-4y}{y-3}\right)}\cdot\frac{y(y-3)}{y(y-3)}$$
$$= \frac{y(y-3)-(y-3)}{y(1-4y)}$$
$$= \frac{y^2-3y-y+3}{y-4y^2}$$
$$= \frac{y^2-4y+3}{-y(-1+4y)}$$
$$= -\frac{(y-3)(y-1)}{y(4y-1)}, \quad y\neq 3$$

39. $$\frac{\left(\frac{10}{x+1}\right)}{\left(\frac{1}{2x+2}+\frac{3}{x+1}\right)} = \frac{\frac{10}{x+1}}{\frac{1}{2(x+1)}+\frac{3}{x+1}}\cdot\frac{2(x+1)}{2(x+1)}$$
$$= \frac{10(2)}{1+3(2)} = \frac{20}{7}, \quad x\neq -1$$

41. $$\frac{\left(\frac{1}{x}-\frac{1}{x+1}\right)}{\left(\frac{1}{x+1}\right)} = \frac{\frac{1}{x}-\frac{1}{x+1}}{\frac{1}{x+1}}\cdot\frac{x(x+1)}{x(x+1)}$$
$$= \frac{x+1-x}{x} = \frac{1}{x}, \quad x\neq -1$$

43. $$\frac{\left(\frac{x}{x-3}-\frac{2}{3}\right)}{\left(\frac{10}{3x}+\frac{x^2}{x-3}\right)} = \frac{\left(\frac{x}{x-3}-\frac{2}{3}\right)}{\left(\frac{10}{3x}+\frac{x^2}{x-3}\right)}\cdot\frac{3x(x-3)}{3x(x-3)}$$
$$= \frac{3x^2-2x(x-3)}{10(x-3)+3x^3}$$
$$= \frac{3x^2-2x^2+6x}{10x-30+3x^3}$$
$$= \frac{x^2+6x}{3x^3+10x-30}, \quad x\neq 0, x\neq 3$$

45. $$\frac{2y-y^{-1}}{10-y^{-2}} = \frac{2y-\frac{1}{y}}{10-\frac{1}{y^2}}\cdot\frac{y^2}{y^2}$$
$$= \frac{2y^3-y}{10y^2-1} = \frac{y(2y^2-1)}{10y^2-1}, \quad y\neq 0$$

47. $$\frac{7x^2+2x^{-1}}{5x^{-3}+x} = \frac{7x^2+\frac{2}{x}}{\frac{5}{x^3}+x}\cdot\frac{x^3}{x^3}$$
$$= \frac{7x^5+2x^2}{5+x^4} = \frac{x^2(7x^3+2)}{x^4+5}, \quad x\neq 0$$

49. $$\frac{x^{-1}+y^{-1}}{x^{-1}-y^{-1}} = \frac{\frac{1}{x}+\frac{1}{y}}{\frac{1}{x}-\frac{1}{y}}\cdot\frac{xy}{xy}$$
$$= \frac{y+x}{y-x}, \quad x\neq 0, y\neq 0$$

51. $$\frac{x^{-2}-y^{-2}}{(x+y)^2} = \frac{\frac{1}{x^2}-\frac{1}{y^2}}{\frac{(x+y)^2}{1}} \cdot \frac{x^2y^2}{x^2y^2}$$
$$= \frac{y^2-x^2}{x^2y^2(x+y)(x+y)}$$
$$= \frac{(y-x)(y+x)}{x^2y^2(x+y)(x+y)}$$
$$= \frac{y-x}{x^2y^2(x+y)}$$

53. $$\frac{f(2+h)-f(2)}{h} = \frac{\frac{1}{2+h}-\frac{1}{2}}{h}$$
$$= \frac{\frac{1}{2+h}-\frac{1}{2}}{h} \cdot \frac{2(2+h)}{2(2+h)}$$
$$= \frac{2-(2+h)}{2h(2+h)}$$
$$= \frac{2-2-h}{2h(2+h)}$$
$$= \frac{-h}{2h(2+h)}$$
$$= \frac{-1}{2(2+h)}, \quad h \neq 0$$

55. $$\frac{\frac{x}{5}+\frac{x}{6}}{2} = \frac{\frac{x}{5}+\frac{x}{6}}{2} \cdot \frac{30}{30} = \frac{6x+5x}{60} = \frac{11x}{60}$$

57. $$\frac{\frac{2x}{3}+\frac{x}{4}}{2} = \frac{\frac{2x}{3}+\frac{x}{4}}{2} \cdot \frac{12}{12}$$
$$= \frac{\frac{2x}{3}(12)+\frac{x}{4}(12)}{2(12)}$$
$$= \frac{8x+3x}{24}$$
$$= \frac{11x}{24}$$

59. $$\frac{\frac{b+5}{4}+\frac{2}{b}}{2} = \frac{\frac{b+5}{4}+\frac{2}{b}}{2} \cdot \frac{4b}{4b}$$
$$= \frac{\frac{b+5}{4}(4b)+\frac{2}{b}(4b)}{2(4b)}$$
$$= \frac{(b+5)b+8}{8b}$$
$$= \frac{b^2+5b+8}{8b}$$

61. $$\frac{\frac{x}{6}-\frac{x}{9}}{4} = \frac{\frac{x}{6}-\frac{x}{9}}{\frac{4}{1}} \cdot \frac{18}{18} = \frac{3x-2x}{72} = \frac{x}{72}$$

$$x_1 = \frac{x}{9}+\frac{x}{72} = \frac{8x}{72}+\frac{x}{72} = \frac{9x}{72} = \frac{x}{8}$$
$$x_2 = \frac{x}{8}+\frac{x}{72} = \frac{9x}{72}+\frac{x}{72} = \frac{10x}{72} = \frac{5x}{36}$$
$$x_3 = \frac{5x}{36}+\frac{x}{72} = \frac{10x}{72}+\frac{x}{72} = \frac{11x}{72}$$

63. $$\frac{1}{\left(\frac{1}{R_1}+\frac{1}{R_2}\right)} = \frac{1}{\left(\frac{1}{R_1}+\frac{1}{R_2}\right)} \cdot \frac{R_1R_2}{R_1R_2}$$
$$= \frac{R_1R_2}{\frac{1}{R_1}(R_1R_2)+\frac{1}{R_2}(R_1R_2)}$$
$$= \frac{R_1R_2}{R_2+R_1}, \quad R_1, R_2 \neq 0$$

65. (a)

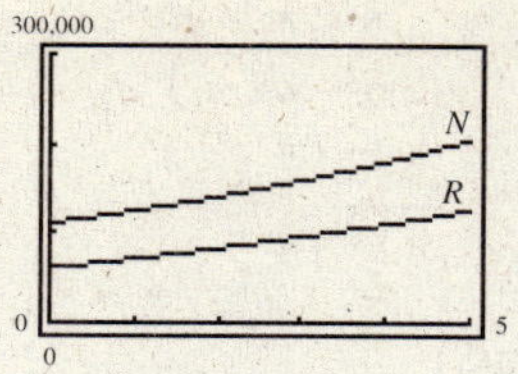

Keystrokes:

N [Y=] [(] 6433.62 [X,T,θ] [+] 111 039.2 [)] [÷] [(] − 0.06 [X,T,θ] [+] 1 [)] [ENTER]

R [(] 8123.73 [X,T,θ] [+] 60 227.5 [)] [÷] [(] − 0.04 [X,T,θ] [+] 1 [)] [GRAPH]

(b) *Verbal Model:*

[Average bill per subscriber] = [Revenue] ÷ [Number of subscribers]

Labels: Revenue = R

Number of subscribers = N

Equation: Average bill per subscriber $= \dfrac{R}{N}$

$$B = \frac{R}{N}$$

$$= \frac{\dfrac{8123.73t + 60{,}227.5}{-0.04t + 1} \cdot 1{,}000{,}000}{\dfrac{6433.62t + 111{,}039.2}{-0.06t + 1} \cdot 1{,}000} \cdot \frac{1}{12}$$

$$= \frac{(8123.73t + 60{,}227.5) \cdot (-0.06t + 1)}{(-0.04t + 1)(6433.62t + 111{,}039.2)} \cdot \frac{1{,}000}{12}$$

$$= \frac{250(-487.42t^2 + 4510.08t + 60{,}227.5)}{3(-257.34t^2 + 1992.05t + 111{,}039.2)}, \quad 0 \le t \le 5$$

67. No. A complex fraction can be written as the division of two rational expressions, so the simplified form will be a rational expression.

69. In the second step, the set of parentheses cannot be moved because division is not associative.

$$\frac{\left(\frac{a}{b}\right)}{b} = \frac{a}{b} \cdot \frac{1}{b} = \frac{a}{b^2}$$

71. $(2y)^3(3y)^2 = 8y^3 \cdot 9y^2 = 72y^5$

73. $3x^2 + 5x - 2 = (3x - 1)(x + 2)$

75. $\dfrac{x^2}{2} \div 4x = \dfrac{x^2}{2} \cdot \dfrac{1}{4x} = \dfrac{x^2}{8x} = \dfrac{x}{8}, \quad x \ne 0$

77. $\dfrac{(x+1)^2}{x+2} \div \dfrac{x+1}{(x+2)^3} = \dfrac{(x+1)^2(x+2)^3}{(x+2)(x+1)} = (x+1)(x+2)^2, \quad x \ne -2, \quad x \ne -1$

Mid-Chapter Quiz for Chapter 6

1. $f(x) = \dfrac{x}{x^2 + x}$

Domain: $x^2 + x \ne 0$

$x(x + 1) \ne 0$

$x \ne 0 \quad x + 1 \ne 0$

$x \ne -1$

$(-\infty, -1) \cup (-1, 0) \cup (0, \infty)$

2. $h(x) = \dfrac{x^2 - 9}{x^2 - x - 2}$

(a) $h(-3) = \dfrac{(-3)^2 - 9}{(-3)^2 - (-3) - 2}$

$= \dfrac{9 - 9}{9 + 3 - 2} = \dfrac{0}{10} = 0$

(b) $h(0) = \dfrac{0^2 - 9}{0^2 - 0 - 2} = \dfrac{-9}{-2} = \dfrac{9}{2}$

(c) $h(-1) = \dfrac{(-1)^2 - 9}{(-1)^2 - (-1) - 2}$

$= \dfrac{1 - 9}{1 + 1 - 2} = \dfrac{-8}{0} =$ undefined

(d) $h(5) = \dfrac{5^2 - 9}{5^2 - 5 - 2} = \dfrac{25 - 9}{25 - 5 - 2} = \dfrac{16}{18} = \dfrac{8}{9}$

3. $\dfrac{9y^2}{6y} = \dfrac{3y}{2}, \quad y \neq 0$

4. $\dfrac{6u^4v^3}{15uv^3} = \dfrac{3 \cdot 2 \cdot u^4 \cdot v^3}{3 \cdot 5 \cdot u \cdot v^3} = \dfrac{2u^3}{5}, \quad u \neq 0, v \neq 0$

5. $\dfrac{4x^2 - 1}{x - 2x^2} = \dfrac{(2x - 1)(2x + 1)}{x(1 - 2x)} = \dfrac{(2x - 1)(2x + 1)}{-x(2x - 1)}$

$= \dfrac{2x + 1}{-x}, \quad x \neq \dfrac{1}{2}$

6. $\dfrac{(z + 3)^2}{2z^2 + 5z - 3} = \dfrac{(z + 3)(z + 3)}{(2z - 1)(z + 3)}$

$= \dfrac{z + 3}{2z - 1}, \quad z \neq -3$

7. $\dfrac{5a^2b + 3ab^3}{a^2b^2} = \dfrac{ab(5a + 3b^2)}{a^2b^2} = \dfrac{5a + 3b^2}{ab}$

8. $\dfrac{2mn^2 - n^3}{2m^2 + mn - n^2} = \dfrac{n^2(2m - n)}{(2m - n)(m + n)}$

$= \dfrac{n^2}{m + n}, \quad 2m - n \neq 0$

9. $\dfrac{11t^2}{6} \cdot \dfrac{9}{33t} = \dfrac{11t^2(9)}{6(33t)} = \dfrac{t}{2}, \quad t \neq 0$

10. $(x^2 + 2x) \cdot \dfrac{5}{x^2 - 4} = \dfrac{x(x + 2)5}{(x - 2)(x + 2)}$

$= \dfrac{5x}{x - 2}, \quad x \neq -2$

11. $\dfrac{4}{3(x - 1)} \cdot \dfrac{12x}{6(x^2 + 2x - 3)} = \dfrac{4(12x)}{3(x - 1)6(x + 3)(x - 1)}$

$= \dfrac{8x}{3(x - 1)^2(x + 3)}$

$= \dfrac{8x}{3(x - 1)(x^2 + 2x - 3)}$

$= \dfrac{8x}{3(x - 1)(x + 3)(x - 1)}$

12. $\dfrac{32z^4}{5x^5y^5} \div \dfrac{80z^5}{25x^8y^6} = \dfrac{32z^4 \cdot 25x^8y^6}{5x^5y^5 \cdot 80z^5}$

$= \dfrac{16 \cdot 2 \cdot 5 \cdot 5 \cdot x^8 \cdot y^6 \cdot z^4}{16 \cdot 5 \cdot 5 \cdot x^5 \cdot y^5 \cdot z^5}$

$= \dfrac{2x^3y}{z}, \quad x \neq 0, y \neq 0$

13. $\dfrac{a - b}{9a + 9b} \div \dfrac{a^2 - b^2}{a^2 + 2a + 1} = \dfrac{a - b}{9(a + b)} \cdot \dfrac{(a + 1)(a + 1)}{(a - b)(a + b)}$

$= \dfrac{(a - b)(a + 1)^2}{9(a + b)^2(a - b)}$

$= \dfrac{(a + 1)^2}{9(a + b)^2}, \quad a \neq b$

14. $\dfrac{5u}{3(u + v)} \cdot \dfrac{2(u^2 - v^2)}{3v} \div \dfrac{25u^2}{18(u - v)} = \dfrac{5u \cdot 2(u - v)(u + v) \cdot 18(u - v)}{3(u + v)(3v)(25u^2)} = \dfrac{4(u - v)^2}{5uv}, \quad u \neq \pm v$

15. $\dfrac{5x - 6}{x - 2} + \dfrac{2x - 5}{x - 2} = \dfrac{5x - 6 + 2x - 5}{x - 2} = \dfrac{7x - 11}{x - 2}$

16. $\dfrac{x}{x^2 - 9} - \dfrac{4(x - 3)}{x + 3} = \dfrac{x}{(x - 3)(x + 3)} - \dfrac{4(x - 3)^2}{(x - 3)(x + 3)}$

$= \dfrac{x - 4(x - 3)^2}{(x - 3)(x + 3)} = \dfrac{x - 4(x^2 - 6x + 9)}{(x - 3)(x + 3)} = \dfrac{x - 4x^2 + 24x - 36}{(x - 3)(x + 3)} = \dfrac{-4x^2 + 25x - 36}{(x - 3)(x + 3)} = -\dfrac{4x^2 - 25x + 36}{(x - 3)(x + 3)}$

17. $\dfrac{x^2+2}{x^2-x-2}+\dfrac{1}{x+1}-\dfrac{x}{x-2}=\dfrac{x^2+2}{(x-2)(x+1)}+\dfrac{1(x-2)}{(x-2)(x+1)}-\dfrac{x(x+1)}{(x-2)(x+1)}$

$$=\frac{x^2+2+x-2-x^2-x}{(x-2)(x+1)}$$

$$=\frac{0}{(x-2)(x+1)}=0,\qquad x\neq 2,\ x\neq -1$$

18. $\dfrac{\dfrac{9t^2}{3-t}}{\dfrac{6t}{t-3}}\cdot\dfrac{t-3}{t-3}=\dfrac{-9t^2}{6t}=-\dfrac{3t}{2},\qquad t\neq 3$

19. $\dfrac{\dfrac{10}{x^2+2x}}{\dfrac{15}{x^2+3x+2}}=\dfrac{\dfrac{10}{x(x+2)}}{\dfrac{15}{(x+2)(x+1)}}\cdot\dfrac{x(x+2)(x+1)}{x(x+2)(x+1)}=\dfrac{10(x+1)}{15x}=\dfrac{2(x+1)}{3x},\quad x\neq -2,\ x\neq -1$

20. $\dfrac{3x^{-1}-y^{-1}}{(x-y)^{-1}}=\dfrac{\left(\dfrac{3}{x}-\dfrac{1}{y}\right)}{\left(\dfrac{1}{x-y}\right)}=\dfrac{\left(\dfrac{3}{x}-\dfrac{1}{y}\right)}{\left(\dfrac{1}{x-y}\right)}\cdot\dfrac{xy(x-y)}{xy(x-y)}$

$$=\frac{3y(x-y)-x(x-y)}{xy}$$

$$=\frac{3xy-3y^2-x^2+xy}{xy}$$

$$=\frac{-3y^2+4xy-x^2}{xy}$$

$$=\frac{-1(3y^2-4xy+x^2)}{xy}$$

$$=\frac{-1(3y-x)(y-x)}{xy}$$

$$=\frac{(3y-x)(x-y)}{xy},\qquad x\neq y$$

21. (a) *Verbal Model:* [Total cost] = [Set up cost] + 144 · [Number of arrangements]

Labels: Total cost = C

Number of arrangements = x

Equation: $C = 25{,}000 + 144x$

(b) *Verbal Model:* [Average cost] = [Total cost] ÷ [Number of arrangements]

Labels: Average cost = $\overline{C}$

Total cost = C

Number of arrangements = x

Equation: $\overline{C}=\dfrac{25{,}000+144x}{x}$

(c) $\overline{C}(500)=\dfrac{25{,}000+144(500)}{500}=\dfrac{97000}{500}=\$194$

22. $\dfrac{x + \dfrac{x}{2} + \dfrac{2x}{3}}{3} = \dfrac{x + \dfrac{x}{2} + \dfrac{2x}{3}}{3} \cdot \dfrac{6}{6} = \dfrac{6x + 3x + 4x}{18} = \dfrac{13x}{18}$

Section 6.5 Dividing Polynomials and Synthetic Division

1. $\begin{aligned}(7x^3 - 2x^2) \div x &= \frac{7x^3 - 2x^2}{x}\\ &= \frac{7x^3}{x} - \frac{2x^2}{x}\\ &= 7x^2 - 2x, \quad x \neq 0\end{aligned}$

3. $\begin{aligned}(4x^2 - 2x) \div (-x) &= \frac{4x^2 - 2x}{-x}\\ &= \frac{4x^2}{-x} - \frac{2x}{-x}\\ &= -4x + 2, \quad x \neq 0\end{aligned}$

5. $\begin{aligned}\frac{m^4 + 2m^2 - 7}{m} &= \frac{m^4}{m} + \frac{2m^2}{m} - \frac{7}{m}\\ &= m^3 + 2m - \frac{7}{m}\end{aligned}$

7. $\begin{aligned}\frac{50z^3 + 30z}{-5z} &= \frac{50z^3}{-5z} + \frac{30z}{-5z}\\ &= -10z^2 - 6, \; z \neq 0\end{aligned}$

9. $\begin{aligned}\frac{4v^4 + 10v^3 - 8v^2}{4v^2} &= \frac{4v^4}{4v^2} + \frac{10v^3}{4v^2} - \frac{8v^2}{4v^2}\\ &= v^2 + \frac{5}{2}v - 2, \quad v \neq 0\end{aligned}$

11. $\begin{aligned}\frac{4x^5 - 6x^4 + 12x^3 - 8x^2}{4x^2} &= \frac{4x^5}{4x^2} - \frac{6x^4}{4x^2} + \frac{12x^3}{4x^2} - \frac{8x^2}{4x^2}\\ &= x^3 - \frac{3}{2}x^2 + 3x - 2, \quad x \neq 0\end{aligned}$

13. $(5x^2y - 8xy + 7xy^2) \div 2xy = \dfrac{5x^2y - 8xy + 7xy^2}{2xy} = \dfrac{5x^2y}{2xy} - \dfrac{8xy}{2xy} + \dfrac{7xy^2}{2xy} = \dfrac{5x}{2} - 4 + \dfrac{7}{2}y, \quad x \neq 0, y \neq 0$

15. $\begin{array}{r} 112\\ 9\overline{)1013}\\ \underline{9}\\ 11\\ \underline{9}\\ 23\\ \underline{18}\\ 5\end{array}$

So $1013 \div 9 = 112 + \frac{5}{9}$

17. $\begin{array}{r} 215\\ 15\overline{)3235}\\ \underline{30}\\ 23\\ \underline{15}\\ 85\\ \underline{75}\\ 10\end{array}$

So $3235 \div 15 = 215 + \frac{10}{15} = 215 + \frac{2}{3}$

19. $\dfrac{x^2 - 8x + 15}{x - 3} = \begin{array}{r} x - 5, \; x \neq 3\\ x - 3\overline{)x^2 - 8x + 15}\\ \underline{x^2 - 3x}\\ -5x + 15\\ \underline{-5x + 15}\end{array}$

21. $(x^2 + 15x + 50) \div (x + 5) = \begin{array}{r} x + 10, \; x \neq -5\\ x + 5\overline{)x^2 + 15x + 50}\\ \underline{x^2 + 5x}\\ 10x + 50\\ \underline{10x + 50}\end{array}$

23.
$$\begin{array}{r} x - 3 + \dfrac{2}{x - 2} \\ x - 2\overline{)x^2 - 5x + 8} \\ \underline{x^2 - 2x} \\ -3x + 8 \\ \underline{-3x + 6} \\ 2 \end{array}$$

25.
$$\begin{array}{r} x + 7,\ x \neq 3 \\ -x + 3\overline{)-x^2 - 4x + 21} \\ \underline{-x^2 + 3x} \\ -7x + 21 \\ \underline{-7x + 21} \end{array}$$

27.
$$\begin{array}{r} 5x - 8 + \dfrac{19}{x + 2} \\ x + 2\overline{)5x^2 + 2x + 3} \\ \underline{5x^2 + 10x} \\ -8x + 3 \\ \underline{-8x - 16} \\ 19 \end{array}$$

29.
$$\begin{array}{r} 4x + 3 + \dfrac{-11}{3x + 2} \\ 3x + 2\overline{)12x^2 + 17x - 5} \\ \underline{12x^2 + 8x} \\ 9x - 5 \\ \underline{9x + 6} \\ -11 \end{array}$$

31.
$$\begin{array}{r} 3t - 4,\ t \neq \tfrac{3}{2} \\ 2t - 3\overline{)6t^2 - 17t + 12} \\ \underline{6t^2 - 9t} \\ -8t + 12 \\ \underline{-8t + 12} \end{array}$$

33.
$$\begin{array}{r} y + 3,\ y \neq -\tfrac{1}{2} \\ 2y + 1\overline{)2y^2 + 7y + 3} \\ \underline{2y^2 + y} \\ 6y + 3 \\ \underline{6y + 3} \end{array}$$

35.
$$\begin{array}{r} x^2 \qquad + 4,\ x \neq 2 \\ x - 2\overline{)x^3 - 2x^2 + 4x - 8} \\ \underline{x^3 - 2x^2} \\ 4x - 8 \\ \underline{4x - 8} \end{array}$$

37.
$$\begin{array}{r} 3x^2 - 3x + 1 + \dfrac{2}{3x + 2} \\ 3x + 2\overline{)9x^3 - 3x^2 - 3x + 4} \\ \underline{9x^3 + 6x^2} \\ -9x^2 - 3x \\ \underline{-9x^2 - 6x} \\ 3x + 4 \\ \underline{3x + 2} \\ 2 \end{array}$$

39.
$$\begin{array}{r} 2 + \dfrac{5}{x + 2} \\ x + 2\overline{)2x + 9} \\ \underline{2x + 4} \\ 5 \end{array}$$

41.
$$\begin{array}{r} x - 4 + \dfrac{32}{x + 4} \\ x + 4\overline{)x^2 + 0x + 16} \\ \underline{x^2 + 4x} \\ -4x + 16 \\ \underline{-4x - 16} \\ 32 \end{array}$$

43.
$$\begin{array}{r} \dfrac{6}{5}z + \dfrac{41}{25} + \dfrac{41}{25(5z - 1)} \\ 5z - 1\overline{)6z^2 + 7z + 0} \\ \underline{6z^2 - \dfrac{6}{5}z} \\ \dfrac{41}{5}z + 0 \\ \underline{\dfrac{41}{5}z - \dfrac{41}{25}} \\ \dfrac{41}{25} \end{array}$$

45.
$$\begin{array}{r} 4x - 1,\ x \neq -\tfrac{1}{4} \\ 4x + 1\overline{)16x^2 + 0x - 1} \\ \underline{16x^2 + 4x} \\ -4x - 1 \\ \underline{-4x - 1} \end{array}$$

47.
$$\begin{array}{r} x^2 - 5x + 25,\ x \neq -5 \\ x + 5\overline{)x^3 + 0x^2 + 0x + 125} \\ \underline{x^3 + 5x^2} \\ -5x^2 + 0x \\ \underline{-5x^2 - 25x} \\ 25x + 125 \\ \underline{25x + 125} \end{array}$$

49.
$$\begin{array}{r} x + 2 + \dfrac{1}{x^2 + 2x + 3} \\ x^2 + 2x + 3\overline{)x^3 + 4x^2 + 7x + 7} \\ \underline{x^3 + 2x^2 + 3x} \\ 2x^2 + 4x + 7 \\ \underline{2x^2 + 4x + 6} \\ 1 \end{array}$$

51.
$$\begin{array}{r}
4x^2 + 12x + 25 + \dfrac{52x - 55}{x^2 - 3x + 2} \\
x^2 - 3x + 2\overline{)4x^4 + 0x^3 - 3x^2 + x - 5} \\
\underline{4x^4 - 12x^3 + 8x^2} \\
12x^3 - 11x^2 + x \\
\underline{12x^3 - 36x^2 + 24x} \\
25x^2 - 23x - 5 \\
\underline{25x^2 - 75x + 50} \\
52x - 55
\end{array}$$

53.
$$\begin{array}{r}
x^3 + x^2 + x + 1, \quad x \neq 1 \\
x - 1\overline{)x^4 + x^3 + x^2 + x - 1} \\
\underline{x^4 - x^3} \\
x^3 + x^2 \\
\underline{x^3 - x^2} \\
x^2 + x \\
\underline{x^2 - x} \\
x - 1 \\
\underline{x - 1}
\end{array}$$

55.
$$\begin{array}{r}
x^3 - x + \dfrac{x}{x^2 + 1} \\
x^2 + 1\overline{)x^5} \\
\underline{x^5 + x^3} \\
-x^3 \\
\underline{-x^3 - x} \\
x
\end{array}$$

57.
$$\begin{aligned}
\frac{8u^2v}{2u} + \frac{3(uv)^2}{uv} &= 4uv + \frac{3u^2v^2}{uv} \\
&= 4uv + 3uv \\
&= 7uv, \quad u \neq 0, v \neq 0
\end{aligned}$$

59.
$$\begin{aligned}
\frac{x^2 + 3x + 2}{x + 2} + (2x + 3) &= \frac{(x + 2)(x + 1)}{x + 2} + (2x + 3) \\
&= (x + 1) + (2x + 3) \\
&= (x + 2x) + (1 + 3) \\
&= 3x + 4, \qquad x \neq -2
\end{aligned}$$

61. $(x^2 + x - 6) \div (x - 2)$

$$\begin{array}{r|rrr}
2 & 1 & 1 & -6 \\
 & & 2 & 6 \\
\hline
 & 1 & 3 & 0
\end{array}$$

$(x^2 + x - 6) \div (x - 2) = x + 3, \ x \neq 2$

63. $\dfrac{x^3 + 3x^2 - 1}{x + 4}$

$$\begin{array}{r|rrrr}
-4 & 1 & 3 & 0 & -1 \\
 & & -4 & 4 & -16 \\
\hline
 & 1 & -1 & 4 & -17
\end{array}$$

$$\frac{x^3 + 3x^2 - 1}{x + 4} = x^2 - x + 4 + \frac{-17}{x + 4}$$

65. $\dfrac{x^4 - 4x^3 + x + 10}{x - 2}$

$$\begin{array}{r|rrrrr}
2 & 1 & -4 & 0 & 1 & 10 \\
 & & 2 & -4 & -8 & -14 \\
\hline
 & 1 & -2 & -4 & -7 & -4
\end{array}$$

$$\frac{x^4 - 4x^3 + x + 10}{x - 2} = x^3 - 2x^2 - 4x - 7 + \frac{-4}{x - 2}$$

67. $\dfrac{5x^3 - 6x^2 + 8}{x - 4}$

$$\begin{array}{r|rrrr}
4 & 5 & -6 & 0 & 8 \\
 & & 20 & 56 & 224 \\
\hline
 & 5 & 14 & 56 & 232
\end{array}$$

$$\frac{5x^3 - 6x^2 + 8}{x - 4} = 5x^2 + 14x + 56 + \frac{232}{x - 4}$$

69. $\dfrac{10x^4 - 50x^3 - 800}{x - 6}$

$$\begin{array}{r|rrrrr} 6 & 10 & -50 & 0 & 0 & -800 \\ & & 60 & 60 & 360 & 2160 \\ \hline & 10 & 10 & 60 & 360 & 1360 \end{array}$$

$$\frac{10x^4 - 50x^3 - 800}{x - 6} = 10x^3 + 10x^2 + 60x + 360 + \frac{1360}{x - 6}$$

71. $\dfrac{0.1x^2 + 0.8x + 1}{x - 0.2}$

$$\begin{array}{r|rrr} 0.2 & 0.1 & 0.8 & 1 \\ & & 0.02 & 0.164 \\ \hline & 0.1 & 0.82 & 1.164 \end{array}$$

$$\frac{0.1x^2 + 0.8x + 1}{x - 0.2} = 0.1x + 0.82 + \frac{1.164}{x - 0.2}$$

73.

$$\begin{array}{r|rrrr} 3 & 1 & -1 & -14 & 24 \\ & & 3 & 6 & -24 \\ \hline & 1 & 2 & -8 & 0 \end{array}$$

$$x^2 + 2x - 8 = (x + 4)(x - 2)$$
$$x^3 - x^2 - 14x + 24 = (x - 3)(x + 4)(x - 2)$$

75.

$$\begin{array}{r|rrrr} 1 & 4 & 0 & -3 & -1 \\ & & 4 & 4 & 1 \\ \hline & 4 & 4 & 1 & 0 \end{array}$$

$$4x^2 + 4x + 1 = (2x + 1)(2x + 1)$$
$$4x^3 - 3x^2 - 1 = (x - 1)(2x + 1)^2$$

77.

$$\begin{array}{r|rrrrr} -4 & 1 & 7 & 3 & -63 & -108 \\ & & -4 & -12 & 36 & 108 \\ \hline & 1 & 3 & -9 & -27 & 0 \end{array}$$

$$\begin{aligned} x^3 + 3x^2 - 9x - 27 &= x^2(x + 3) - 9(x + 3) \\ &= (x^2 - 9)(x + 3) \\ &= (x + 3)(x - 3)(x + 3) \end{aligned}$$

$$x^4 + 7x^3 + 3x^2 - 63x - 108 = (x + 3)^2(x - 3)(x + 4)$$

79. $\dfrac{15x^2 - 2x - 8}{x - \frac{4}{5}}$

$$\begin{array}{r|rrr} \frac{4}{5} & 15 & -2 & -8 \\ & & 12 & 8 \\ \hline & 15 & 10 & 0 \end{array}$$

$$\begin{aligned} 15x^2 - 2x - 8 &= (15x + 10)\left(x - \frac{4}{5}\right) \\ &= 5(3x + 2)\left(x - \frac{4}{5}\right) \end{aligned}$$

81. $\dfrac{x^3 + 2x^2 - 4x + c}{x - 2}$

$$\begin{array}{r|rrrr} 2 & 1 & 2 & -4 & c \\ & & 2 & 8 & 8 \\ \hline & 1 & 4 & 4 & 0 \end{array}$$

$$c + 8 = 0$$
$$c = -8$$

83. *Keystrokes:*

y_1 [Y=] [(] [X,T,θ] [+] 4 [)] [÷] 2 [X,T,θ] [ENTER]

y_2 [(] [1] [÷] 2 [)] [+] [(] 2 [÷] [X,T,θ] [)] [GRAPH]

$$\frac{x+4}{2x} = \frac{x}{2x} + \frac{4}{2x} = \frac{1}{2} + \frac{2}{x}$$

So, $y_1 = y_2$.

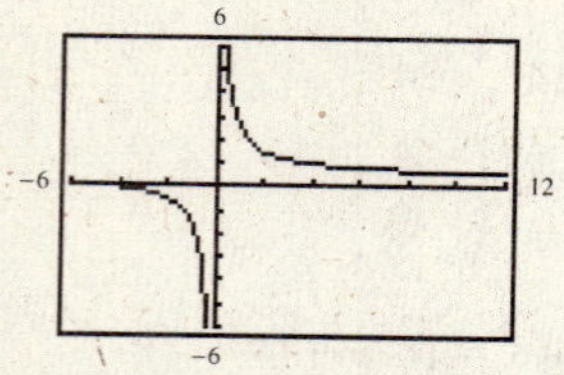

85.

$$\begin{array}{r} x^{2n} + x^n + 4, \quad x^n \neq -2 \\ x^n + 2\overline{\big)\, x^{3n} + 3x^{2n} + 6x^n + 8} \\ \underline{x^{3n} + 2x^{2n}} \qquad\qquad \\ x^{2n} + 6x^n \qquad \\ \underline{x^{2n} + 2x^n} \qquad \\ 4x^n + 8 \\ \underline{4x^n + 8} \end{array}$$

87. Dividend = Divisor · Quotient + Remainder

$$= (x-6)\cdot(x^2 - x + 1) - 4$$

$$= x^3 + x^2 + x - 6x^2 - 6x - 6 - 4$$

$$= x^3 - 5x^2 - 5x - 10$$

89.

k	$f(k)$	Divisor $(x-k)$	Remainder
-2	-8	$x+2$	-8
-1	0	$x+1$	0
0	0	x	0
$\frac{1}{2}$	$-\frac{9}{8}$	$x-\frac{1}{2}$	$-\frac{9}{8}$
1	-2	$x-1$	-2
2	0	$x-2$	0

$$f(-2) = (-2)^3 - (-2)^2 - 2(-2) = -8 - 4 + 4 = -8$$

$$\begin{array}{r|rrrr} -2 & 1 & -1 & -2 & 0 \\ & & -2 & 6 & -8 \\ \hline & 1 & -3 & 4 & -8 \end{array}$$

$$f(-1) = (-1)^3 - (-1)^2 - 2(-1) = -1 - 1 + 2 = 0$$

$$\begin{array}{r|rrrr} -1 & 1 & -1 & -2 & 0 \\ & & -1 & 2 & 0 \\ \hline & 1 & -2 & 0 & 0 \end{array}$$

$$f(0) = 0^3 - 0^2 - 2(0) = 0$$

$$\begin{array}{r|rrrr} 0 & 1 & -1 & -2 & 0 \\ & & 0 & 0 & 0 \\ \hline & 1 & -1 & -2 & 0 \end{array}$$

$$f\left(\tfrac{1}{2}\right) = \left(\tfrac{1}{2}\right)^3 - \left(\tfrac{1}{2}\right)^2 - 2\left(\tfrac{1}{2}\right) = \tfrac{1}{8} - \tfrac{1}{4} - 1 = \tfrac{1}{8} - \tfrac{2}{8} - \tfrac{8}{8} = -\tfrac{9}{8}$$

$$\begin{array}{r|rrrr} \frac{1}{2} & 1 & -1 & -2 & 0 \\ & & \frac{1}{2} & -\frac{1}{4} & -\frac{9}{8} \\ \hline & 1 & -\frac{1}{2} & -\frac{9}{4} & -\frac{9}{8} \end{array}$$

$$f(1) = 1^3 - 1^2 - 2(1) = 1 - 1 - 2 = -2$$

$$\begin{array}{r|rrrr} 1 & 1 & -1 & -2 & 0 \\ & & 1 & 0 & -2 \\ \hline & 1 & 0 & -2 & -2 \end{array}$$

$$f(2) = 2^3 - 2^2 - 2(2) = 8 - 4 - 4 = 0$$

$$\begin{array}{r|rrrr} 2 & 1 & -1 & -2 & 0 \\ & & 2 & 2 & 0 \\ \hline & 1 & 1 & 0 & 0 \end{array}$$

The polynomial values equal the remainders.

91. *Verbal Model:* $\boxed{\text{Area of base}} = \boxed{\text{Volume}} \div \boxed{\text{Height}}$

$$= \left(x^3 + 3x^2 + 3x + 1\right) \div (x + 1)$$

$$\begin{array}{r|rrrr} -1 & 1 & 3 & 3 & 1 \\ & & -1 & -2 & -1 \\ \hline & 1 & 2 & 1 & 0 \end{array}$$

Area of base $= x^2 + 2x + 1$

93. Volume = Area of triangle · Height (of prism)

$$\text{Area of triangle} = \frac{\text{Volume}}{\text{Height (of prism)}}$$

$$= \frac{x^3 + 18x^2 + 80x + 96}{x + 12}$$

$$= x^2 + 6x + 8$$

$$\text{Area of triangle} = \frac{1}{2} \cdot \text{Base} \cdot \text{Height}$$

$$\text{Height} = \frac{2(\text{Area of triangle})}{\text{Base}}$$

$$= \frac{2\left(x^2 + 6x + 8\right)}{x + 2}$$

$$= 2x + 8 \text{ or } 2(x + 4)$$

95. x is not a factor of the numerator.

97. $\dfrac{x^2 + 4}{x + 1} = x - 1 + \dfrac{5}{x + 1}$

Divisor: $x + 1$

Dividend: $x^2 + 4$

Quotient: $x - 1$

Remainder: 5

99. (a) $z = (x - 4)(x - 2)(x + 1)$

Keystrokes: y_1: [Y=] [(] [X,T,θ] [−] 4 [)] [(] [X,T,θ] [−] 2 [)] [(] [X,T,θ] [+] 1 [)] [ENTER]

(b) $y = \left(x^2 - 6x + 8\right)(x + 1)$

Keystrokes: y_2: [(] [X,T,θ] [x^2] [−] 6 [X,T,θ] [+] 8 [)] [(] [X,T,θ] [+] 1 [)] [ENTER]

(c) $y = x^3 - 5x^2 + 2x + 8$

Keystrokes: y_3: [X,T,θ] [^] 3 [−] 5 [X,T,θ] [x^2] [+] 2 [X,T,θ] [+] 8 [GRAPH]

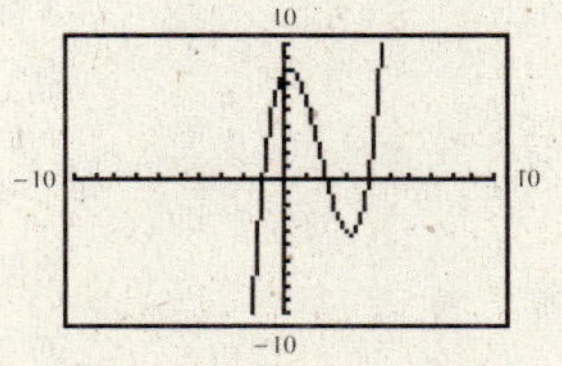

The polynomials in parts (a), (b), and (c) are all equivalent. The x-intercepts are $(-1, 0)$, $(2, 0)$, and $(4, 0)$.

101. $7 - 3x > 4 - x$

$-3x + x > 4 - 7$

$-2x > -3$

$x < \frac{3}{2}$

103. $|x - 3| < 2$

$-2 < x - 3 < 2$

$3 - 2 < x < 2 + 3$

$1 < x < 5$

105. $\left|\frac{1}{4}x - 1\right| \geq 3$

$\frac{1}{4}x - 1 \leq -3$ or $\frac{1}{4}x - 1 \geq 3$

$\frac{1}{4}x \leq -2$ $\qquad$ $\frac{1}{4}x \geq 4$

$x \leq -8$ or $x \geq 16$

107. $(-3, y)$, y is a real number

This point is in Quadrant II or III because the x-coordinate is a negative number which means the point is to the left of the y-axis.

109. The set of points whose x-coordinates are 0 are located on the y-axis.

Section 6.6 Solving Rational Equations

1. (a) $x = 0$

$$\frac{0}{3} - \frac{0}{5} \stackrel{?}{=} \frac{4}{3}$$

$$0 \neq \frac{4}{3}$$

Not a solution

(b) $x = -2$

$$\frac{-2}{3} - \frac{-2}{5} \stackrel{?}{=} \frac{4}{3}$$

$$\frac{-10}{15} + \frac{6}{15} \stackrel{?}{=} \frac{20}{15}$$

$$\frac{-4}{15} \neq \frac{20}{15}$$

Not a solution

(c) $x = \frac{1}{8}$

$$\frac{\frac{1}{8}}{3} - \frac{\frac{1}{8}}{5} \stackrel{?}{=} \frac{4}{3}$$

$$\frac{1}{24} - \frac{1}{40} \stackrel{?}{=} \frac{4}{3}$$

$$\frac{5}{120} - \frac{3}{120} \stackrel{?}{=} \frac{160}{120}$$

$$\frac{2}{120} \neq \frac{160}{120}$$

Not a solution

(d) $x = 10$

$$\frac{10}{3} - \frac{10}{5} \stackrel{?}{=} \frac{4}{3}$$

$$\frac{50}{15} - \frac{30}{15} \stackrel{?}{=} \frac{20}{15}$$

$$\frac{20}{15} = \frac{20}{15}$$

Solution

3. (a) $x = 0$

$$0 \stackrel{?}{=} 4 + \frac{21}{0}$$

$\frac{21}{0}$ is undefined.

Not a solution

(b) $x = -3$

$$-3 \stackrel{?}{=} 4 + \frac{21}{-3}$$

$$-3 \stackrel{?}{=} 4 + (-7)$$

$$-3 = -3$$

Solution

(c) $x = 7$

$$7 \stackrel{?}{=} 4 + \frac{21}{7}$$

$$7 \stackrel{?}{=} 4 + 3$$

$$7 = 7$$

Solution

(d) $x = -1$

$$-1 \stackrel{?}{=} 4 + \frac{21}{-1}$$

$$-1 \stackrel{?}{=} 4 - 21$$

$$-1 \neq -17$$

Not a solution

5.
$$\frac{x}{6} - 1 = \frac{2}{3}$$
$$6\left(\frac{x}{6} - 1\right) = \left(\frac{2}{3}\right)6$$
$$x - 6 = 4$$
$$x = 10$$

Check:
$$\frac{10}{6} - 1 \stackrel{?}{=} \frac{2}{3}$$
$$\frac{5}{3} - \frac{3}{3} \stackrel{?}{=} \frac{2}{3}$$
$$\frac{2}{3} = \frac{2}{3}$$

7.
$$\frac{1}{4} = \frac{z+1}{8}$$
$$4(z+1) = 1(8)$$
$$4z + 4 = 8$$
$$4z = 4$$
$$z = 1$$

Check:
$$\frac{1}{4} \stackrel{?}{=} \frac{1+1}{8}$$
$$\frac{1}{4} \stackrel{?}{=} \frac{2}{8}$$
$$\frac{1}{4} = \frac{1}{4}$$

9.
$$\frac{x}{4} + \frac{x}{2} = \frac{2x}{3}$$
$$12\left(\frac{x}{4} + \frac{x}{2}\right) = \left(\frac{2x}{3}\right)12$$
$$3x + 6x = 8x$$
$$9x = 8x$$
$$x = 0$$

Check:
$$\frac{0}{4} + \frac{0}{2} \stackrel{?}{=} \frac{2(0)}{3}$$
$$0 = 0$$

11.
$$\frac{z+2}{3} = 4 - \frac{z}{12}$$
$$12\left(\frac{z+2}{3}\right) = \left(4 - \frac{z}{12}\right)12$$
$$4(z+2) = 48 - z$$
$$4z + 8 = 48 - z$$
$$5z = 40$$
$$z = 8$$

Check:
$$\frac{8+2}{3} \stackrel{?}{=} 4 - \frac{8}{12}$$
$$\frac{10}{3} \stackrel{?}{=} \frac{12}{3} - \frac{2}{3}$$
$$\frac{10}{3} = \frac{10}{3}$$

13.
$$\frac{x-5}{5} + 3 = -\frac{x}{4}$$
$$20\left(\frac{x-5}{5} + 3\right) = \left(-\frac{x}{4}\right)20$$
$$4(x-5) + 60 = -5x$$
$$4x - 20 + 60 = -5x$$
$$4x + 40 = -5x$$
$$40 = -9x$$
$$-\frac{40}{9} = x$$

Check:
$$\frac{-\frac{40}{9} - 5}{5} + 3 \stackrel{?}{=} -\frac{-\frac{40}{9}}{4}$$
$$\frac{-\frac{40}{9} - \frac{45}{9}}{5} + 3 \stackrel{?}{=} \frac{40}{9} \cdot \frac{1}{4}$$
$$\frac{1}{5}\left(-\frac{85}{9}\right) + 3 \stackrel{?}{=} \frac{10}{9}$$
$$-\frac{17}{9} + \frac{27}{9} \stackrel{?}{=} \frac{10}{9}$$
$$\frac{10}{9} = \frac{10}{9}$$

15.
$$\frac{t}{2} = 12 - \frac{3t^2}{2}$$
$$2\left(\frac{t}{2}\right) = \left(12 - \frac{3t^2}{2}\right)2$$
$$t = 24 - 3t^2$$
$$3t^2 + t - 24 = 0$$
$$(3t - 8)(t + 3) = 0$$
$$3t - 8 = 0 \qquad t + 3 = 0$$
$$t = \frac{8}{3} \qquad t = -3$$

Check:
$$\frac{\frac{8}{3}}{2} \stackrel{?}{=} 12 - \frac{3\left(\frac{8}{3}\right)^2}{2}$$
$$\frac{8}{6} \stackrel{?}{=} 12 - \frac{3\left(\frac{64}{9}\right)}{2}$$
$$\frac{4}{3} \stackrel{?}{=} 12 - \frac{\frac{64}{3}}{2}$$
$$\frac{4}{3} \stackrel{?}{=} 12 - \frac{64}{6}$$
$$\frac{4}{3} \stackrel{?}{=} \frac{36}{3} - \frac{32}{3}$$
$$\frac{4}{3} = \frac{4}{3}$$

Check:
$$\frac{-3}{2} \stackrel{?}{=} 12 - \frac{3(-3)^2}{2}$$
$$\frac{-3}{2} \stackrel{?}{=} 12 - \frac{27}{2}$$
$$\frac{-3}{2} \stackrel{?}{=} \frac{24}{2} - \frac{27}{2}$$
$$\frac{-3}{2} = \frac{-3}{2}$$

17.
$$\frac{5y-1}{12}+\frac{y}{3}=-\frac{1}{4}$$
$$12\left(\frac{5y-1}{12}+\frac{y}{3}\right)=\left(-\frac{1}{4}\right)12$$
$$5y-1+4y=-3$$
$$9y=-2$$
$$y=-\frac{2}{9}$$

Check:
$$\frac{5\left(-\frac{2}{9}\right)-1}{12}+\frac{-\frac{2}{9}}{3}\stackrel{?}{=}-\frac{1}{4}$$
$$\frac{-\frac{10}{9}-\frac{9}{9}}{12}-\frac{2}{27}\stackrel{?}{=}-\frac{1}{4}$$
$$-\frac{19}{108}-\frac{8}{108}\stackrel{?}{=}-\frac{1}{4}$$
$$-\frac{27}{108}\stackrel{?}{=}-\frac{1}{4}$$
$$-\frac{1}{4}=-\frac{1}{4}$$

19.
$$\frac{h+2}{5}-\frac{h-1}{9}=\frac{2}{3}$$
$$45\left(\frac{h+2}{5}-\frac{h-1}{9}\right)=\left(\frac{2}{3}\right)45$$
$$9(h+2)-5(h-1)=30$$
$$9h+18-5h+5=30$$
$$4h+23=30$$
$$4h=7$$
$$h=\frac{7}{4}$$

Check:
$$\frac{\frac{7}{4}+2}{5}-\frac{\frac{7}{4}-1}{9}\stackrel{?}{=}\frac{2}{3}$$
$$\frac{1}{5}\left(\frac{7}{4}+\frac{8}{4}\right)-\frac{1}{9}\left(\frac{7}{4}-\frac{4}{4}\right)\stackrel{?}{=}\frac{2}{3}$$
$$\frac{1}{5}\left(\frac{15}{4}\right)-\frac{1}{9}\left(\frac{3}{4}\right)\stackrel{?}{=}\frac{2}{3}$$
$$\frac{3}{4}-\frac{1}{12}\stackrel{?}{=}\frac{2}{3}$$
$$\frac{9}{12}-\frac{1}{12}\stackrel{?}{=}\frac{2}{3}$$
$$\frac{8}{12}\stackrel{?}{=}\frac{2}{3}$$
$$\frac{2}{3}=\frac{2}{3}$$

21.
$$\frac{x+5}{4}-\frac{3x-8}{3}=\frac{4-x}{12}$$
$$12\left(\frac{x+5}{4}-\frac{3x-8}{3}\right)=\left(\frac{4-x}{12}\right)12$$
$$3(x+5)-4(3x-8)=4-x$$
$$3x+15-12x+32=4-x$$
$$-9x+47=4-x$$
$$-8x=-43$$
$$x=\frac{43}{8}$$

Check:
$$\frac{\frac{43}{8}+5}{4}-\frac{3\left(\frac{43}{8}\right)-8}{3}\stackrel{?}{=}\frac{4-\left(\frac{43}{8}\right)}{12}$$
$$\frac{1}{4}\left(\frac{43}{8}+\frac{40}{8}\right)-\frac{1}{3}\left(\frac{129}{8}-\frac{64}{8}\right)\stackrel{?}{=}\frac{1}{12}\left(\frac{32}{8}-\frac{43}{8}\right)$$
$$\frac{1}{4}\left(\frac{83}{8}\right)-\frac{1}{3}\left(\frac{65}{8}\right)\stackrel{?}{=}\frac{1}{12}\left(-\frac{11}{8}\right)$$
$$\frac{1}{8}\left(\frac{83}{4}-\frac{65}{3}\right)\stackrel{?}{=}\frac{1}{8}\left(-\frac{11}{12}\right)$$
$$\frac{1}{8}\left(\frac{249}{12}-\frac{260}{12}\right)\stackrel{?}{=}\frac{1}{8}\left(-\frac{11}{12}\right)$$
$$\frac{1}{8}\left(-\frac{11}{12}\right)=\frac{1}{8}\left(-\frac{11}{12}\right)$$

23.
$$\frac{9}{25-y}=-\frac{1}{4}$$
$$4(25-y)\left(\frac{9}{25-y}\right)=\left(-\frac{1}{4}\right)4(25-y)$$
$$36=-(25-y)$$
$$36=-25+y$$
$$61=y$$

Check:
$$\frac{9}{25-61}\stackrel{?}{=}-\frac{1}{4}$$
$$-\frac{9}{36}=-\frac{1}{4}$$
$$-\frac{1}{4}=-\frac{1}{4}$$

25.
$$5 - \frac{12}{a} = \frac{5}{3}$$
$$3a\left(5 - \frac{12}{a}\right) = \left(\frac{5}{3}\right)3a$$
$$15a - 36 = 5a$$
$$10a = 36$$
$$a = \frac{36}{10}$$
$$a = \frac{18}{5}$$

Check: $5 - \dfrac{12}{\frac{18}{5}} \stackrel{?}{=} \dfrac{5}{3}$
$$5 - \frac{60}{18} = \frac{5}{3}$$
$$\frac{15}{3} - \frac{10}{3} = \frac{5}{3}$$
$$\frac{5}{3} = \frac{5}{3}$$

27.
$$\frac{4}{x} - \frac{7}{5x} = -\frac{1}{2}$$
$$10x\left(\frac{4}{x} - \frac{7}{5x}\right) = \left(-\frac{1}{2}\right)10x$$
$$40 - 14 = -5x$$
$$26 = -5x$$
$$-\frac{26}{5} = x$$

Check: $\dfrac{4}{-\frac{26}{5}} - \dfrac{7}{5\left(-\frac{26}{5}\right)} \stackrel{?}{=} -\dfrac{1}{2}$
$$-\frac{20}{26} + \frac{7}{26} \stackrel{?}{=} -\frac{1}{2}$$
$$-\frac{13}{26} \stackrel{?}{=} -\frac{1}{2}$$
$$-\frac{1}{2} = -\frac{1}{2}$$

29.
$$\frac{12}{y+5} + \frac{1}{2} = 2$$
$$2(y+5)\left(\frac{12}{y+5} + \frac{1}{2}\right) = (2)2(y+5)$$
$$24 + y + 5 = 4(y+5)$$
$$y + 29 = 4y + 20$$
$$9 = 3y$$
$$3 = y$$

Check: $\dfrac{12}{3+5} + \dfrac{1}{2} \stackrel{?}{=} 2$
$$\frac{3}{2} + \frac{1}{2} = 2$$
$$\frac{4}{2} = 2$$
$$2 = 2$$

31.
$$\frac{5}{x} = \frac{25}{3(x+2)}$$
$$3x(x+2)\left(\frac{5}{x}\right) = \left(\frac{25}{3(x+2)}\right)3x(x+2)$$
$$15(x+2) = 25x$$
$$15x + 30 = 25x$$
$$30 = 10x$$
$$3 = x$$

Check: $\dfrac{5}{3} \stackrel{?}{=} \dfrac{25}{3(3+2)}$
$$\frac{5}{3} = \frac{25}{15}$$
$$\frac{5}{3} = \frac{5}{3}$$

33.
$$\frac{8}{3x+5} = \frac{1}{x+2}$$
$$(3x+5)(x+2)\left(\frac{8}{3x+5}\right) = \left(\frac{1}{x+2}\right)(3x+5)(x+2)$$
$$8(x+2) = 3x+5$$
$$8x+16 = 3x+5$$
$$5x = -11$$
$$x = -\frac{11}{5}$$

Check:
$$\frac{8}{3\left(-\frac{11}{5}\right)+5} \stackrel{?}{=} \frac{1}{-\frac{11}{5}+2}$$
$$\frac{8}{-\frac{33}{5}+\frac{25}{5}} = \frac{1}{-\frac{11}{5}+\frac{10}{5}}$$
$$\frac{8}{-\frac{8}{5}} = \frac{1}{-\frac{1}{5}}$$
$$-5 = -5$$

35.
$$\frac{3}{x+2} - \frac{1}{x} = \frac{1}{5x}$$
$$5x(x+2)\left(\frac{3}{x+2} - \frac{1}{x}\right) = \left(\frac{1}{5x}\right)5x(x+2)$$
$$15x - 5(x+2) = x+2$$
$$15x - 5x - 10 = x+2$$
$$10x - 10 = x+2$$
$$9x = 12$$
$$x = \frac{12}{9}$$
$$x = \frac{4}{3}$$

Check:
$$\frac{1}{\frac{4}{3}+2} - \frac{1}{\frac{4}{3}} \stackrel{?}{=} \frac{1}{5\left(\frac{4}{3}\right)}$$
$$\frac{\frac{3}{10}}{\frac{10}{3}} - \frac{1}{\frac{4}{3}} = \frac{1}{\frac{20}{3}}$$
$$\frac{9}{10} - \frac{3}{4} = \frac{3}{20}$$
$$\frac{18}{20} - \frac{15}{20} = \frac{3}{20}$$
$$\frac{3}{20} = \frac{3}{20}$$

37.
$$\frac{1}{2} = \frac{18}{x^2}$$
$$2x^2\left(\frac{1}{2}\right) = \left(\frac{18}{x^2}\right)2x^2$$
$$x^2 = 36$$
$$x^2 - 36 = 0$$
$$(x-6)(x+6) = 0$$
$$x = 6 \quad x = -6$$

Check:
$$\frac{1}{2} \stackrel{?}{=} \frac{18}{6^2} \qquad \frac{1}{2} \stackrel{?}{=} \frac{18}{(-6)^2}$$
$$\frac{1}{2} = \frac{18}{36} \qquad \frac{1}{2} = \frac{18}{36}$$
$$\frac{1}{2} = \frac{1}{2} \qquad \frac{1}{2} = \frac{1}{2}$$

39.
$$\frac{t}{4} = \frac{4}{t}$$
$$4t\left(\frac{t}{4}\right) = \left(\frac{4}{t}\right)4t$$
$$t^2 = 16$$
$$t = \pm 4$$

Check: $\frac{4}{4} \stackrel{?}{=} \frac{4}{4}$, $1 = 1$ **Check:** $\frac{-4}{4} = \frac{4}{-4}$, $-1 = -1$

41.
$$x + 1 = \frac{72}{x}$$
$$x(x+1) = \left(\frac{72}{x}\right)x$$
$$x^2 + x = 72$$
$$x^2 + x - 72 = 0$$
$$(x+9)(x-8) = 0$$
$$x = -9 \quad x = 8$$

Check:
$$-9 + 1 \stackrel{?}{=} \frac{72}{-9} \qquad 8 + 1 \stackrel{?}{=} \frac{72}{8}$$
$$-8 = -8 \qquad 9 = 9$$

43.
$$\begin{aligned} y + \frac{18}{y} &= 9 \\ y\left(y + \frac{18}{y}\right) &= (9)y \\ y^2 + 18 &= 9y \\ y^2 - 9y + 18 &= 0 \\ (y-3)(y-6) &= 0 \end{aligned}$$

$y - 3 = 0 \quad y - 6 = 0$

$y = 3 \quad y = 6$

Check: $3 + \frac{18}{3} \stackrel{?}{=} 9$

$3 + 6 \stackrel{?}{=} 9$

$9 = 9$

Check: $6 + \frac{18}{6} \stackrel{?}{=} 9$

$6 + 3 \stackrel{?}{=} 9$

$9 = 9$

45.
$$\begin{aligned} \frac{4}{x(x-1)} + \frac{3}{x} &= \frac{4}{x-1} \\ x(x-1)\left(\frac{4}{x(x-1)} + \frac{3}{x}\right) &= \left(\frac{4}{x-1}\right)x(x-1) \\ 4 + 3(x-1) &= 4x \\ 4 + 3x - 3 &= 4x \\ 1 &= x \end{aligned}$$

Check: $\frac{4}{1(1-1)} + \frac{3}{1} \stackrel{?}{=} \frac{4}{1-1}$

$\frac{4}{0} + 3 = \frac{4}{0}$

Division by zero is undefined. Solution $x = 1$ extraneous. No solution.

47.
$$\begin{aligned} \frac{2x}{5} &= \frac{x^2 - 5x}{5x} \\ 5x\left(\frac{2x}{5}\right) &= \left(\frac{x^2 - 5x}{5x}\right)5x \\ 2x^2 &= x^2 - 5x \\ x^2 + 5x &= 0 \\ x(x+5) &= 0 \\ x = 0 \quad x + 5 &= 0 \\ x &= -5 \end{aligned}$$

Check:

$x = 0$

$\frac{2(0)}{5} \stackrel{?}{=} \frac{0^2 - 5(0)}{5(0)}$

$0 \neq$ undefined

so $x = 0$ is extraneous.

$x = -5$

$\frac{2(-5)}{5} \stackrel{?}{=} \frac{(-5)^2 - 5(-5)}{5(-5)}$

$\frac{-10}{5} = \frac{25 + 25}{-25}$

$-2 = -2$

49.
$$\begin{aligned} \frac{y+1}{y+10} &= \frac{y-2}{y+4} \\ (y+1)(y+4) &= (y+10)(y-2) \\ y^2 + 5y + 4 &= y^2 + 8y - 20 \\ 5y + 4 &= 8y - 20 \\ -3y &= -24 \\ y &= 8 \end{aligned}$$

Check: $\frac{8+1}{8+10} \stackrel{?}{=} \frac{8-2}{8+4}$

$\frac{9}{18} \stackrel{?}{=} \frac{6}{12}$

$\frac{1}{2} = \frac{1}{2}$

51.
$$\begin{aligned} \frac{15}{x} + \frac{9x-7}{x+2} &= 9 \\ x(x+2)\left(\frac{15}{x} + \frac{9x-7}{x+2}\right) &= (9)x(x+2) \\ 15(x+2) + x(9x-7) &= 9x(x+2) \\ 15x + 30 + 9x^2 - 7x &= 9x^2 + 18x \\ 8x + 30 &= 18x \\ -10x &= -30 \\ x &= 3 \end{aligned}$$

Check: $\frac{15}{3} + \frac{9(3)-7}{3+2} \stackrel{?}{=} 9$

$5 + \frac{20}{5} \stackrel{?}{=} 9$

$5 + 4 \stackrel{?}{=} 9$

$9 = 9$

53.

$$\frac{2}{6q+5} - \frac{3}{4(6q+5)} = \frac{1}{28}$$
$$28(6q+5)\left(\frac{2}{6q+5} - \frac{3}{4(6q+5)}\right) = \left(\frac{1}{28}\right)28(6q+5)$$
$$28(2) - 7(3) = 6q + 5$$
$$56 - 21 = 6q + 5$$
$$35 = 6q + 5$$
$$30 = 6q$$
$$5 = q$$

Check:
$$\frac{2}{6(5)+5} - \frac{3}{4[6(5)+5]} \stackrel{?}{=} \frac{1}{28}$$
$$\frac{2}{30+5} - \frac{3}{4(30+5)} \stackrel{?}{=} \frac{1}{28}$$
$$\frac{2}{35} - \frac{3}{4(35)} \stackrel{?}{=} \frac{1}{28}$$
$$\frac{8}{140} - \frac{3}{140} \stackrel{?}{=} \frac{1}{28}$$
$$\frac{5}{140} \stackrel{?}{=} \frac{1}{28}$$
$$\frac{1}{28} = \frac{1}{28}$$

55.

$$\frac{4}{2x+3} + \frac{17}{5x-3} = 3$$
$$(5x-3)(2x+3)\left(\frac{4}{2x+3} + \frac{17}{5x-3}\right) = (3)(2x+3)(5x-3)$$
$$4(5x-3) + 17(2x+3) = 3(10x^2 + 9x - 9)$$
$$20x - 12 + 34x + 51 = 30x^2 + 27x - 27$$
$$0 = 30x^2 - 27x - 66$$
$$0 = 10x^2 - 9x - 22$$
$$0 = (10x + 11)(x - 2)$$
$$x = -\frac{11}{10} \quad x = 2$$

Check:
$$\frac{4}{2\left(-\frac{11}{10}\right)+3} + \frac{17}{5\left(-\frac{11}{10}\right)-3} \stackrel{?}{=} 3$$
$$\frac{4}{-\frac{22}{10}+\frac{30}{10}} + \frac{17}{-\frac{55}{10}-\frac{30}{10}} \stackrel{?}{=} 3$$
$$\frac{4}{\frac{8}{10}} + \frac{17}{-\frac{85}{10}} \stackrel{?}{=} 3$$
$$\frac{4}{\frac{4}{5}} + \frac{17}{-\frac{17}{2}} \stackrel{?}{=} 3$$
$$5 + -2 \stackrel{?}{=} 3$$
$$3 = 3$$

Check:
$$\frac{4}{2(2)+3} + \frac{17}{5(2)-3} \stackrel{?}{=} 3$$
$$\frac{4}{7} + \frac{17}{7} \stackrel{?}{=} 3$$
$$3 = 3$$

57.

$$\frac{2}{x-10} - \frac{3}{x-2} = \frac{6}{x^2 - 12x + 20}$$

$$\frac{2}{x-10} - \frac{3}{x-2} = \frac{6}{(x-10)(x-2)}$$

$$(x-10)(x-2)\left(\frac{2}{x-10} - \frac{3}{x-2}\right) = \left(\frac{6}{(x-10)(x-2)}\right)(x-10)(x-2)$$

$$2(x-2) - 3(x-10) = 6$$

$$2x - 4 - 3x + 30 = 6$$

$$-x + 26 = 6$$

$$-x = -20$$

$$x = 20$$

Check: $\frac{2}{20-10} - \frac{3}{20-2} \stackrel{?}{=} \frac{6}{(20)^2 - 12(20) + 20}$

$$\frac{2}{10} - \frac{3}{18} \stackrel{?}{=} \frac{6}{400 - 240 + 20}$$

$$\frac{1}{5} - \frac{1}{6} \stackrel{?}{=} \frac{6}{180}$$

$$\frac{6}{30} - \frac{5}{30} = \frac{1}{30}$$

$$\frac{1}{30} = \frac{1}{30}$$

59.

$$\frac{x+3}{x^2-9} + \frac{4}{3-x} - 2 = 0$$

$$\frac{x+3}{(x-3)(x+3)} - \frac{4}{x-3} - 2 = 0$$

$$(x-3)\left(\frac{1}{x-3} - \frac{4}{x-3} - 2\right) = 0(x-3)$$

$$1 - 4 - 2(x-3) = 0$$

$$-3 - 2x + 6 = 0$$

$$-2x = -3$$

$$x = \frac{3}{2}$$

Check: $\dfrac{\frac{3}{2} + 3}{\left(\frac{3}{2}\right)^2 - 9} + \dfrac{4}{3 - \frac{3}{2}} - 2 \stackrel{?}{=} 0$

$$\frac{\frac{3}{2} + \frac{6}{2}}{\frac{9}{4} - \frac{36}{4}} + \frac{4}{\frac{6}{2} - \frac{3}{2}} - 2 \stackrel{?}{=} 0$$

$$\frac{\frac{9}{2}}{-\frac{27}{4}} + \frac{4}{\frac{3}{2}} - 2 \stackrel{?}{=} 0$$

$$-\frac{2}{3} + \frac{8}{3} - \frac{6}{3} \stackrel{?}{=} 0$$

$$0 = 0$$

61.

$$\frac{x}{x-2}+\frac{3x}{x-4}=\frac{-2(x-6)}{x^2-6x+8}$$

$$(x-2)(x-4)\left(\frac{x}{x-2}+\frac{3x}{x-4}\right)=\left(\frac{-2(x-6)}{(x-4)(x-2)}\right)(x-2)(x-4)$$

$$x(x-4)+3x(x-2)=-2(x-6)$$

$$x^2-4x+3x^2-6x=-2x+12$$

$$4x^2-8x-12=0$$

$$x^2-2x-3=0$$

$$(x-3)(x+1)=0$$

$$x=3 \quad x=-1$$

Check: $\frac{3}{3-2}+\frac{3(3)}{3-4}\stackrel{?}{=}\frac{-2(3-6)}{3^2-6(3)+8}$

$$3+\frac{9}{-1}\stackrel{?}{=}\frac{6}{-1}$$

$$-6=-6$$

Check: $\frac{-1}{-1-2}+\frac{3(-1)}{-1-4}\stackrel{?}{=}\frac{-2(-1-6)}{(-1)^2-6(-1)+8}$

$$\frac{1}{3}+\frac{3}{5}\stackrel{?}{=}\frac{14}{15}$$

$$\frac{5}{15}+\frac{9}{15}\stackrel{?}{=}\frac{14}{15}$$

$$\frac{14}{15}=\frac{14}{15}$$

63.

$$\frac{5}{x^2+4x+3}+\frac{2}{x^2+x-6}=\frac{3}{x^2-x-2}$$

$$(x+3)(x+1)(x-2)\left[\frac{5}{(x+3)(x+1)}+\frac{2}{(x+3)(x-2)}=\frac{3}{(x-2)(x+1)}\right](x+3)(x+1)(x-2)$$

$$5(x-2)+2(x+1)=3(x+3)$$

$$5x-10+2x+2=3x+9$$

$$7x-8=3x+9$$

$$4x=17$$

$$x=\frac{17}{4}$$

Check: $\frac{5}{\left(\frac{17}{4}\right)^2+4\left(\frac{17}{4}\right)+3}+\frac{2}{\left(\frac{17}{4}\right)^2+\left(\frac{17}{4}\right)-6}\stackrel{?}{=}\frac{3}{\left(\frac{17}{4}\right)^2-\left(\frac{17}{4}\right)-2}$

$$\frac{5}{\frac{289}{16}+17+3}+\frac{2}{\frac{289}{16}+\frac{68}{16}-\frac{96}{16}}\stackrel{?}{=}\frac{3}{\frac{289}{16}-\frac{68}{16}-\frac{32}{16}}$$

$$\frac{5}{\frac{289}{16}+\frac{320}{16}}+\frac{2}{\frac{261}{16}}\stackrel{?}{=}\frac{3}{\frac{189}{16}}$$

$$\frac{5}{\frac{609}{16}}+\frac{32}{261}\stackrel{?}{=}\frac{48}{189}$$

$$\frac{80}{609}+\frac{32}{261}\stackrel{?}{=}\frac{48}{189}$$

$$\frac{240}{1827}+\frac{224}{1827}\stackrel{?}{=}\frac{48}{189}$$

$$\frac{464}{1827}\stackrel{?}{=}\frac{48}{189}$$

$$\frac{16}{63}=\frac{16}{63}$$

65. $\dfrac{x}{3} = \dfrac{1 + \dfrac{4}{x}}{1 + \dfrac{2}{x}}$

$$\frac{x}{3} = \frac{1 + \frac{4}{x}}{1 + \frac{2}{x}} \cdot \frac{x}{x}$$

$$3(x + 2)\left(\frac{x}{3}\right) = \left(\frac{x + 4}{x + 2}\right)3(x + 2)$$

$$x(x + 2) = 3(x + 4)$$

$$x^2 + 2x = 3x + 12$$

$$x^2 - x - 12 = 0$$

$$(x - 4)(x + 3) = 0$$

$$x - 4 = 0 \qquad x + 3 = 0$$

$$x = 4 \qquad x = -3$$

Check: $\dfrac{4}{3} \stackrel{?}{=} \dfrac{1 + \frac{4}{4}}{1 + \frac{2}{4}}$

$$\frac{4}{3} \stackrel{?}{=} \frac{1 + 1}{1 + \frac{1}{2}}$$

$$\frac{4}{3} \stackrel{?}{=} \frac{2}{\frac{3}{2}}$$

$$\frac{4}{3} = \frac{4}{3}$$

Check: $\dfrac{-3}{3} \stackrel{?}{=} \dfrac{1 + \frac{4}{-3}}{1 + \frac{2}{-3}}$

$$-1 \stackrel{?}{=} \frac{-\frac{1}{3}}{\frac{1}{3}}$$

$$-1 = -1$$

67. x-intercept: $(-2, 0)$

$$0 = \frac{x + 2}{x - 2}$$

$$(x - 2)(0) = \left(\frac{x + 2}{x - 2}\right)(x - 2)$$

$$0 = x + 2$$

$$-2 = x$$

69. x-intercepts: $(-1, 0)$ and $(1, 0)$

$$0 = x - \frac{1}{x}$$

$$x(0) = \left(x - \frac{1}{x}\right)x$$

$$0 = x^2 - 1$$

$$0 = (x - 1)(x + 1)$$

$$x - 1 = 0 \qquad x + 1 = 0$$

$$x = 1 \qquad x = -1$$

71. (a) *Keystrokes:*

[Y=] [(] [X,T,θ] [–] 4 [)] [÷] [(] [X,T,θ] [+] 5 [)] [GRAPH]

x-intercept: $(4, 0)$

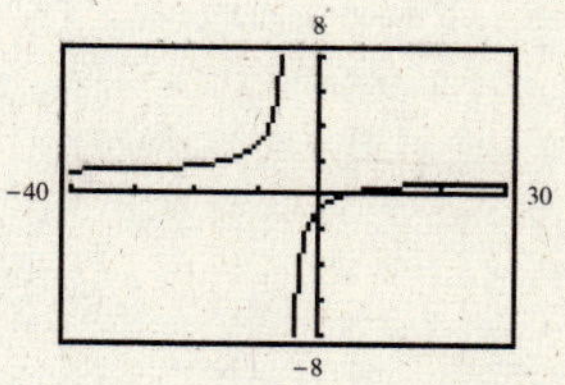

(b) $0 = \dfrac{x - 4}{x + 5}$

$$0 = x - 4$$

$$4 = x$$

73. (a) *Keystrokes:* [Y=] [X,T,θ] [+] 3 [+] 7 [÷] [X,T,θ] [GRAPH]

No x-intercepts

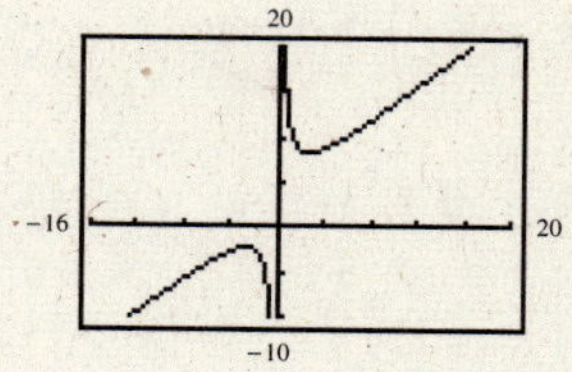

(b) $0 = x + 3 + \dfrac{7}{x}$

$$x(0) = \left(x + 3 + \frac{7}{x}\right)x$$

$$0 = x^2 + 3x + 7$$

Prime

75. (a) *Keystrokes:* Y= (X,T,θ + 1) − 6 ÷ X,T,θ GRAPH

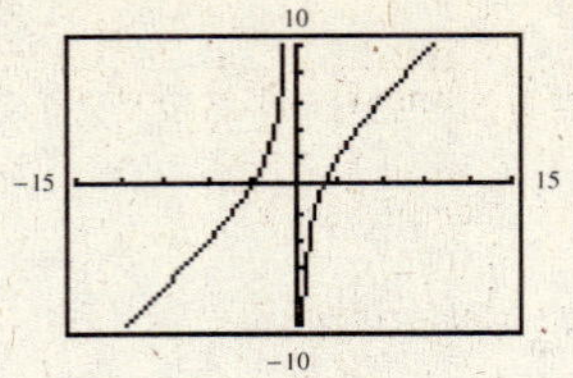

x-intercepts: $(-3, 0)$ and $(2, 0)$

(b)
$$0 = (x + 1) - \frac{6}{x}$$
$$x(0) = \left[(x + 1) - \frac{6}{x}\right]x$$
$$0 = x^2 + x - 6$$
$$0 = (x + 3)(x - 2)$$
$$x + 3 = 0 \qquad x - 2 = 0$$
$$x = -3 \qquad x = 2$$

77.
$$\frac{16}{x^2 - 16} + \frac{x}{2x - 8} = \frac{1}{2} \rightarrow \text{equation}$$
$$2(x - 4)(x + 4)\left(\frac{16}{x^2 - 16} + \frac{x}{2(x - 4)}\right) = \left(\frac{1}{2}\right)2(x - 4)(x + 4)$$
$$32 + x(x + 4) = x^2 - 16$$
$$32 + x^2 + 4x = x^2 - 16$$
$$4x = -48$$
$$x = -12$$

79. $\frac{16}{x^2 - 16} + \frac{x}{2x - 8} + \frac{1}{2} \rightarrow$ expression

$$\frac{16}{(x - 4)(x + 4)} + \frac{x}{2(x - 4)} + \frac{1}{2} = \frac{16(2)}{2(x - 4)(x + 4)} + \frac{x(x + 4)}{2(x - 4)(x + 4)} + \frac{1(x - 4)(x + 4)}{2(x - 4)(x + 4)}$$
$$= \frac{32 + x^2 + 4x + x^2 - 16}{2(x - 4)(x + 4)}$$
$$= \frac{2x^2 + 4x + 16}{2(x - 4)(x + 4)}$$
$$= \frac{2(x^2 + 2x + 8)}{2(x - 4)(x + 4)}$$
$$= \frac{x^2 + 2x + 8}{(x - 4)(x + 4)}$$

81. *Verbal Model*: $\boxed{\text{Number}} + \boxed{\text{Reciprocal}} = \frac{37}{6}$

Labels: *Number* $= x$

Reciprocal $= \frac{1}{x}$

Equation:

$$x + \frac{1}{x} = \frac{37}{6}$$

$$6x\left(x + \frac{1}{x}\right) = \left(\frac{37}{6}\right)6x$$

$$6x^2 + 6 = 37x$$

$$6x^2 - 37x + 6 = 0$$

$$(6x - 1)(x - 6) = 0$$

$$6x - 1 = 0 \qquad x - 6 = 0$$

$$x = \frac{1}{6} \qquad x = 6$$

83. *Verbal Model*:

$\boxed{\text{Rate painter 1}} + \boxed{\text{Rate painter 2}} = \boxed{\text{Rate together}}$

Labels: Painter 1's rate $= \frac{1}{4}$

Painter 2's rate $= \frac{1}{6}$

Rate together $= \frac{1}{x}$

Equation:

$$\frac{1}{4} + \frac{1}{6} = \frac{1}{x}$$

$$12x\left(\frac{1}{4} + \frac{1}{6}\right) = \left(\frac{1}{x}\right)12x$$

$$3x + 2x = 12$$

$$5x = 12$$

$$x = \frac{12}{5} = 2.4 \text{ hours}$$

$$= 2 \text{ hours } 24 \text{ minutes}$$

85. *Verbal Model:* $\boxed{\text{Distance}} \div \boxed{\text{Rate}} = \boxed{\text{Time}}$

$$\frac{\boxed{\text{Distance}}}{\boxed{\text{Rate}}} = \boxed{\text{Time}}$$

$\boxed{\text{Time traveled with wind}} = \boxed{\text{Time traveled without wind}}$

Label: Speed of the wind $= x$

Equation:

$$\frac{680}{300 + x} = \frac{520}{300 - x}$$

$$(300 + x)(300 - x)\left(\frac{680}{300 + x}\right) = \left(\frac{520}{300 - x}\right)(300 + x)(300 - x)$$

$$680(300 - x) = 520(300 + x)$$

$$204{,}000 - 680x = 156{,}000 + 520x$$

$$-1200x = -48{,}000$$

$$x = 40 \text{ miles per hour}$$

87. *Verbal Model:* $\frac{\boxed{\text{Saves}}}{\boxed{\text{Total shots}}} = \boxed{\text{Percent saved shots}}$

Label: $x =$ additional saves

Equation:

$$\frac{707 + x}{799 + x} = 0.900$$

$$(799 + x)\left(\frac{707 + x}{799 + x}\right) = 0.900(799 + x)$$

$$707 + x = 719.10 + 0.900x$$

$$0.10x = 12.10$$

$$x = 121 \text{ saves}$$

89. An extraneous solution is a "trial solution" that does not satisfy the original equation.

91. When the equation is solved, the solution is $x = 0$. However, if $x = 0$, then there is division by zero, so the equation has no solution.

93. $x^2 - 81 = x^2 - (9)^2 = (x + 9)(x - 9)$

95. $4x^2 - \frac{1}{4} = (2x)^2 - \left(\frac{1}{2}\right)^2 = \left(2x + \frac{1}{2}\right)\left(2x - \frac{1}{2}\right)$

97. $f(x) = \frac{2x^2}{5}$

$5 \neq 0$

Domain: $(-\infty, \infty)$

Section 6.7 Applications and Variation

1. $I = kV$

3. $V = kt$

5. $u = kv^2$

7. $p = \frac{k}{d}$

9. $A = \frac{k}{t^4}$

11. $A = klw$

13. $P = \frac{k}{V}$

15. The area of a triangle varies jointly as the base and the height.

17. The volume of a right circular cylinder varies jointly as the square of the radius and the height.

19. The average speed varies directly as the distance and inversely as the time.

21.
$$\begin{aligned} s &= kt \\ 20 &= k(4) \\ 5 &= k \\ s &= 5t \end{aligned}$$

23.
$$\begin{aligned} F &= kx^2 \\ 500 &= k(40)^2 \\ \tfrac{500}{1600} &= k \\ \tfrac{5}{16} &= k \\ F &= \tfrac{5}{16}x^2 \end{aligned}$$

25.
$$\begin{aligned} n &= \frac{k}{m} \\ 32 &= \frac{k}{1.5} \\ 48 &= k \\ n &= \frac{48}{m} \end{aligned}$$

27.
$$\begin{aligned} g &= \frac{k}{\sqrt{z}} \\ \frac{4}{5} &= \frac{k}{\sqrt{25}} \\ 4 &= k \\ g &= \frac{4}{\sqrt{z}} \end{aligned}$$

29.
$$\begin{aligned} F &= kxy \\ 500 &= k(15)(8) \\ \tfrac{500}{120} &= k \\ \tfrac{25}{6} &= k \\ F &= \tfrac{25}{6}xy \end{aligned}$$

31.
$$\begin{aligned} d &= k\left(\frac{x^2}{r}\right) \\ 3000 &= k\left(\frac{10^2}{4}\right) \\ 3000 &= k(25) \\ 120 &= k \\ d &= \frac{120x^2}{r} \end{aligned}$$

33.

x	2	4	6	8	10
$y = kx^2$	4	16	36	64	100

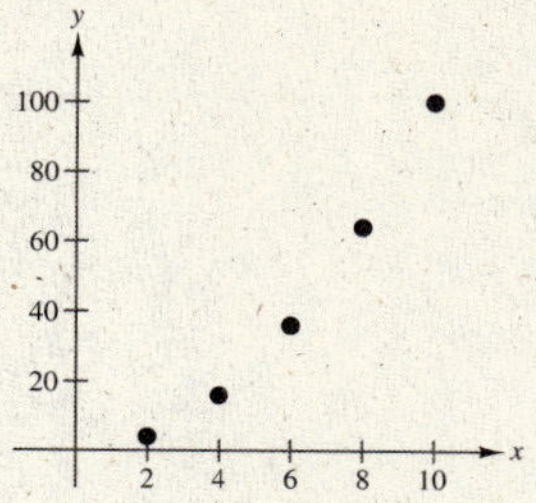

35.

x	2	4	6	8	10
$y = kx^2$	2	8	18	32	50

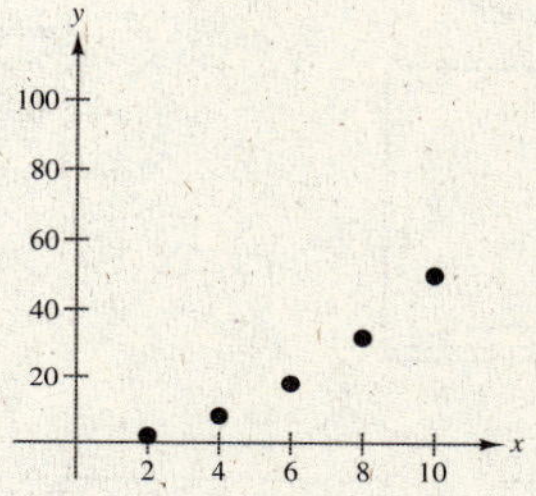

37.

x	2	4	6	8	10
$y = \dfrac{k}{x^2}$	$\dfrac{1}{2}$	$\dfrac{1}{8}$	$\dfrac{1}{18}$	$\dfrac{1}{32}$	$\dfrac{1}{50}$

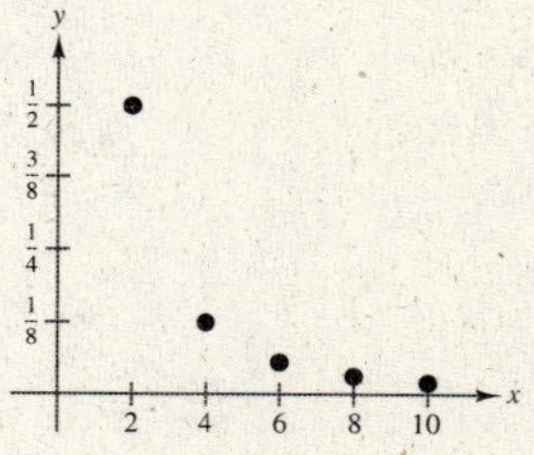

39.

x	2	4	6	8	10
$y = \dfrac{k}{x^2}$	$\dfrac{5}{2}$	$\dfrac{5}{8}$	$\dfrac{5}{18}$	$\dfrac{5}{32}$	$\dfrac{1}{10}$

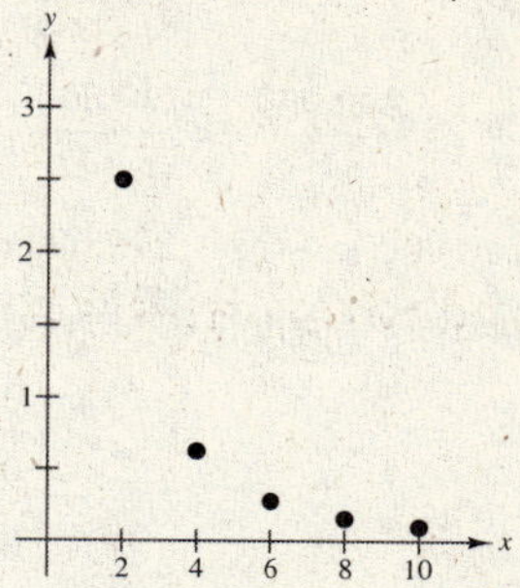

41.

x	10	20	30	40	50
y	$\dfrac{2}{5}$	$\dfrac{1}{5}$	$\dfrac{2}{15}$	$\dfrac{1}{10}$	$\dfrac{2}{25}$

$$\frac{2}{5} = \frac{k}{10} \qquad \frac{1}{5} = \frac{k}{20}$$
$$4 = k \qquad 4 = k$$

Using any two pairs of numbers, k is 4. So, $y = \dfrac{k}{x}$ with $k = 4$.

43. *Verbal Model*: $\boxed{\text{Your time}} = \boxed{\text{Your friend's time}}$

$$\frac{\boxed{\text{Your distance}}}{\boxed{\text{Your rate}}} = \frac{\boxed{\text{Friend's distance}}}{\boxed{\text{Friend's rate}}}$$

Labels: Your distance $= 10$

Your rate $= r$

Friend's distance $= 12$

Friend's rate $= r + 1.5$

Equation:
$$\frac{10}{r} = \frac{12}{r + 1.5}$$
$$10(r + 1.5) = 12r$$
$$10r + 15 = 12r$$
$$15 = 2r$$
$$\frac{15}{2} = r = 7.5 \text{ miles per hour}$$
$$r + 1.5 = 9 \text{ miles per hour}$$

45. *Verbal Model:* $\boxed{\text{Cost per person current group}} - \boxed{\text{Cost per person new group}} = \boxed{4000}$

Labels: Persons in current group $= x$

Persons in new group $= x + 2$

Equation:

$$\begin{aligned}
\frac{240{,}000}{x} - \frac{240{,}000}{x+2} &= 4000 \\
x(x+2)\left(\frac{240{,}000}{x} - \frac{240{,}000}{x+2}\right) &= (4000)x(x+2) \\
240{,}000(x+2) - 240{,}000x &= 4000\left(x^2 + 2x\right) \\
240{,}000x + 480{,}000 - 240{,}000x &= 4000x^2 + 8000x \\
0 &= 4000x^2 + 8000x - 480{,}000 \\
0 &= x^2 + 2x - 120 \\
0 &= (x+12)(x-10) \\
x + 12 &= 0 \qquad x - 10 = 0 \\
x &= -12 \qquad x = 10 \text{ people}
\end{aligned}$$

47. *Verbal Model:* $\boxed{\begin{matrix}\text{Riding}\\ \text{mower}\\ \text{rate}\end{matrix}} + \boxed{\begin{matrix}\text{Push}\\ \text{mower}\\ \text{rate}\end{matrix}} = \boxed{\begin{matrix}\text{Rate}\\ \text{together}\end{matrix}}$

Labels: Riding mower rate $= \dfrac{1}{60}$

Push mower rate $= \dfrac{1}{x}$

Rate together $= \dfrac{1}{45}$

Equation:

$$\begin{aligned}
\frac{1}{60} + \frac{1}{x} &= \frac{1}{45} \\
180x\left(\frac{1}{60} + \frac{1}{x}\right) &= \left(\frac{1}{45}\right)180x \\
3x + 180 &= 4x \\
180 &= x \text{ minutes}
\end{aligned}$$

49. *Verbal Model:* $\boxed{\text{Cost}} = \boxed{\dfrac{120{,}000p}{100 - p}}$

Equation:

$$\begin{aligned}
680{,}000 &= \frac{120{,}000p}{100 - p} \\
(100 - p)(680{,}000) &= \left(\frac{120{,}000p}{100 - p}\right)(100 - p) \\
68{,}000{,}000 - 680{,}000p &= 120{,}000p \\
68{,}000{,}000 &= 800{,}000p \\
85\% &= p
\end{aligned}$$

51. (a) Domain $= \{1, 2, 3, 4, \ldots\}$

(b)

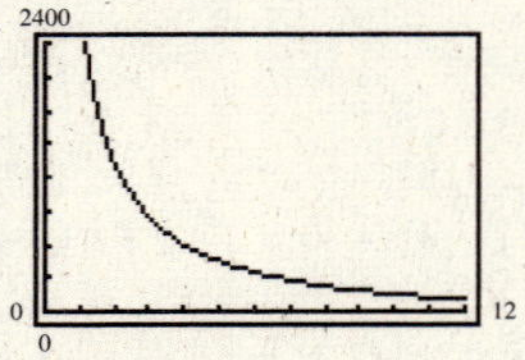

Keystrokes:

$\boxed{\text{Y=}}$ $\boxed{(-)}$ 139.1 $\boxed{+}$ 2921 $\boxed{\div}$ $\boxed{\text{X,T,}\theta}$ $\boxed{\text{GRAPH}}$

(c) $x \approx 10d$

(d)

$$\begin{aligned}
153 &= -139.1 + \frac{2921}{x} \\
292.1 &= \frac{2921}{x} \\
x &= \frac{2921}{292.1} \\
x &= 10
\end{aligned}$$

53.

$$\begin{aligned}
R &= kx \\
4825 &= k(500) \\
\frac{4825}{500} &= k \\
9.65 &= k \\
R &= 9.65x \\
R &= 9.65(620) \\
R &= \$5983
\end{aligned}$$

K is the price per unit.

55. (a)
$$\begin{aligned} d &= kF \\ 5 &= k(50) \\ \tfrac{5}{50} &= k \\ \tfrac{1}{10} &= k \end{aligned}$$
$$\begin{aligned} d &= \tfrac{1}{10}F \\ d &= \tfrac{1}{10}(20) \\ d &= 2 \text{ inches} \end{aligned}$$

(b)
$$\begin{aligned} d &= \tfrac{1}{10}F \\ 1.5 &= \tfrac{1}{10}F \\ 15 \text{ pounds} &= F \end{aligned}$$

57.
$$\begin{aligned} d &= kF \\ 7 &= k(10.5) \\ \tfrac{7}{10.5} &= k \\ \tfrac{70}{105} &= k \\ \tfrac{2}{3} &= k \\ 12 &= \tfrac{2}{3}F \\ 18 \text{ pounds} &= F \end{aligned}$$

59.
$$\begin{aligned} v &= kt \\ -64 &= k(2) \\ -32 &= k \\ v &= -32t \\ v &= -32(4) \\ v &= -128 \text{ feet per second} \end{aligned}$$

61.
$$\begin{aligned} d &= ks^2 \\ 75 &= k(30)^2 \\ 75 &= k(900) \\ \tfrac{75}{900} &= k \\ \tfrac{1}{12} &= k \\ d &= \tfrac{1}{12}s^2 \\ d &= \tfrac{1}{12}(48)^2 \\ d &= \tfrac{1}{12}(2304) \\ d &= 192 \text{ feet} \end{aligned}$$

63.
$$\begin{aligned} P &= kw^3 \\ 400 &= k(20)^3 \\ 400 &= k(8000) \\ \frac{400}{8000} &= k \\ \frac{1}{20} &= k \\ P &= \frac{1}{20}w^3 \\ &= \frac{1}{20}(30)^3 \\ &= \frac{1}{20}(27{,}000) \\ &= 1350 \text{ watts} \end{aligned}$$

65.
$$\begin{aligned} x &= \frac{k}{p} \\ 800 &= \frac{k}{5} \\ 4000 &= k \\ x &= \frac{4000}{6} \\ x &= 666.\overline{6} \approx 667 \text{ boxes} \end{aligned}$$

67.
$$\begin{aligned} p &= \frac{k}{t} \\ 38 &= \frac{k}{3} \\ 114 &= k \\ \text{So, } p &= \frac{114}{t}. \\ p &= \frac{114}{6.5} \\ p &= 17.5\% \end{aligned}$$

69.
$$\begin{aligned} D &= \frac{ka}{p} \\ 2000 &= \frac{k(500)}{5} \\ \frac{5(2000)}{500} &= k \\ 20 &= k \\ 2000 &= \frac{20(600)}{p} \\ p &= \frac{20(600)}{2000} \\ p &= \$6 \end{aligned}$$

71. $i = kpt$

$10 = k(600)\left(\frac{1}{8}\right)$

$\frac{10}{200} = k$

$\frac{1}{20} = k$

$i = \frac{1}{20}(900)\left(\frac{1}{2}\right)$

$i = \frac{900}{40}$

$i = \$22.50$

73. (a) $P = \dfrac{kWD^2}{L}$

(b) Unchanged

(c) Increases by a factor of 8.

(d) Increases by a factor of 4.

(e) Increases by a factor of $\dfrac{1}{4}$.

(f) $2000 = \dfrac{k(3)8^2}{120}$

$2000 = \dfrac{k(192)}{120}$

$1250 = k$

$L = \dfrac{1250(3)10^2}{120}$

$L = 3125$ pounds

75. In a joint variation problem where z varied jointly as x and y, if x increases, then z and y do not both necessarily increase.

77. $y = \dfrac{k}{x^2}$

$y = \dfrac{k}{(2x)^2}$

$y = \dfrac{k}{4x^2}$

y will be $\dfrac{1}{4}$ as great.

79. $(6)(6)(6)(6) = 6^4$

81. $\left(\frac{1}{5}\right)\left(\frac{1}{5}\right)\left(\frac{1}{5}\right)\left(\frac{1}{5}\right)\left(\frac{1}{5}\right) = \left(\frac{1}{5}\right)^5$

83.

$$\begin{array}{r|rrr} -2 & 1 & -5 & -14 \\ & & -2 & 14 \\ \hline & 1 & -7 & 0 \end{array}$$

$(x^2 - 5x - 14) \div (x + 2) = x - 7,\ x \neq -2$

85.

$$\begin{array}{r|rrr} 3 & 4 & -14 & 6 \\ & & 12 & -6 \\ \hline & 4 & -2 & 0 \end{array}$$

$\dfrac{4x^5 - 14x^4 + 6x^3}{x - 3} = 4x^4 - 2x^3,\ x \neq 3$

Review Exercises for Chapter 6

1. $y - 8 \neq 0$

$y \neq 8$

Domain: $(-\infty, 8) \cup (8, \infty)$

3. $f(x) = \dfrac{2x}{x^2 + 1}$

$x^2 + 1 \neq 0$

Domain: $(-\infty, \infty)$

5. $u^2 - 7u + 6 \neq 0$

$(u - 6)(u - 1) \neq 0$

$u \neq 6, \quad u \neq 1$

Domain: $(-\infty, 1) \cup (1, 6) \cup (6, \infty)$

7. Domain: $(0, \infty]$

$P = 2\left(w + \dfrac{36}{w}\right), \quad w \neq 0$

9. $\dfrac{6x^4y^2}{15xy^2} = \dfrac{2 \cdot 3x \cdot x^3 \cdot y^2}{5 \cdot 3x \cdot y^2}$

$= \dfrac{2x^3}{5}, \quad x \neq 0,\ y \neq 0$

11. $\dfrac{5b - 15}{30b - 120} = \dfrac{5(b - 3)}{30(b - 4)}$

$= \dfrac{5(b - 3)}{5 \cdot 6(b - 4)}$

$= \dfrac{b - 3}{6(b - 4)}$

13. $\dfrac{9x - 9y}{y - x} = \dfrac{9(x - y)}{-1(x - y)}$

$= -9, \quad x \neq y$

15. $\dfrac{x^2 - 5x}{2x^2 - 50} = \dfrac{x(x - 5)}{2(x^2 - 25)}$

$= \dfrac{x(x - 5)}{2(x - 5)(x + 5)}$

$= \dfrac{x}{2(x + 5)}, \quad x \neq 5$

17. $\dfrac{\text{Area of shaded region}}{\text{Area of whole figure}} = \dfrac{x(x + 2)}{(x + 2)2x}$

$= \dfrac{x}{2x} = \dfrac{1}{2}, \quad x > 0$

19. $\dfrac{4}{x} \cdot \dfrac{x^2}{12} = \dfrac{4x^2}{4 \cdot 3x} = \dfrac{x}{3}, \quad x \neq 0$

21. $\dfrac{7}{8} \cdot \dfrac{2x}{y} \cdot \dfrac{y^2}{14x^2} = \dfrac{7 \cdot 2 \cdot x \cdot y \cdot y}{2 \cdot 2 \cdot 2 \cdot y \cdot 7 \cdot 2 \cdot x \cdot x}$

$= \dfrac{y}{8x}, \quad y \neq 0$

23. $\dfrac{60z}{z + 6} \cdot \dfrac{z^2 - 36}{5} = \dfrac{5 \cdot 12z(z - 6)(z + 6)}{(z + 6)5}$

$= 12z(z - 6), \quad z \neq -6$

25. $\dfrac{u}{u - 3} \cdot \dfrac{3u - u^2}{4u^2} = \dfrac{u}{u - 3} \cdot \dfrac{-u(u - 3)}{4u^2} = -\dfrac{1}{4}, \quad u \neq 0, u \neq 3$

27. $24x^4 \div \dfrac{6x}{5} = 24x^4 \cdot \dfrac{5}{6x} = \dfrac{6 \cdot 4 \cdot 5 \cdot x^4}{6x} = 20x^3, \quad x \neq 0$

29. $25y^2 \div \dfrac{xy}{5} = 25y \cdot y \cdot \dfrac{5}{xy} = \dfrac{125y}{x}, \quad y \neq 0$

31. $\dfrac{x^2 + 3x + 2}{3x^2 + x - 2} \div (x + 2) = \dfrac{(x + 2)(x + 1)}{(3x - 2)(x + 1)} \cdot \dfrac{1}{(x + 2)} = \dfrac{(x + 2)(x + 1)}{(3x - 2)(x + 1)(x + 2)} = \dfrac{1}{3x - 2}, \quad x \neq -2, x \neq -1$

33. $\dfrac{x^2 - 7x}{x + 1} \div \dfrac{x^2 - 14x + 49}{x^2 - 1} = \dfrac{x(x - 7)}{x + 1} \cdot \dfrac{(x - 1)(x + 1)}{(x - 7)(x - 7)} = \dfrac{x(x - 1)}{x - 7}, \quad x \neq -1, x \neq 1$

35. $\dfrac{4x}{5} + \dfrac{11x}{5} = \dfrac{15x}{5} = 3x$

37. $\dfrac{15}{3x} - \dfrac{3}{3x} = \dfrac{12}{3x} = \dfrac{4}{x}$

39. $\dfrac{8 - x}{4x} + \dfrac{5}{4x} = \dfrac{8 - x + 5}{4x} = \dfrac{13 - x}{4x} = -\dfrac{x - 13}{4x}$

41. $\dfrac{2(3y + 4)}{2y + 1} + \dfrac{3 - y}{2y + 1} = \dfrac{2(3y + 4) + 3 - y}{2y + 1}$

$= \dfrac{6y + 8 + 3 - y}{2y + 1}$

$= \dfrac{5y + 11}{2y + 1}$

43. $\dfrac{4x}{x + 2} + \dfrac{3x - 7}{x + 2} - \dfrac{9}{x + 2} = \dfrac{4x + 3x - 7 - 9}{x + 2}$

$= \dfrac{7x - 16}{x + 2}$

45. $\dfrac{3}{5x^2} + \dfrac{4}{10x} = \dfrac{6}{10x^2} + \dfrac{4x}{10x^2} = \dfrac{6 + 4x}{10x^2}$

$= \dfrac{2(3 + 2x)}{10x^2} = \dfrac{2x + 3}{5x^2}$

47. $\dfrac{1}{x + 5} + \dfrac{3}{x - 12} = \dfrac{1}{x + 5}\left(\dfrac{x - 12}{x - 12}\right) + \dfrac{3}{x - 12}\left(\dfrac{x + 5}{x + 5}\right)$

$= \dfrac{x - 12}{(x + 5)(x - 12)} + \dfrac{3(x + 5)}{(x - 12)(x + 5)}$

$= \dfrac{x - 12 + 3x + 15}{(x + 5)(x - 12)}$

$= \dfrac{4x + 3}{(x + 5)(x - 12)}$

49. $$5x + \frac{2}{x-3} - \frac{3}{x+2} = \frac{5x(x-3)(x+2)}{(x-3)(x+2)} + \frac{2}{(x-3)}\left(\frac{x+2}{x+2}\right) - \frac{3}{(x+2)}\left(\frac{x-3}{x-3}\right)$$
$$= \frac{5x^3 - 5x^2 - 30x + 2x + 4 - 3x + 9}{(x-3)(x+2)}$$
$$= \frac{5x^3 - 5x^2 - 31x + 13}{(x-3)(x+2)}$$

51. $$\frac{6}{x-5} - \frac{4x+7}{x^2-x-20} = \frac{6}{x-5} - \frac{4x+7}{(x-5)(x+4)}$$
$$= \frac{6(x+4)}{(x-5)(x+4)} - \frac{4x+7}{(x-5)(x+4)}$$
$$= \frac{6x + 24 - 4x - 7}{(x-5)(x+4)}$$
$$= \frac{2x+17}{(x-5)(x+4)}$$

53. $$\frac{5}{x+3} - \frac{4x}{(x+3)^2} - \frac{1}{x-3} = \frac{5}{x+3}\left(\frac{(x+3)(x-3)}{(x+3)(x-3)}\right) - \frac{4x}{(x+3)^2}\left(\frac{x-3}{x-3}\right) - \frac{1}{x-3}\left(\frac{(x+3)^2}{(x+3)^2}\right)$$
$$= \frac{5x^2 - 45 - 4x^2 + 12x - x^2 - 6x - 9}{(x+3)^2(x-3)}$$
$$= \frac{6x - 54}{(x+3)^2(x-3)}$$

55. *Keystrokes:*

y_1: [Y=] 1 [÷] [X,T,θ] [−] 3 [÷] [(] [X,T,θ] [+] 3 [)] [ENTER]

y_2: [(] 3 [−] 2 [X,T,θ] [)] [÷] [(] [X,T,θ] [(] [X,T,θ] [+] 3 [)] [)] [GRAPH]

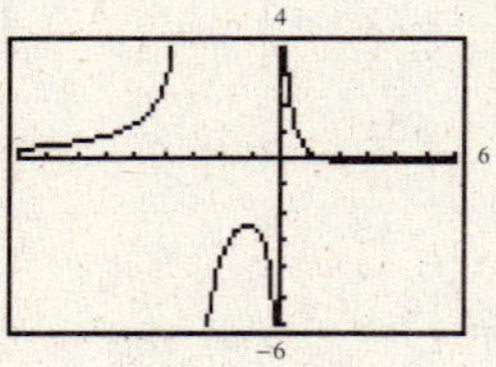

$$\frac{1}{x} - \frac{3}{x+3} = \frac{x+3}{x(x+3)} - \frac{3x}{x(x+3)} = \frac{x+3-3x}{x(x+3)} = \frac{3-2x}{x(x+3)}$$

57. $$\frac{\left(\frac{6}{x}\right)}{\left(\frac{2}{x^3}\right)} = \frac{6}{x} \div \frac{2}{x^3} = \frac{3 \cdot 2}{x} \cdot \frac{x \cdot x^2}{2} = 3x^2, \quad x \neq 0$$

59. $$\frac{\left(\frac{x}{x-2}\right)}{\left(\frac{2x}{2-x}\right)} = \frac{x}{x-2} \cdot \frac{2-x}{2x} = \frac{x(-1)(x-2)}{(x-2)(2x)}$$
$$= -\frac{1}{2}, \quad x \neq 0, \quad x \neq 2$$

61. $$\frac{\left(\frac{6x^2}{x^2+2x-35}\right)}{\left(\frac{x^3}{x^2-25}\right)} = \frac{\frac{6x^2}{(x+7)(x-5)}}{\frac{x^3}{(x-5)(x+5)}} \cdot \frac{(x+7)(x-5)(x+5)}{(x+7)(x-5)(x+5)} = \frac{6x^2(x+5)}{x^3(x+7)} = \frac{6(x+5)}{x(x+7)}, \quad x \neq 5, x \neq -5$$

63. $$\frac{3t}{\left(5 - \frac{2}{t}\right)} \cdot \frac{t}{t} = \frac{3t^2}{5t-2}, t \neq 0$$

65. $$\frac{\left(x - 3 + \frac{2}{x}\right)}{\left(1 - \frac{2}{x}\right)} = \frac{\left(x - 3 + \frac{2}{x}\right)}{\left(1 - \frac{2}{x}\right)} \cdot \frac{x}{x}$$
$$= \frac{x^2 - 3x + 2}{x-2}, \quad x \neq 0$$
$$= \frac{(x-2)(x-1)}{x-2}, \quad x \neq 0$$
$$= x - 1, \quad x \neq 0, \quad x \neq 2$$

67. $$\frac{\left(\dfrac{1}{a^2-16}-\dfrac{1}{a}\right)}{\left(\dfrac{1}{a^2+4a}+4\right)}\cdot\frac{a(a-4)(a+4)}{a(a-4)(a+4)}=\frac{a-(a-4)(a+4)}{a-4+4a(a-4)(a+4)}$$

$$=\frac{a-(a^2-16)}{a-4+4a(a^2-16)}$$

$$=\frac{a-a^2+16}{a-4+4a^3-64a}$$

$$=\frac{-a^2+a+16}{4a^3-63a-4},\ a\neq 0,\ a\neq -4$$

69. $$(4x^3-x)\div 2x=\frac{4x^3-x}{2x}$$

$$=\frac{4x^3}{2x}-\frac{x}{2x}$$

$$=2x^2-\frac{1}{2},\ x\neq 0$$

71. $$\frac{3x^3y^2-x^2y^2+x^2y}{x^2y}=\frac{3x^3y^2}{x^2y}-\frac{x^2y^2}{x^2y}+\frac{x^2y}{x^2y}$$

$$=3xy-y+1,\quad x\neq 0,\ y\neq 0$$

73.
$$\begin{array}{r|l}
 & 2x^2+\frac{4}{3}x-\frac{8}{9}+\frac{10}{9(3x-1)} \\ \hline
3x-1 & 6x^3+2x^2-4x+2 \\
 & \underline{6x^3-2x^2} \\
 & 4x^2-4x \\
 & \underline{4x^2-\frac{4}{3}x} \\
 & -\frac{8}{3}x+2 \\
 & \underline{-\frac{8}{3}x+\frac{8}{9}} \\
 & \frac{10}{9}
\end{array}$$

75.
$$\begin{array}{r|l}
 & x^2-2,\ x\neq\pm 1 \\ \hline
x^2-1 & x^4+0x^3-3x^2+2 \\
 & \underline{x^4-x^2} \\
 & -2x^2+2 \\
 & \underline{-2x^2+2}
\end{array}$$

77.
$$\begin{array}{r|l}
 & x^2-x-3+\frac{-3x^2+2x+3}{x^3-2x^2+x-1} \\ \hline
x^3-2x^2+x-1 & x^5-3x^4+0x^3+x^2+0x+6 \\
 & \underline{x^5-2x^4+x^3-x^2} \\
 & -x^4-x^3+2x^2+0x \\
 & \underline{-x^4+2x^3-x^2+x} \\
 & -3x^3+3x^2-x+6 \\
 & \underline{-3x^3+6x^2-3x+3} \\
 & -3x^2+2x+3
\end{array}$$

79.
$$\begin{array}{r|rrr}
-1 & 1 & 3 & 5 \\
 & & -1 & -2 \\ \hline
 & 1 & 2 & 3
\end{array}$$

$$\frac{x^2+3x+5}{x+1}=x+2+\frac{3}{x+1}$$

81.
$$\begin{array}{r|rrrr}
-2 & 1 & 7 & 3 & -14 \\
 & & -2 & -10 & 14 \\ \hline
 & 1 & 5 & -7 & 0
\end{array}$$

$$\frac{x^3-7x^2+3x-14}{x+2}=x^2+5x-7,\ x\neq -2$$

83.
$$\begin{array}{r|rrrrr}
3 & 1 & 0 & -3 & 0 & -25 \\
 & & 3 & 9 & 18 & 54 \\ \hline
 & 1 & 3 & 6 & 18 & 29
\end{array}$$

$$(x^4-3x^2-25)\div(x-3)=x^3+3x^2+6x+18+\frac{29}{x-3}$$

85.

$$\begin{array}{r|rrrr} 2 & 1 & 2 & -5 & -6 \\ & & 2 & 8 & 6 \\ \hline & 1 & 4 & 3 & 0 \end{array}$$

$x^3 + 2x^2 - 5x - 6 = (x - 2)(x + 1)(x + 3)$

$x^2 + 4x + 3$

$(x + 3)(x + 1)$

87.

$$\frac{x}{15} + \frac{3}{5} = 1$$
$$15\left(\frac{x}{15} + \frac{3}{5}\right) = (1)15$$
$$x + 9 = 15$$
$$x = 6$$

89.

$$\frac{3x}{8} = -15 + \frac{x}{4}$$
$$8\left(\frac{3}{8}x\right) = (-15)8 + \left(\frac{x}{4}\right)8$$
$$3x = -120 + 2x$$
$$x = -120$$

Check: $\frac{3(-120)}{8} \stackrel{?}{=} -15 + \frac{-120}{4}$

$$\frac{-360}{8} \stackrel{?}{=} -15 + -30$$
$$-45 = -45$$

91.

$$\frac{x^2}{6} - \frac{x}{12} = \frac{1}{2}$$
$$12\left(\frac{x^2}{6} - \frac{x}{12}\right) = \left(\frac{1}{2}\right)12$$
$$2x^2 - x = 6$$
$$2x^2 - x - 6 = 0$$
$$(2x + 3)(x - 2) = 0$$

$2x + 3 = 0 \qquad x - 2 = 0$

$x = -\frac{3}{2} \qquad x = 2$

93. $(3t)\left(8 - \frac{12}{t}\right) = \frac{1}{3}(3t)$

$$24t - 36 = t$$
$$23t = 36$$
$$t = \frac{36}{23}$$

Check: $8 - \frac{12}{\left(\frac{36}{23}\right)} \stackrel{?}{=} \frac{1}{3}$

$$8 - \frac{23}{3} \stackrel{?}{=} \frac{1}{3}$$
$$\frac{24}{3} - \frac{23}{3} \stackrel{?}{=} \frac{1}{3}$$
$$\frac{1}{3} = \frac{1}{3}$$

95.

$$\frac{2}{y} - \frac{1}{3y} = \frac{1}{3}$$
$$3y\left(\frac{2}{y} - \frac{1}{3y}\right) = \left(\frac{1}{3}\right)3y$$
$$6 - 1 = y$$
$$5 = y$$

Check: $\frac{2}{5} - \frac{1}{3(5)} \stackrel{?}{=} \frac{1}{3}$

$$\frac{2}{5} - \frac{1}{15} \stackrel{?}{=} \frac{1}{3}$$
$$\frac{6}{15} - \frac{1}{15} \stackrel{?}{=} \frac{1}{3}$$
$$\frac{5}{15} \stackrel{?}{=} \frac{1}{3}$$
$$\frac{1}{3} = \frac{1}{3}$$

97.

$$r = 2 + \frac{24}{r}$$
$$r(r) = \left(2 + \frac{24}{r}\right)r$$
$$r^2 = 2r + 24$$
$$r^2 - 2r - 24 = 0$$
$$(r - 6)(r + 4) = 0$$
$$r = 6, r = -4$$

Check: $6 \stackrel{?}{=} 2 + \frac{24}{6}$

$$6 \stackrel{?}{=} 2 + 4$$
$$6 = 6$$

Check: $-4 \stackrel{?}{=} 2 + \frac{24}{-4}$

$$-4 \stackrel{?}{=} 2 - 6$$
$$-4 = -4$$

99.
$$
\begin{aligned}
8\left(\frac{6}{x} - \frac{1}{x+5}\right) &= 15\\
\left(\frac{6}{x} - \frac{1}{x+5}\right) &= \frac{15}{8}\\
8x(x+5)\left(\frac{6}{x} - \frac{1}{x+5}\right) &= \left(\frac{15}{8}\right)8x(x+5)\\
48(x+5) - 8x &= 15x(x+5)\\
48x + 240 - 8x &= 15x^2 + 75x\\
240 + 40x &= 15x^2 + 75x\\
0 &= 15x^2 + 35x - 240\\
0 &= 5\left(3x^2 + 7x - 48\right)\\
0 &= 5(3x+16)(x-3)
\end{aligned}
$$

$$3x + 16 = 0 \qquad x - 3 = 0$$
$$3x = -16 \qquad x = 3$$
$$x = -\frac{16}{3}$$

Check:
$$
\begin{aligned}
8\left(\frac{6}{-\frac{16}{3}} - \frac{1}{-\frac{16}{3} + 5}\right) &\stackrel{?}{=} 15\\
8\left(-\frac{18}{16} - \frac{1}{-\frac{16}{3} + \frac{15}{3}}\right) &\stackrel{?}{=} 15\\
8\left(-\frac{9}{8} - \frac{1}{-\frac{1}{3}}\right) &\stackrel{?}{=} 15\\
8\left(-\frac{9}{8} + 3\right) &\stackrel{?}{=} 15\\
8\left(-\frac{9}{8} + \frac{24}{8}\right) &\stackrel{?}{=} 15\\
8\left(\frac{15}{8}\right) &\stackrel{?}{=} 15\\
15 &= 15
\end{aligned}
$$

Check:
$$
\begin{aligned}
8\left(\frac{6}{3} - \frac{1}{3+5}\right) &\stackrel{?}{=} 15\\
8\left(2 - \frac{1}{8}\right) &\stackrel{?}{=} 15\\
8\left(\frac{16}{8} - \frac{1}{8}\right) &\stackrel{?}{=} 15\\
8\left(\frac{15}{8}\right) &\stackrel{?}{=} 15\\
15 &= 15
\end{aligned}
$$

101.
$$
\begin{aligned}
\frac{3}{y+1} - \frac{8}{y} &= 1\\
y(y+1)\left(\frac{3}{y+1} - \frac{8}{y}\right) &= (1)y(y+1)\\
3y - 8(y+1) &= y(y+1)\\
3y - 8y - 8 &= y^2 + y\\
-5y - 8 &= y^2 + y\\
0 &= y^2 + 6y + 8\\
0 &= (y+4)(y+2)
\end{aligned}
$$

$$y + 4 = 0 \qquad y + 2 = 0$$
$$y = -4 \qquad y = -2$$

103.
$$
\begin{aligned}
\frac{2x}{x-3} - \frac{3}{x} &= 0\\
x(x-3)\left(\frac{2x}{x-3} - \frac{3}{x}\right) &= (0)x(x-3)\\
2x(x) - 3(x-3) &= 0\\
2x^2 - 3x + 9 &= 0
\end{aligned}
$$

No real solution

105.

$$\frac{12}{x^2 + x - 12} - \frac{1}{x - 3} = -1$$

$$(x - 3)(x + 4)\left(\frac{12}{x^2 + x - 12} - \frac{1}{x - 3}\right) = (-1)(x - 3)(x + 4)$$

$$12 - (x + 4) = -1(x^2 + x - 12)$$

$$12 - x - 4 = -x^2 - x + 12$$

$$(x^2 - 4) = 0$$

$$(x - 2)(x + 2) = 0$$

$$x - 2 = 0 \qquad x + 2 = 0$$

$$x = 2 \qquad x = -2$$

Check: $\frac{12}{2^2 + 2 - 12} - \frac{1}{2 - 3} \stackrel{?}{=} -1$

$$\frac{12}{-6} - \frac{1}{-1} \stackrel{?}{=} -1$$

$$-2 + 1 \stackrel{?}{=} -1$$

$$-1 = -1$$

Check: $\frac{12}{(-2)^2 + (-2) - 12} - \frac{1}{(-2) - 3} \stackrel{?}{=} -1$

$$\frac{12}{-10} - \frac{1}{-5} \stackrel{?}{=} -1$$

$$-\frac{6}{5} + \frac{1}{5} \stackrel{?}{=} -1$$

$$-1 = -1$$

107.

$$\frac{5}{x^2 - 4} - \frac{6}{x - 2} = -5$$

$$(x - 2)(x + 2)\left(\frac{5}{x^2 - 4} - \frac{6}{x - 2}\right) = (-5)(x - 2)(x + 2)$$

$$5 - 6(x + 2) = -5(x^2 - 4)$$

$$5 - 6x - 12 = -5x^2 + 20$$

$$5x^2 - 6x - 27 = 0$$

$$(5x + 9)(x - 3) = 0$$

$$5x + 9 = 0 \qquad x - 3 = 0$$

$$x = -\frac{9}{5} \qquad x = 3$$

Check: $\frac{5}{\left(-\frac{9}{5}\right)^2 - 4} - \frac{6}{-\frac{9}{5} - 2} \stackrel{?}{=} -5$

$$\frac{5}{-\frac{19}{25}} - \frac{6}{-\frac{19}{5}} \stackrel{?}{=} -5$$

$$-\frac{125}{19} + \frac{30}{19} \stackrel{?}{=} -5$$

$$-\frac{95}{19} \stackrel{?}{=} -5$$

$$-5 = -5$$

Check: $\frac{5}{3^2 - 4} - \frac{6}{3 - 2} \stackrel{?}{=} -5$

$$\frac{5}{5} - \frac{6}{1} \stackrel{?}{=} -5$$

$$-5 = -5$$

109. *Verbal Model*: $\boxed{\text{Your time}} = \boxed{\text{Your friend's time}}$

$$\boxed{\frac{\text{Your distance}}{\text{Your rate}}} = \boxed{\frac{\text{Friend's distance}}{\text{Friend's rate}}}$$

Labels: Your distance $= 24$

Your rate $= r$

Friend's distance $= 15$

Friend's rate $= r - 6$

Equation:

$$\frac{24}{r} = \frac{15}{r-6}$$

$$r(r-6)\left(\frac{24}{r}\right) = \left(\frac{15}{r-6}\right)r(r-6)$$

$$24(r-6) = 15r$$

$$24r - 144 = 15r$$

$$9r = 144$$

$$r = 16 \text{ miles per hour} \qquad r - 6 = 10 \text{ miles per hour}$$

111. *Verbal Model:* $\boxed{\text{Share per person now}} = \boxed{\text{Share per person later}} + 5000$

Labels: People presently in group $= x$

People in new group $= x + 2$

Equation:

$$\frac{60{,}000}{x} = \frac{60{,}000}{x+2} + 5000$$

$$x(x+2)\left(\frac{60{,}000}{x}\right) = \left(\frac{60{,}000}{x+2} + 5000\right)x(x+2)$$

$$60{,}000(x+2) = 60{,}000x + 5000x(x+2)$$

$$60{,}000x + 120{,}000 = 60{,}000x + 5000x^2 + 10{,}000x$$

$$0 = 5000x^2 + 10{,}000x - 120{,}000$$

$$0 = x^2 + 2x - 24$$

$$0 = (x+6)(x-4)$$

$$x = -6, \quad x = 4 \text{ people}$$

113.

$$400 = \frac{20(4 + 3t)}{1 + 0.05t}$$

$$400 + 20t = 80 + 60t$$

$$320 = 40t$$

$$8 \text{ years} = t$$

115.

$$d = kF$$

$$4 = k(100)$$

$$\frac{4}{100} = k$$

$$\frac{1}{25} = k$$

$$6 = \frac{1}{25}F$$

$$150 \text{ pounds} = F$$

117.

$$t = \frac{k}{r}$$

$$3 = \frac{k}{65}$$

$$3(65) = k$$

$$195 = k$$

$$t = \frac{195}{80}$$

$$t \approx 2.44 \text{ hours}$$

119. $D = \dfrac{ka}{p^2}$

$900 = \dfrac{k(20{,}000)}{55^2}$

$\dfrac{900(55^2)}{20{,}000} = k$

$136.125 = k$

$900 = \dfrac{136.125(25{,}000)}{p^2}$

$p^2 = \dfrac{136.125(25{,}000)}{900}$

$p^2 = 3781.25$

$p = \$61.49$

121. $i = ktp$

$675 = k\left(\dfrac{9}{12}\right)(12{,}000)$

$675 = k(9{,}000)$

$\dfrac{675}{9000} = k$

$0.075 = k$

$i = 0.075\left(\dfrac{18}{12}\right)(8200)$

$i = \$922.50$

Chapter Test for Chapter 6

1. $f(x) = \dfrac{x+1}{x^2 - 6x + 5}$

$x^2 - 6x + 5 \neq 0$

$(x-1)(x-5) \neq 0$

$x - 1 \neq 0 \quad x - 5 \neq 0$

$x \neq 1 \quad x \neq 5$

Domain: $(-\infty, 1) \cup (1, 5) \cup (5, \infty)$

2. $\dfrac{4-2x}{x-2} = \dfrac{-2(x-2)}{x-2} = -2, \quad x \neq 2$

3. $\dfrac{2a^2 - 5a - 12}{5a - 20} = \dfrac{(2a+3)(a-4)}{5(a-4)}$

$= \dfrac{2a+3}{5}, \quad a \neq 4$

4. Least common multiple of x^2, $3x^3$, and $(x+4)^2$:

$3x^3(x+4)^2$

5. $\dfrac{4z^3}{5} \cdot \dfrac{25}{12z^2} = \dfrac{4 \cdot z^2 \cdot z \cdot 5 \cdot 5}{5 \cdot 4 \cdot 3 \cdot z^2}$

$= \dfrac{5z}{3}, \quad z \neq 0$

6. $\dfrac{y^2 + 8y + 16}{2(y-2)} \cdot \dfrac{8y - 16}{(y+4)^3} = \dfrac{(y+4)^2 \cdot 8(y-2)}{2(y-2)(y+4)^2(y+4)}$

$= \dfrac{4}{y+4}, \quad y \neq 2$

7. $\dfrac{(2xy^2)^3}{15} \div \dfrac{12x^3}{21} = \dfrac{(2xy^2)^3}{15} \cdot \dfrac{21}{12x^3}$

$= \dfrac{8x^3y^6 \cdot 7 \cdot 3}{5 \cdot 3 \cdot 4 \cdot 3x^3}$

$= \dfrac{14y^6}{15}, \quad x \neq 0$

8. $(4x^2 - 9) \div \dfrac{2x+3}{2x^2 - x - 3} = (2x-3)(2x+3) \cdot \dfrac{(2x-3)(x+1)}{2x+3}$

$= (2x-3)^2(x+1), \quad x \neq -\dfrac{3}{2}, \quad x \neq -1$

9. $\dfrac{3}{x-3} + \dfrac{x-2}{x-3} = \dfrac{3 + x - 2}{x-3} = \dfrac{x+1}{x-3}$

10. $2x + \dfrac{1-4x^2}{x+1} = 2x\left(\dfrac{x+1}{x+1}\right) + \dfrac{1-4x^2}{x+1} = \dfrac{2x^2 + 2x}{x+1} + \dfrac{1-4x^2}{x+1} = \dfrac{-2x^2 + 2x + 1}{x+1}$

11. $\dfrac{5x}{x+2} - \dfrac{2}{x^2 - x - 6} = \dfrac{5x}{x+2} - \dfrac{2}{(x-3)(x+2)} = \dfrac{5x}{x+2}\left(\dfrac{x-3}{x-3}\right) - \dfrac{2}{(x-3)(x+2)} = \dfrac{5x^2 - 15x - 2}{(x+2)(x-3)}$

12. $$\frac{3}{x}-\frac{5}{x^2}+\frac{2x}{x^2+2x+1}=\frac{3}{x}-\frac{5}{x^2}+\frac{2x}{(x+1)^2}$$
$$=\frac{3}{x}\left[\frac{x(x+1)^2}{x(x+1)^2}\right]-\frac{5}{x^2}\left[\frac{(x+1)^2}{(x+1)^2}\right]+\frac{2x}{(x+1)^2}\left(\frac{x^2}{x^2}\right)$$
$$=\frac{3x(x^2+2x+1)-5(x^2+2x+1)+2x^3}{x^2(x+1)^2}$$
$$=\frac{3x^3+6x^2+3x-5x^2-10x-5+2x^3}{x^2(x+1)^2}$$
$$=\frac{5x^3+x^2-7x-5}{x^2(x+1)^2}$$

13. $$\frac{\left(\frac{3x}{x+2}\right)}{\left(\frac{12}{x^3+2x^2}\right)}=\frac{3x}{x+2}\div\frac{12}{x^3+2x^2}=\frac{3x}{x+2}\cdot\frac{x^2(x+2)}{12}=\frac{x^3}{4},\quad x\neq 0,-2$$

14. $$\frac{\left(9x-\frac{1}{x}\right)}{\left(\frac{1}{x}-3\right)}=\frac{\left(9x-\frac{1}{x}\right)}{\left(\frac{1}{x}-3\right)}\cdot\frac{x}{x}$$
$$=\frac{9x(x)-\frac{1}{x}(x)}{\frac{1}{x}(x)-3(x)}$$
$$=\frac{9x^2-1}{1-3x}$$
$$=\frac{(3x-1)(3x+1)}{-1(-1+3x)}$$
$$=-(3x+1),\quad x\neq 0,\frac{1}{3}$$

15. $$\frac{3x^{-2}+y^{-1}}{(x+y)^{-1}}=\frac{\frac{3}{x^2}+\frac{1}{y}}{\frac{1}{(x+y)}}\cdot\frac{x^2y(x+y)}{x^2y(x+y)}$$
$$=\frac{3y(x+y)+x^2(x+y)}{x^2y}$$
$$=\frac{(3y+x^2)(x+y)}{x^2y},\quad x\neq -y$$

16. $$\frac{6x^2-4x+8}{2x}=\frac{6x^2}{2x}-\frac{4x}{2x}+\frac{8}{2x}=3x-2+\frac{4}{x}$$

17. $$\frac{t^4+t^2-6t}{t^2-2}=t^2+3-\frac{6t-6}{t^2-2}$$

$$\begin{array}{r} t^2 \qquad\quad + 3 \\ t^2-2\overline{)\,t^4+0t^3+t^2-6t+0} \\ \underline{t^4 \qquad\quad -2t^2 \qquad\qquad} \\ 3t^2-6t \qquad \\ \underline{3t^2 \qquad -6} \\ -6t+6 \end{array}$$

18. $$\frac{2x^4-15x^2-7}{x-3}$$

3	2	0	−15	0	−7
		6	18	9	27
	2	6	3	9	20

$$\frac{2x^4-15x^2-7}{x-3}=2x^3+6x^2+3x+9+\frac{20}{x-3}$$

19. $$\frac{3}{h+2}=\frac{1}{6}$$
$$6(h+2)\left(\frac{3}{h+2}\right)=\left(\frac{1}{6}\right)6(h+2)$$
$$18=h+2$$
$$16=h$$

20. $$\frac{2}{x+5} - \frac{3}{x+3} = \frac{1}{x}$$

$$2x(x+3) - 3x(x+5) = (x+5)(x+3)$$

$$2x^2 + 6x - 3x^2 - 15x = x^2 + 3x + 5x + 15$$

$$-2x^2 - 17x - 15 = 0$$

$$2x^2 + 17x + 15 = 0$$

$$(2x+15)(x+1) = 0$$

$$2x + 15 = 0 \qquad x + 1 = 0$$

$$2x = 15 \qquad x = -1$$

$$x = -\frac{15}{2}$$

Check: $$\frac{2}{-\frac{15}{2}+5} - \frac{3}{-\frac{15}{2}+3} \stackrel{?}{=} \frac{1}{-\frac{15}{2}}$$

$$-\frac{12}{15} + \frac{10}{15} \stackrel{?}{=} -\frac{2}{15}$$

$$-\frac{2}{15} = -\frac{2}{15}$$

Check: $$\frac{2}{-1+5} - \frac{3}{-1+3} \stackrel{?}{=} -\frac{1}{1}$$

$$\frac{2}{4} - \frac{3}{2} \stackrel{?}{=} 1$$

$$\frac{1}{2} - \frac{3}{2} \stackrel{?}{=} -1$$

$$-1 = -1$$

21. $$\frac{1}{x+1} + \frac{1}{x-1} = \frac{2}{x^2-1}$$

$$x - 1 + x + 1 = 2$$

$$2x = 2$$

$$x = 1$$

Check: $$\frac{1}{1+1} + \frac{1}{1-1} \neq \frac{2}{1-1}$$

Division by zero is undefined. Solution is extraneous, so equation has no solution.

22. $$v = k\sqrt{u}$$

$$\tfrac{3}{2} = k\sqrt{36}$$

$$\tfrac{3}{2} = k \cdot 6$$

$$\tfrac{1}{6} \cdot \tfrac{3}{2} = k$$

$$\tfrac{1}{4} = k$$

$$v = \tfrac{1}{4}\sqrt{u}$$

23. $$P = \frac{k}{V}$$

$$1 = \frac{k}{180}$$

$$180 = k$$

$$0.75 = \frac{180}{V}$$

$$V = \frac{180}{0.75}$$

$V = 240$ cubic meters

CHAPTER 7
Radicals and Complex Numbers

CHAPTER 7
Radicals and Complex Numbers

Section 7.1 Radicals and Rational Exponents

1. $\sqrt{64} = 8$ because $8 \cdot 8 = 64$.

3. $-\sqrt{49} = -7$ because $7 \cdot 7 = 49$.

5. $\sqrt[3]{-27} = -3$ because $-3 \cdot -3 \cdot -3 = -27$.

7. $\sqrt{-1}$ is not a real number because no real number multiplied by itself yields -1.

9. 49: Perfect square because $7^2 = 49$.

11. 1728: Perfect cube because $12^3 = 1728$.

13. 96: Neither because there is no integer that can be squared or cubed and yield 96.

15. The square roots of 25 are -5 and 5 because $(-5)(-5) = 25$ and $(5)(5) = 25$.

17. The square roots of $\frac{9}{16}$ are $-\frac{3}{4}$ and $\frac{3}{4}$ because $\left(-\frac{3}{4}\right)\left(-\frac{3}{4}\right) = \frac{9}{16}$ and $\left(\frac{3}{4}\right)\left(\frac{3}{4}\right) = \frac{9}{16}$.

19. The square roots of $\frac{1}{49}$ are $-\frac{1}{7}$ and $\frac{1}{7}$ because $\left(-\frac{1}{7}\right)\left(-\frac{1}{7}\right) = \frac{1}{49}$ and $\left(\frac{1}{7}\right)\left(\frac{1}{7}\right) = \frac{1}{49}$.

21. The square roots of 0.16 are -0.4 and 0.4 because $(-0.4)(-0.4) = 0.16$ and $(0.4)(0.4) = 0.16$.

23. The cube root of 8 is 2 because $(2)(2)(2) = 8$.

25. The cube root of $-\frac{1}{27}$ is $-\frac{1}{3}$ because $\left(-\frac{1}{3}\right)\left(-\frac{1}{3}\right)\left(-\frac{1}{3}\right) = -\frac{1}{27}$.

27. The cube root of $\frac{1}{1000}$ is $\frac{1}{10}$ because $\left(\frac{1}{10}\right)\left(\frac{1}{10}\right)\left(\frac{1}{10}\right) = \frac{1}{1000}$.

29. The cube root of 0.001 is 0.1 because $(0.1)(0.1)(0.1) = 0.001$.

31. $\sqrt{3}$ is an irrational number because 3 is not a perfect square.

33. $\sqrt{16}$ is a rational number because 16 is a perfect square.

35. $\sqrt{\frac{4}{25}}$ is a rational number because $\frac{4}{25}$ is a perfect square.

37. $\sqrt{\frac{3}{16}}$ is an irrational number because $\frac{3}{16}$ is not a perfect square.

39. $\sqrt{8^2} = |8| = 8$

(index is even)

41. $\sqrt{(-10)^2} = |-10| = 10$

(index is even)

43. $\sqrt{-9^2}$ is not a real number

(even root of a negative number)

45. $-\sqrt{\left(\frac{2}{3}\right)^2} = -\frac{2}{3}$

(index is even)

47. $\sqrt{-\left(\frac{3}{10}\right)^2}$ is not a real number

(even root of a negative number)

49. $\left(\sqrt{5}\right)^2 = 5$

(inverse property of powers and roots)

51. $-\left(\sqrt{23}\right)^2 = -23$

(inverse property of powers and roots)

55. $\sqrt[3]{10^3} = 10$

(index is odd)

57. $-\sqrt[3]{(-6)^3} = 6$

(index is odd)

59. $\sqrt[3]{\left(-\frac{1}{4}\right)^3} = -\frac{1}{4}$

(index is odd)

61. $\left(\sqrt[3]{11}\right)^3 = 11$

(inverse property of powers and roots)

63. $\left(-\sqrt[3]{24}\right)^3 = -24$

(inverse property of powers and roots)

65. $\sqrt[4]{3^4} = 3$
(inverse property of powers and roots)

67. $-\sqrt[4]{2^4} = -2$
(inverse property of powers and roots)

69. *Radical Form* — *Rational Exponent Form*
$\sqrt{36} = 6$ — $36^{1/2} = 6$

71. *Radical Form* — *Rational Exponent Form*
$\sqrt[4]{256^3} = 64$ — $256^{3/4} = 64$

73. $25^{1/2} = \sqrt{25} = 5$
Root is 2. Power is 1.

75. $-36^{1/2} = -\sqrt{36} = -6$
Root is 2. Power is 1.

77. $32^{-2/5} = \dfrac{1}{\left(\sqrt[5]{32}\right)^2} = \dfrac{1}{2^2} = \dfrac{1}{4}$
Root is 5. Power is 2.

79. $(-27)^{-2/3} = \dfrac{1}{(-27)^{2/3}} = \dfrac{1}{\left(\sqrt[3]{-27}\right)^2} = \dfrac{1}{9}$
Root is 3. Power is 2.

81. $\left(\frac{8}{27}\right)^{2/3} = \left(\sqrt[3]{\frac{8}{27}}\right)^2 = \left(\frac{2}{3}\right)^2 = \frac{4}{9}$
Root is 3. Power is 2.

83. $\left(\frac{121}{9}\right)^{-1/2} = \left(\frac{9}{121}\right)^{1/2} = \sqrt{\frac{9}{121}} = \frac{3}{11}$
Root is 2. Power is 1.

85. $\left(-3^3\right)^{2/3} = (-27)^{2/3} = \left(\sqrt[3]{-27}\right)^2 = (-3)^2 = 9$

87. $-\left(4^4\right)^{3/4} = -4^{4\cdot 3/4} = -4^3 = -64$
Root is 4. Power is 3.

89. $\sqrt{t} = t^{1/2}$
Root is 2. Power is 1.

91. $x\sqrt[3]{x^6} = x \cdot x^{6/3} = x \cdot x^2 = x^3$
Root is 3. Power is 6.

93. $u^2\sqrt[3]{u} = u^2 \cdot u^{1/3} = u^{2+1/3} = u^{7/3}$
Root is 3. Power is 7.

95. $\dfrac{\sqrt{x}}{\sqrt{x^3}} = \dfrac{x^{1/2}}{x^{3/2}} = x^{1/2-3/2} = x^{-1} = \dfrac{1}{x}$

97. $\dfrac{\sqrt[4]{t}}{\sqrt{t^5}} = \dfrac{t^{1/4}}{t^{5/2}} = t^{1/4-5/2} = t^{1/4-10/4} = t^{-9/4} = \dfrac{1}{t^{9/4}}$

99. $\sqrt[3]{x^2} \cdot \sqrt[3]{x^7} = x^{2/3} \cdot x^{7/3} = x^{2/3+7/3} = x^{9/3} = x^3$

101. $\sqrt[4]{y^3} \cdot \sqrt[3]{y} = y^{3/4} \cdot y^{1/3} = y^{3/4+1/3}$
$= y^{9/12+4/12} = y^{13/12}$

103. $\sqrt[4]{x^3 y} = \left(x^3 y\right)^{1/4} = x^{3/4}y^{1/4}$

105. $z^2\sqrt{y^5 z^4} = z^2 \cdot \left(y^5 z^4\right)^{1/2}$
$= z^2 y^{5/2} z^2$
$= z^{2+2} y^{5/2}$
$= z^4 y^{5/2}$

107. $3^{1/4} \cdot 3^{3/4} = 3^{1/4+3/4} = 3^{4/4} = 3^1$

109. $\left(2^{1/2}\right)^{2/3} = 2^{1/2\cdot 2/3} = 2^{1/3} = \sqrt[3]{2}$

111. $\dfrac{2^{1/5}}{2^{6/5}} = 2^{1/5-6/5} = 2^{-5/5} = 2^{-1} = \dfrac{1}{2}$

113. $\left(c^{3/2}\right)^{1/3} = c^{3/2\cdot 1/3} = c^{1/2} = \sqrt{c}$

115. $\dfrac{18y^{4/3}z^{-1/3}}{24y^{-2/3}z} = \dfrac{6\cdot 3y^{4/3-(-2/3)}z^{-1/3-1}}{6\cdot 4} = \dfrac{3y^{6/3}z^{-4/3}}{4} = \dfrac{3y^2}{4z^{4/3}}$

117. $\left(3x^{-1/3}y^{3/4}\right)^2 = 3^2x^{-2/3}y^{3/2} = \dfrac{9y^{3/2}}{x^{2/3}}$

119. $\left(\dfrac{x^{1/4}}{x^{1/6}}\right)^3 = \left(x^{1/4-1/6}\right)^3$
$= \left(x^{3/12-2/12}\right)^3$
$= \left(x^{1/12}\right)^3$
$= x^{3/12}$
$= x^{1/4}$

121. $\sqrt{\sqrt[4]{y}} = \left(y^{1/4}\right)^{1/2} = y^{1/4\cdot 1/2} = y^{1/8} = \sqrt[8]{y}$

123. $\sqrt[4]{\sqrt{x^3}} = \sqrt[4]{x^{3/2}} = \left(x^{3/2}\right)^{1/4} = x^{3/2\cdot 1/4} = x^{3/8}$

125. $\dfrac{(x+y)^{3/4}}{\sqrt[4]{x+y}} = \dfrac{(x+y)^{3/4}}{(x+y)^{1/4}}$

$= (x+y)^{3/4-1/4}$

$= (x+y)^{2/4}$

$= (x+y)^{1/2}$

$= \sqrt{x+y}$

127. $\dfrac{(3u-2v)^{2/3}}{\sqrt{(3u-2v)^3}} = \dfrac{(3u-2v)^{2/3}}{(3u-2v)^{3/2}}$

$= (3u-2v)^{2/3-3/2}$

$= (3u-2v)^{4/6-9/6}$

$= (3u-2v)^{-5/6}$

$= \dfrac{1}{(3u-2v)^{5/6}}$

129. $\sqrt{35} \approx 5.9161$

Scientific: 35 [√]

Graphing: [√] 35 [)] [ENTER]

131. $315^{2/5} = \left(\sqrt[5]{315}\right)^2 \approx 9.9845$

Scientific: 315 [y^x] [(] 2 [÷] 5 [)] [=]

Graphing: 315 [^] [(] 2 [÷] 5 [)] [ENTER]

133. $82^{-3/4} \approx 0.0367$

Scientific: 82 [y^x] [(] 3 [÷] 4 [+/−] [)] [=]

Graphing: 82 [^] [(] [(−)] 3 [÷] 4 [)] [ENTER]

135. $\sqrt[4]{212} \approx 3.8158$

$\sqrt[4]{212} = 212^{1/4}$

Scientific: 212 [y^x] [(] 1 [÷] 4 [)] [=]

Graphing: 212 [^] [(] 1 [÷] 4 [)] [ENTER]

137. $\sqrt[3]{545^2} \approx 66.7213$

$\sqrt[3]{545^2} = 545^{2/3}$

Scientific: 545 [y^x] [(] 2 [÷] 3 [)] [=]

Graphing: 545 [^] [(] 2 [÷] 3 [)] [ENTER]

139. $\dfrac{8-\sqrt{35}}{2} \approx 1.0420$

Scientific: [(] 8 [−] 35 [√] [)] [÷] 2 [=]

Graphing: [(] 8 [−] [√] 35 [)] [÷] 2 [ENTER]

141. $\dfrac{3+\sqrt{17}}{9} \approx 0.7915$

Scientific: [(] 3 [+] 17 [√] [)] [÷] 9 [=]

Graphing: [(] 3 [+] [√] 17 [)] [÷] 9 [ENTER]

143. $f(x) = \sqrt{2x+9}$

(a) $f(0) = \sqrt{2(0)+9} = \sqrt{9} = 3$

(b) $f(8) = \sqrt{2(8)+9} = \sqrt{25} = 5$

(c) $f(-6) = \sqrt{2(-6)+9} = \sqrt{-3} =$ not a real number

(d) $f(36) = \sqrt{2(36)+9} = \sqrt{81} = 9$

145. $g(x) = \sqrt[3]{x+1}$

(a) $g(7) = \sqrt[3]{7+1} = \sqrt[3]{8} = 2$

(b) $g(26) = \sqrt[3]{26+1} = \sqrt[3]{27} = 3$

(c) $g(-9) = \sqrt[3]{-9+1} = \sqrt[3]{-8} = -2$

(d) $g(-65) = \sqrt[3]{-65+1} = \sqrt[3]{-64} = -4$

147. $f(x) = \sqrt[4]{x-3}$

(a) $f(19) = \sqrt[4]{19-3} = \sqrt[4]{16} = 2$

(b) $f(1) = \sqrt[4]{1-3} = \sqrt[4]{-2} =$ not a real number

(c) $f(84) = \sqrt[4]{84-3} = \sqrt[4]{81} = 3$

(d) $f(4) = \sqrt[4]{4-3} = \sqrt[4]{1} = 1$

149. $f(x) = 3\sqrt{x},\ x \ge 0$

Domain: $[0, \infty)$

151. The domain of $g(x) = \sqrt{4-9x}$ is $\left(-\infty, \frac{4}{9}\right]$.

$4 - 9x \ge 0$

$-9x \ge -4$

$x \le \frac{4}{9}$

153. The domain of $f(x) = \sqrt[3]{x^4}$ is all reals $(-\infty, \infty)$.

155. $h(x) = \sqrt{2x + 9}$

$2x + 9 \geq 0$

$2x \geq -9$

$x \geq -\frac{9}{2}$

Domain: $\left[-\frac{9}{2}, \infty\right)$

157. The domain of $g(x) = \dfrac{2}{\sqrt[4]{x}}$ is the set of all nonnegative real numbers or $(0, \infty)$.

159. *Keystrokes:*

[Y=] 5 [÷] 4 [MATH 5] [X,T,θ] [MATH 3] [GRAPH]

Domain is $(0, \infty)$ so graphing utility did complete the graph.

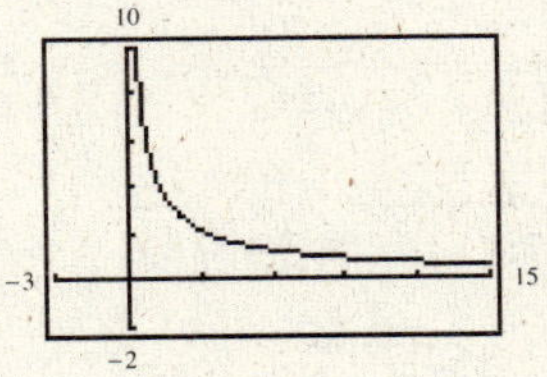

161. *Keystrokes:*

[Y=] 2 [X,T,θ] [^] [(] 3 [÷] 5 [)] [GRAPH]

Domain is $(-\infty, \infty)$ so graphing utility did complete the graph.

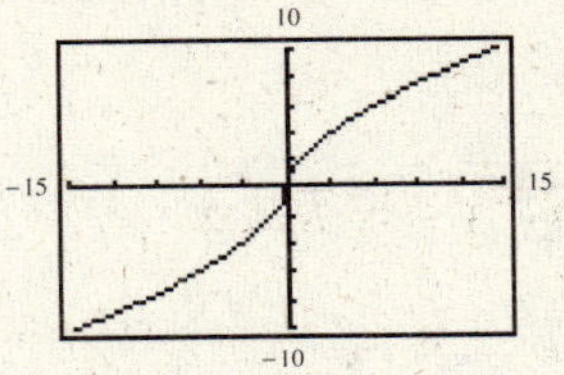

163. $x^{1/2}(2x - 3) = 2x^{3/2} - 3x^{1/2}$

165. $y^{-1/3}\left(y^{1/3} + 5y^{4/3}\right) = y^0 + 5y^{3/3} = 1 + 5y$

167. $r = 1 - \left(\dfrac{25,000}{75,000}\right)^{1/8} = 1 - \left(\dfrac{1}{3}\right)^{1/8} \approx 0.128 \approx 12.8\%$

169. *Verbal Model:* [Area] = [Side] · [Side]

Labels: Area = 529

Side = x

Equation: $529 = x \cdot x$

$529 = x^2$

$\sqrt{529} = x$

$23 = x$

23 feet × 23 feet

171. $D = \sqrt{l^2 + w^2 + h^2}$

$= \sqrt{79^2 + 65^2 + 22^2}$

$= \sqrt{6241 + 4225 + 484}$

$= \sqrt{10,950}$

≈ 104.64 inches

173. If a and b are real numbers, n is an integer greater than or equal to 2, and $a = b^n$, then b is the nth root of a.

175. No. $\sqrt{2}$ is an irrational number. Its decimal representation is a nonterminating, nonrepeating decimal.

177. (a) "Last digits:"

1 (Perfect square 81)

4 (Perfect square 64)

5 (Perfect square 25)

6 (Perfect square 36)

9 (Perfect square 49)

0 (Perfect square 100)

(b) Yes, 4,322,788,986 ends in a 6, but it is not a perfect square.

179. $\dfrac{a}{5} = \dfrac{a-3}{2}$

$10\left(\dfrac{a}{5}\right) = \left(\dfrac{a-3}{2}\right)^{10}$

$2a = 5(a - 3)$

$2a = 5a - 15$

$-3a = -15$

$a = 5$

181. $\dfrac{2}{u+4} = \dfrac{5}{8}$

$8(u + 4)\left(\dfrac{2}{u+4}\right) = \left(\dfrac{5}{8}\right)8(u + 4)$

$16 = 5(u + 4)$

$16 = 5u + 20$

$-4 = 5u$

$-\dfrac{4}{5} = u$

182. $\frac{6}{b} + 22 = 24$

$\frac{6}{b} = 2$

$b\left(\frac{6}{b}\right) = (2)b$

$6 = 2b$

$3 = b$

183. $s = kt^2$

184. $r = \frac{k}{x^4}$

185. $a = kbc$

186. $x = \frac{ky}{z}$

Section 7.2 Simplifying Radical Expressions

1. $\sqrt{18} = \sqrt{9 \cdot 2} = \sqrt{3^2 \cdot 2} = 3\sqrt{2}$

3. $\sqrt{45} = \sqrt{9 \cdot 5} = \sqrt{3^2 \cdot 5} = 3\sqrt{5}$

5. $\sqrt{96} = \sqrt{16 \cdot 6} = \sqrt{4^2 \cdot 6} = 4\sqrt{6}$

7. $\sqrt{153} = \sqrt{9 \cdot 17} = \sqrt{3^2 \cdot 17} = 3\sqrt{17}$

9. $\sqrt{1183} = \sqrt{169 \cdot 7} = \sqrt{13^2 \cdot 7} = 13\sqrt{7}$

11. $\sqrt{0.04} = \sqrt{4 \cdot 0.01} = \sqrt{4}\sqrt{0.01} = 2 \cdot 0.1 = 0.2$

13. $\sqrt{0.0072} = \sqrt{36 \cdot 2 \cdot 0.0001}$

$= \sqrt{36} \cdot \sqrt{2} \cdot \sqrt{0.0001}$

$= 6 \cdot 0.01\sqrt{2}$

$= 0.06\sqrt{2}$

15. $\sqrt{\frac{60}{3}} = \sqrt{20} = \sqrt{4 \cdot 5} = \sqrt{2^2 \cdot 5} = 2\sqrt{5}$

17. $\sqrt{\frac{13}{25}} = \frac{\sqrt{13}}{5}$

19. $\sqrt{9x^5} = \sqrt{3^2x^4 \cdot x} = 3 \cdot x^2 \cdot \sqrt{x} = 3x^2\sqrt{x}$

21. $\sqrt{48y^4} = \sqrt{16 \cdot 3 \cdot y^4} = 4y^2\sqrt{3}$

23. $\sqrt{117y^6} = \sqrt{9 \cdot 13 \cdot y^6} = 3\left|y^3\right|\sqrt{13}$

25. $\sqrt{120x^2y^3} = \sqrt{4 \cdot 30 \cdot x^2 \cdot y^2 \cdot y} = 2|x|y\sqrt{30y}$

27. $\sqrt{192a^5b^7} = \sqrt{64 \cdot 3 \cdot a^4 \cdot a \cdot b^6 \cdot b} = 8a^2b^3\sqrt{3ab}$

29. $\sqrt[3]{48} = \sqrt[3]{16 \cdot 3} = \sqrt[3]{2^4 \cdot 3} = 2\sqrt[3]{3 \cdot 2} = 2\sqrt[3]{6}$

31. $\sqrt[3]{112} = \sqrt[3]{8 \cdot 14} = \sqrt[3]{8} \cdot \sqrt[3]{14} = 2\sqrt[3]{14}$

33. $\sqrt[3]{40x^5} = \sqrt[3]{8 \cdot 5 \cdot x^3 \cdot x^2} = 2x\sqrt[3]{5x^2}$

35. $\sqrt[4]{324y^6} = \sqrt[4]{81 \cdot 4 \cdot y^4 \cdot y^2}$

$= 3|y|\sqrt[4]{4y^2}$

$= 3|y|\sqrt[4]{2^2y^2}$

$= 3|y|\sqrt{2y}$

37. $\sqrt[3]{x^4y^3} = \sqrt[3]{x^3 \cdot x \cdot y^3} = xy\sqrt[3]{x}$

39. $\sqrt[4]{4x^4y^6} = \sqrt[4]{4x^4 \cdot y^4 \cdot y^2} = |xy|\sqrt[4]{4y^2}$

41. $\sqrt[5]{32x^5y^6} = \sqrt[5]{2^5 \cdot x^5 \cdot y^5 \cdot y} = 2xy\sqrt[5]{y}$

43. $\sqrt[3]{\frac{35}{64}} = \frac{\sqrt[3]{35}}{4}$

45. $\frac{\sqrt{39y^2}}{\sqrt{3}} = \sqrt{\frac{39}{3}y^2} = \sqrt{13y^2} = |y|\sqrt{13}$

47. $\sqrt{\frac{32a^4}{b^2}} = \frac{\sqrt{16 \cdot 2 \cdot a^4}}{\sqrt{b^2}} = \frac{4a^2\sqrt{2}}{|b|}$

49. $\sqrt[5]{\frac{32x^2}{y^5}} = \sqrt[5]{\frac{2^5x^2}{y^5}} = \frac{2}{y}\sqrt[5]{x^2}$

51. $\sqrt[3]{\frac{54a^4}{b^9}} = \sqrt[3]{\frac{3^3 \cdot 2 \cdot a^3 \cdot a}{b^9}} = \frac{3a}{b^3}\sqrt[3]{2a}$

53. $-\sqrt[3]{\frac{3w^4}{8z^3}} = -\frac{\sqrt[3]{3 \cdot w^3 \cdot w}}{\sqrt[3]{8z^3}} = -\frac{w}{2z}\sqrt[3]{3w}$

55. $\sqrt{\frac{1}{3}} = \frac{1}{\sqrt{3}} \cdot \frac{\sqrt{3}}{\sqrt{3}} = \frac{\sqrt{3}}{3}$

57. $\frac{1}{\sqrt{7}} = \frac{1}{\sqrt{7}} \cdot \frac{\sqrt{7}}{\sqrt{7}} = \frac{\sqrt{7}}{7}$

59. $\sqrt[4]{\frac{5}{4}} = \frac{\sqrt[4]{5}}{\sqrt[4]{2^2}} \cdot \frac{\sqrt[4]{2^2}}{\sqrt[4]{2^2}} = \frac{\sqrt[4]{5 \cdot 2^2}}{\sqrt[4]{2^4}} = \frac{\sqrt[4]{20}}{2}$

61. $\dfrac{6}{\sqrt[3]{32}} = \dfrac{6}{\sqrt[3]{2^3 \cdot 2^2}}$

$= \dfrac{6}{2\sqrt[3]{2^2}} \cdot \dfrac{\sqrt[3]{2}}{\sqrt[3]{2}} = \dfrac{6\sqrt[3]{2}}{2\sqrt[3]{2^3}} = \dfrac{6\sqrt[3]{2}}{4} = \dfrac{3\sqrt[3]{2}}{2}$

63. $\dfrac{1}{\sqrt{y}} = \dfrac{1}{\sqrt{y}} \cdot \dfrac{\sqrt{y}}{\sqrt{y}} = \dfrac{\sqrt{y}}{\sqrt{y^2}} = \dfrac{\sqrt{y}}{y}$

65. $\sqrt{\dfrac{4}{x}} = \dfrac{\sqrt{4}}{\sqrt{x}} = \dfrac{2}{\sqrt{x}} \cdot \dfrac{\sqrt{x}}{\sqrt{x}} = \dfrac{2\sqrt{x}}{x}$

67. $\dfrac{1}{x\sqrt{2}} = \dfrac{1}{x\sqrt{2}} \cdot \dfrac{\sqrt{2}}{\sqrt{2}} = \dfrac{\sqrt{2}}{2x}$

69. $\dfrac{6}{\sqrt{3b^3}} = \dfrac{6}{b\sqrt{3b}} \cdot \dfrac{\sqrt{3b}}{\sqrt{3b}} = \dfrac{6\sqrt{3b}}{3b^2} = \dfrac{2\sqrt{3b}}{b^2}$

71. $\sqrt[3]{\dfrac{2x}{3y}} = \dfrac{\sqrt[3]{2x}}{\sqrt[3]{3y}} \cdot \dfrac{\sqrt[3]{3^2 y^2}}{\sqrt[3]{3^2 y^2}} = \dfrac{\sqrt[3]{2x \cdot 3^2 y^2}}{\sqrt[3]{3^3 y^3}} = \dfrac{\sqrt[3]{18xy^2}}{3y}$

73. $c = \sqrt{a^2 + b^2}$

$= \sqrt{6^2 + 3^2}$

$= \sqrt{36 + 9}$

$= \sqrt{45}$

$= \sqrt{9 \cdot 5}$

$= 3\sqrt{5}$

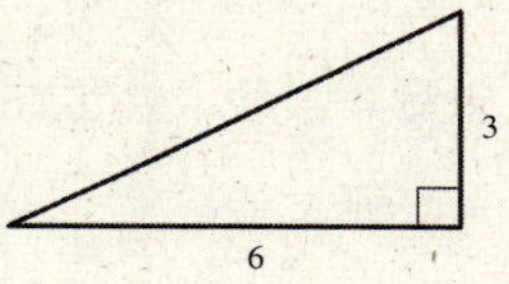

75. $f = \dfrac{1}{100}\sqrt{\dfrac{400 \times 10^6}{5}}$

$\approx 8.9443 \times 10^1$

≈ 89.443

≈ 89.44 cycles per second

77. *Verbal Model:* $\boxed{\text{Hypotenuse}}^2 = \boxed{\text{Leg 1}}^2 + \boxed{\text{Leg 2}}^2$

Labels: Hypotenuse $= c$

Leg 1 $= 26$

Leg 2 $= 10$

Equation: $c^2 = 26^2 + 10^2$

$c^2 = 676 + 100$

$c^2 = 776$

$c = \sqrt{776} \approx 27.86$ feet

79. Example: $\sqrt{6} \cdot \sqrt{15} = \sqrt{6 \cdot 15}$

$= \sqrt{3 \cdot 2 \cdot 3 \cdot 5}$

$= 3\sqrt{10}$

81. $\left(\dfrac{5}{\sqrt{3}}\right)^2 = \dfrac{5}{\sqrt{3}} \cdot \dfrac{5}{\sqrt{3}} = \dfrac{25}{3}$

No. Rationalizing the denominator produces an expression equivalent to the original expression. Squaring a number does not.

83. To find a perfect *n*th root factor, first factor the radicand completely. If the same factor appears at least *n* times, the perfect *n*th root factor is the common factor to the *n*th power.

85. $\begin{cases} 2x + 3y = 12 \\ 4x - y = 10 \end{cases} \Rightarrow \begin{matrix} 3y = -2x + 12 \\ -y = -4x + 10 \end{matrix} \Rightarrow \begin{matrix} y = -\frac{2}{3}x + 4 \\ y = 4x - 10 \end{matrix}$

Solution: $(3, 2)$

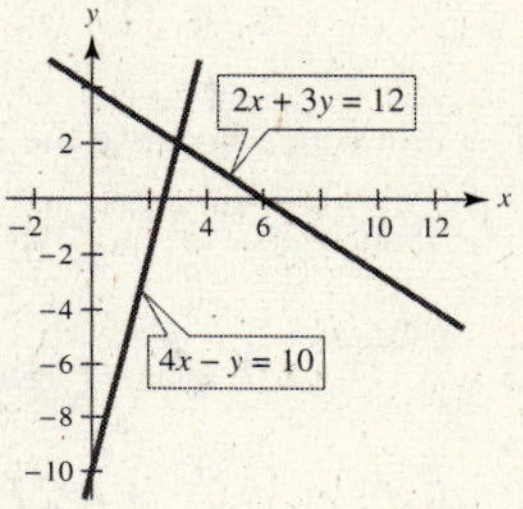

87. $\begin{cases} y = x + 2 \\ y - x = 8 \end{cases}$

$(x + 2) - x = 8$

$2 \neq 8$

No solution

89. $\begin{cases} x + 4y + 3z = 2 \\ 2x + y + z = 10 \\ -x + y + 2z = 8 \end{cases}$

$$\begin{aligned} x + 4y + 3z &= 2 \\ -7y - 5z &= 6 \\ 5y + 5z &= 10 \end{aligned}$$

$$\begin{aligned} x + 4y + 3z &= 2 \\ -7y - 5z &= 6 \\ -2y &= 16 \\ y &= -8 \end{aligned} \qquad \begin{aligned} -7(-8) - 5z &= 6 \\ 56 - 5z &= 6 \\ -5z &= -50 \\ z &= 10 \end{aligned} \qquad \begin{aligned} x + 4(-8) + 3(10) &= 2 \\ x - 32 + 30 &= 2 \\ x - 2 &= 2 \\ x &= 4 \end{aligned}$$

$(4, -8, 10)$

Section 7.3 Adding and Subtracting Radical Expressions

1. $3\sqrt{2} - \sqrt{2} = 2\sqrt{2}$

3. $2\sqrt{6} + 5\sqrt{6} = (2 + 5)\sqrt{6} = 7\sqrt{6}$

5. $8\sqrt{5} + 9\sqrt[3]{5}$ cannot be combined because the indices are not the same.

7. $9\sqrt[3]{5} - 6\sqrt[3]{5} = 3\sqrt[3]{5}$

9. $4\sqrt[3]{y} + 9\sqrt[3]{y} = 13\sqrt[3]{y}$

11. $15\sqrt[4]{s} - \sqrt[4]{s} = 14\sqrt[4]{s}$

13. $8\sqrt{2} + 6\sqrt{2} - 5\sqrt{2} = (8 + 6 - 5)\sqrt{2} = 9\sqrt{2}$

15. $\sqrt[4]{5} - 6\sqrt[4]{13} + 3\sqrt[4]{5} - \sqrt[4]{13} = (1 + 3)\sqrt[4]{5} + (-6 - 1)\sqrt[4]{13} = 4\sqrt[4]{5} - 7\sqrt[4]{13}$

17. $9\sqrt[3]{7} - \sqrt{3} + 4\sqrt[3]{7} + 2\sqrt{3} = (9\sqrt[3]{7} + 4\sqrt[3]{7}) + (-\sqrt{3} + 2\sqrt{3}) = 13\sqrt[3]{7} + \sqrt{3}$

19. $$\begin{aligned} 8\sqrt{27} - 3\sqrt{3} &= 8\sqrt{9 \cdot 3} - 3\sqrt{3} \\ &= 8(3)\sqrt{3} - 3\sqrt{3} \\ &= 24\sqrt{3} - 3\sqrt{3} \\ &= 21\sqrt{3} \end{aligned}$$

21. $$\begin{aligned} 3\sqrt{45} + 7\sqrt{20} &= 3\sqrt{9 \cdot 5} + 7\sqrt{4 \cdot 5} \\ &= 3(3)\sqrt{5} + 7(2)\sqrt{5} \\ &= 9\sqrt{5} + 14\sqrt{5} \\ &= 23\sqrt{5} \end{aligned}$$

23. $$\begin{aligned} 2\sqrt[3]{54} + 12\sqrt[3]{16} &= 2\sqrt[3]{27 \cdot 2} + 12\sqrt[3]{8 \cdot 2} \\ &= 6\sqrt[3]{2} + 24\sqrt[3]{2} = 30\sqrt[3]{2} \end{aligned}$$

25. $5\sqrt{9x} - 3\sqrt{x} = 15\sqrt{x} - 3\sqrt{x} = 12\sqrt{x}$

27. $3\sqrt{x+1} + 10\sqrt{x+1} = 13\sqrt{x+1}$

29. $\sqrt{25y} + \sqrt{64y} = 5\sqrt{y} + 8\sqrt{y} = 13\sqrt{y}$

31. $$\begin{aligned} 10\sqrt[3]{z} - \sqrt[3]{z^4} &= 10\sqrt[3]{z} - \sqrt[3]{z^3 \cdot z} = 10\sqrt[3]{z} - z\sqrt[3]{z} \\ &= (10 - z)\sqrt[3]{z} \end{aligned}$$

33. $$\begin{aligned} \sqrt{5a} + 2\sqrt{45a^3} &= \sqrt{5a} + 2\sqrt{9 \cdot 5 \cdot a^2 \cdot a} \\ &= \sqrt{5a} + 2(3)(a)\sqrt{5a} \\ &= \sqrt{5a} + 6a\sqrt{5a} \\ &= (1 + 6a)\sqrt{5a} \\ &= (6a + 1)\sqrt{5a} \end{aligned}$$

35. $$\begin{aligned} \sqrt[3]{6x^4} + \sqrt[3]{48x} &= \sqrt[3]{6 \cdot x^3 \cdot x} + \sqrt[3]{8 \cdot 6 \cdot x} \\ &= x\sqrt[3]{6x} + 2\sqrt[3]{6x} \\ &= (x + 2)\sqrt[3]{6x} \end{aligned}$$

37. $$\begin{aligned} \sqrt{9x - 9} + \sqrt{x - 1} &= \sqrt{9(x - 1)} + \sqrt{x - 1} \\ &= 3\sqrt{x - 1} + \sqrt{x - 1} \\ &= 4\sqrt{x - 1} \end{aligned}$$

39. $$\begin{aligned} \sqrt{x^3 - x^2} + \sqrt{4x - 4} &= \sqrt{x^2(x - 1)} + \sqrt{4(x - 1)} \\ &= x\sqrt{x - 1} + 2\sqrt{x - 1} \\ &= (x + 2)\sqrt{x - 1} \end{aligned}$$

41. $$\begin{aligned} 2\sqrt[3]{a^4b^2} + 3a\sqrt[3]{ab^2} &= 2\sqrt[3]{a^3 \cdot a \cdot b^2} + 3a\sqrt[3]{ab^2} \\ &= 2a\sqrt[3]{ab^2} + 3a\sqrt[3]{ab^2} \\ &= 5a\sqrt[3]{ab^2} \end{aligned}$$

43. $\sqrt{4r^7s^5} + 3r^2\sqrt{r^3s^5} - 2rs\sqrt{r^5s^3} = \sqrt{4r^6 \cdot r \cdot s^4 \cdot s} + 3r^2\sqrt{r^2 \cdot r \cdot s^4 \cdot s} - 2rs\sqrt{r^4 \cdot r \cdot s^2 \cdot s}$

$= 2r^3s^2\sqrt{rs} + 3r^3s^2\sqrt{rs} - 2r^3s^2\sqrt{rs}$

$= 3r^3s^2\sqrt{rs}$

45. $\sqrt[3]{128x^9y^{10}} - 2x^2y\sqrt[3]{16x^3y^7} = \sqrt[3]{64 \cdot 2 \cdot x^9 \cdot y^9 \cdot y} - 2x^2y\sqrt[3]{8 \cdot 2 \cdot x^3 \cdot y^6 \cdot y}$

$= 4x^3y^3\sqrt[3]{2y} - 4x^3y^3\sqrt[3]{2y} = 0$

47. $\sqrt{5} - \dfrac{3}{\sqrt{5}} = \sqrt{5} - \left(\dfrac{3}{\sqrt{5}} \cdot \dfrac{\sqrt{5}}{\sqrt{5}}\right) = \sqrt{5} - \dfrac{3\sqrt{5}}{5}$

$= \left(1 - \dfrac{3}{5}\right)\sqrt{5}$

$= \dfrac{2}{5}\sqrt{5}$

49. $\sqrt{32} + \sqrt{\dfrac{1}{2}} = \sqrt{16 \cdot 2} + \dfrac{\sqrt{1}}{\sqrt{2}} \cdot \dfrac{\sqrt{2}}{\sqrt{2}}$

$= 4\sqrt{2} + \dfrac{\sqrt{2}}{2}$

$= \dfrac{8\sqrt{2}}{2} + \dfrac{1\sqrt{2}}{2}$

$= \dfrac{9\sqrt{2}}{2}$

51. $\sqrt{12y} - \dfrac{y}{\sqrt{3y}} = \sqrt{4 \cdot 3y} - \dfrac{y}{\sqrt{3y}} \cdot \dfrac{\sqrt{3y}}{\sqrt{3y}}$

$= 2\sqrt{3y} - \dfrac{y\sqrt{3y}}{3y}$

$= \dfrac{6}{3}\sqrt{3y} - \dfrac{1}{3}\sqrt{3y}$

$= \dfrac{5}{3}\sqrt{3y}$

53. $\dfrac{2}{\sqrt{3x}} + \sqrt{3x} = \dfrac{2}{\sqrt{3x}} \cdot \dfrac{\sqrt{3x}}{\sqrt{3x}} + \sqrt{3x}$

$= \dfrac{2\sqrt{3x}}{3x} + \dfrac{3x\sqrt{3x}}{3x}$

$= \dfrac{(3x + 2)\sqrt{3x}}{3x}$

55. $\sqrt{7y^3} - \sqrt{\dfrac{9}{7y^3}} = \sqrt{7y^2 \cdot y} - \dfrac{3}{\sqrt{7y^2 \cdot y}}$

$= y\sqrt{7y} - \dfrac{3}{y\sqrt{7y}} \cdot \dfrac{\sqrt{7y}}{\sqrt{7y}}$

$= y\sqrt{7y} - \dfrac{3\sqrt{7y}}{y \cdot 7y}$

$= y\sqrt{7y} - \dfrac{3\sqrt{7y}}{7y^2}$

$= \dfrac{7y^2}{7y^2}\left(\dfrac{y\sqrt{7y}}{1}\right) - \dfrac{3\sqrt{7y}}{7y^2}$

$= \dfrac{7y^3\sqrt{7y}}{7y^2} - \dfrac{3\sqrt{7y}}{7y^2}$

$= \dfrac{\sqrt{7y}(7y^3 - 3)}{7y^2}$

57. $\sqrt{7} + \sqrt{18} > \sqrt{7 + 18}$

59. $5 < \sqrt{9^2 - 4^2}$

$5 < \sqrt{81 - 16}$

$5 < \sqrt{65}$

$5 < 8.06$

61. $P = \sqrt{54x} + \sqrt{96x} + \sqrt{150x}$

$= \sqrt{9 \cdot 6x} + \sqrt{16 \cdot 6x} + \sqrt{25 \cdot 6x}$

$= 3\sqrt{6x} + 4\sqrt{6x} + 5\sqrt{6x}$

$= 12\sqrt{6x}$

63. $P = \sqrt{12x} + \sqrt{48}x + \sqrt{27x} + \sqrt{75}x$

$= \sqrt{4 \cdot 3x} + \sqrt{16 \cdot 3}x + \sqrt{9 \cdot 3x} + \sqrt{25 \cdot 3}x$

$= 2\sqrt{3x} + 4\sqrt{3}x + 3\sqrt{3x} + 5\sqrt{3}x$

$= (2\sqrt{3x} + 3\sqrt{3x}) + (4\sqrt{3} + 5\sqrt{3})x$

$= 5\sqrt{3x} + (9\sqrt{3})x$

$= 9x\sqrt{3} + 5\sqrt{3x}$

65. (a) $c = \sqrt{a^2 + b^2}$

$c = \sqrt{(15)^2 + (5)^2}$

$c = \sqrt{225 + 25}$

$c = \sqrt{250}$

$c = \sqrt{25 \cdot 10}$

$c = 5\sqrt{10}$

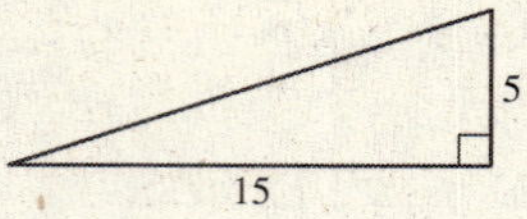

(b) Area of roof $= 2 \cdot$ Length $\cdot$ Width

$A = 2 \cdot 40 \cdot 5\sqrt{10}$

$A = 400\sqrt{10}$

Thus, the total area of the roof is $400\sqrt{10} \approx 1264.9$ square feet.

67. *Verbal Model:* [Total immigrants] = [South America immigrants] − [Colombia immigrants]

$T = S - C$

Radical Expression: $T = \left(-182 - 409.2t + 94.6\sqrt{t^3} + 565.5\sqrt{t}\right) - \left(-118 - 195.1t + 42.8\sqrt{t^3} + 286.7\sqrt{t}\right)$

$T = -182 - 409.2t + 94.6t\sqrt{t} + 565.5\sqrt{t} + 118 + 195.1t - 42.8t\sqrt{t} - 286.7\sqrt{t}$

$T = (-182 + 118) + (-409.2 + 195.1)t + (94.6 - 42.8)t\sqrt{t} + (565.5 - 286.7)\sqrt{t}$

$T = -64 - 214.1t + 51.8t\sqrt{t} + 278.8\sqrt{t}$

T in 2003 $= -64 - 214.1(3) + 51.8(3)(\sqrt{3}) + 278.8(\sqrt{3}) = 45.75646 \times 1000 = 45{,}756$ people

69. No; $\sqrt{5} + \left(-\sqrt{5}\right) = 0$.

71. $\sqrt{2x} + \sqrt{2x} = (1 + 1)\sqrt{2x}$

$= 2\sqrt{2x} = \sqrt{4} \cdot \sqrt{2x} = \sqrt{8x}$

73. (a) The student combined terms with unlike radicands; can be simplified no further.

(b) The student combined terms with unlike indices; can be simplified no further.

75. $\dfrac{7z - 2}{2z} - \dfrac{4z + 1}{2z} = \dfrac{7z - 2 - 4z - 1}{2z} = \dfrac{3z - 3}{2z}$

$= \dfrac{3(z - 1)}{2z}$

77. $\dfrac{2x + 3}{x - 3} + \dfrac{6 - 5x}{x - 3} = \dfrac{2x + 3 + 6 - 5x}{x - 3} = \dfrac{-3x + 9}{x - 3}$

$= \dfrac{-3(x - 3)}{x - 3} = -3, \quad x \neq 3$

79. $\dfrac{2v}{v - 5} - \dfrac{3}{5 - v} = \dfrac{2v}{v - 5} - \dfrac{3}{-1(v - 5)}$

$= \dfrac{2v}{v - 5} + \dfrac{3}{v - 5} = \dfrac{2v + 3}{v - 5}$

81. $\dfrac{\left(\dfrac{27a^3}{4b^2c}\right)}{\left(\dfrac{9ac^2}{10b^2}\right)} = \dfrac{27a^3}{4b^2c} \cdot \dfrac{10b^2}{9ac^2}$

$= \dfrac{(27)(10)a^3 \cdot b^2}{(4)(9)a \cdot b^2 \cdot c^3} = \dfrac{15a^2}{2c^3}, \quad a \neq 0, \quad b \neq 0$

83. $\dfrac{3w - 9}{\left(\dfrac{w^2 - 10w + 21}{w + 1}\right)} = 3w - 9 \cdot \dfrac{w + 1}{w^2 - 10w + 21}$

$= \dfrac{3(w - 3)(w + 1)}{(w - 7)(w - 3)}$

$= \dfrac{3(w + 1)}{w - 7}, \quad w \neq -1, w \neq 3$

Mid-Chapter Quiz for Chapter 7

1. $\sqrt{225} = 15$ because $15 \cdot 15 = 225$.

2. $\sqrt[4]{\frac{81}{16}} = \frac{3}{2}$ because $\frac{3}{2} \cdot \frac{3}{2} \cdot \frac{3}{2} \cdot \frac{3}{2} = \frac{81}{16}$.

3. $49^{1/2} = \sqrt{49} = 7$ because $7 \cdot 7 = 49$.

4. $(-27)^{2/3} = \sqrt[3]{(-27)^2} = \left(\sqrt[3]{-27}\right)^2 = (-3)^2 = 9$

5. $f(x) = \sqrt{3x - 5}$

(a) $f(0) = \sqrt{3(0) - 5} = \sqrt{-5}$ = not a real number

(b) $f(2) = \sqrt{3(2) - 5} = \sqrt{1} = 1$

(c) $f(10) = \sqrt{3(10) - 5} = \sqrt{25} = 5$

6. $g(x) = \sqrt{9 - x}$

(a) $g(-7) = \sqrt{9 - (-7)} = \sqrt{9 + 7} = \sqrt{16} = 4$

(b) $g(5) = \sqrt{9 - 5} = \sqrt{4} = 2$

(c) $g(9) = \sqrt{9 - 9} = \sqrt{0} = 0$

7. $g(x) = \dfrac{12}{\sqrt[3]{x}}$

Domain: $(-\infty, 0) \cup (0, \infty)$

$\sqrt[3]{x} \neq 0$

$x \neq 0$

8. $h(x) = \sqrt{3x + 10}$; Domain: $\left[-\frac{10}{3}, \infty\right)$

$3x + 10 \geq 0$

$3x \geq -10$

$x \geq -\frac{10}{3}$

9. $\sqrt{27x^2} = \sqrt{9 \cdot 3 \cdot x^2} = 3|x|\sqrt{3}$

10. $\sqrt[4]{32x^8} = \sqrt[4]{16 \cdot 2 \cdot (x^2)^4} = 2x^2\sqrt[4]{2}$

11. $\sqrt{\dfrac{4u^3}{9}} = \dfrac{\sqrt{4 \cdot u^2 \cdot u}}{\sqrt{9}} = \dfrac{2u\sqrt{u}}{3}$

12. $\sqrt[3]{\dfrac{16}{u^6}} = \dfrac{\sqrt[3]{16}}{\sqrt[3]{u^6}} = \dfrac{\sqrt[3]{16}}{u^2} = \dfrac{\sqrt[3]{8 \cdot 2}}{u^2} = \dfrac{2\sqrt[3]{2}}{u^2}$

13. $\sqrt{125x^3y^2z^4} = \sqrt{25 \cdot 5 \cdot x^2 \cdot x \cdot y^2 \cdot z^4}$

$= 5x|y|z^2\sqrt{5x}$

15. $\dfrac{24}{\sqrt{12}} = \dfrac{24}{\sqrt{4 \cdot 3}} = \dfrac{24}{2\sqrt{3}} = \dfrac{12}{\sqrt{3}} \cdot \dfrac{\sqrt{3}}{\sqrt{3}} = \dfrac{12\sqrt{3}}{3} = 4\sqrt{3}$

17. $2\sqrt{3} - 4\sqrt{7} + \sqrt{3} = (2\sqrt{3} + \sqrt{3}) - 4\sqrt{7} = 3\sqrt{3} - 4\sqrt{7}$

19. $5\sqrt{12} + 2\sqrt{3} - \sqrt{75} = 5\sqrt{4 \cdot 3} + 2\sqrt{3} - \sqrt{25 \cdot 3} = 10\sqrt{3} + 2\sqrt{3} - 5\sqrt{3} = 7\sqrt{3}$

21. $6x\sqrt[3]{5x^2} + 2\sqrt[3]{40x^4} = 6x\sqrt[3]{5x^2} + 2\sqrt[3]{8 \cdot 5 \cdot x^3 \cdot x} = 6x\sqrt[3]{5x^2} + 4x\sqrt[3]{5x}$

23. $C = \sqrt{2^2 + 2^2} = \sqrt{4 + 4} = \sqrt{8}$

Equation:

$P = 2(7) + 2\left(4\frac{1}{2}\right) + 4(\sqrt{8}) = 14 + 9 + 8\sqrt{2} = 23 + 8\sqrt{2}$ inches

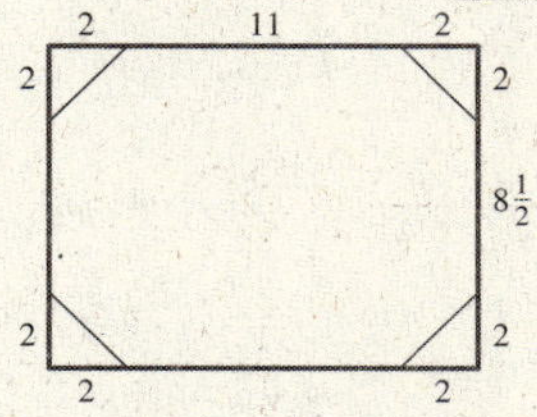

Section 7.4 Multiplying and Dividing Radical Expressions

1. $\sqrt{2} \cdot \sqrt{8} = \sqrt{2 \cdot 8} = \sqrt{16} = 4$

3. $\sqrt{3} \cdot \sqrt{15} = \sqrt{3 \cdot 15} = \sqrt{45} = \sqrt{9 \cdot 5} = 3\sqrt{5}$

5. $\sqrt[3]{12} \cdot \sqrt[3]{6} = \sqrt[3]{12 \cdot 6} = \sqrt[3]{8 \cdot 9} = 2\sqrt[3]{9}$

7. $\sqrt[4]{8} \cdot \sqrt[4]{2} = \sqrt[4]{8 \cdot 2} = \sqrt[4]{16} = 2$

9. $\sqrt{7}(3 - \sqrt{7}) = 3\sqrt{7} - 7$

11. $\sqrt{2}(\sqrt{20} + 8) = \sqrt{2}\sqrt{20} + 8\sqrt{2}$

$= \sqrt{40} + 8\sqrt{2}$

$= 2\sqrt{10} + 8\sqrt{2}$

13. $\sqrt{6}(\sqrt{12}-\sqrt{3}) = \sqrt{6}\sqrt{12}-\sqrt{6}\sqrt{3}$
$= \sqrt{72}-\sqrt{18}$
$= \sqrt{36\cdot 2}-\sqrt{9\cdot 2}$
$= 6\sqrt{2}-3\sqrt{2}$
$= 3\sqrt{2}$

15. $4\sqrt{3}(\sqrt{3}-\sqrt{5}) = 4\sqrt{3}\cdot\sqrt{3}-4\sqrt{3}\cdot\sqrt{5}$
$= 4\cdot 3-4\sqrt{3\cdot 5}$
$= 12-4\sqrt{15}$

17. $\sqrt{y}(\sqrt{y}+4) = (\sqrt{y})^2+4\sqrt{y} = y+4\sqrt{y}$

19. $\sqrt{a}(4-\sqrt{a}) = \sqrt{a}\cdot 4-\sqrt{a}\sqrt{a} = 4\sqrt{a}-a$

21. $\sqrt[3]{4}(\sqrt[3]{2}-7) = \sqrt[3]{4}\sqrt[3]{2}-7\sqrt[3]{4}$
$= \sqrt[3]{8}-7\sqrt[3]{4}$
$= 2-7\sqrt[3]{4}$

23. $(\sqrt{5}+3)(\sqrt{3}-5) = \sqrt{15}-5\sqrt{5}+3\sqrt{3}-15$

25. $(\sqrt{20}+2)^2 = (\sqrt{20})^2+2\cdot\sqrt{20}\cdot 2+2^2$
$= 20+4\sqrt{20}+4$
$= 24+4\sqrt{4\cdot 5}$
$= 24+8\sqrt{5}$

27. $(\sqrt[3]{6}-3)(\sqrt[3]{4}+3) = \sqrt[3]{6}\sqrt[3]{4}+3\sqrt[3]{6}-3\sqrt[3]{4}-9$
$= \sqrt[3]{24}+3\sqrt[3]{6}-3\sqrt[3]{4}-9$
$= \sqrt[3]{8\cdot 3}+3\sqrt[3]{6}-3\sqrt[3]{4}-9$
$= 2\sqrt[3]{3}+3\sqrt[3]{6}-3\sqrt[3]{4}-9$

29. $(\sqrt{3}+2)(\sqrt{3}-2) = (\sqrt{3})^2-2^2 = 3-4 = -1$

31. $(6-\sqrt{7})(6+\sqrt{7}) = 6^2-(\sqrt{7})^2 = 36-7 = 29$

33. $(\sqrt{5}-\sqrt{3})(\sqrt{5}-\sqrt{3}) = (\sqrt{5})^2-2\cdot\sqrt{5}\cdot\sqrt{3}+(\sqrt{3})^2 = 5-2\sqrt{15}+3 = 8-2\sqrt{15}$

35. $(10+\sqrt{2x})^2 = 10^2+2\cdot 10\cdot\sqrt{2x}+(\sqrt{2x})^2 = 100+20\sqrt{2x}+2x$

37. $(9\sqrt{x}+2)(5\sqrt{x}-3) = (9\sqrt{x})(5\sqrt{x})-27\sqrt{x}+10\sqrt{x}-6 = 45x-17\sqrt{x}-6$

39. $(2\sqrt{2x}-\sqrt{5})(2\sqrt{2x}+\sqrt{5}) = (2\sqrt{2x})^2-(\sqrt{5})^2 = 4\cdot 2x-5 = 8x-5$

41. $(\sqrt[3]{2x}+5)^2 = (\sqrt[3]{2x})^2+2\cdot 5\sqrt[3]{2x}+5^2 = \sqrt[3]{(2x)^2}+10\sqrt[3]{2x}+25 = \sqrt[3]{4x^2}+10\sqrt[3]{2x}+25$

43. $(\sqrt[3]{y}+2)(\sqrt[3]{y^2}-5) = \sqrt[3]{y}\cdot\sqrt[3]{y^2}-5\sqrt[3]{y}+2\sqrt[3]{y^2}-10 = \sqrt[3]{y^3}-5\sqrt[3]{y}+2\sqrt[3]{y^2}-10 = y-5\sqrt[3]{y}+2\sqrt[3]{y^2}-10$

45. $(\sqrt[3]{t}+1)(\sqrt[3]{t^2}+4\sqrt[3]{t}-3) = \sqrt[3]{t}\sqrt[3]{t^2}+\sqrt[3]{t}\cdot 4\sqrt[3]{t}-3\sqrt[3]{t}+\sqrt[3]{t^2}+4\sqrt[3]{t}-3$
$= \sqrt[3]{t^3}+4\sqrt[3]{t^2}-3\sqrt[3]{t}+\sqrt[3]{t^2}+4\sqrt[3]{t}-3 = t+5\sqrt[3]{t^2}+\sqrt[3]{t}-3$

47. $\sqrt{x^3y^4}\left(2\sqrt{xy^2}-\sqrt{x^3y}\right) = \sqrt{x^3y^4}\cdot 2\sqrt{xy^2}-\sqrt{x^3y^4}\cdot\sqrt{x^3y}$
$= 2\sqrt{x^4y^6}-\sqrt{x^6y^5}$
$= 2x^2y^3-x^3\sqrt{y^4\cdot y}$
$= 2x^2y^3-x^3y^2\sqrt{y}$
$= x^2y^2\left(2y-|x|\sqrt{y}\right)$

49. $2\sqrt[3]{x^4y^5}\left(\sqrt[3]{8x^{12}y^4}+\sqrt[3]{16xy^9}\right) = 2\sqrt[3]{x^4y^5}\cdot\sqrt[3]{8x^{12}y^4}+2\sqrt[3]{x^4y^5}\cdot\sqrt[3]{16xy^9}$
$= 2\sqrt[3]{8x^{16}y^9}+2\sqrt[3]{16x^5y^{14}}$
$= 2\sqrt[3]{8x^{15}\cdot x\cdot y^9}+2\sqrt[3]{8\cdot 2x^3\cdot x^2\cdot y^{12}\cdot y^2}$
$= 4x^5y^3\sqrt[3]{x}+4xy^4\sqrt[3]{2x^2y^2}$
$= 4xy^3\left(x^4\sqrt[3]{x}+y\sqrt[3]{2x^2y^2}\right)$

51. $5x\sqrt{3} + 15\sqrt{3} = 5\sqrt{3}(x + 3)$

53. $4\sqrt{12} - 2x\sqrt{27} = 4\sqrt{4 \cdot 3} - 2x\sqrt{9 \cdot 3}$
$= 8\sqrt{3} - 6x\sqrt{3}$
$= 2\sqrt{3}(4 - 3x)$

55. $6u^2 + \sqrt{18u^3} = 6u^2 + \sqrt{9 \cdot 2u^2 \cdot u}$
$= 6u^2 + 3u\sqrt{2u}$
$= 3u(2u + \sqrt{2u})$

57. $2 + \sqrt{5}$, conjugate $= 2 - \sqrt{5}$

Product $= (2 + \sqrt{5})(2 - \sqrt{5})$
$= 2^2 - (\sqrt{5})^2$
$= 4 - 5 = -1$

59. $\sqrt{11} - \sqrt{3}$, conjugate $= \sqrt{11} + \sqrt{3}$

Product $= (\sqrt{11} - \sqrt{3})(\sqrt{11} + \sqrt{3})$
$= (\sqrt{11})^2 - (\sqrt{3})^2$
$= 11 - 3 = 8$

61. $\sqrt{15} + 3$, conjugate $= \sqrt{15} - 3$

Product $= (\sqrt{15} + 3)(\sqrt{15} - 3)$
$= \sqrt{15} \cdot \sqrt{15} - 3\sqrt{15} + 3\sqrt{15} - 9$
$= 15 - 9 = 6$

63. $\sqrt{x} - 3$, conjugate $= \sqrt{x} + 3$

Product $= (\sqrt{x} - 3)(\sqrt{x} + 3)$
$= (\sqrt{x})^2 - 3^2$
$= x - 9$

65. $\sqrt{2u} - \sqrt{3}$, conjugate $= \sqrt{2u} + \sqrt{3}$

Product $= (\sqrt{2u} - \sqrt{3})(\sqrt{2u} + \sqrt{3})$
$= (\sqrt{2u})^2 - (\sqrt{3})^2$
$= 2u - 3$

67. $2\sqrt{2} + \sqrt{4}$, conjugate $= 2\sqrt{2} - \sqrt{4}$

Product $= (2\sqrt{2} + \sqrt{4})(2\sqrt{2} - \sqrt{4})$
$= (2\sqrt{2})^2 - (\sqrt{4})^2$
$= 4 \cdot 2 - 4$
$= 8 - 4 = 4$

69. $\sqrt{x} + \sqrt{y}$, conjugate $= \sqrt{x} - \sqrt{y}$

Product $= (\sqrt{x} + \sqrt{y})(\sqrt{x} - \sqrt{y})$
$= (\sqrt{x})^2 - (\sqrt{y})^2$
$= x - y$

71. $f(x) = x^2 - 6x + 1$

(a) $f(2 - \sqrt{3}) = (2 - \sqrt{3})^2 - 6(2 - \sqrt{3}) + 1$
$= 4 - 4\sqrt{3} + 3 - 12 + 6\sqrt{3} + 1$
$= 2\sqrt{3} - 4$

(b) $f(3 - 2\sqrt{2}) = (3 - 2\sqrt{2})^2 - 6(3 - 2\sqrt{2}) + 1$
$= 9 - 12\sqrt{2} + 8 - 18 + 12\sqrt{2} + 1$
$= 0$

73. $f(x) = x^2 - 2x - 2$

(a) $f(1 + \sqrt{3}) = (1 + \sqrt{3})^2 - 2(1 + \sqrt{3}) - 2$
$= 1 + 2\sqrt{3} + 3 - 2 - 2\sqrt{3} - 2$
$= (1 + 3 - 2 - 2) + (2 - 2)\sqrt{3}$
$= 0$

(b) $f(3 - \sqrt{3}) = (3 - \sqrt{3})^2 - 2(3 - \sqrt{3}) - 2$
$= 9 - 6\sqrt{3} + 3 - 6 + 2\sqrt{3} - 2$
$= (9 + 3 - 6 - 2) + (-6 + 2)\sqrt{3}$
$= 4 - 4\sqrt{3}$

75. $\dfrac{6}{\sqrt{11} - 2} = \dfrac{6}{\sqrt{11} - 2} \cdot \dfrac{\sqrt{11} + 2}{\sqrt{11} + 2}$
$= \dfrac{6(\sqrt{11} + 2)}{(\sqrt{11})^2 - 2^2}$
$= \dfrac{6(\sqrt{11} + 2)}{11 - 4}$
$= \dfrac{6(\sqrt{11} + 2)}{7}$

77. $\dfrac{7}{\sqrt{3} + 5} = \dfrac{7}{\sqrt{3} + 5} \cdot \dfrac{\sqrt{3} - 5}{\sqrt{3} - 5}$
$= \dfrac{7(\sqrt{3} - 5)}{(\sqrt{3})^2 - 5^2}$
$= \dfrac{7(\sqrt{3} - 5)}{3 - 25}$
$= \dfrac{7(\sqrt{3} - 5)}{-22}$
$= \dfrac{7(5 - \sqrt{3})}{22}$

79. $\dfrac{3}{2\sqrt{10}-5} = \dfrac{3}{2\sqrt{10}-5}\cdot\dfrac{2\sqrt{10}+5}{2\sqrt{10}+5}$

$= \dfrac{3(2\sqrt{10}+5)}{(2\sqrt{10})^2 - 5^2}$

$= \dfrac{3(2\sqrt{10}+5)}{40-25}$

$= \dfrac{3(2\sqrt{10}+5)}{15}$

$= \dfrac{2\sqrt{10}+5}{5}$

81. $\dfrac{2}{\sqrt{6}+\sqrt{2}} = \dfrac{2}{\sqrt{6}+\sqrt{2}}\cdot\dfrac{\sqrt{6}-\sqrt{2}}{\sqrt{6}-\sqrt{2}}$

$= \dfrac{2(\sqrt{6}-\sqrt{2})}{6-2}$

$= \dfrac{2(\sqrt{6}-\sqrt{2})}{4}$

$= \dfrac{\sqrt{6}-\sqrt{2}}{2}$

83. $\dfrac{10}{2\sqrt{3}-\sqrt{7}} = \dfrac{10}{2\sqrt{3}-\sqrt{7}}\cdot\dfrac{2\sqrt{3}+\sqrt{7}}{2\sqrt{3}+\sqrt{7}}$

$= \dfrac{10(2\sqrt{3}+\sqrt{7})}{12-7}$

$= \dfrac{10(2\sqrt{3}+\sqrt{7})}{5}$

$= 2(2\sqrt{3}+\sqrt{7})$

85. $(\sqrt{7}+2)\div(\sqrt{7}-2) = \dfrac{\sqrt{7}+2}{\sqrt{7}-2}\cdot\dfrac{\sqrt{7}+2}{\sqrt{7}+2}$

$= \dfrac{(\sqrt{7})^2 + 2\sqrt{7} + 2\sqrt{7} + 4}{(\sqrt{7})^2 - 2^2}$

$= \dfrac{7 + 4\sqrt{7} + 4}{7-4}$

$= \dfrac{11 + 4\sqrt{7}}{3}$

87. $(\sqrt{x}-5)\div(2\sqrt{x}-1) = \dfrac{\sqrt{x}-5}{2\sqrt{x}-1}\cdot\dfrac{2\sqrt{x}+1}{2\sqrt{x}+1}$

$= \dfrac{2x + \sqrt{x} - 10\sqrt{x} - 5}{(2\sqrt{x})^2 - 1^2}$

$= \dfrac{2x - 9\sqrt{x} - 5}{4x-1}$

89. $\dfrac{3x}{\sqrt{15}-\sqrt{3}} = \dfrac{3x}{\sqrt{15}-\sqrt{3}}\cdot\dfrac{\sqrt{15}+\sqrt{3}}{\sqrt{15}+\sqrt{3}}$

$= \dfrac{3x(\sqrt{15}+\sqrt{3})}{(\sqrt{15})^2 - (\sqrt{3})^2}$

$= \dfrac{3x(\sqrt{15}+\sqrt{3})}{15-3}$

$= \dfrac{3x(\sqrt{15}+\sqrt{3})}{12}$

$= \dfrac{x\sqrt{15}+x\sqrt{3}}{4}$ or $\dfrac{(\sqrt{15}+\sqrt{3})x}{4}$

91. $\dfrac{\sqrt{5t}}{\sqrt{5}-\sqrt{t}} = \dfrac{\sqrt{5t}}{\sqrt{5}-\sqrt{t}}\cdot\dfrac{\sqrt{5}+\sqrt{t}}{\sqrt{5}+\sqrt{t}}$

$= \dfrac{\sqrt{5t}\cdot\sqrt{5} + \sqrt{5t}\cdot\sqrt{t}}{5-t}$

$= \dfrac{5\sqrt{t} + t\sqrt{5}}{5-t}$

93. $\dfrac{8a}{\sqrt{3a}+\sqrt{a}} = \dfrac{8a}{\sqrt{3a}+\sqrt{a}}\cdot\dfrac{\sqrt{3a}-\sqrt{a}}{\sqrt{3a}-\sqrt{a}}$

$= \dfrac{8a(\sqrt{3a}-\sqrt{a})}{(\sqrt{3a})^2 - (\sqrt{a})^2}$

$= \dfrac{8a(\sqrt{3a}-\sqrt{a})}{3a-a}$

$= \dfrac{8a(\sqrt{3a}-\sqrt{a})}{2a}$

$= 4(\sqrt{3a}-\sqrt{a}),\quad a \neq 0$

95. $\dfrac{3(x-4)}{x^2-\sqrt{x}} = \dfrac{3(x-4)}{x^2-\sqrt{x}}\cdot\dfrac{x^2+\sqrt{x}}{x^2+\sqrt{x}}$

$= \dfrac{3(x-4)(x^2+\sqrt{x})}{(x^2)^2 - (\sqrt{x})^2}$

$= \dfrac{3(x-4)(x^2+\sqrt{x})}{x^4 - x}$

$= \dfrac{3(x-4)(x^2+\sqrt{x})}{x(x^3-1)}$

$= \dfrac{3(x-4)(x^2+\sqrt{x})}{x(x-1)(x^2+x+1)}$

97. $\dfrac{\sqrt{u+v}}{\sqrt{u-v}-\sqrt{u}} = \dfrac{\sqrt{u+v}}{\sqrt{u-v}-\sqrt{u}}\cdot\dfrac{\sqrt{u-v}+\sqrt{u}}{\sqrt{u-v}+\sqrt{u}}$

$= \dfrac{\sqrt{u+v}(\sqrt{u-v}+\sqrt{u})}{u-v-u}$

$= \dfrac{\sqrt{u+v}(\sqrt{u-v}+\sqrt{u})}{-v}$

$= \dfrac{-\sqrt{u+v}(\sqrt{u-v}+\sqrt{u})}{v}$

99. *Keystrokes:*

y_1: [Y=] 10 [÷] [(] [√] [X,T,θ] [+] 1 [)] [ENTER]

y_2: [(] 10 [(] [√] [X,T,θ] [−] 1 [)] [)] [÷] [(] [X,T,θ] [−] 1 [)] [GRAPH]

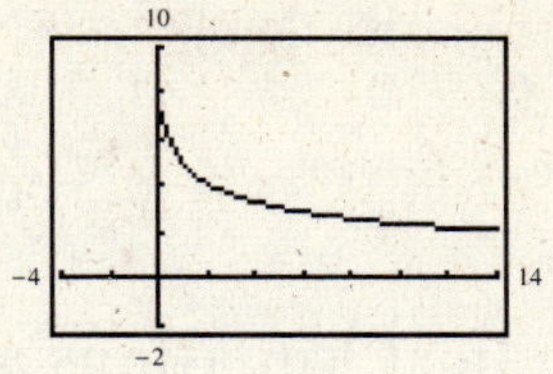

$y_1 = y_2$, except at $x = 1$

$$\frac{10}{\sqrt{x}+1} = \frac{10}{\sqrt{x}+1}\cdot\frac{\sqrt{x}-1}{\sqrt{x}-1} = \frac{10(\sqrt{x}-1)}{x-1},\ x \neq 1$$

101. *Keystrokes:*

y_1: [Y=] [(] 2 [√] [(] 3 [X,T,θ] [)] [÷] [(] 2 [−] [√] [(] 3 [X,T,θ] [)] [)] [ENTER]

y_2: [(] 2 [(] 2 [√] [(] 3 [X,T,θ] [)] [+] 3 [X,T,θ] [)] [)] [÷] [(] 4 [−] 3 [X,T,θ] [)] [GRAPH]

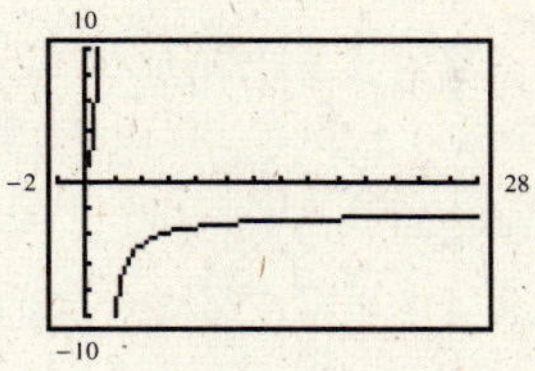

$y_1 = y_2$

$$\frac{2\sqrt{3x}}{2-\sqrt{3x}} = \frac{2\sqrt{3x}}{2-\sqrt{3x}}\cdot\frac{2+\sqrt{3x}}{2+\sqrt{3x}} = \frac{4\sqrt{3x}+2\cdot 3x}{4-3x} = \frac{2(2\sqrt{3x}+3x)}{4-3x}$$

103. $$\frac{\sqrt{2}}{7} = \frac{\sqrt{2}}{7}\cdot\frac{\sqrt{2}}{\sqrt{2}} = \frac{2}{7\sqrt{2}}$$

105. $$\frac{\sqrt{10}}{\sqrt{3x}} = \frac{\sqrt{10}}{\sqrt{3x}}\cdot\frac{\sqrt{10}}{\sqrt{10}} = \frac{10}{\sqrt{30x}}$$

107. $$\frac{\sqrt{7}+\sqrt{3}}{5} = \frac{\sqrt{7}+\sqrt{3}}{5}\cdot\frac{\sqrt{7}-\sqrt{3}}{\sqrt{7}-\sqrt{3}}$$
$$= \frac{(\sqrt{7})^2-(\sqrt{3})^2}{5(\sqrt{7}-\sqrt{3})}$$
$$= \frac{7-3}{5(\sqrt{7}-\sqrt{3})}$$
$$= \frac{4}{5(\sqrt{7}-\sqrt{3})}$$

109. $$\frac{\sqrt{y}-5}{\sqrt{3}} = \frac{\sqrt{y}-5}{\sqrt{3}}\cdot\frac{\sqrt{y}+5}{\sqrt{y}+5}$$
$$= \frac{(\sqrt{y})^2-5^2}{\sqrt{3}(\sqrt{y}+5)}$$
$$= \frac{y-25}{\sqrt{3y}+5\sqrt{3}} \text{ or } \frac{y-25}{\sqrt{3}(\sqrt{y}+\sqrt{5})}$$

111. Area $= h \cdot w$
$$= \sqrt{24^2-(8\sqrt{3})^2}\cdot 8\sqrt{3}$$
$$= \sqrt{576-192}\cdot 8\sqrt{3}$$
$$= \sqrt{384}\cdot 8\sqrt{3}$$
$$= 8\sqrt{1152}$$
$$= 8\sqrt{2^7\cdot 3^2}$$
$$= 8\cdot 2^3\cdot 3\sqrt{2}$$
$$= 192\sqrt{2} \text{ square inches}$$

113. $$\frac{500k}{\dfrac{1}{\sqrt{k^2+1}}+\dfrac{k^2}{\sqrt{k^2+1}}} = \frac{500k}{\dfrac{1+k^2}{\sqrt{k^2+1}}} = \frac{500k\sqrt{k^2+1}}{1+k^2}$$

115. (a) If either a or b (or both) equal zero, the expression $a\sqrt{b}$ is zero and therefore rational.

(b) If the product of a and b is a perfect square, then the expression is rational.

117. Conjugate of $\sqrt{a}+\sqrt{b}$ is $\sqrt{a}-\sqrt{b}$.

Product $= (\sqrt{a}+\sqrt{b})(\sqrt{a}-\sqrt{b})$
$= (\sqrt{a})^2-(\sqrt{b})^2 = a-b$

Conjugate of $\sqrt{b}+\sqrt{a}$ is $\sqrt{b}-\sqrt{a}$.

Product $= (\sqrt{b}+\sqrt{a})(\sqrt{b}-\sqrt{a})$
$= (\sqrt{b})^2-(\sqrt{a})^2 = b-a$

By changing the order of the terms, the conjugate and the product both change by a factor of -1.

119. $3x - 18 = 0$
$3x = 18$
$x = 6$

121. $3x - 4 = 3x$
$-4 = 0$
no solution

123. $x^2 - 144 = 0$
$(x+12)(x-12) = 0$
$x + 12 = 0$ $\quad$ $x - 12 = 0$
$x = -12$ $\quad$ $x = 12$

125. $x^2 + 2x - 15 = 0$

$(x + 5)(x - 3) = 0$

$x + 5 = 0 \qquad x - 3 = 0$

$x = -5 \qquad x = 3$

127. $\sqrt{32x^2y^5} = \sqrt{16 \cdot 2 \cdot x^2 \cdot y^4 \cdot y} = 4|x|y^2\sqrt{2y}$

129. $\sqrt[4]{32x^2y^5} = \sqrt[4]{16 \cdot 2 \cdot x^2 \cdot y^4 \cdot y} = 2y\sqrt[4]{2x^2y}$

Section 7.5 Radical Equations and Applications

1. (a) $x = -4$ $\quad \sqrt{-4} - 10 \neq 0$ $\quad$ Not a solution

(b) $x = -100$ $\quad \sqrt{-100} - 10 \neq 0$ $\quad$ Not a solution

(c) $x = \sqrt{10}$ $\quad \sqrt{\sqrt{10}} - 10 \neq 0$ $\quad$ Not a solution

(d) $x = 100$ $\quad \sqrt{100} - 10 = 0$ $\quad$ A solution

3. (a) $x = -60$ $\quad \sqrt[3]{-60 - 4} \neq 4$ $\quad$ Not a solution

(b) $x = 68$ $\quad \sqrt[3]{68 - 4} = 4$ $\quad$ A solution

(c) $x = 20$ $\quad \sqrt[3]{20 - 4} \neq 4$ $\quad$ Not a solution

(d) $x = 0$ $\quad \sqrt[3]{0 - 4} \neq 4$ $\quad$ Not a solution

5. $\sqrt{x} = 12$

$\left(\sqrt{x}\right)^2 = 12^2$

$x = 144$

Check: $\sqrt{144} \stackrel{?}{=} 12$

$12 = 12$

7. $\sqrt{y} = 7$

$\left(\sqrt{y}\right)^2 = 7^2$

$y = 49$

Check: $\sqrt{49} \stackrel{?}{=} 7$

$7 = 7$

9. $\sqrt[3]{z} = 3$

$\left(\sqrt[3]{z}\right)^3 = 3^3$

$z = 27$

Check: $\sqrt[3]{27} \stackrel{?}{=} 3$

$3 = 3$

11. $\sqrt{y} - 7 = 0$

$\sqrt{y} = 7$

$\left(\sqrt{y}\right)^2 = 7^2$

$y = 49$

Check: $\sqrt{49} - 7 \stackrel{?}{=} 0$

$7 - 7 \stackrel{?}{=} 0$

$0 = 0$

13. $\sqrt{u} + 13 = 0$

$\sqrt{u} = -13$

$\left(\sqrt{u}\right)^2 = (-13)^2$

$u = 169$

Check: $\sqrt{169} + 13 \stackrel{?}{=} 0$

$13 + 13 \neq 0$

No solution

15. $\sqrt{x} - 8 = 0$

$\sqrt{x} = 8$

$\left(\sqrt{x}\right)^2 = 8^2$

$x = 64$

Check: $\sqrt{64} - 8 \stackrel{?}{=} 0$

$8 - 8 \stackrel{?}{=} 0$

$0 = 0$

17. $\sqrt{10x} = 30$

$\left(\sqrt{10x}\right)^2 = 30^2$

$10x = 900$

$x = 90$

Check: $\sqrt{10 \cdot 90} \stackrel{?}{=} 30$

$\sqrt{900} \stackrel{?}{=} 30$

$30 = 30$

19. $\sqrt{-3x} = 9$

$$\left(\sqrt{-3x}\right)^2 = 9^2$$
$$-3x = 81$$
$$x = -27$$

Check: $\sqrt{-3(-27)} \stackrel{?}{=} 9$

$$\sqrt{81} \stackrel{?}{=} 9$$
$$9 = 9$$

21. $\sqrt{5t} - 2 = 0$

$$\sqrt{5t} = 2$$
$$\left(\sqrt{5t}\right)^2 = 2^2$$
$$5t = 4$$
$$t = \tfrac{4}{5}$$

Check: $\sqrt{5\left(\tfrac{4}{5}\right)} - 2 \stackrel{?}{=} 0$

$$\sqrt{4} - 2 \stackrel{?}{=} 0$$
$$2 - 2 \stackrel{?}{=} 0$$
$$0 = 0$$

23. $\sqrt{3y + 1} = 4$

$$\left(\sqrt{3y + 1}\right)^2 = 4^2$$
$$3y + 1 = 16$$
$$3y = 15$$
$$y = 5$$

Check: $\sqrt{3(5) + 1} \stackrel{?}{=} 4$

$$\sqrt{16} \stackrel{?}{=} 4$$
$$4 = 4$$

25. $\sqrt{9 - 2x} = -9$

$$\left(\sqrt{9 - 2x}\right)^2 = (-9)^2$$
$$9 - 2x = 81$$
$$-2x = 72$$
$$x = -36$$

Check: $\sqrt{9 - 2(-36)} \stackrel{?}{=} -9$

$$\sqrt{9 + 72} \stackrel{?}{=} -9$$
$$\sqrt{81} \stackrel{?}{=} -9$$
$$9 \neq -9$$

No solution

27. $\sqrt[3]{y - 3 + 4} = 6$

$$\left(\sqrt[3]{y - 3 + 4}\right)^3 = (6)^3$$
$$y - 3 + 4 = 216$$
$$y = 215$$

Check: $\sqrt[3]{215 - 3 + 4} \stackrel{?}{=} 6$

$$\sqrt[3]{216} \stackrel{?}{=} 6$$
$$6 = 6$$

29. $6\sqrt[4]{x + 3} = 15$

$$\left(6\sqrt[4]{x + 3}\right)^4 = (15)^4$$
$$1296(x + 3) = 50{,}625$$
$$1296x + 3888 = 50{,}625$$
$$1296x = 46{,}737$$
$$x = \frac{46{,}737}{1296}$$
$$x = \frac{577}{16}$$

Check: $6\sqrt[4]{\frac{577}{16} + 3} \stackrel{?}{=} 15$

$$6\sqrt[4]{\frac{577}{16} + \frac{48}{16}} \stackrel{?}{=} 15$$
$$6\sqrt[4]{\frac{625}{16}} \stackrel{?}{=} 15$$
$$6\left(\frac{5}{2}\right) \stackrel{?}{=} 15$$
$$15 = 15$$

31. $\sqrt{x + 3} = \sqrt{2x - 1}$

$$\left(\sqrt{x + 3}\right)^2 = \left(\sqrt{2x - 1}\right)^2$$
$$x + 3 = 2x - 1$$
$$4 = x$$

Check: $\sqrt{4 + 3} \stackrel{?}{=} \sqrt{2(4) - 1}$

$$\sqrt{7} = \sqrt{7}$$

33. $\sqrt{3y - 5} - 3\sqrt{y} = 0$

$$\sqrt{3y - 5} = 3\sqrt{y}$$
$$\left(\sqrt{3y - 5}\right)^2 = \left(3\sqrt{y}\right)^2$$
$$3y - 5 = 9y$$
$$-5 = 6y$$
$$-\tfrac{5}{6} = y$$

Check: $\sqrt{3\left(-\tfrac{5}{6}\right) - 5} - 3\sqrt{-\tfrac{5}{6}} \stackrel{?}{=} 0$

No solution

35. $\sqrt[3]{3x-4} = \sqrt[3]{x+10}$

$\left(\sqrt[3]{3x-4}\right)^3 = \left(\sqrt[3]{x+10}\right)^3$

$3x - 4 = x + 10$

$2x = 14$

$x = 7$

Check: $\sqrt[3]{3(7)-4} \stackrel{?}{=} \sqrt[3]{7+10}$

$\sqrt[3]{17} = \sqrt[3]{17}$

37. $\sqrt[3]{2x+15} - \sqrt[3]{x} = 0$

$\sqrt[3]{2x+15} = \sqrt[3]{x}$

$\left(\sqrt[3]{2x+15}\right)^3 = \left(\sqrt[3]{x}\right)^3$

$2x + 15 = x$

$x = -15$

Check: $\sqrt[3]{2(-15)+15} - \sqrt[3]{-15} \stackrel{?}{=} 0$

$\sqrt[3]{-15} - \sqrt[3]{-15} \stackrel{?}{=} 0$

$0 = 0$

39. $\sqrt{x^2-2} = x + 4$

$\left(\sqrt{x^2-2}\right)^2 = (x+4)^2$

$x^2 - 2 = x^2 + 8x + 16$

$-2 = 8x + 16$

$-18 = 8x$

$-\frac{18}{8} = x$

$-\frac{9}{4} = x$

Check: $\sqrt{\left(-\frac{9}{4}\right)^2 - 2} \stackrel{?}{=} -\frac{9}{4} + 4$

$\sqrt{\frac{81}{16} - \frac{32}{16}} \stackrel{?}{=} -\frac{9}{4} + \frac{16}{4}$

$\sqrt{\frac{49}{16}} \stackrel{?}{=} \frac{7}{4}$

$\frac{7}{4} = \frac{7}{4}$

41. $\sqrt{2x} = x - 4$

$\left(\sqrt{2x}\right)^2 = (x-4)^2$

$2x = x^2 - 8x + 16$

$0 = x^2 - 10x + 16$

$0 = (x-8)(x-2)$

$8 = x, \quad x = 2$, Not a solution

Check: $\sqrt{2(8)} \stackrel{?}{=} 8 - 4$

$\sqrt{16} \stackrel{?}{=} 4$

$4 = 4$

$\sqrt{2(2)} \stackrel{?}{=} 2 - 4$

$\sqrt{4} \stackrel{?}{=} -2$

$2 \neq -2$

43. $\sqrt{8x+1} = x + 2$

$\left(\sqrt{8x+1}\right)^2 = (x+2)^2$

$8x + 1 = x^2 + 4x + 4$

$0 = x^2 - 4x + 3$

$0 = (x-3)(x-1)$

$3 = x, \quad x = 1$

Check: $\sqrt{8(3)+1} \stackrel{?}{=} 3 + 2$

$\sqrt{25} \stackrel{?}{=} 5$

$5 = 5$

$\sqrt{8(1)+1} \stackrel{?}{=} 1 + 2$

$\sqrt{9} \stackrel{?}{=} 3$

$3 = 3$

45. $\sqrt{3x+4} = \sqrt{4x+3}$

$\left(\sqrt{3x+4}\right)^2 = \left(\sqrt{4x+3}\right)^2$

$3x + 4 = 4x + 3$

$1 = x$

Check: $\sqrt{3(1)+4} \stackrel{?}{=} \sqrt{4(1)+3}$

$\sqrt{7} = \sqrt{7}$

47.
$$\sqrt{z+2} = 1 + \sqrt{z}$$
$$\left(\sqrt{z+2}\right)^2 = \left(1+\sqrt{z}\right)^2$$
$$z + 2 = 1 + 2\sqrt{z} + z$$
$$1 = 2\sqrt{z}$$
$$1^2 = \left(2\sqrt{z}\right)^2$$
$$1 = 4z$$
$$\tfrac{1}{4} = z$$

Check: $\sqrt{\frac{1}{4}+2} \stackrel{?}{=} 1 + \sqrt{\frac{1}{4}}$
$$\sqrt{\tfrac{9}{4}} \stackrel{?}{=} 1 + \tfrac{1}{2}$$
$$\tfrac{3}{2} = \tfrac{3}{2}$$

49.
$$\sqrt{2t+3} = 3 - \sqrt{2t}$$
$$\left(\sqrt{2t+3}\right)^2 = \left(3 - \sqrt{2t}\right)^2$$
$$2t + 3 = 9 - 6\sqrt{2t} + 2t$$
$$-6 = -6\sqrt{2t}$$
$$1 = \sqrt{2t}$$
$$1^2 = \left(\sqrt{2t}\right)^2$$
$$1 = 2t$$
$$\tfrac{1}{2} = t$$

Check: $\sqrt{2\left(\frac{1}{2}\right)+3} \stackrel{?}{=} 3 - \sqrt{2\left(\frac{1}{2}\right)}$
$$\sqrt{1+3} \stackrel{?}{=} 3 - \sqrt{1}$$
$$\sqrt{4} \stackrel{?}{=} 3 - 1$$
$$2 = 2$$

51.
$$\sqrt{x+5} - \sqrt{x} = 1$$
$$\sqrt{x+5} = 1 + \sqrt{x}$$
$$\left(\sqrt{x+5}\right)^2 = \left(1+\sqrt{x}\right)^2$$
$$x + 5 = 1 + 2\sqrt{x} + x$$
$$4 = 2\sqrt{x}$$
$$2 = \sqrt{x}$$
$$2^2 = \left(\sqrt{x}\right)^2$$
$$4 = x$$

Check: $\sqrt{4+5} - \sqrt{4} \stackrel{?}{=} 1$
$$\sqrt{9} - \sqrt{4} \stackrel{?}{=} 1$$
$$3 - 2 \stackrel{?}{=} 1$$
$$1 = 1$$

53.
$$\sqrt{x-6} + 3 = \sqrt{x+9}$$
$$\left(\sqrt{x-6}+3\right)^2 = \left(\sqrt{x+9}\right)^2$$
$$x - 6 + 6\sqrt{x-6} + 9 = x + 9$$
$$6\sqrt{x-6} + 3 = 9$$
$$6\sqrt{x-6} = 6$$
$$\sqrt{x-6} = 1$$
$$\left(\sqrt{x-6}\right)^2 = 1^2$$
$$x - 6 = 1$$
$$x = 7$$

Check: $\sqrt{7-6} + 3 \stackrel{?}{=} \sqrt{7+9}$
$$1 + 3 \stackrel{?}{=} \sqrt{16}$$
$$4 = 4$$

55.
$$t^{3/2} = 8$$
$$\sqrt{t^3} = 8$$
$$\left(\sqrt{t^3}\right)^2 = 8^2$$
$$t^3 = 64$$
$$t = 4$$

Check: $4^{3/2} \stackrel{?}{=} 8$
$$\left(\sqrt{4}\right)^3 \stackrel{?}{=} 8$$
$$2^3 \stackrel{?}{=} 8$$
$$8 = 8$$

57.
$$3y^{1/3} = 18$$
$$y^{1/3} = 6$$
$$\sqrt[3]{y} = 6$$
$$\left(\sqrt[3]{y}\right)^3 = 6^3$$
$$y = 216$$

Check: $3(216)^{1/3} \stackrel{?}{=} 18$
$$3\sqrt[3]{216} \stackrel{?}{=} 18$$
$$3 \cdot 6 \stackrel{?}{=} 18$$
$$18 = 18$$

59.
$$\begin{aligned}(x+4)^{2/3} &= 4\\ \sqrt[3]{(x+4)^2} &= 4\\ \left(\sqrt[3]{(x+4)^2}\right)^3 &= (4)^3\\ (x+4)^2 &= 64\\ x+4 &= \pm\sqrt{64}\\ x &= -4 \pm 8\\ &= 4, -12\end{aligned}$$

Check:
$$\begin{aligned}(4+4)^{2/3} &\stackrel{?}{=} 4\\ 8^{2/3} &\stackrel{?}{=} 4\\ 2^2 &= 4\\ (-12+4)^{2/3} &\stackrel{?}{=} 4\\ (-8)^{2/3} &\stackrel{?}{=} 4\\ (-2)^2 &= 4\end{aligned}$$

61.
$$\begin{aligned}(2x+5)^{1/3} + 3 &= 0\\ \sqrt[3]{(2x+5)} &= -3\\ \left(\sqrt[3]{2x+5}\right)^3 &= (-3)^3\\ 2x+5 &= -27\\ 2x &= -32\\ x &= -16\end{aligned}$$

Check:
$$\begin{aligned}(2(-16)+5)^{1/3} + 3 &\stackrel{?}{=} 0\\ (-32+5)^{1/3} + 3 &\stackrel{?}{=} 0\\ (-27)^{1/3} + 3 &\stackrel{?}{=} 0\\ -3+3 &\stackrel{?}{=} 0\\ 0 &= 0\end{aligned}$$

63. $\sqrt{x} = 2(2 - x)$

Keystrokes:

y_1: [Y=] [√] [X,T,θ] [ENTER]

y_2: 2 [(] 2 [−] [X,T,θ] [)] [GRAPH]

Approximate solution: $x \approx 1.407$

Check algebraically: $\sqrt{1.407} \stackrel{?}{=} 2(2 - 1.407)$

$1.186 = 1.186$

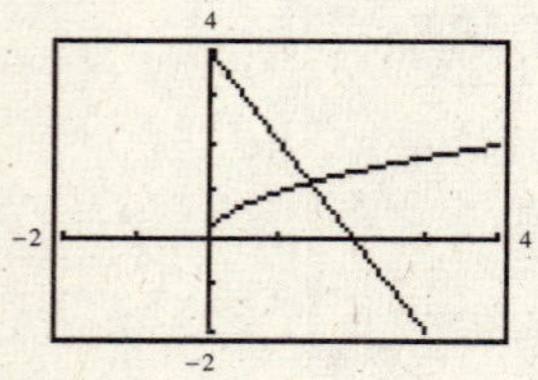

65. $\sqrt{x^2 + 1} = 5 - 2x$

Keystrokes:

y_1: [Y=] [√] [(] [X,T,θ] [x^2] [+] 1 [)] [ENTER]

y_2: 5 [−] 2 [X,T,θ] [GRAPH]

Approximate solution: $x \approx 1.569$

Check algebraically: $\sqrt{1.569^2 + 1} \stackrel{?}{=} 5 - 2(1.569)$

$1.86 = 1.86$

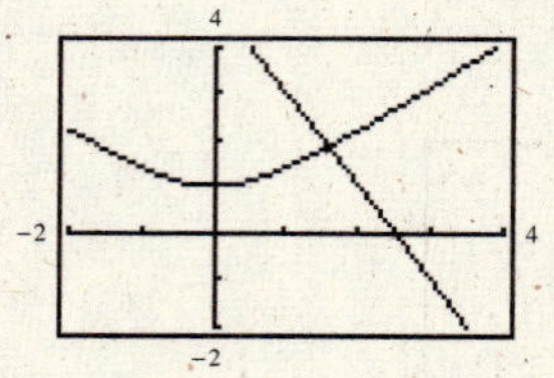

67. $\sqrt{x + 3} = 5 - \sqrt{x}$

Keystrokes:

y_1: [Y=] [√] [(] [X,T,θ] [+] 3 [)] [ENTER]

y_2: 5 [−] [√] [X,T,θ] [GRAPH]

Approximate solution: $x \approx 4.840$

Check algebraically: $\sqrt{4.840 + 3} \stackrel{?}{=} 5 - \sqrt{4.840}$

$2.8 = 2.8$

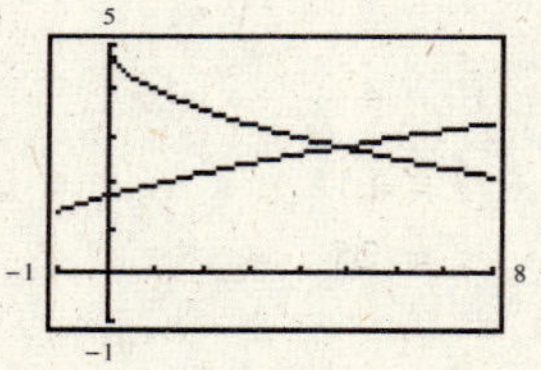

69. $3\sqrt[4]{x} = 9 - x$

Keystrokes:

y_1: [Y=] [3] [X,T,θ] [^] [(] 1 [÷] 4 [)] [ENTER]

y_2: 9 [−] [X,T,θ] [GRAPH]

Approximate solution: $x \approx 4.605$

Check algebraically: $3\sqrt[4]{4.605} \stackrel{?}{=} 9 - 4.605$

$4.395 = 4.395$

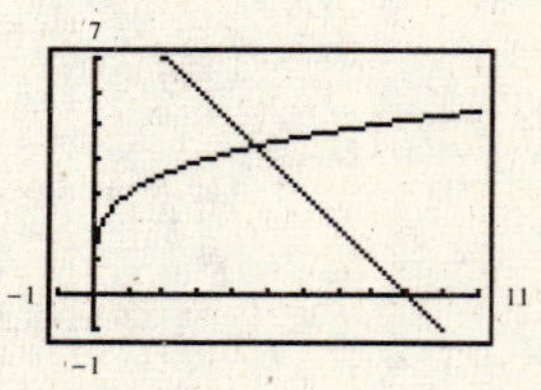

71. $\sqrt{15 - 4x} = 2x$

Keystrokes:

y_1: [Y=] [√] [(] 15 [−] 4 [X,T,θ] [)] [ENTER]

y_2: [2] [X,T,θ] [GRAPH]

Solution: $x = 1.5$

Check algebraically: $\sqrt{15 - 4(1.5)} \stackrel{?}{=} 2(1.5)$

$$\sqrt{9} \stackrel{?}{=} 3$$
$$3 = 3$$

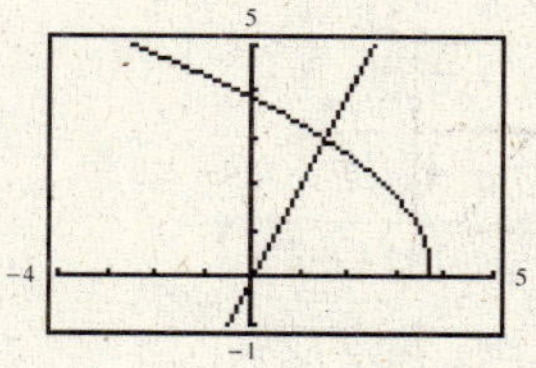

73.

$$\begin{aligned}
1 &= \sqrt{x} - \sqrt{x - 9} \\
1 + \sqrt{x - 9} &= \sqrt{x} \\
\left(1 + \sqrt{x - 9}\right)^2 &= \left(\sqrt{x}\right)^2 \\
1 + 2\sqrt{x - 9} + x - 9 &= x \\
2\sqrt{x - 9} - 8 &= 0 \\
2\sqrt{x - 9} &= 8 \\
\sqrt{x - 9} &= 4 \\
\left(\sqrt{x - 9}\right)^2 &= 4^2 \\
x - 9 &= 16 \\
x &= 25
\end{aligned}$$

Check: $1 \stackrel{?}{=} \sqrt{25} - \sqrt{25 - 9}$

$$1 \stackrel{?}{=} 5 - \sqrt{16}$$
$$1 \stackrel{?}{=} 5 - 4$$
$$1 = 1$$

75.

$$\begin{aligned}
-3 &= \sqrt{x - 2} - \sqrt{4x + 1} \\
\sqrt{4x + 1} - 3 &= \sqrt{x - 2} \\
\left(\sqrt{4x + 1} - 3\right)^2 &= \left(\sqrt{x - 2}\right)^2 \\
4x + 1 - 6\sqrt{4x + 1} + 9 &= x - 2 \\
-6\sqrt{4x + 1} + 10 &= -3x - 2 \\
-6\sqrt{4x + 1} &= -3x - 12 \\
2\sqrt{4x + 1} &= x + 4 \\
\left(2\sqrt{4x + 1}\right)^2 &= (x + 4)^2 \\
4(4x + 1) &= x^2 + 8x + 16 \\
16x + 4 &= x^2 + 8x + 16 \\
0 &= x^2 - 8x + 12 \\
0 &= (x - 6)(x - 2) \\
x = 6, &\quad x = 2
\end{aligned}$$

Check: $-3 \stackrel{?}{=} \sqrt{6 - 2} - \sqrt{4(6) + 1}$

$$-3 \stackrel{?}{=} \sqrt{4} - \sqrt{25}$$
$$-3 \stackrel{?}{=} 2 - 5$$
$$-3 = -3$$
$$-3 \stackrel{?}{=} \sqrt{2 - 2} - \sqrt{4(2) + 1}$$
$$-3 \stackrel{?}{=} \sqrt{0} - \sqrt{9}$$
$$-3 \stackrel{?}{=} 0 - 3$$
$$-3 = -3$$

77.

$$\begin{aligned}
0 &= \sqrt{x + 5} - 3 + \sqrt{x} \\
3 - \sqrt{x} &= \sqrt{x + 5} \\
\left(3 - \sqrt{x}\right)^2 &= \left(\sqrt{x + 5}\right)^2 \\
9 - 6\sqrt{x} + x &= x + 5 \\
-6\sqrt{x} &= -4 \\
3\sqrt{x} &= 2 \\
\left(3\sqrt{x}\right)^2 &= (2)^2 \\
9x &= 4 \\
x &= \tfrac{4}{9}
\end{aligned}$$

Check: $0 \stackrel{?}{=} \sqrt{\frac{4}{9} + 5} - 3 + \sqrt{\frac{4}{9}}$

$$0 \stackrel{?}{=} \sqrt{\frac{4}{9} + \frac{45}{9}} - 3 + \frac{2}{3}$$
$$0 \stackrel{?}{=} \sqrt{\frac{49}{9}} - 3 + \frac{2}{3}$$
$$0 \stackrel{?}{=} \frac{7}{3} - \frac{9}{3} + \frac{2}{3}$$
$$0 = 0$$

79.

$$0 = \sqrt{3x-2} - 1 - \sqrt{2x-3}$$
$$\sqrt{2x-3} = \sqrt{3x-2} - 1$$
$$\left(\sqrt{2x-3}\right)^2 = \left(\sqrt{3x-2} - 1\right)^2$$
$$2x - 3 = 3x - 2 - 2\sqrt{3x-2} + 1$$
$$-x - 2 = -2\sqrt{3x-2}$$
$$x + 2 = 2\sqrt{3x-2}$$
$$(x+2)^2 = \left(2\sqrt{3x-2}\right)^2$$
$$x^2 + 4x + 4 = 4(3x-2)$$
$$x^2 + 4x + 4 = 12x - 8$$
$$x^2 - 8x + 12 = 0$$
$$(x-2)(x-6) = 0$$
$$x - 2 = 0 \qquad x - 6 = 0$$
$$x = 2 \qquad x = 6$$

Check: $0 \stackrel{?}{=} \sqrt{3(2)-2} - 1 - \sqrt{2(2)-3}$

$0 \stackrel{?}{=} \sqrt{4} - 1 - \sqrt{1}$

$0 \stackrel{?}{=} 2 - 1 - 1$

$0 = 0$

Check: $0 \stackrel{?}{=} \sqrt{3(6)-2} - 1 - \sqrt{2(6)-3}$

$0 \stackrel{?}{=} \sqrt{16} - 1 - \sqrt{9}$

$0 \stackrel{?}{=} 4 - 1 - 3$

$0 = 0$

81.

$$c^2 = a^2 + b^2$$
$$15^2 = x^2 + 12^2$$
$$225 = x^2 + 144$$
$$81 = x^2$$
$$\sqrt{81} = x^2$$
$$9 = x$$

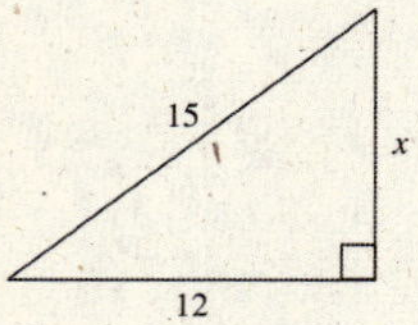

83.

$$c^2 = a^2 + b^2$$
$$13^2 = x^2 + 5^2$$
$$169 = x^2 + 25$$
$$144 = x^2$$
$$\sqrt{144} = x$$
$$12 = x$$

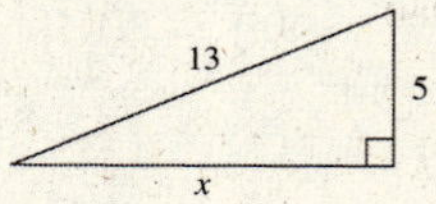

85.

$$c = \sqrt{a^2 + b^2}$$
$$50 = \sqrt{a^2 + 43.75^2}$$
$$50 = \sqrt{a^2 + 1914.0625}$$
$$(50)^2 = \left(\sqrt{a^2 + 1914.0625}\right)^2$$
$$2500 = a^2 + 1914.0625$$
$$585.9375 = a^2$$
$$24.21 \text{ inches} \approx a$$

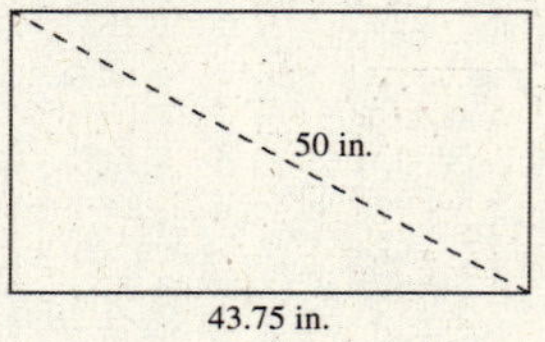

87. $x^2 = 6^2 + 2^2$

$x^2 = 36 + 4$

$x^2 = 40$

$x = \sqrt{40}$

$x = \sqrt{4 \cdot 10}$

$x = 2\sqrt{10} \approx 6.32$ meters

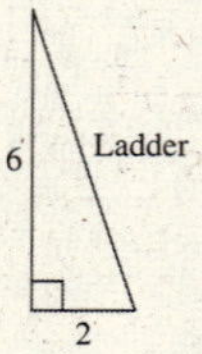

89. $17^2 = x^2 + 8^2$

$x^2 = 17^2 - 8^2$

$x = \sqrt{289 - 64}$

$x = \sqrt{225}$

$x = 15$ feet

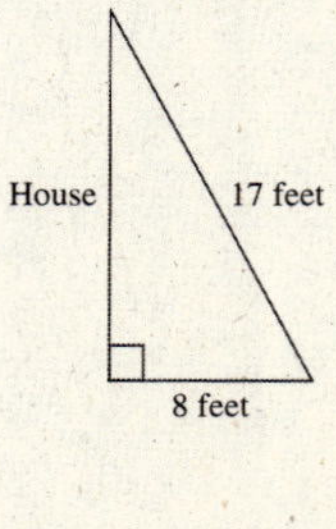

91.
$$\begin{aligned} P &= 2l + 2w \\ 92 &= 2l + 2w \\ 46 &= l + w \\ 46 - w &= l \\ 34^2 &= w^2 + (46 - w)^2 \\ 1156 &= w^2 + 2116 - 92w + w^2 \\ 0 &= 2w^2 - 92w + 960 \\ 0 &= w^2 - 46w + 480 \\ 0 &= (w - 30)(w - 16) \end{aligned}$$

$w = 30 \qquad w = 16$

$l = 16 \qquad l = 30$

30 inches × 16 inches

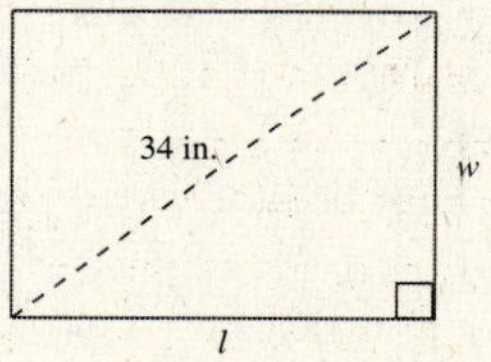

93.
$$\begin{aligned} S &= \pi r\sqrt{r^2 + h^2} \\ \frac{S}{\pi r} &= \sqrt{r^2 + h^2} \\ \left(\frac{S}{\pi r}\right)^2 &= \left(\sqrt{r^2 + h^2}\right)^2 \\ \frac{S^2}{\pi^2 r^2} &= r^2 + h^2 \\ \frac{S^2}{\pi^2 r^2} - r^2 &= h^2 \\ \frac{S^2 - \pi^2 r^4}{\pi^2 r^2} &= h^2 \\ \sqrt{\frac{S^2 - \pi^2 r^4}{\pi^2 r^2}} &= h \\ \frac{\sqrt{S^2 - \pi^2 r^4}}{\pi r} &= h \end{aligned}$$

$$\begin{aligned} h &= \frac{\sqrt{\left(364\pi\sqrt{2}\right)^2 - \pi^2(14)^4}}{\pi(14)} \\ &= \frac{\sqrt{264{,}992\pi^2 - 38{,}416\pi^2}}{14\pi} \\ &= \frac{\sqrt{226{,}576\pi^2}}{14\pi} \\ &= \frac{476\pi}{14\pi} \\ &= 34 \text{ cm} \end{aligned}$$

95. Area of circle $= \pi r^2$
$$\begin{aligned} A &= \pi r^2 \\ \frac{A}{\pi} &= r^2 \\ \sqrt{\frac{A}{\pi}} &= r \end{aligned}$$

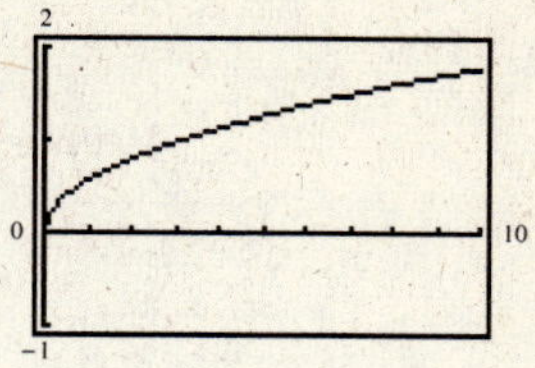

Keystrokes:

[Y=] [√] [(] [X,T,θ] [÷] [π] [)] [GRAPH]

97.
$$\begin{aligned} t &= \sqrt{\frac{d}{16}} \\ 2 &= \sqrt{\frac{d}{16}} \\ 2^2 &= \left(\sqrt{\frac{d}{16}}\right)^2 \\ 4 &= \frac{d}{16} \\ 64 \text{ feet} &= d \end{aligned}$$

99.
$$\begin{aligned} v &= \sqrt{2gh} \\ v &= \sqrt{2(32)(80)} \\ v &= \sqrt{5120} \\ v &= \sqrt{1024 \cdot 5} \\ v &= 32\sqrt{5} \approx 71.55 \text{ feet per second} \end{aligned}$$

101.
$$\begin{aligned} v &= \sqrt{2gh} \\ 50 &= \sqrt{2(32)h} \\ 50 &= \sqrt{64h} \\ 2500 &= 64h \\ \tfrac{2500}{64} &= h \approx 39.06 \text{ feet} \end{aligned}$$

103.
$$\begin{aligned} t &= 2\pi\sqrt{\frac{L}{32}} \\ 1.5 &= 2\pi\sqrt{\frac{L}{32}} \\ \left(\frac{1.5}{2\pi}\right)^2 &= \left(\sqrt{\frac{L}{32}}\right)^2 \\ \frac{2.25}{4\pi^2} &= \frac{L}{32} \\ \frac{2.25}{4\pi^2}(32) &= L \\ 1.82 \text{ feet} &\approx L \end{aligned}$$

105.
$$\begin{aligned} p &= 50 - \sqrt{0.8(x-1)} \\ 30.02 &= 50 - \sqrt{0.8(x-1)} \\ \sqrt{0.8(x-1)} &= 19.98 \\ \left(\sqrt{0.8(x-1)}\right)^2 &= (19.98)^2 \\ 0.8(x-1) &= 399.2004 \\ 0.8x - 0.8 &= 399.2004 \\ 0.8x &= 400.0004 \\ x &= 500.0005 \\ &\approx 500 \text{ units} \end{aligned}$$

107. (a) *Keystrokes:*

[Y=] 280 [+] 325 [√] [X,T,θ] [GRAPH]

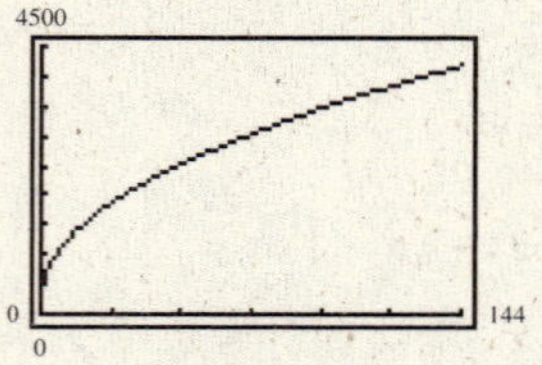

(b)
$$\begin{aligned} 3400 &= 280 + 325\sqrt{t} \\ 3120 &= 325\sqrt{t} \\ 9.6 &= \sqrt{t} \\ (9.6)^2 &= \left(\sqrt{t}\right)^2 \\ 92.16 &= t \approx 92 \text{ months} \end{aligned}$$

117.
$$\begin{aligned} 4x - y &= 10 \\ -7x - 2y &= -25 \end{aligned} \Rightarrow \begin{bmatrix} 4 & -1 & \vdots & 10 \\ -7 & -2 & \vdots & -25 \end{bmatrix} \tfrac{1}{4}R_1 \begin{bmatrix} 1 & -\tfrac{1}{4} & \vdots & \tfrac{5}{2} \\ -7 & -2 & \vdots & -25 \end{bmatrix}$$

$$7R_1 + R_2 \begin{bmatrix} 1 & -\tfrac{1}{4} & \vdots & \tfrac{5}{2} \\ 0 & -\tfrac{15}{4} & \vdots & -\tfrac{15}{2} \end{bmatrix} -\tfrac{4}{15}R_2 \begin{bmatrix} 1 & -\tfrac{1}{4} & \vdots & \tfrac{5}{2} \\ 0 & 1 & \vdots & 2 \end{bmatrix}$$

$$\tfrac{1}{4}R_2 + R_1 \begin{bmatrix} 1 & 0 & \vdots & 3 \\ 0 & 1 & \vdots & 2 \end{bmatrix}$$

(3, 2)

119. $a^{3/5} \cdot a^{1/5} = a^{3/5+1/5} = a^{4/5}$

109. $\left(\sqrt{x} + \sqrt{6}\right)^2 \neq \left(\sqrt{x}\right)^2 + \left(\sqrt{6}\right)^2$

$\left(\sqrt{x} + \sqrt{6}\right)^2$ must be multiplied by FOIL.

111. Substitute $x = 20$ into the equation, then choose any value of a such that $a \leq 20$ and solve the resulting equation for b.

Example:

$$\begin{aligned} 20 + \sqrt{20-4} &= b \quad \text{let } a = 4 \\ 20 + \sqrt{16} &= b \\ 20 + 4 &= b \\ 24 &= b \end{aligned}$$

113. L_1 and L_2 are parallel because $m_1 = 4$ and $m_2 = 4$. $m_1 = m_2$

115. L_1 and L_2 are perpendicular because $m_1 = -1$ and $m_2 = 1$. $m_1 \cdot m_2 = -1$

121.
$$\left(\frac{x^{1/2}}{x^{1/8}}\right)^4 = \left(x^{1/2-1/8}\right)^4 = \left(x^{4/8-1/8}\right)^4 = \left(x^{3/8}\right)^4 = x^{3/8 \cdot 4/1} = x^{3/2}$$

Section 7.6 Complex Numbers

1. $\sqrt{-4} = \sqrt{-1 \cdot 4} = \sqrt{-1} \cdot \sqrt{4} = 2i$

3. $-\sqrt{-144} = -\sqrt{144 \cdot -1} = -\sqrt{144} \cdot \sqrt{-1} = -12i$

5. $\sqrt{-\frac{4}{25}} = \sqrt{\frac{4}{25} \cdot -1} = \sqrt{\frac{4}{25}} \cdot \sqrt{-1} = \frac{2}{5}i$

7. $-\sqrt{-\frac{36}{121}} = -\sqrt{\frac{36}{121} \cdot -1} = -\sqrt{\frac{36}{121}} \cdot \sqrt{-1} = -\frac{6}{11}i$

9. $\sqrt{-8} = \sqrt{4 \cdot 2 \cdot -1} = \sqrt{4} \cdot \sqrt{2} \cdot \sqrt{-1} = 2\sqrt{2}i$

11. $\sqrt{-7} = \sqrt{7 \cdot -1} = \sqrt{7} \cdot \sqrt{-1} = \sqrt{7}i$

13. $\dfrac{\sqrt{-12}}{\sqrt{-3}} = \dfrac{\sqrt{4\cdot 3\cdot -1}}{\sqrt{3\cdot -1}}$

$= \dfrac{\sqrt{4}\cdot\sqrt{3}\cdot\sqrt{-1}}{\sqrt{3}\cdot\sqrt{-1}}$

$= \sqrt{4} = 2$ or

$\dfrac{\sqrt{-12}}{\sqrt{-3}} = \sqrt{\dfrac{-12}{-3}} = \sqrt{4} = 2$

15. $\sqrt{-\dfrac{18}{25}} = \sqrt{-1}\cdot\dfrac{\sqrt{18}}{\sqrt{25}} = i\cdot\dfrac{\sqrt{9\cdot 2}}{5} = \dfrac{3\sqrt{2}}{5}i$

17. $\sqrt{-0.09} = \sqrt{0.09\cdot -1} = \sqrt{0.09}\cdot\sqrt{-1} = 0.3i$

19. $\sqrt{-16} + \sqrt{-36} = 4i + 6i = (4+6)i = 10i$

21. $\sqrt{-9} - \sqrt{-1} = 3i - i = 2i$

23. $\sqrt{-50} - \sqrt{-8} = 5i\sqrt{2} - 2i\sqrt{2}$

$= \left(5\sqrt{2} - 2\sqrt{2}\right)i$

$= 3\sqrt{2}i$

25. $\sqrt{-48} + \sqrt{-12} - \sqrt{-27} = \sqrt{16\cdot 3\cdot -1} + \sqrt{4\cdot 3\cdot -1} - \sqrt{9\cdot 3\cdot -1}$

$= 4i\sqrt{3} + 2i\sqrt{3} - 3i\sqrt{3}$

$= (4i + 2i - 3i)\sqrt{3}$

$= 3\sqrt{3}i$

27. $\sqrt{-12}\sqrt{-2} = 2\sqrt{3}i\cdot\sqrt{2}i = 2\sqrt{6}i^2 = -2\sqrt{6}$

29. $\sqrt{-18}\sqrt{-3} = \left(3i\sqrt{2}\right)\left(i\sqrt{3}\right)$

$= 3\sqrt{6}\cdot i^2$

$= -3\sqrt{6}$

31. $\sqrt{-0.16}\sqrt{-1.21} = (0.4i)(1.1i) = 0.44i^2 = -0.44$

33. $\sqrt{-3}\left(\sqrt{-3} + \sqrt{-4}\right) = i\sqrt{3}\left(i\sqrt{3} + 2i\right)$

$= \left(i\sqrt{3}\right)^2 + 2\sqrt{3}i^2$

$= -3 - 2\sqrt{3}$

35. $\sqrt{-5}\left(\sqrt{-16} - \sqrt{-10}\right) = i\sqrt{5}\left(4i - i\sqrt{10}\right)$

$= i^2 4\sqrt{5} - i^2\sqrt{50}$

$= -4\sqrt{5} + 5\sqrt{2}$

$= 5\sqrt{2} - 4\sqrt{5}$

37. $\sqrt{-2}\left(3 - \sqrt{-8}\right) = i\sqrt{2}\left(3 - 2i\sqrt{2}\right)$

$= 3\sqrt{2}i - 2i^2(2)$

$= 4 + 3\sqrt{2}i$

39. $\left(\sqrt{-16}\right)^2 = (4i)^2 = 16i^2 = -16$

41. $\left(\sqrt{-4}\right)^3 = (2i)^3 = 8i^3 = -8i$

43. $\sqrt{1} + \sqrt{-25} = 1 + 5i$

45. $\sqrt{27} - \sqrt{-8} = 3\sqrt{3} - 2\sqrt{2}i \neq 3\sqrt{3} + 2\sqrt{2}i$

47. $3 - 4i = a + bi$

$a = 3, \quad b = -4$

49. $5 - 4i = (a+3) + (b-1)i$

$a + 3 = 5 \qquad b - 1 = -4$

$a = 2 \qquad b = -3$

51. $-4 - \sqrt{-8} = a + bi$

$-4 - 2i\sqrt{2} = a + bi$

$-4 = a \qquad -2i\sqrt{2} = bi$

$-2\sqrt{2} = b$

53. $\sqrt{a} + \sqrt{-49} = 8 + bi$

$\sqrt{a} = 8 \qquad \sqrt{-49} = bi$

$a = 64 \qquad 7i = bi$

$7 = b$

55. $(4 - 3i) + (6 + 7i) = (4 + 6) + (-3 + 7)i$

$= 10 + 4i$

57. $(-4 - 7i) + (-10 - 33i) = (-4 - 10) + (-7 - 33)i$

$= -14 - 40i$

59. $13i - (14 - 7i) = (-14) + (13 + 7)i$

$= -14 + 20i$

61. $(30 - i) - (18 + 6i) + 3i^2 = 30 - i - 18 - 6i - 3$

$= 9 - 7i$

63. $6 - (3 - 4i) + 2i = 6 - 3 + 4i + 2i = 3 + 6i$

65. $\left(\frac{4}{3} + \frac{1}{3}i\right) + \left(\frac{5}{6} + \frac{7}{6}i\right) = \left(\frac{4}{3} + \frac{5}{6}\right) + \left(\frac{1}{3} + \frac{7}{6}\right)i = \left(\frac{8}{6} + \frac{5}{6}\right) + \left(\frac{2}{6} + \frac{7}{6}\right)i = \frac{13}{6} + \frac{9}{6}i = \frac{13}{6} + \frac{3}{2}i$

67. $(0.05 + 2.50i) - (6.2 + 11.8i) = (0.05 - 6.2) + (2.50 - 11.8)i = -6.15 - 9.3i$

69. $15i - (3 - 25i) + \sqrt{-81} = 15i - 3 + 25i + 9i = -3 + (15 + 25 + 9)i = -3 + 49i$

71. $(3i)(12i) = 36i^2 = -36$

73. $(3i)(-8i) = -24i^2 = -24(-1) = 24$

75. $(-5i)(-i)(\sqrt{-49}) = -5i \cdot -i \cdot 7i = 35i^3$
$= 35 \cdot i^2 \cdot i = 35 \cdot (-1) \cdot i = -35i$

77. $(-3i)^3 = -27i^3 = 27i$

79. $(-3i)^2 = 9i^2 = 9(-1) = -9$

81. $-5(13 + 2i) = -65 - 10i$

83. $4i(-3 - 5i) = -12i - 20i^2 = 20 - 12i$

85. $(9 - 2i)(\sqrt{-4}) = (9 - 2i)(2i)$
$= 18i - 4i^2$
$= 18i + 4$
$= 4 + 18i$

87. $(4 + 3i)(-7 + 4i) = -28 + 16i - 21i + 12i^2$
$= -28 - 12 - 5i = -40 - 5i$

89. $(-7 + 7i)(4 - 2i) = -28 + 14i + 28i - 14i^2$
$= -28 + 42i + 14$
$= -14 + 42i$

91. $(-2 + \sqrt{-5})(-2 - \sqrt{-5}) = (-2 + i\sqrt{5})(-2 - i\sqrt{5})$
$= 4 + 2i\sqrt{5} - 2i\sqrt{5} - 5i^2$
$= 4 + 5$
$= 9$

93. $(3 - 4i)^2 = 3^2 - 2(3)(4i) + (4i)^2$
$= 9 - 24i + 16i^2$
$= 9 - 16 - 24i$
$= -7 - 24i$

95. $(2 + 5i)^2 = 2^2 + 2(2)(5i) + (5i)^2$
$= 4 + 20i + 25i^2$
$= 4 - 25 + 20i$
$= -21 + 20i$

97. $(3 + i)^3 = (3 + i)^2(3 + i)$
$= (9 + 6i + i^2)(3 + i)$
$= (9 + 6i - 1)(3 + i)$
$= (8 + 6i)(3 + i)$
$= 24 + 8i + 18i + 6i^2$
$= 24 + (8 + 18)i + 6(-1)$
$= (24 - 6) + 26i$
$= 18 + 26i$

99. $i^7 = (i^4)(i^2)i = (1)(-1)i = -i$

101. $i^{24} = (i^4)^6 = 1^6 = 1$

103. $i^{42} = (i^4)^{10}(i^2) = (1)^{10}(-1) = -1$

105. $i^9 = (i^4)^2 i = (1)^2 i = i$

107. $(-i)^6 = i^6 = i^4 \cdot i^2 = 1 \cdot -1 = -1$

109. $2 + i$, conjugate $= 2 - i$
Product $= (2 + i)(2 - i)$
$= 2^2 - i^2 = 4 + 1 = 5$

111. $-2 - 8i$, conjugate $= -2 + 8i$
Product $= (-2 - 8i)(-2 + 8i)$
$= (-2)^2 - (8i)^2$
$= 4 - 64i^2 = 4 + 64 = 68$

113. $5 - \sqrt{6}i$, conjugate $= 5 + \sqrt{6}i$
Product $= (5 - \sqrt{6}i)(5 + \sqrt{6}i)$
$= 5^2 - (\sqrt{6}i)^2$
$= 25 - 6i^2 = 25 + 6 = 31$

115. $10i$, conjugate $= -10i$
Product $= (10i)(-10i) = -(10i)^2 = -100i^2 = 100$

117. -12, conjugate: $-12 - 0i$
Product: $(-12 + 0i)(-12 - 0i) = 144$

119. $1 + \sqrt{-3} = 1 + i\sqrt{3}$, conjugate $= 1 - i\sqrt{3}$
Product $= (1 + i\sqrt{3})(1 - i\sqrt{3})$
$= 1^2 - (i\sqrt{3})^2$
$= 1 - 3i^2 = 1 + 3 = 4$

121. $1.5 + \sqrt{-0.25} = 1.5 + 0.5i$, conjugate $= 1.5 - 0.5i$

$$\begin{aligned}\text{Product} &= (1.5 + 0.5i)(1.5 - 0.5i)\\ &= 1.5^2 - (0.5i)^2\\ &= 2.25 + 0.25 = 2.5\end{aligned}$$

123. $\dfrac{20}{2i} = \dfrac{10}{i} \cdot \dfrac{-i}{-i} = \dfrac{-10i}{1} = -10i$

125. $\dfrac{2+i}{-5i} = \dfrac{2+i}{-5i} \cdot \dfrac{i}{i}$

$$= \frac{(2+i)i}{-5 \cdot i^2} = \frac{2i + i^2}{-5(-1)} = \frac{-1+2i}{5} = -\frac{1}{5} + \frac{2}{5}i$$

127. $\dfrac{4}{1-i} = \dfrac{4}{1-i} \cdot \dfrac{1+i}{1+i}$

$$\begin{aligned}&= \frac{4(1+i)}{1+1}\\ &= \frac{4(1+i)}{2}\\ &= 2(1+i)\\ &= 2 + 2i\end{aligned}$$

129. $\dfrac{7i+14}{7i} = \dfrac{7i+14}{7i} \cdot \dfrac{-i}{-i}$

$$\begin{aligned}&= \frac{-i(7i+14)}{-7i^2}\\ &= \frac{-7i^2 - 14i}{-7(-1)}\\ &= \frac{7 - 14i}{7}\\ &= \frac{7}{7} - \frac{14}{7}i\\ &= 1 - 2i\end{aligned}$$

131. $\dfrac{-12}{2+7i} = \dfrac{-12}{2+7i} \cdot \dfrac{2-7i}{2-7i}$

$$\begin{aligned}&= \frac{-12(2-7i)}{4+49}\\ &= \frac{-12(2-7i)}{53}\\ &= \frac{-24+84i}{53}\\ &= -\frac{24}{53} + \frac{84}{53}i\end{aligned}$$

133. $\dfrac{3i}{5+2i} = \dfrac{3i}{5+2i} \cdot \dfrac{5-2i}{5-2i}$

$$\begin{aligned}&= \frac{3i(5-2i)}{25+4}\\ &= \frac{15i+6}{29}\\ &= \frac{6}{29} + \frac{15}{29}i\end{aligned}$$

135. $\dfrac{5-i}{5+i} = \dfrac{5-i}{5+i} \cdot \dfrac{5-i}{5-i} = \dfrac{25 - 2(5i) + i^2}{25 - i^2} = \dfrac{25 - 10i - 1}{25 - (-1)}$

$$= \frac{24 - 10i}{26} = \frac{24}{26} - \frac{10}{26}i = \frac{12}{13} - \frac{5}{13}i$$

137. $\dfrac{4+5i}{3-7i} = \dfrac{4+5i}{3-7i} \cdot \dfrac{3+7i}{3+7i}$

$$\begin{aligned}&= \frac{(4+5i)(3+7i)}{9+49}\\ &= \frac{12 + 28i + 15i - 35}{58}\\ &= \frac{-23+43i}{58}\\ &= -\frac{23}{58} + \frac{43}{58}i\end{aligned}$$

139. $\dfrac{5}{3+i} + \dfrac{1}{3-i} = \dfrac{5}{3+i} \cdot \dfrac{3-i}{3-i} + \dfrac{1}{3-i} \cdot \dfrac{3+i}{3+i}$

$$\begin{aligned}&= \frac{5(3-i)}{9+1} + \frac{3+i}{9+1}\\ &= \frac{15 - 5i + 3 + i}{10}\\ &= \frac{18 - 4i}{10}\\ &= \frac{18}{10} - \frac{4}{10}i\\ &= \frac{9}{5} - \frac{2}{5}i\end{aligned}$$

141. $\dfrac{3i}{1+i} + \dfrac{2}{2+3i} = \dfrac{3i}{1+i}\cdot\dfrac{1-i}{1-i} + \dfrac{2}{2+3i}\cdot\dfrac{2-3i}{2-3i}$

$$= \frac{3i(1-i)}{1+1} + \frac{2(2-3i)}{4+9}$$

$$= \frac{3i(1-i)}{2} + \frac{2(2-3i)}{13}$$

$$= \frac{3i \cdot 13(1-i) + 2 \cdot 2(2-3i)}{26}$$

$$= \frac{39i(1-i) + 4(2-3i)}{26}$$

$$= \frac{39i + 39 + 8 - 12i}{26}$$

$$= \frac{47 + 27i}{26}$$

$$= \frac{47}{26} + \frac{27}{26}i$$

143. $\dfrac{1+i}{i} - \dfrac{3}{5-2i} = \dfrac{1+i}{i}\cdot\dfrac{-i}{-i} - \dfrac{3}{5-2i}\cdot\dfrac{5+2i}{5+2i}$

$$= \frac{-i(1+i)}{1} - \frac{3(5+2i)}{25+4}$$

$$= \frac{-i(1+i)}{1} - \frac{3(5+2i)}{29}$$

$$= \frac{-29i(1+i) - 3(5+2i)}{29}$$

$$= \frac{-29i + 29 - 15 - 6i}{29}$$

$$= \frac{14 - 35i}{29}$$

$$= \frac{14}{29} - \frac{35}{29}i$$

145. $x^2 + 2x + 5 = 0$

(a) $x = -1 + 2i$

$$(-1+2i)^2 + 2(-1+2i) + 5 \stackrel{?}{=} 0$$

$$1 - 4i + 4i^2 - 2 + 4i + 5 \stackrel{?}{=} 0$$

$$1 - 4 - 2 + 5 \stackrel{?}{=} 0$$

$$0 = 0$$

Solution

(b) $x = -1 - 2i$

$$(-1-2i)^2 + 2(-1-2i) + 5 \stackrel{?}{=} 0$$

$$1 + 4i + 4i^2 - 2 - 4i + 5 \stackrel{?}{=} 0$$

$$1 - 4 - 2 + 5 \stackrel{?}{=} 0$$

$$0 = 0$$

Solution

147. $x^3 + 4x^2 + 9x + 36 = 0$

(a) $x = -4$

$$(-4)^3 + 4(-4)^2 + 9(-4) + 36 \stackrel{?}{=} 0$$

$$-64 + 64 - 36 + 36 \stackrel{?}{=} 0$$

$$0 = 0$$

Solution

(b) $x = -3i$

$$(-3i)^3 + 4(-3i)^2 + 9(-3i) + 36 \stackrel{?}{=} 0$$

$$-27i^3 + 36i^2 - 27i + 36 \stackrel{?}{=} 0$$

$$27i - 36 - 27i + 36 \stackrel{?}{=} 0$$

$$0 = 0$$

Solution

149. (a) $\left(\dfrac{-5 + 5\sqrt{3}i}{2}\right)^3 = \left(\dfrac{-5}{2} + \dfrac{5}{2}\sqrt{3}i\right)^2\left(\dfrac{-5}{2} + \dfrac{5}{2}\sqrt{3}i\right)$

$$= \left(\frac{25}{4} - \frac{25}{2}\sqrt{3}i + \frac{25}{4}(3)i^2\right)\left(\frac{-5}{2} + \frac{5}{2}\sqrt{3}i\right)$$

$$= \left(\frac{25}{4} - \frac{25}{2}\sqrt{3}i - \frac{75}{4}\right)\left(\frac{-5}{2} + \frac{5}{2}\sqrt{3}i\right)$$

$$= \left(\frac{-50}{4} - \frac{25}{2}\sqrt{3}i\right)\left(\frac{-5}{2} + \frac{5}{2}\sqrt{3}i\right)$$

$$= \left(\frac{-25}{2} - \frac{25}{2}\sqrt{3}i\right)\left(\frac{-5}{2} + \frac{5}{2}\sqrt{3}i\right)$$

$$= \frac{125}{4} - \frac{125}{4}\sqrt{3}i + \frac{125}{4}\sqrt{3}i - \frac{125}{4}(3)i^2$$

$$= \frac{125}{4} + \frac{375}{4} = \frac{500}{4} = 125$$

(b) Use the same method as part (a).

151. (a) $1, \dfrac{-1+\sqrt{3}i}{2}, \dfrac{-1-\sqrt{3}i}{2}$

(b) $2, \dfrac{-2+2\sqrt{3}i}{2} = -1+\sqrt{3}i,$

$\dfrac{-2-2\sqrt{3}i}{2} = -1-\sqrt{3}i$

(c) $4, \dfrac{-4+4\sqrt{3}i}{2} = -2+2\sqrt{3}i,$

$\dfrac{-4-4\sqrt{3}i}{2} = -2-2\sqrt{3}i$

153. $(a+bi)+(a-bi) = (a+a)+(b-b)i = 2a+0i$

155. $(a+bi)-(a-bi) = (a-a)+(b+b)i = 2bi$

157. Exercise 153: The sum of complex conjugates of the form $a+bi$ and $a-bi$ is twice the real number a, or $2a$.

Exercise 154: The product of complex conjugates of the form $a+bi$ and $a-bi$ is the sum of squares of a and b, or a^2+b^2.

Exercise 155: The difference of complex conjugates of the form $a+bi$ and $a-bi$ is twice the imaginary number bi, or $2bi$.

Exercise 156: The sum of the squares of complex conjugates of the form $a+bi$ and $a-bi$ is the difference of twice the squares of a and b, or $2a^2-2b^2$.

159. The numbers must be written in i-form first.

$\sqrt{-3}\sqrt{-3} = (\sqrt{3}i)(\sqrt{3}i) = 3i^2 = 3(-1) = -3$

161. To simplify the quotient, multiply the numerator and the denominator by $-bi$. This will yield a positive real number in the denominator. The number i can also be used to simplify the quotient. The denominator will be the opposite of b, but the resulting number will be the same.

163. $(x-5)(x+7) = 0$

$x-5=0 \qquad x+7=0$

$x=5 \qquad x=-7$

165. $3y(y-3)(y+4) = 0$

$3y=0 \qquad y-3=0 \qquad y+4=0$

$y=0 \qquad y=3 \qquad y=-4$

167. $\sqrt{x} = 9$

$(\sqrt{x})^2 = (9)^2$

$x = 81$

Check: $\sqrt{81} \stackrel{?}{=} 9$

$9 = 9$

169. $\sqrt{x} - 5 = 0$

$\sqrt{x} = 5$

$(\sqrt{x})^2 = (5)^2$

$x = 25$

Check: $\sqrt{25} - 5 \stackrel{?}{=} 0$

$5 - 5 \stackrel{?}{=} 0$

$0 = 0$

Review Exercises for Chapter 7

1. $-\sqrt{81} = -9$ because $9 \cdot 9 = 81$.

3. $-\sqrt[3]{64} = -4$ because $4 \cdot 4 \cdot 4 = 64$.

5. $-\sqrt{\left(\frac{3}{4}\right)^2} = -\frac{3}{4}$

7. $\sqrt[3]{-\left(\frac{1}{5}\right)^3} = -\frac{1}{5}$ (inverse property of powers and roots)

9. $\sqrt{-2^2} = 2i$ (not a real number)

11.

Radical Form	Rational Exponent Form
$\sqrt[3]{27} = 3$	$27^{1/3} = 3$

13. $\sqrt[3]{216} = 6$

15. $27^{4/3} = \left(\sqrt[3]{27}\right)^4 = 3^4 = 81$

17. $(-25)^{3/2} = \left(\sqrt{-25}\right)^3$ (not a real number)

19. $8^{-4/3} = \dfrac{1}{8^{4/3}} = \dfrac{1}{\left(\sqrt[3]{8}\right)^4} = \dfrac{1}{2^4} = \dfrac{1}{16}$

21. $x^{3/4} \cdot x^{-1/6} = x^{3/4+(-1/6)} = x^{9/12+(-2/12)} = x^{7/12}$

23. $z\sqrt[3]{z^2} = z \cdot z^{2/3} = z^{1+2/3} = z^{5/3}$

25. $\dfrac{\sqrt[4]{x^3}}{\sqrt{x^4}} = \dfrac{x^{3/4}}{x^{4/2}} = x^{3/4-2} = x^{3/4-8/4} = x^{-5/4} = \dfrac{1}{x^{5/4}}$

27. $\sqrt[3]{a^3b^2} = a\sqrt[3]{b^2} = ab^{2/3}$

29. $\sqrt[4]{\sqrt{x}} = \sqrt[4]{x^{1/2}} = \left(x^{1/2}\right)^{1/4} = x^{1/8}$

31. $\dfrac{(3x+2)^{2/3}}{\sqrt[3]{3x+2}} = \dfrac{(3x+2)^{2/3}}{(3x+2)^{1/3}}$
$= (3x+2)^{2/3-1/3}$
$= (3x+2)^{1/3}$
$= \sqrt[3]{3x+2},\ x \neq -\dfrac{2}{3}$

33. $75^{-3/4} = 0.0392377 \approx 0.0392$

35. $\sqrt{13^2 - 4(2)(7)} = 10.630146 \approx 10.6301$

37. $f(x) = \sqrt{x-2}$

(a) $f(-7) = \sqrt{-7-2} = \sqrt{-9}$ (not a real number)

(b) $f(51) = \sqrt{51-2} = \sqrt{49} = 7$

39. $g(x) = \sqrt[3]{2x-1}$

(a) $g(0) = \sqrt[3]{2(0)-1} = \sqrt[3]{-1} = -1$

(b) $g(14) = \sqrt[3]{2(14)-1} = \sqrt[3]{27} = 3$

41. $f(x) = \sqrt{9-2x}$

Domain: $\left(-\infty, \frac{9}{2}\right]$

$9 - 2x \geq 0$
$-2x \geq -9$
$x \leq \frac{9}{2}$

43. $\sqrt{36u^5v^2} = \sqrt{6^2 \cdot u^4 \cdot u \cdot v^2} = 6u^2|v|\sqrt{u}$

45. $\sqrt{0.25x^4y} = \sqrt{25 \times 10^{-2}x^4y}$
$= 5 \times 10^{-1}x^2\sqrt{y}$
$= 0.5x^2\sqrt{y}$

47. $\sqrt[3]{48a^3b^4} = \sqrt[3]{8 \cdot 6a^3b^3b} = 2ab\sqrt[3]{6b}$

49. $\sqrt{\dfrac{5}{6}} = \sqrt{\dfrac{5}{6}} \cdot \dfrac{\sqrt{6}}{\sqrt{6}} = \dfrac{\sqrt{30}}{6}$

51. $\dfrac{2}{\sqrt[3]{2x}} = \dfrac{2}{\sqrt[3]{2x}} \cdot \dfrac{\sqrt[3]{2^2x^2}}{\sqrt[3]{2^2x^2}} = \dfrac{2\sqrt[3]{4x^2}}{\sqrt[3]{8x^3}} = \dfrac{2\sqrt[3]{4x^2}}{2x} = \dfrac{\sqrt[3]{4x^2}}{x}$

53. $c = \sqrt{a^2 + b^2}$
$c = \sqrt{9^2 + 8^2}$
$c = \sqrt{81 + 64}$
$c = \sqrt{145}$

55. $2\sqrt{24} + 7\sqrt{6} - \sqrt{54} = 2\sqrt{4 \cdot 6} + 7\sqrt{6} - \sqrt{9 \cdot 6} = 4\sqrt{6} + 7\sqrt{6} - 3\sqrt{6} = (4 + 7 - 3)\sqrt{6} = 8\sqrt{6}$

57. $5\sqrt{x} - \sqrt[3]{x} + 9\sqrt{x} - 8\sqrt[3]{x} = 5\sqrt{x} + 9\sqrt{x} - \sqrt[3]{x} - 8\sqrt[3]{x} = (5 + 9)\sqrt{x} + (-1 - 8)\sqrt[3]{x} = 14\sqrt{x} - 9\sqrt[3]{x}$

59. $10\sqrt[4]{y+3} - 3\sqrt[4]{y+3} = (10-3)\sqrt[4]{y+3} = 7\sqrt[4]{y+3}$

61. $2x\sqrt[3]{24x^2y} - \sqrt[3]{3x^5y} = 2x\sqrt[3]{8 \cdot 3 \cdot x^2y} - \sqrt[3]{3 \cdot x^3 \cdot x^2 \cdot y} = 4x\sqrt[3]{3x^2y} - x\sqrt[3]{3x^2y} = 3x\sqrt[3]{3x^2y}$

63. $c = \sqrt{a^2 + b^2}$
$c = \sqrt{\left(2\sqrt{x}\right)^2 + \left(2\sqrt{3x}\right)^2}$
$c = \sqrt{4x + 12x}$
$c = \sqrt{16x}$
$c = 4\sqrt{x}$

Perimeter $= 2\sqrt{x} + 2\sqrt{3x} + 4\sqrt{x} = 6\sqrt{x} + 2\sqrt{3x}$

65. $\sqrt{15} \cdot \sqrt{20} = \sqrt{15 \cdot 20}$
$= \sqrt{300}$
$= \sqrt{100 \cdot 3}$
$= 10\sqrt{3}$

67. $\sqrt{10}\left(\sqrt{2} + \sqrt{5}\right) = \sqrt{10}\sqrt{2} + \sqrt{10}\sqrt{5}$
$= \sqrt{20} + \sqrt{50}$
$= \sqrt{4 \cdot 5} + \sqrt{25 \cdot 2}$
$= 2\sqrt{5} + 5\sqrt{2}$

69. $\left(\sqrt{3}-\sqrt{x}\right)\left(\sqrt{3}+\sqrt{x}\right) = 3 - \sqrt{3x} + \sqrt{3x} - x$

$= 3 - x$

71. $3 - \sqrt{7}$

Conjugate: $3 + \sqrt{7}$

$\left(3-\sqrt{7}\right)\left(3+\sqrt{7}\right) = 3^2 - \left(\sqrt{7}\right)^2$

$= 9 - 7$

$= 2$

73. $\sqrt{x} + 20$

Conjugate: $\sqrt{x} - 20$

$\left(\sqrt{x}+20\right)\left(\sqrt{x}-20\right) = \left(\sqrt{x}\right)^2 - 20^2$

$= x - 400$

75. $$\frac{\sqrt{2}-1}{\sqrt{3}-4} = \frac{\sqrt{2}-1}{\sqrt{3}-4}\cdot\frac{\sqrt{3}+4}{\sqrt{3}+4} = \frac{\sqrt{6}+4\sqrt{2}-\sqrt{3}-4}{\left(\sqrt{3}\right)^2-4^2} = \frac{\sqrt{6}+4\sqrt{2}-\sqrt{3}-4}{3-16} = \frac{\sqrt{6}+4\sqrt{2}-\sqrt{3}-4}{-13} = -\frac{\sqrt{6}+4\sqrt{2}-\sqrt{3}-4}{13}$$

or $-\dfrac{\left(\sqrt{2}-1\right)\left(\sqrt{3}+4\right)}{13}$

77. $$\frac{\sqrt{x}+10}{\sqrt{x}-10} = \frac{\sqrt{x}+10}{\sqrt{x}-10}\cdot\frac{\sqrt{x}+10}{\sqrt{x}+10} = \frac{x+10\sqrt{x}+10\sqrt{x}+100}{\left(\sqrt{x}\right)^2-10^2} = \frac{x+20\sqrt{x}+100}{x-100}$$

79. $\sqrt{2x} - 8 = 0$

$\sqrt{2x} = 8$

$\left(\sqrt{2x}\right)^2 = 8^2$

$2x = 64$

$x = 32$

Check: $\sqrt{2(32)} - 8 \stackrel{?}{=} 0$

$\sqrt{64} - 8 \stackrel{?}{=} 0$

$8 - 8 \stackrel{?}{=} 0$

$0 = 0$

81. $\sqrt[4]{3x-1} + 6 = 3$

$\sqrt[4]{3x-1} = -3$

$\left(\sqrt[4]{3x-1}\right)^4 = (-3)^4$

$3x - 1 = 81$

$3x = 82$

$x = \frac{82}{3}$

Not a real solution

Check: $\sqrt[4]{3\left(\frac{82}{3}\right) - 1} + 6 \stackrel{?}{=} 3$

$\sqrt[4]{82-1} + 6 \stackrel{?}{=} 3$

$\sqrt[4]{81} + 6 \stackrel{?}{=} 3$

$3 + 6 \stackrel{?}{=} 3$

$9 \neq 3$

83. $\sqrt[3]{5x+2} - \sqrt[3]{7x-8} = 0$

$\sqrt[3]{5x+2} = \sqrt[3]{7x-8}$

$\left(\sqrt[3]{5x+2}\right)^3 = \left(\sqrt[3]{7x-8}\right)^3$

$5x + 2 = 7x - 8$

$10 = 2x$

$5 = x$

Check: $\sqrt[3]{5(5)+2} - \sqrt[3]{7(5)-8} \stackrel{?}{=} 0$

$\sqrt[3]{27} - \sqrt[3]{27} \stackrel{?}{=} 0$

$0 = 0$

85. $\sqrt{2(x+5)} = x + 5$

$\left(\sqrt{2(x+5)}\right)^2 = (x+5)^2$

$2(x+5) = x^2 + 10x + 25$

$2x + 10 = x^2 + 10x + 25$

$0 = x^2 + 8x + 15$

$0 = (x+5)(x+3)$

$-5 = x, \quad x = -3$

Check: $\sqrt{2(-5+5)} \stackrel{?}{=} -5 + 5$

$\sqrt{0} \stackrel{?}{=} 0$

$0 = 0$

$\sqrt{2(-3+5)} \stackrel{?}{=} -3 + 5$

$\sqrt{4} \stackrel{?}{=} 2$

$2 = 2$

87.
$$\sqrt{1+6x} = 2-\sqrt{6x}$$
$$\left(\sqrt{1+6x}\right)^2 = \left(2-\sqrt{6x}\right)^2$$
$$1+6x = 4-4\sqrt{6x}+6x$$
$$1 = 4-4\sqrt{6x}$$
$$-3 = -4\sqrt{6x}$$
$$(3)^2 = \left(4\sqrt{6x}\right)^2$$
$$9 = 16(6x)$$
$$\frac{9}{96} = x$$
$$\frac{3}{32} = x$$

Check:
$$\sqrt{1+6\left(\frac{3}{32}\right)} \stackrel{?}{=} 2-\sqrt{6\left(\frac{3}{32}\right)}$$
$$\sqrt{\frac{32}{32}+\frac{18}{32}} \stackrel{?}{=} 2-\sqrt{\frac{18}{32}}$$
$$\sqrt{\frac{50}{32}} \stackrel{?}{=} 2-\sqrt{\frac{9\cdot 2}{16\cdot 2}}$$
$$\sqrt{\frac{25\cdot 2}{16\cdot 2}} \stackrel{?}{=} 2-\sqrt{\frac{9\cdot 2}{16\cdot 2}}$$
$$\sqrt{\frac{25}{16}} \stackrel{?}{=} 2-\sqrt{\frac{9}{16}}$$
$$\frac{5}{4} \stackrel{?}{=} 2-\frac{3}{4}$$
$$\frac{5}{4} \stackrel{?}{=} \frac{8}{4}-\frac{3}{4}$$
$$\frac{5}{4} = \frac{5}{4}$$

89.
$$P = 2l+2w$$
$$46 = 2l+2w$$
$$23 = l+w$$
$$23-w = l$$
$$c^2 = a^2+b^2$$
$$17^2 = w^2+(23-w)^2$$
$$289 = w^2+529-46w+w^2$$
$$0 = 2w^2-46w+240$$
$$0 = w^2-23w+120$$
$$0 = (w-8)(w-15)$$

$w-8=0$ $\quad$ $w-15=0$

$w=8$ $\quad$ $w=15$

$l=23-8=15$ $\quad$ $l=23-15=8$

8 inches × 15 inches

91.
$$t = 2\pi\sqrt{\frac{L}{32}}$$
$$1.9 = 2\pi\sqrt{\frac{L}{32}}$$
$$\frac{1.9}{2\pi} = \sqrt{\frac{L}{32}}$$
$$\left(\frac{1.9}{2\pi}\right)^2 = \left(\sqrt{\frac{L}{32}}\right)^2$$
$$\frac{3.61}{4\pi^2} = \frac{L}{32}$$
$$\frac{3.61(32)}{4\pi^2} = L$$
$$2.93 \text{ feet} \approx L$$

93.
$$v = \sqrt{2gh}$$
$$64 = \sqrt{2gh}$$
$$64^2 = \left(\sqrt{2gh}\right)^2$$
$$4096 = 2(32)h$$
$$\tfrac{4096}{64} = h$$
$$64 \text{ feet} = h$$

95. $\sqrt{-48} = \sqrt{16\cdot 3\cdot -1} = 4\sqrt{3}i$

97.
$$10-3\sqrt{-27} = 10-3\sqrt{-1\cdot 9\cdot 3}$$
$$= 10-3\sqrt{-1}\cdot\sqrt{9}\cdot\sqrt{3}$$
$$= 10-9\sqrt{3}i$$

99.
$$\tfrac{3}{4}-5\sqrt{-\tfrac{3}{25}} = \tfrac{3}{4}-5\sqrt{\tfrac{3}{25}\cdot -1}$$
$$= \tfrac{3}{4}-\tfrac{5}{5}i\sqrt{3}$$
$$= \tfrac{3}{4}-\sqrt{3}i$$

101. $\sqrt{-81}+\sqrt{-36} = 9i+6i = 15i$

103.
$$\sqrt{-10}\left(\sqrt{-4}-\sqrt{-7}\right) = i\sqrt{10}\left(2i-i\sqrt{7}\right)$$
$$= 2i^2\sqrt{10}-i^2\sqrt{70}$$
$$= -2\sqrt{10}+\sqrt{70}$$

105. $12-5i = (a+2)+(b-1)i$

$a+2=12$ $\quad$ $b-1=-5$

$a=10$ $\quad$ $b=-4$

107.
$$\sqrt{-49}+4 = a+bi$$
$$7i+4 = a+bi$$
$$a=4,\quad b=7$$

109. $(-4 + 5i) - (-12 + 8i) = (-4 + 12) + (5 - 8)i$
$= 8 - 3i$

111. $(4 - 3i)(4 + 3i) = 4^2 - (3i)^2 = 16 + 9 = 25$

113. $(6 - 5i)^2 = 6^2 - 2(6)(5i) + (5i)^2$
$= 36 - 60i - 25$
$= 11 - 60i$

115. $\dfrac{7}{3i} = \dfrac{7}{3i} \cdot \dfrac{-i}{-i} = \dfrac{-7i}{-3i^2} = -\dfrac{7i}{3}$

117. $\dfrac{-3i}{4 - 6i} = \dfrac{-3i}{4 - 6i} \cdot \dfrac{4 + 6i}{4 + 6i} = \dfrac{-12i - 18i^2}{16 - 36i^2}$
$= \dfrac{-12i + 18}{16 + 36}$
$= \dfrac{18 - 12i}{52}$
$= \dfrac{18}{52} - \dfrac{12}{52}i$
$= \dfrac{9}{26} - \dfrac{3}{13}i$

119. $\dfrac{3 - 5i}{6 + i} = \dfrac{3 - 5i}{6 + i} \cdot \dfrac{6 - i}{6 - i}$
$= \dfrac{18 - 3i - 30i + 5i^2}{6^2 - i^2}$
$= \dfrac{18 - 33i - 5}{36 + 1}$
$= \dfrac{13 - 33i}{37}$
$= \dfrac{13}{37} - \dfrac{33}{37}i$

Chapter Test for Chapter 7

1. (a) $16^{3/2} = (\sqrt{16})^3 = 4^3 = 64$

(b) $\sqrt{5}\sqrt{20} = \sqrt{5 \cdot 20} = \sqrt{100} = 10$

2. (a) $125^{-2/3} = \dfrac{1}{125^{2/3}} = \dfrac{1}{(\sqrt[3]{125})^2} = \dfrac{1}{5^2} = \dfrac{1}{25}$

(b) $\sqrt{3}\sqrt{12} = \sqrt{3 \cdot 12} = \sqrt{36} = 6$

3. $f(x) = \sqrt{9 - 5x}$

(a) $f(-8) = \sqrt{9 - 5(-8)} = \sqrt{9 + 40} = \sqrt{49} = 7$

(b) $f(0) = \sqrt{9 - 5(0)} = \sqrt{9} = 3$

4. $g(x) = \sqrt{7x - 3}$
$7x - 3 \ge 0$
$7x \ge 3$
$x \ge \frac{3}{7}$
$\left[\frac{3}{7}, \infty\right)$

5. (a) $\left(\dfrac{x^{1/2}}{x^{1/3}}\right)^2 = \dfrac{x}{x^{2/3}} = x^{1-2/3} = x^{1/3},\ x \ne 0$

(b) $5^{1/4} \cdot 5^{7/4} = 5^{1/4+7/4} = 5^{8/4} = 5^2 = 25$

6. (a) $\sqrt{\dfrac{32}{9}} = \sqrt{\dfrac{16 \cdot 2}{9}} = \dfrac{4}{3}\sqrt{2}$

(b) $\sqrt[3]{24} = \sqrt[3]{8 \cdot 3} = 2\sqrt[3]{3}$

7. (a) $\sqrt{24x^3} = \sqrt{4 \cdot 6 \cdot x^2 \cdot x} = 2x\sqrt{6x}$

(b) $\sqrt[4]{16x^5y^8} = \sqrt[4]{16x^4xy^8} = 2xy^2\sqrt[4]{x}$

8. $\dfrac{2}{\sqrt[3]{9y}} = \dfrac{2}{\sqrt[3]{9y}} \cdot \dfrac{\sqrt[3]{3y^2}}{\sqrt[3]{3y^2}} = \dfrac{2\sqrt[3]{3y^2}}{\sqrt[3]{27y^3}} = \dfrac{2\sqrt[3]{3y^2}}{3y}$

9. $\dfrac{10}{\sqrt{6} - \sqrt{2}} = \dfrac{10}{\sqrt{6} - \sqrt{2}} \cdot \dfrac{\sqrt{6} + \sqrt{2}}{\sqrt{6} + \sqrt{2}}$
$= \dfrac{10(\sqrt{6} + \sqrt{2})}{(\sqrt{6})^2 - (\sqrt{2})^2}$
$= \dfrac{10(\sqrt{6} + \sqrt{2})}{6 - 2}$
$= \dfrac{10(\sqrt{6} + \sqrt{2})}{4}$
$= \dfrac{5(\sqrt{6} + \sqrt{2})}{2}$

10. $6\sqrt{18x} - 3\sqrt{32x} = 6\sqrt{9 \cdot 2x} - 3\sqrt{16 \cdot 2x}$
$= 18\sqrt{2x} - 12\sqrt{2x}$
$= 6\sqrt{2x}$

11. $\sqrt{5}\left(\sqrt{15x}+3\right) = \sqrt{75x}+3\sqrt{5}$
$= \sqrt{25\cdot 3x}+3\sqrt{5}$
$= 5\sqrt{3x}+3\sqrt{5}$

12. $\left(4-\sqrt{2x}\right)^2 = 16-8\sqrt{2x}+2x$

13. $7\sqrt{27}+14y\sqrt{12} = 7\sqrt{9\cdot 3}+14y\sqrt{4\cdot 3}$
$= 21\sqrt{3}+28y\sqrt{3}$
$= 7\sqrt{3}(3+4y)$

14. $\sqrt{6z}+5 = 17$
$\sqrt{6z} = 12$
$\left(\sqrt{6z}\right)^2 = 12^2$
$6z = 144$
$z = \frac{144}{6} = 24$

Check: $\sqrt{6(24)}+5 \stackrel{?}{=} 17$
$\sqrt{144}+5 \stackrel{?}{=} 17$
$12+5 \stackrel{?}{=} 17$
$17 = 17$

15. $\sqrt{x^2-1} = x-2$
$\left(\sqrt{x^2-1}\right)^2 = (x-2)^2$
$x^2-1 = x^2-4x+4$
$4x = 5$
$x = \frac{5}{4}$

No solution

Check: $\sqrt{\left(\frac{5}{4}\right)^2-1} \stackrel{?}{=} \frac{5}{4}-2$
$\sqrt{\frac{25}{16}-\frac{16}{16}} \stackrel{?}{=} \frac{5}{4}-\frac{8}{4}$
$\sqrt{\frac{9}{16}} \stackrel{?}{=} -\frac{3}{4}$
$\frac{3}{4} \neq -\frac{3}{4}$

16. $\sqrt{x}-x+6 = 0$
$\left(\sqrt{x}\right)^2 = (x-6)^2$
$x = x^2-12x+36$
$0 = x^2-13x+36$
$0 = (x-9)(x-4)$

$0 = x-9$ $\quad$ $0 = x-4$
$9 = x$ $\quad$ $4 = x$

Not a solution

Check: $\sqrt{9}-9+6 \stackrel{?}{=} 0$
$3-9+6 \stackrel{?}{=} 0$
$0 = 0$
$\sqrt{4}-4+6 \stackrel{?}{=} 0$
$2-4+6 \stackrel{?}{=} 0$
$4 \neq 0$

17. $(2+3i)-\sqrt{-25} = 2+3i-5i = 2-2i$

18. $(3-5i)^2 = 3^2-2(3)(5i)+(5i)^2$
$= 9-30i+25i^2$
$= 9-30i-25$
$= -16-30i$

19. $\sqrt{-16}\left(1+\sqrt{-4}\right) = 4i(1+2i) = 4i+8i^2 = -8+4i$

20. $(3-2i)(1+5i) = 3+13i-10i^2$
$= 3+13i+10$
$= 13+13i$

21. $\frac{5-2i}{3+i} = \frac{5-2i}{3+i}\cdot\frac{3-i}{3-i}$
$= \frac{(5-2i)(3-i)}{9+1}$
$= \frac{15-5i-6i+2i^2}{10}$
$= \frac{15-11i-2}{10}$
$= \frac{13}{10}-\frac{11}{10}i$

22. $v = \sqrt{2gh}$
$96 = \sqrt{2(32)h}$
$96^2 = \left(\sqrt{64h}\right)^2$
$9216 = 64h$
$\frac{9216}{64} = h$
144 feet $= h$

Cumulative Test for Chapters 5–7

1. $(-2x^5y^{-2}z^0)^{-1} = -2^{-1}x^{-5}y^2 = -\dfrac{y^2}{2x^5},\quad z \neq 0, y \neq 0$

2. $\dfrac{12s^5t^{-2}}{20s^{-2}t^{-1}} = \dfrac{3s^{5+2}t^{-2+1}}{5} = \dfrac{3s^7}{5t},\quad s \neq 0$

3. $\left(\dfrac{2x^{-4}y^3}{3x^5y^{-3}z^0}\right)^{-2} = \left(\dfrac{2x^{(-4)+(-5)}y^{3+3}}{3}\right)^{-2}$

$= \left(\dfrac{2x^{-9}y^6}{3}\right)^{-2} = \left(\dfrac{2y^6}{3x^9}\right)^{-2} = \left(\dfrac{3x^9}{2y^6}\right)^2$

$= \dfrac{9x^{18}}{4y^{12}},\ x \neq 0,\ z \neq 0$

4. $(5 \times 10^3)^2 = 5^2 \times 10^{3\cdot 2} = 25 \times 10^6 = 2.5 \times 10^7$

5. $(x^5 + 2x^3 + x^2 - 10x) - (2x^3 - x^2 + x - 4) = x^5 + 2x^3 + x^2 - 10x - 2x^3 + x^2 - x + 4$

$= x^5 + (2x^3 - 2x^3) + (x^2 + x^2) + (-10x - x) + 4$

$= x^5 + 2x^2 - 11x + 4$

6. $-3(3x^3 - 4x^2 + x) + 3x(2x^2 + x - 1) = -9x^3 + 12x^2 - 3x + 6x^3 + 3x^2 - 3x$

$= (-9x^3 + 6x^3) + (12x^2 + 3x^2) + (-3x - 3x)$

$= -3x^3 + 15x^2 - 6x$

7. $(x + 8)(3x - 2) = 3x^2 - 2x + 24x - 16 = 3x^2 + 22x - 16$

8. $(3x + 2)(3x^2 - x + 1) = 9x^3 - 3x^2 + 3x + 6x^2 - 2x + 2$

$= 9x^3 + (-3x^2 + 6x^2) + (3x - 2x) + 2$

$= 9x^3 + 3x^2 + x + 2$

9. $2x^2 - 11x + 15 = (2x - 5)(x - 3)$

10. $9x^2 - 144 = (3x - 12)(3x + 12)$

$= 3(x - 4)3(x + 4)$

$= 9(x - 4)(x + 4)$

11. $y^3 - 3y^2 - 9y + 27 = (y^3 - 3y^2) + (-9y + 27)$

$= y^2(y - 3) - 9(y - 3)$

$= (y - 3)(y^2 - 9)$

$= (y - 3)(y - 3)(y + 3)$

$= (y - 3)^2(y + 3)$

12. $8t^3 - 40t^2 + 50t = 2t(4t^2 - 20t + 25)$

$= 2t(2t - 5)(2t - 5)$

$= 2t(2t - 5)^2$

13. $3x^2 + x - 24 = 0$

$(3x - 8)(x + 3) = 0$

$3x - 8 = 0 \qquad x + 3 = 0$

$3x = 8 \qquad x = -3$

$x = \frac{8}{3}$

14. $6x^3 - 486x = 0$

$6x(x^2 - 81) = 0$

$6x(x + 9)(x - 9) = 0$

$6x = 0 \qquad x + 9 = 0 \qquad x - 9 = 0$

$x = 0 \qquad x = -9 \qquad x = 9$

15. $\dfrac{x^2 + 8x + 16}{18x^2} \cdot \dfrac{2x^4 + 4x^3}{x^2 - 16} = \dfrac{(x + 4)^2}{18x^2} \cdot \dfrac{2x^3(x + 2)}{(x - 4)(x + 4)} = \dfrac{x(x + 4)(x + 2)}{9(x - 4)},\ x \neq -4,\ x \neq 0$

16. $\dfrac{x^2+4x}{2x^2-7x+3} \div \dfrac{x^2-16}{x-3} = \dfrac{x(x+4)}{(2x-1)(x-3)} \cdot \dfrac{x-3}{(x-4)(x+4)}$

$= \dfrac{x(x+4)(x-3)}{(2x-1)(x-3)(x-4)(x+4)}$

$= \dfrac{x}{(2x-1)(x-4)}, \quad x \neq -4, x \neq 3$

17. $\dfrac{5x}{x+2} - \dfrac{2}{x^2-x-6} = \dfrac{5x}{x+2} - \dfrac{2}{(x-3)(x+2)} = \dfrac{5x(x-3)}{(x+2)(x-3)} - \dfrac{2}{(x-3)(x+2)} = \dfrac{5x^2-15x-2}{(x+2)(x-3)}$

18. $\dfrac{2}{x} - \dfrac{x}{x^3+3x^2} + \dfrac{1}{x+3} = \dfrac{2}{x} - \dfrac{x}{x^2(x+3)} + \dfrac{1}{x+3}$

$= \dfrac{2}{x} - \dfrac{1}{x(x+3)} + \dfrac{1}{x+3}$

$= \dfrac{2}{x}\left(\dfrac{x+3}{x+3}\right) - \dfrac{1}{x(x+3)}\left(\dfrac{1}{1}\right) + \dfrac{1}{x+3}\left(\dfrac{x}{x}\right)$

$= \dfrac{2x+6}{x(x+3)} - \dfrac{1}{x(x+3)} + \dfrac{x}{x(x+3)}$

$= \dfrac{2x+6-1+x}{x(x+3)}$

$= \dfrac{3x+5}{x(x+3)}$

19. $\dfrac{\left(\dfrac{3x}{x+2}\right)}{\left(\dfrac{12}{x^3+2x^2}\right)} = \dfrac{3x}{x+2} \div \dfrac{12}{x^3+2x^2} = \dfrac{3x}{x+2} \cdot \dfrac{x^2(x+2)}{12} = \dfrac{(3x)(x^2)(x+2)}{(x+2)12} = \dfrac{x^3}{4}, \quad x \neq -2, x \neq 0$

20. $\dfrac{\left(\dfrac{x}{y} - \dfrac{y}{x}\right)}{\left(\dfrac{x-y}{xy}\right)} = \dfrac{\left(\dfrac{x}{y} - \dfrac{y}{x}\right)}{\left(\dfrac{x-y}{xy}\right)} \cdot \dfrac{xy}{xy} = \dfrac{x^2-y^2}{x-y} = \dfrac{(x-y)(x+y)}{x-y} = x+y, \quad x \neq 0, y \neq 0, x \neq y$

21.

−4	2	7	0	−5
		−8	4	−16
	2	−1	4	−21

$(2x^3+7x^2-5) \div (x+4) = 2x^2 - x + 4 - \dfrac{21}{x+4}$

22.

$$\begin{array}{r} 2x^3 - 2x^2 - x - \dfrac{4}{2x-1} \\ 2x-1 \overline{\big)\, 4x^4 - 6x^3 + 0x^2 + x - 4} \\ \underline{4x^4 - 2x^3} \qquad\qquad\qquad \\ -4x^3 + 0x^2 \qquad\quad \\ \underline{-4x^3 + 2x^2} \qquad\quad \\ -2x^2 + x \\ \underline{-2x^2 + x} \end{array}$$

23.

$\dfrac{1}{x} + \dfrac{4}{10-x} = 1$

$x(10-x)\left(\dfrac{1}{x} + \dfrac{4}{10-x}\right) = (1)x(10-x)$

$10 - x + 4x = 10x - x^2$

$x^2 - 7x + 10 = 0$

$(x-5)(x-2) = 0$

$x = 5, \quad x = 2$

Check: $\dfrac{1}{5} + \dfrac{4}{10-5} \stackrel{?}{=} 1$ $\qquad$ $\dfrac{1}{2} + \dfrac{4}{10-2} \stackrel{?}{=} 1$

$\dfrac{1}{5} + \dfrac{4}{5} \stackrel{?}{=} 1$ $\qquad$ $\dfrac{1}{2} + \dfrac{4}{8} \stackrel{?}{=} 1$

$\dfrac{5}{5} \stackrel{?}{=} 1$ $\qquad$ $\dfrac{1}{2} + \dfrac{1}{2} \stackrel{?}{=} 1$

$1 = 1$ $\qquad$ $1 = 1$

24.
$$\frac{x-3}{x}+1=\frac{x-4}{x-6}$$
$$x(x-6)\left(\frac{x-3}{x}+1\right)=\left(\frac{x-4}{x-6}\right)x(x-6)$$
$$(x-6)(x-3)+x(x-6)=x(x-4)$$
$$x^2-9x+18+x^2-6x=x^2-4x$$
$$x^2-11x+18=0$$
$$(x-9)(x-2)=0$$
$$x=9,\quad x=2$$

Check:

$$\frac{9-3}{9}+1\stackrel{?}{=}\frac{9-4}{9-6}$$
$$\frac{6}{9}+1\stackrel{?}{=}\frac{5}{3}$$
$$\frac{2}{3}+\frac{3}{3}\stackrel{?}{=}\frac{5}{3}$$
$$\frac{5}{3}=\frac{5}{3}$$

$$\frac{2-3}{2}+1\stackrel{?}{=}\frac{2-4}{2-6}$$
$$\frac{-1}{2}+\frac{2}{2}\stackrel{?}{=}\frac{-2}{-4}$$
$$\frac{1}{2}=\frac{1}{2}$$

25. $\sqrt{24x^2y^3}=\sqrt{4\cdot 6\cdot x^2\cdot y^2\cdot y}=2|x|y\sqrt{6y}$

26. $\sqrt[3]{80a^{15}b^8}=\sqrt[3]{8\cdot 10\cdot a^{15}\cdot b^6\cdot b^2}=2a^5b^2\sqrt[3]{10b^2}$

27. $\left(12a^{-4}b^6\right)^{1/2}=\sqrt{12}a^{-2}\left|b^3\right|=\dfrac{\sqrt{4\cdot 3}\left|b^3\right|}{a^2}=\dfrac{2\sqrt{3}\left|b^3\right|}{a^2}$

28. $\left(\dfrac{t^{1/2}}{t^{1/4}}\right)^2=\dfrac{t}{t^{1/2}}=t^{1-1/2}=t^{1/2}=\sqrt{t},\ t\neq 0$

29.
$$\begin{aligned}10\sqrt{20x}+3\sqrt{125x}&=10\sqrt{4\cdot 5x}+3\sqrt{25\cdot 5x}\\&=20\sqrt{5x}+15\sqrt{5x}\\&=35\sqrt{5x}\end{aligned}$$

30. $\left(\sqrt{2x}-3\right)^2=2x-6\sqrt{2x}+9$

31.
$$\begin{aligned}\frac{3}{\sqrt{10}-\sqrt{x}}&=\frac{3}{\sqrt{10}-\sqrt{x}}\cdot\frac{\sqrt{10}+\sqrt{x}}{\sqrt{10}+\sqrt{x}}\\&=\frac{3\left(\sqrt{10}+\sqrt{x}\right)}{\left(\sqrt{10}\right)^2-\left(\sqrt{x}\right)^2}\\&=\frac{3\left(\sqrt{10}+\sqrt{x}\right)}{10-x}\end{aligned}$$

32.
$$\sqrt{x-5}-6=0$$
$$\sqrt{x-5}=6$$
$$\left(\sqrt{x-5}\right)^2=6^2$$
$$x-5=36$$
$$x=41$$

Check:
$$\sqrt{41-5}-6\stackrel{?}{=}0$$
$$\sqrt{36}-6\stackrel{?}{=}0$$
$$6-6\stackrel{?}{=}0$$
$$0=0$$

33.
$$\sqrt{3-x}+10=11$$
$$\sqrt{3-x}=1$$
$$\left(\sqrt{3-x}\right)^2=1^2$$
$$3-x=1$$
$$2=x$$

Check:
$$\sqrt{3-2}+10\stackrel{?}{=}11$$
$$\sqrt{1}+10\stackrel{?}{=}11$$
$$1+10\stackrel{?}{=}11$$
$$11=11$$

34.
$$\sqrt{x+5}-\sqrt{x-7}=2$$
$$\sqrt{x+5}=2+\sqrt{x-7}$$
$$\left(\sqrt{x+5}\right)^2=\left(2+\sqrt{x-7}\right)^2$$
$$x+5=4+4\sqrt{x-7}+x-7$$
$$5=-3+4\sqrt{x-7}$$
$$8=4\sqrt{x-7}$$
$$2=\sqrt{x-7}$$
$$2^2=\left(\sqrt{x-7}\right)^2$$
$$4=x-7$$
$$11=x$$

Check:
$$\sqrt{11+5}-\sqrt{11-7}\stackrel{?}{=}2$$
$$\sqrt{16}-\sqrt{4}\stackrel{?}{=}2$$
$$4-2\stackrel{?}{=}2$$
$$2=2$$

35.
$$\begin{aligned}
\sqrt{x-4} &= \sqrt{x+7}-1\\
\left(\sqrt{x-4}\right)^2 &= \left(\sqrt{x+7}-1\right)^2\\
x-4 &= x+7-2\sqrt{x+7}+1\\
-4 &= 8-2\sqrt{x+7}\\
-12 &= -2\sqrt{x+7}\\
6 &= \sqrt{x+7}\\
6^2 &= \left(\sqrt{x+7}\right)^2\\
36 &= x+7\\
29 &= x
\end{aligned}$$

Check:
$$\begin{aligned}
\sqrt{29-4} &\stackrel{?}{=} \sqrt{29+7}-1\\
\sqrt{25} &\stackrel{?}{=} \sqrt{36}-1\\
5 &\stackrel{?}{=} 6-1\\
5 &= 5
\end{aligned}$$

36.
$$\begin{aligned}
\sqrt{-2}\left(\sqrt{-8}+3\right) &= i\sqrt{2}\left(2i\sqrt{2}+3\right)\\
&= 2i^2\cdot 2+3i\sqrt{2}\\
&= -4+3\sqrt{2}i
\end{aligned}$$

37.
$$\begin{aligned}
(-4+11i)-(3-5i) &= -4+11i-3+5i\\
&= \left[-4+(-3)\right]+(11i+5i)\\
&= -7+16i
\end{aligned}$$

38.
$$\begin{aligned}
(5+2i)^2 &= 5^2+2(5)(2i)+(2i)^2\\
&= 25+20i+4i^2\\
&= 25+20i+4(-1)\\
&= 21+20i
\end{aligned}$$

39.
$$\begin{aligned}
\frac{2+3i}{6-2i} &= \frac{2+3i}{6-2i}\cdot\frac{6+2i}{6+2i} = \frac{12+4i+18i+6i^2}{6^2-(2i)^2}\\
&= \frac{12+22i+6(-1)}{36-4(-1)} = \frac{6+22i}{40}\\
&= \frac{6}{40}+\frac{22}{40}i = \frac{3}{20}+\frac{11}{20}i
\end{aligned}$$

40.
$$\begin{aligned}
P &= (x+1)+(x+5)+x+x+(2x+1)+(2x+5)\\
&= 8x+12\\
&= 4(2x+3)
\end{aligned}$$

41. Your time $= 2$ hours, Your rate $= \frac{1}{2}$

Friend's time $= 3.5$ hours,

Friend's rate $= \frac{1}{3.5} = \frac{1}{7/2} = \frac{2}{7}$

Verbal Model: [Your rate] + [Friend's rate] = [Rate together]

Labels: Time together $= x$

Rate together $= \frac{1}{x}$

Equation:
$$\begin{aligned}
\frac{1}{2}+\frac{2}{7} &= \frac{1}{x}\\
7x+4x &= 14\\
11x &= 14\\
x &= \frac{14}{11}\text{ hours}
\end{aligned}$$

42.
$$\begin{aligned}
250 &= \frac{10(5+3t)}{1+0.004t}\\
250(1+0.004t) &= 10(5+3t)\\
250+t &= 50+30t\\
200 &= 29t\\
\frac{200}{29} &= t\\
6.9\text{ years} &\approx t
\end{aligned}$$

or

$$t \approx 6.9\text{ years}$$

43.
$$\begin{aligned}
c^2 &= a^2+b^2\\
c &= \sqrt{180^2+90^2}\\
c &= \sqrt{32{,}400+8100}\\
c &= \sqrt{40{,}500}\\
c &= \sqrt{81\cdot 100\cdot 5}\\
c &= 90\sqrt{5} \approx 201.25\text{ feet}
\end{aligned}$$

44.
$$\begin{aligned}
t &= \sqrt{\frac{d}{16}}\\
5 &= \sqrt{\frac{d}{16}}\\
(5)^2 &= \left(\sqrt{\frac{d}{16}}\right)^2\\
25 &= \frac{d}{16}\\
25(16) &= d\\
400\text{ feet} &= d
\end{aligned}$$

CHAPTER 8
Quadratic Equations, Functions, and Inequalities

CHAPTER 8
Quadratic Equations, Functions, and Inequalities

Section 8.1 Solving Quadratic Equations: Factoring and Special Forms

1. $x^2 - 15x + 54 = 0$

$(x - 6)(x - 9) = 0$

$x - 6 = 0 \qquad x - 9 = 0$

$x = 6 \qquad x = 9$

3. $x^2 - x - 30 = 0$

$(x - 6)(x + 5) = 0$

$x = 6, \qquad x = -5$

5. $x^2 + 4x = 45$

$x^2 + 4x - 45 = 0$

$(x + 9)(x - 5) = 0$

$x = -9, \qquad x = 5$

7. $x^2 - 16x + 64 = 0$

$(x - 8)(x - 8) = 0$

$x - 8 = 0 \qquad x - 8 = 0$

$x = 8 \qquad x = 8$

9. $9x^2 - 10x - 16 = 0$

$(9x + 8)(x - 2) = 0$

$9x + 8 = 0 \qquad x - 2 = 0$

$9x = -8 \qquad x = 2$

$x = -\frac{8}{9}$

11. $4x^2 - 12x = 0$

$4x(x - 3) = 0$

$4x = 0 \qquad x - 3 = 0$

$x = 0 \qquad x = 3$

13. $u(u - 9) - 12(u - 9) = 0$

$(u - 9)(u - 12) = 0$

$u - 9 = 0 \qquad x - 12 = 0$

$u = 9 \qquad u = 12$

15. $2x(x - 5) + 9(x - 5) = 0$

$(x - 5)(2x + 9) = 0$

$x - 5 = 0 \qquad 2x + 9 = 0$

$x = 5 \qquad 2x = -9$

$x = -\frac{9}{2}$

17. $(y - 4)(y - 3) = 6$

$y^2 - 7y + 12 - 6 = 0$

$y^2 - 7y + 6 = 0$

$(y - 6)(y - 1) = 0$

$y - 6 = 0 \qquad y - 1 = 0$

$y = 6 \qquad y = 1$

19. $2x(3x + 2) = 5 - 6x^2$

$6x^2 + 4x = 5 - 6x^2$

$12x^2 + 4x - 5 = 0$

$(6x + 5)(2x - 1) = 0$

$6x + 5 = 0 \qquad 2x - 1 = 0$

$x = -\frac{5}{6} \qquad x = \frac{1}{2}$

21. $x^2 = 49$

$x = \pm\sqrt{49}$

$x = \pm 7$

23. $6x^2 = 54$

$x^2 = 9$

$x = \pm\sqrt{9}$

$x = \pm 3$

25. $25x^2 = 16$

$x^2 = \frac{16}{25}$

$x = \pm\sqrt{\frac{16}{25}}$

$x = \pm\frac{4}{5}$

27. $\frac{w^2}{4} = 49$

$w^2 = 196$

$w = \pm\sqrt{196}$

$w = \pm 14$

29. $4x^2 - 25 = 0$

$4x^2 = 25$

$x^2 = \frac{25}{4}$

$x = \pm\sqrt{\frac{25}{4}}$

$x = \pm\frac{5}{2}$

31. $4u^2 - 225 = 0$

$u^2 = \frac{225}{4}$

$u = \pm\sqrt{\frac{225}{4}}$

$u = \pm\frac{15}{2}$

33. $(x + 4)^2 = 64$

$x + 4 = \pm\sqrt{64}$

$x = -4 \pm 8$

$x = 4, -12$

35. $(x - 3)^2 = 0.25$

$x - 3 = \pm\sqrt{0.25}$

$x = 3 \pm 0.5$

$x = 3.5, 2.5$

37. $(x - 2)^2 = 7$

$x - 2 = \pm\sqrt{7}$

$x = 2 \pm \sqrt{7}$

39. $(2x + 1)^2 = 50$

$2x + 1 = \pm\sqrt{50}$

$2x = -1 \pm 5\sqrt{2}$

$x = \dfrac{-1 \pm 5\sqrt{2}}{2}$

41. $(9m - 2)^2 - 108 = 0$

$(9m - 2)^2 = 108$

$9m - 2 = \pm\sqrt{108}$

$9m = 2 \pm 6\sqrt{3}$

$m = \dfrac{2 \pm 6\sqrt{3}}{9}$

$m = \dfrac{2}{9} \pm \dfrac{2\sqrt{3}}{3}$

43. $z^2 = -36$

$z = \pm\sqrt{-36}$

$z = \pm 6i$

45. $x^2 + 4 = 0$

$x^2 = -4$

$x = \pm\sqrt{-4}$

$x = \pm 2i$

47. $9u^2 + 17 = 0$

$9u^2 = -17$

$u = \pm\sqrt{-\dfrac{17}{9}}$

$= \pm\dfrac{\sqrt{17}}{3}i$

49. $(t - 3)^2 = -25$

$t - 3 = \pm\sqrt{-25}$

$t = 3 \pm 5i$

51. $(3z + 4)^2 + 144 = 0$

$(3z + 4)^2 = -144$

$3z + 4 = \pm\sqrt{-144}$

$3z + 4 = \pm 12i$

$3z = -4 \pm 12i$

$z = \dfrac{-4 \pm 12i}{3}$

$z = -\dfrac{4}{3} \pm 4i$

53. $(4m + 1)^2 = -80$

$4m + 1 = \pm\sqrt{-80}$

$4m = -1 \pm 4\sqrt{5}i$

$m = \dfrac{-1 \pm 4\sqrt{5}i}{4}$

$m = -\dfrac{1}{4} \pm \sqrt{5}i$

55. $36(t + 3)^2 = -100$

$(t + 3)^2 = -\frac{100}{36}$

$t + 3 = \pm\sqrt{-\frac{100}{36}}$

$t = -3 \pm \frac{10}{6}i$

$t = -3 \pm \frac{5}{3}i$

57. $(x - 1)^2 = -27$

$x - 1 = \pm\sqrt{-27}$

$x = 1 \pm 3\sqrt{3}i$

59. $(x + 1)^2 + 0.04 = 0$

$$(x + 1)^2 = -0.04$$
$$x + 1 = \pm\sqrt{-0.04}$$
$$x = -1 \pm 0.2i$$

61. $\left(c - \frac{2}{3}\right)^2 + \frac{1}{9} = 0$

$$\left(c - \frac{2}{3}\right)^2 = -\frac{1}{9}$$
$$c - \frac{2}{3} = \pm\sqrt{-\frac{1}{9}}$$
$$c = \frac{2}{3} \pm \frac{1}{3}i$$

63. $\left(x + \frac{7}{3}\right)^2 = -\frac{38}{9}$

$$x + \frac{7}{3} = \pm\sqrt{-\frac{38}{9}}$$
$$x = -\frac{7}{3} \pm \frac{\sqrt{38}}{3}i$$

65. $2x^2 - 5x = 0$

$$x(2x - 5) = 0$$
$$x = 0, \qquad 2x - 5 = 0$$
$$x = \frac{5}{2}$$

67. $2x^2 + 5x - 12 = 0$

$$(2x - 3)(x + 4) = 0$$
$$x = \frac{3}{2}, \qquad x = -4$$

69. $x^2 - 900 = 0$

$$x^2 = 900$$
$$x = \pm 30$$

71. $x^2 + 900 = 0$

$$x^2 = -900$$
$$x = \pm\sqrt{-900}$$
$$x = \pm 30i$$

73. $\frac{2}{3}x^2 = 6$

$$\frac{3}{2} \cdot \frac{2}{3}x^2 = 6 \cdot \frac{3}{2}$$
$$x^2 = 9$$
$$x = \pm 3$$

75. $(p - 2)^2 - 108 = 0$

$$(p - 2)^2 = 108$$
$$p - 2 = \pm\sqrt{108}$$
$$p = 2 \pm 6\sqrt{3}$$

77. $(p - 2)^2 + 108 = 0$

$$(p - 2)^2 = -108$$
$$p - 2 = \pm\sqrt{-108}$$
$$p = 2 \pm 6\sqrt{3}i$$

79. $(x + 2)^2 + 18 = 0$

$$(x + 2)^2 = -18$$
$$x + 2 = \pm\sqrt{-18}$$
$$x = -2 \pm 3\sqrt{2}i$$

81. $y = x^2 - 9$

Keystrokes:

[Y=] [X,T,θ] [x^2] [−] 9 [GRAPH]

x-intercepts are −3 and 3.

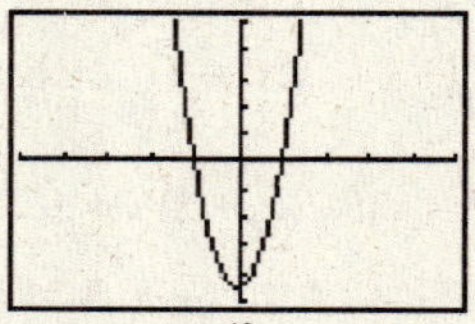

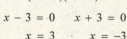

$$0 = x^2 - 9$$
$$= (x - 3)(x + 3)$$
$$x - 3 = 0 \qquad x + 3 = 0$$
$$x = 3 \qquad x = -3$$

83. $y = x^2 - 2x - 15$

Keystrokes:

[Y=] [X,T,θ] [x^2] [−] 2 [X,T,θ] [−] 15 [GRAPH]

x-intercepts are −3 and 5.

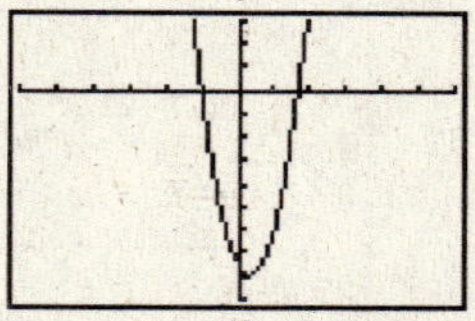

$$0 = x^2 - 2x - 15$$
$$0 = (x - 5)(x + 3)$$
$$x - 5 = 0 \qquad x + 3 = 0$$
$$x = 5 \qquad x = -3$$

85. $y = 4 - (x - 3)^2$

Keystrokes:

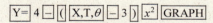

[Y=] 4 [−] [(] [X,T,θ] [−] 3 [)] [x^2] [GRAPH]

x-intercepts are 1 and 5.

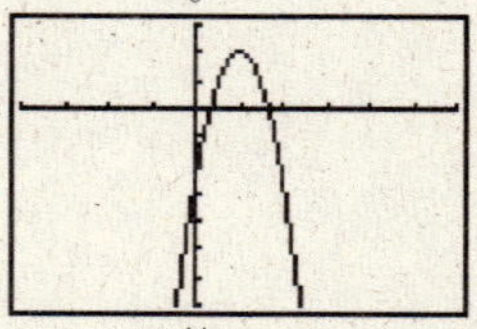

$$0 = 4 - (x - 3)^2$$
$$(x - 3)^2 = 4$$
$$x - 3 = \pm 2$$
$$x = 5, 1$$

87. $y = 2x^2 - x - 6$

Keystrokes:

[Y=] 2 [X,T,θ] [x^2] [−] [X,T,θ] [−] 6 [GRAPH]

x-intercepts are $-\frac{3}{2}$ and 2.

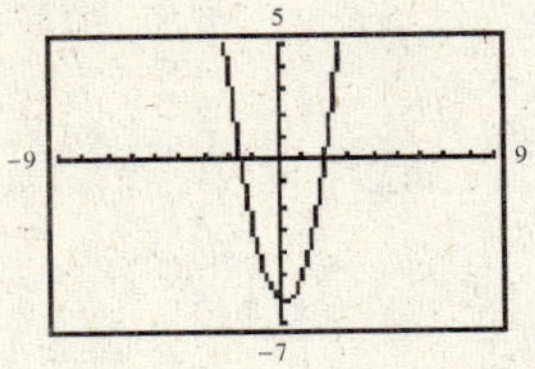

$0 = 2x^2 - x - 6$

$0 = (2x + 3)(x - 2)$

$x = -\frac{3}{2}, \quad x = 2$

89. $y = 3x^2 - 13x - 10$

Keystrokes:

[Y=] 3 [X,T,θ] [x^2] [−] 13 [X,T,θ] [−] 10 [GRAPH]

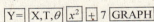

x-intercepts are $-\frac{2}{3}$ and 5.

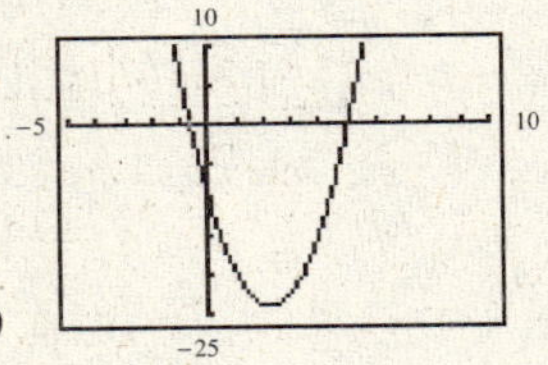

$0 = 3x^2 - 13x - 10$

$0 = (3x + 2)(x - 5)$

$0 = 3x + 2 \qquad x - 5 = 0$

$-2 = 3x \qquad x = 5$

$-\frac{2}{3} = x \qquad \left(-\frac{2}{3}, 0\right), (5, 0)$

91. $y = x^2 + 7$

Keystrokes:

[Y=] [X,T,θ] [x^2] [+] 7 [GRAPH]

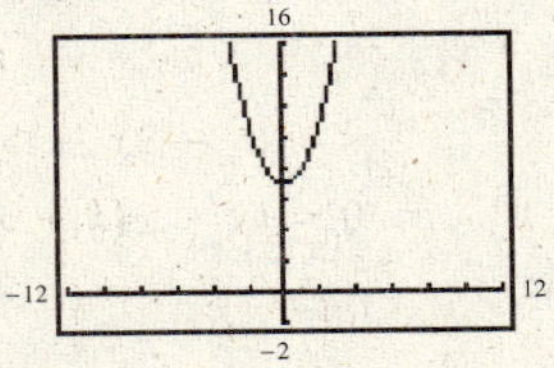

$0 = x^2 + 7$

$-7 = x^2$

$\pm\sqrt{-7} = x$

$\pm\sqrt{7}i = x$

The equation has complex solutions.

93. $y = (x - 4)^2 + 2$

Keystrokes:

[Y=] [(] [X,T,θ] [−] 4 [)] [x^2] [+] 2 [GRAPH]

No x-intercepts

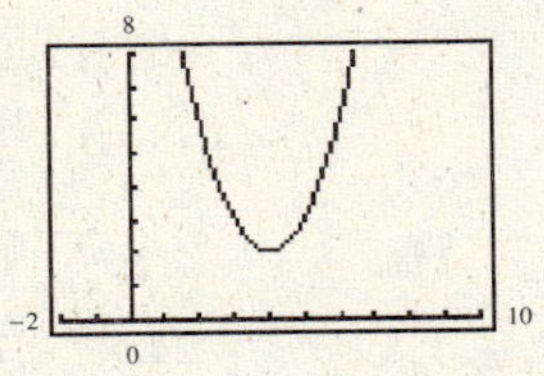

$0 = (x - 4)^2 + 2$

$-2 = (x - 4)^2$

$\pm\sqrt{-2} = x - 4$

$4 \pm \sqrt{2}i = x$

The equation has complex solutions.

95. $y = (x + 3)^2 + 5$

Keystrokes:

[Y=] [(] [X,T,θ] [+] 3 [)] [x^2] [+] 5 [GRAPH]

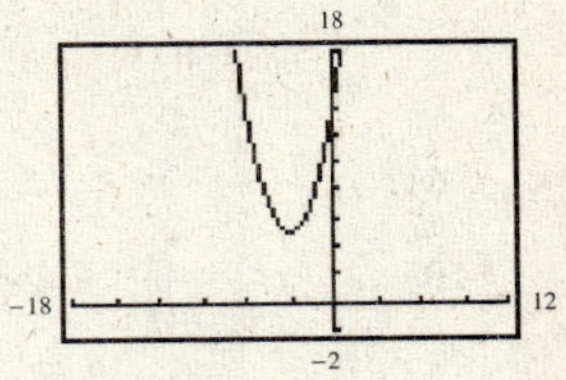

$0 = (x + 3)^2 + 5$

$-5 = (x + 3)^2$

$\pm\sqrt{-5} = x + 3$

$\pm\sqrt{5}i = x + 3$

$-3 \pm \sqrt{5}i = x$

The equation has complex solutions.

97. $x^2 + y^2 = 4$

$y^2 = 4 - x^2$

$y = \pm\sqrt{4 - x^2}$

Keystrokes:

y_1: [Y=] [√] [(] 4 [−] [X,T,θ] [x^2] [)] [ENTER]

y_2: [(−)] [√] [(] 4 [−] [X,T,θ] [x^2] [)] [GRAPH]

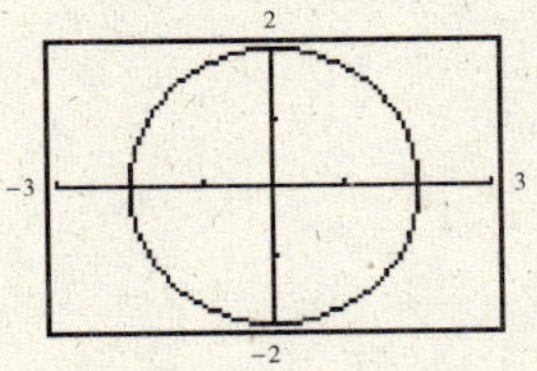

99. $x^2 + 4y^2 = 4$

$4y^2 = 4 - x^2$

$y^2 = \frac{4 - x^2}{4}$

$y = \pm\sqrt{\frac{4 - x^2}{4}} = \pm\frac{\sqrt{4 - x^2}}{2}$

Keystrokes:

y_1: [Y=] [√] [(] 4 [−] [X,T,θ] [x^2] [)] [÷] 2 [ENTER]

y_2: [(−)] [√] [(] 4 [−] [X,T,θ] [x^2] [)] [÷] 2 [GRAPH]

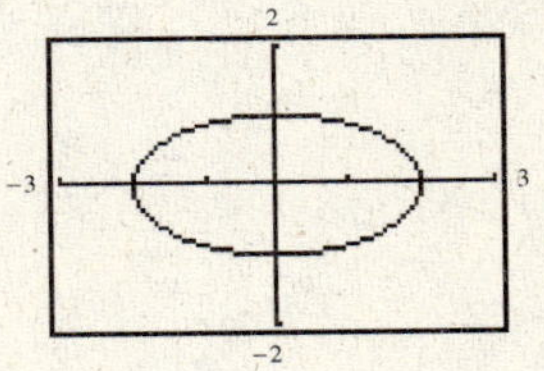

101.

$x^4 - 5x^2 + 4 = 0$

$(x^2 - 4)(x^2 - 1) = 0$

$(x - 2)(x + 2)(x - 1)(x + 1) = 0$

$x - 2 = 0 \qquad x + 2 = 0 \qquad x - 1 = 0 \qquad x + 1 = 0$

$x = 2 \qquad x = -2 \qquad x = 1 \qquad x = -1$

103. $x^4 - 5x^2 + 6 = 0$

$(x^2 - 3)(x^2 - 2) = 0$

$x^2 - 3 = 0 \qquad x^2 - 2 = 0$

$x^2 = 3 \qquad x^2 = 2$

$x = \pm\sqrt{3} \qquad x = \pm\sqrt{2}$

105. $(x^2 - 4)^2 + 2(x^2 - 4) - 3 = 0$

$[(x^2 - 4) + 3][(x^2 - 4) - 1] = 0$

$(x^2 - 1)(x^2 - 5) = 0$

$x^2 - 1 = 0 \qquad x^2 - 5 = 0$

$x^2 = 1 \qquad x^2 = 5$

$x = \pm 1 \qquad x = \pm\sqrt{5}$

107. $x - 3\sqrt{x} - 4 = 0$

Let $u = \sqrt{x}$.

$(\sqrt{x})^2 - 3\sqrt{x} - 4 = 0$

$u^2 - 3u - 4 = 0$

$(u - 4)(u + 1) = 0$

$u = 4 \qquad u = -1$

$\sqrt{x} = 4 \qquad \sqrt{x} = -1$

$(\sqrt{x})^2 = 4^2 \qquad (\sqrt{x})^2 = (-1)^2$

$x = 16 \qquad x = 1$

Check: $16 - 3\sqrt{16} - 4 \stackrel{?}{=} 0$

$16 - 12 - 4 \stackrel{?}{=} 0$

$0 = 0$

Check: $1 - 3\sqrt{1} - 4 \stackrel{?}{=} 0$

$1 - 3 - 4 \stackrel{?}{=} 0$

$-6 \neq 0$

109. $x - 7\sqrt{x} + 10 = 0$

Let $u = \sqrt{x}$.

$(\sqrt{x})^2 - 7(\sqrt{x}) + 10 = 0$

$u^2 - 7u + 10 = 0$

$(u - 5)(u - 2) = 0$

$u = 5 \qquad u = 2$

$\sqrt{x} = 5 \qquad \sqrt{x} = 2$

$x = 25 \qquad x = 4$

Check: $25 - 7\sqrt{25} + 10 \stackrel{?}{=} 0$

$25 - 35 + 10 \stackrel{?}{=} 0$

$0 = 0$

Check: $4 - 7\sqrt{4} + 10 \stackrel{?}{=} 0$

$4 - 14 + 10 \stackrel{?}{=} 0$

$0 = 0$

111. $x^{2/3} - x^{1/3} - 6 = 0$

$(x^{1/3} - 3)(x^{1/3} + 2) = 0$

$x^{1/3} - 3 = 0 \qquad x^{1/3} + 2 = 0$

$x^{1/3} = 3 \qquad x^{1/3} = -2$

$x = 27 \qquad x = -8$

113. $2x^{2/3} - 7x^{1/3} + 5 = 0$

Let $u = x^{1/3}$.

$2(x^{1/3})^2 - 7x^{1/3} + 5 = 0$

$2u^2 - 7u + 5 = 0$

$(2u - 5)(u - 1) = 0$

$2u - 5 = 0 \qquad u - 1 = 0$

$2u = 5 \qquad u = 1$

$u = \frac{5}{2} \qquad x^{1/3} = 1$

$x^{1/3} = \frac{5}{2} \qquad (x^{1/3}) = 1^3$

$(x^{1/3}) = \left(\frac{5}{2}\right)^3 \qquad x = 1$

$x = \frac{125}{8}$

115. $x^{2/5} - 3x^{1/5} + 2 = 0$

$(x^{1/5} - 2)(x^{1/5} - 1) = 0$

$x^{1/5} = 2 \qquad x^{1/5} = 1$

$x = 2^5 \qquad x = 1^5$

$x = 32 \qquad x = 1$

117. $2x^{2/5} - 7x^{1/5} + 3 = 0$

$(2x^{1/5} - 1)(x^{1/5} - 3) = 0$

$x^{1/5} = \frac{1}{2}$ $\quad$ $x^{1/5} = 3$

$x = \left(\frac{1}{2}\right)^5$ $\quad$ $x = 3^5$

$x = \frac{1}{32}$ $\quad$ $x = 243$

119. $x^{1/3} - x^{1/6} - 6 = 0$

$(x^{1/6} - 3)(x^{1/6} + 2) = 0$

$x^{1/6} = 3$ $\quad$ $x^{1/6} = -2$

$(x^{1/6})^6 = 3^6$ $\quad$ $(x^{1/6})^6 = (-2)^6$

$x = 729$ $\quad$ $x = 64$

Check: $729^{1/3} - 729^{1/6} - 6 \stackrel{?}{=} 0$

$9 - 3 - 6 \stackrel{?}{=} 0$

$0 = 0$

Check: $64^{1/3} - 64^{1/6} - 6 \stackrel{?}{=} 0$

$4 - 2 - 6 \stackrel{?}{=} 0$

$-4 \neq 0$

121. $x^{1/2} - 3x^{1/4} + 2 = 0$

$(x^{1/4} - 2)(x^{1/4} - 1) = 0$

$x^{1/4} = 2$ $\quad$ $x^{1/4} = 1$

$(x^{1/4})^4 = 2^4$ $\quad$ $(x^{1/4})^4 = 1^4$

$x = 16$ $\quad$ $x = 1$

Check: $16^{1/2} - 3(16)^{1/4} + 2 \stackrel{?}{=} 0$

$4 - 3(2) + 2 \stackrel{?}{=} 0$

$0 = 0$

Check: $1^{1/2} - 3(1)^{1/4} + 2 \stackrel{?}{=} 0$

$1 - 3 + 2 \stackrel{?}{=} 0$

$0 = 0$

123. $\frac{1}{x^2} - \frac{3}{x} + 2 = 0$

$1 - 3x + 2x^2 = 0$

$2x^2 - 3x + 1 = 0$

$(2x - 1)(x - 1) = 0$

$2x - 1 = 0$ $\quad$ $x - 1 = 0$

$x = \frac{1}{2}$ $\quad$ $x = 1$

125. $4x^{-2} - x^{-1} - 5 = 0$

Let $u = x^{-1}$.

$4u^2 - u - 5 = 0$

$(u + 1)(4u - 5) = 0$

$u = -1$ $\quad$ $u = \frac{5}{4}$

$x^{-1} = -1$ $\quad$ $x^{-1} = \frac{5}{4}$

$x = -1$ $\quad$ $x = \frac{4}{5}$

127. $(x^2 - 3x)^2 - 2(x^2 - 3x) - 8 = 0$

Let $u = x^2 - 3x$.

$u^2 - 2u - 8 = 0$

$(u - 4)(u + 2) = 0$

$u - 4 = 0$ $\quad$ $u + 2 = 0$

$u = 4$ $\quad$ $u = -2$

$x^2 - 3x = 4$

$x^2 - 3x - 4 = 0$

$(x - 4)(x + 1) = 0$

$x - 4 = 0$ $\quad$ $x + 1 = 0$

$x = 4$ $\quad$ $x = -1$

$x^2 - 3x = -2$

$x^2 - 3x + 2 = 0$

$(x - 2)(x - 1) = 0$

$x - 2 = 0$ $\quad$ $x - 1 = 0$

$x = 2$ $\quad$ $x = 1$

129. $16\left(\frac{x - 1}{x - 8}\right)^2 + 8\left(\frac{x - 1}{x - 8}\right) + 1 = 0$

Let $u = \left(\frac{x - 1}{x - 8}\right)$.

$16u^2 + 8u + 1 = 0$

$(4u + 1)(4u + 1) = 0$

$u = -\frac{1}{4}$

$\frac{x - 1}{x - 8} = -\frac{1}{4}$

$4x - 4 = -x + 8$

$5x = 12$

$x = \frac{12}{5}$

131. Surface area of a sphere $= 4\pi r^2$

$$SA = 4\pi\left(\frac{d}{2}\right)^2, \qquad r = \frac{d}{2}$$
$$= 4\pi\frac{d^2}{4}$$
$$SA = \pi d^2$$
$$45{,}239 = \pi d^2$$
$$\sqrt{\frac{45{,}239}{\pi}} = d$$
$$d \approx 120 \text{ feet}$$

133.
$$0 = -16t^2 + 256$$
$$16t^2 = 256$$
$$t^2 = 16$$
$$t = 4 \text{ seconds}$$

135.
$$0 = -16t^2 + 128$$
$$16t^2 = 128$$
$$t^2 = 8$$
$$t = \pm\sqrt{8}$$
$$t = \pm 2\sqrt{2}$$
$$t = 2\sqrt{2} \approx 2.83 \text{ seconds}$$

137.
$$0 = 144 + 128 - 16^2$$
$$0 = -16t^2 + 128t + 144$$
$$0 = -16(t^2 - 8t - 9)$$
$$0 = -16(t - 9)(t + 1)$$
$$t - 9 = 0 \qquad t + 1 = 0$$
$$t = 9 \text{ seconds} \qquad t = -1$$

139. $P = \$1500$, $A = \$1685.40$
$$A = P(1 + r)^2$$
$$1685.40 = 1500(1 + r)^2$$
$$1.1236 = (1 + r)^2$$
$$1.06 = 1 + r$$
$$0.06 = r$$
$$6\% = r$$

141.
$$y = 4.95t^2 + 876$$
$$1500 = 4.95t^2 + 876$$
$$624 = 4.95t^2$$
$$\frac{624}{4.92} = t^2$$
$$t = \sqrt{\frac{624}{4.95}} \approx 11$$
$$1997 + 4 = 2001$$

143. If $a = 0$, the equation would not be quadratic because it would be of degree 1, not 2.

145. Write the equation in the form $u^2 = d$, where u is an algebraic expression and d is a positive constant. Take the square roots of each side to obtain the solutions $u = \pm\sqrt{d}$.

147.
$$3x - 8 > 4$$
$$3x > 12$$
$$x > 4$$

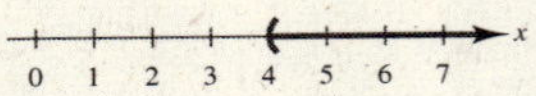

149.
$$2x - 6 \le 9 - x$$
$$3x \le 15$$
$$x \le 5$$

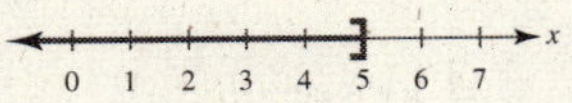

151.
$$\begin{aligned} x + y - z &= 4 \\ 2x + y + 2z &= 10 \\ x - 3y - 4z &= -7 \end{aligned} \Rightarrow \begin{bmatrix} 1 & 1 & -1 & \vdots & 4 \\ 2 & 1 & 2 & \vdots & 10 \\ 1 & -3 & -4 & \vdots & -7 \end{bmatrix} \begin{matrix} \\ -2R_1 + R_2 \\ -R_1 + R_3 \end{matrix} \begin{bmatrix} 1 & 1 & -1 & \vdots & 4 \\ 0 & -1 & 4 & \vdots & 2 \\ 0 & -4 & -3 & \vdots & -11 \end{bmatrix}$$

$$\begin{matrix} \\ -R_2 \\ \\ \end{matrix} \begin{bmatrix} 1 & 1 & -1 & \vdots & 4 \\ 0 & 1 & -4 & \vdots & -2 \\ 0 & -4 & -3 & \vdots & -11 \end{bmatrix} \begin{matrix} -R_2 + R_1 \\ \\ 4R_2 + R_3 \end{matrix} \begin{bmatrix} 1 & 0 & 3 & \vdots & 6 \\ 0 & 1 & -4 & \vdots & -2 \\ 0 & 0 & -19 & \vdots & -19 \end{bmatrix} -\tfrac{1}{19}R_3$$

$$\begin{bmatrix} 1 & 0 & 3 & \vdots & 6 \\ 0 & 1 & -4 & \vdots & -2 \\ 0 & 0 & 1 & \vdots & 1 \end{bmatrix} \begin{matrix} -3R_3 + R_1 \\ 4R_3 + R_2 \\ \end{matrix} \begin{bmatrix} 1 & 0 & 0 & \vdots & 3 \\ 0 & 1 & 0 & \vdots & 2 \\ 0 & 0 & 1 & \vdots & 1 \end{bmatrix} \qquad \begin{aligned} x &= 3 \\ y &= 2 \\ z &= 1 \end{aligned} \qquad (3, 2, 1)$$

153. $5\sqrt{3} - 2\sqrt{3} = (5 - 2)\sqrt{3} = 3\sqrt{3}$

155. $16\sqrt[3]{y} - 9\sqrt[3]{x} = 16\sqrt[3]{y} - 9\sqrt[3]{x}$

Radical expressions are not alike so cannot be combined.

157. $\sqrt{16m^4n^3} + m\sqrt{m^2n} = 4m^2n\sqrt{n} + m^2\sqrt{n} = (4m^2n + m^2)\sqrt{n} = (4n + 1)m^2\sqrt{n}$

Section 8.2 Completing the Square

1. $x^2 + 8x + 16$

$\left[16 = \left(\frac{8}{2}\right)^2\right]$

3. $y^2 - 20y + 100$

$\left[100 = \left(-\frac{20}{2}\right)^2\right]$

5. $x^2 + 14x + 49$

$\left[49 = \left(\frac{14}{2}\right)^2\right]$

7. $t^2 + 5t + \frac{25}{4}$

$\left[\frac{25}{4} = \left(\frac{5}{2}\right)^2\right]$

9. $x^2 - 9x + \frac{81}{4}$

$\left[\frac{81}{4} = \left(-\frac{9}{2}\right)^2\right]$

11. $a^2 - \frac{1}{3}a + \frac{1}{36}$

$\left[\frac{1}{36} = \left(-\frac{1}{3}\right)\left(\frac{1}{2}\right)^2\right]$

13. $y^2 + \frac{8}{5}y + \frac{16}{25}$

$\left[\frac{16}{25} = \left(\frac{1}{2} \cdot \frac{8}{5}\right)^2\right]$

15. $r^2 - 0.4r + 0.04$

$\left[0.04 = \left(-\frac{0.4}{2}\right)^2\right]$

17. (a)
$$\begin{aligned} x^2 - 20x + 100 &= 100 \\ (x - 10)^2 &= 100 \\ x - 10 &= \pm 10 \\ x &= 10 \pm 10 \\ x &= 20, 0 \end{aligned}$$

(b)
$$\begin{aligned} x^2 - 20x &= 0 \\ x(x - 20) &= 0 \\ x &= 0, 20 \end{aligned}$$

19. (a)
$$\begin{aligned} x^2 + 6x + 9 &= 0 + 9 \\ (x + 3)^2 &= 9 \\ x + 3 &= \pm 3 \\ x &= -3 \pm 3 \\ x &= -6, 0 \end{aligned}$$

(b)
$$\begin{aligned} x^2 + 6x &= 0 \\ x(x + 6) &= 0 \end{aligned}$$
$x = 0, \quad x + 6 = 0$

$x = -6, 0$

21. (a)
$$\begin{aligned} y^2 - 5y &= 0 \\ y^2 - 5y + \tfrac{25}{4} &= \tfrac{25}{4} \\ \left(y - \tfrac{5}{2}\right)^2 &= \tfrac{25}{4} \\ y - \tfrac{5}{2} &= \pm\tfrac{5}{2} \\ y &= \tfrac{5}{2} \pm \tfrac{5}{2} \\ &= 0, 5 \end{aligned}$$

(b)
$$\begin{aligned} y^2 - 5y &= 0 \\ y(y - 5) &= 0 \end{aligned}$$
$y = 0, \quad y - 5 = 0$

$y = 5$

23. (a)
$$\begin{aligned} t^2 - 8t + 16 &= -7 + 16 \\ (t - 4)^2 &= 9 \\ t - 4 &= \pm 3 \\ t &= 4 \pm 3 \\ t &= 7, 1 \end{aligned}$$

(b)
$$\begin{aligned} t^2 - 8t + 7 &= 0 \\ (t - 7)(t - 1) &= 0 \end{aligned}$$
$t = 7, \quad t = 1$

25. (a) $x^2 + 7x + \frac{49}{4} = -12 + \frac{49}{4}$

$\left(x + \frac{7}{2}\right)^2 = \frac{1}{4}$

$x + \frac{7}{2} = \pm\frac{1}{2}$

$x = -\frac{7}{2} \pm \frac{1}{2}$

$x = -\frac{6}{2}, -\frac{8}{2}$

$x = -3, -4$

(b) $x^2 + 7x + 12 = 0$

$(x + 4)(x + 3) = 0$

$x = -4, \quad x = -3$

27. (a) $x^2 - 3x + \frac{9}{4} = 18 + \frac{9}{4}$

$\left(x - \frac{3}{2}\right)^2 = \frac{81}{4}$

$x - \frac{3}{2} = \pm\frac{9}{2}$

$x = \frac{3}{2} \pm \frac{9}{2}$

$x = \frac{12}{2}, -\frac{6}{2}$

$x = 6, -3$

(b) $x^2 - 3x - 18 = 0$

$(x - 6)(x + 3) = 0$

$x = 6, -3$

29. (a) $2u^2 - 12u + 18 = 0$

$u^2 - 6u + 9 = 0$

$(u - 3)^2 = 0$

$u - 3 = \pm 0$

$u = 3$

(b) $2u^2 - 12u + 18 = 0$

$u^2 - 6u + 9 = 0$

$(u - 3)(u - 3) = 0$

$u - 3 = 0 \quad u - 3 = 0$

$u = 3 \quad u = 3$

31. (a) $4x^2 + 4x - 15 = 0$

$x^2 + x - \frac{15}{4} = 0$

$x^2 + x = \frac{15}{4}$

$x^2 + x + \frac{1}{4} = \frac{15}{4} + \frac{1}{4}$

$\left(x + \frac{1}{2}\right)^2 = \frac{16}{4}$

$x + \frac{1}{2} = \pm\sqrt{4}$

$x = -\frac{1}{2} \pm 2$

$x = \frac{3}{2}, -\frac{5}{2}$

(b) $4x^2 + 4x - 15 = 0$

$(2x - 3)(2x + 5) = 0$

$x = \frac{3}{2}, -\frac{5}{2}$

33. $x^2 - 4x - 3 = 0$

$x^2 - 4x + 4 = 3 + 4$

$(x - 2)^2 = 7$

$x - 2 = \sqrt{7}$

$x = 2 \pm \sqrt{7}$

$x \approx 4.65, -0.65$

35. $x^2 + 4x - 3 = 0$

$x^2 + 4x + 4 = 3 + 4$

$(x + 2)^2 = 7$

$x + 2 = \pm\sqrt{7}$

$x = -2 \pm \sqrt{7}$

$x \approx 0.65, -4.65$

37. $x^2 + 6x = 7$

$x^2 + 6x + 9 = 7 + 9$

$(x + 3)^2 = 16$

$x + 3 = \pm 4$

$x = -3 \pm 4$

$x = 1, -7$

39. $x^2 - 12x = -10$

$x^2 - 12x + 36 = -10 + 36$

$(x - 6)^2 = 26$

$x - 6 = \pm\sqrt{26}$

$x = 6 \pm \sqrt{26}$

$x = 6 + \sqrt{26} \approx 11.10$

$x = 6 - \sqrt{26} \approx 0.90$

41. $x^2 + 8x + 7 = 0$

$$x^2 + 8x + 16 = -7 + 16$$
$$(x + 4)^2 = 9$$
$$x + 4 = \pm 3$$
$$x = -4 \pm 3$$
$$x = -1, -7$$

43. $x^2 - 10x + 21 = 0$

$$x^2 - 10x + 25 = -21 + 25$$
$$(x - 5)^2 = 4$$
$$x - 5 = \pm 2$$
$$x = 5 \pm 2$$
$$x = 7, 3$$

45. $y^2 + 5y + 3 = 0$

$$y^2 + 5y + \frac{25}{4} = -3 + \frac{25}{4}$$
$$\left(y + \frac{5}{2}\right)^2 = -\frac{12}{4} + \frac{25}{4}$$
$$\left(y + \frac{5}{2}\right)^2 = \frac{13}{4}$$
$$y + \frac{5}{2} = \pm\sqrt{\frac{13}{4}}$$
$$y = -\frac{5}{2} \pm \frac{\sqrt{13}}{2}$$
$$y \approx -0.70, -4.30$$

47. $x^2 + 10 = 6x$

$$x^2 - 6x + 9 = -10 + 9$$
$$(x - 3)^2 = -1$$
$$x - 3 = \pm\sqrt{-1}$$
$$x = 3 \pm i$$

49. $z^2 + 4z + 13 = 0$

$$z^2 + 4z + 4 = -13 + 4$$
$$(z + 2)^2 = -9$$
$$z + 2 = \pm\sqrt{-9}$$
$$z = -2 \pm 3i$$

51. $-x^2 + x - 1 = 0$

$$x^2 - x + 1 = 0$$
$$x^2 - x + \frac{1}{4} = -1 + \frac{1}{4}$$
$$\left(x - \frac{1}{2}\right)^2 = -\frac{3}{4}$$
$$x - \frac{1}{2} = \pm\sqrt{-\frac{3}{4}}$$
$$x = \frac{1}{2} \pm \frac{\sqrt{3}}{2}i$$
$$x = \frac{1 \pm \sqrt{3}i}{2}$$
$$x \approx 0.5 + 0.87i$$
$$x \approx 0.5 - 0.87i$$

53. $a^2 + 7a + 11 = 0$

$$a^2 + 7a + \frac{49}{4} = -11 + \frac{49}{4}$$
$$\left(a + \frac{7}{2}\right)^2 = \frac{5}{4}$$
$$a + \frac{7}{2} = \pm\sqrt{\frac{5}{4}}$$
$$a = -\frac{7}{2} \pm \frac{\sqrt{5}}{2}$$
$$a = \frac{-7 + \sqrt{5}}{2} \approx -2.38$$
$$a = \frac{-7 - \sqrt{5}}{2} \approx -4.62$$

55. $x^2 - \frac{2}{3}x - 3 = 0$

$$x^2 - \frac{2}{3}x + \frac{1}{9} = 3 + \frac{1}{9}$$
$$\left(x - \frac{1}{3}\right)^2 = \frac{28}{9}$$
$$x - \frac{1}{3} = \pm\sqrt{\frac{28}{9}}$$
$$x = \frac{1}{3} \pm \frac{2}{3}\sqrt{7}$$
$$x = \frac{1 \pm 2\sqrt{7}}{3}$$
$$x \approx 2.10, -1.43$$

57. $v^2 + \frac{3}{4}v - 2 = 0$

$$v^2 + \frac{3}{4}v + \frac{9}{64} = 2 + \frac{9}{64}$$

$$\left(v + \frac{3}{8}\right)^2 = \frac{128}{64} + \frac{9}{64}$$

$$\left(v + \frac{3}{8}\right)^2 = \frac{137}{64}$$

$$v + \frac{3}{8} = \pm\sqrt{\frac{137}{64}}$$

$$v = -\frac{3}{8} \pm \frac{\sqrt{137}}{8}$$

$$v \approx 1.09, -1.84$$

59. $2x^2 + 8x + 3 = 0$

$$x^2 + 4x + 4 = -\frac{3}{2} + 4$$

$$(x + 2)^2 = \frac{5}{2}$$

$$x + 2 = \pm\sqrt{\frac{5}{2}} \cdot \frac{\sqrt{2}}{\sqrt{2}}$$

$$x = -2 \pm \frac{\sqrt{10}}{2}$$

$$x \approx -0.42, -3.58$$

61. $3x^2 + 9x + 5 = 0$

$$x^2 + 3x + \frac{9}{4} = -\frac{5}{3} + \frac{9}{4}$$

$$\left(x + \frac{3}{2}\right)^2 = \frac{-20 + 27}{12}$$

$$\left(x + \frac{3}{2}\right)^2 = \frac{7}{12}$$

$$x + \frac{3}{2} = \pm\sqrt{\frac{7}{12}} \cdot \frac{\sqrt{3}}{\sqrt{3}}$$

$$x = -\frac{3}{2} \pm \frac{\sqrt{21}}{6}$$

$$x = \frac{-9 \pm \sqrt{21}}{6}$$

$$x \approx -0.74, -2.26$$

63. $4y^2 + 4y - 9 = 0$

$$y^2 + y + \frac{1}{4} = \frac{9}{4} + \frac{1}{4}$$

$$\left(y + \frac{1}{2}\right)^2 = \frac{10}{4}$$

$$y + \frac{1}{2} = \pm\sqrt{\frac{10}{4}}$$

$$y = -\frac{1}{2} \pm \frac{\sqrt{10}}{2}$$

$$y = \frac{-1 \pm \sqrt{10}}{2}$$

$$y \approx 1.08, -2.08$$

65. $5x^2 - 3x + 10 = 0$

$$x^2 - \frac{3}{5}x = -2$$

$$x^2 - \frac{3}{5}x + \frac{9}{100} = -2 + \frac{9}{100}$$

$$\left(x - \frac{3}{10}\right)^2 = \frac{-200}{100} + \frac{9}{100}$$

$$\left(x - \frac{3}{10}\right)^2 = -\frac{191}{100}$$

$$x - \frac{3}{10} = \pm\sqrt{-\frac{191}{100}}$$

$$x = \frac{3}{10} \pm \frac{\sqrt{191}}{10}i$$

$$x \approx 0.30 + 1.38i, 0.30 - 1.38i$$

67. $x\left(x - \frac{2}{3}\right) = 14$

$$x^2 - \frac{2}{3}x + \frac{1}{9} = 14 + \frac{1}{9}$$

$$\left(x - \frac{1}{3}\right)^2 = \frac{127}{9}$$

$$x - \frac{1}{3} = \pm\sqrt{\frac{127}{9}}$$

$$x = \frac{1}{3} \pm \frac{\sqrt{127}}{3}$$

$$x = \frac{1 + \sqrt{127}}{3} \approx 4.09$$

$$x = \frac{1 - \sqrt{127}}{3} \approx -3.42$$

69.
$$\begin{aligned} 0.1x^2 + 0.5x &= -0.2 \\ 0.1x^2 + 0.5x + 0.2 &= 0 \\ x^2 + 5x + 2 &= 0 \\ x^2 + 5x + \frac{25}{4} &= -2 + \frac{25}{4} \\ \left(x + \frac{5}{2}\right)^2 &= \frac{-8 + 25}{4} \\ \left(x + \frac{5}{2}\right)^2 &= \frac{17}{4} \\ x + \frac{5}{2} &= \pm\sqrt{\frac{17}{4}} \\ x &= -\frac{5}{2} \pm \frac{\sqrt{17}}{2} \\ x &= \frac{-5 \pm \sqrt{17}}{2} \\ x &\approx -0.44, -4.56 \end{aligned}$$

71.
$$\begin{aligned} 0.75x^2 + 1.25x + 1.5 &= 0 \\ x^2 + \frac{5}{3}x + \frac{25}{36} &= -2 + \frac{25}{36} \\ \left(x + \frac{5}{6}\right)^2 &= -\frac{47}{36} \\ x + \frac{5}{6} &= \pm\sqrt{-\frac{47}{36}} \\ x &= -\frac{5}{6} \pm \frac{\sqrt{47}}{6}i \\ x &= -\frac{5}{6} + \frac{\sqrt{47}}{6}i \approx -0.83 + 1.14i \\ x &= -\frac{5}{6} - \frac{\sqrt{47}}{6}i \approx -0.83 - 1.14i \end{aligned}$$

73.
$$\begin{aligned} \frac{x}{2} - \frac{1}{x} &= 1 \\ 2x\left(\frac{x}{2} - \frac{1}{x}\right) &= (1)2x \\ x^2 - 2 &= 2x \\ x^2 - 2x + 1 &= 2 + 1 \\ (x - 1)^2 &= 3 \\ x - 1 &= \pm\sqrt{3} \\ x &= 1 \pm \sqrt{3} \end{aligned}$$

75.
$$\begin{aligned} \frac{x^2}{8} &= \frac{x + 3}{2} \\ 8\left(\frac{x^2}{8}\right) &= \left(\frac{x + 3}{2}\right)8 \\ x^2 &= 4(x + 3) \\ x^2 &= 4x + 12 \\ x^2 - 4x + 4 &= 12 + 4 \\ (x - 2)^2 &= 16 \\ x - 2 &= \pm 4 \\ x &= 6, -2 \end{aligned}$$

77.
$$\begin{aligned} \sqrt{2x + 1} &= x - 3 \\ \left(\sqrt{2x + 1}\right)^2 &= (x - 3)^2 \\ 2x + 1 &= x^2 - 6x + 9 \\ 0 &= x^2 - 8x + 8 \\ 16 - 8 &= x^2 - 8x + 16 \\ 8 &= (x - 4)^2 \\ \pm\sqrt{8} &= x - 4 \\ 4 \pm \sqrt{8} &= x \\ 4 \pm 2\sqrt{2} &= x \end{aligned}$$

79. $y = x^2 + 4x - 1$

Keystrokes:

[Y=] [X,T,θ] [x²] [+] 4 [X,T,θ] [−] 1 [GRAPH]

$$\begin{aligned} 0 &= x^2 + 4x - 1 \\ 1 &= x^2 + 4x \\ 1 + 4 &= x^2 + 4x + 4 \\ 5 &= (x + 2)^2 \\ \pm\sqrt{5} &= x + 2 \\ -2 \pm \sqrt{5} &= x \\ x &\approx 0.236, -4.236 \end{aligned}$$

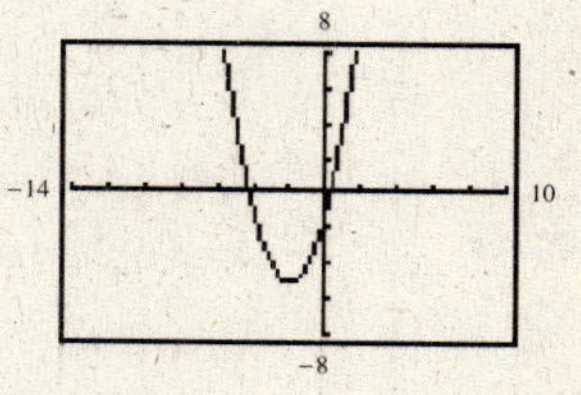

81. $y = x^2 - 2x - 5$

Keystrokes:

[Y=] [X,T,θ] [x²] [−] 2 [X,T,θ] [−] 5 [GRAPH]

$$\begin{aligned} 0 &= x^2 - 2x - 5 \\ 5 &= x^2 - 2x \\ 1 + 5 &= x^2 - 2x + 1 \\ 6 &= (x - 1)^2 \\ \pm\sqrt{6} &= x - 1 \\ 1 \pm \sqrt{6} &= x \\ x &\approx 3.449, -1.449 \end{aligned}$$

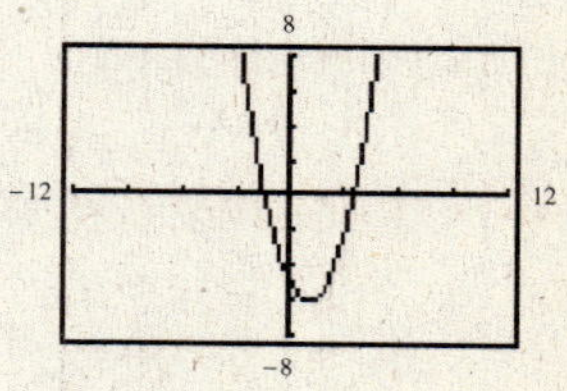

83. $y = \frac{1}{3}x^2 + 2x - 6$

Keystrokes:

[Y=] [(] 1 [÷] 3 [)] [X,T,θ] [x²] [+] 2 [X,T,θ] [−] 6 [GRAPH]

$$\begin{aligned} 0 &= \tfrac{1}{3}x^2 + 2x - 6 \\ 0 &= x^2 + 6x - 18 \\ 18 &= x^2 + 6x \\ 9 + 18 &= x^2 + 6x + 9 \\ 27 &= (x + 3)^2 \\ \pm\sqrt{27} &= x + 3 \\ -3 \pm 3\sqrt{3} &= x \\ x &\approx 2.20,\ -8.20 \end{aligned}$$

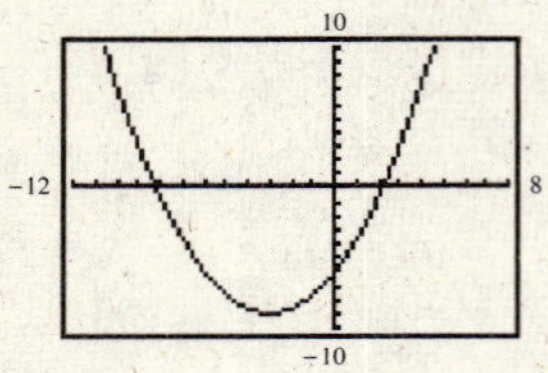

85. $y = x - 2\sqrt{x} + 1$

Keystrokes:

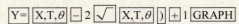

[Y=] [X,T,θ] [−] 2 [√] [X,T,θ] [)] [+] 1 [GRAPH]

x-intercept: $(1, 0)$

$$\begin{aligned} 0 &= x - 2\sqrt{x} + 1 \\ 2\sqrt{x} &= x + 1 \\ \left(2\sqrt{x}\right)^2 &= (x + 1)^2 \\ 4x &= x^2 + 2x + 1 \\ 0 &= x^2 - 2x + 1 \\ 0 &= (x - 1)^2 \\ 0 &= x - 1 \\ 1 &= x \end{aligned}$$

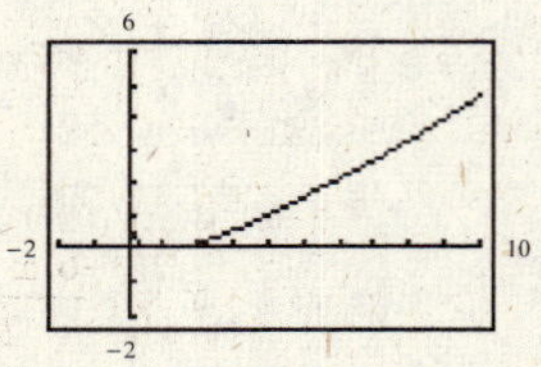

87. (a) Area of square $= x \cdot x = x^2$

Area of vertical rectangle $= 4 \cdot x = 4x$

Area of horizontal rectangle $= 4 \cdot x = 4x$

Total area $= x^2 + 4x + 4x = x^2 + 8x$

(b) Area of small square $= 4 \cdot 4 = 16$

Total area $= x^2 + 8x + 16$

(c) $(x + 4)(x + 4) = x^2 + 8x + 16$

89. *Verbal Model:* [Area] = [Length] · [Width]

Labels: Length $= x$

Width $= \dfrac{200 - 4x}{3}$

Equation:

$$\begin{aligned} 1400 &= 2\left[x \cdot \left(\frac{200 - 4x}{3}\right)\right] \\ 1400 &= 2\left[\frac{200}{3}x - \frac{4x^2}{3}\right] \\ 1400 &= \frac{400x}{3} - \frac{8x^2}{3} \\ 4200 &= 400x - 8x^2 \\ 8x^2 - 400x + 4200 &= 0 \\ x^2 - 50x + 525 &= 0 \\ (x - 35)(x - 15) &= 0 \end{aligned}$$

$x - 35 = 0$ | $x - 15 = 0$

$x = 35$ meters | $x = 15$ meters

$\dfrac{200 - 4x}{3} = 20$ meters | $\dfrac{200 - 4x}{3} = 46\frac{2}{3}$ meters

91. *Verbal Model:*

[Volume] = [Length] · [Width] · [Height]

Labels: Length $= x + 4$

Width $= x$

Height $= 6$

Equation:

$$\begin{aligned} 840 &= (x + 4)(x)(6) \\ 140 &= x(x + 4) \\ 140 &= x^2 + 4x \\ 0 &= x^2 + 4x - 140 \\ 0 &= (x + 14)(x - 10) \end{aligned}$$

$x + 14 = 0$ | $x - 10 = 0$

$x = -14$ | $x = 10$

Not a solution

Thus, the dimensions are:
length $= 10 + 4 = 14$ inches,
width $= x = 10$ inches, height $= 6$ inches

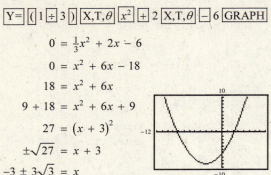

93. $R = x\left(80 - \frac{1}{2}x\right)$

$2750 = 80x - \frac{1}{2}x^2$

$0 = -\frac{1}{2}x^2 + 80x - 2750$

$0 = x^2 - 160x + 5500$

$0 = (x - 50)(x - 110)$

$0 = x - 50 \qquad x - 110 = 0$

$50 = x \qquad x = 110$

50 pairs, 110 pairs

95. Use the method of completing the square to write the quadratic equation in the form $u^2 = d$. Then use the Square Root Property to simplify.

97. (a) $d = 0$

(b) $d > 0$, and d is a perfect square.

(c) $d > 0$, and d is not a perfect square.

(d) $d < 0$

99. $3\sqrt{5}\sqrt{500} = 3 \cdot \sqrt{5 \cdot 500}$

$= 3 \cdot \sqrt{25 \cdot 100} = 3 \cdot 5 \cdot 10 = 150$

101. $\left(3 + \sqrt{2}\right)\left(3 - \sqrt{2}\right) = 3^2 - \left(\sqrt{2}\right)^2 = 9 - 2 = 7$

103. $\left(3 + \sqrt{2}\right)^2 = 3^2 + 2(3)\left(\sqrt{2}\right) + \left(\sqrt{2}\right)^2$

$= 9 + 6\sqrt{2} + 2$

$= 11 + 6\sqrt{2}$

105. $\dfrac{8}{\sqrt{10}} = \dfrac{8}{\sqrt{10}} \cdot \dfrac{\sqrt{10}}{\sqrt{10}} = \dfrac{8\sqrt{10}}{10} = \dfrac{4\sqrt{10}}{5}$

107. Product Rule: $\sqrt{ab} = \sqrt{a} \cdot \sqrt{b}$

Section 8.3 The Quadratic Formula

1. $2x^2 = 7 - 2x$

$2x^2 + 2x - 7 = 0$

3. $x(10 - x) = 5$

$10x - x^2 = 5$

$-x^2 + 10x - 5 = 0$

$x^2 - 10x + 5 = 0$

5. (a) $x^2 - 11x + 28 = 0$

$$x = \frac{11 \pm \sqrt{(-11)^2 - 4(1)(28)}}{2(1)}$$

$$x = \frac{11 \pm \sqrt{121 - 112}}{2}$$

$$x = \frac{11 \pm \sqrt{9}}{2}$$

$$x = \frac{11 \pm 3}{2}$$

$x = 7, 4$

(b) $(x - 7)(x - 4) = 0$

$x - 7 = 0 \qquad x - 4 = 0$

$x = 7 \qquad x = 4$

7. (a) $x^2 + 6x + 8 = 0$

$$x = \frac{-6 \pm \sqrt{6^2 - 4(1)(8)}}{2(1)}$$

$$x = \frac{6 \pm \sqrt{36 - 32}}{2}$$

$$x = \frac{-6 \pm \sqrt{4}}{2}$$

$$x = \frac{-6 \pm 2}{2}$$

$x = -2, -4$

(b) $(x + 4)(x + 2) = 0$

$x + 4 = 0 \qquad x + 2 = 0$

$x = -4 \qquad x = -2$

9. (a) $16x^2 + 8x + 1 = 0$

$$x = \frac{-8 \pm \sqrt{8^2 - 4(16)(1)}}{2(16)}$$

$$x = \frac{-8 \pm \sqrt{64 - 64}}{32}$$

$$x = \frac{-8 \pm \sqrt{0}}{32}$$

$$x = -\frac{8}{32} = -\frac{1}{4}$$

(b) $(4x + 1)(4x + 1) = 0$

$$4x + 1 = 0 \qquad 4x + 1 = 0$$

$$4x = -1 \qquad 4x = -1$$

$$x = -\frac{1}{4} \qquad x = -\frac{1}{4}$$

11. (a) $4x^2 + 12x + 9 = 0$

$$x = \frac{-12 \pm \sqrt{12^2 - 4(4)(9)}}{2(4)}$$

$$x = \frac{-12 \pm \sqrt{144 - 144}}{8}$$

$$x = \frac{-12 \pm 0}{8}$$

$$x = -\frac{12}{8} = -\frac{3}{2}$$

(b) $(2x + 3)(2x + 3) = 0$

$$2x + 3 = 0 \qquad 2x + 3 = 0$$

$$x = -\frac{3}{2} \qquad x = -\frac{3}{2}$$

13. (a) $x^2 - 5x - 300 = 0$

$$x = \frac{-(-5) \pm \sqrt{(-5)^2 - 4(1)(-300)}}{2(1)}$$

$$x = \frac{5 \pm \sqrt{25 + 1200}}{2}$$

$$x = \frac{5 \pm \sqrt{1225}}{2}$$

$$x = \frac{5 \pm 35}{2}$$

$$x = 20, -15$$

(b) $(x - 20)(x + 15) = 0$

$$x - 20 = 0 \qquad x + 15 = 0$$

$$x = 20 \qquad x = -15$$

15. $x^2 - 2x - 4 = 0$

$$x = \frac{-(-2) \pm \sqrt{(-2)^2 - 4(1)(-4)}}{2(1)}$$

$$x = \frac{2 \pm \sqrt{4 + 16}}{2}$$

$$x = \frac{2 \pm \sqrt{20}}{2}$$

$$x = \frac{2 \pm 2\sqrt{5}}{2}$$

$$x = \frac{2(1 \pm \sqrt{5})}{2}$$

$$x = 1 \pm \sqrt{5}$$

17. $t^2 + 4t + 1 = 0$

$$t = \frac{-4 \pm \sqrt{4^2 - 4(1)(1)}}{2(1)}$$

$$t = \frac{-4 \pm \sqrt{16 - 4}}{2}$$

$$t = \frac{4 \pm \sqrt{12}}{2}$$

$$t = \frac{-4 \pm 2\sqrt{3}}{2}$$

$$t = \frac{2(-2 \pm \sqrt{3})}{2}$$

$$t = -2 \pm \sqrt{3}$$

19. $x^2 - 10x + 23 = 0$

$$x = \frac{-(-10) \pm \sqrt{(-10)^2 - 4(1)(23)}}{2(1)}$$

$$x = \frac{10 \pm \sqrt{100 - 92}}{2}$$

$$x = \frac{10 \pm \sqrt{8}}{2}$$

$$x = \frac{10 \pm 2\sqrt{2}}{2}$$

$$x = \frac{2(5 \pm \sqrt{2})}{2}$$

$$x = 5 \pm \sqrt{2}$$

21. $2x^2 + 3x + 3 = 0$

$$x = \frac{-3 \pm \sqrt{3^2 - 4(2)(3)}}{2(2)}$$

$$x = \frac{-3 \pm \sqrt{9 - 24}}{4}$$

$$x = \frac{-3 \pm \sqrt{-15}}{4}$$

$$x = \frac{-3 \pm i\sqrt{15}}{4}$$

$$x = -\frac{3}{4} \pm \frac{\sqrt{15}}{4}i$$

23. $3v^2 - 2v - 1 = 0$

$$v = \frac{-(-2) \pm \sqrt{(-2)^2 - 4(3)(-1)}}{2(3)}$$

$$v = \frac{2 \pm \sqrt{4 + 12}}{6}$$

$$v = \frac{2 \pm \sqrt{16}}{6}$$

$$v = \frac{2 \pm 4}{6}$$

$$v = \frac{6}{6}, -\frac{2}{6}$$

$$v = 1, -\frac{1}{3}$$

25. $2x^2 + 4x - 3 = 0$

$$x = \frac{-4 \pm \sqrt{4^2 - 4(2)(-3)}}{2(2)}$$

$$x = \frac{-4 \pm \sqrt{16 + 24}}{4}$$

$$x = \frac{-4 \pm \sqrt{40}}{4}$$

$$x = \frac{-4 \pm 2\sqrt{10}}{4}$$

$$x = \frac{2\left(-2 \pm \sqrt{10}\right)}{4}$$

$$x = \frac{-2 \pm \sqrt{10}}{2}$$

27. $-4x^2 - 6x + 3 = 0$

$$x = \frac{-(-6) \pm \sqrt{(-6)^2 - 4(-4)(3)}}{2(-4)}$$

$$x = \frac{6 \pm \sqrt{36 + 48}}{-8}$$

$$x = \frac{6 \pm \sqrt{84}}{-8}$$

$$x = \frac{6 \pm 2\sqrt{21}}{-8}$$

$$x = \frac{-3 \pm \sqrt{21}}{4}$$

29. $8x^2 - 6x + 2 = 0$ (Divide by 2.)

$4x^2 - 3x + 1 = 0$

$$x = \frac{-(-3) \pm \sqrt{(-3)^2 - 4(4)(1)}}{2(4)}$$

$$x = \frac{3 \pm \sqrt{9 - 16}}{8}$$

$$x = \frac{3 \pm \sqrt{-7}}{8} = \frac{3}{8} \pm \frac{\sqrt{7}}{8}i$$

31. $-4x^2 + 10x + 12 = 0$ (Divide by -2.)

$2x^2 - 5x - 6 = 0$

$$x = \frac{-(-5) \pm \sqrt{(-5)^2 - 4(2)(-6)}}{2(2)}$$

$$x = \frac{5 \pm \sqrt{25 + 48}}{4}$$

$$x = \frac{5 \pm \sqrt{73}}{4}$$

33. $9x^2 = 1 + 9x$

$9x^2 - 9x - 1 = 0$

$$x = \frac{-(-9) \pm \sqrt{(-9)^2 - 4(9)(-1)}}{2(9)}$$

$$x = \frac{9 \pm \sqrt{81 + 36}}{18}$$

$$x = \frac{9 \pm \sqrt{117}}{18}$$

$$x = \frac{9}{18} \pm \frac{3\sqrt{13}}{18}$$

$$x = \frac{1}{2} \pm \frac{\sqrt{13}}{6} \text{ or } \frac{3 \pm \sqrt{13}}{6}$$

35. $2x - 3x^2 = 3 - 7x^2$

$4x^2 + 2x - 3 = 0$

$x = \dfrac{-2 \pm \sqrt{2^2 - 4(4)(-3)}}{2(4)}$

$x = \dfrac{-2 \pm \sqrt{4 + 48}}{8}$

$x = \dfrac{-2 \pm \sqrt{52}}{8}$

$x = \dfrac{-2 \pm 2\sqrt{13}}{8}$

$x = -\dfrac{1}{4} \pm \dfrac{\sqrt{13}}{4}$

37. $x^2 - 0.4x - 0.16 = 0$

$x = \dfrac{-(-0.4) \pm \sqrt{(-0.4)^2 - 4(1)(-0.16)}}{2(1)}$

$x = \dfrac{0.4 \pm \sqrt{0.16 + 0.64}}{2}$

$x = \dfrac{0.4 \pm \sqrt{0.80}}{2}$

$x = \dfrac{0.4 \pm 2\sqrt{0.2}}{2}$

$x = 0.2 \pm \sqrt{0.2}$ or $\dfrac{1 \pm \sqrt{5}}{5}$

39. $2.5x^2 + x - 0.9 = 0$

$x = \dfrac{-1 \pm \sqrt{1^2 - 4(2.5)(-0.9)}}{2(2.5)}$

$x = \dfrac{-1 \pm \sqrt{1 + 9}}{5}$

$x = \dfrac{-1 \pm \sqrt{10}}{5}$

41. $b^2 - 4ac = 1^2 - 4(1)(1) = 1 - 4 = -3$

2 distinct complex solutions

43. $b^2 - 4ac = (-2)^2 - 4(8)(-5) = 4 + 160 = 164$

Two distinct rational solutions

45. $b^2 - 4ac = (-24)^2 - 4(9)(16) = 576 - 576 = 0$

1 repeated rational solution

47. $b^2 - 4ac = (-1)^2 - 4(3)(2) = 1 - 24 = -23$

2 distinct complex solutions

49. $z^2 - 169 = 0$

$z^2 = 169$

$z = \pm 13$

51. $5y^2 + 15y = 0$

$5y(y + 3) = 0$

$5y = 0 \qquad y + 3 = 0$

$y = 0 \qquad y = -3$

53. $25(x - 3)^2 - 36 = 0$

$(x - 3)^2 = \frac{36}{25}$

$x - 3 = \pm\sqrt{\frac{36}{25}}$

$x = 3 \pm \frac{6}{5}$

$x = \frac{15}{5} \pm \frac{6}{5}$

$x = \frac{21}{5}, \frac{9}{5}$

55. $2y(y - 18) + 3(y - 18) = 0$

$(y - 18)(2y + 3) = 0$

$y - 18 = 0 \qquad 2y + 3 = 0$

$y = 18 \qquad 2y = -3$

$y = -\frac{3}{2}$

57. $x^2 + 8x + 25 = 0$

$x^2 + 8x + 16 = -25 + 16$

$(x + 4)^2 = -9$

$x + 4 = \pm\sqrt{-9}$

$x = -4 \pm 3i$

59. $3x^2 - 13x - 169 = 0$

$x = \dfrac{-(-13) \pm \sqrt{(-13)^2 - 4(3)(169)}}{2(3)}$

$x = \dfrac{13 \pm \sqrt{169 - 2028}}{6}$

$x = \dfrac{13 \pm \sqrt{-1859}}{6}$

$x = \dfrac{13}{6} \pm \dfrac{13\sqrt{11}}{6}i$

61. $25x^2 + 80x + 61 = 0$

$$x = \frac{-80 \pm \sqrt{80^2 - 4(25)(61)}}{2(25)}$$

$$x = \frac{-80 \pm \sqrt{6400 - 6100}}{50}$$

$$x = \frac{-80 \pm \sqrt{300}}{50}$$

$$x = \frac{-80 \pm 10\sqrt{3}}{50}$$

$$x = -\frac{8}{5} \pm \frac{\sqrt{3}}{5}$$

63. $7x(x + 2) + 5 = 3x(x + 1)$

$7x^2 + 14x + 5 = 3x^2 + 3x$

$4x^2 + 11x + 5 = 0$

$$x = \frac{-11 \pm \sqrt{11^2 - 4(4)(5)}}{2(4)}$$

$$x = \frac{-11 \pm \sqrt{121 - 80}}{8}$$

$$x = \frac{-11 \pm \sqrt{41}}{8}$$

65. $x = 5 \qquad x = -2$

$x - 5 = 0 \qquad x + 2 = 0$

$(x - 5)(x + 2) = 0$

$x^2 - 3x - 10 = 0$

67. $x = 1 \qquad x = 7$

$x - 1 = 0 \qquad x - 7 = 0$

$(x - 1)(x - 7) = 0$

$x^2 - 8x + 7 = 0$

69. $x = 1 + \sqrt{2} \qquad x = 1 - \sqrt{2}$

$x - \left(1 + \sqrt{2}\right) = 0 \qquad x - \left(1 - \sqrt{2}\right) = 0$

$\left[x - \left(1 + \sqrt{2}\right)\right]\left[x - \left(1 - \sqrt{2}\right)\right] = 0$

$\left[(x - 1) - \sqrt{2}\right]\left[(x - 1) + \sqrt{2}\right] = 0$

$(x - 1)^2 - \left(\sqrt{2}\right)^2 = 0$

$x^2 - 2x + 1 - 2 = 0$

$x^2 - 2x - 1 = 0$

71. $x = 5i \qquad x = -5i$

$x - 5i = 0 \qquad x + 5i = 0$

$(x - 5i)(x + 5i) = 0$

$x^2 - 25i^2 = 0$

$x^2 + 25 = 0$

73. $x = 12 \qquad x = 12$

$x - 12 = 0 \qquad x - 12 = 0$

$(x - 12)(x - 12) = 0$

$x^2 - 24x + 144 = 0$

75. $3x^2 - 6x + 1 = 0$

Keystrokes:

[Y=] 3 [X,T,θ] [x²] [−] 6 [X,T,θ] [+] 1 [GRAPH]

$0 = 3x^2 - 6x + 1$

$$x = \frac{-(-6) \pm \sqrt{(-6)^2 - 4(3)(1)}}{2(3)}$$

$$x = \frac{6 \pm \sqrt{36 - 12}}{6}$$

$$x = \frac{6 \pm \sqrt{24}}{6}$$

$x \approx 1.82, 0.18$

$(1.82, 0), (0.18, 0)$

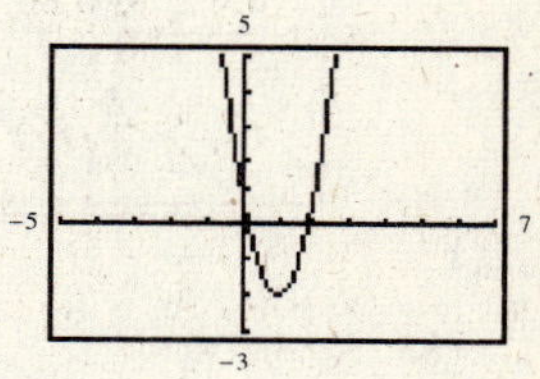

77. $y = x^2 - 4x + 3$

Keystrokes:

[Y=] [X,T,θ] [x²] [−] 4 [X,T,θ] [+] 3 [GRAPH]

$0 = x^2 - 4x + 3$

$0 = (x - 3)(x - 1)$

$x - 3 = 0 \qquad x - 1 = 0$

$x = 3 \qquad x = 1$

$(3, 0), (1, 0)$

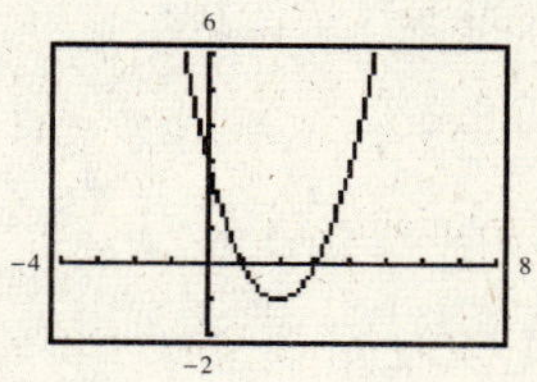

79. $-0.03x^2 + 2x - 0.4 = 0$

Keystrokes:

[Y=] [(-)] 0.03 [X,T,θ] [x^2] [+] 2 [X,T,θ] [−] 0.4 [GRAPH]

$$x = \frac{-2 \pm \sqrt{2^2 - 4(-0.03)(-0.4)}}{2(-0.03)}$$

$$x = \frac{-2 \pm \sqrt{4 - 0.048}}{-0.06}$$

$$x = \frac{-2 \pm \sqrt{3.952}}{-0.06}$$

$x \approx 0.20, 66.47$

$(0.20, 0), (66.47, 0)$

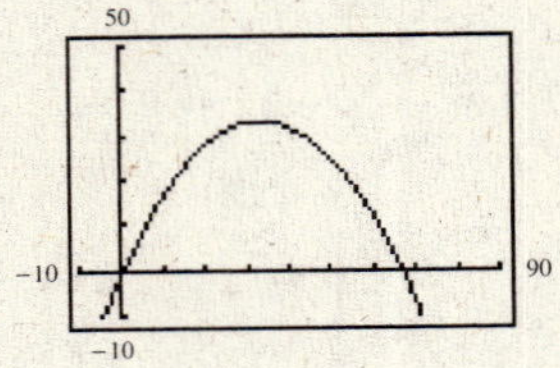

81. *Keystrokes:*

[Y=] 2 [X,T,θ] [x^2] [−] 5 [X,T,θ] [+] 5 [GRAPH]

$b^2 - 4ac = (-5)^2 - 4(2)(5) = 25 - 40 = -15$

No real solutions

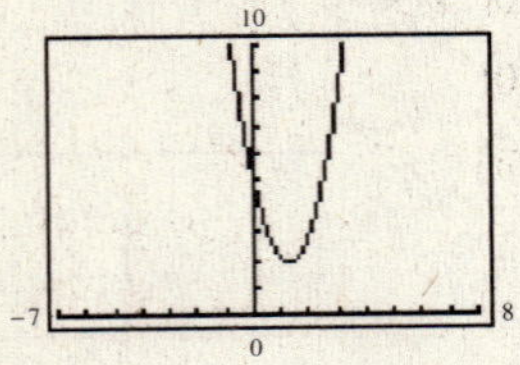

83. *Keystrokes:*

[Y=] [X,T,θ] [x^2] [+] 6 [X,T,θ] [−] 40 [GRAPH]

$b^2 - 4ac = 6^2 - 4(1)(-40) = 36 + 160 = 196$

Two real solutions

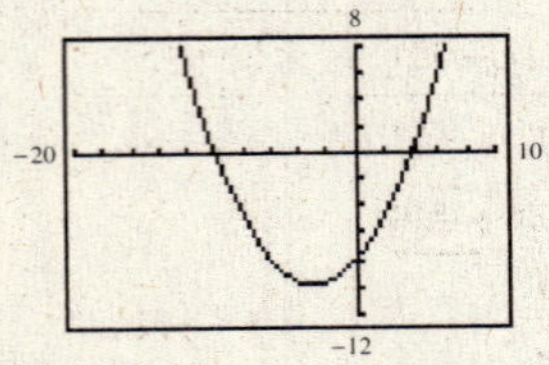

85. $f(x) = 2x^2 - 7x + 1$

$-3 = 2x^2 - 7x + 1$

$0 = 2x^2 - 7x + 4$

$$x = \frac{-(-7) \pm \sqrt{(-7)^2 - 4(2)(4)}}{2(2)}$$

$$x = \frac{7 \pm \sqrt{49 - 32}}{4}$$

$$x = \frac{7 \pm \sqrt{17}}{4}$$

87. $g(x) = 2x^2 - 3x + 16$

$14 = 2x^2 - 3x + 16$

$0 = 2x^2 - 3x + 2$

$$x = \frac{-(-3) \pm \sqrt{(-3)^2 - 4(2)(2)}}{2(2)}$$

$$x = \frac{3 \pm \sqrt{9 - 16}}{4}$$

$$x = \frac{3 \pm \sqrt{-7}}{4}$$

No real values

89. $$\frac{x^2}{4} - \frac{2x}{3} = 1$$

$$12\left(\frac{x^2}{4} - \frac{2x}{3}\right) = (1)12$$

$3x^2 - 8x - 12 = 0$

$$x = \frac{-(-8) \pm \sqrt{(-8)^2 - 4(3)(-12)}}{2(3)}$$

$$x = \frac{8 \pm \sqrt{64 + 144}}{6}$$

$$x = \frac{8 \pm \sqrt{208}}{6}$$

$$x = \frac{8 \pm 4\sqrt{13}}{6}$$

$$x = \frac{4}{3} \pm \frac{2\sqrt{13}}{3}$$

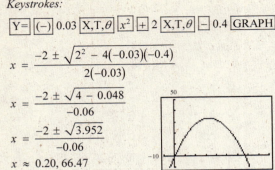

91. $\sqrt{x+3} = x - 1$

$$\left(\sqrt{x+3}\right)^2 = (x-1)^2$$
$$x + 3 = x^2 - 2x + 1$$
$$0 = x^2 - 3x - 2$$
$$x = \frac{-(-3) \pm \sqrt{(-3)^2 - 4(1)(-2)}}{2(1)}$$
$$x = \frac{3 \pm \sqrt{9+8}}{2}$$
$$x = \frac{3 \pm \sqrt{17}}{2}$$
$$x = \frac{3 + \sqrt{17}}{2}$$
$$x = \frac{3 - \sqrt{17}}{2} \text{ does not check.}$$

93. $x^2 - 6x + c = 0$

(a) $$b^2 - 4ac > 0$$
$$(-6)^2 - 4(1)c > 0$$
$$36 - 4c > 0$$
$$-4c > -36$$
$$c < 9$$

(b) $$b^2 - 4ac = 0$$
$$(-6)^2 - 4(1)c = 0$$
$$36 - 4c = 0$$
$$-4c = -36$$
$$c = 9$$

(c) $$b^2 - 4ac < 0$$
$$(-6)^2 - 4(1)c < 0$$
$$36 - 4c < 0$$
$$-4c < -36$$
$$c > 9$$

95. *Verbal Model:* $\boxed{\text{Area}} = \boxed{\text{Length}} \cdot \boxed{\text{Width}}$

Labels: Length $= x + 6.3$

Width $= x$

Equation: $58.14 = (x + 6.3) \cdot x$

$$58.14 = x^2 + 6.3x$$
$$0 = x^2 + 6.3x - 58.14$$
$$x = \frac{-6.3 \pm \sqrt{6.3^2 - 4(1)(-58.14)}}{2(1)}$$
$$x = \frac{-6.3 \pm \sqrt{39.69 + 232.56}}{2}$$
$$x = \frac{-6.3 \pm \sqrt{272.25}}{2}$$
$$x \approx 5.1 \text{ inches}$$
$$x + 6.3 \approx 11.4 \text{ inches}$$

97. $h = -16t^2 + 40t + 50$

(a) $$50 = -16t^2 + 40t + 50$$
$$0 = -16t^2 + 40t$$
$$0 = -8\left(2t^2 - 5t\right)$$
$$0 = 2t^2 - 5t$$
$$0 = t(2t - 5)$$
$$0 = t, \quad 2t - 5 = 0$$
$$t = \frac{5}{2} = 2.5 \text{ seconds}$$

(b) $$0 = -16t^2 + 40t + 50$$
$$0 = -2\left(8t^2 - 20t - 25\right)$$
$$t = \frac{-(-20) \pm \sqrt{(-20)^2 - 4(8)(-25)}}{2(8)}$$
$$t = \frac{20 \pm \sqrt{400 + 800}}{16}$$
$$t = \frac{20 \pm \sqrt{1200}}{16}$$
$$t = \frac{20 \pm 20\sqrt{3}}{16}$$
$$t = \frac{4\left(5 \pm 5\sqrt{3}\right)}{16}$$
$$t = \frac{5 + 5\sqrt{3}}{4}, \frac{5 - 5\sqrt{3}}{4} \text{ reject}$$
$$t \approx 3.415 \text{ seconds}$$

(c) No. In order for the discriminant to be greater than or equal to zero, the value of c must be greater than or equal to -25. Therefore, the height cannot exceed 75 feet, or the value of c would be less than -25 when the equation is set equal to zero.

99. $d = -0.25t^2 + 1.7t + 3.5, \quad 0 \le t \le 7$

$$6 = -0.25t^2 + 1.7t + 3.5$$

$$0 = -0.25t^2 + 1.7t - 2.5$$

$$0 = t^2 - 6.8t + 10$$

$$t = \frac{-(-6.8) \pm \sqrt{6.8^2 - 4(1)(10)}}{2(1)}$$

$$t = \frac{6.8 \pm \sqrt{46.24 - 40}}{2}$$

$$t = \frac{6.8 \pm \sqrt{6.24}}{2}$$

$$t = \frac{6.8 + \sqrt{6.24}}{2} \approx 4.65 \text{ hours after a heavy rain}$$

$$t = \frac{6.8 - \sqrt{6.24}}{2} \approx 2.15 \text{ hours after a heavy rain}$$

101. (a) *Keystrokes:* [Y=] −0.013 [X,T,θ] [x^2] [+] 1.25 [X,T,θ] [+] 5.6 [GRAPH]

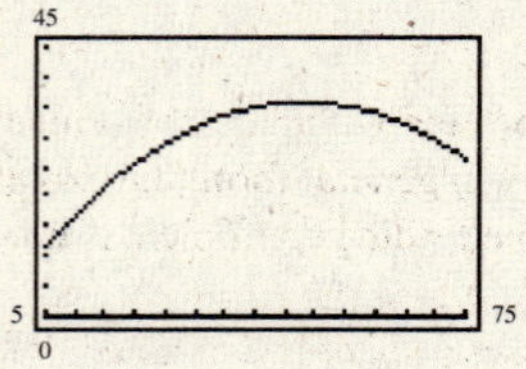

(b) $32 = -0.013x^2 + 1.25x + 5.6$

$$0 = -0.013x^2 + 1.25x - 26.4$$

$$0 = 1.3x^2 - 125x + 2640$$

$$x = \frac{-(-125) \pm \sqrt{(-125)^2 - 4(1.3)(2640)}}{2(1.3)}$$

$$x = \frac{125 \pm \sqrt{15{,}625 - 13{,}728}}{2.6}$$

$$x = \frac{125 + \sqrt{1897}}{2.6}$$

$x \approx 64.8$ miles per hour

$$x = \frac{125 - \sqrt{1897}}{2.6}$$

$x \approx 31.3$ miles per hour

Using the graph, you would have to travel approximately 65 miles per hour or 31 miles per hour to obtain a fuel economy of 32 miles per gallon.

103.

	x_1, x_2	$x_1 + x_2$	$x_1 \cdot x_2$
(a) $x^2 - x - 6 = 0$ $(x - 3)(x + 2) = 0$ $x = 3, \quad x = -2$	$3, -2$	1	-6
(b) $2x^2 + 5x - 3 = 0$ $(2x - 1)(x + 3) = 0$ $x = \frac{1}{2}, \quad x = -3$	$\frac{1}{2}, -3$	$-\frac{5}{2}$	$-\frac{3}{2}$
(c) $4x^2 - 9 = 0$ $(2x - 3)(2x + 3) = 0$ $x = \frac{3}{2}, \quad x = -\frac{3}{2}$	$\frac{3}{2}, -\frac{3}{2}$	0	$-\frac{9}{4}$
(d) $x^2 - 10x + 34 = 0$ $x^2 - 10x + 25 = -34 + 25$ $(x - 5)^2 = -9$ $x - 5 = \pm\sqrt{-9}$ $x = 5 \pm 3i$	$5 + 3i, 5 - 3i$	10	34

105. The Square Root Property would be convenient because the equation is of the form $u^2 = d$.

107. The Quadratic Formula would be convenient because the equation is already in general form, the expression cannot be factored, and the leading coefficient is not equal to 1.

109. When the Quadratic Formula is applied to $ax^2 + bx + c = 0$, the square root of the discriminant is evaluated. When the discriminant is positive, the square root of the discriminant is positive and will yield two real solutions (or x-intercepts). When the discriminant is zero, the equation has one real solution (or x-intercept). When the discriminant is negative, the square root of the discriminant is negative and will yield two complex solutions (or no x-intercepts).

111. $A(-1, 11), B(2, 2), C(1, 5)$

$$AB = \sqrt{(-1 - 2)^2 + (11 - 2)^2} = \sqrt{9 + 81} = \sqrt{90} = 3\sqrt{10}$$

$$BC = \sqrt{(2 - 1)^2 + (2 - 5)^2} = \sqrt{1 + 9} = \sqrt{10}$$

$$AC = \sqrt{(-1 - 1)^2 + (11 - 5)^2} = \sqrt{4 + 36} = \sqrt{40} = 2\sqrt{10}$$

$\sqrt{10} + 2\sqrt{10} = 3\sqrt{10}$, collinear

113. $A(-6, -2), B(-3, -4), C(3, -4)$

$$AB = \sqrt{(-6 - (-3))^2 + (-2 - (-4))^2} = \sqrt{9 + 4} = \sqrt{13}$$

$$BC = \sqrt{(-3 - 3)^2 + (-4 - (-4))^2} = \sqrt{36 + 0} = \sqrt{36} = 6$$

$$AC = \sqrt{(-6 - 3)^2 + (-2 - (-4))^2} = \sqrt{81 + 4} = \sqrt{85}$$

$\sqrt{13} + 6 \neq \sqrt{85}$, not collinear

115. $f(x) = (x - 1)^2$

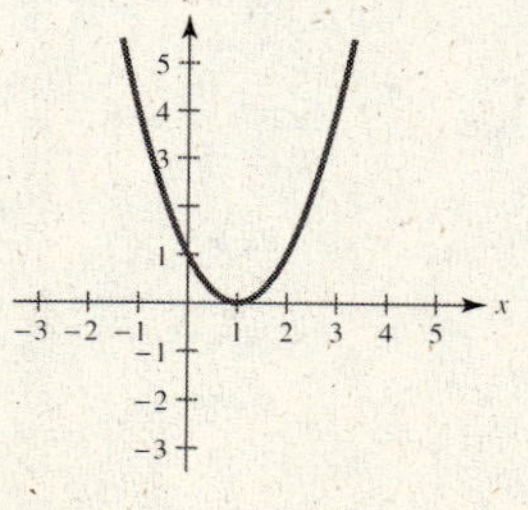

117. $f(x) = (x - 2)^2 + 4$

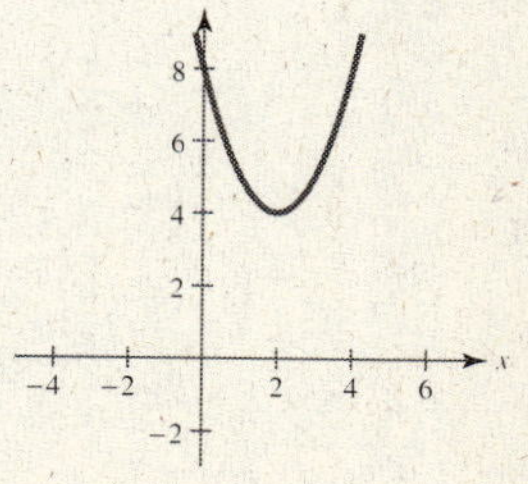

Mid-Chapter Quiz for Chapter 8

1. $2x^2 - 72 = 0$

$2(x^2 - 36) = 0$

$2(x - 6)(x + 6) = 0$

$x - 6 = 0 \qquad x + 6 = 0$

$x = 6 \qquad x = -6$

2. $2x^2 + 3x - 20 = 0$

$(2x - 5)(x + 4) = 0$

$2x - 5 = 0 \qquad x + 4 = 0$

$x = \frac{5}{2} \qquad x = -4$

3. $3x^2 = 36$

$x^2 = 12$

$x = \pm\sqrt{12}$

$x = \pm 2\sqrt{3}$

4. $(u - 3)^2 - 16 = 0$

$(u - 3)^2 = 16$

$u - 3 = \pm 4$

$u = 3 \pm 4 = 7, -1$

5. $m^2 + 7m + 2 = 0$

$$m^2 + 7m + \frac{49}{4} = -2 + \frac{49}{4}$$

$$\left(m + \frac{7}{2}\right)^2 = \frac{41}{4}$$

$$m + \frac{7}{2} = \pm\sqrt{\frac{41}{4}}$$

$$m = -\frac{7}{2} \pm \frac{\sqrt{41}}{2}$$

6. $2y^2 + 6y - 5 = 0$

$$y^2 + 3y = \frac{5}{2}$$

$$y^2 + 3y + \frac{9}{4} = \frac{5}{2} + \frac{9}{4}$$

$$\left(y + \frac{3}{2}\right)^2 = \frac{10}{4} + \frac{9}{4}$$

$$\left(y + \frac{3}{2}\right)^2 = \frac{19}{4}$$

$$y + \frac{3}{2} = \pm\frac{\sqrt{19}}{2}$$

$$y = -\frac{3}{2} \pm \frac{\sqrt{19}}{2}$$

7. $x^2 + 4x - 6 = 0$

$$x = \frac{-4 \pm \sqrt{4^2 - 4(1)(-6)}}{2(1)}$$

$$x = \frac{-4 \pm \sqrt{16 + 24}}{2}$$

$$x = \frac{-4 \pm \sqrt{40}}{2}$$

$$x = \frac{-4 \pm 2\sqrt{10}}{2} = -2 \pm \sqrt{10}$$

8. $6v^2 - 3v - 4 = 0$

$$v = \frac{-(-3) \pm \sqrt{(-3)^2 - 4(6)(-4)}}{2(6)}$$

$$v = \frac{3 \pm \sqrt{9 + 96}}{12}$$

$$v = \frac{3 \pm \sqrt{105}}{12}$$

9. $x^2 + 5x + 7 = 0$

$$x = \frac{-5 \pm \sqrt{5^2 - 4(1)(7)}}{2(1)}$$

$$x = \frac{-5 \pm \sqrt{25 - 28}}{2}$$

$$x = \frac{-5 \pm \sqrt{-3}}{2}$$

$$x = \frac{-5 \pm \sqrt{3}i}{2} = -\frac{5}{2} \pm \frac{\sqrt{3}}{2}i$$

10. $36 = (t - 4)^2$

$\pm 6 = t - 4$

$4 \pm 6 = t$

$10, -2 = t$

11. $(x - 10)(x + 3) = 0$

$(x - 10) = 0 \qquad x + 3 = 0$

$x = 10 \qquad x = -3$

12. $x^2 - 3x - 10 = 0$

$(x - 5)(x + 2) = 0$

$x - 5 = 0 \qquad x + 2 = 0$

$x = 5 \qquad x = -2$

13. $4b^2 - 12b + 9 = 0$

$(2b - 3)(2b - 3) = 0$

$2b - 3 = 0 \qquad 2b - 3 = 0$

$b = \frac{3}{2} \qquad b = \frac{3}{2}$

14. $3m^2 + 10m + 5 = 0$

$$m = \frac{-10 \pm \sqrt{10^2 - 4(3)(5)}}{2(3)}$$

$$m = \frac{-10 \pm \sqrt{100 - 60}}{6}$$

$$m = \frac{10 \pm \sqrt{40}}{6}$$

$$m = \frac{-10 \pm 2\sqrt{10}}{6}$$

$$m = \frac{-5 \pm \sqrt{10}}{3}$$

15. $x - 4\sqrt{x} - 21 = 0$

$x - 21 = 4\sqrt{x}$

$(x - 21)^2 = \left(4\sqrt{x}\right)^2$

$x^2 - 42x + 441 = 16x$

$x^2 - 58x + 441 = 0$

$(x - 9)(x - 49) = 0$

$x - 9 = 0 \qquad x - 49 = 0$

$x = 9 \qquad x = 49$

Check: $9 - 4\sqrt{9} - 21 \stackrel{?}{=} 0$

$9 - 12 - 21 \stackrel{?}{=} 0$

$-24 \neq 0$

Not a solution

Check: $49 - 4\sqrt{49} - 21 \stackrel{?}{=} 0$

$49 - 28 - 21 \stackrel{?}{=} 0$

$0 = 0$

Solution

16. $x^4 + 7x^2 + 12 = 0$

$(x^2 + 4)(x^2 + 3) = 0$

$x^2 = -4 \qquad x^2 = -3$

$x = \pm\sqrt{-4} \qquad x = \pm\sqrt{-3}$

$x = \pm 2i \qquad x = \pm\sqrt{3}i$

17. $x - 4\sqrt{x} + 3 = 0$

Let $u = \sqrt{x}$, $u^2 = x$.

$u^2 - 4u + 3 = 0$

$(u - 3)(u - 1) = 0$

$u - 3 = 0 \qquad u - 1 = 0$

$u = 3 \qquad u = 1$

$\sqrt{x} = 3 \qquad \sqrt{x} = 1$

$x = 9 \qquad x = 1$

Check: $9 - 4\sqrt{9} + 3 \stackrel{?}{=} 0$

$9 - 12 + 3 \stackrel{?}{=} 0$

$0 = 0$

Solution

Check: $1 - 4\sqrt{1} + 3 \stackrel{?}{=} 0$

$1 - 4 + 3 \stackrel{?}{=} 0$

$0 = 0$

Solution

18. $x^4 - 14x^2 + 24 = 0$

Let $u = x^2$, $u^2 = x^4$.

$u^2 - 14u + 24 = 0$

$u^2 - 14u + 49 = -24 + 49$

$(u - 7)^2 = 25$

$u - 7 = \pm 5$

$u = 7 \pm 5$

$u = 12 \qquad u = 2$

$x^2 = 12 \qquad x^2 = 2$

$x = \pm\sqrt{12} \qquad x = \pm\sqrt{2}$

$x = \pm 2\sqrt{3}$

19. *Keystrokes:*

[Y=] .5 [X,T,θ] [x^2] [−] 3 [X,T,θ] [−] 1 [GRAPH]

$0 = 0.5x^2 - 3x - 1$

$0 = x^2 - 6x - 2$

$x = \dfrac{-(-6) \pm \sqrt{(-6)^2 - 4(1)(-2)}}{2(1)}$

$x = \dfrac{6 \pm \sqrt{36 + 8}}{2}$

$x = \dfrac{6 \pm \sqrt{44}}{2}$

$x = \dfrac{6 \pm 2\sqrt{11}}{2}$

$x = 3 \pm \sqrt{11}$

$x \approx 6.32$ and -0.32

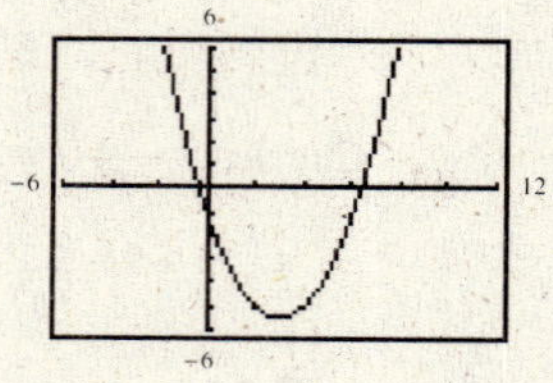

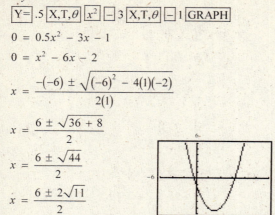

20. *Keystrokes:*

[Y=] [X,T,θ] [x^2] [+] .045 [X,T,θ] [−] 4 [GRAPH]

$0 = x^2 + 0.45x - 4$

$x = \dfrac{-0.45 \pm \sqrt{(0.45)^2 - 4(1)(-4)}}{2(1)}$

$x = \dfrac{-0.45 \pm \sqrt{0.2025 + 16}}{2}$

$x = \dfrac{-0.45 \pm \sqrt{16.2025}}{2}$

$x \approx 1.79$ and -2.24

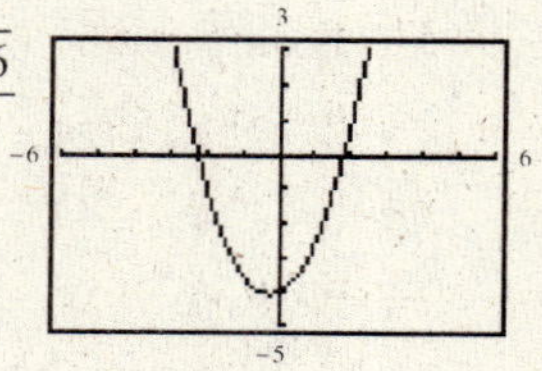

21. $R = x(180 - 1.5x)$

$5400 = 180x - 1.5x^2$

$0 = 1.5x^2 - 180x + 5400$

$0 = x^2 - 120x + 3600$

$0 = (x - 60)(x - 60)$

$0 = x - 60 \qquad x - 60 = 0$

$60 = x \qquad x = 60$

60 video games

22. *Verbal Model:* [Area] = [Length] · [Width]

Equation: $2275 = x \cdot (100 - x)$

$2275 = 100x - x^2$

$0 = x^2 - 100x + 2275$

$0 = (x - 35)(x - 65)$

$x - 35 = 0 \qquad x - 65 = 0$

$x = 35$ meters $\qquad x = 65$ meters

35 meters × 65 meters

Section 8.4 Graphs of Quadratic Functions

1. $y = (x + 1)^2 - 3$ (e)

3. $y = x^2 - 3$ (b)

5. $y = (x - 2)^2$ (d)

7. $y = x^2 - 2x = (x^2 - 2x + 1) - 1 = (x - 1)^2 - 1$

vertex $= (1, -1)$

9. $y = x^2 - 4x + 7$

$= (x^2 - 4x + 4) + 7 - 4$

$= (x - 2)^2 + 3$

vertex $= (2, 3)$

11. $y = x^2 + 6x + 5$

$y = (x^2 + 6x + 9) + 5 - 9$

$y = (x + 3)^2 - 4$

vertex $= (-3, -4)$

13. $y = -x^2 + 6x - 10$

$y = -1(x^2 - 6x) - 10$

$y = -1(x^2 - 6x + 9) - 10 + 9$

$y = -1(x - 3)^2 - 1$

vertex $= (3, -1)$

15. $y = -x^2 - 8x + 5$

$= -1(x^2 + 8x + 16) + 5 + 16$

$= -(x + 4)^2 + 21$

vertex $= (-4, 21)$

17. $y = 2x^2 + 6x + 2$

$= 2\left(x^2 + 3x + \frac{9}{4}\right) + 2 - \frac{9}{2}$

$= 2\left(x + \frac{3}{2}\right)^2 - \frac{5}{2}$

vertex $= \left(-\frac{3}{2}, -\frac{5}{2}\right)$

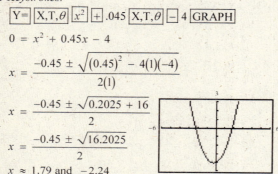

19. $f(x) = x^2 - 8x + 15$

$a = 1, \quad b = -8$

$x = \frac{-b}{2a} = \frac{-(-8)}{2(1)} = 4$

$f\left(\frac{-b}{2a}\right) = 4^2 - 8(4) + 15 = 16 - 32 + 15 = -1$

vertex $= (4, -1)$

21. $g(x) = -x^2 - 2x + 1$

$a = -1, \quad b = -2$

$x = \frac{-b}{2a} = \frac{-(-2)}{2(-1)} = -1$

$g\left(\frac{-b}{2a}\right) = -(-1)^2 - 2(-1) + 1 = -1 + 2 + 1 = 2$

vertex $= (-1, 2)$

23. $y = 4x^2 + 4x + 4$

$a = 4, \quad b = 4$

$x = \frac{-b}{2a} = \frac{-4}{2(4)} = -\frac{1}{2}$

$y = 4\left(-\frac{1}{2}\right)^2 + 4\left(-\frac{1}{2}\right) + 4$

$= 4\left(\frac{1}{4}\right) - 2 + 4$

$= 1 - 2 + 4$

$= 3$

vertex $= \left(-\frac{1}{2}, 3\right)$

25. $y = 2(x - 0)^2 + 2$

$2 > 0$ opens upward.

vertex $= (0, 2)$

27. $y = 4 - (x - 10)^2$

$-1 < 0$ opens downward.

vertex $= (10, 4)$

29. $y = x^2 - 6$

$1 > 0$ opens upward.

vertex $= (0, -6)$

31. $y = -(x - 3)^2$

$-1 < 0$ opens downward.

vertex $= (3, 0)$

33. $y = -x^2 + 6x$

$-1 < 0$ opens downward.

vertex $= (3, 9)$

$x = -\frac{b}{2a}$

$= -\frac{6}{2(-1)} = 3$

$y = -3^2 + 6(3)$

$= -9 + 18 = 9$

35. $y = 25 - x^2$

$0 = 25 - x^2$

$x^2 = 25$

$x = \pm 5$

$(5, 0), (-5, 0)$

$y = 25 - x^2$

$y = 25 - 0^2$

$y = 25$

$(0, 25)$

37. $y = x^2 - 9x$

$0 = x^2 - 9x$

$0 = x(x - 9)$

$(0, 0), (9, 0)$

$y = x^2 - 9x$

$y = 0^2 - 9(0)$

$y = 0$

$(0, 0)$

39. $y = -x^2 - 6x + 7$

$0 = -x^2 - 6x + 7$

$0 = x^2 + 6x - 7$

$0 = (x + 7)(x - 1)$

$(-7, 0), (1, 0)$

$y = -x^2 - 6x + 7$

$y = -0^2 - 6(0) + 7$

$y = 7$

$(0, 7)$

41. $y = 4x^2 - 12x + 9$

$0 = 4x^2 - 12x + 9$

$0 = (2x - 3)^2$

$0 = 2x - 3$

$\frac{3}{2} = x$

$\left(\frac{3}{2}, 0\right)$

$y = 4x^2 - 12x + 9$

$y = 4(0)^2 - 12(0) + 9$

$y = 9$

$(0, 9)$

43. $y = x^2 - 3x + 3$

$0 = x^2 - 3x + 3$

$x = \dfrac{3 \pm \sqrt{9 - 12}}{2} = \dfrac{3 \pm \sqrt{-3}}{2}$

No x-intercepts

$y = x^2 - 3x + 3$

$y = 0^2 - 3(0) + 3$

$y = 3$

$(0, 3)$

45. $y = -2x^2 - 6x + 5$

$0 = -2x^2 - 6x + 5$

$x = \dfrac{6 \pm \sqrt{36 + 40}}{-4}$

$x = \dfrac{-6 \pm \sqrt{76}}{4}$

$x = \dfrac{-6 \pm 2\sqrt{19}}{4} = \dfrac{-3 \pm \sqrt{19}}{2}$

$\left(\dfrac{-3 \pm \sqrt{19}}{2}, 0\right)$

$y = -2x^2 - 6x + 5$

$y = -2(0)^2 - 6(0) + 5$

$y = 5$

$(0, 5)$

47. $g(x) = x^2 - 4$

x-intercepts:

$0 = x^2 - 4$

$0 = (x - 2)(x + 2)$

$x = 2, \quad x = -2$

vertex: $g(x) = (x - 0)^2 - 4$

$(0, -4)$

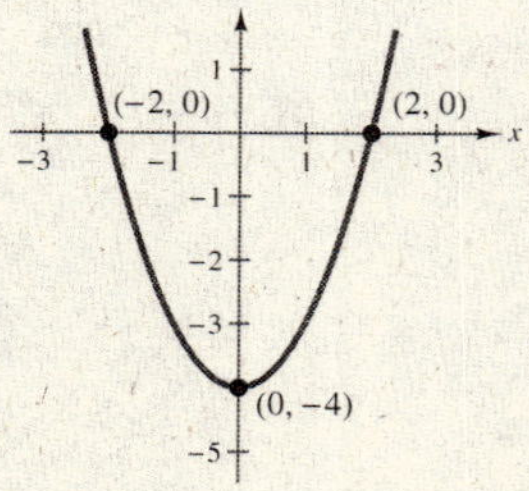

49. $f(x) = -x^2 + 4$

x-intercepts:

$0 = -x^2 + 4$

$x^2 = 4$

$x = \pm 2$

vertex:

$f(x) = -(x - 0)^2 + 4$

$(0, 4)$

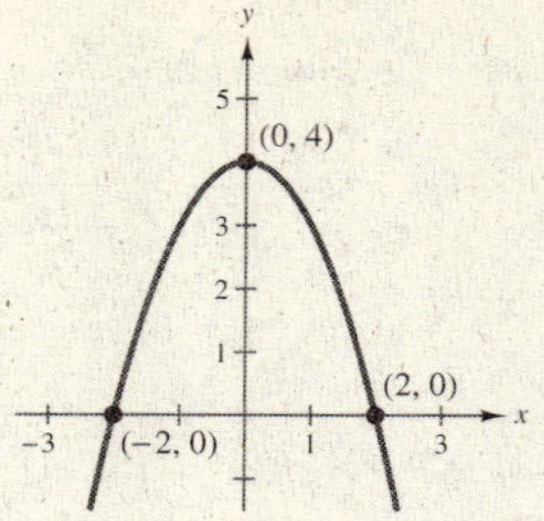

51. $f(x) = x^2 - 3x$

x-intercepts:

$0 = x^2 - 3x$

$0 = x(x - 3)$

$0 = x, \quad x = 3$

vertex:

$f(x) = \left(x^2 - 3x + \frac{9}{4}\right) - \frac{9}{4}$

$f(x) = \left(x - \frac{3}{2}\right)^2 - \frac{9}{4}$

$\left(\frac{3}{2}, -\frac{9}{4}\right)$

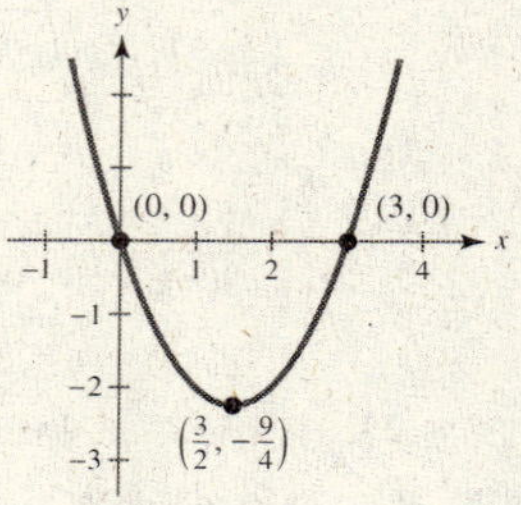

53. $y = -x^2 + 3x$

x-intercepts:

$0 = -x^2 + 3x$

$0 = -x(x - 3)$

$0 = x, \quad x = 3$

vertex:

$y = -1\left(x^2 - 3x + \frac{9}{4}\right) + \frac{9}{4}$

$= -1\left(x - \frac{3}{2}\right)^2 + \frac{9}{4}$

$\left(\frac{3}{2}, \frac{9}{4}\right)$

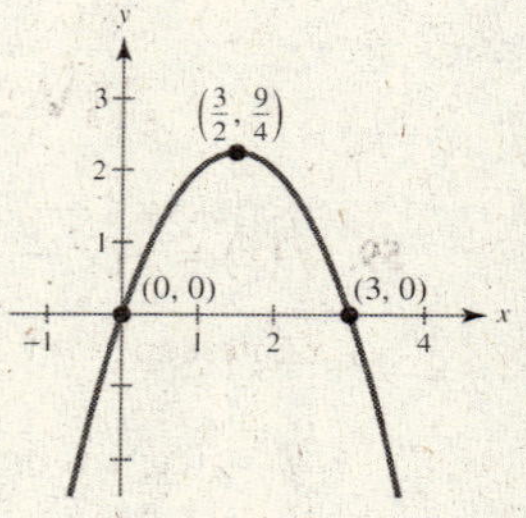

55. $y = (x - 4)^2$

x-intercepts:

$0 = (x - 4)^2$

$0 = x - 4$

$4 = x$

vertex:

$y = (x - 4)^2 + 0$

$(4, 0)$

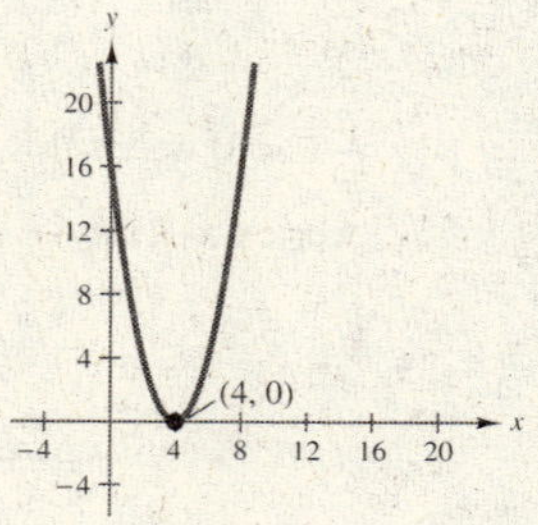

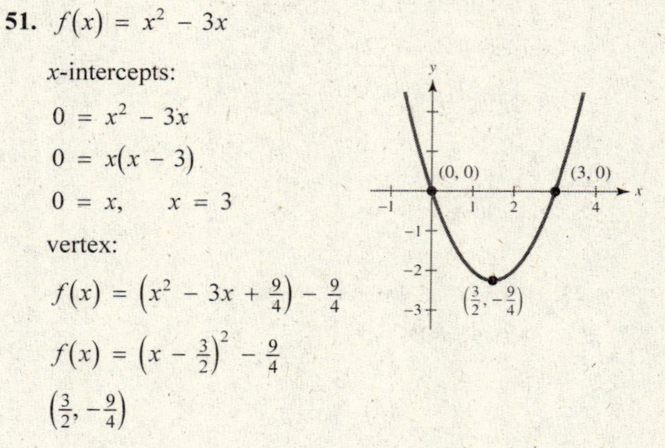

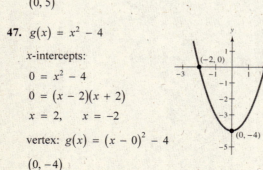

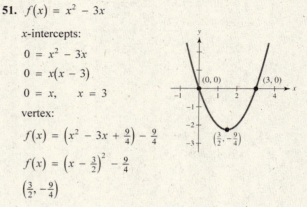

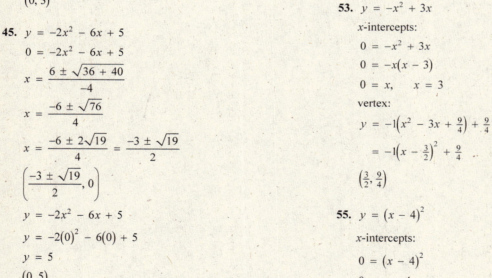

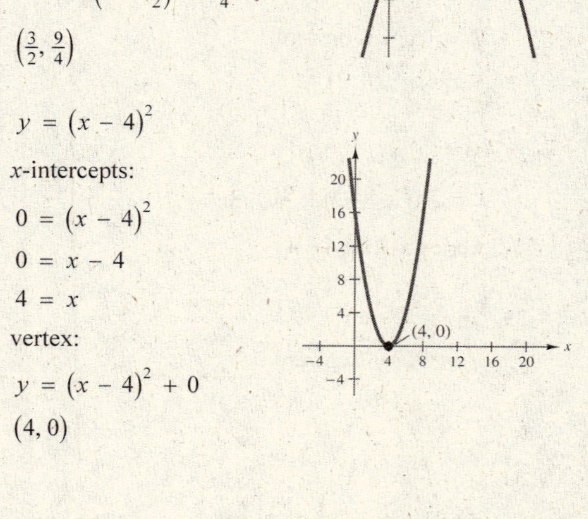

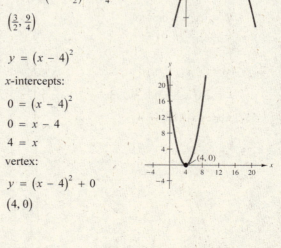

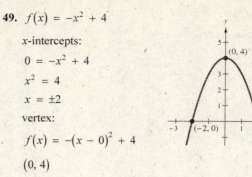

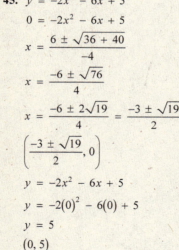

57. $y = x^2 - 9x - 18$

x-intercepts:

$0 = x^2 - 9x - 18$

$x = \dfrac{-(-9) \pm \sqrt{(-9)^2 - 4(1)(-18)}}{2(1)}$

$x = \dfrac{9 \pm \sqrt{81 + 72}}{2} = \dfrac{9 \pm \sqrt{153}}{2}$

$= \dfrac{9 \pm 3\sqrt{17}}{2} = \dfrac{9}{2} \pm \dfrac{3\sqrt{17}}{2}$

$\approx (10.68, 0), (-1.68, 0)$

vertex:

$y = \left(x^2 - 9x + \dfrac{81}{4}\right) - 18 - \dfrac{81}{4}$

$= \left(x - \dfrac{9}{2}\right)^2 - \dfrac{72}{4} - \dfrac{81}{4}$

$= \left(x - \dfrac{9}{2}\right)^2 - \dfrac{153}{4}$

$\left(\dfrac{9}{2}, -\dfrac{153}{4}\right) = (4.5, -38.25)$

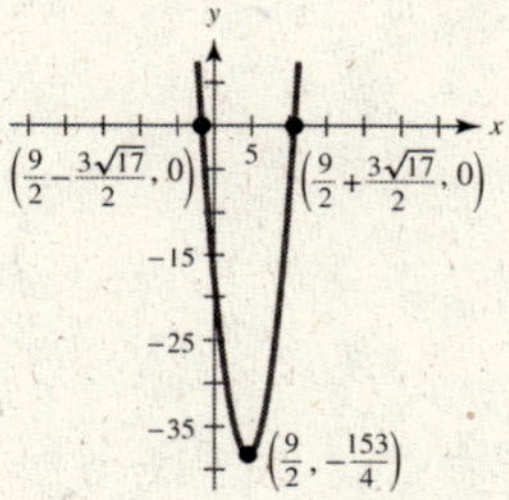

59. $f(x) = -(x^2 + 6x + 5)$

x-intercepts:

$0 = x^2 + 6x + 5$

$0 = (x + 5)(x + 1)$

$-5 = x, \quad x = -1$

vertex:

$y = -(x^2 + 6x + 9) - 5 + 9$

$y = -(x + 3)^2 + 4$

$(-3, 4)$

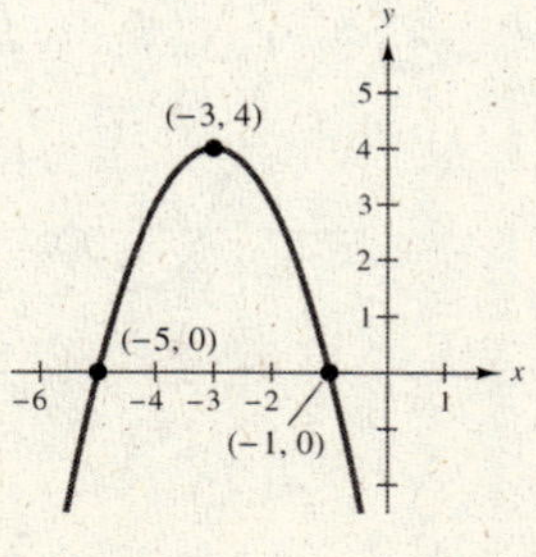

61. $g(x) = -x^2 + 6x - 7$

x-intercepts:

$0 = -x^2 + 6x - 7$

$x = \dfrac{-6 \pm \sqrt{6^2 - 4(-1)(-7)}}{2(-1)}$

$x = \dfrac{-6 \pm \sqrt{36 - 28}}{-2}$

$x = \dfrac{-6 \pm \sqrt{8}}{-2}$

$x = \dfrac{-6 \pm 2\sqrt{2}}{2}$

$x = -3 \pm \sqrt{2}$

vertex:

$g(x) = -x^2 + 6x - 7$

$g(x) = -(x^2 - 6x) - 7$

$g(x) = -(x^2 - 6x + 9 - 9) - 7$

$g(x) = -(x^2 - 6x + 9) + 9 - 7$

$g(x) = -(x - 3)^2 + 2$

$(3, 2)$

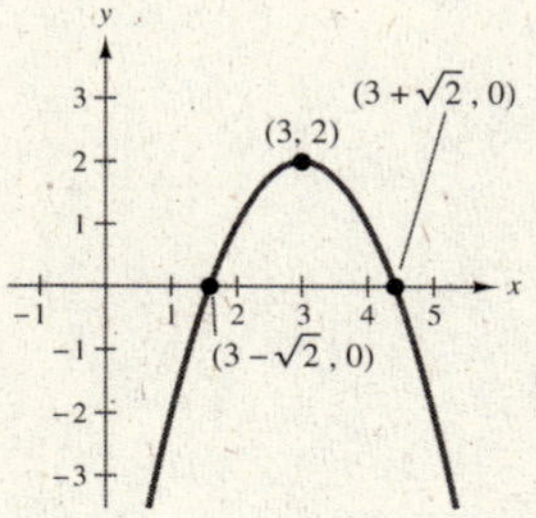

63. $y = -2x^2 - 12x - 21$

x-intercepts:

$$0 = -2x^2 - 12x - 21$$

$$x = \frac{-(-12) \pm \sqrt{(-12)^2 - 4(-2)(-21)}}{2(-2)}$$

$$x = \frac{12 \pm \sqrt{144 - 168}}{-4}$$

$$x = \frac{12 \pm \sqrt{-24}}{-4}$$

not real, so no x-intercepts

vertex:

$$y = -2(x^2 + 6x + 9) - 21 + 18$$

$$= -2(x + 3)^2 - 3$$

$(-3, -3)$

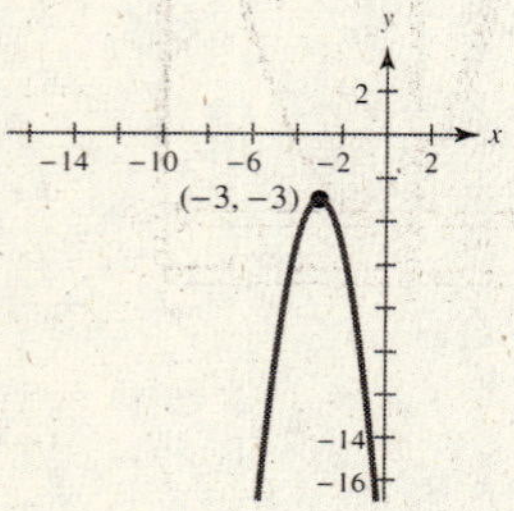

65. $y = \frac{1}{2}(x^2 - 2x - 3)$

vertex:

$$y = \tfrac{1}{2}(x^2 - 2x + 1) - \tfrac{3}{2} - \tfrac{1}{2}$$

$$y = \tfrac{1}{2}(x - 1)^2 - 2$$

$(1, -2)$

x-intercepts:

$$0 = x^2 - 2x - 3$$

$$0 = (x - 3)(x + 1)$$

$$3 = x, \qquad x = -1$$

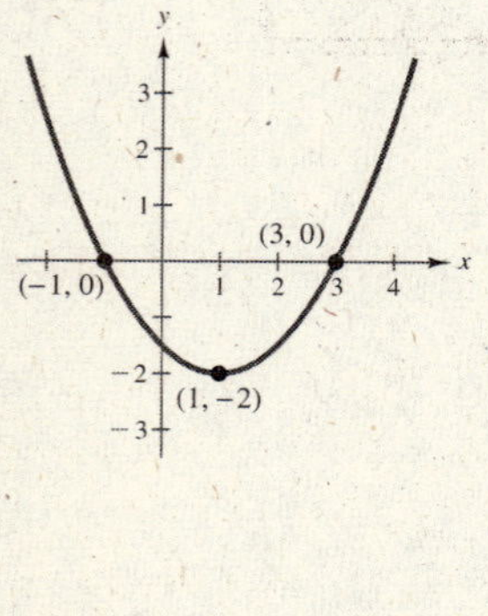

67. $y = \frac{1}{5}(3x^2 - 24x + 38)$

x-intercepts:

$$0 = 3x^2 - 24x + 38$$

$$x = \frac{24 \pm \sqrt{576 - 456}}{6}$$

$$x = \frac{24 \pm \sqrt{120}}{6} = \frac{12 \pm \sqrt{30}}{3} \approx 5.83, 2.17$$

vertex:

$$y = \frac{3}{5}(x^2 - 8x + 16) + \frac{38}{5} - \frac{48}{5}$$

$$y = \frac{3}{5}(x - 4)^2 - 2$$

$(4, -2)$

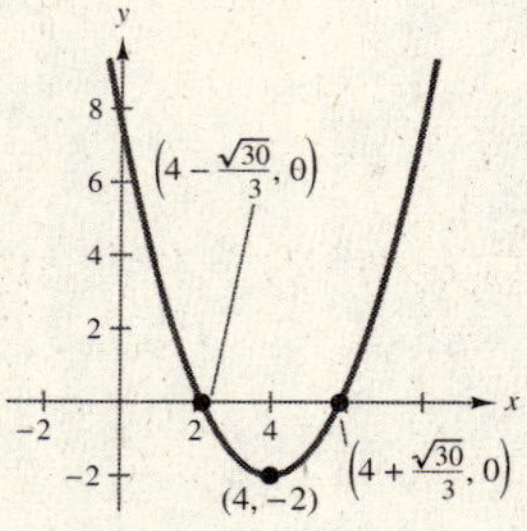

69. $f(x) = 5 - \frac{1}{3}x^2$

x-intercepts:

$$0 = -\tfrac{1}{3}x^2 + 5$$

$$\tfrac{1}{3}x^2 = 5$$

$$x^2 = 15$$

$$x = \pm\sqrt{15}$$

$$x \approx 3.87, -3.87$$

vertex:

$$f(x) = -\tfrac{1}{3}x^2 + 5$$

$$f(x) = -\tfrac{1}{3}(x - 0)^2 + 5$$

$(0, 5)$

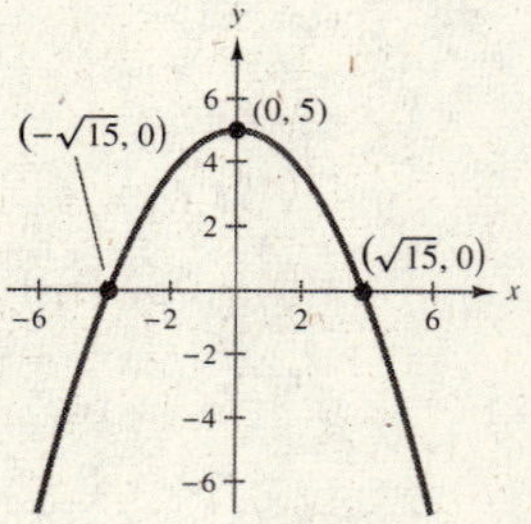

71. $h(x) = x^2 - 1$

Vertical shift 1 unit down

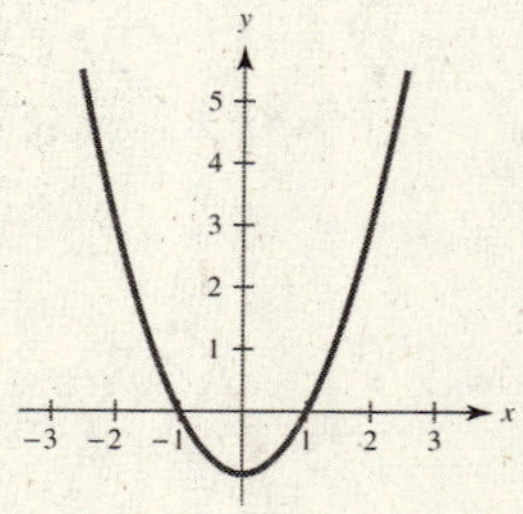

73. $h(x) = (x + 2)^2$

Horizontal shift 2 units left

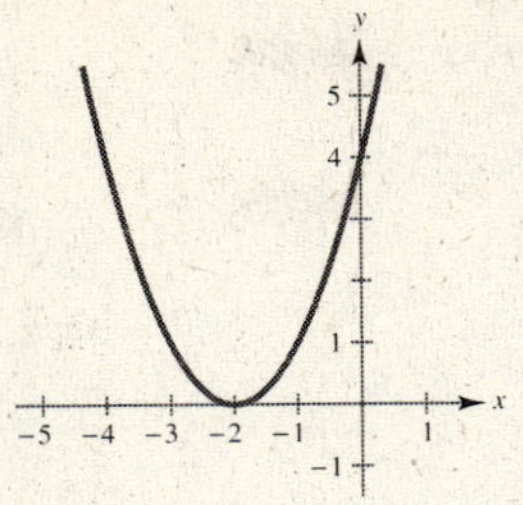

75. $h(x) = -(x + 5)^2$

Horizontal shift 5 units left

Reflection in the x-axis

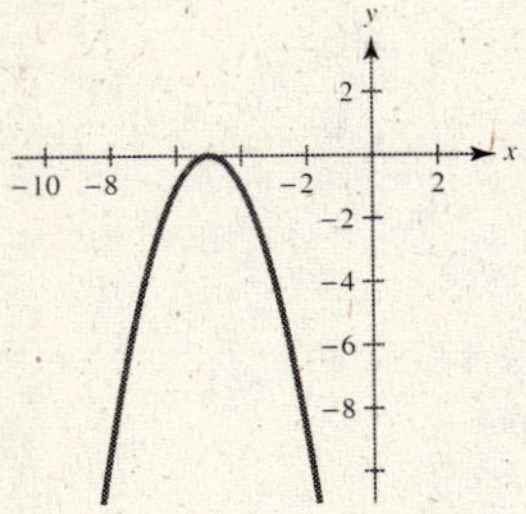

77. $h(x) = -(x - 2)^2 - 3$

Horizontal shift 2 units right

Vertical shift 3 units down

Reflection in the x-axis

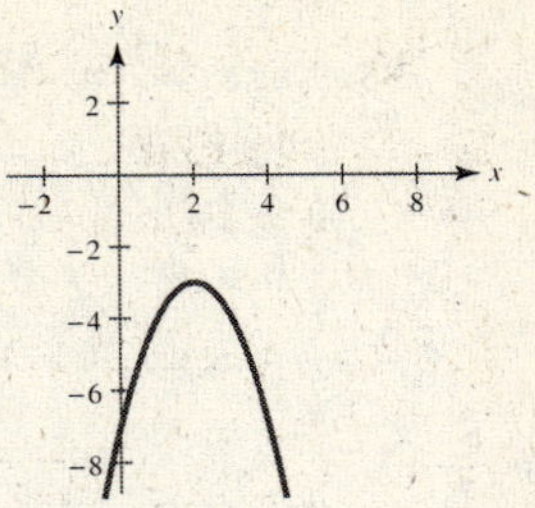

79. $y = \frac{1}{6}(2x^2 - 8x + 11)$

Keystrokes: [Y=] [(] 1 [÷] 6 [)] [(] 2 [X,T,θ] [x²] [−] 8 [X,T,θ] [+] 11 [)] [GRAPH]

vertex: $(2, 0.5)$

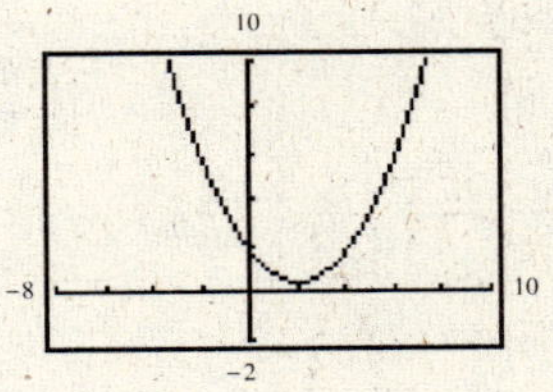

Check:

$$y = \frac{1}{3}x^2 - \frac{4}{3}x + \frac{11}{6}$$

$$a = \frac{1}{3}, \quad b = -\frac{4}{3}$$

$$x = -\frac{b}{2a} = \frac{-\left(-\frac{4}{3}\right)}{2\left(\frac{1}{3}\right)} = \frac{\frac{4}{3}}{\frac{2}{3}} = 2$$

$$y = \frac{1}{3}(2)^2 - \frac{4}{3}(2) + \frac{11}{6} = \frac{4}{3} - \frac{8}{3} + \frac{11}{6} = -\frac{4}{3} + \frac{11}{6} = \frac{-8 + 11}{6} = \frac{3}{6} = \frac{1}{2}$$

81. $y = -0.7x^2 - 2.7x + 2.3$

Keystrokes: [Y=] [(−)] .7 [X,T,θ] [x²] [(−)] 2.7 [X,T,θ] [+] 2.3 [GRAPH]

vertex: $(-1.9, 4.9)$

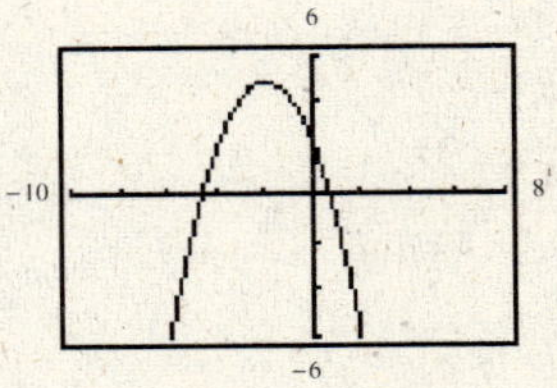

Check:

$a = -0.7, \quad b = -2.7$

$$x = -\frac{b}{2a} = \frac{-(-2.7)}{2(-0.7)} = -\frac{2.7}{1.4} \approx -1.9$$

$$y = -0.7(-1.9)^2 - 2.7(-1.9) + 2.3 = -2.527 + 5.13 + 2.3 \approx 4.9$$

83. vertex = $(0, 4)$, point = $(-2, 0)$

$$y = a(x - 0)^2 + 4$$
$$0 = a(-2 - 0)^2 + 4$$
$$0 = 4a + 4$$
$$-4 = 4a$$
$$-1 = a$$
$$y = -1(x - 0)^2 + 4$$
$$y = -x^2 + 4$$

85. vertex = $(-1, -2)$, point = $(0, 0)$

$$y = a(x - (-1))^2 - 2$$
$$0 = a(0 + 1)^2 - 2$$
$$2 = a$$
$$y = 2(x + 1)^2 - 2$$

87. vertex = $(2, 1)$, $a = 1$

$$y = 1(x - 2)^2 + 1 = x^2 - 4x + 5$$

89. vertex = $(2, -4)$, point = $(0, 0)$

$$0 = a(0 - 2)^2 - 4$$
$$4 = a(4)$$
$$1 = a$$
$$y = 1(x - 2)^2 - 4 = x^2 - 4x$$

91. vertex = $(-2, -1)$, point = $(1, 8)$

$$y = a(x - (-2))^2 - 1$$
$$8 = a(1 + 2)^2 - 1$$
$$9 = a(9)$$
$$1 = a$$
$$y = 1(x + 2)^2 - 1$$
$$y = (x + 2)^2 - 1$$

93. vertex = $(-1, 1)$, point = $(-4, 7)$

$$y = a(x + 1)^2 + 1$$
$$7 = a(-4 + 1)^2 + 1$$
$$6 = a(9)$$
$$\tfrac{2}{3} = \tfrac{6}{9} = a$$
$$y = \tfrac{2}{3}(x + 1)^2 + 1$$

95. $y = -\frac{1}{12}x^2 + 2x + 4$

(a) $y = -\frac{1}{12}(0)^2 + 2(0) + 4$

$y = 4$ feet

(b) $y = -\frac{1}{12}x^2 + 2x + 4$

$y = -\frac{1}{12}(x^2 - 24x + 144) + 4 + 12$

$y = -\frac{1}{12}(x - 12)^2 + 16$

Maximum height = 16 feet

(c) $0 = -\frac{1}{12}x^2 + 2x + 4$

$0 = x^2 - 24x - 48$

$x = \dfrac{24 \pm \sqrt{576 + 192}}{2} \approx 25.9$ feet

97. $y = -\frac{1}{5}x^2 + 6x + 3$

(a) $y = -\frac{1}{5}(0)^2 + 6(0) + 3$

$y = 3$ feet

(b) $y = -\frac{1}{5}(x^2 - 30x + 225) + 3 + 45$

$y = -\frac{1}{5}(x - 15)^2 + 48$

Maximum height = 48 feet

(c) $0 = -\frac{1}{5}x^2 + 6x + 3$

$0 = x^2 - 30x - 15$

$x = \dfrac{-(-30) \pm \sqrt{(-30)^2 - 4(1)(-15)}}{2(1)}$

$x = \dfrac{30 \pm \sqrt{900 + 60}}{2}$

$x = \dfrac{30 \pm \sqrt{960}}{2}$

$x \approx 30.49; -0.49$ reject

$x \approx 30.49$ feet

99. $y = -\frac{1}{480}x^2 + \frac{1}{2}x$

(a) $y = -\frac{1}{480}(0)^2 + \frac{1}{2}(0) = 0$ yards

(b) $y = -\frac{1}{480}(x^2 - 240x + 14{,}400) + 30$

$= -\frac{1}{480}(x - 120)^2 + 30$

Maximum height $= 30$ yards

(c) $0 = -\frac{1}{480}x^2 + \frac{1}{2}x$

$0 = x^2 - 240x$

$0 = x(x - 240)$

$0 = x \qquad x - 240 = 0$

$x = 240$

240 yards

101. $y = -\frac{4}{9}x^2 + \frac{24}{9}x + 10$

$y = -\frac{4}{9}(x^2 - 6x) + 10$

$y = -\frac{4}{9}(x^2 - 6x + 9 - 9) + 10$

$y = -\frac{4}{9}(x^2 - 6x + 9) + 4 + 10$

$y = -\frac{4}{9}(x - 3)^2 + 14$

The maximum height of the diver is 14 feet.

103. $C = 800 - 10x + \frac{1}{4}x^2, 0 < x < 40$

Keystrokes: [Y=] 800 [−] 10 [X,T,θ] [+] [(] 1 [÷] 4 [)] [X,T,θ] [x^2] [GRAPH]

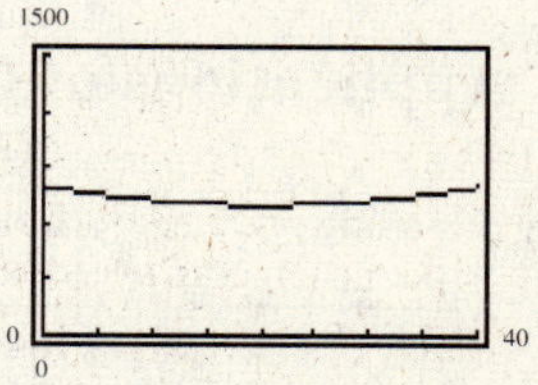

$x = 19.789474$ when $C = 700.01108$

$x \approx 20$ units

105. $y = 6\left[80 - \left(\frac{x^2}{2400}\right)\right]$

Keystrokes: [Y=] 6 [(] 80 [−] [(] [X,T,θ] [x^2] [÷] 2400 [)] [)] [GRAPH]

$x \approx 480$ feet is the maximum height.

$y = 480 - \frac{x^2}{400} = \frac{-1}{400}x^2 + 480$

Vertex is $(0, 480)$ so maximum height is 480 feet.

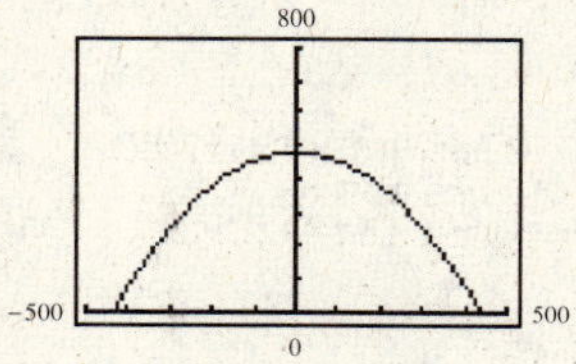

107. $y = a(x - 0)^2 + 0$

$15 = a(30 - 0)^2$

$15 = a(900)$

$\frac{15}{900} = a$

$\frac{1}{60} = a$

$y = \frac{1}{60}x^2$

109. If the discriminant is positive, the parabola has two x-intercepts; if it is zero, the parabola has one x-intercept; and if it is negative, the parabola has no x-intercepts.

111. Find the y-coordinate of the vertex of the graph of the function.

113. $(0, 0), (4, -2)$

$m = \frac{-2 - 0}{4 - 0} = -\frac{2}{4} = -\frac{1}{2}$

$y = mx + b$

$y = -\frac{1}{2}x + 0$

$y = -\frac{1}{2}x$

115. $(-1, -2), (3, 6)$

$m = \frac{6 - (-2)}{3 - (-1)} = \frac{8}{4} = 2$

$y - y_1 = m(x - x_1)$

$y - 6 = 2(x - 3)$

$y - 6 = 2x - 6$

$y = 2x$

117. $\left(\frac{3}{2}, 8\right), \left(\frac{11}{2}, \frac{5}{2}\right)$

$$m = \frac{\frac{5}{2} - 8}{\frac{11}{2} - \frac{3}{2}} = \frac{\frac{5}{2} - \frac{16}{2}}{\frac{8}{2}} = \frac{-\frac{11}{2}}{4} = -\frac{11}{8}$$

$$y - y_1 = m(x - x_1)$$

$$y - 8 = -\frac{11}{8}\left(x - \frac{3}{2}\right)$$

$$y = -\frac{11}{8}x + \frac{33}{16} + 8$$

$$y = -\frac{11}{8}x + \frac{33}{16} + \frac{128}{16}$$

$$y = -\frac{11}{8}x + \frac{161}{16}$$

119. $(0, 8), (5, 8)$

$$m = \frac{8 - 8}{5 - 0} = \frac{0}{5} = 0$$

$$y = 8$$

Horizontal line

121. $\sqrt{-64} = \sqrt{-1 \cdot 64} = 8i$

123. $\sqrt{-0.0081} = \sqrt{-1 \cdot 0.0081} = 0.09i$

Section 8.5 Applications of Quadratic Equations

1. *Verbal Model:* $\boxed{\text{Selling price per dozen eggs}} = \boxed{\text{Cost per dozen eggs}} + \boxed{\text{Profit per dozen eggs}}$

Equation: $\frac{21.60}{x} = \frac{21.60}{x + 6} + 0.30$

Labels: Number of eggs sold $= x$

Number of eggs purchased $= x + 6$

$$21.60(x + 6) = 21.60x + 0.30x(x + 6)$$

$$21.6x + 129.6 = 21.6x + 0.3x^2 + 1.8x$$

$$0 = 0.3x^2 + 1.8x - 129.6$$

$$0 = 3x^2 + 18x - 1296$$

$$0 = x^2 + 6x - 432$$

$$0 = (x + 24)(x - 18)$$

$$x = -24, \quad x = 18 \text{ dozen}$$

$$\text{Selling price} = \frac{21.60}{18} = \$1.20 \text{ per dozen}$$

3. *Verbal Model:* [Selling price per DVD] = [Cost per DVD] + [Profit per DVD]

Labels: Number of DVDs sold $= x$

Number of DVDs purchased $= x + 15$

Equation:

$$\frac{50}{x} = \frac{50}{x+15} + 3$$
$$50(x+15) = 50x + 3x(x+15)$$
$$50x + 750 = 50x + 3x^2 + 45x$$
$$0 = 3x^2 + 45x - 750$$
$$0 = x^2 + 15x - 250$$
$$0 = (x+25)(x-10)$$
$$x = -25, \quad x = 10 \text{ DVDs}$$

$$\text{Selling price} = \frac{50}{10} = \$5$$

5. *Verbal Model:* 2 [Length] + 2 [Width] = [Perimeter]

Labels: Length $= l$

Width $= 1.4l$

Equation:

$$2l + 2(1.4l) = 54$$
$$2l + 2.8l = 54$$
$$4.8l = 54$$
$$l = 11.25 \text{ inches}$$
$$w = 1.4l = 15.75 \text{ inches}$$

Verbal Model: [Length] · [Width] = [Area]

Equation:

$$11.25 \cdot 15.75 = A$$
$$177.1875 \text{ in.}^2 = A$$

7. *Verbal Model:* [Area] = [Length] · [Width]

Labels: Length $= 2.5w$

Width $= w$

Equation:

$$250 = 2.5w \cdot w$$
$$250 = 2.5w^2$$
$$100 = w^2$$
$$10 = w$$
$$25 = 2.5w$$

Verbal Model: 2 [Length] + 2 [Width] = [Perimeter]

Equation:

$$2(25) + 2(10) = P$$
$$70 \text{ feet} = P$$

9. *Verbal Model:* [Length] · [Width] = [Area]

Labels: Length $= l$

Width $= \frac{1}{3}l$

Equation:

$$l \cdot \tfrac{1}{3}l = 192$$
$$\tfrac{1}{3}l^2 = 192$$
$$l^2 = 576$$
$$l = 24 \text{ inches}$$
$$w = \tfrac{1}{3}l = 8 \text{ inches}$$

Verbal Model: 2 [Length] + 2 [Width] = [Perimeter]

Equation:

$$2(24) + 2(8) = P$$
$$48 + 16 = P$$
$$64 \text{ inches} = P$$

11. *Verbal Model:* 2 [Length] + 2 [Width] = [Perimeter]

Labels: Length $= w + 3$

Width $= w$

Equation:

$$2(w+3) + 2w = 54$$
$$2w + 6 + 2w = 54$$
$$4w = 48$$
$$w = 12 \text{ km}$$
$$l = w + 3 = 15 \text{ km}$$

Verbal Model: [Length] · [Width] = [Area]

Equation: $15 \cdot 12 = 180 \text{ km}^2 = A$

13. *Verbal Model:* $\boxed{\text{Length}} \cdot \boxed{\text{Width}} = \boxed{\text{Area}}$

Labels: Length $= l$

Width $= l - 20$

Equation:

$$l \cdot (l - 20) = 12{,}000$$
$$l^2 - 20l = 12{,}000$$
$$l^2 - 20l + 100 = 12{,}000 + 100$$
$$(l - 10)^2 = 12{,}100$$
$$l - 10 = \pm\sqrt{12{,}100}$$
$$l = 10 + 110 = 120 \text{ meters}$$
$$w = l - 20 = 100 \text{ meters}$$

Verbal Model: $2\boxed{\text{Length}} + 2\boxed{\text{Width}} = \boxed{\text{Perimeter}}$

Equation: $2(120) + 2(100) = 440 \text{ meters} = P$

15. *Verbal Model:* $\boxed{\text{Area}} = \boxed{\text{Length}} \cdot \boxed{\text{Width}}$

Labels: Length $= x + 4$

Width $= x$

Equation:

$$192 = (x + 4)x$$
$$192 = x^2 + 4x$$
$$0 = x^2 + 4x - 192$$
$$0 = (x + 16)(x - 12)$$
$$x = -16, \quad x = 12 \text{ inches}$$
$$x + 4 = 16 \text{ inches}$$

17. *Verbal Model:* $\boxed{\text{Length}} \cdot \boxed{\text{Width}} = \boxed{\text{Area}}$

Labels: Length $= 350 - 2x$

Width $= x$

Equation:

$$(350 - 2x) \cdot x = 12{,}500$$
$$350x - 2x^2 = 12{,}500$$
$$2x^2 - 350x + 12{,}500 = 0$$
$$x^2 - 175x + 6250 = 0$$
$$x = \frac{175 \pm \sqrt{175^2 - 4(1)(6250)}}{2(1)}$$
$$x = \frac{175 \pm \sqrt{5625}}{2} = \frac{175 \pm 75}{2}$$
$$x = 125, 50$$
$$350 - 2x = 100, 250$$

100 feet × 125 feet or 50 feet × 250 feet

19. *Verbal Model:* $\boxed{\text{Side 1}} + \boxed{\text{Side 2}} + \boxed{\text{Side 3}} = 550$

Equation:
$$\begin{aligned} x + x + b &= 550 \\ 2x + b &= 550 \\ b &= 550 - 2x \end{aligned}$$

Verbal Model: $\boxed{\tfrac{1}{2}} \cdot \boxed{\text{Height}}\boxed{(}\boxed{\text{Base 1}} + \boxed{\text{Base 2}}\boxed{)} = \boxed{\text{Area}}$

Labels: Height $= x$

Base 1 $= x$

Base 2 $= b$

Equation:
$$\begin{aligned} \tfrac{1}{2}x(x + b) &= 43{,}560 \\ \tfrac{1}{2}x(x + 550 - 2x) &= 43{,}560 \\ \tfrac{1}{2}x(-x + 550) &= 43{,}560 \\ -\tfrac{1}{2}x^2 + 275x &= 43{,}560 \\ -x^2 + 550x &= 87{,}120 \\ 0 &= x^2 - 550x + 87{,}120 \end{aligned}$$

This has no real solution, so it would be impossible to have an area of 43,560 square feet.

21. *Verbal Model:* $\boxed{\text{Height}} \cdot \boxed{\text{Width}} = \boxed{\text{Area}}$

Labels: Height $= x$

Width $= 48 - 2x$

Equation:
$$\begin{aligned} x \cdot (48 - 2x) &= 288 \\ 2x^2 - 48x + 288 &= 0 \\ x^2 - 24x + 144 &= 0 \\ (x - 12)(x - 12) &= 0 \\ x &= 12 \end{aligned}$$

height $= 12$ inches

width $= 48 - 2(12)$

$= 48 - 24 = 24$ inches

23.
$$\begin{aligned} A &= P(1 + r)^2 \\ 11{,}990.25 &= 10{,}000(1 + r)^2 \\ 1.199025 &= (1 + r)^2 \\ 1.095 &= 1 + r \\ 0.095 &= r \text{ or } 9.5\% \end{aligned}$$

25.
$$\begin{aligned} A &= P(1 + r)^2 \\ 572.45 &= 500(1 + r)^2 \\ \frac{572.45}{500} &= (1 + r)^2 \\ 1.1449 &= (1 + r)^2 \\ 1.07 &= 1 + r \\ 0.07 &= r \text{ or } 7\% \end{aligned}$$

27.
$$\begin{aligned} A &= P(1 + r)^2 \\ 7372.46 &= 6500(1 + r)^2 \\ 1.134224615 &\approx (1 + r)^2 \\ 1.064999819 &\approx 1 + r \\ 0.064999819 &\approx r \\ 0.065 &\approx r \text{ or } 6.5\% \end{aligned}$$

29. *Verbal Model:* $\boxed{\text{Cost per ticket}} \cdot \boxed{\text{Number of people going}} = 210$

Equation: $\left(\dfrac{210}{x} - 3.50\right) \cdot (x + 3) = 210$

$$210 + \frac{630}{x} - 3.5x - 10.50 = 210$$

$$210x + 630 - 3.5x^2 - 10.5x = 210x$$

$$-3.5x^2 - 10.5x + 630 = 0$$

$$0 = 3.5x^2 + 10.5x - 630$$

$$0 = 3.5(x^2 + 3x - 180)$$

$$0 = 3.5(x - 12)(x + 15)$$

$x - 12 = 0$ $\quad$ $x + 15 = 0$

$x = 12$ $\quad$ $x = -15$

There are $12 + 3 = 15$ people going to the game.

31. *Common formula:* $a^2 + b^2 = c^2$

Labels: Length of one leg $= x$

Length of other leg $= 12 - x$

Length of diagonal $= 9$

Equation: $x^2 + (12 - x)^2 = 9^2$

$$x^2 + 144 - 24x + x^2 = 81$$

$$2x^2 - 24x + 63 = 0$$

$$x = \frac{-(-24) \pm \sqrt{(-24)^2 - 4(2)(63)}}{2(2)}$$

$$x = \frac{24 \pm \sqrt{576 - 504}}{4}$$

$$x = \frac{24 \pm \sqrt{72}}{4}$$

$$= \frac{24 \pm 6\sqrt{2}}{4}$$

$$= \frac{12 \pm 3\sqrt{2}}{2}$$

$$\approx 3.88 \text{ miles or } 8.12 \text{ miles}$$

33. (a) $d = \sqrt{100^2 + h^2}$

$d = \sqrt{10,000 + h^2}$

(b) *Keystrokes:* [Y=] [√] [(] 10,000 [+] [X,T,θ] [x^2] [)] [GRAPH]

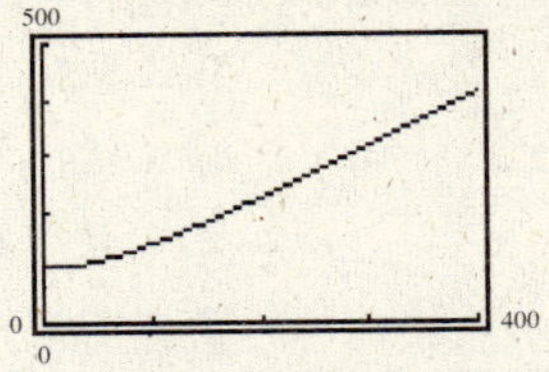

(c) When $d = 200$ feet h is approximately 173.2 feet.

(d)

h	0	100	200	300
d	100	141.4	223.6	316.2

$d = \sqrt{10,000 + 0^2} = \sqrt{10,000} = 100$

$d = \sqrt{10,000 + 100^2} = \sqrt{20,000} \approx 141.4$

$d = \sqrt{10,000 + 200^2} = \sqrt{10,000 + 40,000} = \sqrt{50,000} \approx 223.6$

$d = \sqrt{10,000 + 300^2} = \sqrt{10,000 + 90,000} = \sqrt{100,000} \approx 316.2$

35. *Verbal Model:* [Work done by Machine A] + [Work done by Machine B] = [One complete job]

Equation:

$$\frac{1}{x}(6) + \frac{1}{x+3}(6) = 1$$

$$x(x+3)\left[\frac{1}{x}(6) + \frac{1}{x+3}(6)\right] = (1)x(x+3)$$

$$6(x+3) + 6x = x(x+3)$$

$$6x + 18 + 6x = x^2 + 3x$$

$$-x^2 + 9x + 18 = 0$$

$$x^2 - 9x - 18 = 0$$

$$x = \frac{9 \pm \sqrt{(-9)^2 - 4(1)(-18)}}{2(1)} = \frac{9 \pm \sqrt{81 + 72}}{2} = \frac{9 \pm \sqrt{153}}{2}$$

$x = 10.684658, -1.6846584$ $\qquad x \approx 10.7$ minutes

$x + 3 = 13.684658$ $\qquad x + 3 \approx 13.7$ minutes

37.

$h = h_0 - 16t^2$

$0 = 169 - 16t^2$

$16t^2 = 169$

$t^2 = 10.5625$

$t = 3.25$ or $3\frac{1}{4}$ seconds

39.

$h = h_0 - 16t^2$

$0 = 1454 - 16t^2$

$16t^2 = 1454$

$t^2 = 90.875$

$t = 9.532838$ seconds

≈ 9.5 seconds

41. $h = 3 + 75t - 16t^2$

$0 = 3 + 75t - 16t^2$

$0 = 16t^2 - 75t - 3$

$t = \dfrac{75 \pm \sqrt{(-75)^2 - 4(16)(-3)}}{2(16)}$

$t = \dfrac{75 \pm \sqrt{5625 + 192}}{32}$

$t = \dfrac{75 \pm \sqrt{5817}}{32}$

$t = \dfrac{75 \pm 76.26926}{32}$

$t = 4.7271644, -0.0396644$ reject

≈ 4.7 seconds

43. (a) $336 = -16t^2 + 160t$

$0 = -16t^2 + 160t - 336$

$0 = t^2 - 10t + 21$

$0 = (t - 7)(t - 3)$

at 3 seconds and at 7 seconds

(b) $0 = -16t^2 + 160t$

$0 = -16t(t - 10)$

$t = 0, 10$ after 10 seconds

(c) $h = -16t^2 + 160t$

$= -16(t^2 - 10t + 25) + 400$

$= -16(t - 5)^2 + 400$

Maximum height is 400 feet.

45. *Verbal Model:* [Integer] · [Integer] = [Product]

Labels: First integer $= n$

Second integer $= n + 1$

Equation: $n \cdot (n + 1) = 182$

$n^2 + n - 182 = 0$

$(n + 14)(n - 13) = 0$

$n + 14 = 0 \qquad n - 13 = 0$

reject $\begin{cases} n = -14 \\ n + 1 = -13 \end{cases} \qquad n = 13, \quad n + 1 = 14$

47. *Verbal Model:* [Even integer] · [Even integer] = [Product]

Labels: First even integer $= 2n$

Second even integer $= 2n + 2$

Equation: $2n \cdot (2n + 2) = 168$

$4n^2 + 4n = 168$

$n^2 + n = 42$

$n^2 + n - 42 = 0$

$(n + 7)(n - 6) = 0$

$n + 7 = 0 \qquad n - 6 = 0$

$n = -7 \qquad n = 6$

reject $\begin{cases} 2n = -14 \\ 2n + 2 = -16 \end{cases} \qquad 2n = 12, \quad 2n + 2 = 14$

49. *Verbal Model:* [Odd integer] · [Odd integer] = [Product]

Labels: First odd integer $= 2n + 1$

Second odd integer $= 2n + 3$

Equation: $(2n + 1) \cdot (2n + 3) = 323$

$4n^2 + 8n + 3 = 323$

$4n^2 + 8n - 320 = 0$

$n^2 + 2n - 80 = 0$

$(n + 10)(n - 8) = 0$

reject $\begin{cases} n + 10 = 0 \\ n = -10 \end{cases} \qquad n - 8 = 0, \quad n = 8$

$2n + 1 = 17$

$2n + 3 = 19$

51. *Verbal Model:* [Original time] = [New time] + $\left[\frac{1}{5}\right]$

Labels: Speed $= x$

Increased speed $= x + 40$

Equation: $\dfrac{720}{x} = \dfrac{720}{x + 40} + \dfrac{1}{5}$

$720(5)(x + 40) = 720(5x) + x(x + 40)$

$3600x + 144{,}000 = 3600x + x^2 + 40x$

$0 = x^2 + 40x - 144{,}000$

$x = \dfrac{-40 \pm \sqrt{40^2 - 4(1)(-144{,}000)}}{2(1)}$

$x = \dfrac{40 \pm \sqrt{1600 + 576{,}000}}{2}$

$x = \dfrac{-40 \pm 760}{2}$

$x = 360, -400$, reject

$x + 40 = 400$ miles per hour

53. *Verbal Model:*

$\boxed{\text{Total cost}} = \boxed{\text{Wage cost}} + \boxed{\text{Fuel cost}}$

Label: Time $= x$

Equation:

$$36 = 15x + x\left[\frac{\left(\frac{80}{x}\right)^2}{300}\right] \qquad \left[\textit{Note: } v = \frac{d}{t}\right.$$

$$v = \frac{80}{x}$$

$$\left. v^2 = \left(\frac{80}{x}\right)^2\right]$$

$$36 = 15x + \frac{x}{300} \cdot \frac{6400}{x^2}$$

$$36 = 15x + \frac{64}{3x}$$

$$108x = 45x^2 + 64$$

$$0 = 45x^2 - 108x + 64$$

$$0 = (3x - 4)(15x - 16)$$

$$3x - 4 = 0 \qquad 15x - 16 = 0$$

$$x = \frac{4}{3} \qquad x = \frac{16}{15}$$

$$v = \frac{80}{\frac{4}{3}} = 80 \cdot \frac{3}{4} = 60 \text{ miles per hour}$$

$$v = \frac{80}{\frac{16}{15}} = 80 \cdot \frac{15}{16} = 75 \text{ miles per hour}$$

55.

$$d = \sqrt{(x_1 - x_2)^2 + (y_1 - y_2)^2}$$

$$10 = \sqrt{(x_1 - 2)^2 + (9 - 3)^2}$$

$$100 = (x_1 - 2)^2 + 36$$

$$64 = (x_1 - 2)^2$$

$$\pm 8 = x_1 - 2$$

$$2 \pm 8 = x_1$$

$$x_1 = 10 \qquad x_1 = -6$$

$$(10, 9) \qquad (-6, 9)$$

57. (a) $a + b = 20$ $\qquad A = \pi ab$

$b = 20 - a$ $\qquad A = \pi a(20 - a)$

(b)

a	4	7	10	13	16
A	201.1	285.9	314.2	285.9	201.1

$$A = \pi(4)(20 - 4) \quad A = \pi(7)(20 - 7) \quad A = \pi(10)(20 - 10) \quad A = \pi(13)(20 - 13) \quad A = \pi(16)(20 - 16)$$

$$= \pi(4)(16) \quad = \pi(7)(13) \quad = \pi(10)(10) \quad = \pi(13)(7) \quad = \pi(16)(4)$$

$$= 64\pi \quad = 91\pi \quad = 100\pi \quad = 91\pi \quad = 64\pi$$

$$\approx 201.1 \quad \approx 285.9 \quad \approx 314.2 \quad \approx 285.9 \quad \approx 201.1$$

(c) $300 = \pi a(20 - a)$

$$0 = 20\pi a - \pi a^2 - 300$$

$$0 = \pi a^2 - 20\pi a + 300$$

$$a = \frac{-(-20\pi) \pm \sqrt{(-20\pi)^2 - 4(\pi)(300)}}{2(\pi)}$$

$$a = \frac{20\pi \pm \sqrt{177.9305761}}{2\pi}$$

$$a \approx 12.1, 7.9$$

(d) $A = \pi a(20 - a)$

Keystrokes: [Y=] [π] [X,T,θ] [(] 20 [−] [X,T,θ] [)] [GRAPH]

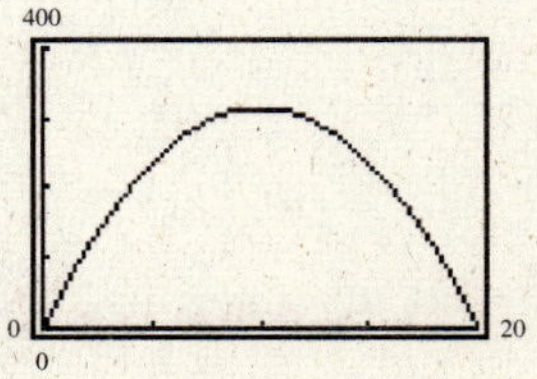

59. To solve a rational equation, each side of the equation is multiplied by the LCD. The resulting equations in this section are quadratic equations.

61. No. For each additional person, the cost-per-person decrease gets smaller because the discount is distributed to more people.

63. $5 - 3x > 17$

$-3x > 12$

$x < -4$

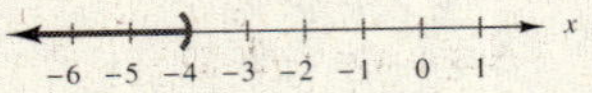

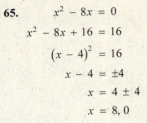

65.
$$x^2 - 8x = 0$$
$$x^2 - 8x + 16 = 16$$
$$(x - 4)^2 = 16$$
$$x - 4 = \pm 4$$
$$x = 4 \pm 4$$
$$x = 8, 0$$

Section 8.6 Quadratic and Rational Inequalities

1. $x(2x - 5) = 0$

$x = 0, \quad 2x - 5 = 0$

$x = \frac{5}{2}$

Critical numbers: $0, \frac{5}{2}$

3. $4x^2 - 81 = 0$

$x^2 = \frac{81}{4}$

$x = \pm\frac{9}{2}$

Critical numbers: $\frac{9}{2}, -\frac{9}{2}$

5. $x(x + 3) - 5(x + 3) = 0$

$(x - 5)(x + 3) = 0$

$x = 5, \quad x = -3$

Critical numbers: $5, -3$

7. $x^2 - 4x + 3 = 0$

$(x - 3)(x - 1) = 0$

$x = 3, \quad x = 1$

Critical numbers: $3, 1$

9. $6x^2 + 13x - 15 = 0$

$(6x - 5)(x + 3) = 0$

$6x - 5 = 0 \quad x + 3 = 0$

$6x = 5 \quad x = -3$

$x = \frac{5}{6}$

Critical numbers: $\frac{5}{6}, -3$

11. $x - 4$

Negative: $(-\infty, 4)$

Positive: $(4, \infty)$

Choose a test value from each interval.

$(-\infty, 4) \Rightarrow x = 0 \Rightarrow 0 - 4 = -4 < 0$

$(4, \infty) \Rightarrow x = 5 \Rightarrow 5 - 4 = 1 > 0$

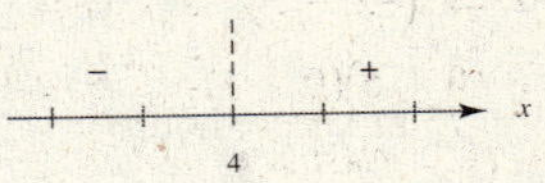

13. $3 - \frac{1}{2}x$

Negative: $(6, \infty)$

Positive: $(-\infty, 6)$

Choose a test value from each interval.

$(-\infty, 6) \Rightarrow x = 0 \Rightarrow 3 - \frac{1}{2}(0) = 3 > 0$

$(6, \infty) \Rightarrow x = 8 \Rightarrow 3 - \frac{1}{2}(8) = -1 < 0$

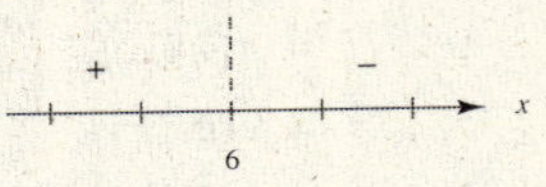

15. $4x(x - 5)$

$4x = 0 \quad x - 5 = 0$

$x = 0 \quad x = 5$ Critical numbers

Negative: $(0, 5)$

Positive: $(-\infty, 0) \cup (5, \infty)$

Choose a test value from each interval.

$(-\infty, 0) \Rightarrow x = -1 \Rightarrow 4(-1)(-1 - 5) = 24 > 0$

$(0, 5) \Rightarrow x = 1 \Rightarrow 4(1)(1 - 5) = -16 < 0$

$(5, \infty) \Rightarrow x = 6 \Rightarrow 4(6)(6 - 5) = 24 > 0$

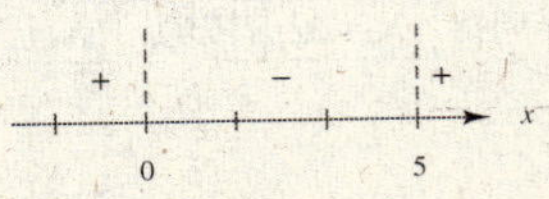

17. $4 - x^2 = (2 - x)(2 + x)$

Negative: $(-\infty, -2) \cup (2, \infty)$

Positive: $(-2, 2)$

Choose a test value from each interval.

$(-\infty, -2) \Rightarrow x = -3 \Rightarrow (2 - (-3))(2 + (-3)) = -5 < 0$

$(-2, 2) \Rightarrow x = 0 \Rightarrow (2 - 0)(2 + 0) = 4 > 0$

$(2, \infty) \Rightarrow x = 3 \Rightarrow (2 - 3)(2 + 3) = -5 < 0$

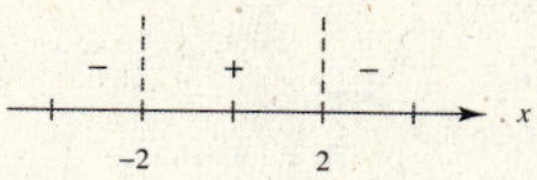

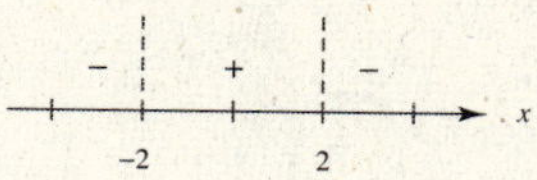

19. $(x - 5)(x + 1)$

Positive: $(-\infty, -1)$

Negative: $(-1, 5)$

Positive: $(5, \infty)$

Choose a test value from each interval.

$(-\infty, -1) \Rightarrow x = -2 \Rightarrow (-2 - 5)(-2 + 1) = 7 > 0$

$(-1, 5) \Rightarrow x = 0 \Rightarrow (0 - 5)(0 + 1) = -5 < 0$

$(5, \infty) \Rightarrow x = 6 \Rightarrow (6 - 5)(6 + 1) = 7 > 0$

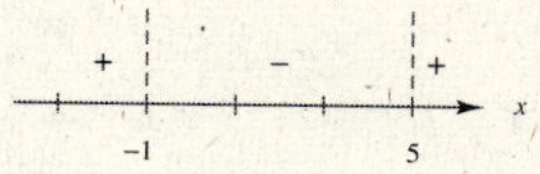

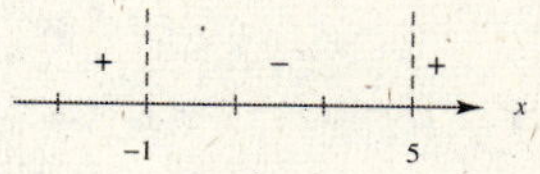

21. $3x(x - 2) < 0$

Critical numbers: $x = 0, 2$

Test intervals:

Positive: $(-\infty, 0)$

Negative: $(0, 2)$

Positive: $(2, \infty)$

Solution: $(0, 2)$

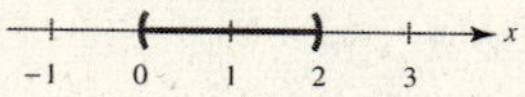

23. $3x(2 - x) \ge 0$

Critical numbers: $x = 0, 2$

Test intervals:

Negative: $(-\infty, 0]$

Positive: $[0, 2]$

Negative: $[2, \infty)$

Solution: $[0, 2]$

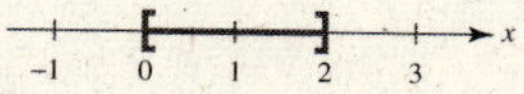

25. $x^2 > 4$

$x^2 - 4 > 0$

$(x - 2)(x + 2) > 0$

Critical numbers: $x = 2, -2$

Test intervals:

Positive: $(-\infty, 2)$

Negative: $(-2, 2)$

Positive: $(2, \infty)$

Solution: $(-\infty, -2) \cup (2, \infty)$

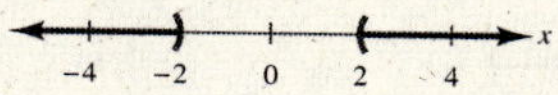

27. $x^2 - 3x - 10 \ge 0$

$(x - 5)(x + 2) \ge 0$

Critical numbers: $x = -2, 5$

Test intervals:

Positive: $(-\infty, -2)$

Negative: $(-2, 5)$

Positive: $(5, \infty)$

Solution: $(-\infty, -2] \cup [5, \infty)$

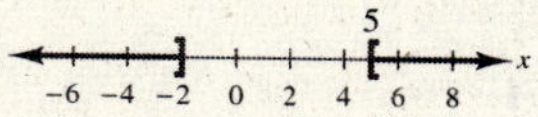

29. $x^2 + 4x > 0$

$x(x + 4) > 0$

Critical numbers: $x = -4, 0$

Test intervals:

Positive: $(-\infty, -4)$

Negative: $(-4, 0)$

Positive: $(0, \infty)$

Solution: $(-\infty, -4) \cup (0, \infty)$

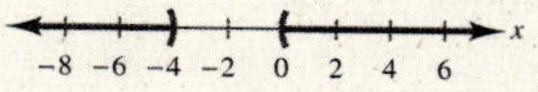

31. $x^2 + 5x \le 36$

$x^2 + 5x - 36 \le 0$

$(x + 9)(x - 4) \le 0$

Critical numbers: $-9, 4$

Test intervals:

Positive: $(-\infty, -9)$

Negative: $(-9, 4)$

Positive: $(4, \infty)$

Solution: $[-9, 4]$

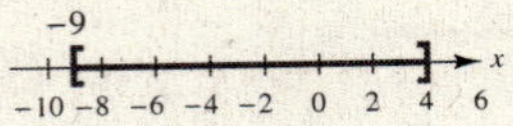

33. $u^2 + 2u - 2 > 1$

$u^2 + 2u - 3 > 0$

$(u + 3)(u - 1) > 0$

Critical numbers: $u = -3, 1$

Test intervals:

Positive: $(-\infty, -3)$

Negative: $(-3, 1)$

Positive: $(1, \infty)$

Solution: $(-\infty, -3) \cup (1, \infty)$

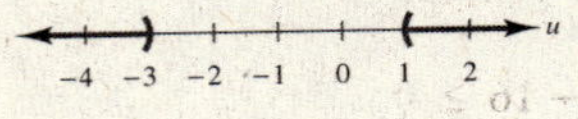

35. $x^2 + 4x + 5 < 0$

$$x = \frac{-4 \pm \sqrt{16 - 20}}{2}$$

No critical numbers

$x^2 + 4x + 5$ is not less than zero for any value of x.

Solution: none

37. $(x + 1)^2 \ge 0$

$(x + 1)^2 \ge 0$ for all real numbers

Solution: $(-\infty, \infty)$

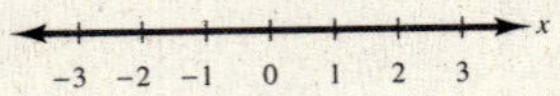

39. $x^2 - 4x + 2 > 0$

$$x = \frac{4 \pm \sqrt{16 - 8}}{2}$$

$$= \frac{4 \pm \sqrt{8}}{2} = \frac{4 \pm 2\sqrt{2}}{2}$$

$$= 2 \pm \sqrt{2}$$

Critical numbers: $x = 2 + \sqrt{2}, 2 - \sqrt{2}$

Test intervals:

Positive: $\left(-\infty, 2 - \sqrt{2}\right)$

Negative: $\left(2 - \sqrt{2}, 2 + \sqrt{2}\right)$

Positive: $\left(2 + \sqrt{2}, \infty\right)$

Solution: $\left(-\infty, 2 - \sqrt{2}\right) \cup \left(2 + \sqrt{2}, \infty\right)$

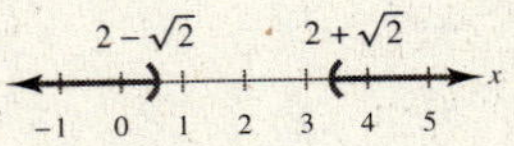

41. $x^2 - 6x + 9 \ge 0$

$(x - 3)^2 \ge 0$

$(x - 3)^2 \ge 0$ for all real numbers

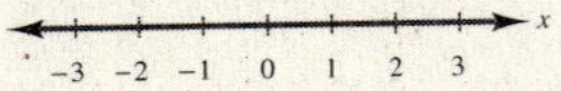

43. $u^2 - 10u + 25 < 0$

$(u - 5)(u - 5) < 0$

Critical number: $u = 5$

Test intervals:

Positive: $(-\infty, 5)$

Positive: $(5, \infty)$

Solution: none

45. $3x^2 + 2x - 8 \le 0$

$(3x - 4)(x + 2) \le 0$

Critical numbers: $x = \frac{4}{3}, -2$

Test intervals:

Positive: $(-\infty, -2]$

Negative: $\left[-2, \frac{4}{3}\right]$

Positive: $\left[\frac{4}{3}, \infty\right)$

Solution: $\left[-2, \frac{4}{3}\right]$

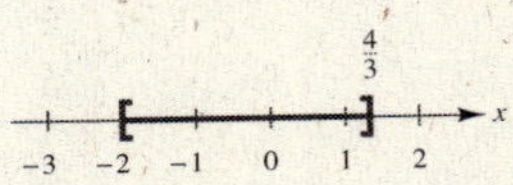

47. $-6u^2 + 19u - 10 > 0$

$6u^2 - 19u + 10 < 0$ (Multiply by -1.)

$(3u - 2)(2u - 5) < 0$

Critical numbers: $u = \frac{2}{3}, \frac{5}{2}$

Test intervals:

Positive: $\left(-\infty, \frac{2}{3}\right)$

Negative: $\left(\frac{2}{3}, \frac{5}{2}\right)$

Positive: $\left(\frac{5}{2}, \infty\right)$

Solution: $\left(\frac{2}{3}, \frac{5}{2}\right)$

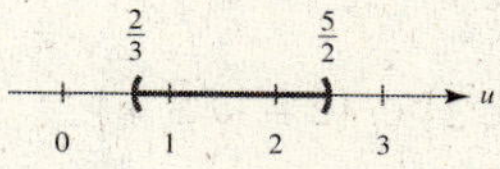

49. $2u^2 - 7u - 4 > 0$

$(2u + 1)(u - 4) > 0$

Critical numbers: $u = -\frac{1}{2}, 4$

Test intervals:

Positive: $\left(-\infty, -\frac{1}{2}\right)$

Negative: $\left(-\frac{1}{2}, 4\right)$

Positive: $(4, \infty)$

Solution: $\left(-\infty, -\frac{1}{2}\right) \cup (4, \infty)$

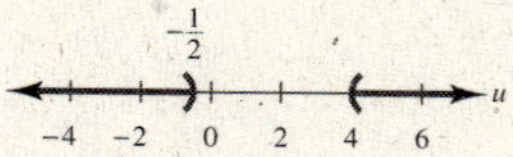

51. $4x^2 + 28x + 49 \le 0$

$(2x + 7)(2x + 7) \le 0$

Critical number: $x = -\frac{7}{2}$

Test intervals:

Positive: $\left(-\infty, -\frac{7}{2}\right)$

Positive: $\left(-\frac{7}{2}, \infty\right)$

Solution: $-\frac{7}{2}$

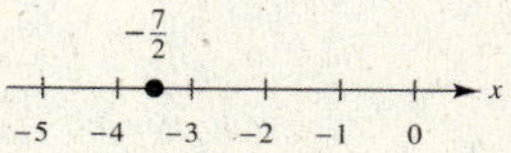

53. $(x - 2)^2 < 0$

Critical number: 2

Test intervals:

Positive: $(-\infty, 2) \cup (2, \infty)$

Negative: none

Solution: none

55. $6 - (x - 2)^2 < 0$

$\left[\sqrt{6} - (x - 2)\right]\left[\sqrt{6} + (x - 2)\right] < 0$

$-\left(x - 2 - \sqrt{6}\right)\left(x - 2 + \sqrt{6}\right) < 0$

Critical numbers: $2 - \sqrt{6}, 2 + \sqrt{6}$

Test intervals:

Negative: $\left(-\infty, 2 - \sqrt{6}\right)$

Positive: $\left(2 - \sqrt{6}, 2 + \sqrt{6}\right)$

Negative: $\left(2 + \sqrt{6}, \infty\right)$

Solution: $\left(-\infty, 2 - \sqrt{6}\right) \cup \left(2 + \sqrt{6}, \infty\right)$

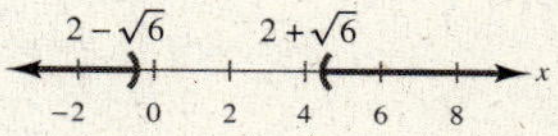

57. $16 \le (u + 5)^2$

$(u + 5)^2 \ge 16$

$u^2 + 10u + 25 - 16 \ge 0$

$u^2 + 10u + 9 \ge 0$

$(u + 9)(u + 1) \ge 0$

Critical numbers: $u = -9, -1$

Test intervals:

Positive: $(-\infty, -9]$

Negative: $(-9, -1]$

Positive: $[-1, \infty)$

Solution: $(-\infty, -9] \cup [-1, \infty)$

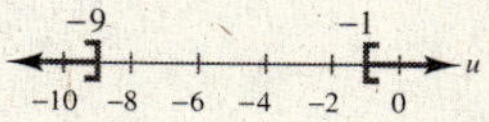

59. $x(x - 2)(x + 2) > 0$

Critical numbers: $x = 0, 2, -2$

Test intervals:

Negative: $(-\infty, -2)$

Positive: $(-2, 0)$

Negative: $(0, 2)$

Positive: $(2, \infty)$

Solution: $(-2, 0) \cup (2, \infty)$

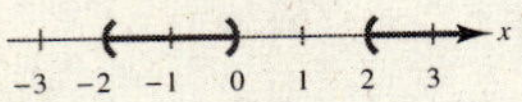

61. *Keystrokes:*

[Y=] [X,T,θ] [x^2] [−] 6 [X,T,θ] [TEST] 5 0 [GRAPH]

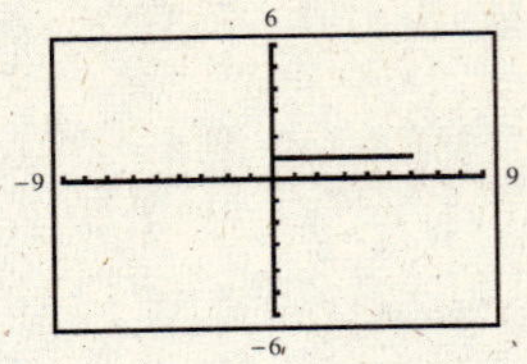

$x^2 - 6x < 0$

$x(x - 6) < 0$

Critical numbers: $x = 0, 6$

Test intervals:

Positive: $(-\infty, 0)$

Negative: $(0, 6)$

Positive: $(6, \infty)$

Solution: $(0, 6)$

63. *Keystrokes:* [Y=] 0.5 [X,T,θ] [x^2] [+] 1.25 [X,T,θ] [−] 3 [TEST] 3 0 [GRAPH]

$0.5x^2 + 1.25x - 3 > 0$

$x^2 + 2.5x - 6 > 0$

$2x^2 + 5x - 12 > 0$

$(2x - 3)(x + 4) > 0$

Critical numbers: $x = -4, \frac{3}{2}$

Test intervals:

Positive: $(-\infty, -4)$

Negative: $\left(-4, \frac{3}{2}\right)$

Positive: $\left(\frac{3}{2}, \infty\right)$

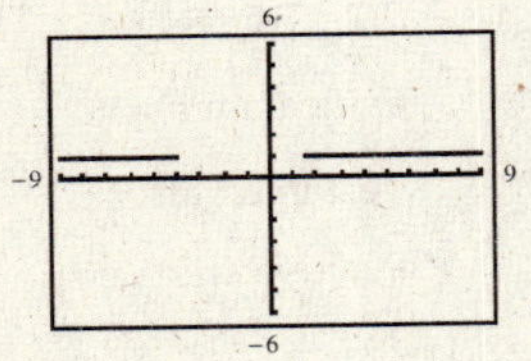

Solution:

$(-\infty, -4) \cup \left(\frac{3}{2}, \infty\right)$

65. *Keystrokes:* [Y=] [X,T,θ] [x^2] [+] 6 [X,T,θ] [+] 5 [TEST] 4 8 [GRAPH]

$x^2 + 6x + 5 \geq 8$

$x^2 + 6x - 3 \geq 0$

$$x = \frac{-6 \pm \sqrt{6^2 - 4(1)(-3)}}{2(1)} = \frac{-6 \pm \sqrt{48}}{2} = \frac{-6 \pm 4\sqrt{3}}{2} = -3 \pm 2\sqrt{3}$$

Critical numbers: $x = -3 - 2\sqrt{3}, -3 + 2\sqrt{3}$

Test intervals:

Positive: $\left(-\infty, -3 - 2\sqrt{3}\right)$

Negative: $\left(-3 - 2\sqrt{3}, -3 + 2\sqrt{3}\right)$

Positive: $\left(-3 + 2\sqrt{3}, \infty\right)$

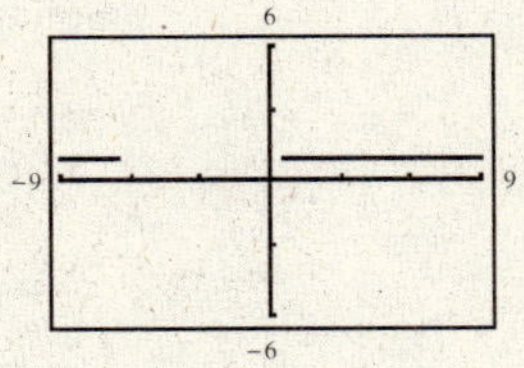

Solution: $\left(-\infty, -3 - 2\sqrt{3}\right] \cup \left[-3 + 2\sqrt{3}, \infty\right)$

67. *Keystrokes:* [Y=] 9 [−] 0.2 [(] [X,T,θ] [−] 2 [)] [x^2] [TEST] 5 4 [GRAPH]

$$9 - 0.2(x - 2)^2 < 4$$

$$9 - 0.2(x^2 - 4x + 4) < 4$$

$$9 - 0.2x^2 + 0.8x - 0.8 - 4 < 0$$

$$-0.2x^2 + 0.8x + 4.2 < 0$$

$$2x^2 - 8x - 42 > 0$$

$$x^2 - 4x - 21 > 0$$

$$(x - 7)(x + 3) > 0$$

Critical numbers: $x = -3, 7$

Test intervals:

Positive: $(-\infty, -3)$

Negative: $(-3, 7)$

Positive: $(7, \infty)$

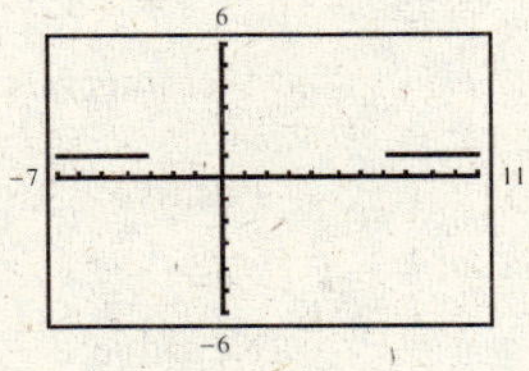

Solution: $(-\infty, -3) \cup (7, \infty)$

69. $f(x) = x^2 - 2x + 3$

Keystrokes: [Y=] [X,T,θ] [x^2] [−] 2 [X,T,θ] [+] 3 [GRAPH]

(a) $f(x) \geq 0, (-\infty, \infty)$

Function is entirely above x-axis.

(b) $f(x) \leq 6, [-1, 3]$

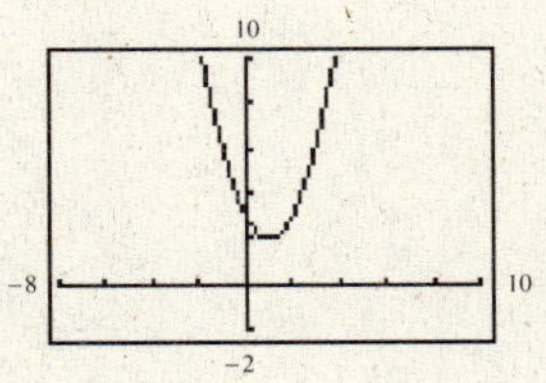

71. $f(x) = -2x^2 + 6x - 9$

Keystrokes: [Y=] [−] 2 [X,T,θ] [x^2] [+] 6 [X,T,θ] [−] 9 [GRAPH]

(a) $f(x) > -11, (-0.303, 3.303)$

(b) $f(x) < 10, (-\infty, \infty)$ Function is entirely below the line $f(x) = 10$.

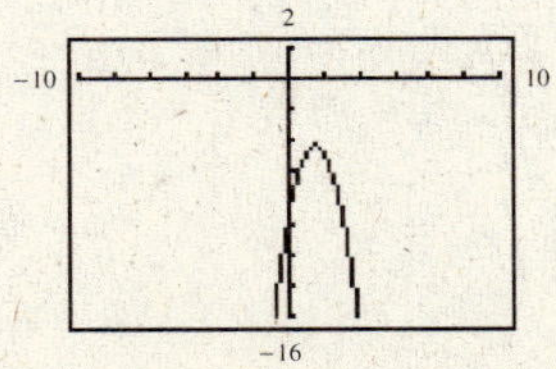

73. Critical number: $x = 3$

75. Critical numbers: $x = 0, -5$

77. $\dfrac{5}{x-3} > 0$

Critical number: $x = 3$

Test intervals:

Negative: $(-\infty, 3)$

Positive: $(3, \infty)$

Solution: $(3, \infty)$

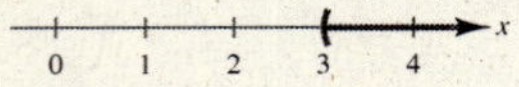

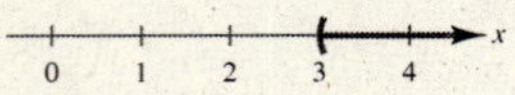

79. $\dfrac{-5}{x-3} > 0$

Critical number: $x = 3$

Test intervals:

Positive: $(-\infty, 3)$

Negative: $(3, \infty)$

Solution: $(-\infty, 3)$

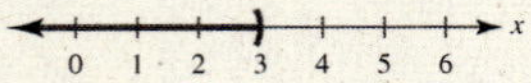

81.
$$\frac{3}{y-1} \le -1$$
$$\frac{3}{y-1} + 1 \le 0$$
$$\frac{3 + (y-1)}{y-1} \le 0$$
$$\frac{y+2}{y-1} \le 0$$

Critical numbers: $x = -2, 1$

Test intervals:

Positive: $(-\infty, -2)$

Negative: $(-2, 1)$

Positive: $(1, \infty)$

Solution: $[-2, 1)$

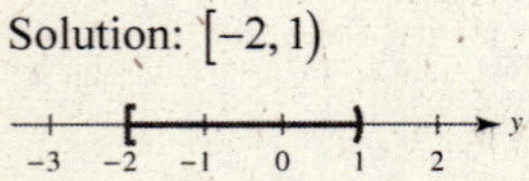

83. $\dfrac{x+4}{x-2} > 0$

Critical numbers: $x = -4, 2$

Test intervals:

Positive: $(-\infty, -4)$

Negative: $(-4, 2)$

Positive: $(2, \infty)$

Solution: $(-\infty, -4) \cup (2, \infty)$

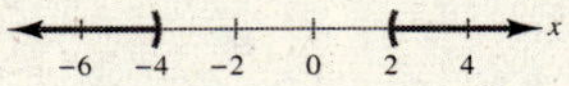

85. $\dfrac{y-4}{y-1} \le 0$

Critical numbers: $y = 1, 4$

Test intervals:

Positive: $(-\infty, 1)$

Negative: $(1, 4)$

Positive: $(4, \infty)$

Solution: $(1, 4]$

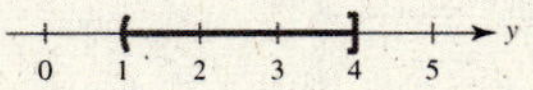

87. $\dfrac{4x-2}{2x-4} > 0$

Critical numbers: $x = \frac{1}{2}, 2$

Test intervals:

Positive: $\left(-\infty, \dfrac{1}{2}\right)$

Negative: $\left(\dfrac{1}{2}, 2\right)$

Positive: $(2, \infty)$

Solution: $\left(-\infty, \dfrac{1}{2}\right) \cup (2, \infty)$

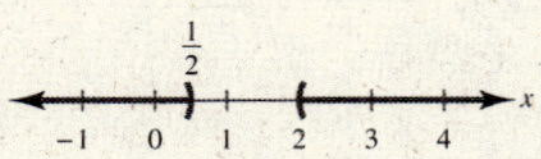

89. $\dfrac{x+2}{4x+6} \le 0$

Critical numbers: $x = -2, -\dfrac{3}{2}$

Test intervals:

Positive: $(-\infty, -2]$

Negative: $\left[-2, -\dfrac{3}{2}\right)$

Positive: $\left(-\dfrac{3}{2}, \infty\right)$

Solution: $\left[-2, -\dfrac{3}{2}\right)$

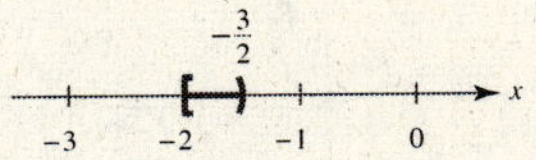

91. $\dfrac{3(u-3)}{u+1} < 0$

Critical numbers: $u = 3, -1$

Test intervals:

Positive: $(-\infty, -1)$

Negative: $(-1, 3)$

Positive: $(3, \infty)$

Solution: $(-1, 3)$

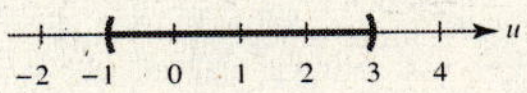

93. $\dfrac{2}{x-5} \ge 3$

$\dfrac{2}{x-5} - 3 \ge 0$

$\dfrac{2-3(x-5)}{x-5} \ge 0$

$\dfrac{2-3x+15}{x-5} \ge 0$

$\dfrac{-3x+17}{x-5} \ge 0$

Critical numbers: $5, \dfrac{17}{3}$

Test intervals:

Negative: $(-\infty, 5)$

Positive: $\left(5, \dfrac{17}{3}\right)$

Negative: $\left(\dfrac{17}{3}, \infty\right)$

Solution: $\left(5, \dfrac{17}{3}\right]$

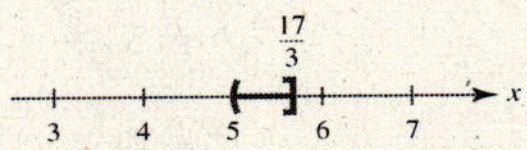

95. $\dfrac{4x}{x+2} < -1$

$\dfrac{4x}{x+2} + 1 < 0$

$\dfrac{4x+(x+2)}{x+2} < 0$

$\dfrac{5x+2}{x+2} < 0$

Critical numbers: $x = -\dfrac{2}{5}, -2$

Test intervals:

Positive: $(-\infty, -2)$

Negative: $\left(-2, -\dfrac{2}{5}\right)$

Positive: $\left(-\dfrac{2}{5}, \infty\right)$

Solution: $\left(-2, -\dfrac{2}{5}\right)$

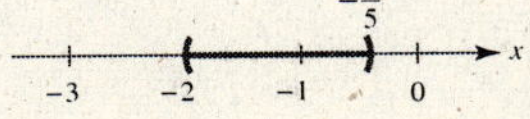

97. $$\frac{x-3}{x-6} \le 4$$
$$\frac{x-3}{x-6} - 4 \le 0$$
$$\frac{x-3-4(x-6)}{x-6} \le 0$$
$$\frac{x-3-4x+24}{x-6} \le 0$$
$$\frac{-3x+21}{x-6} \le 0$$

Critical numbers: 6, 7

Test intervals:

Negative: $(-\infty, 6)$

Positive: $(6, 7)$

Negative: $(7, \infty)$

Solution: $(-\infty, 6) \cup [7, \infty)$

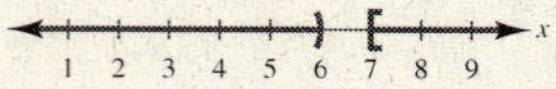

99. *Keystrokes:*

[Y=] 1 [÷] [X,T,θ] [−] [X,T,θ] [TEST] 3 0 [GRAPH]

$$\frac{1}{x} - x > 0$$
$$\frac{1-x^2}{x} > 0$$
$$\frac{(1-x)(1+x)}{x} > 0$$

Critical numbers: $x = -1, 0, 1$

Test intervals:

Positive: $(-\infty, -1)$

Negative: $(-1, 0)$

Positive: $(0, 1)$

Negative: $(1, \infty)$

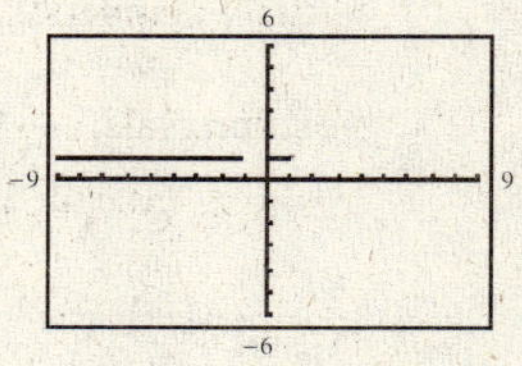

Solution: $(-\infty, -1) \cup (0, 1)$

101. *Keystrokes:* [Y=] [(] [X,T,θ] [+] 6 [)] [÷] [(] [X,T,θ] [+] 1 [)] [−] 2 [TEST] 5 0 [GRAPH]

$$\frac{x+6}{x+1} - 2 < 0$$
$$\frac{x+6}{x+1} - \frac{2(x+1)}{x+1} < 0$$
$$\frac{x+6-2x-2}{x+1} < 0$$
$$\frac{-x+4}{x+1} < 0$$
$$\frac{x-4}{x+1} > 0$$

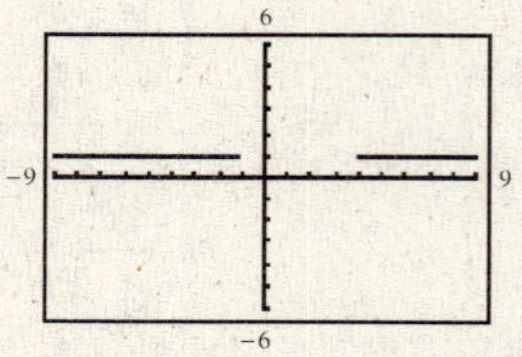

Critical numbers: $x = -1, 4$

Test intervals:

Positive: $(-\infty, -1)$

Negative: $(-1, 4)$

Positive: $(4, \infty)$

Solution: $(-\infty, -1) \cup (4, \infty)$

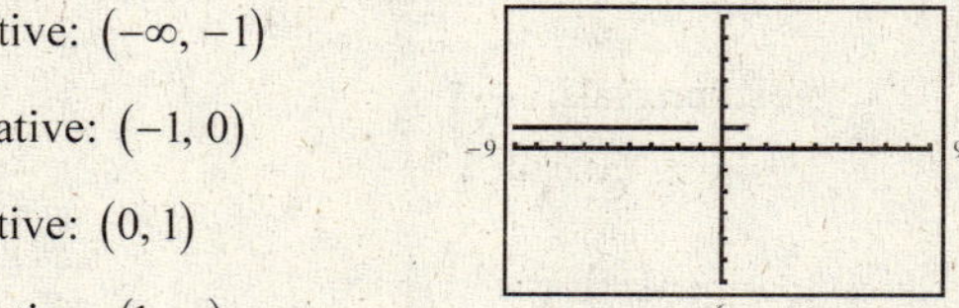

103. *Keystrokes:* [Y=] [(] 6 [X,T,θ] [−] 3 [)] [÷] [(] [X,T,θ] [+] 5 [)] [TEST] 5 2 [GRAPH]

$$\frac{6x-3}{x+5} < 2$$

$$\frac{6x-3}{x+5} - \frac{2(x+5)}{x+5} < 0$$

$$\frac{6x-3-2x-10}{x+5} < 0$$

$$\frac{4x-13}{x+5} < 0$$

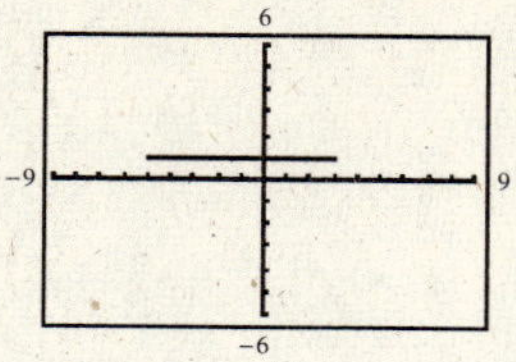

Critical numbers: $x = -5, \frac{13}{4}$

Test intervals:

Positive: $(-\infty, -5)$

Negative: $\left(-5, \frac{13}{4}\right)$

Positive: $\left(\frac{13}{4}, \infty\right)$

Solution: $\left(-5, \frac{13}{4}\right)$

105. *Keystrokes:* [Y=] [X,T,θ] [+] 1 [÷] [X,T,θ] [TEST] 3 3 [GRAPH]

$$x + \frac{1}{x} > 3$$

$$\frac{x^2}{x} + \frac{1}{x} - \frac{3x}{x} > 0$$

$$\frac{x^2-3x+1}{x} > 0$$

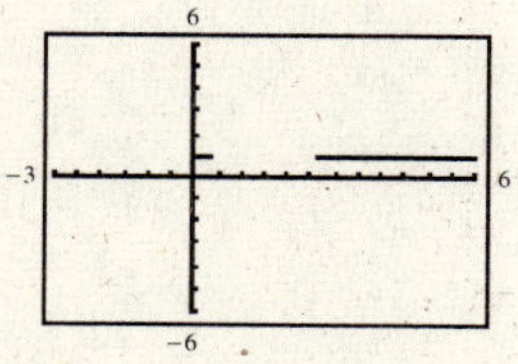

Critical numbers: $x = 0.382, 0, 2.618$

Test intervals:

Negative: $(-\infty, 0)$

Positive: $(0, 0.382)$

Negative: $(0.382, 2.618)$

Positive: $(2.618, \infty)$

Solution: $(0, 0.382) \cup (2.618, \infty)$

107. *Keystrokes:* [Y=] 3 [X,T,θ] ÷ [(] [X,T,θ] [−] 2 [)] [GRAPH]

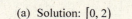

(a) Solution: $[0, 2)$

(Look at x-axis and vertical asymptote $x = 2$.)

(b) $(2, 4]$

(Graph $y = 6$ as y_2 and find the intersection.)

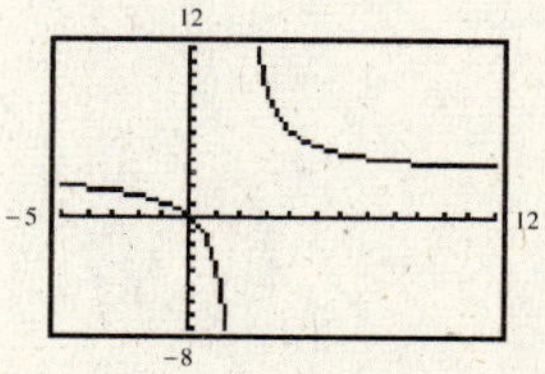

109. *Keystrokes:*

[Y=] 2 [X,T,θ] [x^2] [÷] [(] [X,T,θ] [x^2] [+] 4 [)] [GRAPH]

(a) Solution: $(-\infty, -2] \cup [2, \infty)$

(Graph $y = 1$ as y_2 and find the intersection.)

(b) Solution $(-\infty, \infty)$

(Notice graph stays below line $y = 2$.)

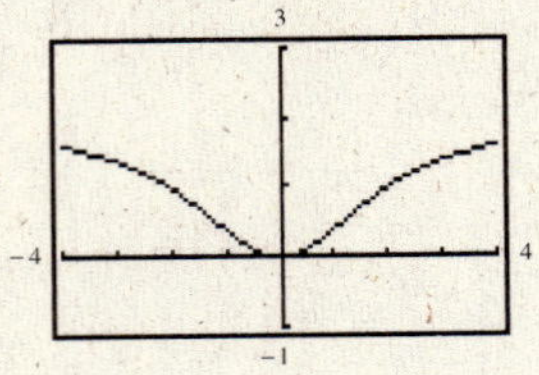

111. $h = -16t^2 + 128t$

$$\text{height} > 240$$
$$-16t^2 + 128t > 240$$
$$-16t^2 + 128t - 240 > 0$$
$$t^2 - 8t + 15 < 0$$
$$(t - 3)(t - 5) < 0$$

Critical numbers: $x = 3, 5$

Test intervals:

Positive: $(-\infty, 3)$

Negative: $(3, 5)$

Positive: $(5, \infty)$

Solution: $(3, 5)$

113.

$$1000(1 + r)^2 > 1150$$
$$1000(1 + 2r + r^2) > 1150$$
$$1000 + 2000r + 1000r^2 > 1150$$
$$1000r^2 + 2000r - 150 > 0$$
$$20r^2 + 40r - 3 > 0$$

Critical numbers: $r = \dfrac{-40 + \sqrt{1840}}{40}, \dfrac{-40 - \sqrt{1840}}{40}$

r cannot be negative.

Test intervals:

Negative: $\left(0, \dfrac{-40 + \sqrt{1840}}{40}\right)$

Positive: $\left(\dfrac{-40 + \sqrt{1840}}{40}, \infty\right)$

Solution: $\left(\dfrac{-40 + \sqrt{1840}}{40}, \infty\right)$

$(0.0724, \infty)$, $r > 7.24\%$

115.

$$\text{Area} > 240$$
$$l(32 - l) > 240$$
$$32l - l^2 > 240$$
$$-l^2 + 32l - 240 > 0$$
$$l^2 - 32l + 240 < 0$$
$$(l - 20)(l - 12) < 0$$

Critical numbers: $l = 20, 12$

Test intervals:

Positive: $(-\infty, 12)$

Negative: $(12, 20)$

Positive: $(20, \infty)$

Solution: $(12, 20)$

117. *Verbal Model:* [Profit] = [Revenue] − [Cost]

Labels: Profit $= P(x)$

Revenue $= x(50 - 0.0002x)$

Cost $= 12x + 150{,}000$

Inequality:

$$P(x) \geq 1{,}650{,}000$$
$$50x - 0.0002x^2 - 12x - 150{,}000 \geq 1{,}650{,}000$$
$$-0.0002x^2 + 38x - 150{,}000 \geq 1{,}650{,}000$$
$$-0.0002x^2 + 38x - 1{,}800{,}000 \geq 0$$
$$2x^2 - 380{,}000x + 18{,}000{,}000{,}000 \leq 0$$
$$x^2 - 190{,}000x + 9{,}000{,}000{,}000 \leq 0$$
$$(x - 90{,}000)(x - 100{,}000) \leq 0$$

Critical numbers: $x = 90{,}000, 100{,}000$

Test intervals:

Positive: $(-\infty, 90{,}000]$

Negative: $[90{,}000, 100{,}000]$

Positive: $[100{,}000, \infty)$

Solution: $[90{,}000, 100{,}000]$

119. (a) *Keystrokes:*

[Y=] [(] 21.9 [−] 0.043 [X,T,θ] [)] [÷] [(] 1 [+] 0.005 [X,T,θ] [)] [GRAPH]

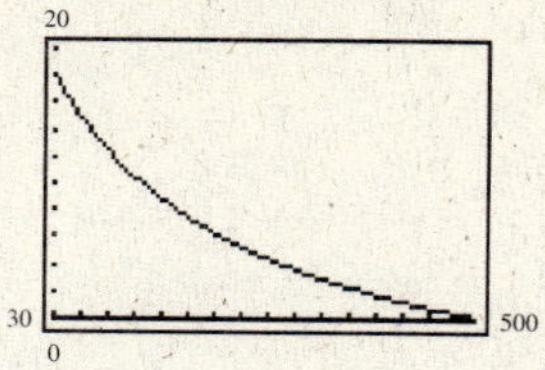

(b)
$$\frac{21.9 - 0.043t}{1 + 0.005t} < 5$$
$$\frac{21.9 - 0.043t}{1 + 0.005t} - 5\left(\frac{1 + 0.005t}{1 + 0.005t}\right) < 0$$
$$\frac{21.9 - 0.043t - 5 - 0.025t}{1 + 0.005t} < 0$$
$$\frac{16.9 - 0.068t}{1 + 0.005t} < 0$$

Critical numbers: 248.5, −200

−200 not in domain.

Test intervals:

Positive: $(30, 248.5)$

Negative: $(248.5, 500)$

It takes about 248.5 minutes for the concentration to fall below 5 milligrams per liter.

121. The critical numbers of a polynomial are its zeros, so the value of the polynomial is zero at its critical numbers.

123. No solution. The value of the polynomial is positive for every real value of x, so there are no values that would make the polynomial negative.

125. $\frac{4xy^3}{x^2y} \cdot \frac{y}{8x} = \frac{4xy^4}{8x^3y} = \frac{y^3}{2x^2}, \quad y \neq 0$

127. $\frac{x^2 - x - 6}{4x^3} \cdot \frac{x + 1}{x^2 + 5x + 6} = \frac{(x - 3)(x + 2)(x + 1)}{4x^3(x + 3)(x + 2)} = \frac{(x - 3)(x + 1)}{4x^3(x + 3)}, \quad x \neq -2$

129. $\frac{x^2 + 8x + 16}{x^2 - 6x} \div (3x - 24) = \frac{(x + 4)(x + 4)}{x(x - 6)} \cdot \frac{1}{3(x - 8)} = \frac{(x + 4)^2}{3x(x - 6)(x - 8)}$

131. $x = -\frac{1}{3}$

$x^2 = \left(-\frac{1}{3}\right)^2 = \frac{1}{9}$

133. $x = 1.06$

$\frac{100}{x^4} = \frac{100}{(1.06)^4} = 79.21$

Review Exercises for Chapter 8

1. $x^2 + 12x = 0$

$x(x + 12) = 0$

$x = 0 \qquad x + 12 = 0$

$x = 0 \qquad x = -12$

3. $3y^2 - 27 = 0$

$3(y^2 - 9) = 0$

$3(y - 3)(y + 3) = 0$

$y - 3 = 0 \qquad y + 3 = 0$

$y = 3 \qquad y = -3$

5. $4y^2 + 20y + 25 = 0$

$(2y + 5)(2y + 5) = 0$

$2y + 5 = 0 \qquad 2y + 5 = 0$

$2y = -5 \qquad 2y = -5$

$y = -\frac{5}{2} \qquad y = -\frac{5}{2}$

7. $2x^2 - 2x - 180 = 0$

$2(x^2 - x - 90) = 0$

$2(x - 10)(x + 9) = 0$

$x - 10 = 0 \qquad x + 9 = 0$

$x = 10 \qquad x = -9$

9. $6x^2 - 12x = 4x^2 - 3x + 18$

$2x^2 - 9x - 18 = 0$

$(2x + 3)(x - 6) = 0$

$x = -\frac{3}{2} \qquad x = 6$

11. $z^2 = 144$

$z = \pm\sqrt{144}$

$z = \pm 12$

13. $y^2 - 12 = 0$

$y^2 = 12$

$y = \pm\sqrt{12}$

$y = \pm 2\sqrt{3}$

15. $(x - 16)^2 = 400$

$x - 16 = \pm\sqrt{400}$

$x = 16 \pm 20$

$x = 36, -4$

17. $z^2 = -121$

$z = \pm\sqrt{-121}$

$z = \pm 11i$

19. $y^2 + 50 = 0$

$y^2 = -50$

$y = \pm\sqrt{-50}$

$y = \pm 5\sqrt{2}i$

21. $(y + 4)^2 + 18 = 0$

$(y + 4)^2 = -18$

$y + 4 = \pm\sqrt{-18}$

$y = -4 \pm 3\sqrt{2}i$

23. $x^4 - 4x^2 - 5 = 0$

$(x^2 - 5)(x^2 + 1) = 0$

$x^2 + 1 = 0$

$x^2 - 5 = 0 \qquad x^2 = -1$

$x^2 = 5 \qquad x = \pm\sqrt{-1}$

$x = \pm\sqrt{5} \qquad x = \pm i$

25. $x - 4\sqrt{x} + 3 = 0$

$(\sqrt{x} - 3)(\sqrt{x} - 1) = 0$

$(\sqrt{x} - 3) = 0 \qquad (\sqrt{x} - 1) = 0$

$\sqrt{x} = 3 \qquad \sqrt{x} = 1$

$(\sqrt{x})^2 = 3^2 \qquad (\sqrt{x})^2 = 1^2$

$x = 9 \qquad x = 1$

Check: $9 - 4\sqrt{9} + 3 \stackrel{?}{=} 0$

$9 - 12 + 3 \stackrel{?}{=} 0$

$0 = 0$

Check: $1 - 4\sqrt{1} + 3 \stackrel{?}{=} 0$

$1 - 4 + 3 \stackrel{?}{=} 0$

$0 = 0$

27. $(x^2 - 2x)^2 - 4(x^2 - 2x) - 5 = 0$

$[(x^2 - 2x) - 5][(x^2 - 2x) + 1] = 0$

$(x^2 - 2x - 5)(x^2 - 2x + 1) = 0$

$x = \dfrac{-(-2) \pm \sqrt{(-2)^2 - 4(1)(-5)}}{2(1)}$

$x = \dfrac{2 \pm \sqrt{4 + 20}}{2}$

$x = \dfrac{2 \pm \sqrt{24}}{2}$

$x = \dfrac{2 \pm 2\sqrt{6}}{2} \qquad (x - 1)^2 = 0$

$x = 1 \pm \sqrt{6} \qquad x = 1$

29. $x^{2/3} + 3x^{1/3} - 28 = 0$

$(x^{1/3} + 7)(x^{1/3} - 4) = 0$

$x^{1/3} + 7 = 0 \qquad x^{1/3} - 4 = 0$

$x^{1/3} = -7 \qquad x^{1/3} = 4$

$\sqrt[3]{x} = -7 \qquad \sqrt[3]{x} = 4$

$\left(\sqrt[3]{x}\right)^3 = (-7)^3 \qquad \left(\sqrt[3]{x}\right)^3 = 4^3$

$x = -343 \qquad x = 64$

31. $z^2 + 18z + 81$

$\left[81 = \left(\frac{18}{2}\right)^2\right]$

33. $x^2 - 15x + \frac{225}{4}$

$\left[\frac{225}{4} = \left(-\frac{15}{2}\right)^2\right]$

35. $y^2 + \frac{2}{5}y + \frac{1}{25}$

$\left[\frac{1}{25} = \left(\frac{2/5}{2}\right)^2\right]$

$\left[\frac{1}{25} = \left(\frac{1}{5}\right)^2\right]$

37. $x^2 - 6x - 3 = 0$

$x^2 - 6x + 9 = 3 + 9$

$(x - 3)^2 = 12$

$x - 3 = \pm\sqrt{12}$

$x = 3 \pm 2\sqrt{3}$

$x \approx 6.46, -0.46$

39. $v^2 + 5v + 4 = 0$

$v^2 + 5v + \frac{25}{4} = -4 + \frac{25}{4}$

$\left(v + \frac{5}{2}\right)^2 = -\frac{16}{4} + \frac{25}{4}$

$\left(v + \frac{5}{2}\right)^2 = \frac{9}{4}$

$v + \frac{5}{2} = \pm\frac{3}{2}$

$v = -\frac{5}{2} \pm \frac{3}{2}$

$v = -\frac{2}{2}, -\frac{8}{2}$

$v = -1, -4$

41. $y^2 - \frac{2}{3}y + 2 = 0$

$y^2 - \frac{2}{3}y = -2$

$y^2 - \frac{2}{3}y + \frac{1}{9} = -2 + \frac{1}{9}$

$\left(y - \frac{1}{3}\right)^2 = \frac{-17}{9}$

$y - \frac{1}{3} = \pm\sqrt{\frac{-17}{9}}$

$y = \frac{1}{3} \pm \frac{\sqrt{17}i}{3}$

$y \approx 0.33 + 1.37i,\ 0.33 - 1.37i$

43. $v^2 + v - 42 = 0$

$v = \frac{-1 \pm \sqrt{1^2 - 4(1)(-42)}}{2(1)}$

$v = \frac{-1 \pm \sqrt{1 + 168}}{2}$

$v = \frac{-1 \pm \sqrt{169}}{2}$

$v = \frac{-1 \pm 13}{2}$

$v = \frac{12}{2}, \frac{-14}{2}$

$v = 6, -7$

45. $2y^2 + y - 21 = 0$

$y = \frac{-1 \pm \sqrt{1^2 - 4(2)(-21)}}{2(2)}$

$y = \frac{-1 \pm \sqrt{1 + 168}}{4}$

$y = \frac{-1 \pm \sqrt{169}}{4}$

$y = \frac{-1 \pm 13}{4}$

$y = 3, -\frac{7}{2}$

47. $5x^2 - 16x + 2 = 0$

$$x = \frac{-(-16) \pm \sqrt{(-16)^2 - 4(5)(2)}}{2(5)}$$

$$x = \frac{16 \pm \sqrt{256 - 40}}{10}$$

$$x = \frac{16 \pm \sqrt{216}}{10}$$

$$x = \frac{16 \pm 6\sqrt{6}}{10}$$

$$x = \frac{8 \pm 3\sqrt{6}}{5}$$

49. $x^2 + 4x + 4 = 0$

$$b^2 - 4ac = 4^2 - 4(1)(4)$$
$$= 16 - 16$$
$$= 0$$

One repeated rational solution

51. $s^2 - s - 20 = 10$

$$b^2 - 4ac = (-1)^2 - 4(1)(-20)$$
$$= 1 + 80$$
$$= 81$$

Two distinct rational solutions

53. $4t^2 + 16t + 10 = 0$

$$b^2 - 4ac = 16^2 - 4(4)(10)$$
$$= 256 - 160$$
$$= 96$$

Two distinct irrational solutions

55. $v^2 - 6v + 21 = 0$

$$b^2 - 4ac = (-6)^2 - 4(1)(21)$$
$$= 36 - 84$$
$$= -48$$

Two distinct complex solutions

57. $x = 3 \qquad x = -7$

$x - 3 = 0 \qquad x + 7 = 0$

$$(x - 3)(x + 7) = 0$$
$$x^2 + 4x - 21 = 0$$

59. $x = 5 + \sqrt{7} \qquad x = 5 - \sqrt{7}$

$x - (5 + \sqrt{7}) = 0 \qquad x - (5 - \sqrt{7}) = 0$

$(x - 5) - \sqrt{7} = 0 \qquad (x - 5) + \sqrt{7} = 0$

$$[(x - 5) - \sqrt{7}][(x - 5) + \sqrt{7}] = 0$$
$$(x - 5)^2 - (\sqrt{7})^2 = 0$$
$$x^2 - 10x + 25 - 7 = 0$$
$$x^2 - 10x + 18 = 0$$

61. $x = 6 + 2i \qquad x = 6 - 2i$

$x - (6 + 2i) = 0 \qquad x - (6 - 2i) = 0$

$(x - 6) - 2i = 0 \qquad (x - 6) + 2i = 0$

$$[(x - 6) - 2i][(x - 6) + 2i] = 0$$
$$(x - 6)^2 - (2i)^2 = 0$$
$$x^2 - 12x + 36 + 4 = 0$$
$$x^2 - 12x + 40 = 0$$

63. $y = x^2 - 8x + 3$

$$= (x^2 - 8x + 16) + 3 - 16$$
$$= (x - 4)^2 - 13$$

Vertex: $(4, -13)$

65. $y = 2x^2 - x + 3$

$$= 2\left(x^2 - \tfrac{1}{2}x\right) + 3$$
$$= 2\left(x^2 - \tfrac{1}{2}x + \tfrac{1}{16}\right) + 3 - \tfrac{1}{8}$$
$$= 2\left(x - \tfrac{1}{4}\right)^2 + \tfrac{23}{8}$$

Vertex: $\left(\frac{1}{4}, \frac{23}{8}\right)$

67. $y = x^2 + 8x$

x-intercepts:

$0 = x^2 + 8x$

$0 = x(x + 8)$

$x = 0, \quad x = -8$

Vertex:

$y = x^2 + 8x + 16 - 16$

$y = (x + 4)^2 - 16$

$(-4, -16)$

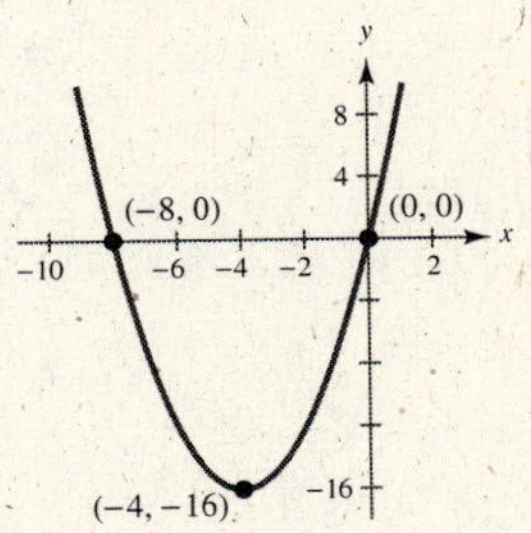

69. $f(x) = -x^2 - 2x + 4$

x-intercepts:

$0 = x^2 + 2x - 4$

$$x = \frac{-2 \pm \sqrt{4 + 16}}{2}$$

$$x = \frac{-2 \pm \sqrt{20}}{2}$$

$$x = \frac{-2 \pm 2\sqrt{5}}{2}$$

$x = -1 \pm \sqrt{5}$

Vertex:

$f(x) = -1(x^2 + 2x) + 4$

$= -(x^2 + 2x + 1) + 4 + 1$

$= -(x + 1)^2 + 5$

$(-1, 5)$

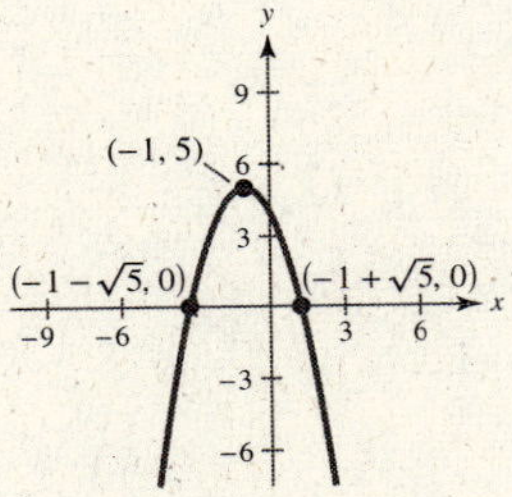

71. Vertex: $(2, -5)$; y-intercept: $(0, 3)$

$y = a(x - h)^2 + k$

$y = a(x - 2)^2 - 5$

$3 = a(0 - 2)^2 - 5$

$3 = a(4) - 5$

$8 = a(4)$

$2 = a$

$y = 2(x - 2)^2 - 5$ or $y = 2x^2 - 8x + 3$

73. Vertex: $(5, 0)$; passes through the point $(1, 1)$

$y = a(x - h)^2 + k$

$1 = a(1 - 5)^2 + 0$

$1 = a(16)$

$\frac{1}{16} = a$

$y = \frac{1}{16}(x - 5)^2 + 0$ or $y = \frac{1}{16}x^2 - \frac{5}{8}x + \frac{25}{16}$

75. (a) *Keystrokes:* [Y=] [(−)] [X,T,θ] [x^2] [÷] 10 [+] 3 [X,T,θ] [+] 6 [GRAPH]

(b) $y = -\frac{1}{10}(0)^2 + 3(0) + 6$

$y = 0 + 0 + 6$

$y = 6$ feet

(c) $x = -\frac{b}{2a} = -\frac{3}{2\left(-\frac{1}{10}\right)} = \frac{-3}{-\frac{1}{5}} = 15$

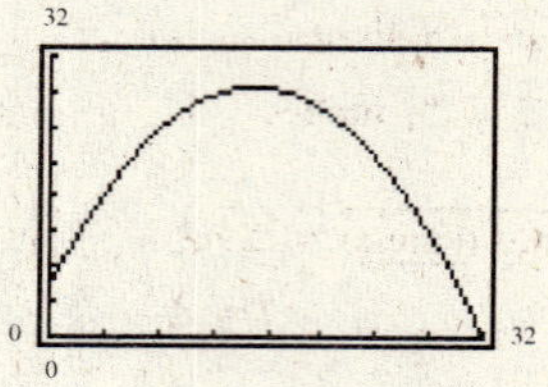

$y = \frac{1}{10}(15)^2 + 3(15) + 6$

$= -\frac{1}{10}(225) + 45 + 6$

$= -22.5 + 45 + 6 = 28.5$ feet

(d) $0 = -\frac{1}{10}x^2 + 3x + 6$

$$x = \frac{-3 \pm \sqrt{3^2 - 4\left(-\frac{1}{10}\right)(6)}}{2\left(-\frac{1}{10}\right)}$$

$$x = \frac{-3 \pm \sqrt{9 + 2.4}}{-\frac{1}{5}}$$

$$x = \frac{-3 \pm \sqrt{11.4}}{-\frac{1}{5}}$$

$x = -5\left(-3 \pm \sqrt{11.4}\right) = 15 \pm 5\sqrt{11.4} = 31.9$

The ball is 31.9 feet from the child when it hits the ground.

77. *Verbal Model:* [Selling price per car] = [Cost per car] + [Profit per car]

Labels: Number cars sold $= x$

Number cars purchased $= x + 4$

Equation:

$$\frac{80{,}000}{x} = \frac{80{,}000}{x + 4} + 1000$$

$$x(x + 4)\left(\frac{80{,}000}{x}\right) = \left(\frac{80{,}000}{x + 4} + 1000\right)x(x + 4)$$

$$80{,}000(x + 4) = 80{,}000x + 1000x(x + 4)$$

$$80{,}000x + 320{,}000 = 80{,}000x + 1000x^2 + 4000x$$

$$0 = 1000x^2 + 4000x - 320{,}000$$

$$0 = x^2 + 4x - 320$$

$$0 = (x + 20)(x - 16)$$

reject $x = -20$, $x = 16$ cars

Average price per car $= \frac{80{,}000}{16} = \$5000$

79. *Verbal Model:* $\boxed{\text{Area}} = \boxed{\text{Length}} \cdot \boxed{\text{Width}}$

Labels: Width $= x$

Length $= x + 12$

Equation:

$$85 = (x + 12)x$$
$$0 = x^2 + 12x - 85$$
$$0 = (x + 17)(x - 5)$$

reject $x = -17$, $\quad x = 5$ inches

$x + 12 = 17$ inches

81. *Verbal Model:* $\boxed{\text{Cost per ticket}} \cdot \boxed{\text{Number of tickets}} = \boxed{\$96}$

Labels: Number in team $= x$

Number going to game $= x + 3$

Equation:

$$\left(\frac{96}{x} - 1.60\right)(x + 3) = 96$$
$$\left(\frac{96 - 1.60x}{x}\right)(x + 3) = 96$$
$$(96 - 1.6x)(x + 3) = 96x$$
$$96x - 1.6x^2 - 4.8x + 288 = 96x$$
$$1.6x^2 + 4.8x - 288 = 0$$
$$x^2 + 3x - 180 = 0$$
$$(x - 12)(x + 15) = 0$$

$x - 12 = 0 \qquad x + 15 = 0$

$x = 12 \qquad x = -15$ reject

$x + 3 = 15$

83. *Verbal Model:* [Work done by Person 1] + [Work done by Person 2] = [One complete job]

Labels: Time Person 1 $= x$

Time Person 2 $= x + 2$

Equation: $\frac{1}{x}(10) + \frac{1}{x+2}(10) = 1$

$$x(x+2)\left[10\left(\frac{1}{x} + \frac{1}{x+2}\right)\right] = [1]x(x+2)$$

$$10(x+2) + 10x = x(x+2)$$

$$10x + 20 + 10x = x^2 + 2x$$

$$0 = x^2 - 18x - 20$$

$$x = \frac{-(-18) \pm \sqrt{(-18)^2 - 4(1)(-20)}}{2(1)}$$

$$x = \frac{18 \pm \sqrt{324 + 80}}{2}$$

$$x = \frac{18 \pm \sqrt{404}}{2}$$

$$x = \frac{18 \pm 2\sqrt{101}}{2}$$

$$x = 9 \pm \sqrt{101}$$

$x \approx 19$, $x = -1$, reject

$x + 2 \approx 21$

19 hours, 21 hours

85. $2x(x + 7)$

$2x = 0 \qquad x + 7 = 0$

$x = 0 \qquad x = -7$

Critical numbers: $x = 0, -7$

87. $x^2 - 6x - 27$

$(x - 9)(x + 3)$

$x - 9 = 0 \qquad x + 3 = 0$

$x = 9 \qquad x = -3$

Critical numbers: $x = 9, -3$

89. $5x(7 - x) > 0$

Critical numbers: $x = 0, 7$

Test intervals:

Negative: $(-\infty, 0)$

Positive: $(0, 7)$

Negative: $(7, \infty)$

Solution: $(0, 7)$

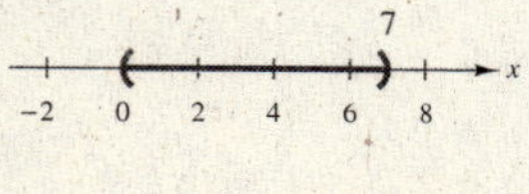

91. $16 - (x - 2)^2 \le 0$

$(4 - x + 2)(4 + x - 2) \le 0$

$(6 - x)(2 + x) \le 0$

Critical numbers: $x = -2, 6$

Test intervals:

Negative: $(-\infty, -2]$

Positive: $[-2, 6]$

Negative: $[6, \infty)$

Solution: $(-\infty, -2] \cup [6, \infty)$

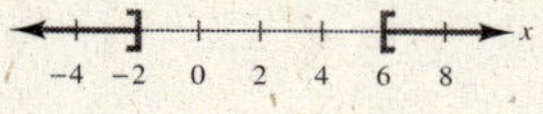

93. $2x^2 + 3x - 20 < 0$

$(2x - 5)(x + 4) < 0$

Critical numbers: $x = -4, \frac{5}{2}$

Test intervals:

Positive: $(-\infty, -4)$

Negative: $\left(-4, \frac{5}{2}\right)$

Positive: $\left(\frac{5}{2}, \infty\right)$

Solution: $\left(-4, \frac{5}{2}\right)$

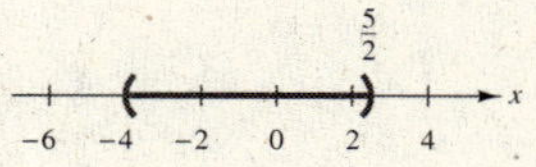

95. $\dfrac{x + 3}{2x - 7} \geq 0$

Critical numbers: $x = -3, \dfrac{7}{2}$

Test intervals:

Positive: $(-\infty, -3)$

Negative: $\left[-3, \dfrac{7}{2}\right]$

Positive: $\left(\dfrac{7}{2}, \infty\right)$

Solution: $(-\infty, -3] \cup \left(\dfrac{7}{2}, \infty\right)$

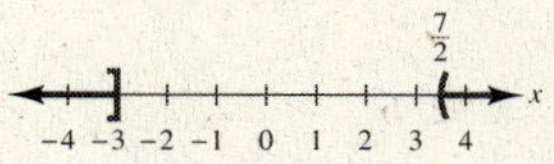

97. $\dfrac{x + 4}{x - 1} < 0$

Critical numbers: $x = -4, 1$

Test intervals:

Positive: $(-\infty, -4)$

Negative: $(-4, 1)$

Positive: $(1, \infty)$

Solution: $(-4, 1)$

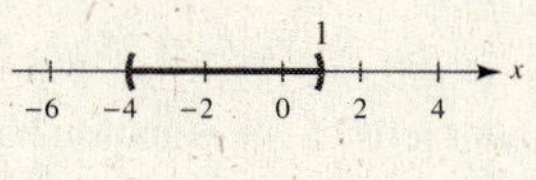

99. $h = -16t^2 + 312t$

$-16t^2 + 312t > 1200$

$-16t^2 + 312t - 1200 > 0$ (Divide by -16.)

$t^2 - 19.5t + 75 < 0$

$$t = \frac{-(-19.5) \pm \sqrt{(-19.5)^2 - 4(1)(75)}}{2(1)}$$

$$t = \frac{19.5 \pm \sqrt{80.25}}{2}$$

$t \approx 14.2, 5.3$

Critical numbers: $t = 14.2, 5.3$

Test intervals:

Positive: $(-\infty, 5.3)$

Negative: $(5.3, 14.2)$

Positive: $(14.2, \infty)$

Solution: $(5.3, 14.2)$

$5.3 < t < 14.2$

Chapter Test for Chapter 8

1. $x(x-3)-10(x-3)=0$

$(x-3)(x-10)=0$

$x-3=0 \qquad x-10=0$

$x=3 \qquad x=10$

2. $6x^2-34x-12=0$

$2(3x^2-17x-6)=0$

$2(3x+1)(x-6)=0$

$3x+1=0 \qquad x-6=0$

$x=-\frac{1}{3} \qquad x=6$

3. $(x-2)^2=0.09$

$x-2=\pm 0.3$

$x=2\pm 0.3$

$x=2.3, 1.7$

4. $(x+4)^2+100=0$

$(x+4)^2=-100$

$x+4=\pm\sqrt{-100}$

$x=-4\pm 10i$

5. $2x^2-6x+3=0$

$x^2-3x+\frac{9}{4}=-\frac{3}{2}+\frac{9}{4}$

$\left(x-\frac{3}{2}\right)^2=\frac{-6+9}{4}$

$\left(x-\frac{3}{2}\right)^2=\frac{3}{4}$

$x-\frac{3}{2}=\pm\sqrt{\frac{3}{4}}$

$x=\frac{3}{2}\pm\frac{\sqrt{3}}{2}$

6. $2y(y-2)=7$

$2y^2-4y-7=0$

$y=\frac{-(-4)\pm\sqrt{(-4)^2-4(2)(-7)}}{2(2)}$

$y=\frac{4\pm\sqrt{16+56}}{4}$

$y=\frac{4\pm\sqrt{72}}{4}$

$y=\frac{4\pm 6\sqrt{2}}{4}$

$y=\frac{2\pm 3\sqrt{2}}{2}\approx 7.41$ and -0.41

7. $\frac{1}{x^2}-\frac{6}{x}+4=0$

$1-6x+4x^2=0$

$4x^2-6x+1=0$

$x=\frac{-(-6)\pm\sqrt{(-6)^2-4(4)(1)}}{2(4)}$

$x=\frac{6\pm\sqrt{36-16}}{8}$

$x=\frac{6\pm\sqrt{20}}{8}$

$x=\frac{6\pm 2\sqrt{5}}{8}$

$x=\frac{3\pm\sqrt{5}}{4}$

8. $x^{2/3}-9x^{1/3}+8=0$

$(x^{1/3}-8)(x^{1/3}-1)=0$

$x^{1/3}-8=0 \qquad x^{1/3}-1=0$

$x^{1/3}=8 \qquad x^{1/3}=1$

$(x^{1/3})^3=8^3 \qquad (x^{1/3})^3=1^3$

$x=512 \qquad x=1$

9. $b^2-4ac=(-12)^2-4(5)(10)$

$=144-200$

$=-56$

A negative discriminant tells us the equation has two complex solutions.

10. $x = -7 \qquad x = -3$

$x + 7 = 0 \qquad x + 3 = 0$

$(x + 7)(x + 3) = 0$

$x^2 + 10x + 21 = 0$

11. $y = -x^2 + 2x - 4$

x-intercepts:

$0 = x^2 - 2x + 4$

$$x = \frac{-(-2) \pm \sqrt{(-2)^2 - 4(1)(4)}}{2(1)}$$

$$x = \frac{2 \pm \sqrt{4 - 16}}{2}$$

$$x = \frac{2 \pm \sqrt{-12}}{2}$$

Not real

No x-intercepts

Keystrokes:

[Y=] [(−)] [X,T,θ] [x^2] [+] 2 [X,T,θ] [−] 4 [GRAPH]

Vertex:

$y = -1(x^2 - 2x) - 4$

$= -(x^2 - 2x + 1) - 4 + 1$

$= -(x - 1)^2 - 3$

$(1, -3)$

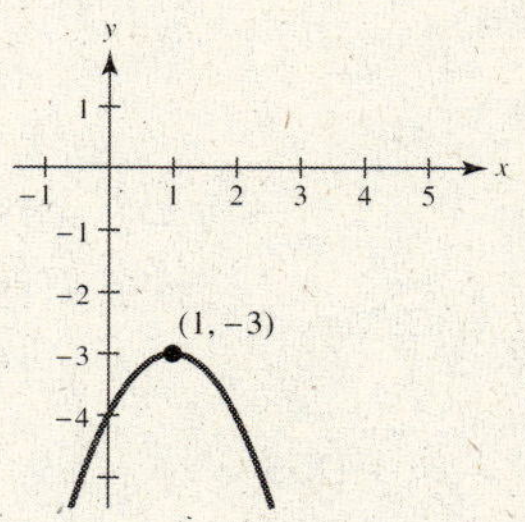

12. $y = x^2 - 2x - 15$

x-intercepts:

$0 = x^2 - 2x - 15$

$0 = (x - 5)(x + 3)$

$x = 5, \quad x = -3$

Keystrokes:

[Y=] [X,T,θ] [x^2] [−] 2 [X,T,θ] [−] 15 [GRAPH]

Vertex:

$y = (x^2 - 2x + 1) - 15 - 1$

$y = (x - 1)^2 - 16$

$(1, -16)$

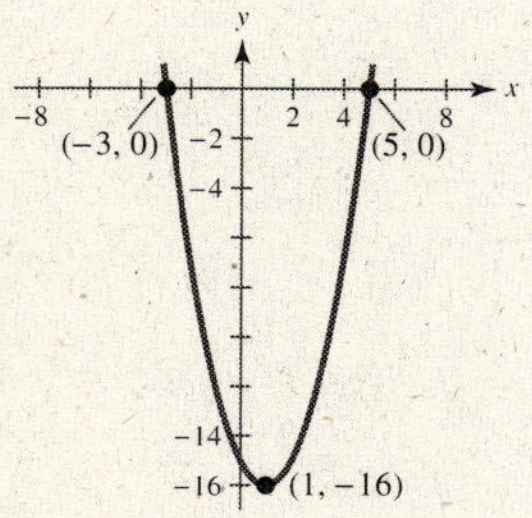

13. $16 \le (x - 2)^2$

$(x - 2)^2 \ge 16$

$x^2 - 4x + 4 \ge 16$

$x^2 - 4x - 12 \ge 0$

$(x - 6)(x + 2) \ge 0$

Critical numbers: $x = -2, 6$

Test intervals:

Positive: $(-\infty, -2]$

Negative: $[-2, 6]$

Positive: $[6, \infty)$

Solution: $(-\infty, -2] \cup [6, \infty)$

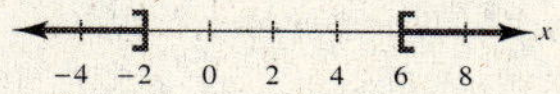

14. $2x(x - 3) < 0$

Critical numbers: $x = 0, 3$

Test intervals:

Positive: $(-\infty, 0)$

Negative: $(0, 3)$

Positive: $(3, \infty)$

Solution: $(0, 3)$

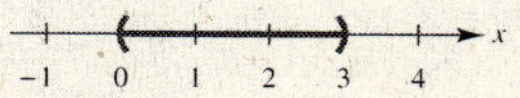

15. $\dfrac{x + 1}{x - 5} \le 0$

Critical numbers: $x = -1, 5$

Test intervals:

Positive: $(-\infty, -1)$

Negative: $(-1, 5)$

Positive: $(5, \infty)$

Solution: $[-1, 5)$

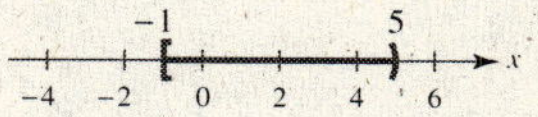

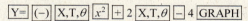

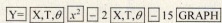

16. *Verbal Model:* $\boxed{\text{Area}} = \boxed{\text{Length}} \cdot \boxed{\text{Width}}$

Labels: Length $= x$

Width $= x - 22$

Equation: $240 = x(x - 22)$

$$0 = x^2 - 22x - 240$$
$$0 = (x - 30)(x + 8)$$
$$0 = x - 30 \qquad \text{reject } x = -8$$
$$30 \text{ feet} = x \qquad x - 22 = 8 \text{ feet}$$

8 feet × 30 feet

17. *Verbal Model:* $\boxed{\begin{matrix}\text{Cost per person}\\\text{Current Group}\end{matrix}} - \boxed{\begin{matrix}\text{Cost per person}\\\text{New Group}\end{matrix}} = 6.25$

Labels: Number Current Group $= x$

Number New Group $= x + 10$

Equation:

$$\frac{1250}{x} - \frac{1250}{x + 10} = 6.25$$
$$x(x + 10)\left(\frac{1250}{x} - \frac{1250}{x + 10}\right) = (6.25)x(x + 10)$$
$$1250(x + 10) - 1250x = 6.25x(x + 10)$$
$$1250x + 12500 - 1250x = 6.25x^2 + 62.5x$$
$$0 = 6.25x^2 + 62.5x - 12{,}500$$
$$0 = x^2 + 10x - 2000$$
$$0 = (x + 50)(x - 40)$$

reject $x = -50$, $\quad x = 40$ club members

18.
$$35 = -16t^2 + 75$$
$$16t^2 = 40$$
$$t^2 = \frac{40}{16} = \frac{5}{2}$$
$$t = \sqrt{\frac{5}{2}}$$
$$t = \frac{\sqrt{10}}{2} \approx 1.5811388$$
$$t \approx 1.58 \text{ seconds}$$

19. *Formula:* $c^2 = a^2 + b^2$

Labels: $c = 125 \qquad a + b = 155$

$a = x \qquad x + b = 155$

$b = 155 - x \qquad b = 155 - x$

Equation:

$$125^2 = x^2 + (155 - x)^2$$
$$15{,}625 = x^2 + 24025 - 310x + x^2$$
$$0 = 2x^2 - 310x + 8400$$
$$0 = x^2 - 155x + 4200$$
$$0 = (x - 35)(x - 120)$$
$$0 = x - 35 \qquad x - 120 = 0$$
$$35 \text{ feet} = x \qquad x = 120 \text{ feet}$$

CHAPTER 9
Exponential and Logarithmic Functions

CHAPTER 9
Exponential and Logarithmic Functions

Section 9.1 Exponential Functions

1. $3^x \cdot 3^{x+2} = 3^{x+(x+2)} = 3^{2x+2}$

3. $\dfrac{e^{x+2}}{e^x} = e^{x+2-x} = e^2$

5. $3(e^x)^{-2} = 3 \cdot \dfrac{1}{(e^x)^2} = \dfrac{3}{e^{2x}}$

7. $\sqrt[3]{-8e^{3x}} = -2e^x$ because

$-2 \cdot -2 \cdot -2 \cdot e^x \cdot e^x \cdot e^x = -8e^{3x}$.

9. $5^{\sqrt{2}} \approx 9.739$

Keystrokes:

Scientific: 5 [y^x] 2 [$\sqrt{\ }$] [=]

Graphing: 5 [^] [$\sqrt{\ }$] 2 [)] [ENTER]

11. $e^{1/3} \approx 1.396$

Keystrokes:

Scientific: [(] 1 [÷] 3 [)] [INV] [ln x] [=]

Graphing: [e^x] [(] 1 [÷] 3 [)] [ENTER]

13. $3(2e^{1/2})^3 = 3 \cdot 8 \cdot e^{3/2} = 24e^{1.5} \approx 107.561$

Keystrokes:

Scientific: 24 [×] 1.5 [e^x] [=]

Graphing: 24 [e^x] 1.5 [)] [ENTER]

15. $\dfrac{4e^3}{12e^2} = \dfrac{e}{3} \approx 0.906$

Keystrokes:

Scientific: 1 [INV] [ln x] [÷] 3 [=]

Graphing: [e] [÷] 3 [ENTER]

17. $f(x) = 3^x$

(a) $f(-2) = 3^{-2} = \frac{1}{9}$

(b) $f(0) = 3^0 = 1$

(c) $f(1) = 3^1 = 3$

19. $g(x) = 2.2^{-x}$

(a) $g(1) = 2 \cdot 2^{-1} \approx 0.455$

(b) $g(3) = 2 \cdot 2^{-3} \approx 0.094$

(c) $g(\sqrt{6}) = 2 \cdot 2^{-\sqrt{6}} \approx 0.145$

21. $f(t) = 500\left(\frac{1}{2}\right)^t$

(a) $f(0) = 500\left(\frac{1}{2}\right)^0 = 500$

(b) $f(1) = 500\left(\frac{1}{2}\right)^1 = 250$

(c) $f(\pi) = 500\left(\frac{1}{2}\right)^\pi \approx 56.657$

23. $f(x) = 1000(1.05)^{2x}$

(a) $f(0) = 1000(1.05)^{(2)(0)} = 1000$

(b) $f(5) = 1000(1.05)^{2(5)} \approx 1628.895$

(c) $f(10) = 1000(1.05)^{2(10)} \approx 2653.298$

25. $h(x) = \dfrac{5000}{(1.06)^{8x}}$

(a) $h(5) = \dfrac{5000}{(1.06)^{8(5)}} \approx 486.111$

(b) $h(10) = \dfrac{5000}{(1.06)^{8(10)}} \approx 47.261$

(c) $h(20) = \dfrac{5000}{(1.06)^{8(20)}} \approx 0.447$

27. $g(x) = 10e^{-0.5x}$

(a) $g(-4) = 10e^{-0.5(-4)} = 10e^2 \approx 73.891$

(b) $g(4) = 10e^{-0.5(4)} = 10e^{-2} \approx 1.353$

(c) $g(8) = 10e^{-0.5(8)} = 10e^{-4} \approx 0.183$

29. $g(x) = \dfrac{1000}{2 + e^{-0.12x}}$

(a) $g(0) = \dfrac{1000}{2 + e^{-0.12(0)}} \approx 333.333$

(b) $g(10) = \dfrac{1000}{2 + e^{-0.12(10)}} \approx 434.557$

(c) $g(50) = \dfrac{1000}{2 + e^{-0.12(50)}} \approx 499.381$

31. $f(x) = 3^x$

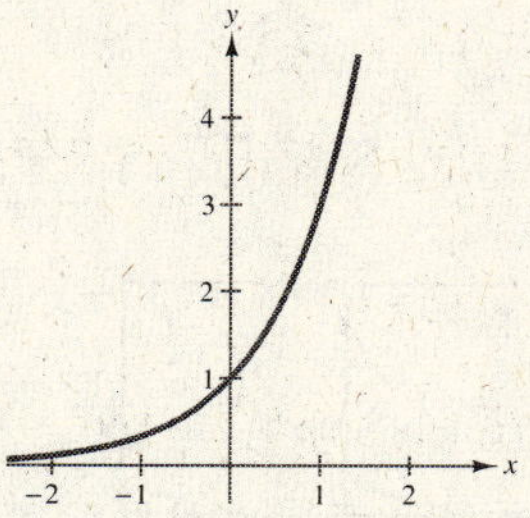

Horizontal asymptote: $y = 0$

Table of values:

x	-2	-1	0	1	2
$f(x)$	0.1	0.3	1	3	9

33. $f(x) = 3^{-x} = \left(\frac{1}{3}\right)^x$

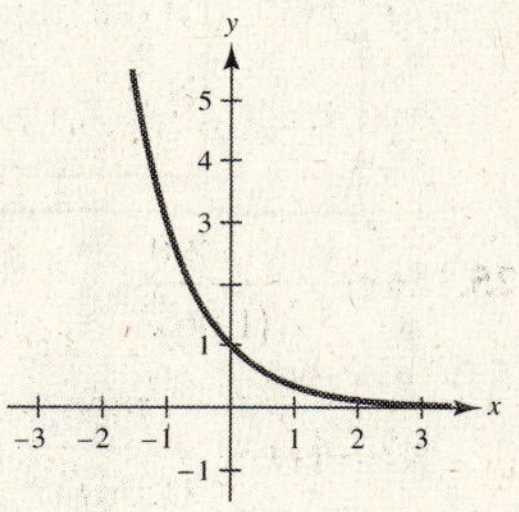

Horizontal asymptote: $y = 0$

35. $g(x) = 3^x - 2$

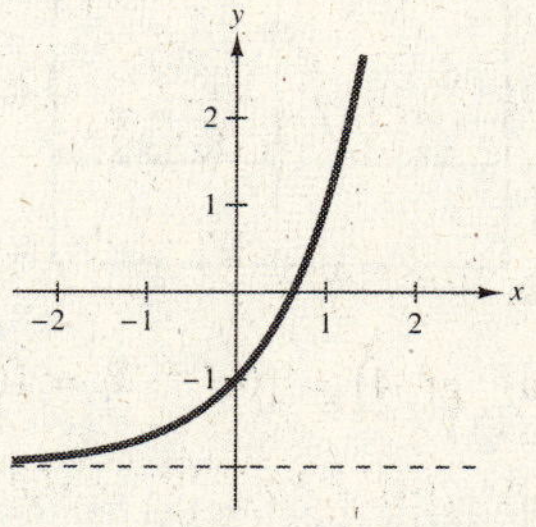

Horizontal asymptote: $y = -2$

Table of values:

x	-2	-1	0	1	2
$g(x)$	-1.9	-1.7	-1	1	7

37. $g(x) = 5^{x-1}$

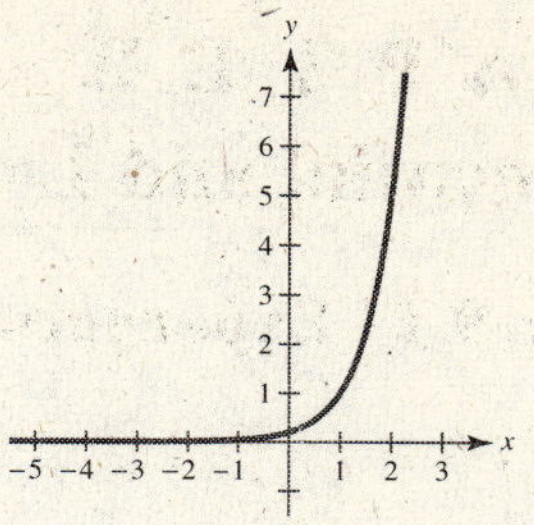

Horizontal asymptote: $y = 0$

Table of values:

x	-2	-1	0	1	2
$g(x)$	0.008	0.04	0.2	1	5

39. $f(t) = 2^{-t^2}$

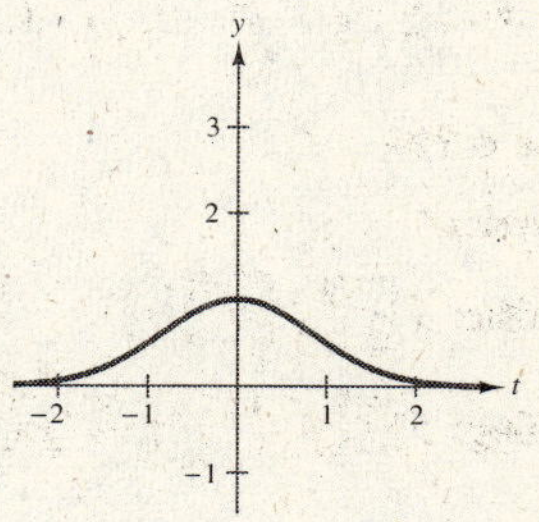

Horizontal asymptote: $y = 0$

Table of values:

t	-2	-1	0	1	2
$f(t)$	0.1	0.5	1	0.5	0.1

41. $f(x) = -2^{0.5x}$

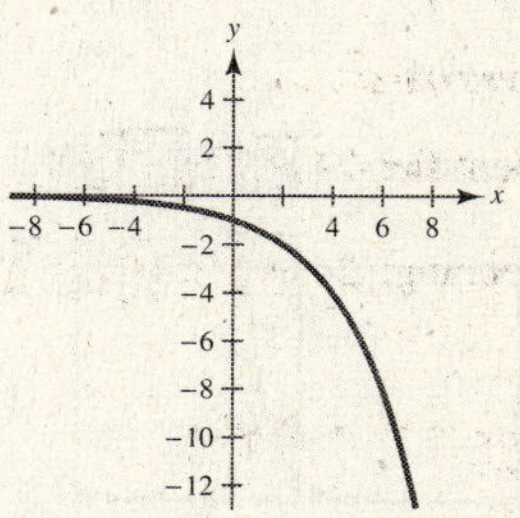

Horizontal asymptote: $y = 0$

Table of values:

x	-2	-1	0	1	2
$f(x)$	-5	-0.7	-1	-1.4	-2

43. $f(x) = -\left(\frac{1}{3}\right)^x$

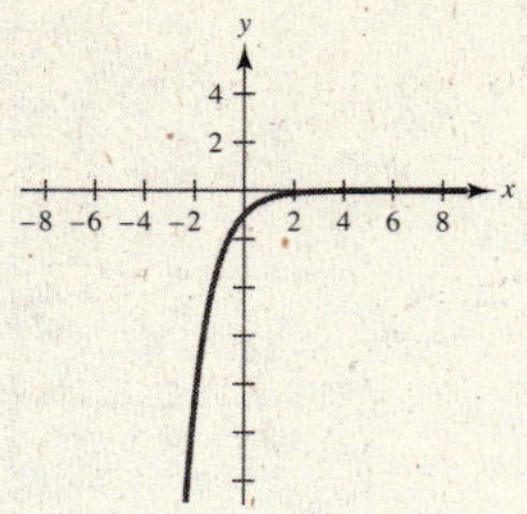

Horizontal asymptote: $y = 0$

Table of values:

x	-2	-1	0	1	2
$f(x)$	-9	-3	-1	-0.3	-0.1

45. $g(t) = 200\left(\frac{1}{2}\right)^t$

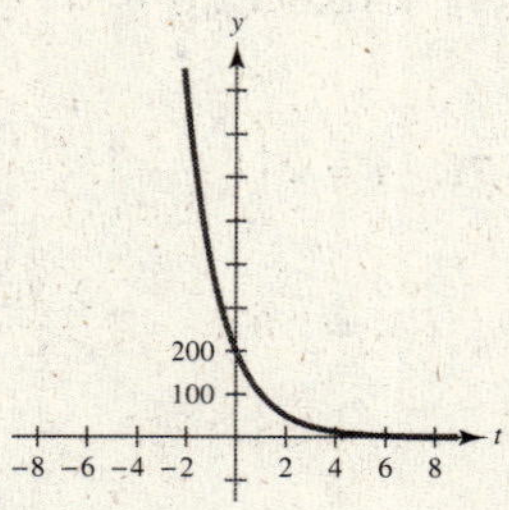

Horizontal asymptote: $y = 0$

Table of values:

t	-2	-1	0	1	2
$g(t)$	800	400	200	100	50

47. $y = 7^{x/2}$

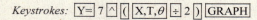

Keystrokes: [Y=] 7 [^] [(] [X,T,θ] [÷] 2 [)] [GRAPH]

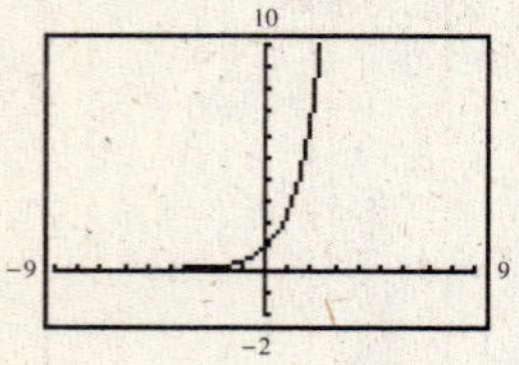

49. $y = 7^{-x/2} + 5$

Keystrokes:

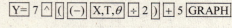

[Y=] 7 [^] [(] [(−)] [X,T,θ] [÷] 2 [)] [+] 5 [GRAPH]

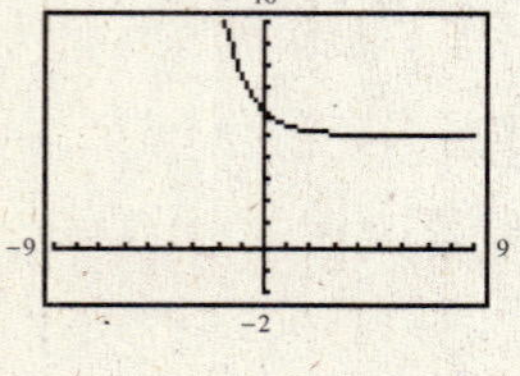

51. $y = 500(1.06)^t$

Keystrokes: [Y=] 500 [(] 1.06 [)] [^] [X,T,θ] [GRAPH]

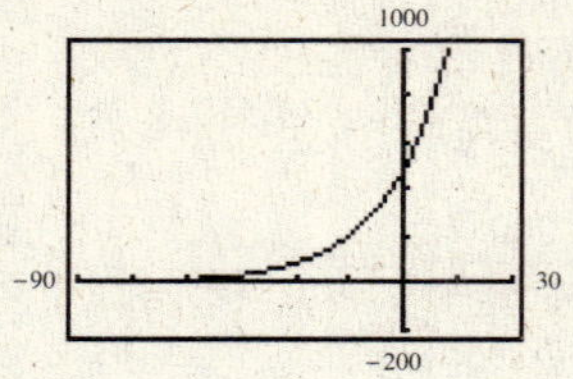

53. $y = 3e^{0.2x}$

Keystrokes: [Y=] 3 [e^x] 0.2 [X,T,θ] [GRAPH]

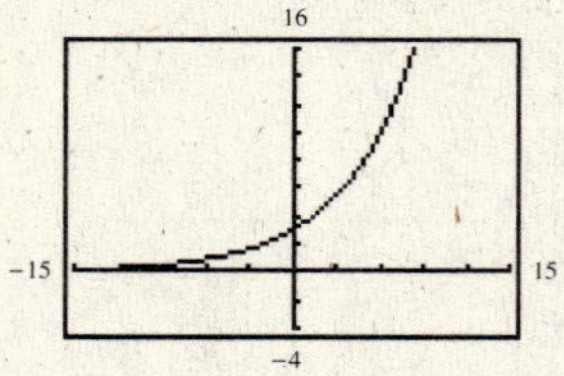

55. $P(t) = 100e^{-0.1t}$

Keystrokes: [Y=] 100 [e^x] [(−)] 0.1 [X,T,θ] [GRAPH]

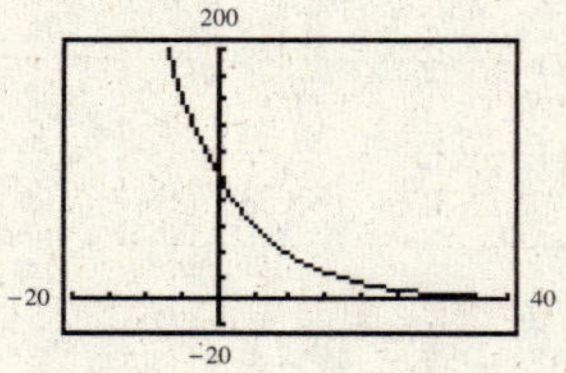

57. $y = 6e^{-x^2/y}$

Keystrokes:

[Y=] 6 [e^x] [(−)] [(] [X,T,θ] [x^2] [÷] 3 [)] [GRAPH]

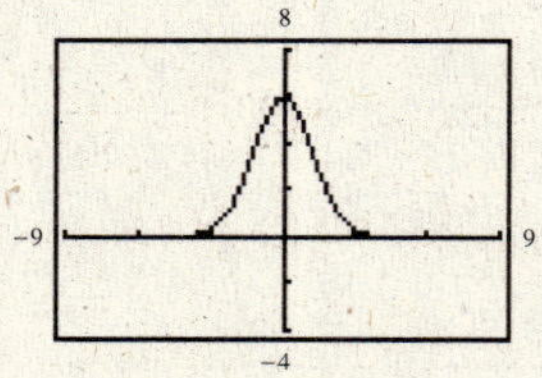

59. $f(x) = 2^{-x}$

(b) Basic graph reflected in the y-axis

61. $f(x) = 2^{x-1}$

(c) Basic graph shifted 1 unit right

63. $h(x) = 4^x - 1$

Vertical shift 1 unit down

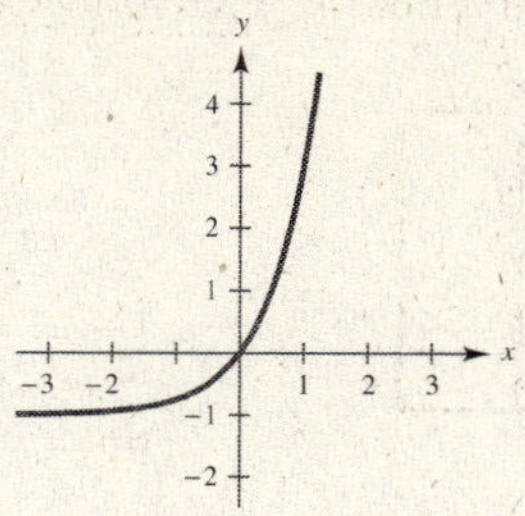

65. $h(x) = 4^{x+6}$

Horizontal shift 2 units left

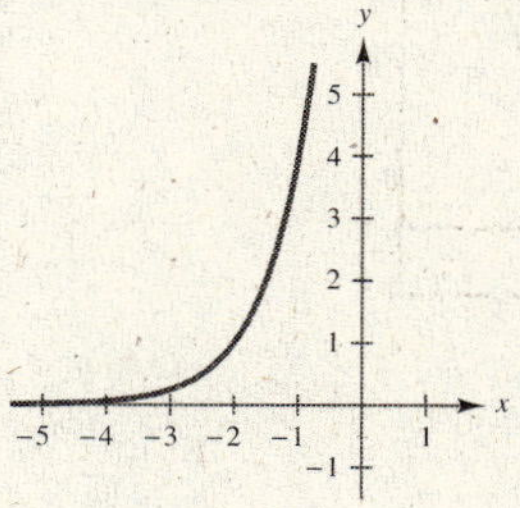

67. $h(x) = -4^x$

Reflection in the x-axis

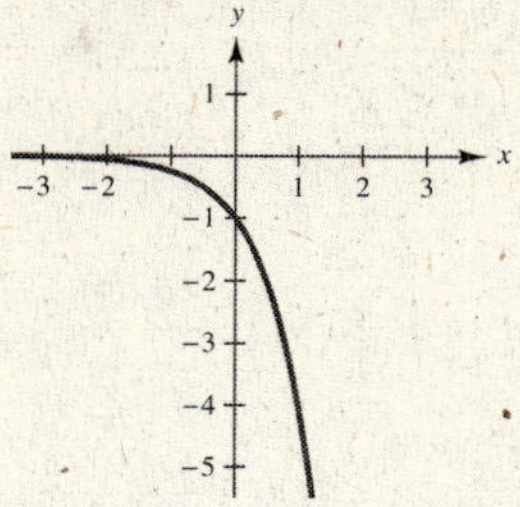

69. $y = 16\left(\frac{1}{2}\right)^{80/30} \approx 2.520$ grams

Keystrokes:

Scientific: 16 [×] 0.5 [y^x] [(] 8 [÷] 3 [)] [=]

Graphing: 16 [×] 0.5 [^] [(] 8 [÷] 3 [)] [ENTER]

71. $A = P\left(1 + \frac{r}{n}\right)^{nt}$

$A = 5000\left(1 + \frac{0.06}{12}\right)^{12(5)}$

$A \approx \$6744.25$

73.

n	1	4	12	365	Continuous
A	\$275.90	\$283.18	\$284.89	\$285.74	\$285.77

Compounded 1 time: $A = 100\left(1 + \frac{0.07}{1}\right)^{1(15)} = \$275.90$

Compounded 4 times: $A = 100\left(1 + \frac{0.07}{4}\right)^{4(15)} = \$283.18$

Compounded 12 times: $A = 100\left(1 + \frac{0.07}{12}\right)^{12(15)} = \$284.89$

Compounded 365 times: $A = 100\left(1 + \frac{0.07}{365}\right)^{365(15)} = \$285.74$

Compounded continuously: $A = Pe^{rt} = 100e^{0.07(15)} = \$285.77$

75.

n	1	4	12	365	Continuous
A	\$4956.46	\$5114.30	\$5152.11	\$5170.78	\$5171.42

Compounded 1 time: $A = 2000\left(1 + \frac{0.095}{1}\right)^{1(10)} = \$4956.46$

Compounded 4 times: $A = 2000\left(1 + \frac{0.095}{4}\right)^{4(10)} = \$5114.30$

Compounded 12 times: $A = 2000\left(1 + \frac{0.095}{12}\right)^{12(10)} = \$5152.11$

Compounded 365 times: $A = 2000\left(1 + \frac{0.095}{365}\right)^{365(10)} = \$5170.78$

Compounded continuously: $A = 2000e^{0.095(10)} = \$5171.42$

77.

n	1	4	12	365	Continuous
P	\$2541.75	\$2498.00	\$2487.98	\$2483.09	\$2482.93

Compounded 1 time:

$$5000 = P\left(1 + \frac{0.07}{1}\right)^{1(10)}$$
$$\frac{5000}{(1.07)^{10}} = P$$
$$\$2541.75 = P$$

Compounded 4 times:

$$5000 = \left(1 + \frac{0.07}{4}\right)^{4(10)}$$
$$\frac{5000}{(1.0175)^{40}} = P$$
$$\$2498.00 = P$$

Compounded 12 times:

$$5000 = P\left(1 + \frac{0.07}{12}\right)^{12(10)}$$
$$\frac{5000}{\left(1.0058\overline{3}\right)^{120}} = P$$
$$\$2487.98 = P$$

Compounded 365 times:

$$5000 = P\left(1 + \frac{0.07}{365}\right)^{365(10)}$$
$$\frac{5000}{(1.0001918)^{3650}} = P$$
$$\$2483.09 = P$$

Compounded continuously:

$$5000 = Pe^{0.07(10)}$$
$$\frac{5000}{e^{0.7}} = P$$
$$\$2482.93 = P$$

79.

n	1	4	12	365	Continuous
P	\$18,429.30	\$15,830.43	\$15,272.04	\$15,004.64	\$14,995.58

Compounded 1 time:

$$1{,}000{,}000 = P\left(1 + \frac{0.105}{1}\right)^{1(40)}$$
$$\frac{1{,}000{,}000}{(1.105)^{40}} = P$$
$$\$18{,}429.30 = P$$

Compounded 4 times:

$$1{,}000{,}000 = P\left(1 + \frac{0.105}{4}\right)^{4(40)}$$
$$\frac{1{,}000{,}000}{(1.02625)^{160}} = P$$
$$\$15{,}830.43 = P$$

Compounded 12 times:

$$1{,}000{,}000 = P\left(1 + \frac{0.105}{12}\right)^{12(40)}$$
$$\frac{1{,}000{,}000}{(1.00875)^{480}} = P$$
$$\$15{,}272.04 = P$$

Compounded 365 times:

$$1{,}000{,}000 = P\left(1 + \frac{0.105}{365}\right)^{365(40)}$$
$$\frac{1{,}000{,}000}{(1.002877)^{14{,}600}} = P$$
$$\$15{,}004.64 = P$$

Compounded continuously:

$$1{,}000{,}000 = Pe^{0.105(40)}$$
$$\frac{1{,}000{,}000}{e^{4.2}} = P$$
$$\$14{,}995.58 = P$$

81. $p = 25 - 0.4^{0.02x}$

(a) $p = 25 - 0.4e^{0.02(100)} = 25 - 0.4e^2 \approx \$22.04$

(b) $p = 25 - 0.4e^{0.02(125)} = 25 - 0.4e^{2.5} \approx \$20.13$

83. $v(t) = 64{,}000(2)^{t/15}$

(a) $v(5) = 64{,}000(2)^{5/15} = 64{,}000(2)^{1/3} \approx \$80{,}634.95$

(b) $v(20) = 64{,}000(2)^{20/15}$

$= 64{,}000(2)^{4/3}$

$\approx \$161{,}269.89$

85. $V(t) = 16{,}000\left(\frac{3}{4}\right)^t$

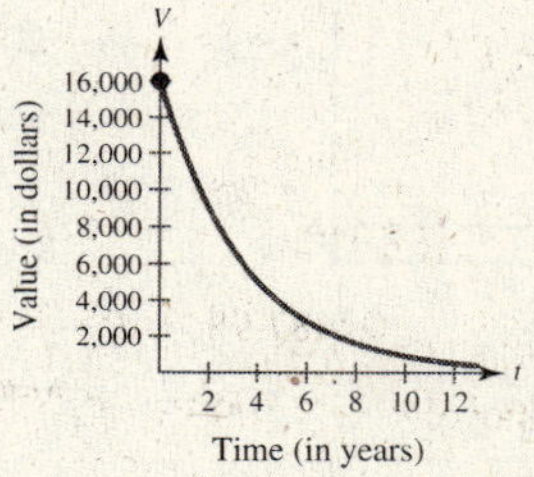

$V(2) = 16{,}000\left(\frac{3}{4}\right)^2 = \$9000$

$V(4) = 16{,}000\left(\frac{3}{4}\right)^4 = \$5062.50$

87. $f(x) = 1024\left(\frac{1}{2}\right)^x$

$f(8) = 1024\left(\frac{1}{2}\right)^8 = 4$ golfers

89. $h = 1950 + 50e^{-0.4433t} - 22t$

(a) *Keystrokes:* [Y=] 1950 [+] 50 [e^x] [(] [–] 0.4433 [X,T,θ] [)] [–] 22 [X,T,θ] [GRAPH]

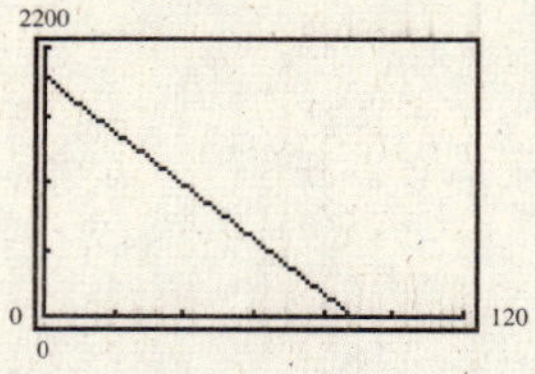

(b)

Time (in seconds)	0	10	20	30	40	50	60	70	80	90
Height (in feet)	2000	1731	1510	1290	1070	850	630	410	190	0

(c) The height changes the most within the first 10 seconds because after the parachute is released, a few seconds pass before the descent is constant.

91. (a) *Graph model:*

Plot data:

Keystrokes: [STAT] [EDIT 1]

Enter each x entry in L 1 followed by [ENTER].

Enter each y entry in L 2 followed by [ENTER].

[STAT PLOT] [ENTER] [ENTER] [ZOOM 9] or set window.

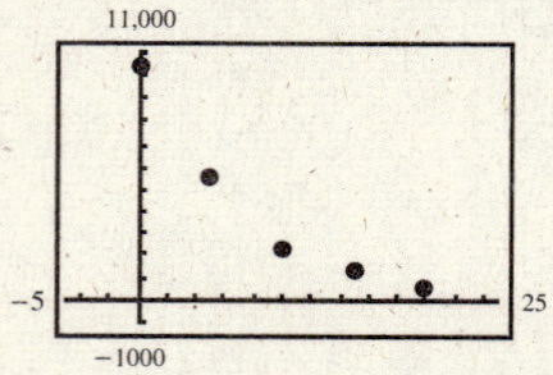

(b) *Keystrokes:* [Y=] 10,958 [e^x] [(–)] .15 [X,T,θ]

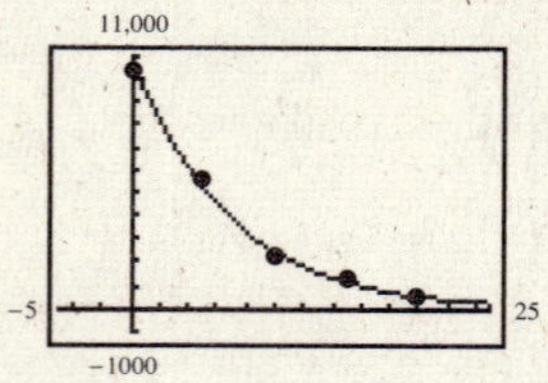

The model fits the data.

(c)

h	0	5	10	15	20
P	10,332	5583	2376	1240	517
Approx.	10,958	5176	2445	1155	546

(d) At an altitude of 8 kilometers, P is approximately 3300 kilograms per square meter. Use table.

Keystrokes: [TABLE] 8 [ENTER]

(e) If P is approximately 2000 kilograms per square meter, altitude is 11.3 kilometers. Graph $y_2 = 2000$ and find the intersection point.

93. (a)

x	1	10	100	1000	10,000
$\left(1+\frac{1}{x}\right)^x$	2	2.5937	2.7048	2.7169	2.7181

(b) *Keystrokes:* [Y=] [(] 1 [+] 1 [÷] [X,T,θ] [)] [^] [X,T,θ] [GRAPH]

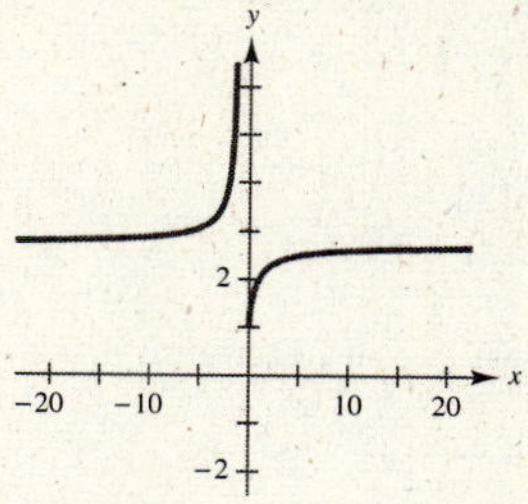

Yes, the graph is approaching a horizontal asymptote.

(c) As x gets larger and larger, $\left(1+\frac{1}{x}\right)^x$ approaches e.

95. By definition, the base of an exponential function must be positive and not equal to 1. If the base is 1, the function simplifies to the constant function $y = 1$.

97. False. e is an irrational number.

$\frac{271,801}{99,990}$ is rational because its equivalent decimal form is a repeating decimal.

99. When $k > 1$, the values of f will increase. When $0 < k < 1$, the values of f will decrease. When $k = 1$, the values of f remain constant.

101. $g(s) = \sqrt{s-4}$, Domain: $[4, \infty)$

$s - 4 \geq 0$

$s \geq 4$

103. $y^2 = x - 1$

y is not a function of x.

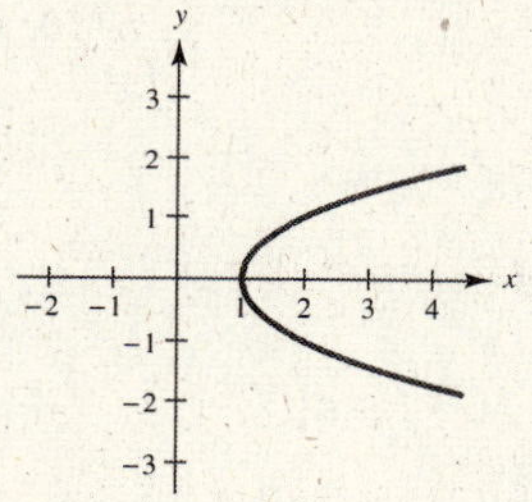

Section 9.2 Composite and Inverse Functions

1. $f(x) = 2x + 3,\ g(x) = x - 6$

(a) $(f \circ g)(x) = 2(x-6) + 3 = 2x - 12 + 3 = 2x - 9$

(b) $(g \circ f)(x) = (2x+3) - 6 = 2x - 3$

(c) $(f \circ g)(4) = 2(4) - 9 = 8 - 9 = -1$

(d) $(g \circ f)(7) = 2(7) - 3 = 14 - 3 = 11$

3. $f(x) = x^2 + 3,\ g(x) = x + 2$

(a) $(f \circ g)(x) = (x+2)^2 + 3$

$= x^2 + 4x + 4 + 3 = x^2 + 4x + 7$

(b) $(g \circ f)(x) = (x^2 + 3) + 2 = x^2 + 5$

(c) $(f \circ g)(2) = 2^2 + 4(2) + 7 = 4 + 8 + 7 = 19$

(d) $(g \circ f)(-3) = (-3)^2 + 5 = 9 + 5 = 14$

5. $f(x) = |x - 3|,\ g(x) = 3x$

(a) $(f \circ g)(x) = |3x - 3|$

(b) $(g \circ f)(x) = 3|x - 3|$

(c) $(f \circ g)(1) = |3 - 3| = 0$

(d) $(g \circ f)(2) = 3|2 - 3| = 3$

7. $f(x) = \sqrt{x-4},\ g(x) = x + 5$

(a) $(f \circ g)(x) = \sqrt{x+5-4} = \sqrt{x+1}$

(b) $(g \circ f)(x) = \sqrt{x-4} + 5$

(c) $(f \circ g)(3) = \sqrt{3+1} = 2$

(d) $(g \circ f)(8) = \sqrt{8-4} + 5 = 2 + 5 = 7$

9. $f(x) = \dfrac{1}{x-3},\ g(x) = \dfrac{2}{x^2}$

(a) $(f \circ g)(x) = \dfrac{1}{\dfrac{2}{x^2} - 3} \cdot \dfrac{x^2}{x^2} = \dfrac{x^2}{2 - 3x^2},\quad x \neq 0$

(b) $(g \circ f)(x) = \dfrac{2}{\left(\dfrac{1}{x-3}\right)^2} = 2(x-3)^2,\quad x \neq 3$

(c) $(f \circ g)(-1) = \dfrac{(-1)^2}{2 - 3(-1)^2} = \dfrac{1}{2-3} = \dfrac{1}{-1} = -1$

(d) $(g \circ f)(2) = 2(2-3)^2 = 2(-1)^2 = 2$

11. (a) $f(1) = -1$

(b) $g(-1) = -2$

(c) $(g \circ f)(1) = g[f(1)] = g[-1] = -2$

13. (a) $(f \circ g)(-3) = f[g(-3)] = f[1] = -1$

(b) $(g \circ f)(-2) = g[f(-2)] = g[3] = 1$

15. (a) $f(2) = 5$

(b) $g(10) = 1$

(c) $(g \circ f)(1) = g[f(1)] = g(2) = 3$

17. (a) $(g \circ f)(4) = g[f(4)] = g[17] = 0$

(b) $(f \circ g)(2) = f[g(2)] = f[3] = 10$

19. $f(x) = 3x + 4,\quad g(x) = x - 7$

(a) $f \circ g = 3(x-7) + 4 = 3x - 21 + 4 = 3x - 17$

Domain: $(-\infty, \infty)$

(b) $g \circ f = (3x + 4) - 7 = 3x - 3$

Domain: $(-\infty, \infty)$

21. $f(x) = \sqrt{x+2},\ g(x) = x - 4$

(a) $(f \circ g)(x) = \sqrt{(x-4)+2} = \sqrt{x-2}$

Domain: $x - 2 \geq 0$

$x \geq 2,\quad [2, \infty)$

(b) $(g \circ f)(x) = \sqrt{x+2} - 4$

Domain: $x + 2 \geq 0$

$x \geq -2,\quad [-2, \infty)$

23. $f(x) = x^2 + 3,\ g(x) = \sqrt{x-1}$

(a) $f \circ g = \left(\sqrt{x-1}\right)^2 + 3 = x - 1 + 3 = x + 2$

Domain: $[1, \infty)$

(b) $g \circ f = \sqrt{(x^2+3) - 1} = \sqrt{x^2 + 2}$

Domain: $(-\infty, \infty)$

25. $f(x) = \dfrac{x}{x+5},\ g(x) = \sqrt{x-1}$

(a) $f \circ g = \dfrac{\sqrt{x-1}}{\sqrt{x-1}+5}$, Domain: $[1, \infty)$

(b) $g \circ f = \sqrt{\dfrac{x}{x+5} - 1}$

$= \sqrt{-\dfrac{5}{x+5}}$, Domain: $(-\infty, -5)$

27. $f(x) = x^3 - 1$

Keystrokes: [Y=] [X,T,θ] [^] 3 [−] 1 [GRAPH]

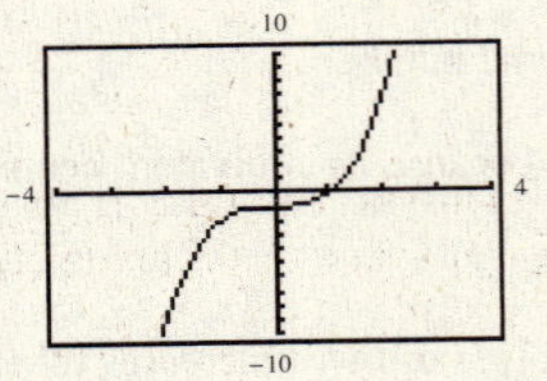

One-to-one

29. $f(t) = \sqrt[3]{5-t}$

Keystrokes:

[Y=] [MATH] 4 [(] 5 [−] [X,T,θ] [)] [GRAPH]

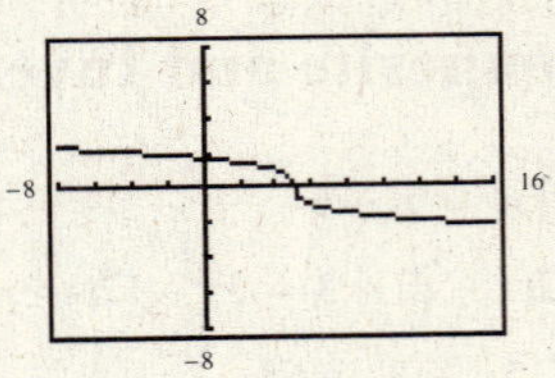

One-to-one

31. $g(x) = (x - 3)^4$

Keystrokes: [Y=] [(] [X,T,θ] [−] 3 [)] [^] 4 [GRAPH]

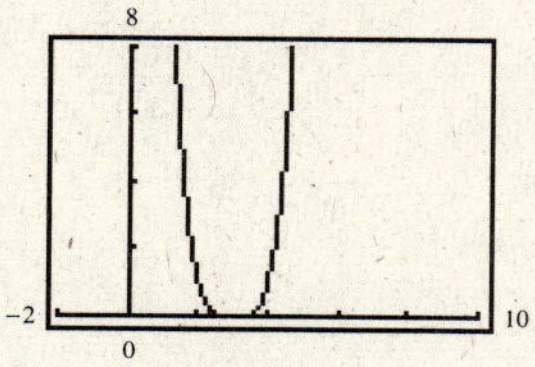

Not one-to-one

33. $h(t) = \frac{5}{t}$

Keystrokes: [Y=] 5 [÷] [X,T,θ] [GRAPH]

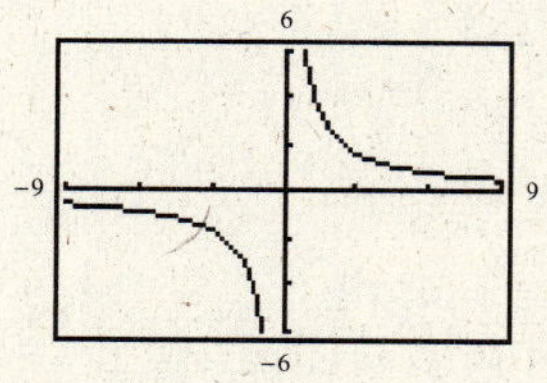

One-to-one

35. $f(x) = x^2 - 2$

No, it does not have an inverse because it is possible to find a horizontal line that intersects the graph of f at more than one point.

37. $f(x) = x^2, x \geq 0$

Yes, it does have an inverse because no horizontal line intersects the graph of f at more than one point.

39. $g(x) = \sqrt{25 - x^2}$

No, it does not have an inverse because it is possible to find a horizontal line that intersects the graph of g at more than one point.

41. $f(x) = -6x, g(x) = -\frac{1}{6}x$

$$f(g(x)) = f\left(-\tfrac{1}{6}x\right) = -6\left(-\tfrac{1}{6}x\right) = x$$
$$g(f(x)) = g(-6x) = -\tfrac{1}{6}(-6x) = x$$

43. $f(x) = 1 - 2x, g(x) = \frac{1}{2}(1 - x)$

$$\begin{aligned} f(g(x)) &= f\left[\tfrac{1}{2}(1 - x)\right] \\ &= 1 - 2\left[\tfrac{1}{2}(1 - x)\right] \\ &= 1 - (1 - x) = 1 - 1 + x = x \\ g(f(x)) &= g(1 - 2x) \\ &= \tfrac{1}{2}[1 - (1 - 2x)] \\ &= \tfrac{1}{2}[1 - 1 + 2x] = \tfrac{1}{2}[2x] = x \end{aligned}$$

45. $f(x) = \sqrt[3]{x + 1}, g(x) = x^3 - 1$

$$\begin{aligned} f(g(x)) &= f(x^3 - 1) = \sqrt[3]{x^3 - 1 + 1} = \sqrt[3]{x^3} = x \\ g(f(x)) &= g\left(\sqrt[3]{x + 1}\right) \\ &= \left(\sqrt[3]{x + 1}\right)^3 - 1 = x + 1 - 1 = x \end{aligned}$$

47. $f(x) = \frac{1}{x}, g(x) = \frac{1}{x}$

$$f(g(x)) = f\left(\frac{1}{x}\right) = \frac{1}{\frac{1}{x}} = x$$
$$g(f(x)) = g\left(\frac{1}{x}\right) = \frac{1}{\frac{1}{x}} = x$$

49. $f(x) = 5x$

$$f^{-1}(x) = \frac{x}{5}$$
$$f(f^{-1}(x)) = f\left(\frac{x}{5}\right) = 5\left(\frac{x}{5}\right) = x$$
$$f^{-1}(f(x)) = f^{-1}(5x) = \frac{5x}{5} = x$$

51. $f(x) = -\frac{2}{5}x$

$$f^{-1}(x) = -\tfrac{5}{2}x$$
$$f(f^{-1}(x)) = f\left(-\tfrac{5}{2}x\right) = -\tfrac{2}{5}\left(-\tfrac{5}{2}x\right)$$
$$f^{-1}(f(x)) = f^{-1}\left(-\tfrac{2}{5}x\right) = -\tfrac{2}{5}\left(-\tfrac{5}{2}x\right)$$

53. $f(x) = x + 10$

$$f^{-1}(x) = x - 10$$
$$f(f^{-1}(x)) = f(x - 10) = x - 10 + 10 = x$$
$$f^{-1}(f(x)) = f^{-1}(x + 10) = x + 10 - 10 = x$$

55. $f(x) = 5 - x$

$$\begin{aligned} f^{-1}(x) &= 5 - x \\ f(f^{-1}(x)) &= f(5 - x) \\ &= 5 - (5 - x) = 5 - 5 + x = x \\ f^{-1}(f(x)) &= f^{-1}(5 - x) \\ &= 5 - (5 - x) = 5 - 5 + x = x \end{aligned}$$

57. $f(x) = x^9$

$$f^{-1}(x) = \sqrt[9]{x}$$
$$f(f^{-1}(x)) = f\left(\sqrt[9]{x}\right) = \left(\sqrt[9]{x}\right)^9 = x$$
$$f^{-1}(f(x)) = f^{-1}(x^9) = \sqrt[9]{x^9} = x$$

59. $f(x) = \sqrt[3]{x}$

$f^{-1}(x) = x^3$

$f(f^{-1}(x)) = f(x^3) = \sqrt[3]{x^3} = x$

$f^{-1}(f(x)) = f^{-1}(\sqrt[3]{x}) = (\sqrt[3]{x})^3 = x$

61. $g(x) = x + 25$

$y = x + 25$

$x = y + 25$

$x - 25 = y$

$g^{-1}(x) = x - 25$

63. $g(x) = 3 - 4x$

$y = 3 - 4x$

$x = 3 - 4y$

$x - 3 = -4y$

$\dfrac{x-3}{-4} = y$

$\dfrac{3-x}{4}$ or $\dfrac{x-3}{-4} = g^{-1}(x)$

65. $g(t) = \frac{1}{4}t + 2$

$y = \frac{1}{4}t + 2$

$t = \frac{1}{4}y + 2$

$t - 2 = \frac{1}{4}y$

$4(t - 2) = y$

$4t - 8 = g^{-1}(t)$

67. $g(x) = x^2 + 4$

$g(x)$ is not one-to-one so an inverse does not exist.

69. $h(x) = \sqrt{x}$

$y = \sqrt{x}$

$x = \sqrt{y}$

$x^2 = y$

$x^2 = h^{-1}(x), \quad x \geq 0$

71. $f(t) = t^3 - 1$

$y = t^3 - 1$

$t = y^3 - 1$

$t + 1 = y^3$

$\sqrt[3]{t+1} = y$

$\sqrt[3]{t+1} = f^{-1}(t)$

73. $f(x) = \sqrt{x+3}$

$y = \sqrt{x+3}$

$x = \sqrt{y+3}$

$x^2 = y + 3$

$x^2 - 3 = y$

$x^2 - 3 = f^{-1}(x), \quad x \geq 0$

75. (b)

76. (c)

77. (d)

78. (a)

79. $f(x) = x + 4, \quad f^{-1}(x) = x - 4$

(0, 4) (4, 0)

(−4, 0) (0, −4)

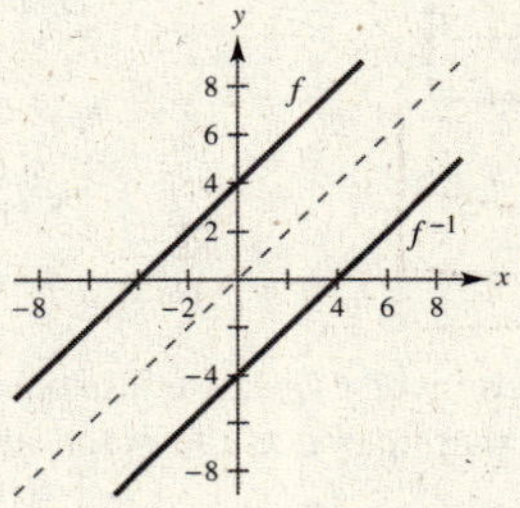

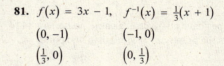

81. $f(x) = 3x - 1, \quad f^{-1}(x) = \frac{1}{3}(x + 1)$

(0, −1) (−1, 0)

$(\frac{1}{3}, 0)$ $(0, \frac{1}{3})$

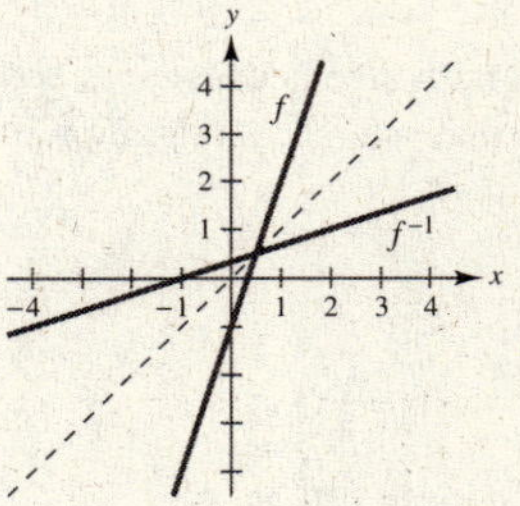

83. $f(x) = x^2 - 1, x \geq 0 \quad f^{-1}(x) = \sqrt{x+1}$

(0, −1) (−1, 0)

(1, 0) (0, 1)

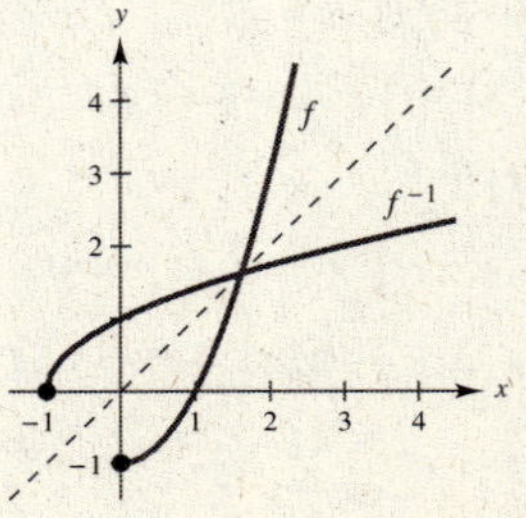

85. $f(x) = \frac{1}{3}x,\ g(x) = 3x$

Keystrokes:

y_1: [Y=] [(] 1 [÷] 3 [)] [X,T,θ] [ENTER]

y_2: 3 [X,T,θ] [GRAPH]

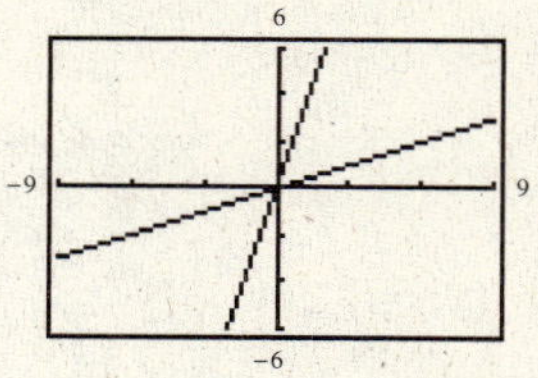

87. $f(x) = \sqrt{x - 4},\ g(x) = x^2 + 4, \quad x \ge 0$

Keystrokes:

y_1: [Y=] [√] [X,T,θ] [−] 4 [)] [ENTER]

y_2: [X,T,θ] [x²] [+] 4 [÷] [(] [X,T,θ] [TEST] 4 0 [)] [GRAPH]

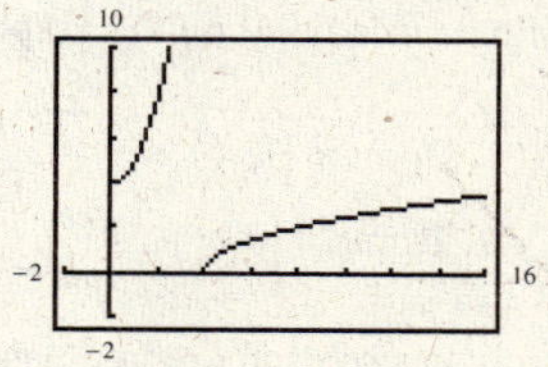

89. $f(x) = \frac{1}{8}x^3,\ g(x) = 2\sqrt[3]{x}$

Keystrokes:

y_1: [Y=] [(] 1 [÷] 8 [)] [X,T,θ] [MATH] 3 [ENTER]

y_2: 2 [MATH] 4 [X,T,θ] [GRAPH]

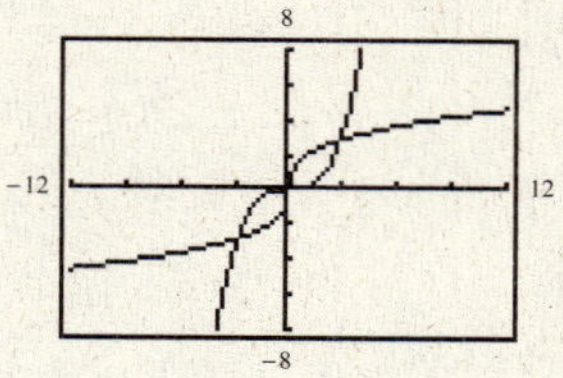

91. $f(x) = |3 - x|, x \ge 3,\ g(x) = 3 + x, x \ge 0$

Keystrokes: y_1: [Y=] [ABS] [(] 3 [−] [X,T,θ] [)] [÷] [(] [X,T,θ] [TEST] 4 3 [)] [ENTER]

y_2: 3 [+] [X,T,θ] [÷] [(] [X,T,θ] [TEST] 4 0 [)] [GRAPH]

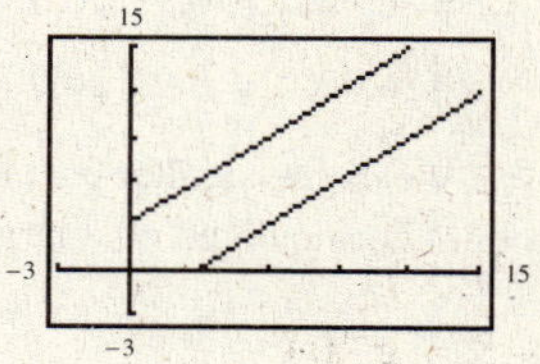

93.

$$
\begin{aligned}
f(x) &= (x - 2)^2, \quad x \ge 2 \\
y &= (x - 2)^2 \\
x &= (y - 2)^2 \\
\sqrt{x} &= y - 2 \\
\sqrt{x} + 2 &= y \\
\sqrt{x} + 2 &= f^{-1}(x), \quad x \ge 0
\end{aligned}
$$

95.

$$
\begin{aligned}
f(x) &= |x| + 1, \quad x \ge 0 \\
y &= |x| + 1 \\
x &= |y| + 1 \\
x - 1 &= |y| \\
x - 1 &= y \\
x - 1 &= f^{-1}(x), \quad x \ge 1
\end{aligned}
$$

97.

$$
\begin{aligned}
f(x) &= 3 - 2x \\
y &= 3 - 2x \\
x &= 3 - 2y \\
2y &= 3 - x \\
y &= \frac{3 - x}{2} \\
f^{-1}(x) &= \frac{3 - x}{2}
\end{aligned}
$$

99. $f^{-1}(a) = c$

101. $f^{-1}\left(f^{-1}(c)\right) = a$

103. $s(x) = x - 300{,}000, \quad p(s) = 0.03s$

$$
\begin{aligned}
p(s(x)) &= p(x - 300{,}000) \\
&= 0.03(x - 300{,}000) \\
&= 0.03x - 9000, \quad x > 300{,}000
\end{aligned}
$$

This function represents the bonus earned for sales over $300,000.

105. $r(t) = 0.6t, A(r) = \pi r^2$

$A(r(t)) = A(0.6t) = \pi(0.6t)^2 = \pi(0.36t^2) = 0.36\pi t^2$

Input: time

Output: area

$$\begin{aligned} A(r(3)) &= A[0.6(3)] \\ &= A(1.8) \\ &= \pi(1.8)^2 \\ &= \pi(3.24) \\ &\approx 10.2 \text{ square feet} \end{aligned}$$

107. (a) $R = p - 2000$

(b) $S = p - 0.05p$

$S = 0.95p$

(c) $(R \circ S)(p) = R[S(p)]$

$= R(0.95p) = 0.95p - 2000$

5% discount followed by the \$2000 rebate.

$(S \circ R)(p) = S[R(p)]$

$= S(p - 2000) = 0.95(p - 2000)$

5% discount after the price is reduced by the rebate.

(d) $(R \circ S)(26{,}000) = 0.95(26{,}000) - 2000 = \$22{,}700$

$(S \circ R)(26{,}000) = 0.95(26{,}000 - 2000) = \$22{,}800$

$R \circ S$ yields the smaller cost because the dealer discount is calculated on a larger base.

109. (a)

$$\begin{aligned} y &= 9 + 0.65x \\ x &= 9 + 0.65y \\ \frac{x - 9}{0.65} &= y \\ \frac{100}{65}(x - 9) &= y \\ \frac{20}{13}(x - 9) &= y \end{aligned}$$

(b) x: hourly wage

y: number of units produced

(c) Domain: $x \geq 9$

(d) $y = \frac{20}{13}(14.20 - 9)$

$y = \frac{20}{13}(5.20)$

$y = 8$ units

111. $f(x) = 4x, g(x) = x + 6$

(a) $(f \circ g)(x) = f(x + 6) = 4(x + 6) = 4x + 24$

(b) $(f \circ g)^{-1}(x) = \frac{x - 24}{4} = \frac{1}{4}x - 6$

$$\begin{aligned} y &= 4x + 24 \\ x &= 4y + 24 \\ \frac{x - 24}{4} &= y \end{aligned}$$

(c) $f^{-1}(x) = \frac{1}{4}x, g^{-1}(x) = x - 6$

(d) $(g^{-1} \circ f^{-1})(x) = g^{-1}\left(\frac{1}{4}x\right) = \frac{1}{4}x - 6$

The result is the same.

(e) $(f \circ g)^{-1}(x) = (g^{-1} \circ f^{-1})(x)$

113. True. The x-coordinate of a point on the graph of f becomes the y-coordinate of a point on the graph of f^{-1}.

115. False: $f(x) = \sqrt{x - 1}$, Domain $[1, \infty)$

$f^{-1}(x) = x^2 + 1$, Domain $[0, \infty)$

117. Interchange the coordinates of each ordered pair. The inverse of the function defined by $\{(3, 6), (5, -2)\}$ is $\{(6, 3), (-2, 5)\}$.

119. A function can have only one input for every output so the inverse will have one output for every input and is therefore a function.

121. $h(x) = -x^2$

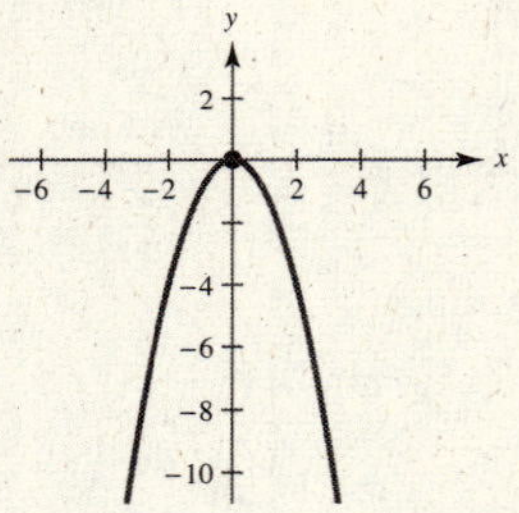

Reflection in the x-axis

123. $k(x) = (x + 3)^2 - 5$

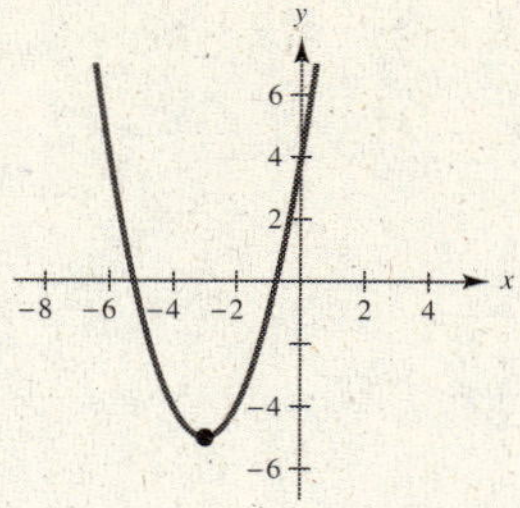

Horizontal shift 3 units left

Vertical shift 5 units down

125. $16 - (y + 2)^2 = [4 - (y + 2)][4 + (y + 2)]$
$= (4 - y - 2)(4 + y + 2)$
$= (2 - y)(6 + y)$

127. $5 - u + 5u^2 - u^3 = 1(5 - u) + u^2(5 - u)$
$= (5 - u)(1 + u^2)$
$= -1(u - 5)(u^2 + 1)$
$= -(u^2 + 1)(u - 5)$

129. $3x - 4y = 6$

Intercepts:

$3(0) - 4y = 6$

$y = -\frac{3}{2}$

$3x - 4(0) = 6$

$3x = 6$

$x = 2$

131. $y = -(x - 2)^2 + 1$

Intercepts:

$y = -(0 - 2)^2 + 1 = -3, (0, -3)$

$0 = -(x - 2)^2 + 1$

$(x - 2)^2 = 1$

$x - 2 = \pm 1$

$x = 2 \pm 1$

$x = 3, 1$

$(3, 0), (1, 0)$

Vertex:

$(2, 1)$

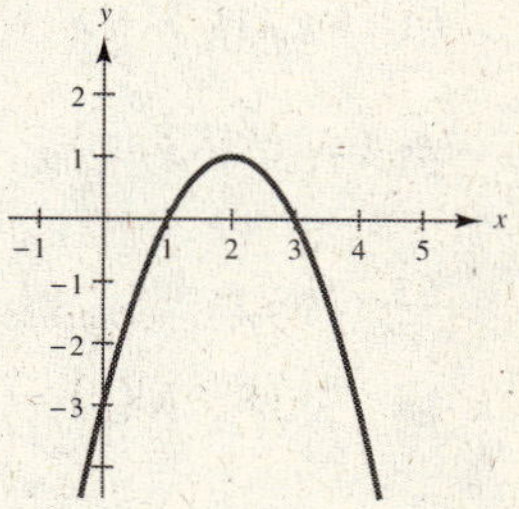

Section 9.3 Logarithmic Functions

1. $\log_7 49 = 2$
$7^2 = 49$

3. $\log_2 \frac{1}{32} = -5$
$2^{-5} = \frac{1}{32}$

5. $\log_3 \frac{1}{243} = -5$
$3^{-5} = \frac{1}{243}$

7. $\log_{36} 6 = \frac{1}{2}$
$36^{1/2} = 6$

9. $\log_8 4 = \frac{2}{3}$
$8^{2/3} = 4$

11. $\log_2 5.278 \approx 2.4$
$2^{2.4} \approx 5.278$

13. $6^2 = 36$
$\log_6 36 = 2$

15. $5^{-3} = \frac{1}{125}$
$\log_5 \frac{1}{125} = -3$

17. $8^{2/3} = 4$
$\log_8 4 = \frac{2}{3}$

19. $25^{-1/2} = \frac{1}{5}$
$\log_{25} \frac{1}{5} = -\frac{1}{2}$

21. $4^0 = 1$
$\log_4 1 = 0$

23. $5^{1.4} \approx 9.518$
$\log_5 9.518 \approx 1.4$

25. $\log_2 8 = 3$ because $2^3 = 8$.

27. $\log_{10} 1000 = 3$ because $10^3 = 1000$.

29. $\log_2 \frac{1}{16} = -4$ because $2^{-4} = \frac{1}{16}$.

31. $\log_4 \frac{1}{64} = -3$ because $4^{-3} = \frac{1}{64}$.

33. $\log_{10} \frac{1}{10{,}000} = -4$ because $10^{-4} = \frac{1}{10{,}000}$.

35. $\log_2(-3)$ is not possible because there is no power to which 2 can be raised to obtain −3.

37. $\log_4 1 = 0$ because $4^0 = 1$.

39. $\log_5(-6)$ is not possible because there is no power to which 5 can be raised to obtain –6.

41. $\log_9 3 = \frac{1}{2}$ because $9^{1/2} = 3$.

43. $\log_{16} 8 = \frac{3}{4}$ because $16^{3/4} = 8$.

45. $\log_7 7^4 = 4$ because $7^4 = 7^4$.

47. $\log_{10} 42 \approx 1.6232$

49. $\log_{10} 0.023 \approx -1.6383$

51. $\log_{10}\left(\sqrt{5} + 3\right) \approx 0.7190$

53. $f(x) = 4 + \log_3 x$ matches graph (c).

55. $f(x) = \log_3(-x)$ matches graph (a).

57. $f(x) = 3^x$; $g(x) = \log_3 x$

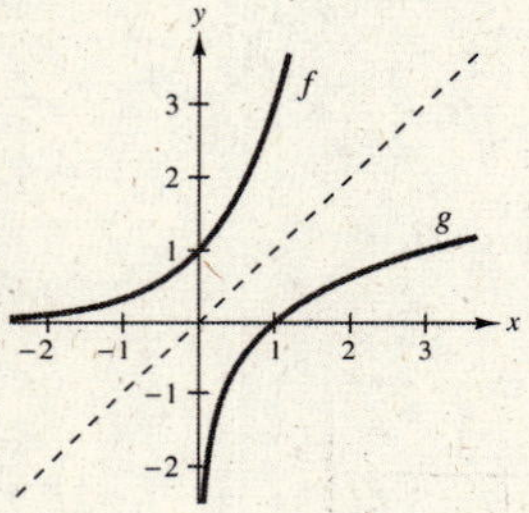

59. $f(x) = 6^x$; $g(x) = \log_6 x$

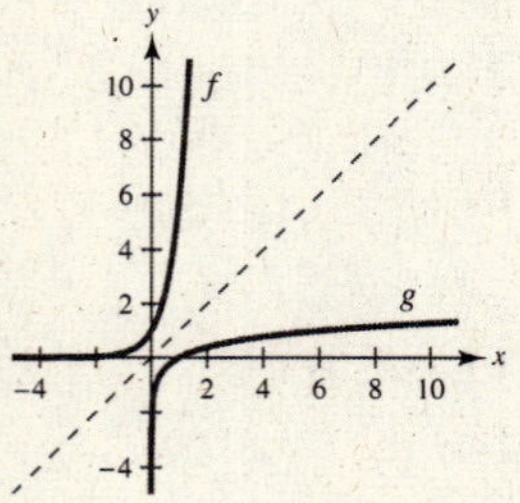

61. $h(x) = 3 + \log_2 x$

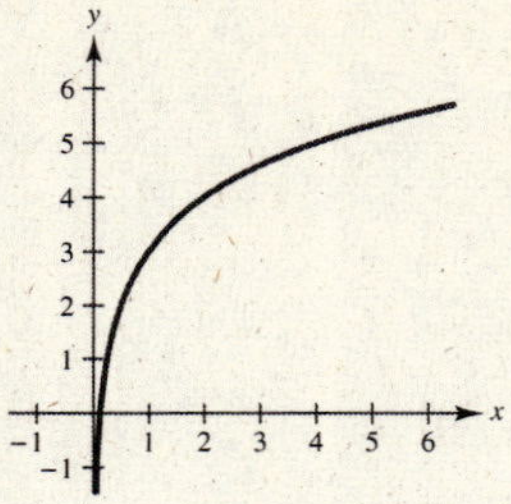

Vertical shift 3 units up

63. $h(x) = \log_2(x - 2)$

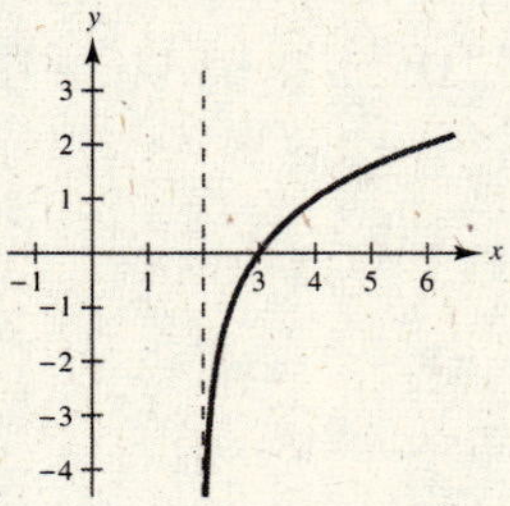

Horizontal shift 2 units right

65. $h(x) = \log_2(-x)$

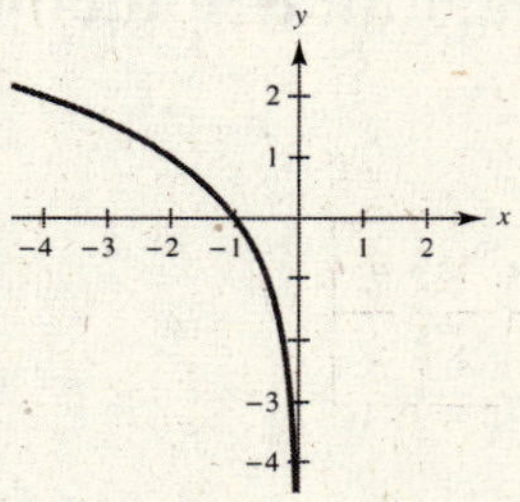

Reflection in the y-axis

67. $f(x) = \log_5 x$

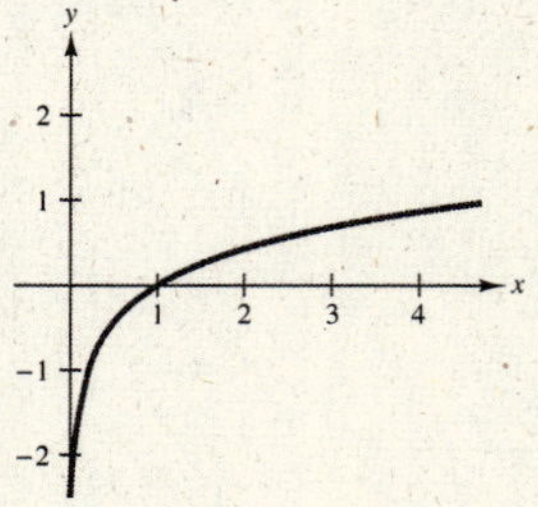

Vertical asymptote: $x = 0$

Table of values:

x	1	5
y	0	1

69. $g(t) = -\log_9 t$

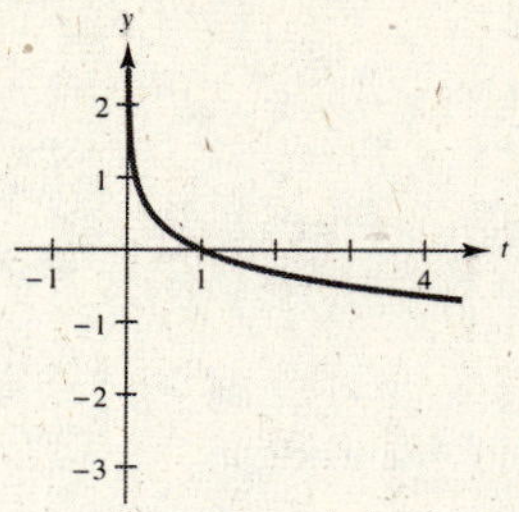

Vertical asymptote: $t = 0$

Table of values:

t	1	9
y	0	−1

71. $f(x) = 2 + \log_4 x$

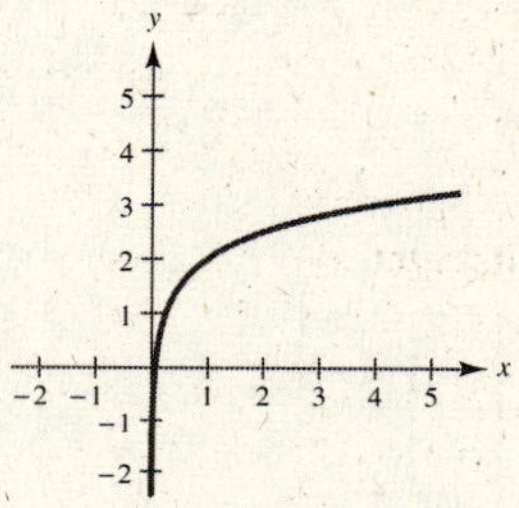

Vertical asymptote: $x = 0$

Table of values:

x	1	4
y	2	3

73. $g(x) = \log_2(x - 3)$

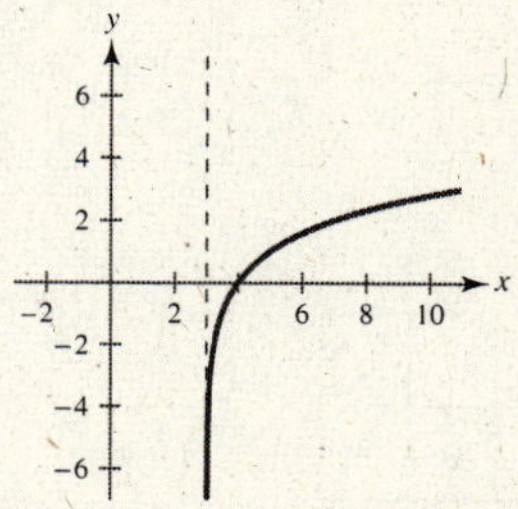

Vertical asymptote: $x = 3$

Table of values:

x	4	7
y	0	2

75. $f(x) = \log_{10}(10x)$

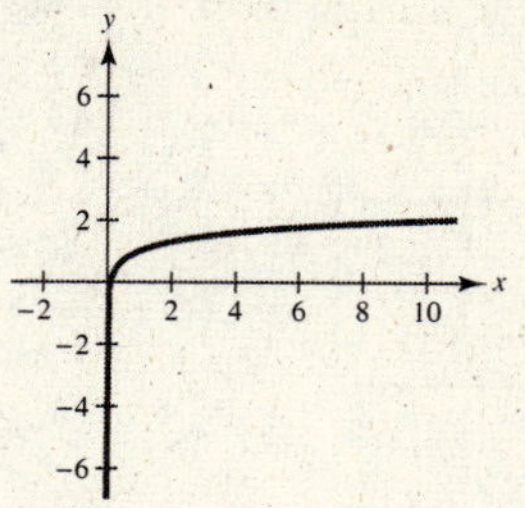

Vertical asymptote: $x = 0$

Table of values:

x	1	10
y	1	2

77. $f(x) = \log_4 x$

Domain: $(0, \infty)$

Vertical asymptote: $x = 0$

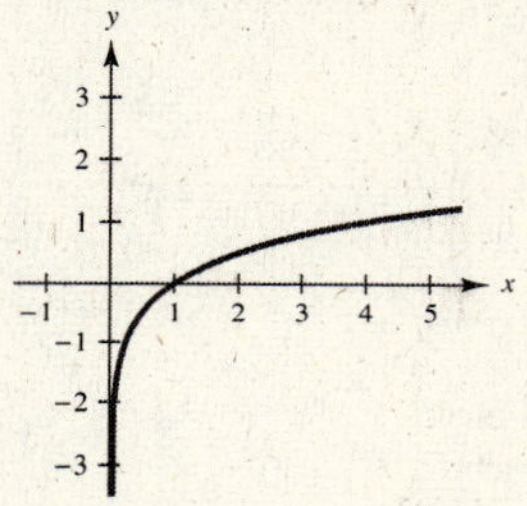

Table of values:

x	1	4
y	0	1

79. $h(x) = \log_5(x - 4)$

Domain: $(4, \infty)$

Vertical asymptote: $x = 4$

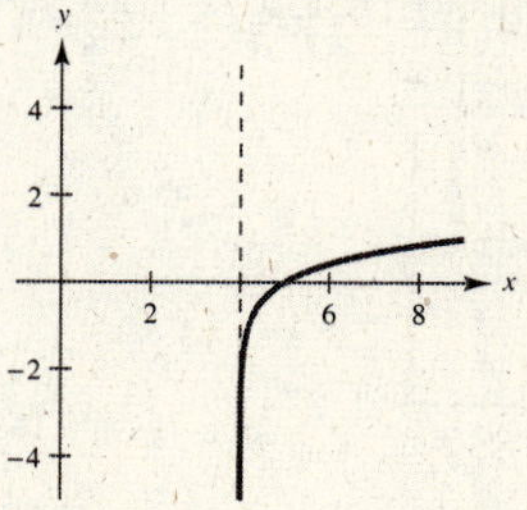

Table of values:

x	5	9
y	0	1

81. $y = -\log_3 x + 2$

Domain: $(0, \infty)$

Vertical asymptote: $x = 0$

Table of values:

x	1	3
y	2	1

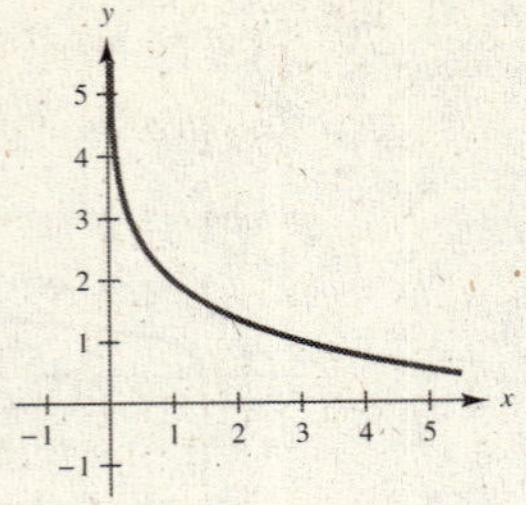

83. $y = 5 \log_{10} x$

Keystrokes: [Y=] 5 [LOG] [X,T,θ] [GRAPH]

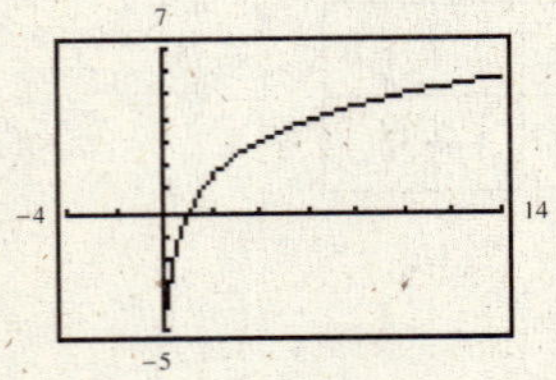

Domain: $(0, \infty)$

Vertical asymptote: $x = 0$

85. $y = -3 + 5 \log_{10} x$

Keystrokes: [Y=] [(−)] 3 [+] 5 [LOG] [X,T,θ] [GRAPH]

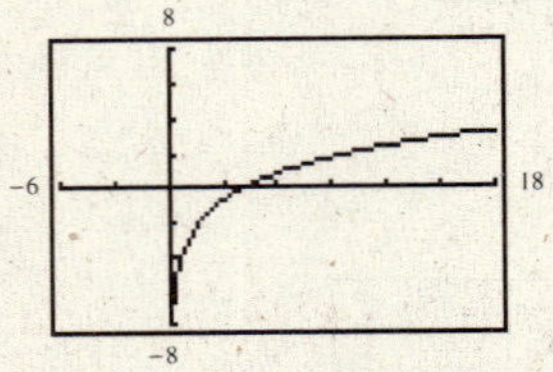

Domain: $(0, \infty)$

Vertical asymptote: $x = 0$

87. $y = \log_{10}\left(\frac{x}{5}\right)$

Keystrokes: [Y=] [(] [LOG] [X,T,θ] [÷] 5 [)] [GRAPH]

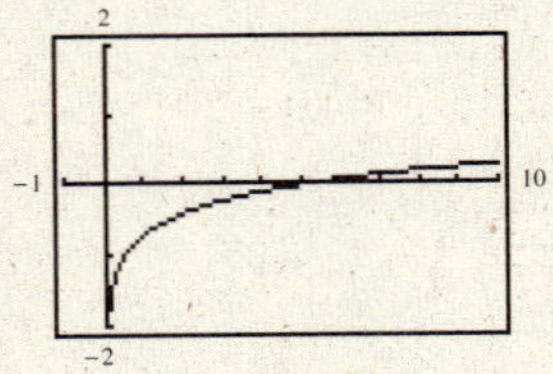

Domain: $(0, \infty)$

Vertical asymptote: $x = 0$

89. $\ln 38 \approx 3.6376$

91. $\ln 0.15 \approx -1.8971$

93. $\ln\left(\frac{3 - \sqrt{2}}{5}\right) \approx -1.1484$

95. $f(x) = \ln(x + 1)$

(b) Basic graph shifted 1 unit left

97. $f(x) = \ln\left(x - \frac{3}{2}\right)$

(d) Basic graph shifted $\frac{3}{2}$ units right

99. $f(x) = -\ln x$

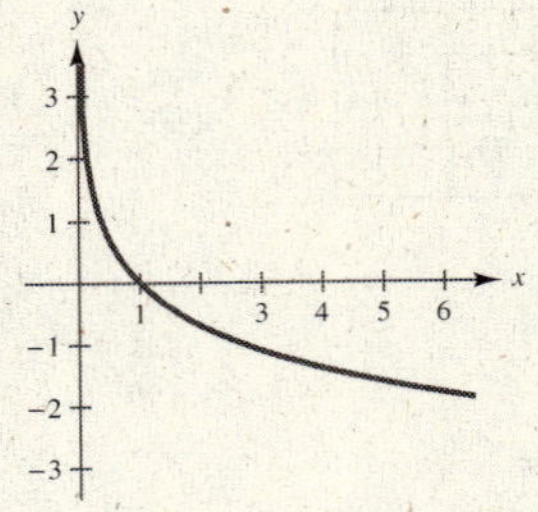

Vertical asymptote: $x = 0$

Table of values:

x	1	e
y	0	−1

101. $f(x) = 3 \ln x$

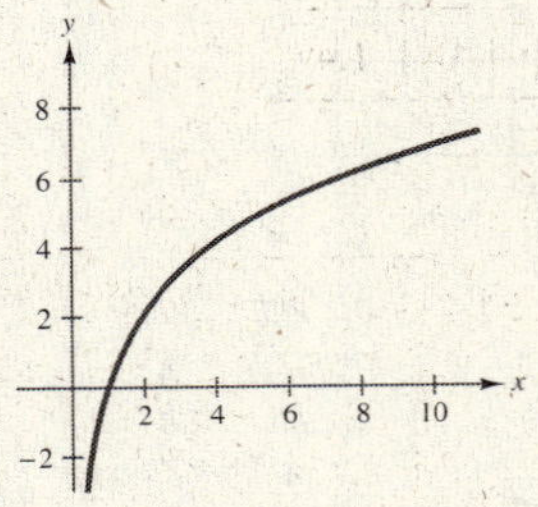

Vertical asymptote: $x = 0$

Table of values:

x	1	e
y	0	3

103. $f(x) = 3 + \ln x$

Vertical asymptote: $x = 0$

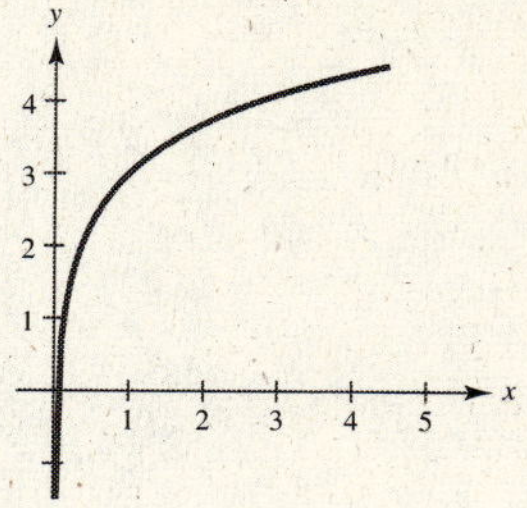

Table of values:

x	1	e
y	3	4

105. $g(t) = 2\ln(t - 4)$

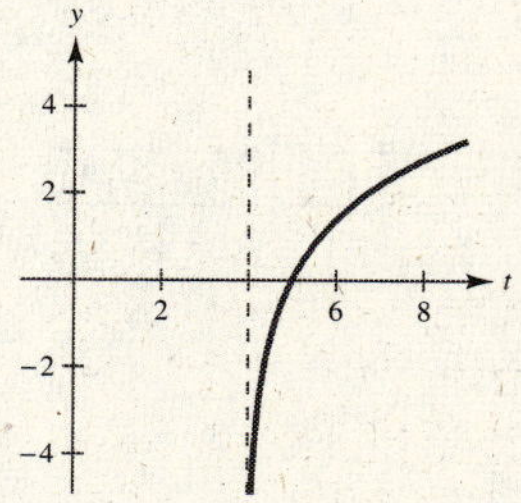

Vertical asymptote: $x = 4$

Table of values:

x	5	6
y	0	1.4

107. $g(x) = -\ln(x + 1)$

Keystrokes: [Y=] [(−)] [LN] [(] [X,T,θ] [+] 1 [)] [GRAPH]

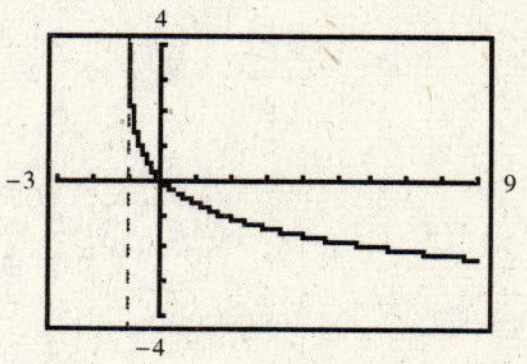

Domain: $(-1, \infty)$

Vertical asymptote: $x = -1$

109. $f(t) = 7 + 3\ln t$

Keystrokes: [Y=] 7 [+] 3 [LN] [X,T,θ] [GRAPH]

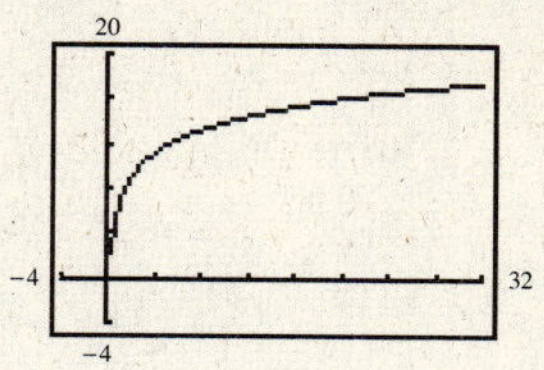

Domain: $(0, \infty)$

Vertical asymptote: $t = 0$

111. $\log_9 36 = \dfrac{\log 36}{\log 9} \approx 1.6309$

$= \dfrac{\ln 36}{\ln 9} \approx 1.6309$

113. $\log_5 14 = \dfrac{\log 14}{\log 5} \approx 1.6397$

$= \dfrac{\ln 14}{\ln 5} \approx 1.6397$

115. $\log_2 0.72 = \dfrac{\log 0.72}{\log 2} \approx -0.4739$

$= \dfrac{\ln 0.72}{\ln 2} \approx -0.4739$

117. $\log_{15} 1250 = \dfrac{\log 1250}{\log 15} \approx 2.6332$

$= \dfrac{\ln 1250}{\ln 15} \approx 2.6332$

119. $\log_{1/4} 16 = \dfrac{\log 16}{\log 1/4} = -2$

$= \dfrac{\ln 16}{\ln 1/4} = -2$

121. $\log_4 \sqrt{42} = \dfrac{\log \sqrt{42}}{\log 4} \approx 1.3481$

$= \dfrac{\ln \sqrt{42}}{\ln 4} \approx 1.3481$

123. $\log_2(1 + e) = \dfrac{\log(1 + e)}{\log 2} \approx 1.8946$

$= \dfrac{\ln(1 + e)}{\ln 2} \approx 1.8946$

125. $h = 116\log_{10}(55 + 40) - 176$

$= 116\log_{10}(95) - 176$

≈ 53.4 inches

127. r of 0.07: $t = \dfrac{\ln 2}{0.07} \approx 9.9021$

r of 0.08: $t = \dfrac{\ln 2}{0.08} \approx 8.6643$

r of 0.09: $t = \dfrac{\ln 2}{0.09} \approx 7.7016$

r of 0.10: $t = \dfrac{\ln 2}{0.10} \approx 6.9315$

r of 0.11: $t = \dfrac{\ln 2}{0.11} \approx 6.3013$

r of 0.12: $t = \dfrac{\ln 2}{0.12} \approx 5.7762$

r	0.07	0.08	0.09	0.10	0.11	0.12
t	9.9	8.7	7.7	6.9	6.3	5.8

129. (a) *Keystrokes:*

[Y=] 10 [LN] [(] [(] 10 [+] [√] [(] 100 [−] [X,T,θ] [x²] [)] [)] [÷] [X,T,θ] [)] [−] [√] [(] 100 [−] [X,T,θ] [x²] [)] [GRAPH]

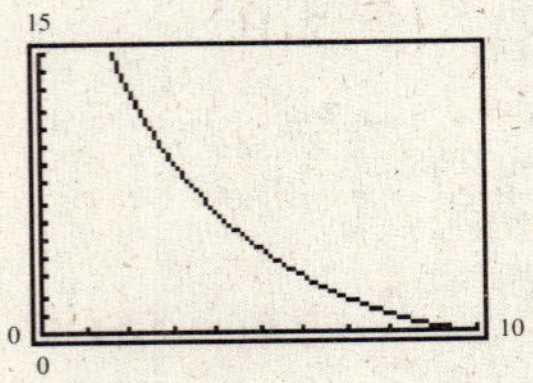

Domain: $(0, 10]$

(b) Vertical asymptote: $x = 0$

(c) $y = 13.126$ when $x = 2$. Trace to $x = 2$.

131. Domain = positive real numbers, $(0, \infty)$

133. If $1000 \le x \le 10{,}000$, then $f(x) = \log_{10} x$ lies $3 \le f(x) \le 4$.

135. When $f(x)$ increases by 1 unit, x increases by a factor of 10.

137. Logarithmic functions with base 10 are common logarithms. Logarithmic functions with base e are natural logarithms.

139. A vertical shift or reflection in the x-axis of a logarithmic graph does not affect the domain or range. A horizontal shift or reflection in the y-axis of a logarithmic graph affects the domain, but the range stays the same.

141. $(-m^6n)(m^4n^3) = -(m^{6+4}n^{1+3}) = -m^{10}n^4$

143. $\dfrac{36x^4y}{8xy^3} = \dfrac{36}{8} \cdot x^{4-1} \cdot y^{1-3} = \dfrac{9x^3}{2y^2}, \quad x \ne 0$

145.
$$\begin{aligned} 25\sqrt{3x} - 3\sqrt{12x} &= 25\sqrt{3x} - 3\sqrt{4 \cdot 3x} \\ &= 25\sqrt{3x} - 6\sqrt{3x} \\ &= (25 - 6)\sqrt{3x} \\ &= 19\sqrt{3x} \end{aligned}$$

147.
$$\begin{aligned} \sqrt{u}\left(\sqrt{20} - \sqrt{5}\right) &= \sqrt{u}\left(\sqrt{4 \cdot 5} - \sqrt{5}\right) \\ &= \sqrt{u}\left(2\sqrt{5} - \sqrt{5}\right) \\ &= \sqrt{u} \cdot \sqrt{5} \\ &= \sqrt{5u} \end{aligned}$$

Mid-Chapter Quiz for Chapter 9

1. (a) $f(2) = \left(\frac{4}{3}\right)^2 = \frac{16}{9}$

(b) $f(0) = \left(\frac{4}{3}\right)^0 = 1$

(c) $f(-1) = \left(\frac{4}{3}\right)^{-1} = \frac{3}{4}$

(d) $f(1.5) = \left(\frac{4}{3}\right)^{1.5} \approx 1.54 = \frac{8\sqrt{3}}{9}$

2. $g(x) = -3^{-0.5x}$

Horizontal asymptote: $y = 0$

3.

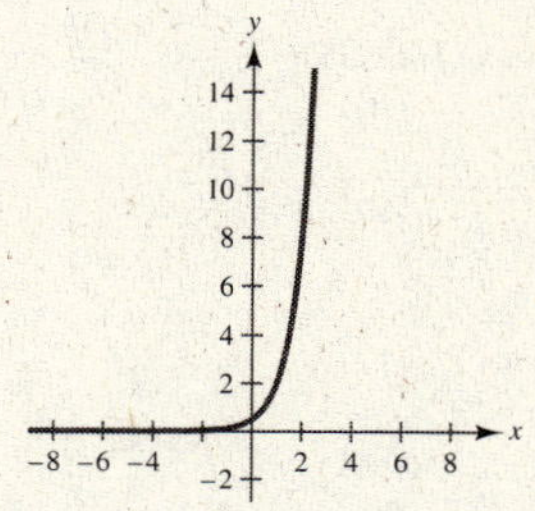

Horizontal asymptote: $y = 0$

4.

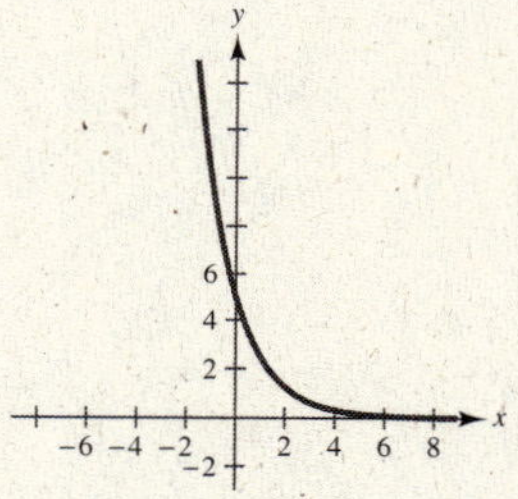

Horizontal asymptote: $y = 0$

5.

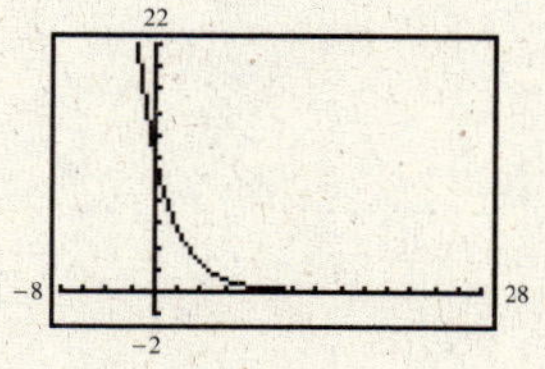

Horizontal asymptote: $y = 0$

6.

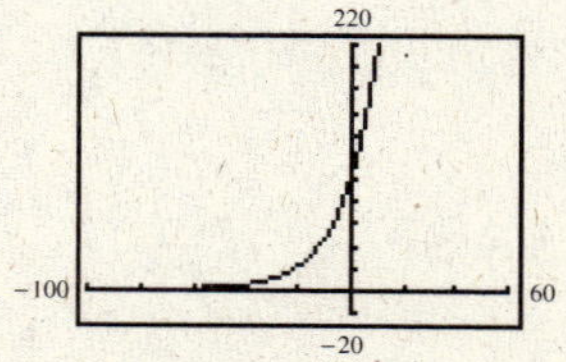

Horizontal asymptote: $y = 0$

7. (a) $(f \circ g)(x) = f[g(x)] = 2x^3 - 3$

(b) $(g \circ f)(x) = g[f(x)] = (2x - 3)^3$

(c) $(f \circ g)(-2) = f[g(-2)]$
$= f[-8]$
$= 2(-8) - 3$
$= -19$

(d) $(g \circ f)(4) = g[f(4)] = g[5] = 5^3 = 125$

8. $(f \circ g)(x) = f\left[\frac{1}{2}(5 - x)\right]$
$= 5 - 2\left[\frac{1}{2}(5 - x)\right]$
$= 5 - (5 - x)$
$= 5 - 5 + x$
$= x$

$(g \circ f)(x) = g[5 - 2x]$
$= \frac{1}{2}[5 - (5 - 2x)]$
$= \frac{1}{2}[5 - 5 + 2x]$
$= \frac{1}{2}(2x)$
$= x$

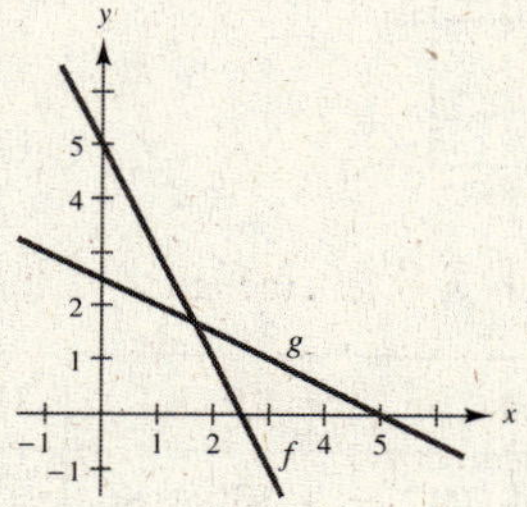

9. $h(x) = 10x + 3$

$y = 10x + 3$

$x = 10y + 3$

$x - 3 = 10y$

$\frac{x - 3}{10} = y$

$\frac{x - 3}{10} = h^{-1}(x)$

10. $g(t) = \frac{1}{2}t^3 + 2$

$$y = \frac{1}{2}t^3 + 2$$
$$t = \frac{1}{2}y^3 + 2$$
$$t - 2 = \frac{1}{2}y^3$$
$$2t - 4 = y^3$$
$$\sqrt[3]{2t - 4} = y$$
$$\sqrt[3]{2t - 4} = g^{-1}(t)$$

11. $\log_9\left(\frac{1}{81}\right) = -2$

$$9^{-2} = \frac{1}{81}$$

12. $2^6 = 64$

$$\log_2 64 = 6$$

13. $\log_5 125 = 3$ because $5^3 = 125$.

14. $f(t) = -2\ln(t + 3)$

Keystrokes:

[Y=] [(−)] 2 [LN] [(] [X,T,θ] [+] 3 [)] [GRAPH]

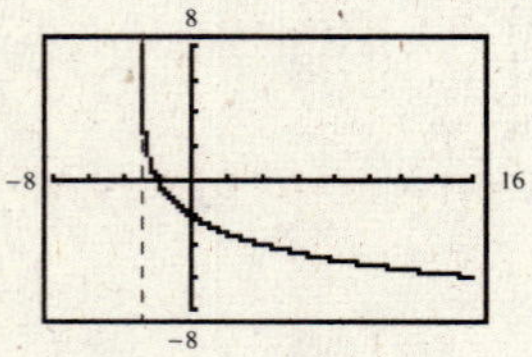

Vertical asymptote: $t = -3$

15. $h(x) = 5 + \frac{1}{2}\ln x$

Keystrokes: [Y=] 5 [+] .5 [LN] [X,T,θ] [GRAPH]

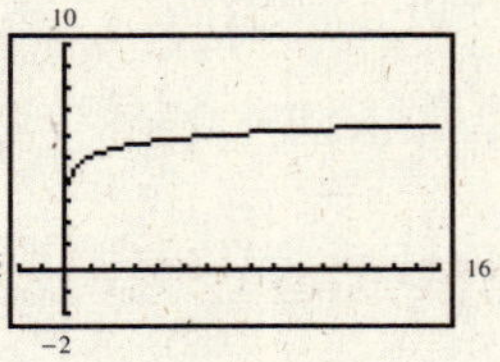

Vertical asymptote: $x = 0$

16. $f(x) = \log_5(x - 2) + 1$

The graph of $f(x) = \log_5 x$ has been shifted 2 units right and 1 unit up, so $h = 2, k = 1$.

17. $\log_3 782 = \dfrac{\log 782}{\log 3} \approx 6.0639$

18.

n	1	4	12	365	Continuous compounding
A	\$2979.31	\$3042.18	\$3056.86	\$3064.06	\$3064.31

Compounded 1 time per year: $A = 1200\left(1 + \dfrac{0.0625}{1}\right)^{1(15)} \approx \$2979.31$

Compounded 4 times per year: $A = 1200\left(1 + \dfrac{0.0625}{4}\right)^{4(15)} \approx \$3042.18$

Compounded 12 times per year: $A = 1200\left(1 + \dfrac{0.0625}{12}\right)^{12(15)} \approx \$3056.86$

Compounded 365 times per year: $A = 1200\left(1 + \dfrac{0.0625}{365}\right)^{365(15)} \approx \$3064.06$

Compounded continuously: $A = 1200e^{(0.0625)(15)} \approx \$3064.31$

19. $y = 14\left(\frac{1}{2}\right)^{t/40}$

$y = 14\left(\frac{1}{2}\right)^{125/40}$

$y = 14\left(\frac{1}{2}\right)^{3.125}$

$y \approx 1.60$ grams

Section 9.4 Properties of Logarithms

1. $\log_{12} 12^3 = 3 \cdot \log_{12} 12 = 3 \cdot 1 = 3$

3. $\log_4\left(\frac{1}{16}\right)^2 = \log_4\left(4^{-2}\right)^2$

$= \log_4 4^{-4}$

$= -4 \log_4 4$

$= -4 \cdot 1$

$= -4$

5. $\log_5 \sqrt[3]{5} = \log_5 5^{1/3} = \frac{1}{3} \cdot \log_5 5 = \frac{1}{3} \cdot 1 = \frac{1}{3}$

7. $\ln 14^0 = 0 \cdot \ln 14 = 0$

9. $\ln e^{-9} = -9 \ln e = -9 \cdot 1 = -9$

11. $\log_4 2 + \log_4 8 = \log_4 2 \cdot 8 = \log_4 16 = 2$ because $4^2 = 16$.

13. $\log_8 4 + \log_8 16 = \log_8 4 \cdot 16 = \log_8 64 = 2$ because $8^2 = 64$.

15. $\log_3 54 - \log_3 2 = \log_3 \frac{54}{2} = \log_3 27 = \log_3 3^3 = 3$

17. $\log_6 72 - \log_6 2 = \log_6 \frac{72}{2} = \log_6 36 = 2$ because $6^2 = 36$.

19. $\log_2 5 - \log_2 40 = \log_2 \frac{5}{40} = \log_2 \frac{1}{8} = \log_2 2^{-3} = -3$ because $2^{-3} = 2^{-3}$.

21. $\ln e^8 + \ln e^4 = \ln e^8 \cdot e^4$

$= \ln e^{12}$

$= 12 \ln e$

$= 12 \cdot 1$

$= 12$

23. $\ln \dfrac{e^3}{e^2} = \ln e = 1$

25. $\log_4 8 = \log_4 2^3 = 3 \log_4 2 \approx 3(0.500) = 1.5000$

27. $\log_4 \frac{3}{2} = \log_4 3 - \log_4 2 = 0.7925 - 0.5000 \approx 0.2925$

29. $\log_4 \sqrt[3]{9} = \log_4(9)^{1/3}$

$= \log_4\left(3^2\right)^{1/3}$

$= \log_4 3^{2/3}$

$= \frac{2}{3} \log_4 3$

$\approx \frac{2}{3}(0.7925)$

≈ 0.5283

31. $\log_4 3^0 = \log_4 1 = 0$

33. $\ln 9 = \ln 3^2$

$= 2 \cdot \ln 3$

$\approx 2 \cdot 1.0986$

≈ 2.1972

35. $\ln \frac{5}{3} = \ln 5 - \ln 3 \approx 1.6094 - 1.0986 \approx 0.5108$

37. $\ln \sqrt{45} = \frac{1}{2} \ln\left(3^2 \cdot 5\right)$

$= \frac{1}{2}[2 \ln 3 + \ln 5]$

$= \ln 3 + \frac{1}{2} \ln 5$

$\approx 1.0986 + \frac{1}{2}(1.6094)$

$\approx 1.0986 + 0.8047$

≈ 1.9033

39. $\ln\left(3^5 \cdot 5^2\right) = 5 \ln 3 + 2 \ln 5$

$\approx 5(1.0986) + 2(1.6094)$

$\approx 5.493 + 3.2188$

≈ 8.7118

41. $-3 \log_4 2 = \log_4 2^{-3} = \log_4 \frac{1}{2^3} = \log_4 \frac{1}{8}$

43. $-3 \log_{10} 3 + \log_{10} \dfrac{3}{2} = \log_{10} 3^{-3} + \log_{10} \dfrac{3}{2}$

$= \log_{10}\left(\dfrac{1}{3^3} \cdot \dfrac{3}{2}\right)$

$= \log_{10}\left(\dfrac{1}{3^2} \cdot \dfrac{1}{2}\right)$

$= \log_{10} \dfrac{1}{18}$

45. $-\ln \frac{1}{7} = \ln\left(\frac{1}{7}\right)^{-1} = \ln 7 = \ln \frac{56}{8} = \ln 56 - \ln 8$

47. $\log_3 11x = \log_3 11 + \log_3 x$

49. $\ln 3y = \ln 3 + \ln y$

51. $\log_7 x^2 = 2 \log_7 x$

53. $\log_4 x^{-3} = -3 \log_4 x$

55. $\log_4 \sqrt{3x} = \log_4(3x)^{1/2}$
$= \frac{1}{2} \log_4(3x)$
$= \frac{1}{2}(\log_4 3 + \log_4 x)$

57. $\log_2 \frac{z}{17} = \log_2 z - \log_2 17$

59. $\log_9 \frac{\sqrt{x}}{12} = \log_9 \sqrt{x} - \log_9 12$
$= \log_9 x^{1/2} - \log_9 12$
$= \frac{1}{2} \log_9 x - \log_9 12$

61. $\ln x^2(y + 2) = \ln x^2 + \ln(y + 2) = 2 \ln x + \ln(y + 2)$

63. $\log_4\left[x^6(x + 7)^2\right] = \log_4 x^6 + \log_4(x + 7)^2$
$= 6 \log_4 x + 2 \log_4(x + 7)$

65. $\log_3 \sqrt[3]{x + 1} = \frac{1}{3} \log_3(x + 1)$

67. $\ln\sqrt{x(x + 2)} = \frac{1}{2}\left[\ln x + \ln(x + 2)\right]$

69. $\ln\left(\frac{x + 1}{x + 4}\right)^2 = 2 \ln\left(\frac{x + 1}{x + 4}\right)$
$= 2\left[\ln(x + 1) - \ln(x + 4)\right]$

71. $\ln \sqrt[3]{\frac{x^2}{x + 1}} = \ln\left(\frac{x^2}{x + 1}\right)^{1/3}$
$= \frac{1}{3} \ln\left(\frac{x^2}{x + 1}\right)$
$= \frac{1}{3}\left[\ln x^2 - \ln(x + 1)\right]$
$= \frac{1}{3}\left[2 \ln x - \ln(x + 1)\right]$

73. $\ln \frac{xy^2}{z^3} = \ln\left(xy^2\right) - \ln z^3$
$= \ln x + \ln y^2 - \ln z^3$
$= \ln x + 2 \ln y - 3 \ln z$

75. $\log_{12} x - \log_{12} 3 = \log_{12} \frac{x}{3}$

77. $\log_3 5 + \log_3 x = \log_3 5x$

79. $\log_{10} 4 - \log_{10} x = \log_{10} \frac{4}{x}$

81. $4 \ln b = \ln b^4, \quad b > 0$

83. $-2 \log_5 2x = \log_5(2x)^{-2} = \log_5 \frac{1}{4x^2}, \quad x > 0$

85. $7 \log_2 x + 3 \log_2 z = \log_2 x^7 + \log_2 z^3 = \log_2 x^7z^3$

87. $\log_3 2 + \frac{1}{2} \log_3 y = \log_3 2 + \log_3 \sqrt{y} = \log_3 2\sqrt{y}$

89. $3 \ln x + \ln y - 2 \ln z = \ln x^3 + \ln y - \ln z^2 = \ln \frac{x^3y}{z^2}$

91. $4(\ln x + \ln y) = \ln(xy)^4$ or $\ln x^4y^4, \quad x > 0, y > 0$

93. $2\left[\ln x - \ln(x + 1)\right] = 2 \ln \frac{x}{x + 1}$
$= \ln\left(\frac{x}{x + 1}\right)^2$
$= \ln \frac{x^2}{(x + 1)^2}, \quad x > 0$

95. $\log_4(x + 8) - 3 \log_4 x = \log_4(x + 8) - \log_4 x^3$
$= \log_4 \frac{(x + 8)}{x^3}, \quad x > 0$

97. $\frac{1}{3} \log_5(x + 3) - \log_5(x - 6) = \log_5(x + 3)^{1/3} - \log_5(x - 6) = \log_5 \frac{\sqrt[3]{x + 3}}{x - 6}$

99. $5 \log_6(c + d) - \frac{1}{2} \log_6(m - n) = \log_6(c + d)^5 - \log_6(m - n)^{1/2} = \log_6 \frac{(c + d)^5}{\sqrt{m - n}}$

101. $\frac{1}{5}(3\log_2 x - 4\log_2 y) = \frac{1}{5}(\log_2 x^3 - \log_2 y^4) = \frac{1}{5}\left(\log_2 \frac{x^3}{y^4}\right) = \log_2 \sqrt[5]{\frac{x^3}{y^4}}, \quad y > 0$

103. $\ln 3e^2 = \ln 3 + \ln e^2 = \ln 3 + 2\ln e = \ln 3 + 2$

105. $\log_5 \sqrt{50} = \frac{1}{2}\left[\log_5(5^2 \cdot 2)\right] = \frac{1}{2}[2\log_5 5 + \log_5 2] = \frac{1}{2}[2 + \log_5 2] = 1 + \frac{1}{2}\log_5 2$

107. $\log_8 \frac{8}{x^3} = \log_8 8 - \log_8 x^3 = 1 - 3\log_8 x$

109. *Keystrokes:* y_1: [Y=] [LN] [(] [(] 10 [÷] [(] [X,T,θ] [x^2] [+] 1 [)] [)] [x^2] [)] [ENTER]

y^2: 2 [(] [LN] 10 [−] [LN] [(] [X,T,θ] [x^2] [+] 1 [)] [)] [GRAPH]

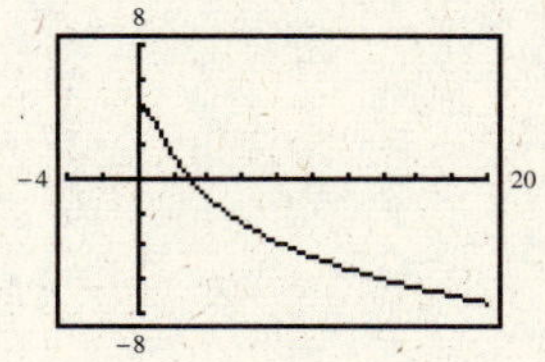

111. *Keystrokes:* y_1: [Y=] [LN] [(] [X,T,θ] [x^2] [(] [X,T,θ] [+] 2 [)] [)] [ENTER]

y_2: 2 [LN] [X,T,θ] [+] [LN] [(] [X,T,θ] [+] 2 [)] [GRAPH]

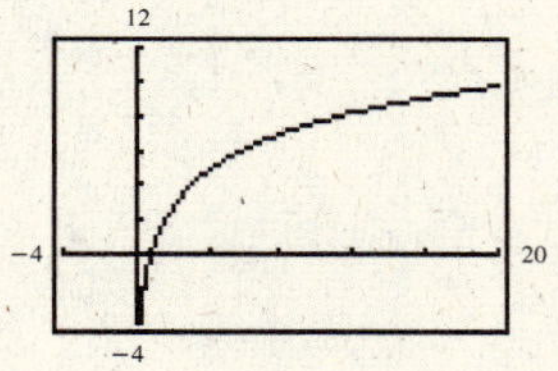

113. (a) $B = 10\log_{10}\left(\frac{I}{10^{-16}}\right)$

$B = 10\left[\log_{10} I - \log_{10} 10^{-16}\right]$

$B = 10[\log_{10} I + 16\log_{10} 10]$

$B = 10(\log_{10} I + 16)$

(b) $B = 10(\log 10^{-3} + 16)$

$= 10(-3 + 16)$

$= 10(13)$

$= 130$ decibels

115. $E = 1.4(\log_{10} C_2 - \log_{10} C_1) = 1.4\left(\log_{10}\frac{C_2}{C_1}\right)$

$= \log_{10}\left(\frac{C_2}{C_1}\right)^{1.4}$

117. True; $\log_2 8x = \log_2 8 + \log_2 x = 3 + \log_2 x$

119. False; $\log_3(u + v)$ does not simplify.

121. True, $f(ax) = \log_a ax = \log_a a + \log_a x$

$= 1 + \log_a x$

$= 1 + f(x)$

123. False; 0 is not in the domain of f.

125. False; $f(x - 3) = \ln(x - 3) \neq \ln x - \ln 3$

127. True; $f(1) = 0$, so when $f(x) > 0$, $x > 1$

129. Choose two values for x and y, such as $x = 3$ and $y = 5$, and show the two expressions are not equal.

$$\frac{\ln 3}{\ln 5} \neq \ln\frac{3}{5} = \ln 3 - \ln 5$$

$$0.6826062 \neq -0.5108256 = -0.5108256$$

or

$$\frac{\ln e}{\ln e} \neq \ln\frac{e}{e}$$

$$1 \neq \ln 1$$

$$1 \neq 0$$

131. $x^2 - 10x + 17 = 0$

$$x = \frac{-(-10) \pm \sqrt{(-10) - 4(1)(17)}}{2(1)}$$

$$x = \frac{10 \pm \sqrt{100 - 68}}{2}$$

$$x = \frac{10 \pm \sqrt{32}}{2}$$

$$x = \frac{10 \pm \sqrt{16 \cdot 2}}{2}$$

$$x = \frac{10 \pm 4\sqrt{2}}{2}$$

$$x = 5 \pm 2\sqrt{2}$$

133. $\dfrac{1}{x} + \dfrac{2}{x - 5} = 0$

$$x - 5 + 2x = 0$$

$$3x = 5$$

$$x = \frac{5}{3}$$

135. $\sqrt{x + 2} = 7$

$$\left(\sqrt{x + 2}\right)^2 = 7^2$$

$$x + 2 = 49$$

$$x = 47$$

Check: $\sqrt{47 + 2} \stackrel{?}{=} 7$

$$\sqrt{49} = 7$$

$$7 = 7$$

137. $f(x) = x^2 - 16$

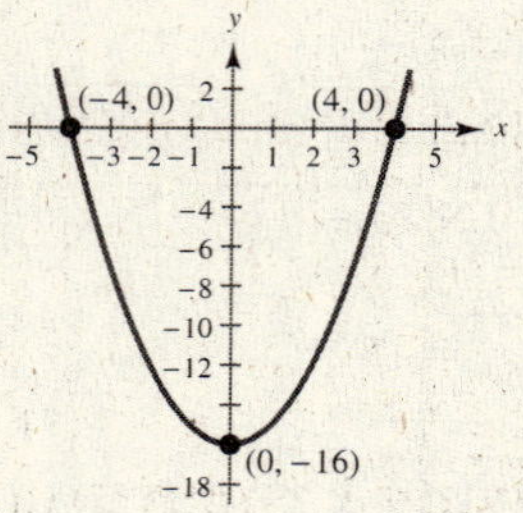

Vertex: $(0, -16)$

x-intercepts: $(4, 0), (-4, 0)$

$$x^2 - 16 = 0$$

$$x^2 = 16$$

$$x = \pm 4$$

139. $h(x) = x^2 + 6x + 14$

$$h(x) = \left(x^2 + 6x + 9\right) + 14 - 9$$

$$= (x + 3)^2 + 5$$

Vertex: $(-3, 5)$

x-intercepts: none

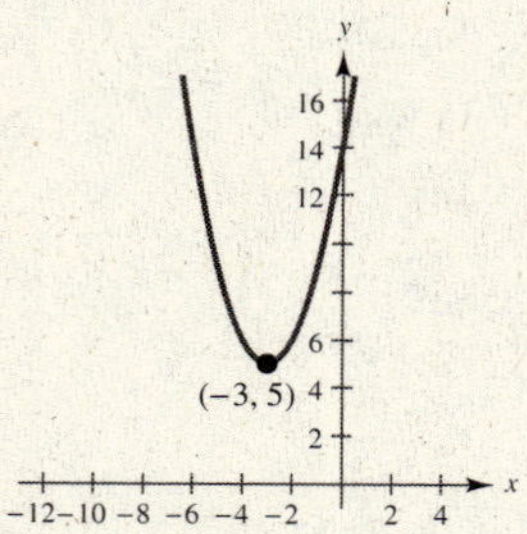

141. $f(x) = \sqrt{x}, g(x) = x - 3$

(a) $(f \circ g)(x) = f(x - 3) = \sqrt{x - 3}$

Domain: $[3, \infty)$

$$x - 3 \geq 0$$

$$x \geq 3$$

(b) $(g \circ f)(x) = g\left(\sqrt{x}\right) = \sqrt{x} - 3$

Domain: $[0, \infty),\ x \geq 0$

143. $f(x) = \dfrac{5}{x^2 - 4}, g(x) = x + 1$

(a) $(f \circ g)(x) = f(x+1) = \dfrac{5}{(x+1)^2 - 4} = \dfrac{5}{x^2 + 2x + 1 - 4} = \dfrac{5}{x^2 + 2x - 3}$

Domain: $(-\infty, -3) \cup (-3, 1) \cup (1, \infty)$

$x^2 + 2x - 3 \neq 0$

$(x + 3)(x - 1) \neq 0$

$x \neq -3, x \neq 1$

(b) $(g \circ f)(x) = g\left(\dfrac{5}{x^2 - 4}\right) = \dfrac{5}{x^2 - 4} + 1 = \dfrac{5 + x^2 - 4}{x^2 - 4} = \dfrac{x^2 + 1}{x^2 - 4}$

Domain: $(-\infty, -2) \cup (-2, 2) \cup (2, \infty)$

$x^2 - 4 \neq 0, \quad (x - 2)(x + 2) \neq 0, \quad x \neq 2, x \neq -2$

Section 9.5 Solving Exponential and Logarithmic Equations

1. (a) $3^{2(1)-5} \stackrel{?}{=} 27$

$3^{-3} \neq 27$

not a solution

(b) $3^{2(4)-5} \stackrel{?}{=} 27$

$3^3 = 27$

solution

3. (a) $e^{-5+\ln 45+5} \stackrel{?}{=} 45$

$e^{\ln 45} \stackrel{?}{=} 45$

$45 = 45$

solution

(b) $e^{-2.1933+5} \stackrel{?}{=} 45$

$e^{2.8067} \stackrel{?}{=} 45$

$16.555 \neq 45$

not a solution

5. (a) $\log_9(6 \cdot 27) \stackrel{?}{=} \frac{3}{2}$

$\log_9 162 \neq \frac{3}{2}$

not a solution

(b) $\log_9\left(6 \cdot \frac{9}{2}\right) \stackrel{?}{=} \frac{3}{2}$

$\log_9 27 = \frac{3}{2}$

solution

7. $7^x = 7^3$

$x = 3$

9. $e^{1-x} = e^4$

$1 - x = 4$

$-x = 3$

$x = -3$

11. $5^{x+6} = 25^5$

$5^{x+6} = (5^2)^5$

$5^{x+6} = 5^{10}$

$x + 6 = 10$

$x = 4$

13. $6^{2x} = 36$

$6^{2x} = 6^2$

$2x = 2$

$x = 1$

15. $3^{2-x} = 81$

$3^{2-x} = 3^4$

$2 - x = 4$

$-x = 2$

$x = -2$

17. $5^x = \frac{1}{125}$

$5^x = 5^{-3}$

so $x = -3$.

19. $2^{x+2} = \frac{1}{16}$

$2^{x+2} = 2^{-4}$

so $x + 2 = -4$

$x = -6$.

21. $4^{x+3} = 32^x$

$(2^2)^{x+3} = (2^5)^x$

so $2(x+3) = 5x$

$2x + 6 = 5x$

$6 = 3x$

$2 = x.$

23. $\ln 5x = \ln 22$

so $5x = 22$

$x = \frac{22}{5}.$

25. $\log_6 3x = \log_6 18$

so $3x = 18$

$x = 6.$

27. $\ln(3 - x) = \ln 10$

so $3 - x = 10$

$-x = 7$

$x = -7.$

29. $\log_2(x + 3) = \log_2 7$

so $x + 3 = 7$

$x = 4.$

31. $\log_5(2x - 3) = \log_5(4x - 5)$

so $2x - 3 = 4x - 5$

$2 = 2x$

$1 = x.$

No solution since expressions on either side are undefined for $x = 1$.

33. $\log_3(2 - x) = 2$

$2 - x = 3^2$

$-x = 7$

$x = -7$

35. $\ln e^{2x-1} = (2x - 1)\ln e$

$= (2x - 1)(1)$

$= 2x - 1$

37. $10^{\log_{10} 2x} = 2x, \quad x > 0$

39. $3^x = 91$

$\log_3 3^x = \log_3 91$

$x = \log_3 91$

$x = \dfrac{\log 91}{\log 3}$

$x \approx 4.11$

41. $5^x = 8.2$

$\log_5 5^x = \log_5 8.2$

$x = \log_5 8.2$

$x = \dfrac{\log 8.2}{\log 5}$

$x \approx 1.31$

43. $6^{2x} = 205$

$\log_6 6^{2x} = \log_6 205$

$2x = \log_6 205$

$2x = \dfrac{\log 205}{\log 6}$

$x = \dfrac{\log 205}{2 \log 6}$

$x \approx 1.49$

45. $7^{3y} = 126$

$\log_7 7^{3y} = \log_7 126$

$3y = \log_7 126$

$y = \dfrac{\log_7 126}{3}$

$y = \dfrac{\log 126}{3 \log 7}$

$y \approx 0.83$

47. $3^{2-x} = 8$

$\log_3 3^{2-x} = \log_3 8$

$2 - x = \log_3 8$

$-x = \log_3 8 - 2$

$x = 2 - \dfrac{\log 8}{\log 3}$

$x \approx 0.11$

49. $10^{x+6} = 250$

$\log 10^{x+6} = \log 250$

$x + 6 = \log 250$

$x = \log 250 - 6$

$x \approx -3.60$

51. $4e^{-x} = 24$

$e^{-x} = 6$

$\ln e^{-x} = \ln 6$

$-x = \ln 6$

$x = -\ln 6$

$x \approx -1.79$

53. $\frac{1}{4}e^x = 5$

$e^x = 20$

$\ln e^x = \ln 20$

$x = \ln 20$

$x \approx 3.00$

55. $\frac{1}{2}e^{-2x} = 9$

$e^{-2x} = 18$

$\ln e^{-2x} = \ln 18$

$-2x = \ln 18$

$x = -\frac{\ln 18}{2}$

$x \approx -1.45$

57. $250(1.04)^x = 1000$

$(1.04)^x = 4$

$\log_{1.04} 1.04^x = \log_{1.04} 4$

$x = \log_{1.04} 4$

$x = \frac{\log 4}{\log 1.04}$

$x \approx 35.35$

59. $300e^{x/2} = 9000$

$e^{x/2} = 30$

$\ln e^{x/2} = \ln 30$

$\frac{x}{2} = \ln 30$

$x = 2 \ln 30$

$x \approx 6.80$

61. $1000^{0.12x} = 25{,}000$

$\log_{1000} 1000^{0.12x} = \log_{1000} 25{,}000$

$0.12x = \log_{1000} 25{,}000$

$x = \frac{\log_{1000} 25{,}000}{0.12}$

$x = \frac{\log 25{,}000}{0.12 \log 1000}$

$x \approx 12.22$

63. $\frac{1}{5}4^{x+2} = 300$

$4^{x+2} = 1500$

$\log_4 4^{x+2} = \log_4 1500$

$x + 2 = \frac{\log 1500}{\log 4}$

$x = \frac{\log 1500}{\log 4} - 2$

$x \approx 3.28$

65. $6 + 2^{x-1} = 1$

$2^{x-1} = -5$

$\log_2 2^{x-1} = \log_2(-5)$

No solution

$\log_2(-5)$ is not possible.

67. $7 + e^{2-x} = 28$

$e^{2-x} = 21$

$\ln e^{2-x} = \ln 21$

$2 - x = \ln 21$

$-x = \ln 21 - 2$

$x = 2 - \ln 21$

$x \approx -1.04$

69. $8 - 12e^{-x} = 7$

$-12e^{-x} = -1$

$e^{-x} = \frac{1}{12}$

$\ln e^{-x} = \ln \frac{1}{12}$

$-x = \ln \frac{1}{12}$

$x = -\ln \frac{1}{12} \approx 2.48$

71. $4 + e^{2x} = 10$

$e^{2x} = 6$

$\ln e^{2x} = \ln 6$

$2x = \ln 6$

$x = \frac{\ln 6}{2}$

$x \approx 0.90$

73. $$\begin{aligned} 17 - e^{x/4} &= 14 \\ -e^{x/4} &= -3 \\ e^{x/4} &= 3 \\ \ln e^{x/4} &= \ln 3 \\ \frac{x}{4} &= \ln 3 \\ x &= 4 \ln 3 \\ x &\approx 4.39 \end{aligned}$$

75. $$\begin{aligned} 23 - 5e^{x+1} &= 3 \\ -5e^{x+1} &= -20 \\ e^{x+1} &= 4 \\ \ln e^{x+1} &= \ln 4 \\ x + 1 &= \ln 4 \\ x &= \ln 4 - 1 \\ x &\approx 0.39 \end{aligned}$$

77. $$\begin{aligned} 4\left(1 + e^{x/3}\right) &= 84 \\ 1 + e^{x/3} &= 21 \\ e^{x/3} &= 20 \\ \ln e^{x/3} &= \ln 20 \\ \frac{x}{3} &= \ln 20 \\ x &= 3 \ln 20 \\ x &\approx 8.99 \end{aligned}$$

79. $$\begin{aligned} \frac{8000}{(1.03)^t} &= 6000 \\ \frac{8000}{6000} &= (1.03)^t \\ \frac{4}{3} &= (1.03)^t \\ \log_{1.03} \frac{4}{3} &= \log_{1.03} 1.03^t \\ \log_{1.03} \frac{4}{3} &= t \\ 9.73 &\approx t \end{aligned}$$

81. $$\begin{aligned} \frac{300}{2 - e^{-0.15t}} &= 200 \\ \frac{300}{200} &= 2 - e^{-0.15t} \\ \frac{3}{2} - 2 &= -e^{-0.15t} \\ -\frac{1}{2} &= -e^{-0.15t} \\ \ln\left(\frac{1}{2}\right) &= \ln e^{-0.15t} \\ \ln\left(\frac{1}{2}\right) &= -0.15t \\ \frac{\ln\left(\frac{1}{2}\right)}{-0.15} &= t \approx 4.62 \end{aligned}$$

83. $$\begin{aligned} \log_{10} x &= -1 \\ 10^{\log_{10} x} &= 10^{-1} \\ x &= 10^{-1} \\ x &= \tfrac{1}{10} - 0.10 \end{aligned}$$

85. $$\begin{aligned} \log_3 x &= 4.7 \\ 3^{\log_3 x} &= 3^{4.7} \\ x &= 3^{4.7} \\ x &\approx 174.77 \end{aligned}$$

87. $$\begin{aligned} 4 \log_3 x &= 28 \\ \log_3 x &= 7 \\ 3^{\log_3 x} &= 3^7 \\ x &= 3^7 \\ x &= 2187.00 \end{aligned}$$

89. $$\begin{aligned} 16 \ln x &= 30 \\ \ln x &= \tfrac{30}{16} \\ e^{\ln x} &= e^{15/8} \\ x &= e^{15/8} \\ x &\approx 6.52 \end{aligned}$$

91. $$\begin{aligned} \log_{10} 4x &= 2 \\ 10^{\log_{10} 4x} &= 10^2 \\ 4x &= 10^2 \\ x &= \frac{10^2}{4} \\ x &= \frac{100}{4} = 25.00 \end{aligned}$$

93. $\ln 2x = \frac{1}{5}$

$$e^{\ln 2x} = e^{1/5}$$
$$2x = e^{1/5}$$
$$x = \frac{e^{1/5}}{2}$$
$$x \approx 0.61$$

95. $\ln x^2 = 6$

$$e^{\ln x^2} = e^6$$
$$x^2 = e^6$$
$$x = \pm\sqrt{e^6}$$
$$x \approx \pm 20.09$$

97. $2\log_4(x+5) = 3$

$$\log_4(x+5) = \tfrac{3}{2}$$
$$4^{\log_4(x+5)} = 4^{1.5}$$
$$x + 5 = 4^{1.5}$$
$$x = 4^{1.5} - 5$$
$$x = 3.00$$

99. $\frac{3}{4}\ln(x+4) = -2$

$$\ln(x+4) = -2 \cdot \tfrac{4}{3}$$
$$\ln(x+4) = -\tfrac{8}{3}$$
$$e^{\ln(x+4)} = e^{-8/3}$$
$$x + 4 = e^{-8/3}$$
$$x = e^{-8/3} - 4$$
$$x \approx -3.93$$

101. $7 - 2\log_2 x = 4$

$$-2\log_2 x = -3$$
$$\log_2 x = \tfrac{3}{2}$$
$$2^{\log_2 x} = 2^{3/2}$$
$$x = 2^{3/2}$$
$$x \approx 2.83$$

103. $-1 + 3\log_{10}\frac{x}{2} = 8$

$$3\log_{10}\frac{x}{2} = 9$$
$$\log_{10}\frac{x}{2} = 3$$
$$10^{\log_{10}(x/2)} = 10^3$$
$$\frac{x}{2} = 10^3$$
$$x = 2(10)^3$$
$$x = 2000.00$$

105. $\log_4 x + \log_4 5 = 2$

$$\log_4 x(5) = 2$$
$$4^{\log_4 5x} = 4^2$$
$$5x = 16$$
$$x = \tfrac{16}{5}$$
$$x = 3.20$$

107. $\log_6(x+8) + \log_6 3 = 2$

$$\log_6(x+8)(3) = 2$$
$$6^{\log_6 3(x+8)} = 6^2$$
$$3x + 24 = 36$$
$$3x = 12$$
$$x = 4.00$$

109. $\log_5(x+3) - \log_5 x = 1$

$$\log_5\left(\frac{x+3}{x}\right) = 1$$
$$5^{\log_5[(x+3)/x]} = 5^1$$
$$\frac{x+3}{x} = 5$$
$$x + 3 = 5x$$
$$3 = 4x$$
$$\frac{3}{4} = x$$
$$0.75 = x$$

111. $\log_{10} x + \log_{10}(x-3) = 1$

$$\log_{10} x(x-3) = 1$$
$$10^{\log_{10} x(x-3)} = 10^1$$
$$x(x-3) = 10$$
$$x^2 - 3x - 10 = 0$$
$$(x-5)(x+2) = 0$$

$x = 5, x = -2$ (which is extraneous)

$x = 5.00$

113.
$$\begin{aligned} \log_2(x-1) + \log_2(x+3) &= 3 \\ \log_2(x-1)(x+3) &= 3 \\ x^2 + 2x - 3 &= 2^3 \\ x^2 + 2x - 11 &= 0 \\ x = \frac{-2 \pm \sqrt{4 - 4(1)(-11)}}{2(1)} &= \frac{-2 \pm \sqrt{4 + 44}}{2} \\ &= \frac{-2 \pm \sqrt{48}}{2} \end{aligned}$$

$x \approx 2.46$ and -4.46 (which is extraneous)

115.
$$\begin{aligned} \log_{10} 4x - \log_{10}(x-2) &= 1 \\ \log_{10}\left(\frac{4x}{x-2}\right) &= 1 \\ \frac{4x}{x-2} &= 10^1 \\ 4x &= 10(x-2) \\ 4x &= 10x - 20 \\ -6x &= -20 \\ x &= \frac{20}{6} \\ x &= \frac{10}{3} \\ x &\approx 3.33 \end{aligned}$$

117.
$$\begin{aligned} \log_2 x + \log_2(x+2) - \log_2 3 &= 4 \\ \log_2 \frac{x(x+2)}{3} &= 4 \\ 2^{\log_2[(x^2+2x)/3]} &= 2^4 \\ \frac{x^2 + 2x}{3} &= 16 \\ x^2 + 2x &= 48 \\ x^2 + 2x - 48 &= 0 \\ (x+8)(x-6) &= 0 \end{aligned}$$

$x = -8$ (which is extraneous)

$x = 6.00$

119. *Keystrokes:*

[Y=] 10 [^] [(] [X,T,θ] [÷] 2 [)] [−] 5 [GRAPH]

x-intercept

$1.3974 \approx 1.40$

$(1.40, 0)$

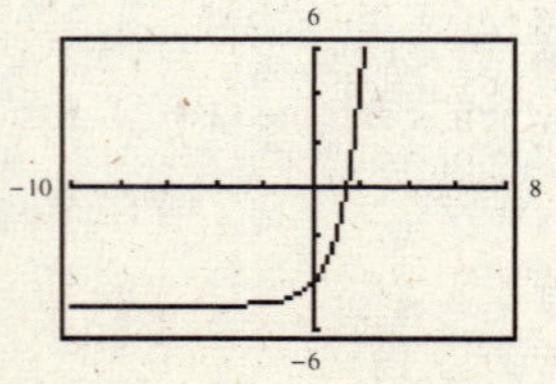

121. *Keystrokes:* [Y=] 6 [LN] [(] .4 [X,T,θ] [)] [−] 13 [GRAPH]

x-intercept

$21.822846 \approx 21.82$

$(21.82, 0)$

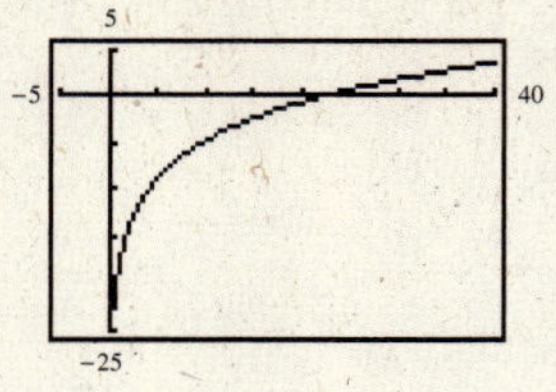

123. *Formula:* $A = Pe^{rt}$

Labels: Principal $= P = \$10{,}000$

Amount $= A = \$11{,}051.71$

Time $= t = 2$ years

Annual interest rate $= r$

Equation:
$$\begin{aligned} 11{,}051.71 &= 10{,}000e^{r(2)} \\ \frac{11{,}051.71}{10{,}000} &= e^{2r} \\ 1.105171 &= e^{2r} \\ \ln 1.105171 &= \ln e^{2r} \\ \ln 1.105171 &= 2r \\ \frac{\ln 1.105171}{2} &= r \\ 0.05 \approx r &\approx 5\% \end{aligned}$$

125.
$$\begin{aligned} 5000 &= 2500e^{0.09t} \\ \frac{5000}{2500} &= e^{0.09t} \\ 2 &= e^{0.09t} \\ \ln 2 &= \ln\left(e^{0.09t}\right) \\ \ln 2 &= 0.09t \\ \frac{\ln 2}{0.09} &= t \\ 7.70 \text{ years} &\approx t \end{aligned}$$

127.
$$\begin{aligned} B &= 10 \log_{10}\left(\frac{I}{10^{-16}}\right) \\ 80 &= 10 \log_{10}\left(\frac{I}{10^{-16}}\right) \\ 8 &= \log_{10}\left(\frac{I}{10^{-16}}\right) \\ 10^8 &= 10 \log_{10}\left(\frac{I}{10^{-16}}\right) \\ 10^8 &= \frac{I}{10^{-16}} \\ 10^8 \cdot 10^{-16} &= I \\ 10^{-8} &= I \end{aligned}$$

watts per square centimeter

129. (a)

$$\begin{aligned} F &= 200e^{-0.5\pi\theta/180} \\ 80 &= 200e^{-0.5\pi\theta/180} \\ 0.4 &= e^{-0.5\pi\theta/180} \\ \ln 0.4 &= \ln e^{-0.5\pi\theta/180} \\ \ln 0.4 &= \frac{-0.5\pi\theta}{180} \\ (\ln 0.4)\left(-\frac{180}{0.5\pi}\right) &= \theta \\ 105^\circ &\approx \theta \end{aligned}$$

(b) *Keystrokes:*

y_1: [Y=] 200 [e^x] [(−)] 0.5 π [X,T,θ] [÷] 180 [)] [ENTER]

y_2: 80 [GRAPH]

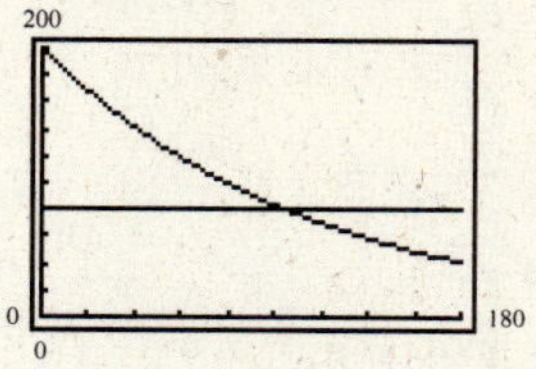

The point of intersection is at $x = 105^\circ$.

131. (a)

$$\begin{aligned} kt &= \ln\frac{T - S}{T_0 - S} \\ k(3) &= \ln\frac{78 - 65}{85 - 65} \\ k &= \frac{1}{3}\ln\frac{13}{20} \\ k &\approx -0.144 \end{aligned}$$

(b)

$$\begin{aligned} -0.144t &= \ln\frac{85 - 65}{98.6 - 65} \\ t &= \frac{\ln\frac{20}{33.6}}{-0.144} \\ t &\approx 3.6 \text{ hours} \\ &\approx 3 \text{ hours } 36 \text{ minutes} \end{aligned}$$

10 P.M. − 3 hr 36 min = 6:24 P.M.

(c)

$$\begin{aligned} -0.144(2) &= \ln\frac{T - 65}{98.6 - 65} \\ -0.288 &= \ln\frac{T - 65}{33.6} \\ e^{-0.288} &= \frac{T - 65}{33.6} \\ 33.6e^{-0.288} + 65 &= T \\ 90.2^\circ\text{F} &\approx T \end{aligned}$$

133. Subantarctic water:

$$\begin{aligned} 8 &= 7.9\ln(1.0245 - d) + 61.84 \\ -53.84 &= 7.9\ln(1.0245 - d) \\ -6.815189873 &= \ln(1.0245 - d) \\ e^{-6.815189873} &= e^{\ln(1.0245 - d)} \\ e^{-6.815189873} &= 1.0245 - d \\ d &= 1.0245 - e^{-6.815189873} \\ d &\approx 1.0234 \text{ grams per cubic centimeter} \end{aligned}$$

Antarctic bottom water:

$$\begin{aligned} 0 &= 7.9\ln(1.0245 - d) + 61.84 \\ -61.84 &= 7.9\ln(1.0245 - d) \\ -7.827848101 &= \ln(1.0245 - d) \\ e^{-7.827848101} &= e^{\ln(1.0245 - d)} \\ e^{-7.827848101} &= 1.0245 - d \\ d &= 1.0245 - d^{-7.827848101} \\ d &\approx 1.0241 \text{ grams per cubic centimeter} \end{aligned}$$

135. Three basic properties of logarithms:

$$\log_a(uv) = \log_a u + \log_a v$$

$$\log_a\left(\frac{u}{v}\right) = \log_a u - \log_a v$$

$$\log_a u^n = n\log_a u$$

137. To solve an exponential equation, first isolate the exponential expression, then take the logarithms of both sides of the equation, and solve for the variable.

To solve a logarithmic equation, first isolate the logarithmic expression, then exponentiate both sides of the equation, and solve for the variable.

139.

$$\begin{aligned} x^2 &= -25 \\ x &= \pm\sqrt{-25} \\ x &= \pm 5i \end{aligned}$$

141.

$$\begin{aligned} 9n^2 - 16 &= 0 \\ n^2 &= \tfrac{16}{9} \\ n &= \pm\sqrt{\tfrac{16}{9}} \\ n &= \pm\tfrac{4}{3} \end{aligned}$$

143.

$$\begin{aligned} t^4 - 13t^2 + 36 &= 0 \\ (t^2 - 9)(t^2 - 4) &= 0 \end{aligned}$$

$$t^2 = 9 \qquad t^2 = 4$$

$$t = \pm 3 \qquad t = \pm 2$$

145. *Verbal Model:* $\boxed{\text{Area}} = \boxed{\text{Length}} \cdot \boxed{\text{Width}}$

Labels: Length $= x$

Width $= 2.5x$

Equation:

$$2x + 2(2.5x) = 42$$
$$7x = 42$$
$$x = 6$$

$$A = 6 \cdot 2.5(6)$$
$$A = 90 \text{ in.}^2$$

147. *Verbal Model:* $\boxed{\text{Perimeter}} = 2 \cdot \boxed{\text{Length}} + 2 \cdot \boxed{\text{Width}}$

Labels: Length $= w + 4$

Width $= w$

Equation:

$$w(w + 4) = 192$$
$$w^2 + 4w - 192 = 0$$
$$(w - 12)(w + 16) = 0$$
$$w = 12, \quad w = -16$$

$$P = 2(12 + 4) + 2(12)$$
$$= 32 + 24$$
$$= 56 \text{ km}$$

Section 9.6 Applications

1.

$$A = P\left(1 + \frac{r}{n}\right)^{nt}$$
$$1004.83 = 500\left(1 + \frac{r}{12}\right)^{12(10)}$$
$$2.00966 = \left(1 + \frac{r}{12}\right)^{120}$$
$$(2.00966)^{1/120} = 1 + \frac{r}{12}$$
$$1.0058333 \approx 1 + \frac{r}{12}$$
$$0.0058333 \approx \frac{r}{12}$$
$$0.07 \approx r$$
$$7\% \approx r$$

3.

$$A = P\left(1 + \frac{r}{n}\right)^{nt}$$
$$36{,}581.00 = 1000\left(1 + \frac{r}{365}\right)^{365(40)}$$
$$36.581 = \left(1 + \frac{r}{365}\right)^{14{,}600}$$
$$(36.581)^{1/14{,}600} = 1 + \frac{r}{365}$$
$$1.0002466 \approx 1 + \frac{r}{365}$$
$$0.0002466 \approx \frac{r}{365}$$
$$0.0899981 \approx r$$
$$9\% \approx r$$

5.

$$A = Pe^{rt}$$
$$8267.38 = 750e^{r(30)}$$
$$11.023173 \approx e^{r(30)}$$
$$\ln 11.023173 \approx \ln e$$
$$\ln 11.023173 \approx 30r$$
$$\frac{\ln 11.023173}{30} \approx r$$
$$0.08 \approx r$$
$$8\% \approx r$$

7.

$$A = P\left(1 + \frac{r}{n}\right)^{nt}$$
$$5000 = 2500\left(1 + \frac{0.075}{12}\right)^{12t}$$
$$2 = (1.00625)^{12t}$$
$$\log_{1.00625} 2 = \log_{1.00625}(1.00625)^{12t}$$
$$\frac{\log 2}{\log 1.00625} = 12t$$
$$\frac{\log 2}{12 \log 1.00625} = t$$
$$t \approx 9.27 \text{ years}$$

9.

$$A = Pe^{rt}$$
$$36{,}000 = 18{,}000e^{0.08t}$$
$$2 = e^{0.08t}$$
$$\ln 2 = \ln e^{0.08t}$$
$$\ln 2 = 0.08t$$
$$\frac{\ln 2}{0.08} = t$$
$$t \approx 8.66 \text{ years}$$

11.
$$A = P\left(1 + \frac{r}{n}\right)^{nt}$$
$$3000 = 1500\left(1 + \frac{0.0725}{12}\right)^{12t}$$
$$2 = \left(1.006041\overline{6}\right)^{12t}$$
$$\log_{1.006041\overline{6}} 2 = \log_{1.006041\overline{6}}\left(1.006041\overline{6}\right)^{12t}$$
$$\frac{\log 2}{\log 1.006041\overline{6}} = 12t$$
$$\frac{\log 2}{12 \log 1.006041\overline{6}} = t$$
$$t \approx 9.59 \text{ years}$$

13. $8954.24 = 5000\left(1 + \frac{0.06}{n}\right)^{n(10)}$

$8954.24 = 5000\left(1 + \frac{0.06}{1}\right)^{1(10)}$

$8954.24 = 8954.24$

Yearly compounding

15. $1587.75 = 750\left(1 + \frac{0.075}{n}\right)^{n(10)}$

$1587.75 = 750e^{0.075(10)}$

$1587.75 = 1587.75$

Continuous compounding

17. $141.48 = 100\left(1 + \frac{0.07}{n}\right)^{n(5)}$

$141.48 = 100\left(1 + \frac{0.07}{4}\right)^{4(5)}$

$141.48 = 141.48$

Quarterly compounding

19. $A = Pe^{rt}$

$A = 1000e^{0.08(1)}$

$A = \$1083.29$

$$\text{Effective yield} = \frac{83.29}{1000} = 0.08329 \approx 8.33\%$$

21. $A = P\left(1 + \frac{r}{n}\right)^{nt}$

$A = 1000\left(1 + \frac{0.07}{12}\right)^{12(1)}$

$A = \$1072.29$

$$\text{Effective yield} = \frac{72.29}{1000} = 0.07229 \approx 7.23\%$$

23. $A = P\left(1 + \frac{r}{n}\right)^{nt}$

$A = 1000\left(1 + \frac{0.06}{4}\right)^{4(1)}$

$A = \$1061.36$

$$\text{Effective yield} = \frac{61.36}{1000} = 0.06136 \approx 6.14\%$$

25. $A = P\left(1 + \frac{r}{n}\right)^{nt}$

$A = 1000\left(1 + \frac{0.08}{12}\right)^{12(1)}$

$A = \$1083.00$

$$\text{Effective yield} = \frac{83.00}{1000} = 0.083 = 8.30\%$$

27. No. Each time the amount is divided by the principal, the result is always 2.

29.
$$A = Pe^{rt}$$
$$10{,}000 = Pe^{0.09(20)}$$
$$\frac{10{,}000}{e^{1.8}} = P$$
$$\$1652.99 \approx P$$

31.
$$A = P\left(1 + \frac{r}{n}\right)^{nt}$$
$$750 = P\left(1 + \frac{0.06}{365}\right)^{365(3)}$$
$$\frac{750}{(1.0001644)^{1095}} = P$$
$$\$626.46 \approx P$$

33.
$$A = P\left(1 + \frac{r}{n}\right)^{nt}$$
$$25{,}000 = P\left(1 + \frac{0.07}{12}\right)^{12(30)}$$
$$\frac{25{,}000}{(1.005833)^{360}} = P$$
$$\$3080.15 \approx P$$

35.
$$A = P\left(1 + \frac{r}{n}\right)^{nt}$$
$$1000 = P\left(1 + \frac{0.05}{365}\right)^{365(1)}$$
$$\frac{1000}{(1.000136986)^{365}} = P$$
$$\$951.23 \approx P$$

37.
$$A = \frac{P(e^{rt} - 1)}{e^{r/12} - 1}$$
$$A = \frac{30(e^{0.08(10)} - 1)}{e^{0.08/12} - 1}$$
$$A \approx \$5496.57$$

39.
$$A = \frac{P(e^{rt} - 1)}{e^{r/12} - 1}$$
$$A = \frac{50(e^{0.10(40)} - 1)}{e^{0.10/12} - 1}$$
$$A \approx \$320{,}250.81$$

41.
$$A = \frac{P(e^{rt} - 1)}{e^{r/12} - 1}$$
$$A = \frac{30(e^{0.08(20)} - 1)}{e^{0.08/12} - 1}$$
$$A \approx \$17{,}729.42$$

Total interest $= \$17{,}729.42 - 7200 \approx \$10{,}529.42$

Total deposits $= \$30 \cdot 12 \cdot 20 = \$7200.00$

43.
$$y = Ce^{kt}$$
$$3 = Ce^{k(0)}$$
$$3 = C$$
$$8 = 3e^{k(2)}$$
$$\frac{8}{3} = e^{2k}$$
$$\ln\frac{8}{3} = \ln e^{2k}$$
$$\ln\frac{8}{3} = 2k$$
$$\frac{\ln\frac{8}{3}}{2} = k \approx 0.4904$$

45.
$$y = Ce^{kt}$$
$$400 = Ce^{k(0)}$$
$$400 = C$$
$$200 = 400e^{k(3)}$$
$$\frac{1}{2} = e^{3k}$$
$$\ln\frac{1}{2} = \ln e^{3k}$$
$$\ln\frac{1}{2} = 3k$$
$$\frac{\ln\frac{1}{2}}{3} = k \approx -0.2310$$

47.
$$y = Ce^{kt}$$
$$90 = Ce^{k(0)}$$
$$90 = C$$
$$106 = 90e^{k(20)}$$
$$\frac{106}{90} = e^{20k}$$
$$\ln\frac{106}{90} = 20k$$
$$\frac{1}{20}\ln\frac{106}{90} = k$$
$$0.0082 \approx k$$
$$y = 90e^{0.0082t}$$
$$y = 90e^{0.0082(25)}$$
$$y \approx 100 \times 10^3$$
$y \approx 110{,}000$ people

49.
$$y = Ce^{kt}$$
$$286 = Ce^{k(0)}$$
$$286 = C$$
$$303 = 286e^{k(20)}$$
$$\frac{303}{286} = e^{20k}$$
$$\ln\frac{303}{286} = 20k$$
$$\frac{1}{20}\ln\frac{303}{286} = k$$
$$0.0029 \approx k$$
$$y = 286e^{0.0029t}$$
$$y = 286e^{0.0029(25)}$$
$$y \approx 308 \times 10^3$$
$y \approx 308{,}000$ people

51.
$$\begin{aligned} y &= Ce^{kt} \\ 2589 &= Ce^{k(0)} \\ 2589 &= C \end{aligned} \qquad \begin{aligned} 2872 &= 2589e^{k(20)} \\ \tfrac{2872}{2589} &= e^{20k} \\ \ln \tfrac{2872}{2589} &= 20k \\ \tfrac{1}{20}\ln \tfrac{2872}{2589} &= k \\ 0.0052 &\approx k \end{aligned} \qquad \begin{aligned} y &= 2589e^{0.0052t} \\ y &= 2589e^{0.0052(25)} \\ y &\approx 2948 \times 10^3 \\ y &\approx 2{,}948{,}000 \text{ people} \end{aligned}$$

53.
$$\begin{aligned} y &= Ce^{kt} \\ 3834 &= Ce^{k(0)} \\ 3834 &= C \end{aligned} \qquad \begin{aligned} 4252 &= 3834e^{k(20)} \\ \tfrac{4252}{3834} &= e^{20k} \\ \ln \tfrac{4252}{3834} &= 20k \\ \tfrac{1}{20}\ln \tfrac{4252}{3834} &= k \\ 0.0052 &\approx k \end{aligned} \qquad \begin{aligned} y &= 3834e^{0.0052t} \\ y &= 3834e^{0.0052(25)} \\ y &\approx 4366 \times 10^3 \\ y &\approx 4{,}366{,}000 \text{ people} \end{aligned}$$

55. K is larger in Exercise 47 because the population of Aruba is increasing faster than the population of Jamaica.

57.
$$P = \frac{11.7}{1 + 1.21e^{-0.0269(35)}}$$
$$P \approx 7.949 \text{ billion people}$$

59. (a)
$$\begin{aligned} y &= Ce^{kt} \\ 1 \times 10^6 &= 73e^{k(20)} \\ \frac{1 \times 10^6}{73} &= e^{20k} \\ \ln \frac{1 \times 10^6}{73} &= 20k \\ \frac{1}{20}\ln \frac{1 \times 10^6}{73} &= k \\ 0.4763 &\approx k \\ y &= 73e^{0.4763t} \end{aligned}$$

(b)
$$\begin{aligned} 5300 &= 73e^{0.4763t} \\ \frac{5300}{73} &= e^{0.4763t} \\ \ln \frac{5300}{73} &= 0.4763t \\ \frac{\ln \frac{5300}{73}}{0.4763} &= t \\ 93 \text{ hours} &\approx t \end{aligned}$$

61. $y = Ce^{kt}$
$$\begin{aligned} 233{,}041{,}000 &= 109{,}478{,}000e^{k(6)} \\ \frac{233{,}041{,}000}{109{,}478{,}000} &= e^{6k} \\ \ln \frac{233{,}041}{109{,}478} &= 6k \\ \frac{1}{6}\ln \frac{233{,}041}{109{,}478} &= k \\ 0.1259 &\approx k \end{aligned} \qquad \begin{aligned} y &= 109{,}478{,}000e^{0.1259t} \\ y &= 109{,}478{,}000e^{0.1259(18)} \\ y &\approx 562{,}518{,}000 \text{ users} \end{aligned}$$

63. $y = Ce^{kt}$

$6 = Ce^{k(0)}$

$6 = C$

$3 = 6e^{k(1620)}$

$0.5 = e^{1620k}$

$\ln 0.5 = \ln e^{1620k}$

$\frac{\ln 0.5}{1620} = k$

$-0.00043 \approx k$

$y = 6e^{-0.00043(1000)}$

$y \approx 3.91$ grams

65. $y = Ce^{kt}$

$0.5C = Ce^{k(5730)}$

$0.5 = e^{5730k}$

$\ln 0.5 = \ln e^{5730k}$

$\ln 0.5 = 5730k$

$\frac{\ln 0.5}{5730} = k$

$-0.00012 \approx k$

$4 = Ce^{-0.00012(1000)}$

$4 = Ce^{-0.12}$

$\frac{4}{e^{-0.12}} = C$

4.51 grams $\approx C$

67. $y = Ce^{kt}$

$4.2 = Ce^{k(0)}$

$4.2 = C$

$2.1 = 4.2e^{k(24,100)}$

$0.5 = e^{24,100k}$

$\ln 0.5 = \ln e^{24,100k}$

$\frac{\ln 0.5}{24,100} = k$

$-0.00003 \approx k$

$y = 4.2e^{-0.00003(1000)}$

$y \approx 4.08$ grams

69. $y = Ce^{kt}$

$5 = Ce^{k(0)}$

$5 = C$

$2.5 = 5e^{k(1620)}$

$0.5 = e^{1620k}$

$\ln 0.5 = \ln e^{1620k}$

$\ln 0.5 = 1620k$

$\frac{\ln 0.5}{1620} = k$

$-0.00043 \approx k$

$y = 5e^{-0.00043(1000)}$

$y \approx 3.3$ grams

71. (a) $y = Ce^{kt}$

$10 = Ce^{k(0)}$

$10 = C$

(b) $5 = 10e^{k(24,360)}$

$0.5 = e^{24,360k}$

$\ln 0.5 = \ln e^{24,360k}$

$\frac{\ln 0.5}{24,360} = k$

$-0.0000285 = k$

(c) $y = 10e^{-0.0000285(10,000)}$

$y \approx 7.5$ grams

73. $26,000 = 34,000e^{k(1)}$

$\frac{26,000}{34,000} = e^k$

$\ln \frac{26,000}{34,000} = \ln e^k$

$\ln \frac{26,000}{34,000} = k$

$-0.26826 \approx k$

$y \approx 34,000e^{-0.26826(3)} \approx \$15,204$

75. $R = \log_{10} I$

Chile (1960): $9.5 = \log_{10} I$

$10^{9.5} = I$

Chile (2008): $5.2 = \log_{10} I$

$10^{5.2} = I$

Ratio of two intensities:

$$\frac{I \text{ for 1960 Chile}}{I \text{ for 2008 Chile}} = \frac{10^{9.5}}{10^{5.2}} = 10^{9.5-5.2} = 10^{4.3} \approx 19{,}953$$

The Chilean earthquake in 1960 was about 19,953 times greater.

77. $R = \log_{10} I$

Fiji: $6.5 = \log_{10} I$

$10^{6.5} = I$

Philippines: $4.7 = \log_{10} I$

$10^{4.7} = I$

Ratio of two intensities:

$$\frac{I \text{ for Fiji}}{I \text{ for Philippines}} = \frac{10^{6.5}}{10^{4.7}} = 10^{6.5-4.7} = 10^{1.8} \approx 63$$

The earthquake in Fiji was about 63 times greater.

79. $\text{pH} = -\log_{10}\left[H^+\right]$

$\text{pH} = -\log_{10}\left(9.2 \times 10^{-8}\right) \approx 7.04$

81. $\text{pH} = -\log_{10}\left[H^+\right]$

Fruit:

$$2.5 = -\log_{10}\left[H^+\right]$$
$$-2.5 = \log_{10}\left[H^+\right]$$
$$10^{-2.5} = 10^{\log_{10}\left[H^+\right]}$$
$$0.0031623 \approx H^+$$

Tablet:

$$9.5 = -\log_{10}\left[H^+\right]$$
$$-9.5 = \log_{10}\left[H^+\right]$$
$$10^{-9.5} = 10^{\log_{10}\left[H^+\right]}$$
$$3.1623 \times 10^{-10} \approx H^+$$

$$\frac{H^+ \text{ of fruit}}{H^+ \text{ of tablet}} \approx \frac{3.1623 \times 10^{-3}}{3.1623 \times 10^{-10}} = 10^7$$

The H^+ of fruit is 10^7 times as great.

83. (a) *Keystrokes:* [Y=] 5000 [÷] [(] 1 [+] 4 [e^x] [(] [(–)] [X,T,θ] [÷] 6 [)] [)] [GRAPH]

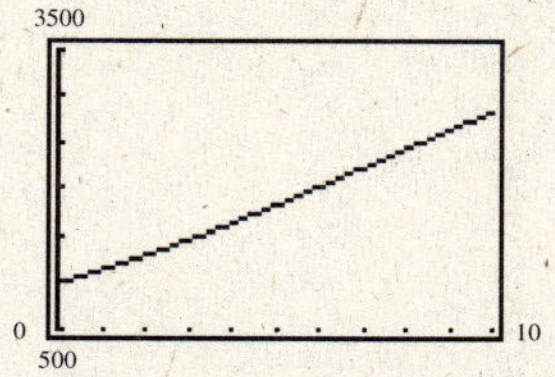

(b) $p(0) = \dfrac{5000}{1 + 4e^{-0/6}} = \dfrac{5000}{5} = 1000$ rabbits

(c) $p(9) = \dfrac{5000}{1 + 4e^{-9/6}} \approx 2642$ rabbits

(d)
$$2000 = \frac{5000}{1 + 4e^{-t/6}}$$
$$1 + 4e^{-t/6} = 2.5$$
$$e^{-t/6} = 0.375$$
$$-\frac{t}{6} = \ln 0.375$$
$$t = -6 \ln 0.375$$
$$t \approx 5.88 \text{ years}$$

85. (a)

$$S = 10\left(1 - e^{kx}\right)$$
$$2.5 = 10\left(1 - e^{k(5)}\right)$$
$$0.25 = 1 - e^{5k}$$
$$-0.75 = -e^{5k}$$
$$0.75 = e^{5k}$$
$$\ln 0.75 = \ln e^{5k}$$
$$\ln 0.75 = 5k$$
$$\frac{\ln 0.75}{5} = k$$
$$-0.0575 \approx k$$
$$S = 10\left(1 - e^{-0.0575x}\right)$$

(b) $S = 10\left(1 - e^{-0.0575(7)}\right)$

$\approx 3.314 \times 1000 \approx 3314$ jeans

87. If $k > 0$, the model represents exponential growth and if $k < 0$, the model represents exponential decay.

89. When the investment is compounded more than once in a year (quarterly, monthly, daily, continuous), the effective yield is greater than the interest rate.

91. $x^2 - 7x - 5 = 0$

$$x = \frac{-(-7) \pm \sqrt{(-7)^2 - 4(1)(-5)}}{2(1)}$$
$$x = \frac{7 \pm \sqrt{49 + 20}}{2}$$
$$x = \frac{7 \pm \sqrt{69}}{2}$$
$$x = \frac{7}{2} \pm \frac{\sqrt{69}}{2}$$

93. $3x^2 + 9x + 4 = 0$

$$x = \frac{-9 \pm \sqrt{9^2 - 4(3)(4)}}{2(3)}$$
$$x = \frac{-9 \pm \sqrt{81 - 48}}{6}$$
$$x = \frac{-9 \pm \sqrt{33}}{6}$$
$$x = -\frac{3}{2} \pm \frac{\sqrt{33}}{6}$$

95. $\dfrac{4}{x - 4} > 0$

Critical number: 4

Test intervals:

Negative: $(-\infty, 4)$

Positive: $(4, \infty)$

Solution: $(4, \infty)$

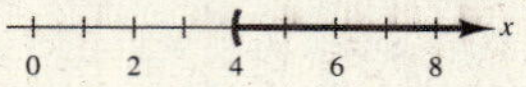

97.

$$\frac{2x}{x - 3} > 1$$
$$\frac{2x}{x - 3} - 1 > 0$$
$$\frac{2x - (x - 3)}{x - 3} > 0$$
$$\frac{x + 3}{x - 3} > 0$$

Critical numbers: $-3, 3$

Test intervals:

Positive: $(-\infty, -3)$

Negative: $(-3, 3)$

Positive: $(3, \infty)$

Solution: $(-\infty, -3) \cup (3, \infty)$

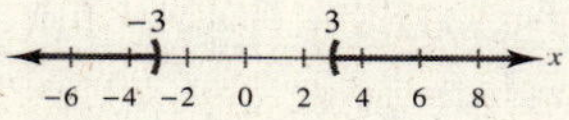

Review Exercises for Chapter 9

1. $f(x) = 4^x$

(a) $f(-3) = 4^{-3} = \frac{1}{4^3} = \frac{1}{64}$

(b) $f(1) = 4^1 = 4$

(c) $f(2) = 4^2 = 16$

3. $g(t) = 5^{-t/3}$

(a) $g(-3) = 5^{-(-3)/3} = 5^1 = 5$

(b) $g(\pi) = 5^{-\pi/3} \approx 0.185$

(c) $g(6) = 5^{-6/3} = 5^{-2} = \frac{1}{25}$

5. $f(x) = 3^x$

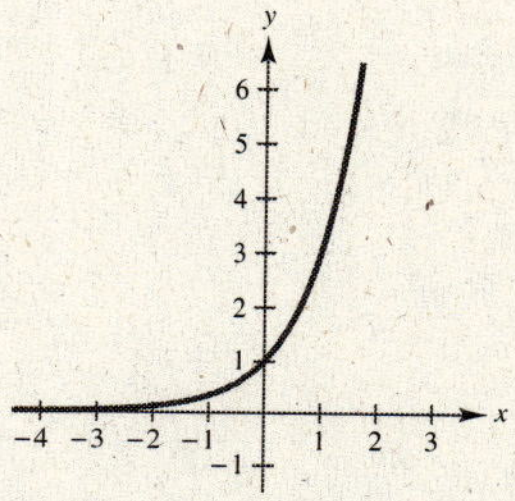

Horizontal asymptote: $y = 0$

Table of values:

x	-1	0	1
y	$\frac{1}{3}$	1	3

7. $f(x) = 3^x - 3$

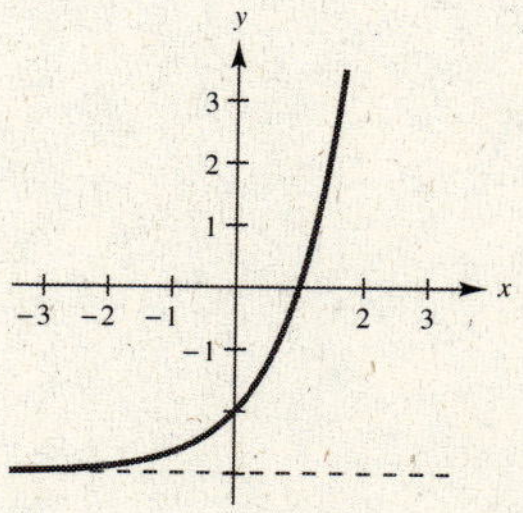

Horizontal asymptote: $y = -3$

9. $f(x) = 3^{(x+1)}$

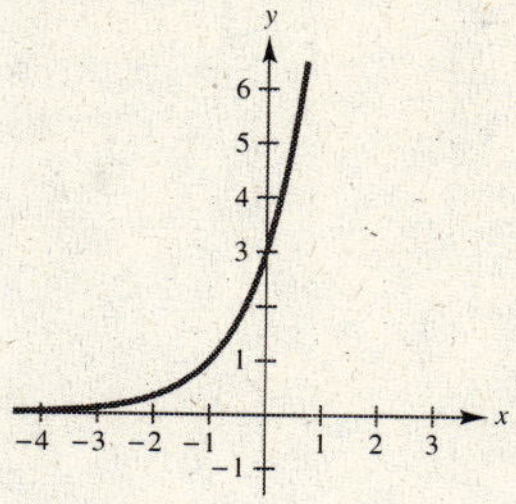

Horizontal asymptote: $y = 0$

Table of values:

x	-1	0	1
y	1	3	9

11. $f(x) = 3^{x/2}$

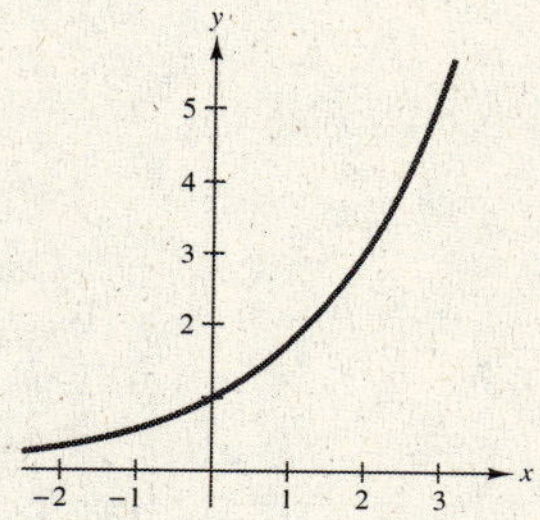

Horizontal asymptote: $y = 0$

Table of values:

x	0	2	-2
y	1	3	$\frac{1}{3}$

13. $f(x) = 3^{x/2} - 2$

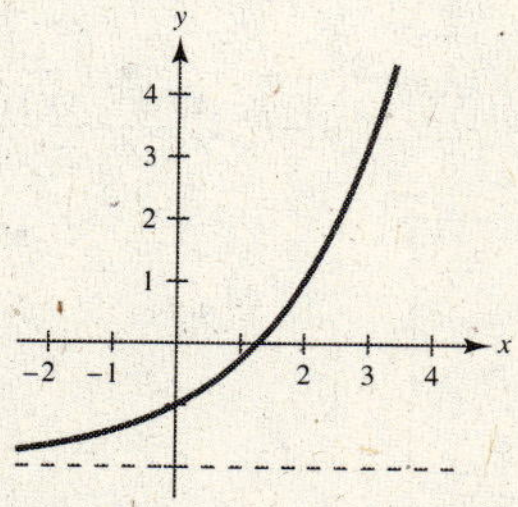

Horizontal asymptote: $y = -2$

Table of values:

x	-2	0	2
y	$-\frac{7}{3}$	-1	1

15. $f(x) = 2^{-x^2}$

Keystrokes: Y= 2 ^ ((−) X,T,θ x²) GRAPH

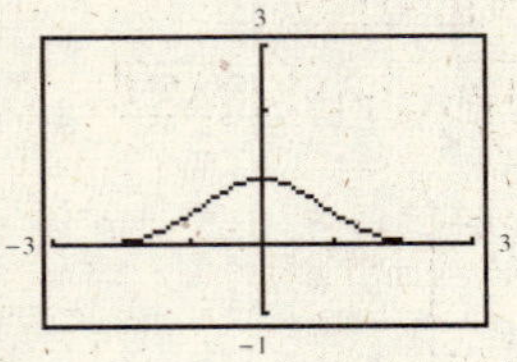

17. $y = 10(1.09)^t$

Keystrokes: Y= 10 (1.09) ^ X,T,θ GRAPH

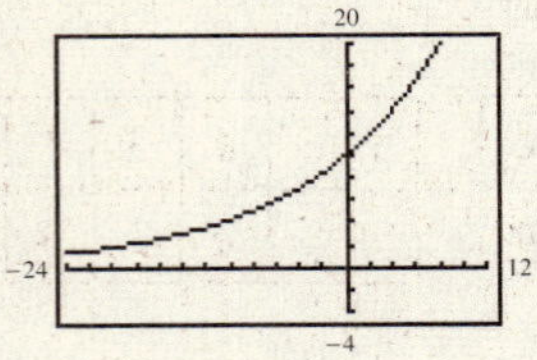

19. $f(x) = 3e^{-2x}$

(a) $f(3) = 3e^{-2(3)} = 3e^{-6} \approx 0.007$

(b) $f(0) = 3e^{-2(0)} = 3e^{0} = 3$

(c) $f(-19) = 3e^{-2(-19)} = 3e^{38} \approx 9.56 \times 10^{16}$

21. $y = 4e^{-x/3}$

Keystrokes: Y= 4 eˣ − X,T,θ ÷ 3) GRAPH

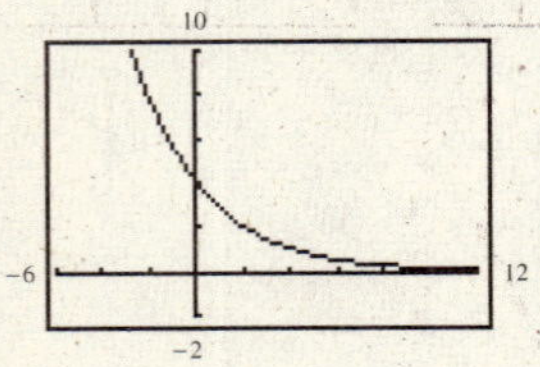

23. $f(x) = e^{x+2}$

Keystrokes: Y= eˣ (X,T,θ + 2) GRAPH

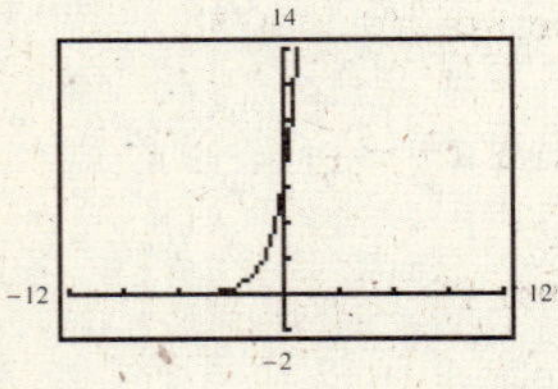

25. (a) $A = P\left(1 + \frac{r}{n}\right)^{nt}$

$= 5000\left(1 + \frac{0.10}{1}\right)^{1(40)}$

$= 5000(1.10)^{40}$

$\approx \$226{,}296.28$

(b) $A = P\left(1 + \frac{r}{n}\right)^{nt}$

$= 5000\left(1 + \frac{0.10}{4}\right)^{4(40)}$

$= 5000(1.025)^{160}$

$\approx \$259{,}889.34$

(c) $A = P\left(1 + \frac{r}{n}\right)^{nt}$

$= 5000\left(1 + \frac{0.10}{12}\right)^{12(40)}$

$= 5000\left(1.008\overline{3}\right)^{480}$

$\approx \$268{,}503.32$

(d) $A = P\left(1 + \frac{r}{n}\right)^{nt}$

$= 5000\left(1 + \frac{0.10}{365}\right)^{365(40)}$

$\approx 5000(1.00027397)^{14600}$

$\approx \$272{,}841.23$

(e) $A = Pe^{rt}$

$= 5000e^{0.10(40)}$

$= 5000e^{4}$

$\approx \$272{,}990.75$

27. $y = 21\left(\frac{1}{2}\right)^{t/25}, \quad t \ge 0$

$y = 21\left(\frac{1}{2}\right)^{58/25}$

$y \approx 4.21$ grams

29. (a) $(f \circ g)(x) = x^2 + 2$

so $(f \circ g)(2) = 2^2 + 2 = 6.$

(b) $(g \circ f)(x) = (x + 2)^2 = x^2 + 4x + 4$

so $(g \circ f)(-1) = (-1)^2 + 4(-1) + 4$

$= 1 - 4 + 4 = 1.$

31. (a) $(f \circ g)(x) = \sqrt{x^2 - 1 + 1} = \sqrt{x^2} = |x|$

so $(f \circ g)(5) = |5| = 5.$

(b) $(g \circ f)(x) = \left(\sqrt{x+1}\right)^2 - 1 = x + 1 - 1 = x$

so $(g \circ f)(-1) = -1.$

33. $f(x) = \sqrt{x+6}, \quad g(x) = 2x$

(a) $(f \circ g)(x) = f(2x) = \sqrt{2x+6}$

Domain: $[-3, \infty)$

$$2x + 6 \ge 0$$
$$x \ge -3$$

(b) $(g \circ f)(x) = g\left(\sqrt{x+6}\right) = 2\sqrt{x+6}$

Domain: $[-6, \infty)$

$$x + 6 \ge 0$$
$$x \ge -6$$

35. No, $f(x)$ does not have an inverse. f is not one-to-one.

37. Yes, $h(x)$ does have an inverse. f is one-to-one.

39.
$$f(x) = 3x + 4$$
$$y = 3x + 4$$
$$x = 3y + 4$$
$$x - 4 = 3y$$
$$\frac{x-4}{3} = y$$
$$\frac{x-4}{3} = f^{-1}(x) = \frac{1}{3}(x - 4)$$

41.
$$h(x) = \sqrt{5x}$$
$$y = \sqrt{5x}$$
$$x = \sqrt{5y}$$
$$x^2 = 5y$$
$$\tfrac{1}{5}x^2 = y$$
$$\tfrac{1}{5}x^2 = h^{-1}(x), \quad x \ge 0$$

43.
$$f(t) = t^3 + 4$$
$$y = t^3 + 4$$
$$t = y^3 + 4$$
$$t - 4 = y^3$$
$$\sqrt[3]{t-4} = y$$
$$\sqrt[3]{t-4} = f^{-1}(t)$$

45. $f(x) = 3x + 4, \; g(x) = \frac{1}{3}(x - 4)$

Keystrokes:

y_1: [Y=] 3 [X,T,θ] [+] 4 [ENTER]

y_2: [(] 1 [÷] 3 [)] [(] [X,T,θ] [−] 4 [)] [GRAPH]

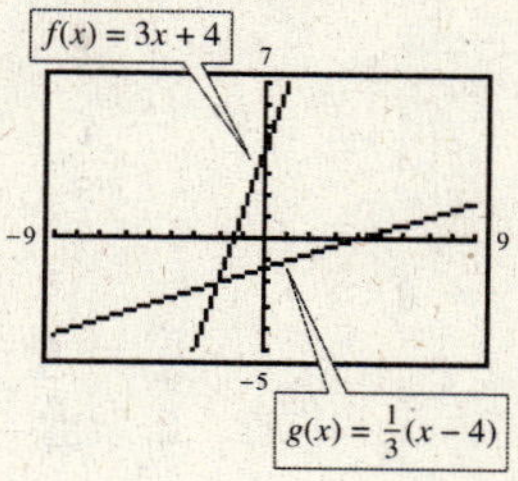

47.

x	0	2	4	6
f	4	3	1	0

x	4	3	1	0
f^{-1}	0	2	4	6

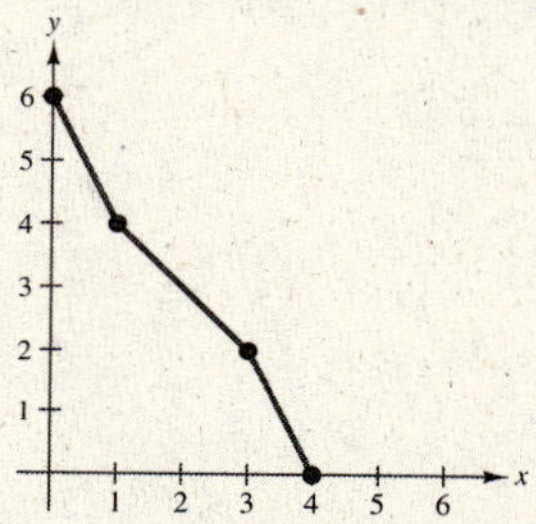

49.

x	−2	−1	1	3
f	−4	−2	2	3

x	−4	−2	2	3
f^{-1}	−2	−1	1	3

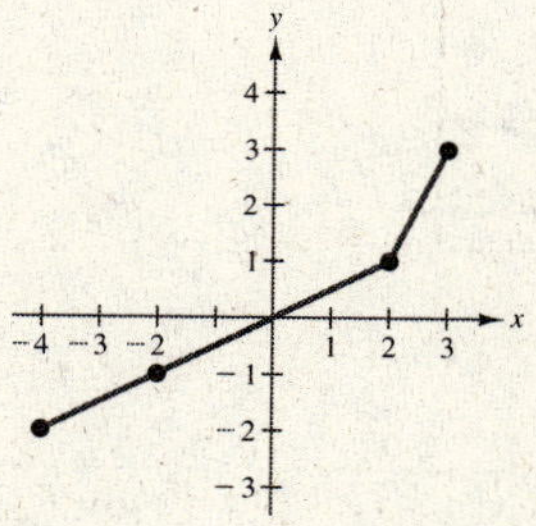

51. $\log_{10} 1000 = 3$ because $10^3 = 1000$.

53. $\log_3 \frac{1}{9} = -2$ because $3^{-2} = \frac{1}{9}$.

55. $\log_2 64 = 6$ because $2^6 = 64$.

57. $\log_3 1 = 0$ because $3^0 = 1$.

59. $f(x) = \log_3 x$

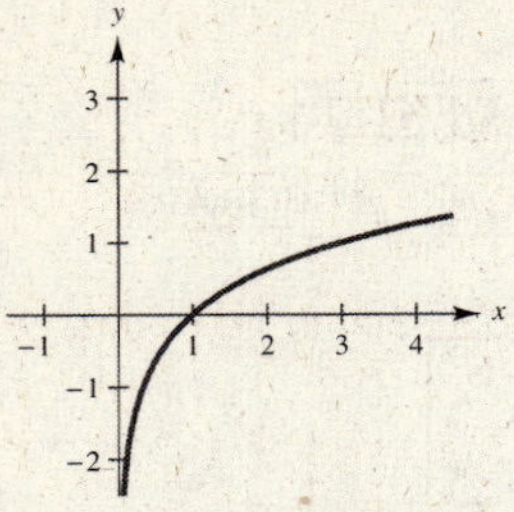

Vertical asymptote: $x = 0$

Table of values:

x	1	3
y	0	1

61. $f(x) = -1 + \log_3 x$

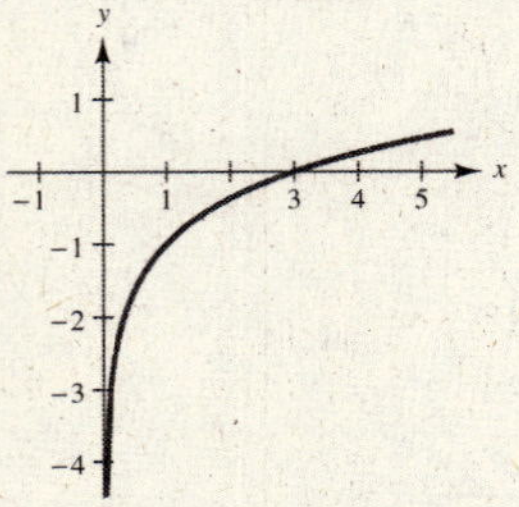

Vertical asymptote: $x = 0$

63. $y = \log_2(x - 4)$

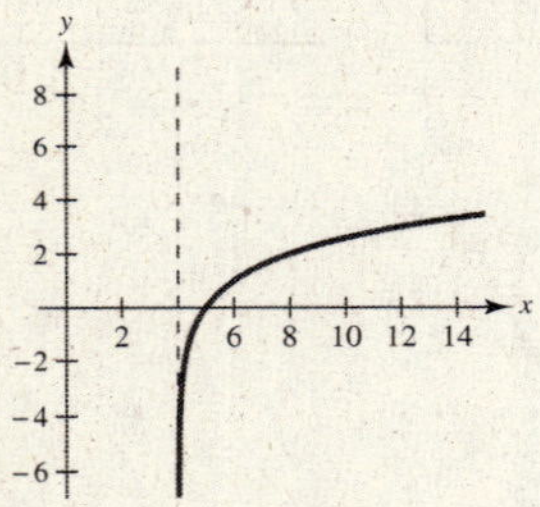

Vertical asymptote: $x = 4$

Table of values:

x	5	6
y	0	1

65. $\ln 50 \approx 3.9120$

67. $y = \ln(x - 3)$

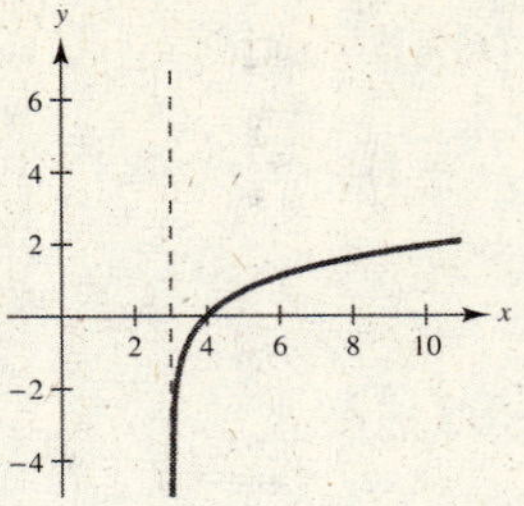

Vertical asymptote: $x = 3$

Table of values:

x	4	5
y	0	0.7

69. $y = 5 - \ln x$

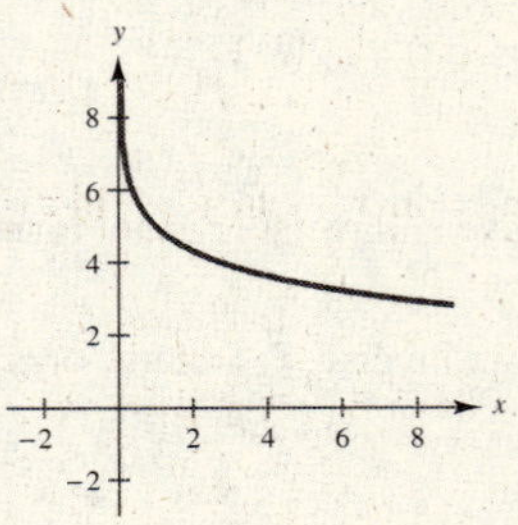

Vertical asymptote: $x = 0$

Table of values:

x	1	e
y	5	4

71. $\log_4 9 = \dfrac{\log 9}{\log 4} \approx 1.5850$

73. $\log_8 160 = \dfrac{\log 160}{\log 8} \approx 2.4406$

$= \dfrac{\ln 160}{\ln 8} \approx 2.4406$

75. $\log_5 18 = \log_5 3^2 + \log_5 2$

$= 2 \log_5 3 + \log 2$

$\approx 2(0.6826) + 0.4307$

≈ 1.7959

77. $\log_5 \frac{1}{2} = \log_5 1 - \log_5 2$

$\approx 0 - (0.4307)$

≈ -0.4307

79. $\log_5(12)^{2/3} = \frac{2}{3}[2\log_5 2 + \log_5 3]$
$\approx \frac{2}{3}[2(0.4307) + 0.6826]$
≈ 1.0293

81. $\log_4 6x^4 = \log_4 6 + 4\log_4 x$

83. $\log_5 \sqrt{x+2} = \frac{1}{2}\log_5(x+2)$

85. $\ln\dfrac{x+2}{x+3} = \ln(x+2) - \ln(x+3)$

87. $\ln\left[\sqrt{2x}(x+3)^5\right] = \ln\sqrt{2x} + \ln(x+3)^5$
$= \ln(2x)^{1/2} + 5\ln(x+3)$
$= \frac{1}{2}[\ln 2 + \ln x] + 5\ln(x+3)$

89. $-\dfrac{2}{3}\ln 3y = \ln(3y)^{-2/3} = \ln\left(\dfrac{1}{3y}\right)^{2/3}, \; y > 0$

91. $\log_8 16x + \log_8 2x^2 = \log_8(16x \cdot 2x^2) = \log_8(32x^3)$

93. $-2(\ln 2x - \ln 3) = \ln\left(\dfrac{2x}{3}\right)^{-2} = \ln\left(\dfrac{3}{2x}\right)^2 = \ln\dfrac{9}{4x^2}, \; x > 0$

95. $4[\log_2 k - \log_2(k-t)] = 4\left[\log_2\left(\dfrac{k}{k-t}\right)\right] = \log_2\left(\dfrac{k}{k-t}\right)^4, \; t < k, k > 0$

97. $3\ln x + 4\ln y + \ln z = \ln x^3 + \ln y^4 + \ln z = \ln x^3y^4z, \; x > 0, \; y > 0, \; z > 0$

99. False

$\log_2 4x = \log_2 4 + \log_2 x = 2 + \log_2 x$

101. True

$\log_{10} 10^{2x} = 2x\log_{10} 10 = 2x$

103. True

$\log_4 \dfrac{16}{x} = \log_4 16 - \log_4 x = 2 - \log_4 x$

105. $y = \ln\left(\dfrac{I_0}{I}\right)^{0.83}$
$y = 0.83\ln\left(\dfrac{I_0}{I}\right)$
$y = 0.83(\ln I_0 - \ln I)$
$y = 0.83(\ln 4.2 - \ln 3.3)$
$y \approx 0.20$

107. $2^x = 64$
$2^x = 2^6$
$x = 6$

109. $4^{x-3} = \frac{1}{16}$
$4^{x-3} = 4^{-2}$
$x - 3 = -2$
$x = 1$

111. $\log_7(x+6) = \log_7 12$
$x + 6 = 12$
$x = 6$

113. $3^x = 500$
$\log_3 3^x = \log_3 500$
$x = \dfrac{\log 500}{\log 3}$
$x \approx 5.66$

115. $2e^{0.5x} = 45$
$e^{0.5x} = 22.5$
$\ln e^{0.5x} = \ln 22.5$
$0.5x = \ln 22.5$
$x = 2\ln 22.5$
$x \approx 6.23$

117. $12(1 - 4^x) = 18$
$1 - 4^x = \frac{18}{12}$
$-4^x = \frac{3}{2} - 1$
$-4^x = \frac{1}{2}$
$4^x = -\frac{1}{2}$

No solution; there is no power that will raise 4 to $-\frac{1}{2}$.

119. $\ln x = 7.25$

$e^{\ln x} = e^{7.25}$

$x = e^{7.25}$

$x \approx 1408.10$

121. $\log_{10} 4x = 2.1$

$4x = 10^{2.1}$

$x = \dfrac{10^{2.1}}{4}$

$x \approx 31.47$

123. $\log_3(2x + 1) = 2$

$3^{\log_3(2x+1)} = 3^2$

$2x + 1 = 9$

$2x = 8$

$x = 4.00$

125. $\frac{1}{3}\log_2 x + 5 = 7$

$\frac{1}{3}\log_2 x = 2$

$\log_2 x = 6$

$2^{\log_2 x} = 2^6$

$x = 2^6 = 64.00$

127. $\log_3 x + \log_3 7 = 4$

$\log_3 7x = 4$

$7x = 3^4$

$x = \dfrac{3^4}{7}$

$x \approx 11.57$

129. $A = Pe^{rt}$

$5751.37 = 5000e^{r(2)}$

$\dfrac{5751.37}{5000} = e^{2r}$

$\ln 1.150274 = \ln e^{2r}$

$\ln 1.150274 = 2r$

$\dfrac{\ln 1.150274}{2} = r \approx 7\%$

131. $A = P\left(1 + \dfrac{r}{n}\right)^{nt}$

$410.90 = 250\left(1 + \dfrac{r}{4}\right)^{4(10)}$

$1.6436 = \left(1 + \dfrac{r}{4}\right)^{40}$

$(1.6436)^{1/40} = 1 + \dfrac{r}{4}$

$1.0124997 \approx 1 + \dfrac{r}{4}$

$0.0124997 \approx \dfrac{r}{4}$

$0.0499 \approx r$

$5\% \approx r$

133. $A = P\left(1 + \dfrac{r}{n}\right)^{nt}$

$15{,}399.30 = 5000\left(1 + \dfrac{r}{365}\right)^{365(15)}$

$3.07986 = \left(1 + \dfrac{r}{365}\right)^{5475}$

$(3.07986)^{1/5475} = 1 + \dfrac{r}{365}$

$1.000205479 \approx 1 + \dfrac{r}{365}$

$0.000205479 \approx \dfrac{r}{365}$

$0.074999 \approx r$

$7.5\% \approx r$

135. $A = Pe^{rt}$

$46{,}422.61 = 1800e^{r(50)}$

$\dfrac{46{,}422.61}{1800} = e^{50r}$

$25.790033\overline{8} = e^{50r}$

$\ln 25.790033\overline{8} = 50r$

$\dfrac{\ln 25.790033\overline{8}}{50} = r$

$0.065 \approx r$

$6.5\% \approx r$

137. $A = P\left(1 + \dfrac{r}{n}\right)^{nt}$

$A = 1000\left(1 + \dfrac{0.055}{365}\right)^{365(1)}$

$A = \$1056.54$

$\text{Effective yield} = \dfrac{56.54}{1000} = 0.05654 \approx 5.65\%$

139. $A = P\left(1 + \frac{r}{n}\right)^{nt}$

$A = 1000\left(1 + \frac{0.075}{4}\right)^{4(1)}$

$A = \$1077.14$

Effective yield $= \frac{77.14}{1000} = 0.07714 \approx 7.71\%$

141. $A = Pe^{rt}$

$A = 1000e^{0.075(1)}$

$A = \$1077.88$

Effective yield $= \frac{77.88}{1000} = 0.07788 \approx 7.79\%$

143.

$y = Ce^{kt}$

$3.5 = Ce^{k(0)}$

$3.5 = C$

$1.75 = 3.5e^{k(1620)}$

$0.5 = e^{1620k}$

$\ln 0.5 = \ln e^{1620k}$

$\ln 0.5 = 1620k$

$\frac{\ln 0.5}{1620} = k$

$-0.00043 \approx k$

$y = 3.5e^{-0.00043(1000)}$

$y \approx 2.282$ grams

145.

$y = Ce^{kt}$

$0.5C = Ce^{k(5730)}$

$0.5 = e^{k(5730)}$

$\ln 0.5 = \ln e^{5730k}$

$\ln 0.5 = 5730k$

$\frac{\ln 0.5}{5730} = k$

$-0.00012 \approx k$

$2.6 = Ce^{-0.00012(1000)}$

$2.6 = Ce^{-0.12}$

$\frac{2.6}{e^{-0.12}} = C$

$2.934 \text{ grams} \approx C$

147.

$y = Ce^{kt}$

$2.5 = 0.5e^{k(24,100)}$

$0.5 = e^{24,100k}$

$\ln 0.5 = 24,100k$

$\frac{\ln 0.5}{24,100} = k$

$-0.0000288 \approx k$

$y = 5e^{-0.0000288(1000)}$

$y \approx 4.858$ grams

149. $R = \log_{10} I$

San Francisco: $8.3 = \log_{10} I$

$10^{8.3} = 10^{\log_{10} I}$

$10^{8.3} = I$

Napa: $4.9 = \log_{10} I$

$10^{4.9} = 10^{\log_{10} I}$

$10^{4.9} = I$

Ratio of two intensities:

$$\frac{I \text{ for San Francisco}}{I \text{ for Napa}} = \frac{10^{8.3}}{10^{4.9}} = 10^{8.3-4.9} = 10^{3.4} \approx 2512$$

The earthquake in San Francisco was about 2512 times greater.

Chapter Test for Chapter 9

1. (a) $f(-1) = 54\left(\frac{2}{3}\right)^{-1}$

$= 54\left(\frac{3}{2}\right)$

$= 81$

(b) $f(0) = 54\left(\frac{2}{3}\right)^{0}$

$= 54$

(c) $f\left(\frac{1}{2}\right) = 54\left(\frac{2}{3}\right)^{1/2}$

≈ 44.09

(d) $f(2) = 54\left(\frac{2}{3}\right)^{2}$

$= 54\left(\frac{4}{7}\right)$

$= 24$

2. $f(x) = 2^{x/3}$

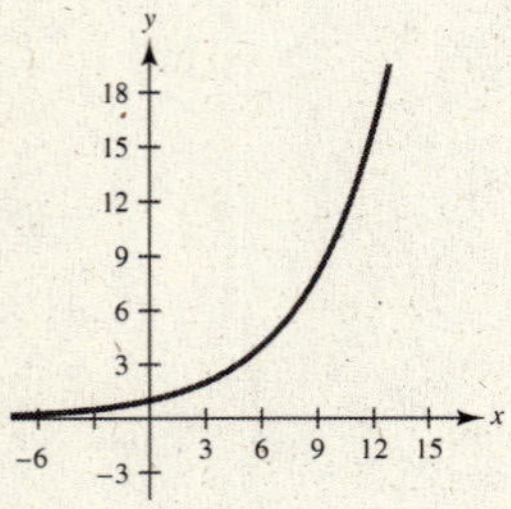

Horizontal asymptote: $y = 0$

3. (a) $f \circ g = f(g(x)) = f(5 - 3x)$

$= 2(5 - 3x)^2 + (5 - 3x)$

$= 2(25 - 30x + 9x^2) + 5 - 3x$

$= 50 - 60x + 18x^2 + 5 - 3x$

$= 18x^2 - 63x + 55$

Domain: $(-\infty, \infty)$

(b) $g \circ f = g(f(x)) = g(2x^2 + x)$

$= 5 - 3(2x^2 + x)$

$= 5 - 6x^2 - 3x$

$= -6x^2 - 3x + 5$

Domain: $(-\infty, \infty)$

4. $f(x) = 9x - 4$

$y = 9x - 4$

$x = 9y - 4$

$x + 4 = 9y$

$\dfrac{x + 4}{9} = y$

$\dfrac{1}{9}(x + 4) = f^{-1}(x)$

5. $f(g(x)) = f(-2x + 6)$

$= -\frac{1}{2}(-2x + 6) + 3$

$= x - 3 + 3$

$= x$

$g(f(x)) = g\left(-\frac{1}{2}x + 3\right)$

$= -2\left(-\frac{1}{2}x + 3\right) + 6$

$= x - 6 + 6$

$= x$

6. $\log_4 \frac{1}{256} = -4$ because $4^{-4} = \frac{1}{256}$.

7. f and g are inverse functions.

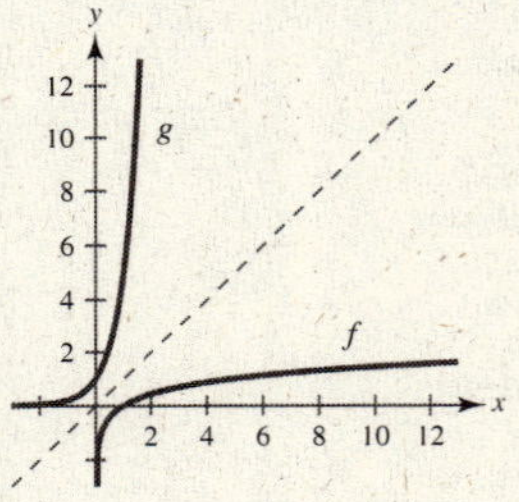

8. $\log_8\left(\dfrac{4\sqrt{x}}{y^4}\right) = \log_8 4\sqrt{x} - \log_8 y^4$

$= \log_8 4 + \log_8 x^{1/2} - \log_8 y^4$

$= \dfrac{2}{3} + \dfrac{1}{2}\log_8 x - 4\log_8 y$

9. $\ln x - \ln y = \ln \dfrac{x}{y^4},\ y > 0$

10. $\log_2 x = 5$

$2^{\log_2 x} = 2^5$

$x = 32$

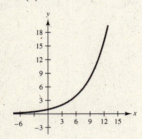

11.
$$\begin{aligned} 9^{2x} &= 182 \\ \log_9 9^{2x} &= \log_9 182 \\ 2x &= \frac{\log 182}{\log 9} \\ x &= \frac{\log 182}{2 \log 9} \\ x &\approx 1.18 \end{aligned}$$

12.
$$\begin{aligned} 400e^{0.08t} &= 1200 \\ e^{0.08t} &= 3 \\ \ln e^{0.08t} &= \ln 3 \\ 0.08t &= \ln 3 \\ t &= \frac{\ln 3}{0.08} \\ t &\approx 13.73 \end{aligned}$$

13.
$$\begin{aligned} 3\ln(2x-3) &= 10 \\ \ln(2x-3) &= \frac{10}{3} \\ e^{\ln(2x-3)} &= e^{10/3} \\ 2x - 3 &= e^{10/3} \\ x &= \frac{e^{10/3}+3}{2} \\ x &\approx 15.52 \end{aligned}$$

14.
$$\begin{aligned} 12(7 - 2^x) &= -300 \\ 7 - 2^x &= -\tfrac{300}{12} \\ 7 - 2^x &= -25 \\ -2^x &= -32 \\ 2^x &= 32 \\ 2^x &= 2^5 \\ x &= 5 \end{aligned}$$

15.
$$\begin{aligned} \log_2 x + \log_2 4 &= 5 \\ \log_2 x(4) &= 5 \\ 2^{\log_2 4x} &= 2^5 \\ 4x &= 32 \\ x &= 8 \end{aligned}$$

16.
$$\begin{aligned} \ln x - \ln 2 &= 4 \\ \ln \frac{x}{2} &= 4 \\ e^{\ln(x/2)} &= e^4 \\ \frac{x}{2} &= e^4 \\ x &= 2e^4 \\ x &\approx 109.20 \end{aligned}$$

17.
$$\begin{aligned} 30(e^x + 9) &= 300 \\ e^x + 9 &= 10 \\ e^x &= 1 \\ e^x &= e^0 \\ \text{so } x &= 0. \end{aligned}$$

18. (a) $A = 2000\left(1 + \frac{0.07}{4}\right)^{4(20)} \approx \$8012.78$

(b) $A = 2000e^{0.07(20)} \approx \$8110.40$

19.
$$\begin{aligned} 100{,}000 &= P\left(1 + \frac{0.09}{4}\right)^{4(25)} \\ \frac{100{,}000}{(1.0225)^{100}} &= P \\ \$10{,}806.08 &\approx P \end{aligned}$$

20.
$$\begin{aligned} 1006.88 &= 500e^{r(10)} \\ 2.01376 &= e^{10r} \\ \ln 2.01376 &= \ln e^{10r} \\ \ln 2.01376 &= 10r \\ \frac{\ln 2.01376}{10} &= r \\ 0.07 &\approx r \\ 7\% &\approx r \end{aligned}$$

21.
$$\begin{aligned} y &= Ce^{kt} \\ 15{,}000 &= 20{,}000e^{k(1)} \\ \tfrac{3}{4} &= e^k \\ \ln \tfrac{3}{4} &= k \\ y &= 20{,}000e^{[\ln(3/4)](5)} \\ y &\approx \$4746.09 \end{aligned}$$

22. $p(0) = \dfrac{2400}{1 + 3e^{-0/4}} = 600$

23. $p(4) = \dfrac{2400}{1 + 3e^{-4/4}} \approx 1141$

24.
$$\begin{aligned} 1200 &= \frac{2400}{1 + 3e^{-t/4}} \\ 1 + 3e^{-t/4} &= \frac{2400}{1200} \\ 3e^{-t/4} &= 1 \\ e^{-t/4} &= \frac{1}{3} \\ \ln e^{-t/4} &= \ln \frac{1}{3} \\ -\frac{t}{4} &= \ln \frac{1}{3} \\ t &= -4 \ln \frac{1}{3} \approx 4.4 \text{ years} \end{aligned}$$

CHAPTER 10
Conics

CHAPTER 10
Conics

Section 10.1 Circles and Parabolas

1. $x^2 + y^2 = 25$ (e)

Center: $(0, 0)$; radius: 5

3. $(x - 2)^2 + (y - 3)^2 = 9$ (d)

Center: $(2, 3)$; radius: 3

5. $y = -\sqrt{4 - x^2}$ (f)

Bottom half of circle

Center: $(0, 0)$; radius: 2

7. Center: $(0, 0)$; radius: 5

$$x^2 + y^2 = r^2$$
$$x^2 + y^2 = 5^2$$
$$x^2 + y^2 = 25$$

9. Center: $(0, 0)$; radius: $\frac{2}{3}$

$$x^2 + y^2 = r^2$$
$$x^2 + y^2 = \left(\tfrac{2}{3}\right)^2$$
$$x^2 + y^2 = \tfrac{4}{9} \text{ or } 9x^2 + 9y^2 = 4$$

11. Passes through $(0, 6)$; center: $(0, 0)$

$$(6 - 0)^2 + (0 - 0)^2 = r^2$$
$$36 = r^2$$
$$x^2 + y^2 = 36$$

13. Center: $(0, 0)$; point: $(5, 2)$

$$r = \sqrt{(5 - 0)^2 + (2 - 0)^2}$$
$$r = \sqrt{25 + 4}$$
$$r = \sqrt{29}$$
$$x^2 + y^2 = r^2$$
$$x^2 + y^2 = \left(\sqrt{29}\right)^2$$
$$x^2 + y^2 = 29$$

15. Center: $(4, 3)$; radius: 10

$$(x - h)^2 + (y - k)^2 = r^2$$
$$(x - 4)^2 + (y - 3)^2 = 10^2$$
$$(x - 4)^2 + (y - 3)^2 = 100$$

17. Center: $(6, -5)$; radius: 3

$$(x - 6)^2 + (y + 5)^2 = 9$$

19. Center: $(-2, 1)$; point: $(0, 1)$

$$r = \sqrt{\left[0 - (-2)^2\right] + (1 - 1)^2}$$
$$r = \sqrt{4 + 0}$$
$$r = 2$$
$$(x - h)^2 + (y - k)^2 = r^2$$
$$\left[x - (-2)\right]^2 + (y - 1)^2 = 2^2$$
$$(x + 2)^2 + (y - 1)^2 = 4$$

21. Center: $(3, 2)$; point: $(4, 6)$

$$r = \sqrt{(4 - 3)^2 + (6 - 2)^2}$$
$$r = \sqrt{1 + 16} = \sqrt{17}$$
$$(x - h)^2 + (y - k)^2 = r^2$$
$$(x - 3)^2 + (y - 2)^2 = \left(\sqrt{17}\right)^2$$
$$(x - 3)^2 + (y - 2)^2 = 17$$

23. $x^2 + y^2 = 16$

Center: $(0, 0)$; radius: 4

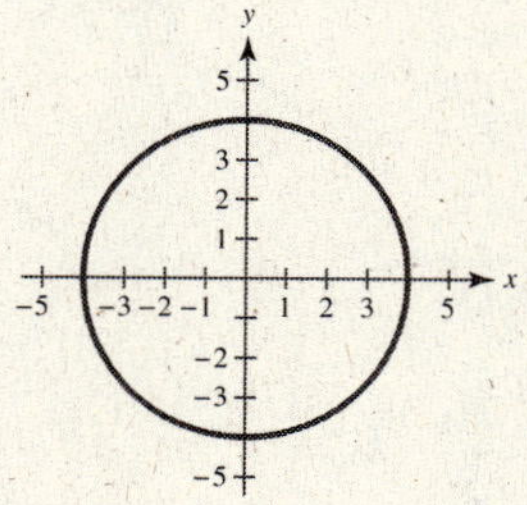

25. $x^2 + y^2 = 36$

Center: $(0, 0)$; radius: 6

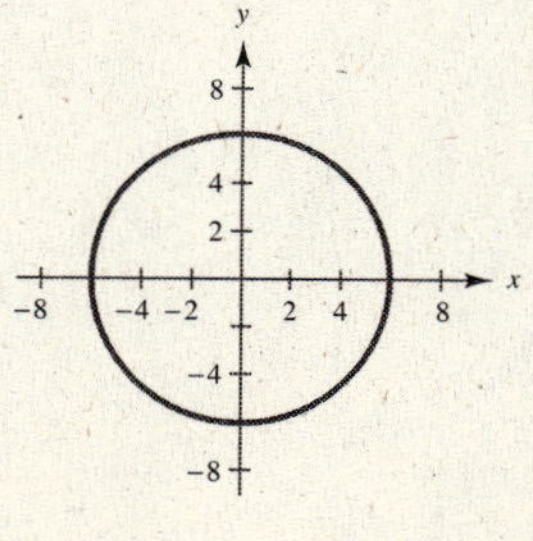

27. $4x^2 + 4y^2 = 1$

$$x^2 + y^2 = \frac{1}{4}$$

$$r = \frac{1}{2}$$

Center: $(0, 0)$; radius: $\frac{1}{2}$

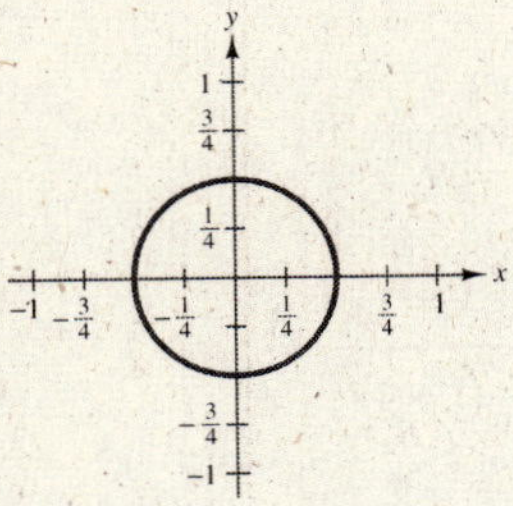

29. $25x^2 + 25y^2 - 144 = 0$

$$25x^2 + 25y^2 = 144$$

$$x^2 + y^2 = \frac{144}{25}$$

Center: $(0, 0)$; radius: $\frac{12}{5}$

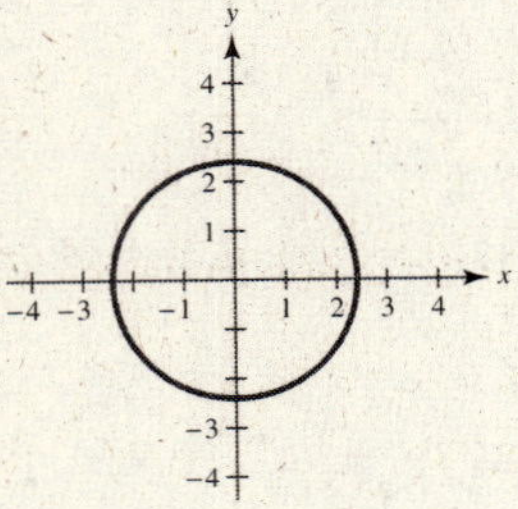

31. $(x + 1)^2 + (y - 5)^2 = 64$

Center: $(-1, 5)$; radius: 8

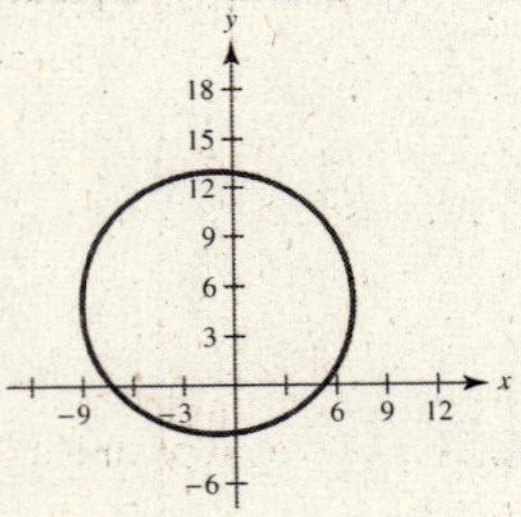

33. $(x - 2)^2 + (y - 3)^2 = 4$

Center: $(2, 3)$; radius: 2

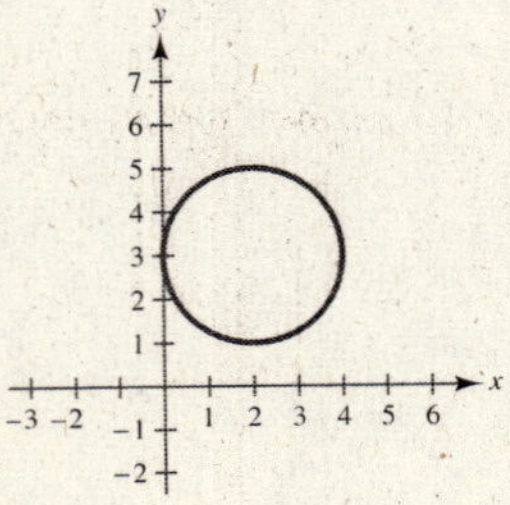

35. $\left(x + \frac{9}{4}\right)^2 + (y - 4)^2 = 16$

Center: $\left(-\frac{9}{4}, 4\right)$; radius: 4

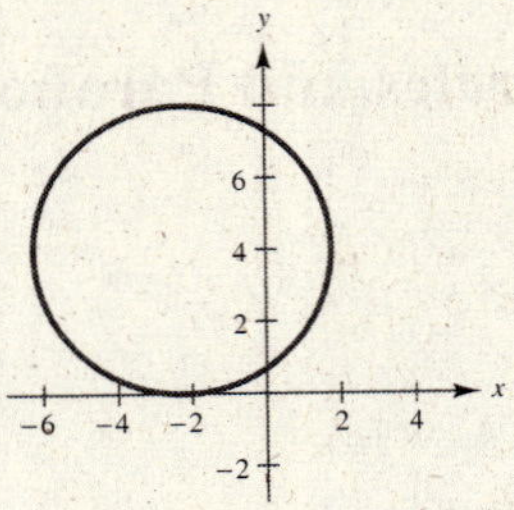

37. $x^2 + y^2 - 4x - 2y + 1 = 0$

$$x^2 - 4x + y^2 - 2y = -1$$

$$\left(x^2 - 4x + 4\right) + \left(y^2 - 2y + 1\right) = -1 + 4 + 1$$

$$(x - 2)^2 + (y - 1)^2 = 4$$

Center: $(2, 1)$; radius: 2

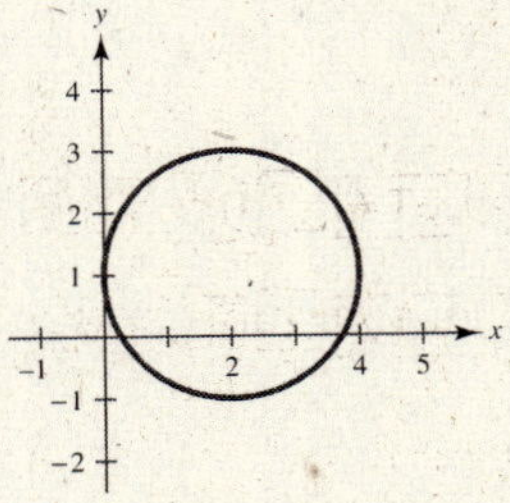

39. $x^2 + y^2 + 2x + 6y + 6 = 0$

$$x^2 + 2x + y^2 + 6y = -6$$

$$\left(x^2 + 2x + 1\right) + \left(y^2 + 6y + 9\right) = -6 + 1 + 9$$

$$(x + 1)^2 + (y + 3)^2 = 4$$

Center: $(-1, -3)$; radius: 2

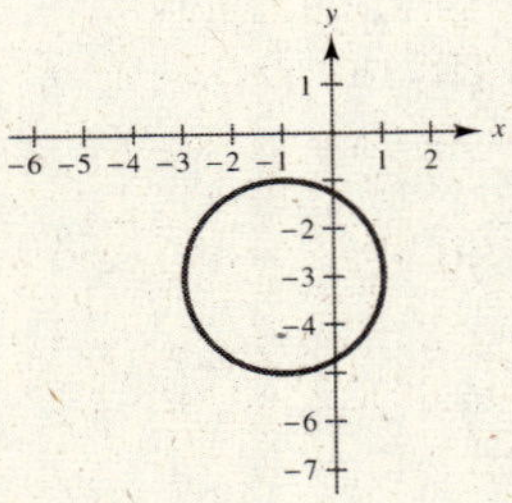

41.
$$x^2 + y^2 + 10x - 4y - 7 = 0$$
$$x^2 + y^2 + 10x - 4y = 7$$
$$(x^2 + 10x + 25) + (y^2 - 4y + 4) = 7 + 25 + 4$$
$$(x + 5)^2 + (y - 2)^2 = 36$$

Center: $(-5, 2)$; radius: 6

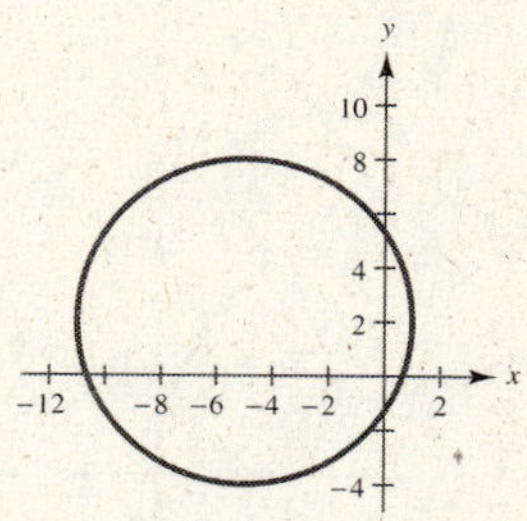

43. $x^2 + y^2 = 30$
$$y^2 = 30 - x^2$$
$$y = \pm\sqrt{30 - x^2}$$

Keystrokes:

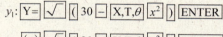

y_1: [Y=] [√] [(] 30 [−] [X,T,θ] [x^2] [)] [ENTER]

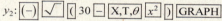

y_2: [(−)] [√] [(] 30 [−] [X,T,θ] [x^2] [)] [GRAPH]

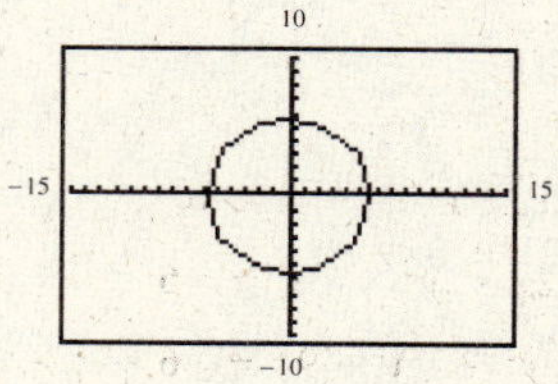

45. $(x - 2)^2 + y^2 = 10$
$$y^2 = 10 - (x - 2)^2$$
$$y = \pm\sqrt{10 - (x - 2)^2}$$

Keystrokes:

y_1: [Y=] [√] [(] 10 [−] [(] [X,T,θ] [−] 2 [)] [x^2] [)] [ENTER]

y_2: [(−)] [√] [(] 10 [−] [(] [X,T,θ] [−] 2 [)] [x^2] [)]

[GRAPH]

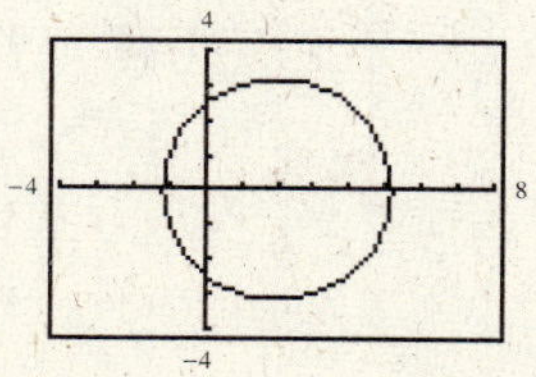

47. $y^2 = -4x$ (f)

Vertex: $(0, 0)$

Hortizontal axis opens left.

49. $x^2 = -8y$ (e)

Vertex: $(0, 0)$

Vertical axis opens down.

51. $(y - 1)^2 = 4(x - 3)$ (b)

Vertex: $(3, 1)$

Horizontal axis opens right.

53. Vertex: $(0, 0)$, point: $(3, 6)$, vertical parabola
$$x^2 = 4py$$
$$3^2 = 4p(6)$$
$$9 = 4p(6)$$
$$\tfrac{9}{6} = 4p$$
$$\tfrac{3}{2} = 4p$$
$$x^2 = \tfrac{3}{2}y$$

55. Vertex: $(0, 0)$, focus: $(0, -\frac{3}{2})$, vertical parabola
$$x^2 = 4py$$
$$p = -\tfrac{3}{2}$$
$$x^2 = 4(-\tfrac{3}{2})y$$
$$x^2 = -6y$$

57. Vertex: $(0, 0)$, focus: $(-2, 0)$, horizontal parabola
$$y^2 = 4px$$
$$p = -2$$
$$y^2 = 4(-2)x$$
$$y^2 = -8x$$

59. Vertex: $(0, 0)$, focus: $(0, 1)$, vertical parabola
$$x^2 = 4py$$
$$p = 1$$
$$x^2 = 4(1)y$$
$$x^2 = 4y$$

61. Vertex: $(0, 0)$, focus: $(6, 0)$, horizontal parabola
$$y^2 = 4px$$
$$p = 6$$
$$y^2 = 4(6)x$$
$$y^2 = 24x$$

63. Vertex: $(0, 0)$, horizontal axis,

Point: $(4, 6)$, horizontal parabola

$$y^2 = 4px$$
$$6^2 = 4p(4)$$
$$36 = 4p(4)$$
$$9 = 4p$$
$$y^2 = 9x$$

65. Vertex: $(3, 1)$, point: $(2, 0)$, vertical parabola

$$(x - h)^2 = 4p(y - k)$$
$$(2 - 3)^2 = 4p(0 - 1)$$
$$1 = 4p(-1)$$
$$-1 = 4p$$
$$(x - 3)^2 = -(y - 1)$$

67. Vertex: $(-2, 0)$, point: $(0, 2)$, horizontal parabola

$$(y - k)^2 = 4p(x - h)$$
$$(2 - 0)^2 = 4p[0 - (-2)]$$
$$4 = 4p(2)$$
$$2 = 4p$$
$$(y - 0)^2 = 2[x - (-2)]$$
$$y^2 = 2(x + 2)$$

69. Vertex: $(3, 2)$, focus: $(1, 2)$, horizontal parabola

$$p = -2$$
$$(y - k)^2 = 4p(x - h)$$
$$(y - 2)^2 = 4(-2)(x - 3)$$
$$(y - 2)^2 = -8(x - 3)$$

71. Vertex: $(0, -4)$, focus: $(0, -1)$, vertical axis

$$p = 3$$
$$(x - 0)^2 = 4(3)(y + 4)$$
$$x^2 = 12(y + 4)$$

73. Vertex: $(0, 2)$, horizontal axis, point: $(1, 3)$

$$(y - k)^2 = 4p(x - h)$$
$$(3 - 2)^2 = 4p(1 - 0)$$
$$1 = 4p$$
$$(y - 2)^2 = 1(x - 0)$$
$$(y - 2)^2 = x$$

75. $y = \frac{1}{2}x^2$

$2y = x^2$

Vertical parabola

$4p = 2$

$p = \frac{2}{4} - \frac{1}{2}$

Vertex: $(0, 0)$

Focus: $\left(0, \frac{1}{2}\right)$

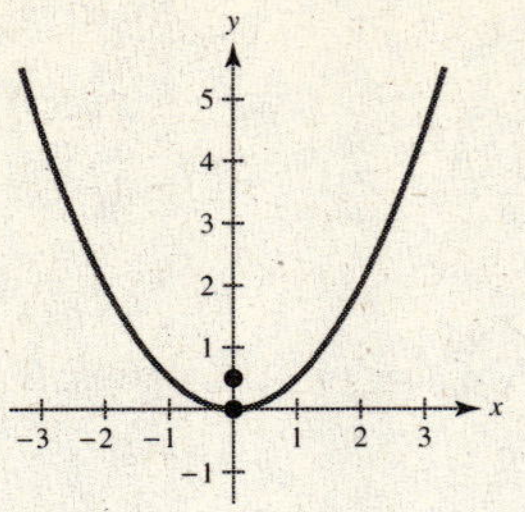

77. $y^2 = -10x$

Horizontal parabola

$4p = -10$

$p = -\frac{5}{2}$

Vertex: $(0, 0)$

Focus: $\left(-\frac{5}{2}, 0\right)$

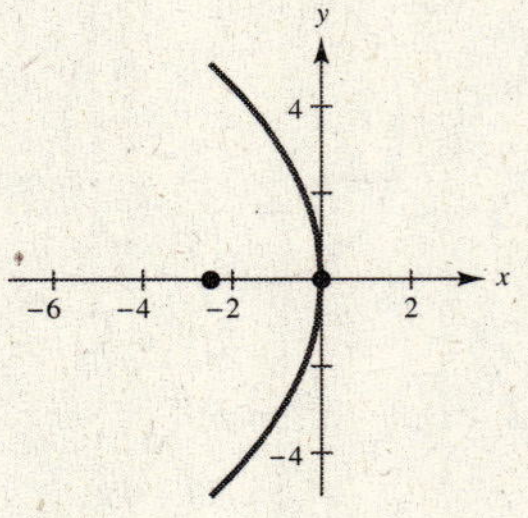

79. $x^2 + 8y = 0$

$x^2 = -8y$

Vertical parabola

$4p = -8$

$p = -\frac{8}{4} = -2$

Vertex: $(0, 0)$

Focus: $(0, -2)$

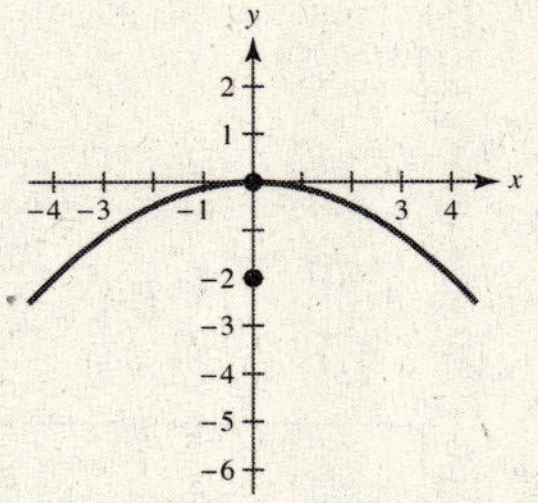

81. $(x - 1)^2 + 8(y + 2) = 0$

$(x - 1)^2 = -8(y + 2)$

Vertical parabola

$4p = -8$

$p = -2$

Vertex: $(1, -2)$

Focus: $(1, -4)$

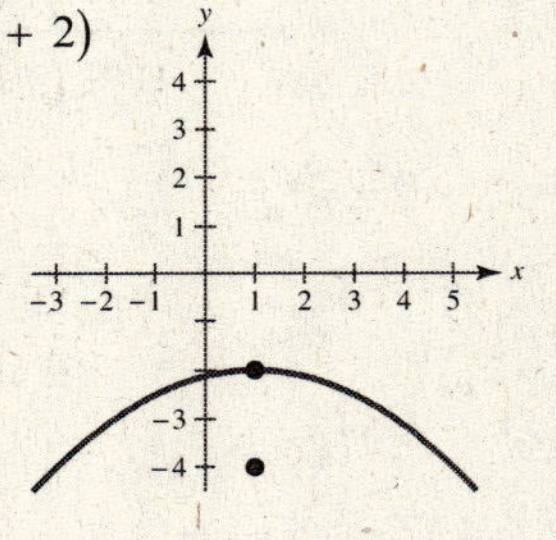

83. $\left(y + \frac{1}{2}\right)^2 = 2(x - 5)$

Horizontal parabola

$4p = 2$

$p = \frac{2}{4} = \frac{1}{2}$

Vertex: $\left(5, -\frac{1}{2}\right)$

Focus: $\left(\frac{11}{2}, -\frac{1}{2}\right)$

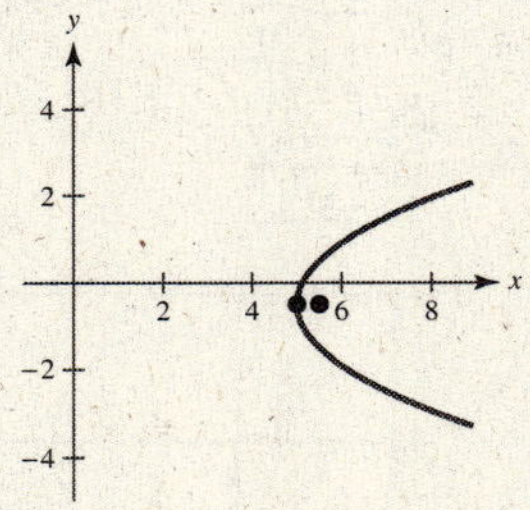

85. $y = \frac{1}{3}(x^2 - 2x + 10)$

$x^2 - 2x + 10 = 3y$

$x^2 - 2x + 1 = 3y - 10 + 1$

$(x - 1)^2 = 3y - 9$

$(x - 1)^2 = 3(y - 3)$

$4p = 3$

$p = \frac{3}{4}$

Vertex: $(1, 3)$

Focus: $\left(1, \frac{15}{4}\right)$

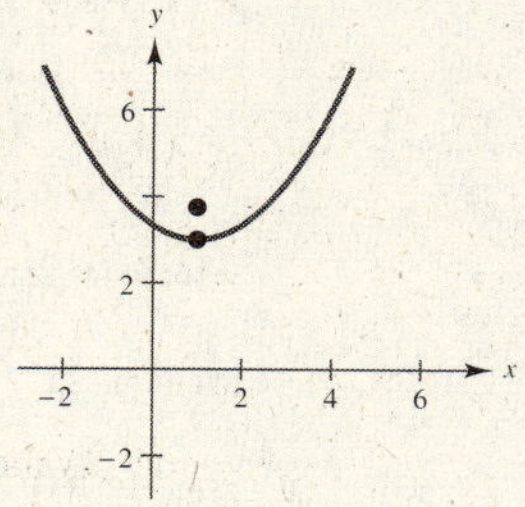

87. $y^2 + 6y + 8x + 25 = 0$

$(y^2 + 6y + 9) = -8x - 25 + 9$

$(y + 3)^2 = -8x - 16$

$(y + 3)^2 = -8(x + 2)$

Horizontal parabola

$4p = -8$

$p = -2$

Vertex: $(-2, -3)$

Focus: $(-4, -3)$

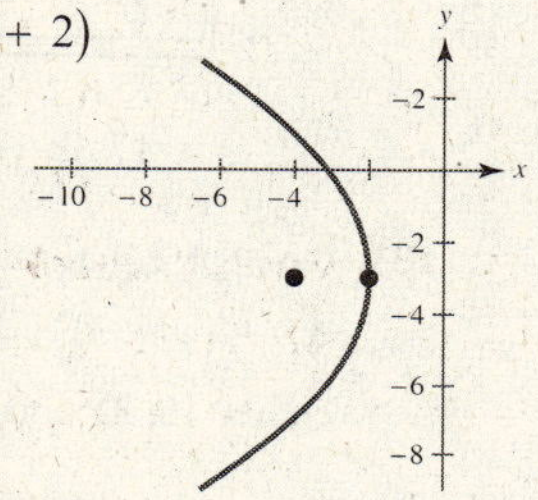

89. $y = -\frac{1}{6}(x^2 + 4x - 2)$

$y = -\frac{1}{6}(x^2 + 4x + 4) + \frac{2}{6} + \frac{4}{6}$

$y = -\frac{1}{6}(x + 2)^2 + 1$

$-6(y - 1) = (x + 2)^2$

$4p = -6$

$p = -\frac{6}{4} = -\frac{3}{2}$

Vertex: $(-2, 1)$

Focus: $\left(-2, -\frac{1}{2}\right)$

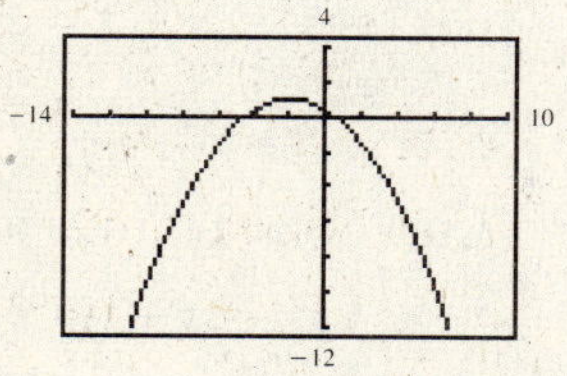

Keystrokes:

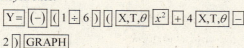

[Y=] [(−)] [(] 1 [÷] 6 [)] [(] [X,T,θ] [x²] [+] 4 [X,T,θ] [−] 2 [)] [GRAPH]

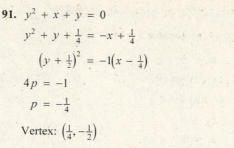

91. $y^2 + x + y = 0$

$y^2 + y + \frac{1}{4} = -x + \frac{1}{4}$

$\left(y + \frac{1}{2}\right)^2 = -1\left(x - \frac{1}{4}\right)$

$4p = -1$

$p = -\frac{1}{4}$

Vertex: $\left(\frac{1}{4}, -\frac{1}{2}\right)$

Focus: $\left(0, -\frac{1}{2}\right)$

Horizontal parabola

$y + \frac{1}{2} = \pm\sqrt{-x + \frac{1}{4}}$

$y = -\frac{1}{2} \pm \sqrt{-x + \frac{1}{4}}$

Keystrokes:

y_1: [Y=] [(−)] [(] 1 [÷] 2 [)] [+] [√] [(] [(−)] [X,T,θ] [+] [(] 1 [÷] 4 [)] [)] [ENTER]

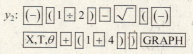

y_2: [(−)] [(] 1 [÷] 2 [)] [−] [√] [(] [(−)] [X,T,θ] [+] [(] 1 [÷] 4 [)] [)] [GRAPH]

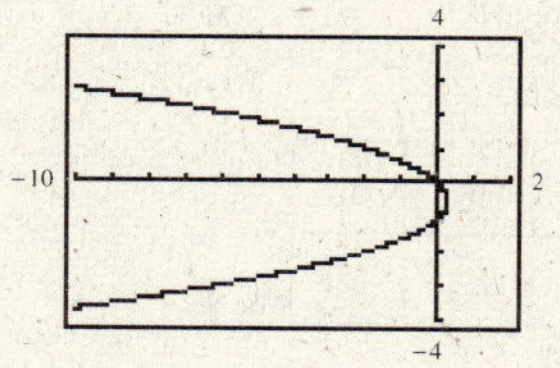

93. $x^2 + y^2 = 4500^2$

$x^2 + y^2 = 20{,}250{,}000$

95. Diameter $= 3$

Radius $= 1.5$, point: $(10, 4)$

Center: $(10, 5.5)$

$(x - 10)^2 + (y - 5.5)^2 = 1.5^2$

$(x - 10)^2 + (y - 5.5)^2 = 2.25$

$(9.25 - 10)^2 + (y - 5.5)^2 = 2.25$

$0.5625 + (y - 5.5)^2 = 2.25$

$(y - 5.5)^2 = 1.6875$

$y - 5.5 \approx 1.30$

$y \approx 6.8$ feet

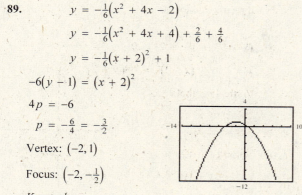

97. (a) Vertex: $(0, 0)$, point: $(60, 20)$

Vertical parabola

$$x^2 = 4py$$
$$60^2 = 4p(20)$$
$$3600 = 4p(20)$$
$$180 = 4p$$
$$x^2 = 180y$$
$$\frac{x^2}{180} = y$$

(b) Let $x = 0$.

$$y = \frac{0^2}{180} = 0$$

Let $x = 20$.

$$y = \frac{20^2}{180} = \frac{400}{180} = \frac{20}{9} = 2\frac{2}{9}$$

Let $x = 40$.

$$y = \frac{40^2}{180} = \frac{1600}{180} = \frac{80}{9} = 8\frac{8}{9}$$

Let $x = 60$.

$$y = \frac{60^2}{180} = \frac{3600}{180} = 20$$

x	0	20	40	60
y	0	$2\frac{2}{9}$	$8\frac{8}{9}$	20

99. $R = 575x - \frac{5}{4}x^2$

(a) *Keystrokes:*

[Y=] 575 [X,T,θ] [−] 5 [÷] 4 [X,T,θ] [x^2] [GRAPH]

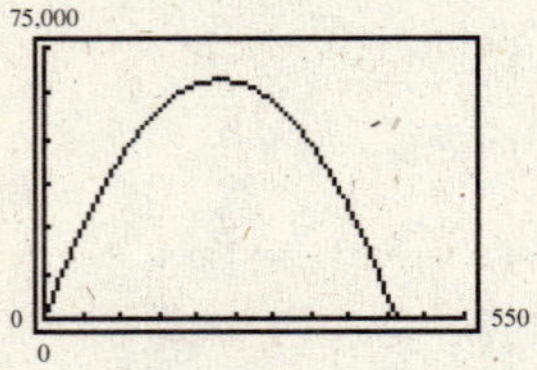

(b) To find the number of sales that will maximize revenue, trace the curve until the highest point is reached. The x-coordinate of that point is 230 so 230 video game systems will maximize revenue.

101. (a) $x^2 + y^2 = 625$ (equation of circle)

(x, y) of the rectangle is also the point on the circle, so y-coordinate equals:

$$x^2 + y^2 = 625$$
$$y^2 = 625 - x^2$$
$$y = \sqrt{625 - x^2}$$
$$\text{width} = 2\left(\sqrt{625 - x^2}\right)$$
$$\text{area} = 2x \cdot 2\left(\sqrt{625} - x^2\right)$$
$$\text{area} = 4x\left(\sqrt{625 - x^2}\right)$$

(b)

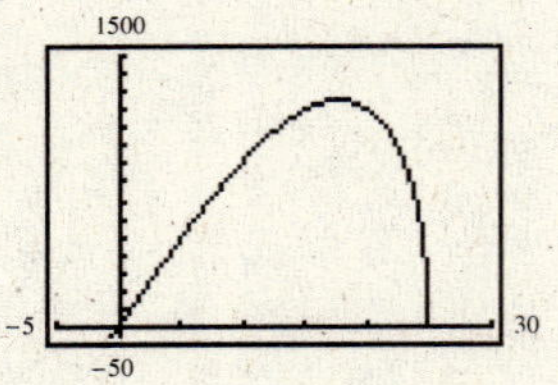

$x \approx 17.68$

103. No. The equation in standard form is $x^2 + (y - 3)^2 = 4$ which does represent a circle with center $(0, 3)$ and radius of 2.

105. All points on the parabola are equidistant from the directrix and focus.

107.
$$x^2 + 4x = 6$$
$$x^2 + 4x + 4 = 6 + 4$$
$$(x + 2)^2 = 10$$
$$x + 2 = \pm\sqrt{10}$$
$$x = -2 \pm \sqrt{10}$$

109.
$$x^2 - 2x - 3 = 0$$
$$x^2 - 2x + 1 = 3 + 1$$
$$(x - 1)^2 = 4$$
$$x - 1 = \pm 2$$
$$x = 1 \pm 2$$
$$x = 3, -1$$

111. $2x^2 + 5x - 8 = 0$

$$x^2 + \frac{5}{2}x = 4$$

$$x^2 + \frac{5}{2}x + \frac{25}{16} = 4 + \frac{25}{16}$$

$$\left(x + \frac{5}{4}\right)^2 = \frac{89}{16}$$

$$x + \frac{5}{4} = \pm\sqrt{\frac{89}{16}}$$

$$x = -\frac{5}{4} \pm \frac{\sqrt{89}}{4}$$

113. $\log_8 x^{10} = 10 \log_9 x$

115. $\ln 5x^2y = \ln 5 + \ln x^2 + \ln y$

$$= \ln 5 + 2\ln x + \ln y$$

117. $\log_{10} x + \log_{10} 6 = \log_{10} 6x$

119. $3\ln x + \ln y - \ln 9 = \ln x^3 + \ln y - \ln 9 = \ln \dfrac{x^3y}{9}$

Section 10.2 Ellipses

1. $\dfrac{x^2}{4} + \dfrac{y^2}{9} = 1$, graph (a)

3. $\dfrac{x^2}{4} + \dfrac{y^2}{25} = 1$, graph (d)

5. $\dfrac{(x-2)^2}{16} + \dfrac{(y+1)^2}{1} = 1$, graph (e)

7. Center: $(0, 0)$

Vertices: $(-4, 0), (4, 0)$

Co-vertices: $(0, -3), (0, 3)$

$$\frac{x^2}{a^2} + \frac{y^2}{b^2} = 1$$

Major axis is x-axis so $a = 4$.

Minor axis is y-axis so $b = 3$.

$$\frac{x^2}{4^2} + \frac{y^2}{3^2} = 1$$

$$\frac{x^2}{16} + \frac{y^2}{9} = 1$$

9. Center: $(0, 0)$

Vertices: $(-2, 0), (2, 0)$

Co-vertices: $(0, -1), (0, 1)$

$$\frac{x^2}{a^2} + \frac{v^2}{b^2} = 1$$

Major axis is x-axis so $a = 2$.

Minor axis is y-axis $b = 1$.

$$\frac{x^2}{2^2} + \frac{y^2}{1^2} = 1$$

$$\frac{x^2}{4} + \frac{y^2}{1} = 1$$

11. Center: $(0, 0)$

Vertices: $(0, -6), (0, 6)$

Co-vertices: $(-3, 0), (3, 0)$

$$\frac{x^2}{b^2} + \frac{y^2}{a^2} = 1$$

Major axis is y-axis so $b = 3$.

Minor axis is x-axis so $a = 6$.

$$\frac{x^2}{3^2} + \frac{y^2}{6^2} = 1$$

$$\frac{x^2}{9} + \frac{y^2}{36} = 1$$

13. Center: $(0, 0)$

Vertices: $(0, -2), (0, 2)$

Co-vertices: $(-1, 0), (1, 0)$

$$\frac{x^2}{b^2} + \frac{y^2}{a^2} = 1$$

Major axis is y-axis so $a = 2$.

Minor axis is x-axis so $b = 1$.

$$\frac{x^2}{1^2} + \frac{y^2}{2^2} = 1$$

$$\frac{x^2}{1} + \frac{y^2}{4} = 1$$

15. Center: $(0, 0)$

Major axis (vertical) 10 units

Minor axis 6 units

$$\frac{x^2}{b^2} + \frac{y^2}{a^2} = 1$$

$$b = 3, \quad a = 5$$

$$\frac{x^2}{3^2} + \frac{y^2}{5^2} = 1$$

$$\frac{x^2}{9} + \frac{y^2}{25} = 1$$

17. Center: $(0, 0)$

Major axis (horizontal) 20 units

Minor axis 12 units

$$\frac{x^2}{a^2} + \frac{y^2}{b^2} = 1$$

$$a = 10, \quad b = 6$$

$$\frac{x^2}{10^2} + \frac{y^2}{6^2} = 1$$

$$\frac{x^2}{100} + \frac{y^2}{36} = 1$$

19. Vertices: $(-4, 0), (4, 0)$

Co-vertices: $(0, 2), (0, -2)$

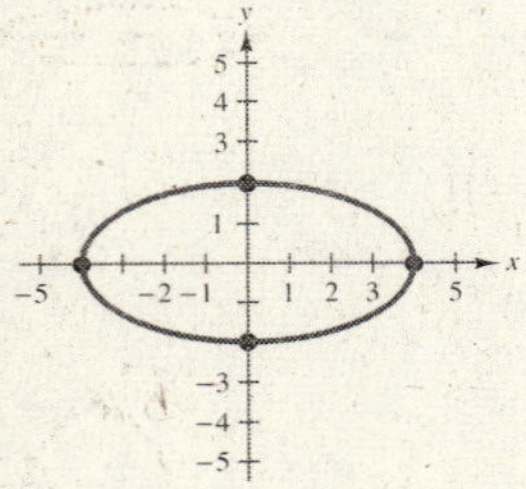

21. Vertices: $(0, 4), (0, -4)$

Co-vertices: $(2, 0), (-2, 0)$

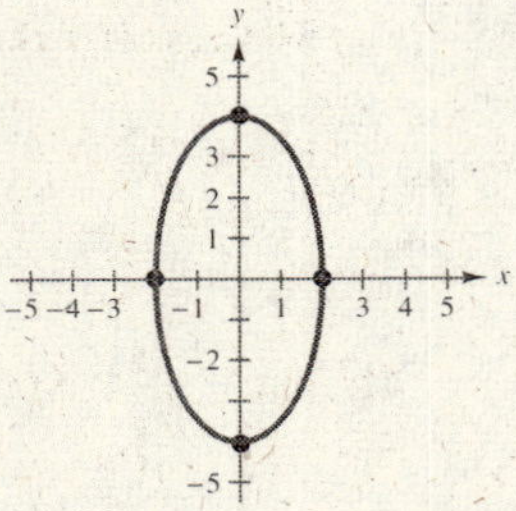

23. Vertices: $\left(-\frac{5}{3}, 0\right), \left(\frac{5}{3}, 0\right)$

Co-vertices: $\left(0, \frac{4}{3}\right), \left(0, -\frac{4}{3}\right)$

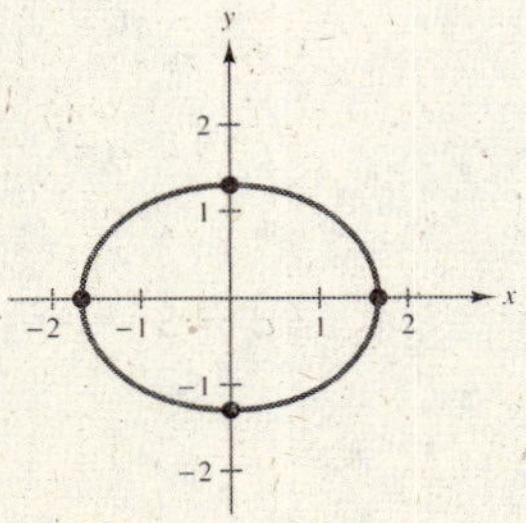

25. $$\frac{9x^2}{4} + \frac{25y^2}{16} = 1$$

$$\frac{x^2}{4/9} + \frac{y^2}{16/25} = 1$$

Vertices: $\left(0, \pm\frac{4}{5}\right)$

Co-vertices: $\left(\pm\frac{2}{3}, 0\right)$

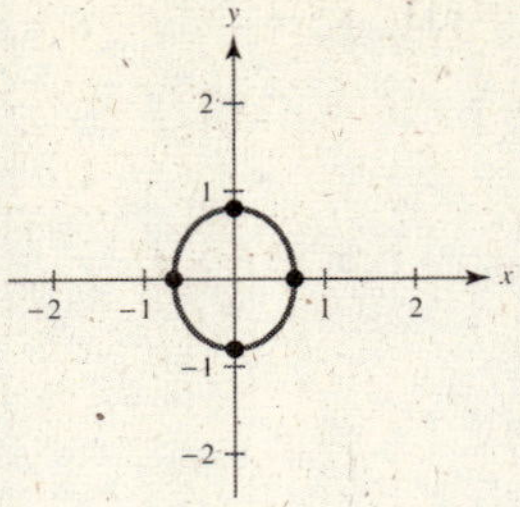

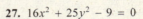

27. $16x^2 + 25y^2 - 9 = 0$

$$\frac{16x^2}{9} + \frac{25y^2}{9} = 1$$

$$\frac{x^2}{9/16} + \frac{y^2}{9/25} = 1$$

Vertices: $\left(\pm\frac{3}{4}, 0\right)$

Co-vertices: $\left(0, \pm\frac{3}{5}\right)$

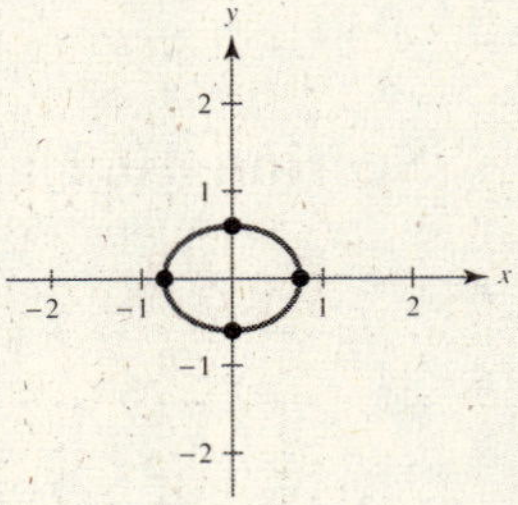

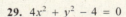

29. $4x^2 + y^2 - 4 = 0$

$$\frac{x^2}{1} + \frac{y^2}{4} = 1$$

Vertices: $(0, 2), (0, -2)$

Co-vertices: $(1, 0), (-1, 0)$

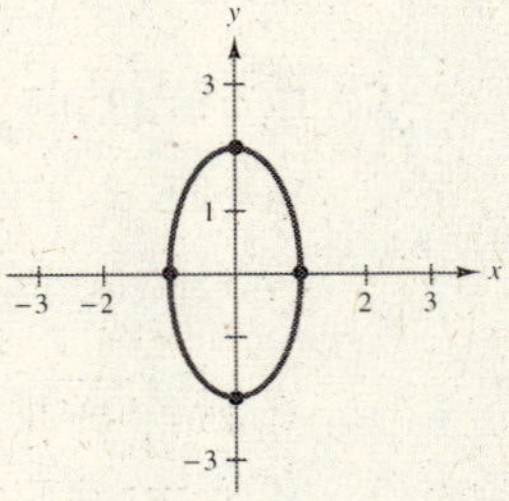

31. $10x^2 + 16y^2 - 160 = 0$

$$\frac{10x^2}{160} + \frac{16y^2}{160} = \frac{160}{160}$$

$$\frac{x^2}{16} + \frac{y^2}{10} = 1$$

Vertices: $(\pm 4, 0)$

Co-vertices: $(0, \pm\sqrt{10})$

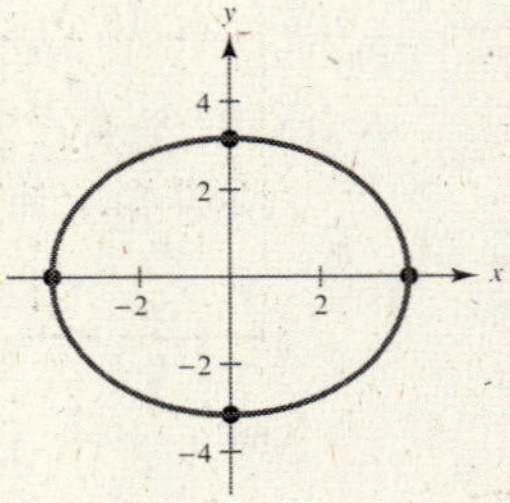

33. $x^2 + 2y^2 = 4$

$$2y^2 = 4 - x^2$$

$$y^2 = \frac{4 - x^2}{2}$$

$$y = \pm\sqrt{\frac{4 - x^2}{2}}$$

Keystrokes:

y_1: [Y=] [√] [(] [(] 4 [−] [X,T,θ] [x^2] [)] [÷] 2 [)] [ENTER]

y_2: [(−)] [√] [(] [(] 4 [−] [X,T,θ] [x^2] [)] [÷] 2 [)] [GRAPH]

Vertices: $(\pm 2, 0)$

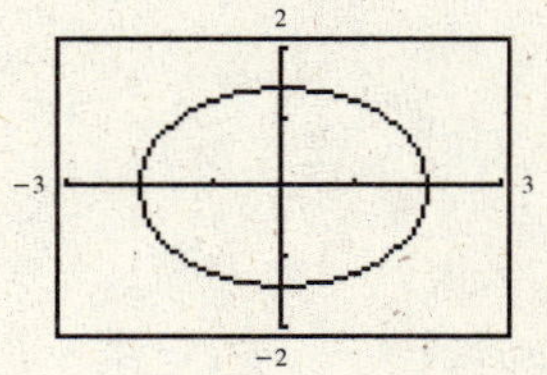

35. $3x^2 + y^2 - 12 = 0$

$$y^2 = 12 - 3x^2$$

$$y = \pm\sqrt{12 - 3x^2}$$

Keystrokes:

y_1: [Y=] [√] [(] 12 [−] 3 [X,T,θ] [x^2] [)] [ENTER]

y_2: [(−)] [√] [(] 12 [−] 3 [X,T,θ] [x^2] [)] [Graph]

Vertices: $\left(0, \pm 2\sqrt{3}\right)$

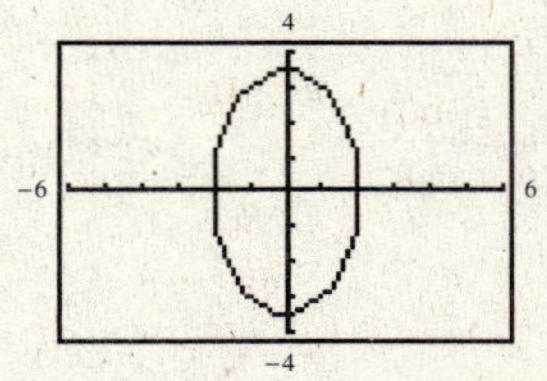

37. Center: $(0, 0)$

Vertices: $(0, \pm 2)$

Co-vertices: $(\pm 1, 0)$

Major axis is y-axis so $a = 2$.

Minor axis is x-axis so $b = 1$.

$$\frac{x^2}{1^2} + \frac{y^2}{2^2} = 1$$

$$\frac{x^2}{1} + \frac{y^2}{4} = 1$$

39. Center: $(4, 0)$

Vertices: $(4, 4), (4, -4)$

Co-vertices: $(1, 0), (7, 0)$

Major axis is vertical so $a = 4$.

Minor axis is horizontal so $b = 3$.

$$\frac{(x - 4)^2}{3^2} + \frac{(y - 0)^2}{4^2} = 1$$

$$\frac{(x - 4)^2}{9} + \frac{y^2}{16} = 1$$

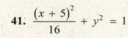

41. $\dfrac{(x + 5)^2}{16} + y^2 = 1$

Center: $(-5, 0)$

Vertices: $(-1, 0), (-9, 0)$

$a^2 = 16 \qquad b^2 = 1$

$a = 4 \qquad b = 1$

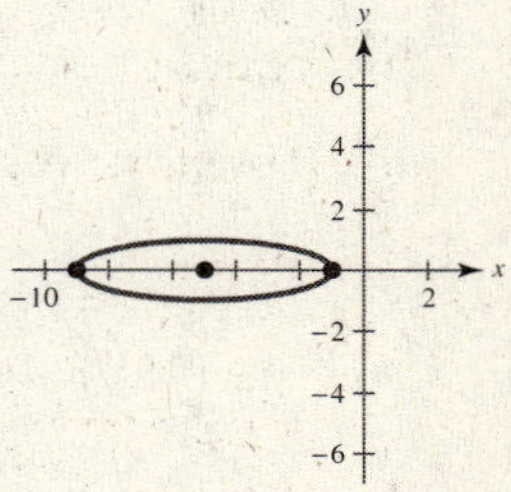

43. $\dfrac{(x - 1)^2}{9} + \dfrac{(y - 5)^2}{25} = 1$

Center: $(1, 5)$

Vertices: $(1, 10), (1, 0)$

$a^2 = 25 \qquad b^2 = 9$

$a = 5 \qquad b = 3$

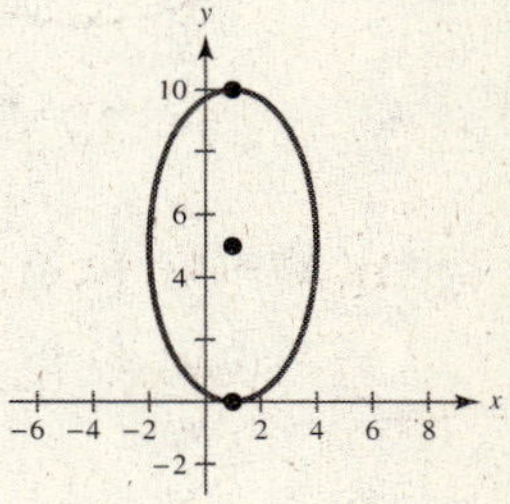

45. $4(x - 2)^2 + 9(y + 2)^2 = 36$

$$\frac{(x - 2)^2}{9} + \frac{(y + 2)^2}{4} = 1$$

Center: $(2, -2)$

Vertices: $(-1, -2), (5, -2)$

$a^2 = 9 \qquad b^2 = 4$

$a = 3 \qquad b = 2$

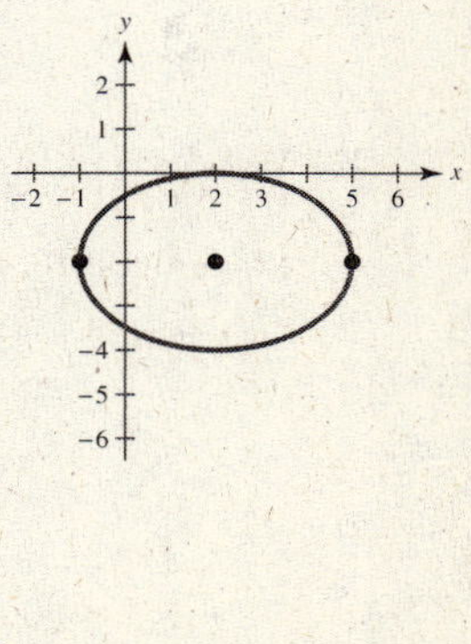

47. $12(x+4)^2 + 3(y-1)^2 = 48$

$$\frac{(x+4)^2}{4} + \frac{(y-1)^2}{16} = 1$$

Center: $(-4, 1)$

Vertices: $(-4, 5), (-4, -3)$

$a^2 = 16$ $\quad b^2 = 4$

$a = 4$ $\quad b = 2$

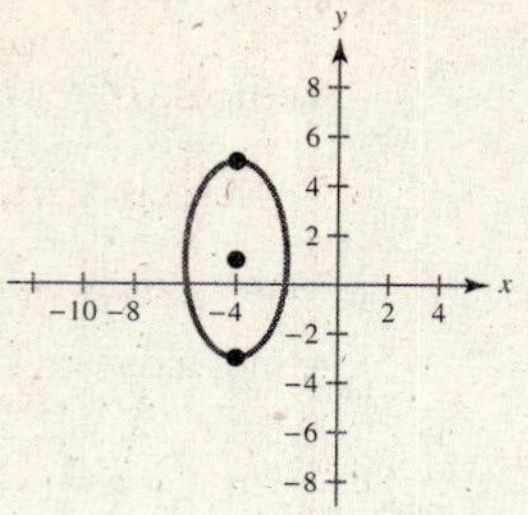

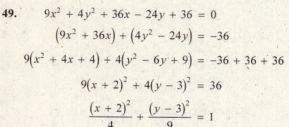

49. $9x^2 + 4y^2 + 36x - 24y + 36 = 0$

$$(9x^2 + 36x) + (4y^2 - 24y) = -36$$

$$9(x^2 + 4x + 4) + 4(y^2 - 6y + 9) = -36 + 36 + 36$$

$$9(x+2)^2 + 4(y-3)^2 = 36$$

$$\frac{(x+2)^2}{4} + \frac{(y-3)^2}{9} = 1$$

Center: $(-2, 3)$

Vertices: $(-2, 0), (-2, 6)$

$a^2 = 9$ $\quad b^2 = 4$

$a = 3$ $\quad b = 2$

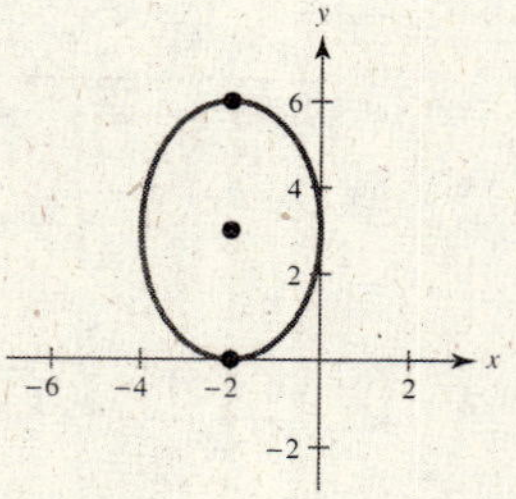

51. $25x^2 + 9y^2 - 200x + 54y + 256 = 0$

$$(25x^2 - 200x) + (9y^2 + 54y) = -256$$

$$25(x^2 - 8x + 16) + 9(y^2 + 6y + 9) = -256 + 400 + 81$$

$$25(x-4)^2 + 9(y+3)^2 = 225$$

$$\frac{(x-4)^2}{9} + \frac{(y+3)^2}{25} = 1$$

Center: $(4, -3)$

Vertices: $(4, 2), (4, -8)$

$a^2 = 25$ $\quad b^2 = 9$

$a = 5$ $\quad b = 3$

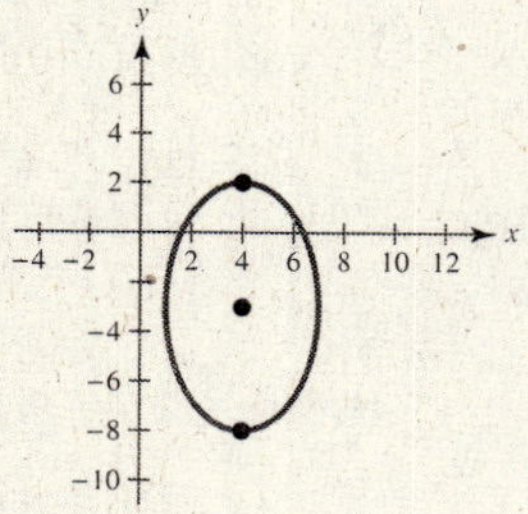

53. $x^2 + 4y^2 - 4x - 8y - 92 = 0$

$$(x^2 - 4x + 4) + (4y^2 - 8y) = 92 + 4$$

$$(x-2)^2 + 4(y^2 - 2y + 1) = 92 + 4 + 4$$

$$(x-2)^2 + 4(y-1)^2 = 100$$

$$\frac{(x-2)^2}{100} + \frac{(y-1)^2}{25} = 1$$

Center: $(2, 1)$

Vertices: $(-8, 1), (12, 1)$

$a^2 = 100$ $\quad b^2 = 25$

$a = 10$ $\quad b = 5$

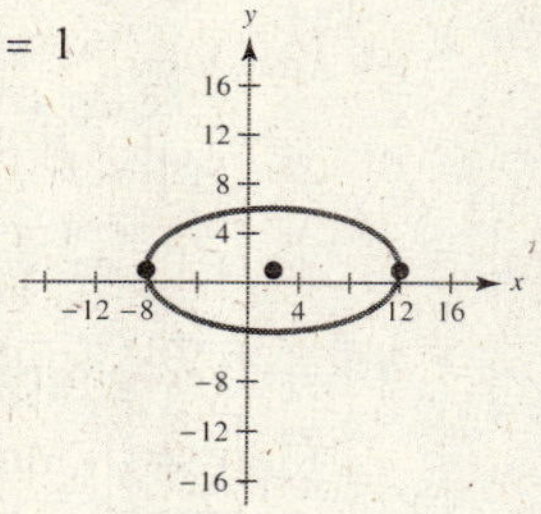

55. $\dfrac{x^2}{324} + \dfrac{y^2}{196} = 1$

$a^2 = 324$ $\quad b^2 = 196$

$a = 18$ $\quad b = 14$

$2a = 36$ $\quad 2b = 28$

36 feet = longest distance

24 feet = shortest distance

57. Equation of ellipse $= \dfrac{x^2}{50^2} + \dfrac{y^2}{40^2} = 1$ or

$$\frac{x^2}{2500} + \frac{y^2}{1600} = 1$$

$$\frac{45^2}{2500} + \frac{y^2}{1600} = 1$$

$$\frac{y^2}{1600} = 0.19$$

$$y^2 = 304$$

$$y = 17.435596 \approx 17.4 \text{ feet}$$

59. Center: $(0, 0)$

Length of major axis: 1230

Length of minor axis: 580

$2a = 1230$ $\quad 2b = 580$

$a = 615$ $\quad b = 290$

$$\frac{x^2}{615^2} + \frac{y^2}{290^2} = 1$$

or

$$\frac{x^2}{290^2} + \frac{y^2}{615^2} = 1$$

61. (a) Every point on the ellipse represents the maximum distance (800 miles) that the plane can safely fly with enough fuel to get from airport A to airport B.

(b) Center: $(0, 0)$ Airport A: $(-250, 0)$

$2c = 500$ Airport B: $(250, 0)$

$c = 250$

(c) Airplane flies maximum of 800 miles without refueling, so $800 - 500 = 300$.

300 = twice the distance from the airport to the vertex.

150 = the distance from the airport to the vertex.

Vertices: $(\pm 400, 0)$

$-250 - 150 = -400$

$250 + 150 = 400$

(d) $c^2 = a^2 - b^2$

$250^2 = 400^2 - b^2$

$400^2 - 250^2 = b^2$ $\qquad \dfrac{x^2}{a^2} + \dfrac{y^2}{b^2} = 1$

$97{,}500 = b^2$

$\sqrt{97{,}500} = b$ $\qquad \dfrac{x^2}{400^2} + \dfrac{y^2}{\left(50\sqrt{39}\right)^2} = 1$

$50\sqrt{39} = 6$

(e) $A = \pi ab$

$A = \pi(400)\left(50\sqrt{39}\right)$

$= 20{,}000\sqrt{39}\pi \approx 392{,}385$ square miles

63. A circle is an ellipse in which the major axis and the minor axis have the same length. Both circles and ellipses have foci; however, in a circle the foci are both at the same point, whereas in an ellipse they are not.

65. The sum of the distances between each point on the ellipse and the two foci is a constant.

67. The graph of an ellipse written in the standard form $\dfrac{(x-h)^2}{a^2} + \dfrac{(y-k)^2}{b^2} = 1$ intersects the y-axis if $|h| > a$ and intersects the x-axis if $|k| > b$. Similarly, the graph of $\dfrac{(x-h)^2}{b^2} + \dfrac{(y-k)^2}{a^2} = 1$ intersects the y-axis if $|h| > b$ and intersects the x-axis if $|k| > a$.

69. $f(x) = 3^{-x}$

(a) $f(-2) = 3^{-(-2)} = 3^2 = 9$

(b) $f(2) = 3^{-2} = \dfrac{1}{3^2} = \dfrac{1}{9}$

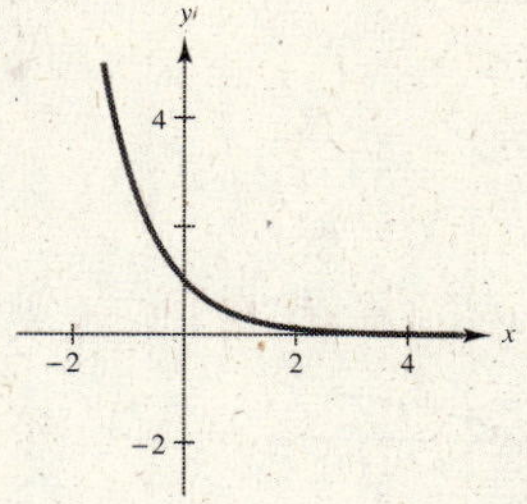

71. $g(x) = 6e^{0.5x}$

(a) $g(-1) = 6e^{0.5(-1)}$

(b) $g(2) = 6e^{0.5(2)} \approx 16.310$

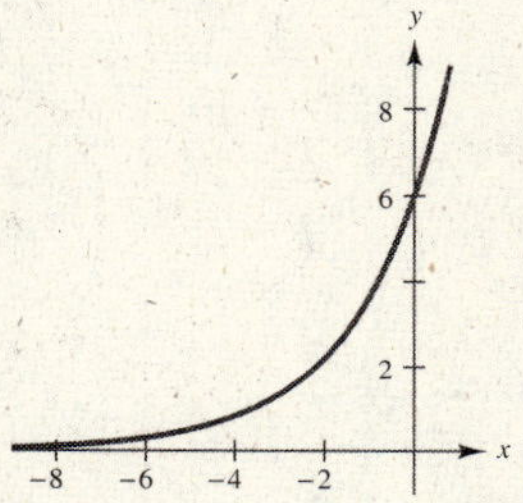

73. $h(x) = \log_{16} 4x$

(a) $h(4) = \log_{16} 4(4) = \log_{16} 16 = 1$

(b) $h(64) = \log_{16} 4(64) = \log_{16} 256 = \log_{16} 16^2 = 2$

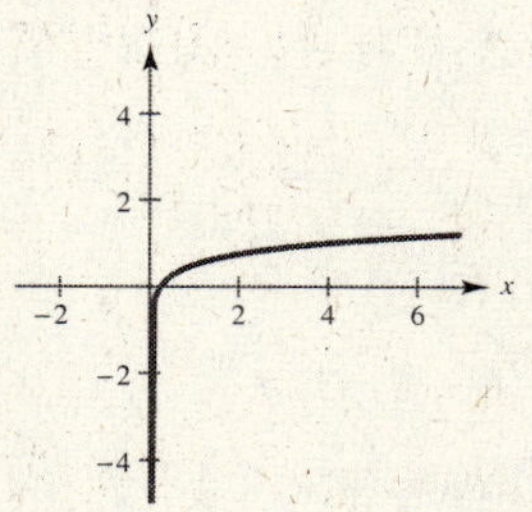

75. $f(x) = \log_4(x - 3)$

(a) $f(3) = \log_4(3 - 3) = \log_4 0 =$ does not exist

(b) $f(35) = \log 4(35 - 3)$

$= \log_4 32$

$= \log_4 2^5$

$= 5 \log_4 2 = 5 \cdot \frac{1}{2} = \frac{5}{2}$

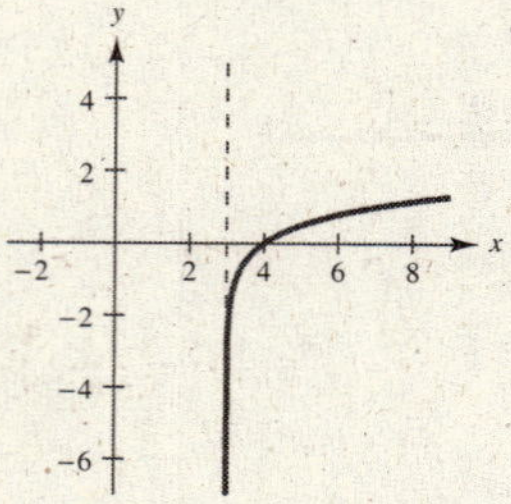

Mid-Chapter Quiz for Chapter 10

1. Center: $(0, 0)$

Radius: 5

$x^2 + y^2 = 5^2$

$x^2 + y^2 = 25$

2. Vertex: $(-2, 1)$, Point: $(6, 9)$

Horizontal parabola

$$(y - k)^2 = 4p(x - h)$$
$$(9 - 1)^2 = 4p[6 - (-2)]$$
$$64 = 4p(8)$$
$$8 = 4p$$
$$(y - 1)^2 = 8(x + 2)$$

3. Center: $(-2, -1)$

Vertices: $(-6, -1), (2, -1)$

Co-vertices: $(-2, 1), (-2, -3)$

Major axis is horizontal so $a = 4$.

Minor axis is vertical so $b = 2$.

$$\frac{[x - (-2)]^2}{4^2} + \frac{[y - (-1)]^2}{2^2} = 1$$
$$\frac{(x + 2)^2}{16} + \frac{(y + 1)^2}{4} = 1$$

4. Center: $(3, -5)$

Point: $(0, -1)$

$$(x - h)^2 + (y - k)^2 = r^2$$
$$(0 - 3)^2 + [-1 - (-5)]^2 = r^2$$
$$(-3)^2 + (4)^2 = r^2$$
$$9 + 16 = r^2$$
$$25 = r^2$$
$$(x - 3) + (y + 5)^2 = 25$$

5. Vertex: $(2, 3)$

Focus: $(2, 1)$

Vertical parabola

$$(x - h)^2 = 4p(y - k)$$
$$p = -2$$
$$(x - 2)^2 = 4(-2)(y - 3)$$
$$(x - 2)^2 = -8(y - 3)$$

6. Vertices: $(0, -10), (0, 10)$

Co-vertices: $(-6, 0), (6, 0)$

Vertical ellipse

$$\frac{x^2}{b^2} + \frac{y^2}{a^2} = 1$$
$$\frac{x^2}{6^2} + \frac{y^2}{10^2} = 1$$
$$\frac{x^2}{36} + \frac{y^2}{100} = 1$$

7. $x^2 + y^2 + 6y - 7 = 0$

$x^2 + y^2 + 6y + 9 = 7 + 9$

$x^2 + (y + 3)^2 = 16$

Center: $(0, -3)$

Radius: 4

8. $x^2 + y^2 + 2x - 4y + 4 = 0$

$(x^2 + 2x + 1) + (y^2 - 4y + 4) = -4 + 1 + 4$

$(x + 1)^2 + (y - 2)^2 = 1$

Center: $(-1, 2)$

Radius: 1

9. $x = y^2 - 6y - 7$

$x + 7 = (y^2 - 6y + 9) - 9$

$x + 16 = (y - 3)^2$

Vertex: $(-16, 3)$

$4p = 1$

$p = \frac{1}{4}$

Focus: $\left(-16 + \frac{1}{4}, 3\right) = \left(-\frac{64}{4} + \frac{1}{4}, 3\right) = \left(-\frac{63}{4}, 3\right)$

10. $x^2 - 8x + y + 12 = 0$

$x^2 - 8x + 16 = -y - 12 + 16$

$(x - 4)^2 = -y + 4$

$(x - 4)^2 = -1(y - 4)$

Vertex: $(4, 4)$

$4p = -1$

$p = -\frac{1}{4}$

Focus: $\left(4, 4 - \frac{1}{4}\right) = \left(4, \frac{16}{4} - \frac{1}{4}\right) = \left(4, \frac{15}{4}\right)$

11. $4x^2 + y^2 - 16x - 20 = 0$

$4x^2 - 16x + y^2 = 20$

$4(x^2 - 4x) + y^2 = 20$

$4(x^2 - 4x + 4) + y^2 = 20 + 16$

$4(x - 2)^2 + y^2 = 36$

$\frac{(x - 2)^2}{9} + \frac{y^2}{36} = 1$

Center: $(2, 0)$

Vertices: $(2, -6), (2, 6)$

$a^2 = 36$

$a = 6$

12. $4x^2 + 9y^2 - 48x + 36y + 144 = 0$

$(4x^2 - 48x) + (9y^2 + 36y) = -144$

$4(x^2 - 12x + 36) + 9(y^2 + 4y + 4) = -144 + 144 + 36$

$4(x - 6)^2 + 9(y + 2)^2 = 36$

$\frac{(x - 6)^2}{9} + \frac{(y + 2)^2}{4} = 1$

Center: $(6, -2)$

Vertices: $(3, -2), (9, -2)$

13. $(x + 5)^2 + (y - 1)^2 = 9$

Circle:

Center: $(-5, 1)$

Radius: 3

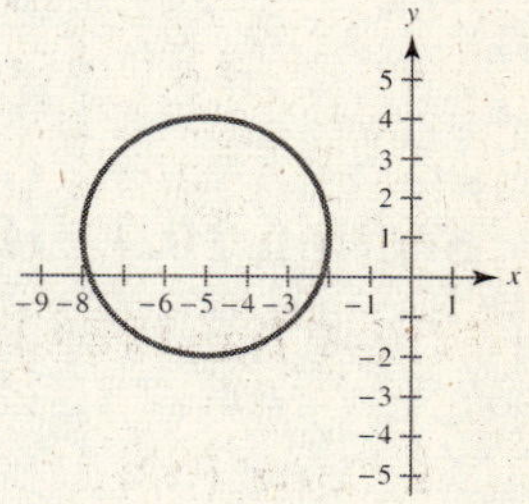

14. $9x^2 + y^2 = 81$

$\frac{x^2}{9} + \frac{y^2}{81} = 1$

Vertical ellipse

$b^2 = 9$ $\quad a^2 = 81$

$b = \pm 3$ $\quad a = \pm 9$

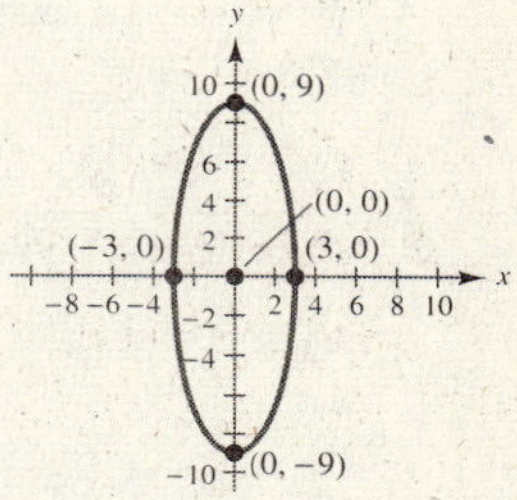

15. $x = -y^2 - 4y$

$x = -1(y^2 + 4y + 4) + 4$

$x - 4 = -1(y + 2)^2$

$-1(x - 4) = (y + 2)^2$

Parabola

Vertex: $(4, -2)$

$4p = -1$

$p = -\frac{1}{4}$

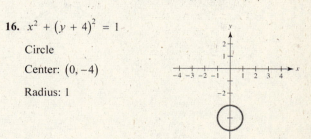

Focus: $\left(4 - \frac{1}{4}, -2\right) = \left(\frac{16}{4} - \frac{1}{4}, -2\right) = \left(\frac{15}{4}, -2\right)$

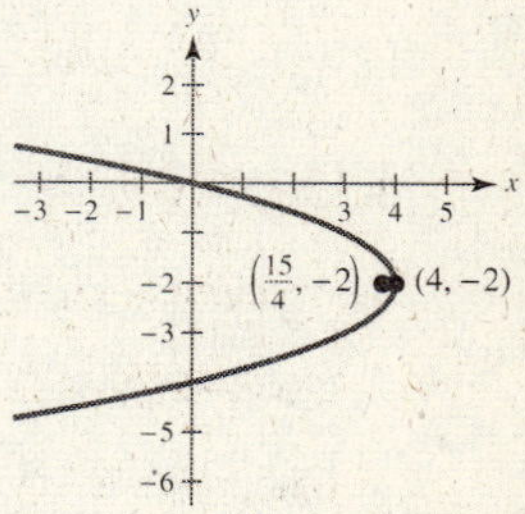

16. $x^2 + (y + 4)^2 = 1$

Circle

Center: $(0, -4)$

Radius: 1

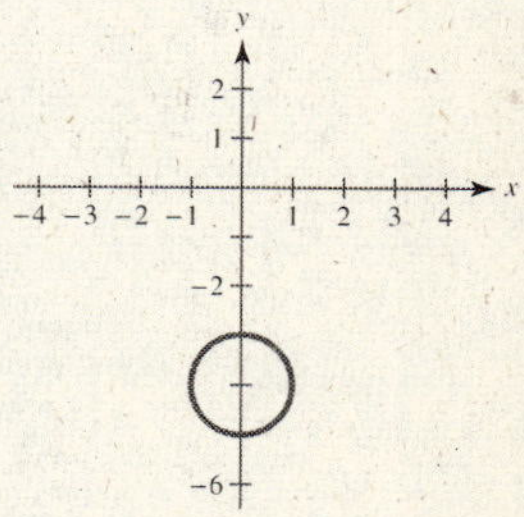

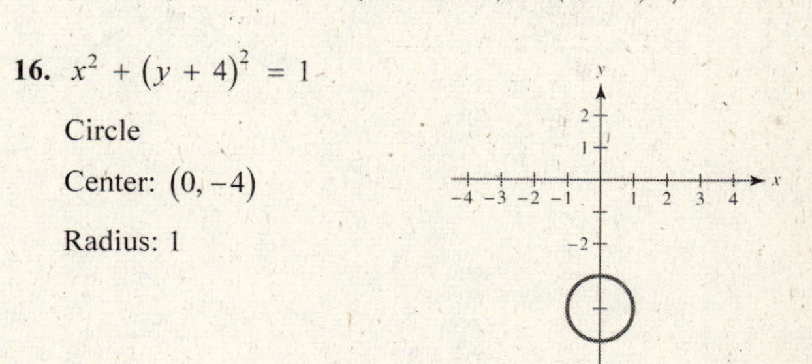

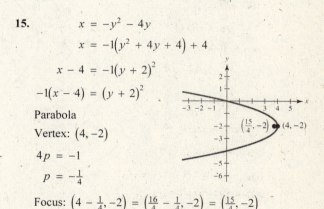

17. $y = x^2 - 2x + 1$

$y = (x - 1)^2$

Parabola

Vertex: $(1, 0)$

$4p = 1$

$p = \frac{1}{4}$

Focus: $\left(1, 0 + \frac{1}{4}\right) = (1, 0.25)$

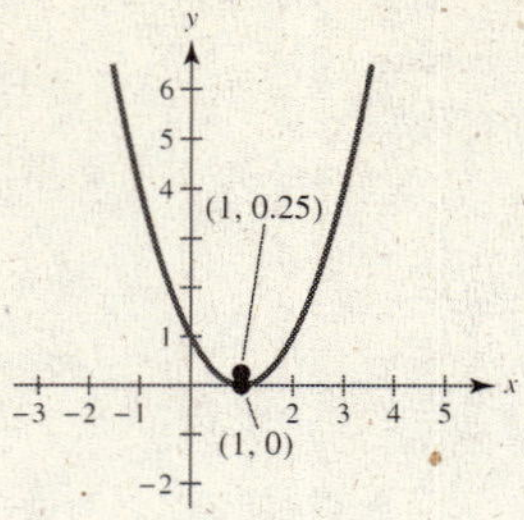

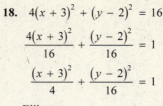

18. $4(x + 3)^2 + (y - 2)^2 = 16$

$\frac{4(x + 3)^2}{16} + \frac{(y - 2)^2}{16} = 1$

$\frac{(x + 3)^2}{4} + \frac{(y - 2)^2}{16} = 1$

Ellipse

$a^2 = 16 \qquad b^2 = 4$

$a = 4 \qquad b = 2$

Center: $(-3, 2)$

Vertices: $(-3, 6), (-3, -2)$

Co-vertices: $(-5, 2), (-1, 2)$

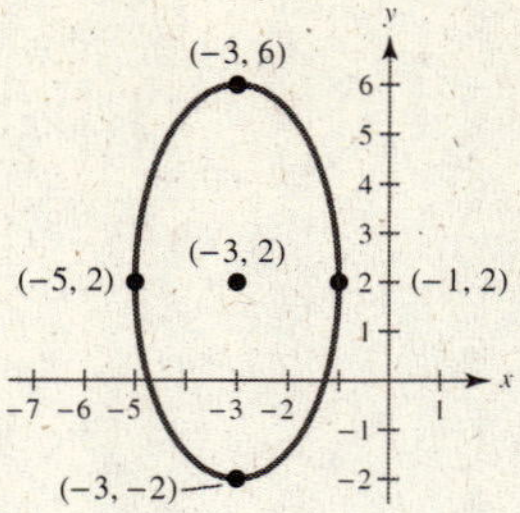

Section 10.3 Hyperbolas

1. $\frac{x^2}{16} - \frac{y^2}{4} = 1$ matches graph (c).

3. $\frac{y^2}{9} - \frac{x^2}{16} = 1$ matches graph (a).

5. $\frac{(x - 1)^2}{16} - \frac{y^2}{4} = 1$ matches graph (b).

7. $x^2 - y^2 = 9$

$\frac{x^2}{9} - \frac{y^2}{9} = 1$

Vertices: $(3, 0), (-3, 0)$

Asymptotes:

$y = \frac{3}{3}x \qquad y = -\frac{3}{3}x$

$y = x \qquad y = -x$

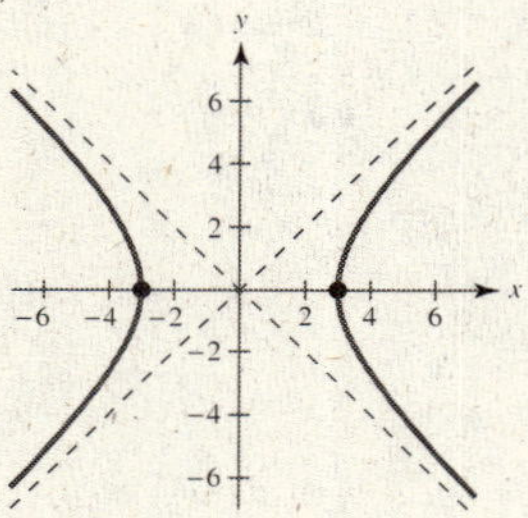

9. $y^2 - x^2 = 9$

$\frac{y^2}{9} - \frac{x^2}{9} = 1$

Vertices: $(0, \pm 3)$

Asymptotes: $y = \pm x$

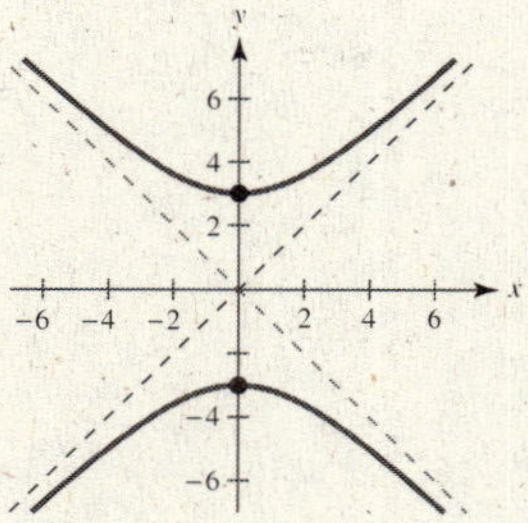

11. $\frac{x^2}{9} - \frac{y^2}{25} = 1$

Vertices: $(3, 0), (-3, 0)$

Asymptotes: $y = \frac{5}{3}x$

$y = -\frac{5}{3}x$

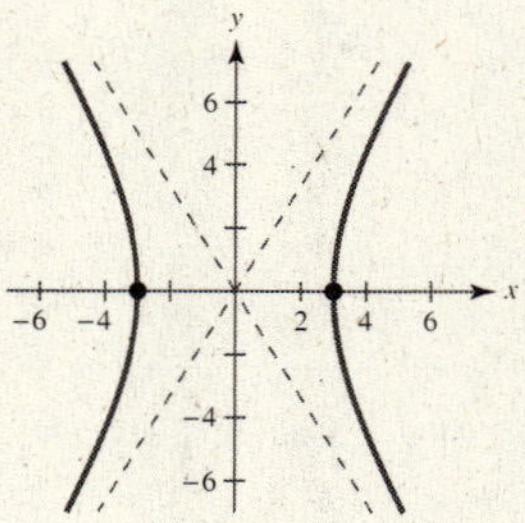

13. $\frac{y^2}{9} - \frac{x^2}{25} = 1$

Vertices: $(0, \pm 3)$

Asymptotes: $y = \pm\frac{3}{5}x$

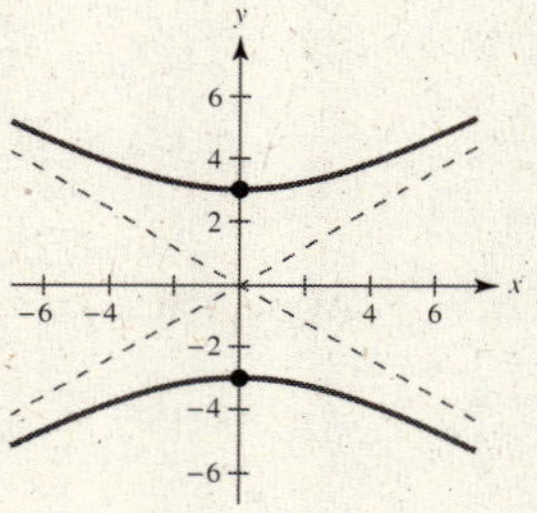

15. $\frac{x^2}{1} - \frac{y^2}{9/4} = 1$

Vertices: $(\pm 1, 0)$

Asymptotes: $y = \pm\frac{3/2}{1}x$

$y = \pm\frac{3}{2}x$

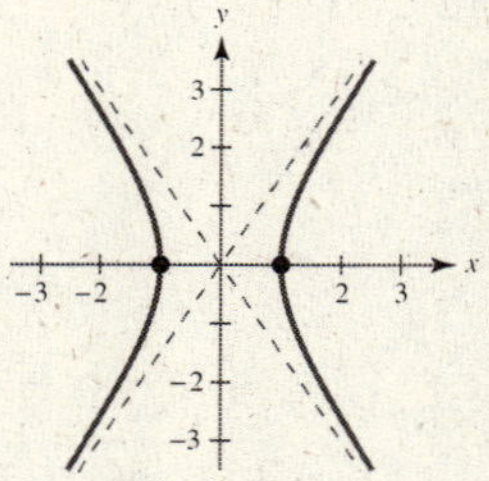

17. $4y^2 - x^2 + 16 = 0$

$$\frac{4y^2}{-16} - \frac{x^2}{-16} = \frac{-16}{-16}$$

$$\frac{-y^2}{4} + \frac{x^2}{16} = 1$$

$$\frac{x^2}{16} - \frac{y^2}{4} = 1$$

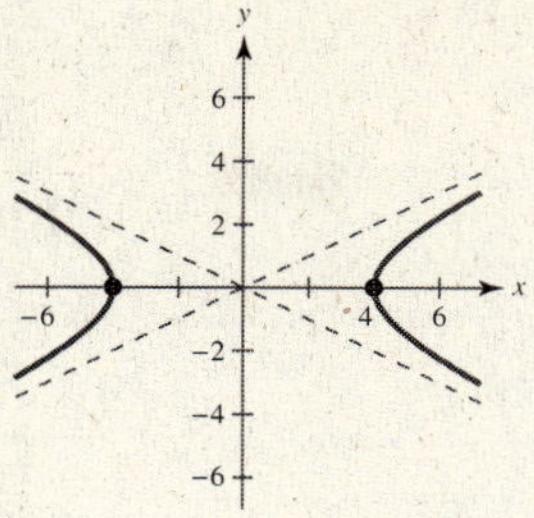

Vertices: $(4, 0), (-4, 0)$

Asymptotes: $y = \frac{2}{4}x = \frac{1}{2}x$

$$y = -\frac{2}{4}x = -\frac{1}{2}x$$

19. Vertices: $(\pm 4, 0)$

Asymptotes: $y = \pm 2x$

$a = 4 \qquad \frac{b}{a} = \pm 2$

$$\frac{b}{4} = \pm 2$$

$$b = \pm 8$$

$$\frac{x^2}{a^2} - \frac{y^2}{b^2} = 1$$

$$\frac{x^2}{4^2} - \frac{y^2}{8^2} = 1$$

$$\frac{x^2}{16} - \frac{y^2}{64} = 1$$

21. Vertices: $(0, \pm 4)$

Asymptotes: $y = \pm\frac{1}{2}x$

$a = 4 \qquad \frac{a}{b} = \pm\frac{1}{2}$

$$\frac{4}{b} = \pm\frac{1}{2}$$

$$b = 8$$

$$\frac{y^2}{a^2} - \frac{x^2}{b^2} = 1$$

$$\frac{y^2}{4^2} - \frac{x^2}{8^2} = 1$$

$$\frac{y^2}{16} - \frac{x^2}{64} = 1$$

23. Vertices: $(\pm 9, 0)$

Asymptotes: $y = \pm\frac{2}{3}x$

$a = 9 \qquad \frac{b}{a} = \pm\frac{2}{3}$

$$\frac{b}{9} = \pm\frac{2}{3}$$

$$b = 6$$

$$\frac{x^2}{a^2} - \frac{y^2}{b^2} = 1$$

$$\frac{x^2}{9^2} - \frac{y^2}{6^2} = 1$$

$$\frac{x^2}{81} - \frac{y^2}{36} = 1$$

25. Vertices: $(0, \pm 1)$

Asymptotes: $y = \pm 2x$

$a = 1 \qquad \frac{a}{b} = \pm 2$

$$\frac{1}{b} = \pm 2$$

$$b = \frac{1}{2}$$

$$\frac{y^2}{a^2} - \frac{x^2}{b^2} = 1$$

$$\frac{y^2}{1^2} - \frac{x^2}{(1/2)^2} = 1$$

$$\frac{y^2}{1} - \frac{x^2}{1/4} = 1$$

27. $\frac{x^2}{16} - \frac{y^2}{4} = 1$

$$x^2 - 4y^2 = 16$$

$$x^2 - 16 = 4y^2$$

$$\frac{x^2 - 16}{4} = y^2$$

$$\pm\sqrt{\frac{x^2 - 16}{4}} = y$$

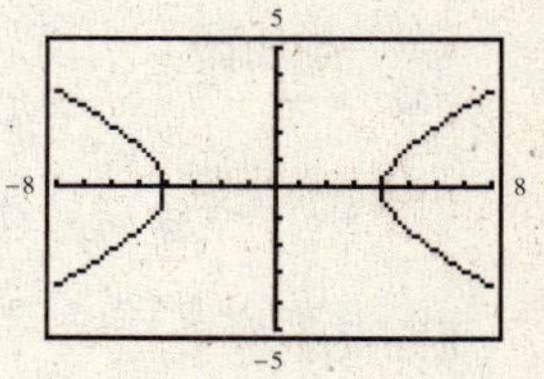

Keystrokes:

y_1: [Y=] [√] [(] [(] [X,T,θ] [x²] [−] 16 [)] [÷] 4 [)] [ENTER]

y_2: [(−)] [√] [(] [(] [X,T,θ] [x²] [−] 16 [)] [÷] 4 [)] [GRAPH]

29. $5x^2 - 2y^2 + 10 = 0$

$$5x^2 + 10 = 2y^2$$

$$\frac{5x^2 + 10}{2} = y^2$$

$$\pm\sqrt{\frac{5x^2 + 10}{2}} = y$$

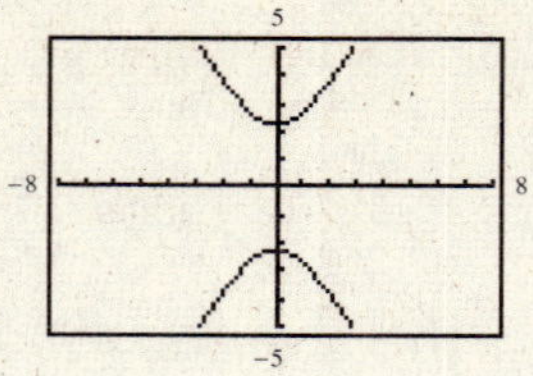

Keystrokes:

y_1: [Y=] [√] [(] [(] 5 [X,T,θ] [x^2] [+] 10 [)] [÷] 2 [)] [ENTER]

y_2: [(−)] [√] [(] [(] 5 [X,T,θ] [x^2] [+] 10 [)] [÷] 2 [)] [GRAPH]

31. $(y + 4)^2 - (x - 3)^2 = 25$

$$\frac{(y + 4)^2}{25} - \frac{(x - 3)^2}{25} = 1$$

Center: $(3, -4)$

Vertices: $(3, 1), (3, -9)$

$a = 5 \qquad b = 5$

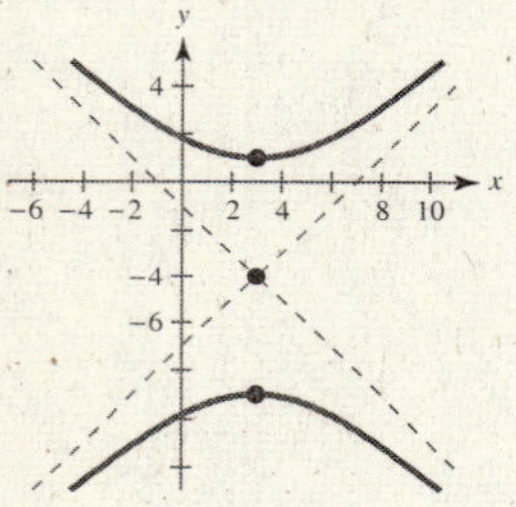

33. $\dfrac{(x - 1)^2}{4} - \dfrac{(y + 2)^2}{1} = 1$

Center: $(1, -2)$

Vertices: $(-1, -2), (3, -2)$

$a^2 = 4 \qquad b^2 = 1$

$a = 2 \qquad b = 1$

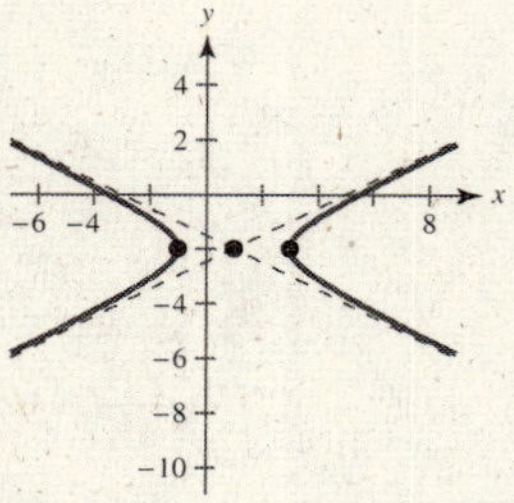

35. $9x^2 - y^2 - 36x - 6y + 18 = 0$

$$(9x^2 - 36x) - (y^2 + 6y) = -18$$

$$9(x^2 - 4x + 4) - (y^2 + 6y + 9) = -18 + 36 - 9$$

$$9(x - 2)^2 - (y + 3)^2 = 9$$

$$\frac{(x - 2)^2}{1} - \frac{(y + 3)^2}{9} = 1$$

Center: $(2, -3)$

Vertices: $(3, -3), (1, -3)$

$a^2 = 1 \qquad b^2 = 9$

$a = 1 \qquad b = 3$

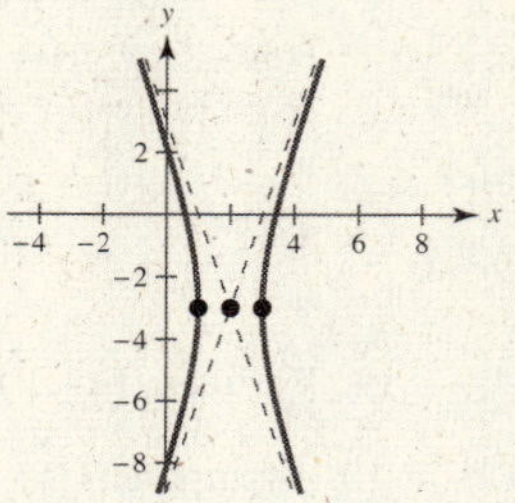

37. $4x^2 - y^2 + 24x + 4y + 28 = 0$

$$(4x^2 + 24x) - (y^2 - 4y) = -28$$

$$4(x^2 + 6x + 9) - (y^2 - 4y + 4) = -28 + 36 - 4$$

$$4(x + 3)^2 - (y - 2)^2 = 4$$

$$\frac{(x + 3)^2}{1} - \frac{(y - 2)^2}{4} = 1$$

Center: $(-3, 2)$

Vertices: $(-4, 2), (-2, 2)$

$a^2 = 1 \qquad b^2 = 4$

$a = 1 \qquad b = 2$

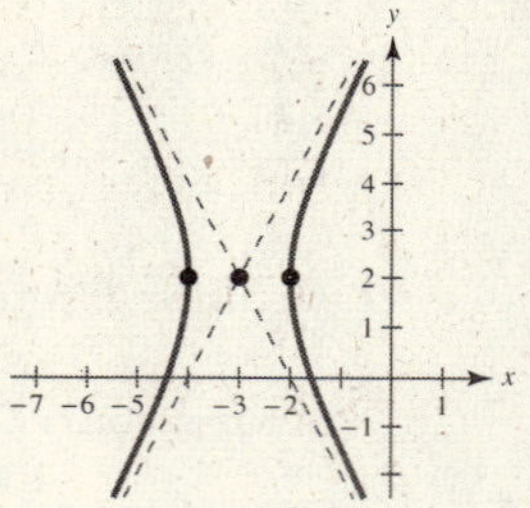

39. Vertices: $(0, \pm 3)$

Vertical axis

Point: $(-2, 5)$

Center: $(0, 0)$

$$\frac{y^2}{a^2} - \frac{x^2}{b^2} = 1$$

$$\frac{(5)^2}{3^2} - \frac{(-2)^2}{b^2} = 1$$

$$\frac{25}{9} - \frac{4}{b^2} = 1$$

$$\frac{25}{9} - \frac{9}{9} = \frac{4}{b^2}$$

$$\frac{16}{9} = \frac{4}{b^2}$$

$$b^2 = \frac{36}{16} = \frac{9}{4}$$

$$\frac{y^2}{9} - \frac{x^2}{9/4} = 1$$

41. Vertices: $(1, 2), (5, 2)$

Horizontal axis

Point: $(0, 0)$

Center: $(3, 2)$

$$\frac{(x-h)^2}{a^2} - \frac{(y-k)^2}{b^2} = 1$$

$$\frac{(0-3)^2}{2^2} - \frac{(0-2)^2}{b^2} = 1$$

$$\frac{9}{4} - \frac{4}{b^2} = 1$$

$$\frac{9}{4} - \frac{4}{4} = \frac{4}{b^2}$$

$$\frac{5}{4} = \frac{4}{b^2}$$

$$b^2 = \frac{16}{5}$$

$$\frac{(x-3)^2}{4} - \frac{(y-2)^2}{16/5} = 1$$

43. $\dfrac{(x-3)^2}{4^2} + \dfrac{(y-4)^2}{6^2} = 1$

Ellipse of the form $\dfrac{(x-h)^2}{b^2} + \dfrac{(y-k)^2}{a^2} = 1$

where $(3, 4)$ is the center.

$a^2 = 6^2$ and $b^2 = 4^2$

45. $x^2 - y^2 = 1$

Hyperbola of the form $\dfrac{x^2}{a^2} - \dfrac{y^2}{b^2} = 1$ where $(0, 0)$ is the center and $a^2 = b^2$.

47.

$$y^2 - x^2 - 2y + 8x - 19 = 0$$

$$(y^2 - 2y) - (x^2 - 8x) = 19$$

$$(y^2 - 2y + 1) - (x^2 - 8x + 16) = 19 + 1 - 16$$

$$(y-1)^2 - (x-4)^2 = 4$$

$$\frac{(y-1)^2}{4} - \frac{(x-4)^2}{4} = 1$$

Hyperbola of the form $\dfrac{(y-k)^2}{a^2} - \dfrac{(x-h)^2}{b^2} = 1$ where $(1, 4)$ is the center and $a^2 = b^2$.

49. $\dfrac{x^2}{93^2} - \dfrac{y^2}{13{,}851} = 1$

$$\frac{x^2}{93^2} - \frac{75^2}{13{,}851} = 1$$

$$x^2 = 93^2\left[1 + \frac{75^2}{13{,}851}\right]$$

$$x^2 \approx 12{,}161.4$$

$$x \approx 110.3$$

51. (a) $\frac{2}{3}\sqrt{9 + (y-1)^2}$ Left half

(b) $\frac{3}{2}\sqrt{(x-3)^2 - 4}$ Top half

53. One. The difference between the given point and the given foci is the same as the difference of the distances between all the other points and the foci.

55. $\begin{cases} -x + 3y = 8 \\ 4x - 12y = -32 \end{cases}$

$3y = x + 8$ $\qquad$ $-12y = -4x - 32$

$y = \frac{1}{3}x + \frac{8}{3}$ $\qquad$ $y = \frac{1}{3}x + \frac{8}{3}$

$m_1 = m_2$ and $b_1 = b_2$; consistent

57. $\begin{cases} x + y = 3 \\ x - y = 2 \end{cases}$

$$\begin{aligned} x + y &= 3 \\ x - y &= 2 \\ \hline 2x &= 5 \\ x &= \tfrac{5}{2} \end{aligned}$$

$$\frac{5}{2} + y = 3$$

$$y = \frac{6}{2} - \frac{5}{2}$$

$$y = \frac{1}{2}$$

$$\left(\frac{5}{2}, \frac{1}{2}\right)$$

Section 10.4 Solving Nonlinear Systems of Equations

1. $\begin{cases} y = 1 \\ x^2 + y = 0 \end{cases}$

Solve for y.

$y = -x^2$

Solution: no real solution

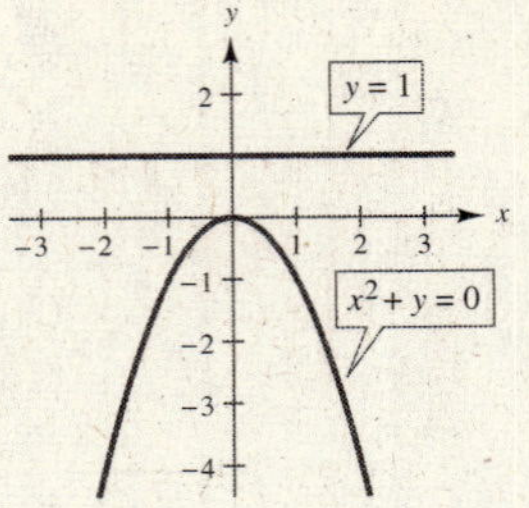

3. $\begin{cases} x = 0 \\ x^2 + y^2 = 9 \end{cases}$

$x = 0$ is the y-axis.

$x^2 + y^2 = 9$ is a circle with center $(0, 0)$ and radius 3.

Solutions: $(0, 3), (0, -3)$

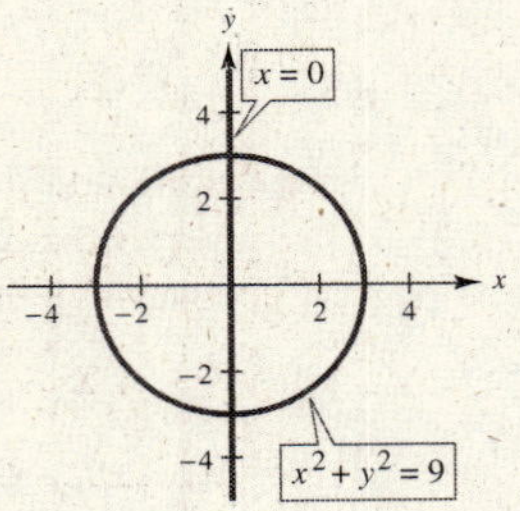

5. $\begin{cases} x + y = 2 \\ x^2 - y = 0 \end{cases}$

Solve for y.

$y = -x + 2$

$y = x^2$

Solutions: $(-2, 4), (1, 1)$

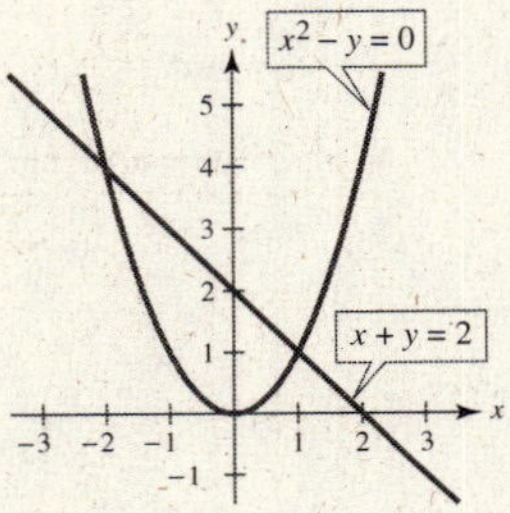

Check:

$-2 + 4 \stackrel{?}{=} 2$

$2 = 2$

$(-2)^2 - 4 \stackrel{?}{=} 0$

$0 = 0$

Check:

$1 + 1 \stackrel{?}{=} 2$

$2 = 2$

$1^2 - 1 \stackrel{?}{=} 0$

$0 = 0$

7. $\begin{cases} x^2 + y = 9 \\ x - y = -3 \end{cases}$

Solve for y.

$y = -x^2 + 9$

$y = x + 3$

Solutions: $(2, 5), (-3, 0)$

Check:

$2^2 + 5 \stackrel{?}{=} 9$

$9 = 9$

$2 - 5 \stackrel{?}{=} -3$

$-3 = -3$

Check:

$(-3)^2 + 0 \stackrel{?}{=} 9$

$9 = 9$

$-3 - 0 \stackrel{?}{=} -3$

$-3 = -3$

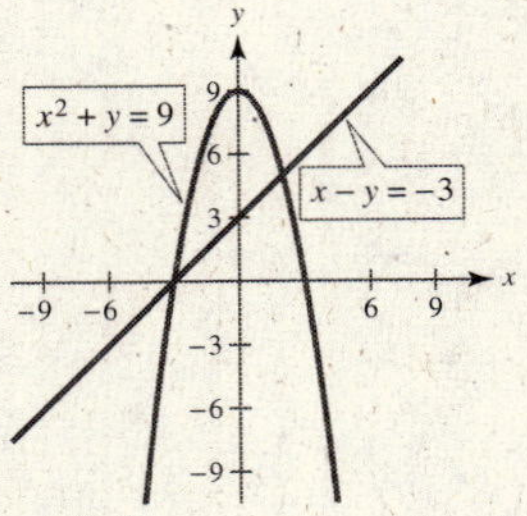

9. $\begin{cases} y = \sqrt{x - 2} \\ x - 2y = 1 \end{cases}$

Solve for y.

$x - 1 = 2y$

$\frac{1}{2}x - \frac{1}{2} = y$

Solution: $(3, 1)$

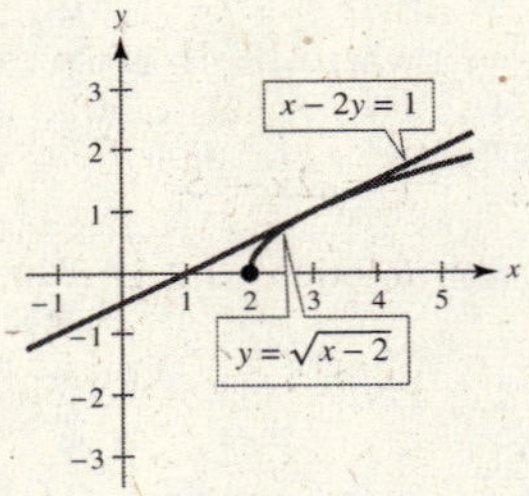

Check:

$1 \stackrel{?}{=} \sqrt{3 - 2}$

$1 = 1$

$3 - 2(1) \stackrel{?}{=} 1$

$1 = 1$

11. $\begin{cases} x^2 + y^2 = 100 \\ x + y = 2 \end{cases}$

Solve for y.

$y = \pm\sqrt{100 - x^2}$

$y = 2 - x$

Solutions: $(-6, 8), (8, -6)$

Check:

$(-6)^2 + 8^2 \stackrel{?}{=} 100$

$100 = 100$

$-6 + 8 \stackrel{?}{=} 2$

$2 = 2$

Check:

$8^2 + (-6)^2 \stackrel{?}{=} 100$

$100 = 100$

$8 + (-6) \stackrel{?}{=} 2$

$2 = 2$

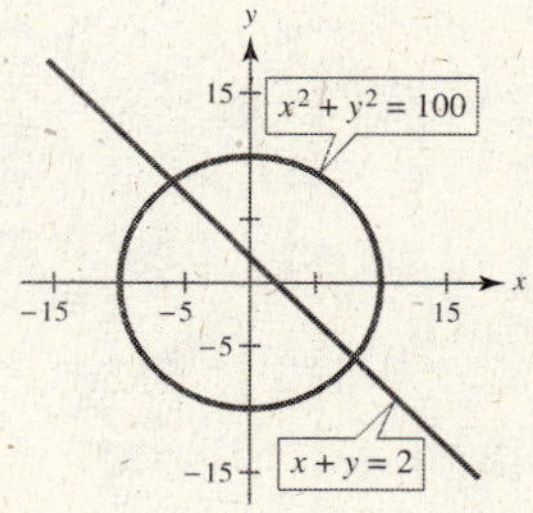

13. $\begin{cases} x^2 + y^2 = 25 \\ 2x - y = -5 \end{cases}$

Solve for y.

$y = \pm\sqrt{25 - x^2}$

$y = 2x + 5$

Solutions: $(0, 5), (-4, -3)$

Check:

$0^2 + 5^2 \stackrel{?}{=} 25$

$25 = 25$

$2(0) - 5 \stackrel{?}{=} -5$

$-5 = -5$

Check:

$(-4)^2 + (-3)^2 \stackrel{?}{=} 25$

$25 = 25$

$2(-4) - (-3) \stackrel{?}{=} -5$

$-5 = -5$

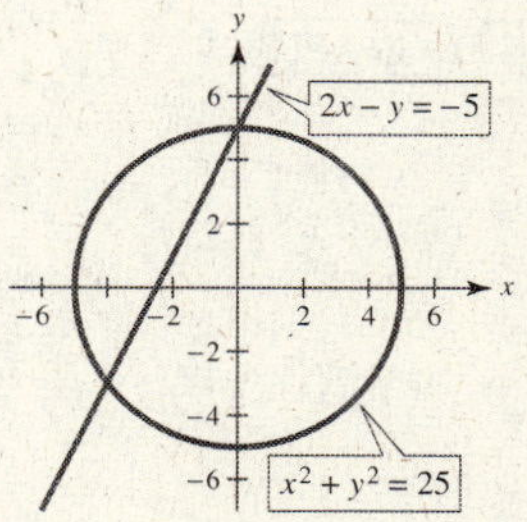

15. $\begin{cases} 9x^2 - 4y^2 = 36 \\ 5x - 2y = 0 \end{cases}$

Rewrite:

$\frac{x^2}{4} - \frac{y^2}{9} = 1$

Solve for y.

$y = \frac{5}{2}x$

No real solution

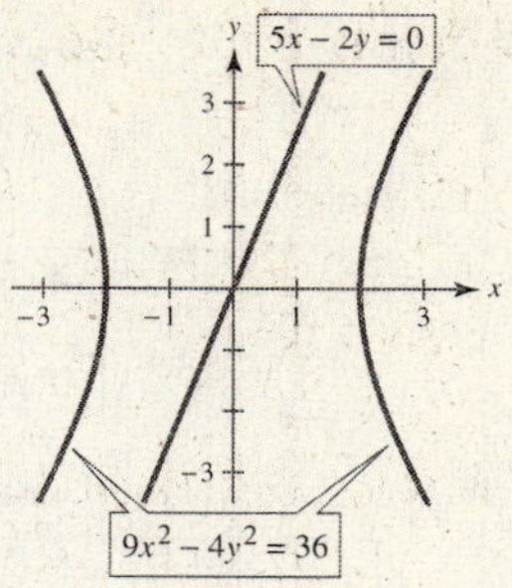

17. $\begin{cases} y = 2x^2 \\ y = -2x + 12 \end{cases}$

Keystrokes:

y_1: [Y=] 2 [X,T,θ] [x^2] [ENTER]

y_2: [(–)] 2 [X,T,θ] [+] 12 [GRAPH]

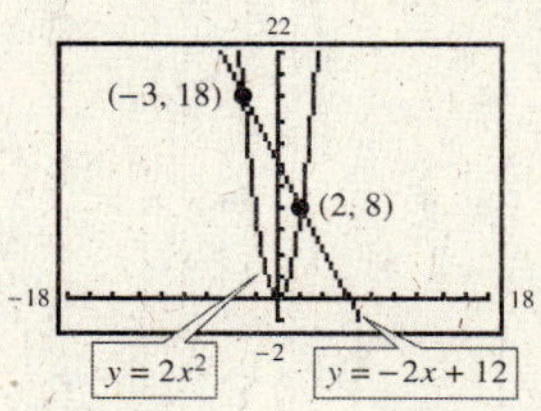

Solutions: $(-3, 18), (2, 8)$

Check:

$18 \stackrel{?}{=} 2(-3)^2$

$18 = 18$

$18 \stackrel{?}{=} -2(-3) + 12$

$18 = 18$

Check:

$8 \stackrel{?}{=} 2(2)^2$

$8 = 8$

$8 \stackrel{?}{=} -2(2) + 12$

$8 = 8$

19. $\begin{cases} y = x \\ y = x^3 \end{cases}$

Keystrokes:

y_1: [Y=] [X,T,θ] [ENTER]

y_2: [X,T,θ] [^] 3 [GRAPH]

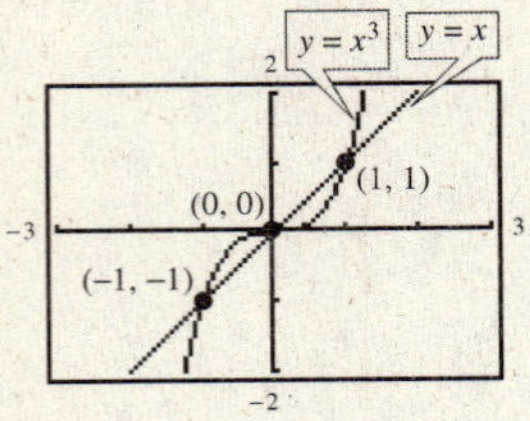

Solutions: $(0, 0), (1, 1), (-1, -1)$

Check:	Check:	Check:
$0 = 0$	$1 = 1$	$-1 = -1$
$0 = 0^3$	$1 = 1^3$	$-1 = (-1)^3$

21. $\begin{cases} y = x^2 \\ y = -x^2 + 4x \end{cases}$

Keystrokes:

y_1: [Y=] [X,T,θ] [x^2] [ENTER]

y_2: [(−)] [X,T,θ] [x^2] [+] 4 [X,T,θ] [Graph]

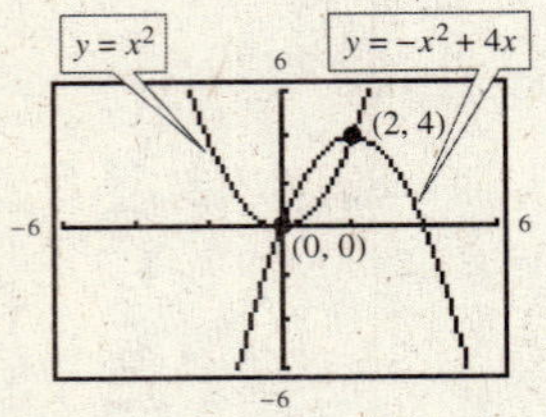

Solutions: $(0, 0), (2, 4)$

Check:

$0 = 0^2$

$0 = -0^2 + 4(0)$

Check:

$4 \stackrel{?}{=} 2^2$

$4 = 4$

$4 \stackrel{?}{=} -2^2 + 4(2)$

$4 = 4$

23. $\begin{cases} x^2 - y = 2 \\ 3x + y = 2 \end{cases} \Rightarrow \begin{cases} y = x^2 - 2 \\ y = -3x + 2 \end{cases}$

Keystrokes:

y_1: [Y=] [X,T,θ] [x^2] [−] 2 [ENTER]

y_2: [(−)] 3 [X,T,θ] [+] 2 [GRAPH]

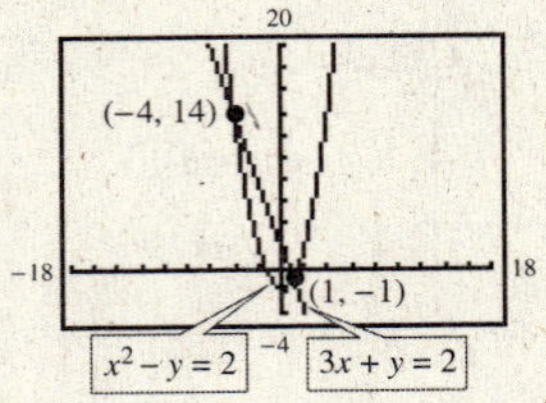

Solutions: $(-4, 14), (1, -1)$

Check:

$(-4)^2 - 14 \stackrel{?}{=} 2$

$2 = 2$

$3(-4) + 14 \stackrel{?}{=} 2$

$2 = 2$

Check:

$1^2 - (-1) \stackrel{?}{=} 2$

$2 = 2$

$3(1) + (-1) \stackrel{?}{=} 2$

$2 = 2$

25. $\begin{cases} x - 3y = 1 \\ \sqrt{x} - 1 = y \end{cases}$

$x - 1 = 3y$

$\frac{1}{3}x - \frac{1}{3}y = y$

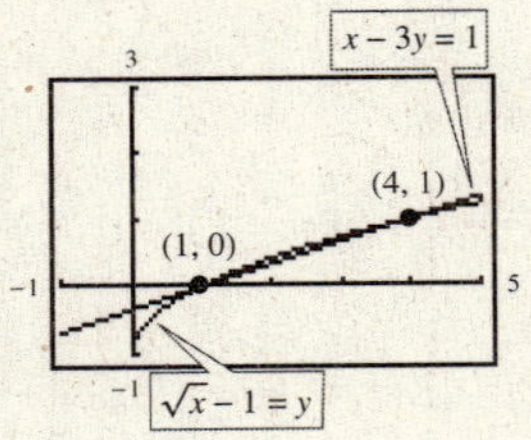

Keystrokes:

y_1: [Y=] 1 [÷] 3 [X,T,θ] [−] 1 [÷] 3 [ENTER]

y_2: [√] [X,T,θ] [)] [−] 1 [GRAPH]

Solutions: $(1, 0), (4, 1)$

Check:

$1 - 3(0) \stackrel{?}{=} 1$

$1 = 1$

$\sqrt{1} - 1 \stackrel{?}{=} 0$

$0 = 0$

Check:

$4 - 3(1) \stackrel{?}{=} 1$

$1 = 1$

$\sqrt{4} - 1 \stackrel{?}{=} 1$

$1 = 1$

27. $\begin{cases} y = x^3 \\ y = x^3 - 3x^2 + 3x \end{cases}$

Keystrokes:

y_1: [Y=] [X,T,θ] [^] 3 [ENTER]

y_2: [X,T,θ] [^] 3 [−] 3 [X,T,θ] [x^2] [+] 3 [X,T,θ] [GRAPH]

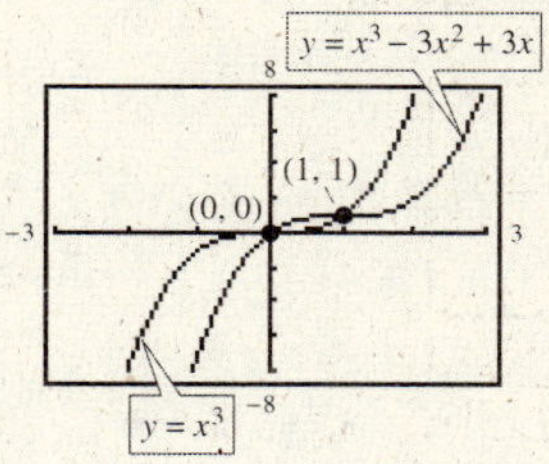

Solutions: $(0, 0), (1, 1)$

Check:

$0 = 0^3$

$0 = 0^3 - 3(0)^2 + 3(0)$

Check:

$1 = 1^3$

$1 \stackrel{?}{=} 1^3 - 3(1)^2 + 3(1)$

$1 = 1$

29. $\begin{cases} y = \ln(x) - 2 \\ y = x - 2 \end{cases}$

Keystrokes:

y_1: [Y=] [LN] [X,T,θ] [−] 2 [ENTER]

y_2: [X,T,θ] [−] 2 [GRAPH]

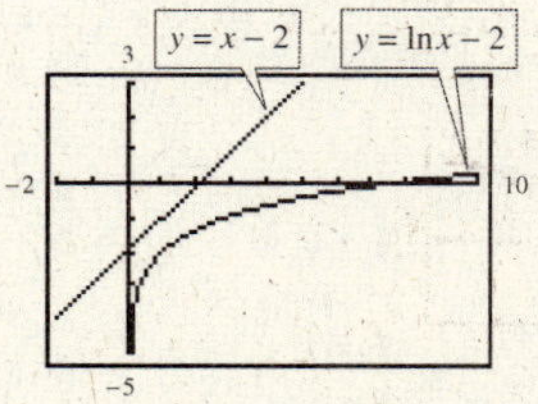

Solution: No real solution

31. $\begin{cases} y = 2x^2 \\ y = 6x - 4 \end{cases}$

$$2x^2 = 6x - 4$$
$$2x^2 - 6x + 4 = 0$$
$$2(x^2 - 3x + 2) = 0$$
$$2(x - 2)(x - 1) = 0$$

$x = 2$ $x = 1$

$y = 2(2)^2 = 8$ $y = 2(1)^2 = 1$

$(2, 8)$ $(1, 2)$

33. $\begin{cases} x^2 + y = 5 \\ 2x + y = 5 \end{cases}$

$$y = 5 - x^2$$
$$2x + (5 - x^2) = 5$$
$$-x^2 + 2x + 5 - 5 = 0$$
$$x^2 - 2x = 0$$
$$x(x - 2) = 0$$

$x = 0$ $x = 2$

$0^2 + y = 5$ $2^2 + y = 5$

$y = 5$ $y = 1$

$(0, 5)$ $(2, 1)$

35. $\begin{cases} x^2 + y^2 = 4 \\ x + y = 2 \end{cases}$

$$y = 2 - x$$
$$x^2 + (2 - x)^2 = 4$$
$$x^2 + 4 - 4x + x^2 = 4$$
$$2x^2 - 4x = 0$$
$$2x(x - 2) = 0$$

$2x = 0$ $x - 2 = 0$

$x = 0$ $x = 2$

$y = 2 - 0$ $y = 2 - 2$

$y = 2$ $y = 0$

$(0, 2)$ $(2, 0)$

37. $\begin{cases} x^2 + y^2 = 25 \\ y = 5 \end{cases}$

$$x^2 + 5^2 = 25$$
$$x^2 = 0$$

$(0, 5)$

39. $\begin{cases} x^2 + y^2 = 64 \\ -3x + y = 8 \end{cases}$

$$y = 3x + 8$$
$$x^2 + (3x + 8)^2 = 64$$
$$x^2 + 9x^2 + 48x + 64 = 64$$
$$10x^2 + 48x = 0$$
$$2x(5x + 24) = 0$$

$2x = 0$, $x = 0$, $y = 3(0) + 8 = 8$ $\qquad$ $5x + 24 = 0$, $x = -\frac{24}{5}$, $y = 3\left(-\frac{24}{5}\right) + 8 = -\frac{72}{5} + \frac{40}{5} = -\frac{32}{5}$

$(0, 8)$ $\qquad$ $\left(-\frac{24}{5}, -\frac{32}{5}\right)$

41. $\begin{cases} 4x - y^2 = 2 \\ 2x - y = -11 \end{cases}$

$$y = 2x + 11$$
$$4x + (2x + 11)^2 = 2$$
$$4x + 4x^2 + 44x + 121 = 2$$
$$4x^2 + 48x + 119 = 0$$
$$(2x + 17)(2x + 7) = 0$$

$2x + 17 = 0$, $x = -\frac{17}{2}$, $y = 2\left(-\frac{17}{2}\right) + 11 = -17 + 11 = -6$ $\qquad$ $2x + 7 = 0$, $x = -\frac{7}{2}$, $y = 2\left(-\frac{7}{2}\right) + 11 = -7 + 11 = 4$

$\left(-\frac{17}{2}, -6\right)$ $\qquad$ $\left(-\frac{7}{2}, 4\right)$

43. $\begin{cases} x^2 + y^2 = 9 \\ x + 2y = 3 \end{cases}$

$$x = 3 - 2y$$
$$(3 - 2y)^2 + y^2 = 9$$
$$9 - 12y + 4y^2 + y^2 = 9$$
$$5y^2 - 12y = 0$$
$$y(5y - 12) = 0$$
$$5y - 12 = 0$$

$y = 0$, $x = 3 - 2(0) = 3$ $\qquad$ $y = \frac{12}{5}$, $x = 3 - 2\left(\frac{12}{5}\right) = \frac{15}{5} - \frac{24}{5} = -\frac{9}{5}$

$(3, 0)$ $\qquad$ $\left(-\frac{9}{5}, \frac{12}{5}\right)$

45. $\begin{cases} 2x^2 - y^2 = -8 \\ x - y = 6 \end{cases}$

$$x = y + 6$$
$$2(y + 6)^2 - y^2 = -8$$
$$2(y^2 + 12y + 36) - y^2 = -8$$
$$2y^2 + 24y + 72 - y^2 = -8$$
$$y^2 + 24y + 80 = 0$$
$$(y + 20)(y + 4) = 0$$

$y + 20 = 0$, $y = -20$, $x = -20 + 6 = -14$ $\qquad$ $y + 4 = 0$, $y = -4$, $x = -4 + 6 = 2$

$(-14, -20)$ $\qquad$ $(2, -4)$

47. $\begin{cases} y = x^2 - 5 \\ 3x + 2y = 10 \end{cases}$

$$3x + 2(x^2 - 5) = 10$$
$$3x + 2x^2 - 10 = 10$$
$$2x^2 + 3x - 20 = 0$$
$$(2x - 5)(x + 4) = 0$$

$2x - 5 = 0$, $x = \frac{5}{2}$, $y = \left(\frac{5}{2}\right)^2 - 5 = \frac{25}{4} - \frac{20}{4} = \frac{5}{4}$ $\qquad$ $x + 4 = 0$, $x = -4$, $y = (-4)^2 - 5 = 16 - 5 = 11$

$\left(\frac{5}{2}, \frac{5}{4}\right)$ $\qquad$ $(-4, 11)$

49. $\begin{cases} y = \sqrt{4 - x} \\ x + 3y = 6 \end{cases}$

$$x + 3\sqrt{4 - x} = 6$$
$$3\sqrt{4 - x} = 6 - x$$
$$9(4 - x) = 36 - 12x + x^2$$
$$36 - 9x = 36 - 12x + x^2$$
$$0 = x^2 - 3x$$
$$0 = x(x - 3)$$

$0 = x$, $y = \sqrt{4 - 0} = 2$ $\qquad$ $x - 3 = 0$, $x = 3$, $y = \sqrt{4 - 3} = 1$

$(0, 2)$ $\qquad$ $(3, 1)$

51. $\begin{cases} x^2 - 4y^2 = 16 \\ x^2 + y^2 = 1 \end{cases}$

$$\begin{aligned} y^2 &= 1 - x^2 \\ x^2 - 4(1 - x^2) &= 16 \\ x^2 - 4 + 4x^2 &= 16 \\ 5x^2 &= 20 \\ x^2 &= 4 \\ x &= \pm 2 \end{aligned}$$

$y^2 = 1 - 4 = -3$

No real solution

53. $\begin{cases} y = x^2 - 3 \\ x^2 + y^2 = 9 \end{cases}$

$$\begin{aligned} x^2 &= y + 3 \\ y + 3 + y^2 &= 9 \\ y^2 + y - 6 &= 0 \\ (y + 3)(y - 2) &= 0 \end{aligned}$$

$$\begin{aligned} y + 3 &= 0 & y - 2 &= 0 \\ y &= -3 & y &= 2 \\ x^2 &= -3 + 3 & x^2 &= 2 + 3 \\ x^2 &= 0 & x^2 &= 5 \\ x &= 0 & x &= \pm\sqrt{5} \end{aligned}$$

$(0, -3)$ $\quad (\pm\sqrt{5}, 2)$

55. $\begin{cases} 16x^2 + 9y^2 = 144 \\ 4x + 3y = 12 \end{cases}$

$$\begin{aligned} 4x &= 12 - 3y \\ y &= \frac{12 - 3y}{4} \\ 16\left(\frac{12 - 3y}{4}\right)^2 + 9y^2 &= 144 \\ 16\left(\frac{144 - 72y + 9y^2}{16}\right) + 9y^2 &= 144 \\ 144 - 72y + 9y^2 + 9y^2 &= 144 \\ 18y^2 - 72y &= 0 \\ 18y(y - 4) &= 0 \end{aligned}$$

$$\begin{aligned} 18y &= 0 & y - 4 &= 0 \\ y &= 0 & y &= 4 \\ x &= \frac{12 - 3(0)}{4} & x &= \frac{12 - 3(4)}{4} \\ x &= 3 & x &= 0 \end{aligned}$$

$(3, 0)$ $\quad (0, 4)$

57. $\begin{cases} x^2 - y^2 = 9 \\ x^2 + y^2 = 1 \end{cases}$

$$\begin{aligned} x^2 &= 1 - y^2 \\ 1 - y^2 - y^2 &= 9 \\ -2y^2 &= 8 \\ y^2 &= -4 \end{aligned}$$

No real solution

59. $\begin{cases} x^2 + 2y = 1 \\ x^2 + y^2 = 4 \end{cases}$

$$\begin{array}{rcrcr} x^2 & + & 2y & = & 1 \\ -x^2 & - & y^2 & = & -4 \\ \hline -y^2 & + & 2y & = & -3 \end{array}$$

$$\begin{aligned} y^2 - 2y - 3 &= 0 \\ (y - 3)(y + 1) &= 0 \end{aligned}$$

$$\begin{aligned} y - 3 &= 0 & y + 1 &= 0 \\ y &= 3 & y &= -1 \\ x^2 + 3^2 &= 4 & x^2 + (-1)^2 &= 4 \\ x^2 &= -5 & x^2 &= 3 \\ & & x &= \pm\sqrt{3} \end{aligned}$$

No real solution $\quad (\pm\sqrt{3}, -1)$

61. $\begin{cases} -x + y^2 = 10 \\ x^2 - y^2 = -8 \end{cases}$

$$\begin{aligned} -x + x^2 &= 2 \\ x^2 - x - 2 &= 0 \\ (x - 2)(x + 1) &= 0 \end{aligned}$$

$$\begin{aligned} x - 2 &= 0 & x + 1 &= 0 \\ x &= 2 & x &= -1 \\ -2 + y^2 &= 10 & -1(-1) + y^2 &= 10 \\ y^2 &= 12 & y^2 &= 9 \\ y &= \pm\sqrt{12} & y &= \pm 3 \\ y &= \pm 2\sqrt{3} \end{aligned}$$

$(2, \pm 2\sqrt{3})$ $\quad (-1, \pm 3)$

63. $\begin{cases} x^2 + y^2 = 7 \\ x^2 - y^2 = 1 \end{cases}$

$$2x^2 = 8$$
$$x^2 = 4$$
$$x = \pm 2$$
$$(\pm 2)^2 + y^2 = 7$$
$$y^2 = 3$$
$$y = \pm\sqrt{3}$$
$$(\pm 2, \pm\sqrt{3})$$

65. $\begin{cases} x^2 - y^2 = 4 \\ x^2 + y^2 = 4 \end{cases}$

$$2x^2 = 8$$
$$x^2 = 4$$
$$x = \pm 2$$
$$(\pm 2)^2 - y^2 = 4$$
$$-y^2 = 0$$
$$y = 0$$
$$(\pm 2, 0)$$

67. $\begin{cases} x^2 + y^2 = 13 \\ 2x^2 + 3y^2 = 30 \end{cases}$

$$-2x^2 - 2y^2 = -26$$
$$2x^2 + 3y^2 = 30$$
$$y^2 = 4$$
$$y = \pm 2$$
$$x^2 + (\pm 2)^2 = 13$$
$$x^2 = 9$$
$$x = \pm 3$$
$$(\pm 3, \pm 2)$$

69. $\begin{cases} 4x^2 + 9y^2 = 36 \\ 2x^2 - 9y^2 = 18 \end{cases}$

$$6x^2 = 54$$
$$x^2 = 9$$
$$x = \pm 3$$
$$4(\pm 3)^2 + 9y^2 = 36$$
$$36 + 9y^2 = 36$$
$$9y^2 = 0$$
$$y^2 = 0$$
$$y = 0$$
$$(\pm 3, 0)$$

71. $\begin{cases} 2x^2 + 3y^2 = 21 \\ x^2 + 2y^2 = 12 \end{cases}$

$$2x^2 + 3y^2 = 21$$
$$-2x^2 - 4y^2 = -24$$
$$-y^2 = -3$$
$$y^2 = 3$$
$$y = \pm\sqrt{3}$$
$$2x^2 + 3(\pm\sqrt{3})^2 = 21$$
$$2x^2 + 9 = 21$$
$$2x^2 = 12$$
$$x^2 = 6$$
$$x = \pm\sqrt{6}$$
$$(\pm\sqrt{6}, \pm\sqrt{3})$$

73. $\begin{cases} -x^2 - 2y^2 = 6 \\ 5x^2 + 15y^2 = 20 \end{cases}$

$$-5x^2 - 10y^2 = 30$$
$$5x^2 + 15y^2 = 20$$
$$5y^2 = 50$$
$$y^2 = 10$$
$$y = \pm\sqrt{10}$$
$$-x^2 - 2(\pm\sqrt{10})^2 = 6$$
$$-x^2 - 20 = 6$$
$$-x^2 = 26$$
$$x^2 = -26$$

No real solution

75. $\begin{cases} x^2 + y^2 = 9 \\ 16x^2 - 4y^2 = 64 \end{cases}$

$$\begin{aligned} 4x^2 + 4y^2 &= 36 \\ 16x^2 - 4y^2 &= 64 \\ 20x^2 &= 100 \\ x^2 &= 5 \\ x &= \pm\sqrt{5} \end{aligned}$$

$$\begin{aligned} \left(\pm\sqrt{5}\right)^2 + y^2 &= 9 \\ y^2 &= 4 \\ y &= \pm 2 \end{aligned}$$

$\left(\pm\sqrt{5}, \pm 2\right)$

77. $\begin{cases} \dfrac{x^2}{4} + y^2 = 1 \\ x^2 + \dfrac{y^2}{4} = 1 \end{cases}$

$$\begin{aligned} x^2 + 4y^2 &= 4 \\ 4x^2 + y^2 &= 4 \\ x^2 + 4y^2 &= 4 \\ -16x^2 - 4y^2 &= -16 \\ -15x^2 &= -12 \\ x^2 &= \frac{12}{15} \\ x^2 &= \frac{4}{5} \\ x &= \pm\sqrt{\frac{4}{5}} \\ x &= \pm\frac{2}{\sqrt{5}} \cdot \frac{\sqrt{5}}{\sqrt{5}} \\ x &= \pm\frac{2\sqrt{5}}{5} \end{aligned}$$

$$\begin{aligned} \left(\pm\frac{2\sqrt{5}}{5}\right)^2 + 4y^2 &= 4 \\ \frac{4}{5} + 4y^2 &= 4 \\ 4y^2 &= \frac{16}{5} \\ y^2 &= \frac{4}{5} \\ y &= \pm\frac{2\sqrt{5}}{5} \end{aligned}$$

$\left(\pm\frac{2\sqrt{5}}{5}, \pm\frac{2\sqrt{5}}{5}\right)$

79. $\begin{cases} y^2 - x^2 = 10 \\ x^2 + y^2 = 16 \end{cases}$

$$\begin{aligned} -x^2 + y^2 &= 10 \\ x^2 + y^2 &= 16 \\ 2y^2 &= 26 \\ y^2 &= 13 \\ y &= \pm\sqrt{13} \end{aligned}$$

$$\begin{aligned} 13 - x^2 &= 10 \\ -x^2 &= -3 \\ x^2 &= 3 \\ x &= \pm\sqrt{3} \end{aligned}$$

$\left(\pm\sqrt{3}, \pm\sqrt{13}\right)$

81. *Verbal Model:* $\boxed{\text{Length}} \times \boxed{\text{Width}} = \boxed{\text{Area}}$

$\boxed{\text{Length}}^2 + \boxed{\text{Width}}^2 = \boxed{\text{Diagonal}}^2$

Labels: Length $= x$

Width $= y$

System:

$$\begin{aligned} xy &= 3000 \\ x^2 + y^2 &= 85^2 \\ y &= \frac{3000}{x} \\ x^2 + \left(\frac{3000}{x}\right)^2 &= 7225 \\ x^2 + \frac{9{,}000{,}000}{x^2} &= 7225 \\ x^4 + 9{,}000{,}000 &= 7225x^2 \\ x^4 - 7225x^2 + 9{,}000{,}000 &= 0 \\ \left(x^2 - 5625\right)\left(x^2 - 1600\right) &= 0 \end{aligned}$$

$$\begin{aligned} x^2 &= 5625 & x^2 &= 1600 \\ x &= 75 & x &= 40 \\ y &= \frac{3000}{75} & y &= \frac{3000}{40} \\ &= 40 & &= 75 \end{aligned}$$

40 feet × 75 feet

83. *Verbal Model:* $\boxed{\text{Length}}^2 + \boxed{\text{Width}}^2 = \boxed{\text{Diagonal}}^2$

$\boxed{\text{Perimeter}} = \boxed{\text{Length}} + \boxed{\text{Width}} + 290$

Labels: Length $= x$

Width $= y$

System: $x^2 + y^2 = 290^2$

$$700 = x + y + 290$$
$$y = 410 - x$$
$$x^2 + (410 - x)^2 = 290^2$$
$$x^2 + 168{,}000 - 820x + x^2 = 84{,}100$$
$$2x^2 - 820x + 84{,}000 = 0$$
$$x^2 - 410x + 42{,}000 = 0$$
$$(x - 200)(x - 210) = 0$$

$x = 200 \qquad x = 210$

85. Find the equation of the line with points $(0, 10)$ and $(5, 0)$.

$$m = \frac{0 - 10}{5 - 0} = -2,\ y = -2x + 10$$

Find the point of intersection of the line and the hyperbola.

$$\begin{cases} y = -2x + 10 \\ \dfrac{x^2}{9} - \dfrac{y^2}{16} = 1 \end{cases} \Rightarrow \begin{matrix} y = -2x + 10 \\ 16x^2 - 9y^2 = 144 \end{matrix}$$

$$16x^2 - 9(-2x + 10)^2 = 144$$
$$16x^2 - 9(4x^2 - 40x + 100) = 144$$
$$16x^2 - 36x^2 + 360x - 900 - 144 = 0$$
$$-20x^2 + 360x - 1044 = 0$$
$$5x^2 - 90x + 261 = 0$$

$$x = \frac{90 \pm \sqrt{90^2 - 4(5)(261)}}{10} \approx 3.633$$

$y = -2(3.633) + 10$

$y \approx 2.733$

$(3.633, 2.733)$

87. Equation of circle: $(x - 0)^2 + (y - 0)^2 = 1^2$

$$x^2 + y^2 = 1$$

Equation of Clarke Street (line): points: $(-2, -1)$ and $(5, 0)$; slope: $m = \dfrac{0 - (-1)}{5 - (-2)} = \dfrac{1}{7}$;

Equation: $y - 0 = \frac{1}{7}(x - 5)$

$$y = \frac{1}{7}x - \frac{5}{7}$$
$$7y = x - 5$$

Find the intersection of the circle and the line by solving the system:

$$x^2 + y^2 = 1$$
$$7y = x - 5$$
$$x = 7y + 5$$
$$(7y + 5)^2 + y^2 = 1$$
$$49y^2 + 70y + 25 + y^2 = 1$$
$$50y^2 + 70y + 24 = 0$$
$$25y^2 + 35y + 12 = 0$$
$$(5y + 4)(5y + 3) = 0$$

$y = -\frac{4}{5}$

$x = 7\left(-\frac{4}{5}\right) + 5 = -\frac{28}{5} + \frac{25}{5}$

$x = -\frac{3}{5}$

$\left(-\frac{3}{5}, -\frac{4}{5}\right)$

$y = -\frac{3}{5}$

$x = 7\left(-\frac{3}{5}\right) + 5 = -\frac{21}{5} + \frac{25}{5}$

$x = \frac{4}{5}$

$\left(\frac{4}{5}, -\frac{3}{5}\right)$

89. Solve one of the equations for one variable in terms of the other. Substitute that expression into the other equation. Solve the equation. Back-substitute the solution into the first equation to find the value of the other variable.

91. Two. The line can intersect a branch of the hyperbola at most twice, and it can intersect only one point on each branch at the same time.

93.
$$\sqrt{6-2x} = 4$$
$$\left(\sqrt{6-2x}\right)^2 = 4^2$$
$$6 - 2x = 16$$
$$-2x = 10$$
$$x = -5$$

Check: $\sqrt{6-2(-5)} \stackrel{?}{=} 4$, $\sqrt{6+10} \stackrel{?}{=} 4$, $\sqrt{16} \stackrel{?}{=} 4$, $4 = 4$

95.
$$\sqrt{x} = x - 6$$
$$\left(\sqrt{x}\right)^2 = (x-6)^2$$
$$x = x^2 - 12x + 36$$
$$0 = x^2 - 13x + 36$$
$$0 = (x-9)(x-4)$$
$$9 = x \qquad x = 4$$

Check: $\sqrt{9} \stackrel{?}{=} 9 - 6$, $3 = 3$ **Check:** $\sqrt{4} \stackrel{?}{=} 4 - 6$, $2 \neq -2$

Solution: $x = 9$

97.
$$3^x = 243$$
$$3^x = 3^5$$
$$x = 5$$

Check: $3^5 \stackrel{?}{=} 243$, $243 = 243$

99.
$$5^{x-1} = 310$$
$$\log 5^{x-1} = \log 310$$
$$(x-1)\log 5 = \log 310$$
$$x\log 5 - \log 5 = \log 310$$
$$x \log 5 = \log 310 + \log 5$$
$$x = \frac{\log 310 + \log 5}{\log 5}$$
$$x \approx 4.564$$

Check: $5^{4.564-1} \stackrel{?}{=} 310$, $5^{3.564} \stackrel{?}{=} 310$, $310 \stackrel{?}{=} 310$

101.
$$\log_{10} x = 0.01$$
$$10^{0.01} = x$$
$$1.023 \approx x$$

Check: $\log_{10} 1.023 \stackrel{?}{=} 0.01$, $0.01 = 0.01$

103.
$$2\ln(x+1) = -2$$
$$\ln(x+1) = -1$$
$$e^{-1} = x + 1$$
$$e^{-1} - 1 = x$$
$$-0.632 \approx x$$

Check: $2\ln(-0.632 + 1) = -2$, $2\ln(0.368) \stackrel{?}{=} -2$, $-2 = -2$

Review Exercises for Chapter 10

1. Ellipse

3. Circle

5. Hyperbola

7. Center: $(0, 0)$

Radius: 6

$$x^2 + y^2 = 6^2$$
$$x^2 + y^2 = 36$$

9. $x^2 + y^2 = 64$

Center: $(0, 0)$

Radius: 8

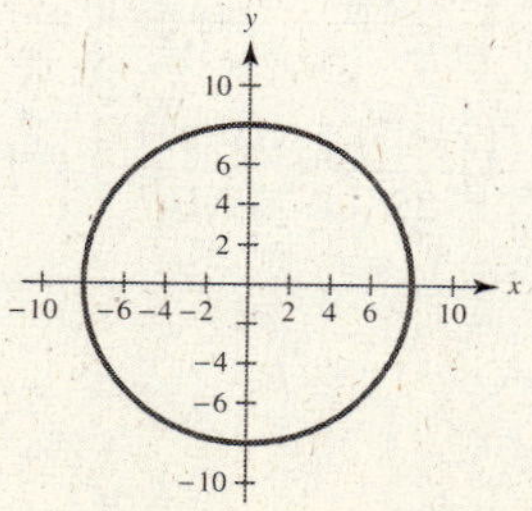

11. Center: $(2, 6)$; Radius: 3

$$(x-h)^2 + (y-k)^2 = r^2$$
$$(x-2)^2 + (y-6)^2 = 9$$

13.
$$x^2 + y^2 + 6x + 8y + 21 = 0$$
$$(x^2 + 6x + 9) + (y^2 + 8y + 16) = -21 + 9 + 16$$
$$(x+3)^2 + (y+4)^2 = 4$$

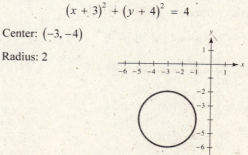

Center: $(-3, -4)$

Radius: 2

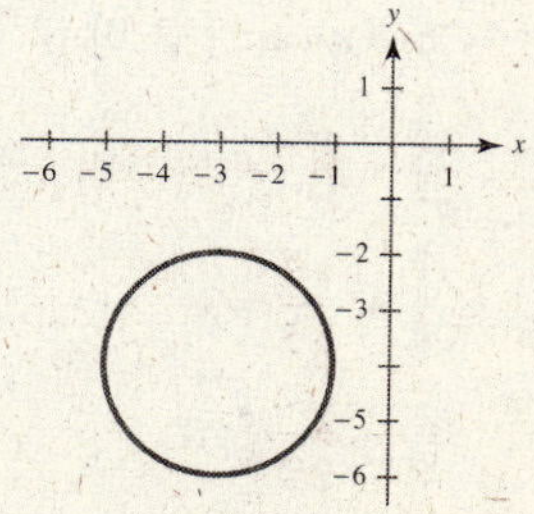

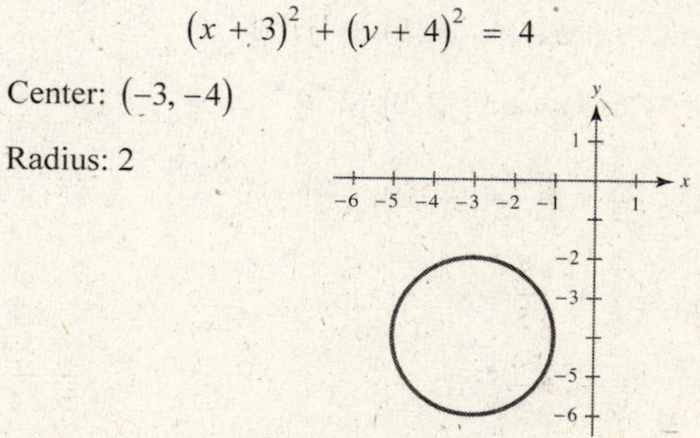

15. Vertex: $(0, 0)$

Focus: $(-2, 0)$

$$y^2 = 4px$$
$$p = -2$$
$$y^2 = 4(-2)x$$
$$y^2 = -8x$$

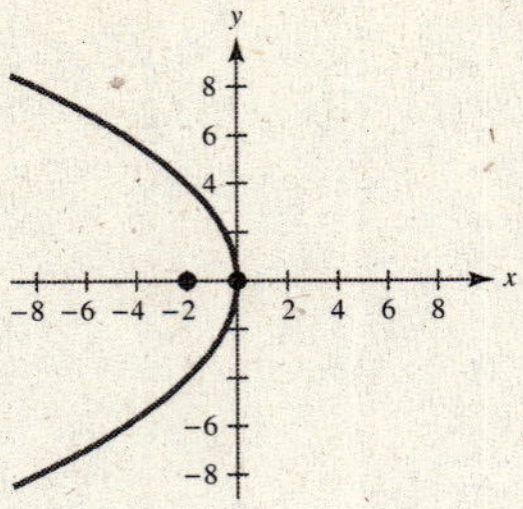

17. Vertex: $(-1, 3)$

Vertical axis

Point: $(-2, 5)$

$$(x - h)^2 = 4p(y - k)$$
$$\left[-2 - (-1)\right]^2 = 4p(5 - 3)$$
$$1 = 4p(2)$$
$$\tfrac{1}{2} = 4p$$
$$\left[x - (-1)\right]^2 = \tfrac{1}{2}(y - 3)$$
$$(x + 1)^2 = \tfrac{1}{2}(y - 3)$$

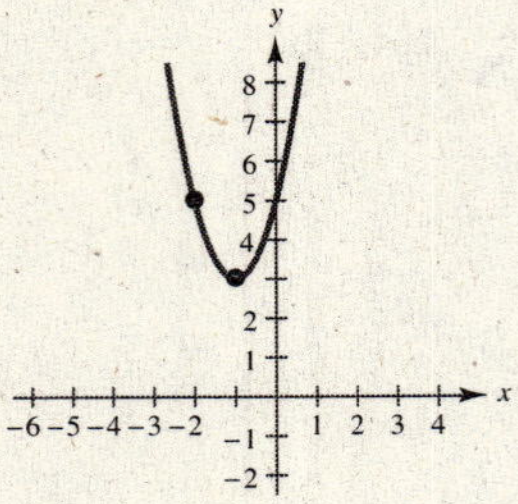

19.
$$y = \tfrac{1}{2}x^2 - 8x + 7$$
$$2y = x^2 - 16x + 14$$
$$2y - 14 = x^2 - 16x$$
$$2y - 14 + 64 = x^2 - 16x + 64$$
$$2y + 50 = (x - 8)^2$$
$$2(y + 25) = (x - 8)^2$$

Vertex: $(8, -25)$

$$4p = 2$$
$$p = \tfrac{1}{2}$$

Focus: $\left(8, -25 + \tfrac{1}{2}\right)$

$$\left(8, -\tfrac{50}{2} + \tfrac{1}{2}\right)$$
$$\left(8, -\tfrac{49}{2}\right)$$

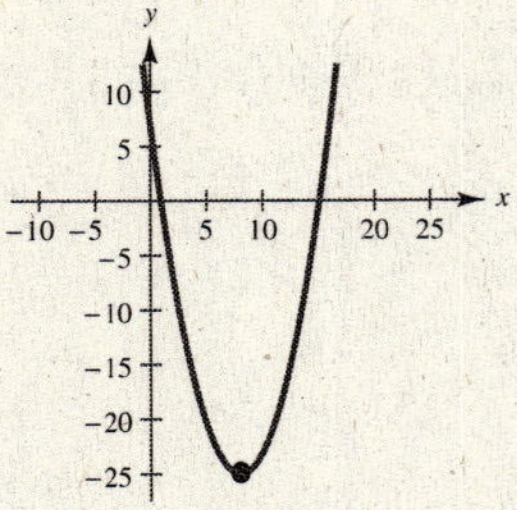

21. Vertices: $(0, -5), (0, 5)$

Co-vertices: $(-2, 0), (2, 0)$

$$\frac{x^2}{a^2} + \frac{y^2}{b^2} = 1$$
$$\frac{x^2}{2^2} + \frac{y^2}{5^2} = 1$$
$$\frac{x^2}{4} + \frac{y^2}{25} = 1$$

23. Major axis (vertical) 6 units

Minor axis 4 units

$$\frac{x^2}{a^2} + \frac{y^2}{b^2} = 1$$

$2b = 6$ $\qquad$ $2a = 4$

$b = 3$ $\qquad$ $a = 2$

$$\frac{x^2}{2^2} + \frac{y^2}{3^2} = 1$$
$$\frac{x^2}{4} + \frac{y^2}{9} = 1$$

25. $\dfrac{x^2}{64} + \dfrac{y^2}{16} = 1$

$a^2 = 64$ $\qquad$ $b^2 = 16$

$a = 8$ $\qquad$ $b = 4$

Vertices: $(\pm 8, 0)$

Co-vertices: $(0, \pm 4)$

Horizontal axis

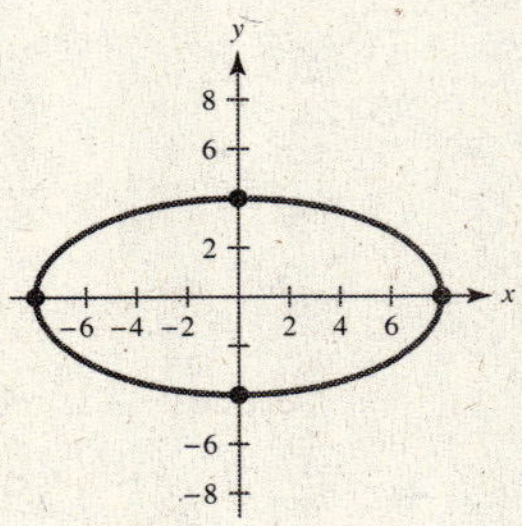

27. $36x^2 + 9y^2 - 36 = 0$

$$x^2 + \frac{y^2}{4} = 1$$

$a^2 = 4$ $\qquad$ $b^2 = 1$

$a = \pm 2$ $\qquad$ $b = \pm 1$

Vertices: $(0, \pm 2)$

Co-vertices: $(\pm 1, 0)$

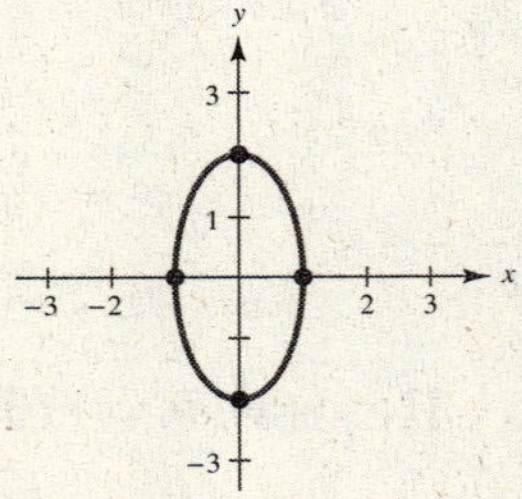

29. Vertices: $(-2, 4), (8, 4)$

Co-vertices: $(3, 0), (3, 8)$

Center: $(3, 4)$

Major axis is horizontal so $a = 5$.

Minor axis is vertical so $b = 4$.

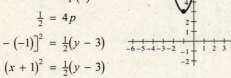

$$\frac{(x - h)^2}{a^2} + \frac{(y - k)^2}{b^2} = 1$$
$$\frac{(x - 3)^2}{5^2} + \frac{(y - 4)^2}{4^2} = 1$$
$$\frac{(x - 3)^2}{25} + \frac{(y - 4)^2}{16} = 1$$

31. Vertices: $(0, 0), (0, 8)$

Co-vertices: $(-3, 4), (3, 4)$

Center: $(0, 4)$

Major axis is vertical so $b = 4$.

Minor axis is horizontal so $a = 3$.

$$\frac{(x-h)^2}{a^2} + \frac{(y-k)^2}{b^2} = 1$$

$$\frac{(x-0)^2}{3^2} + \frac{(y-4)^2}{4^2} = 1$$

$$\frac{x^2}{9} + \frac{(y-4)^2}{16} = 1$$

33. $9(x+1)^2 + 4(y-2)^2 = 144$

$$\frac{(x+1)^2}{16} + \frac{(y-2)^2}{36} = 1$$

Center: $(-1, 2)$

Vertices: $(-1, -4), (-1, 8)$

$a^2 = 36 \qquad b^2 = 16$

$a = 6 \qquad b = 4$

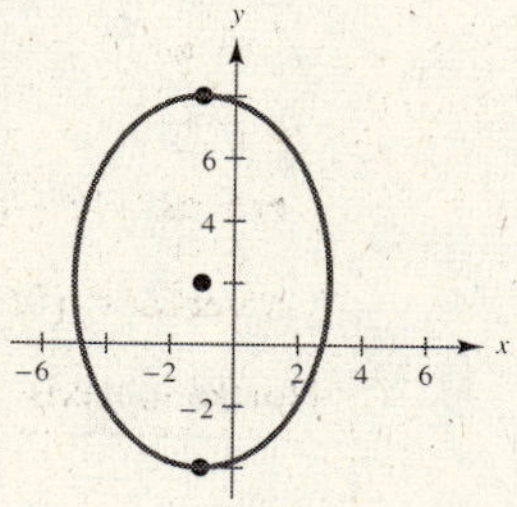

35. $16x^2 + y^2 + 6y - 7 = 0$

$$16x^2 + (y^2 + 6y + 9) = 7 + 9$$

$$16x^2 + (y+3)^2 = 16$$

$$x^2 + \frac{(y+3)^2}{16} = 1$$

Center: $(0, -3)$

Vertices: $(0, -7), (0, 1)$

$a^2 = 16 \qquad b^2 = 1$

$a = 4 \qquad b = 1$

Vertical axis

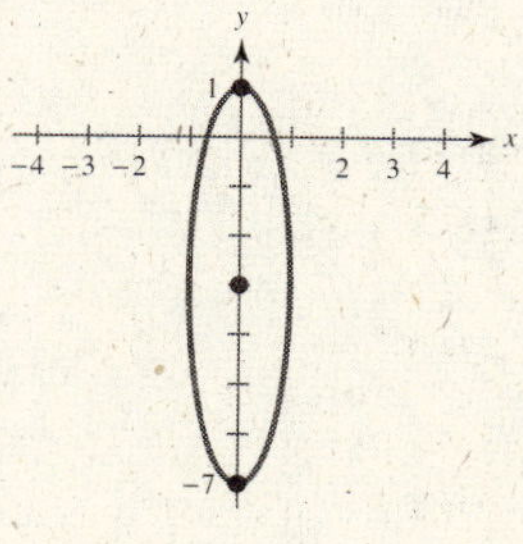

37. $x^2 - y^2 = 25$

$$\frac{x^2}{25} - \frac{y^2}{25} = 1$$

Center: $(0, 0)$

Vertices: $(\pm 5, 0)$

Asymptotes: $y = \pm x$

$a^2 = 25 \qquad b^2 = 25$

$a = 5 \qquad b = 5$

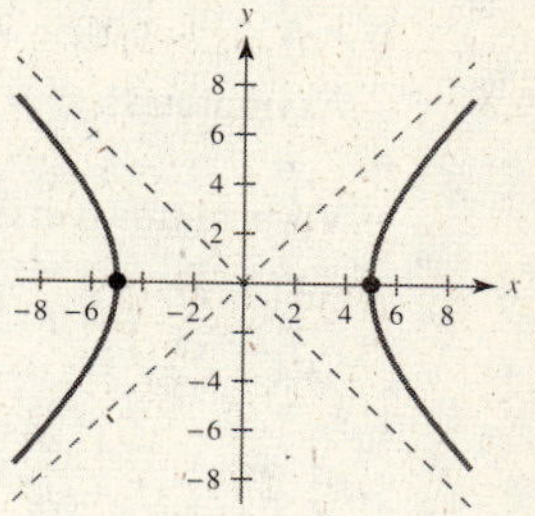

39. $\frac{y^2}{25} - \frac{x^2}{4} = 1$

Center: $(0, 0)$

Vertices: $(0, \pm 5)$

Asymptotes: $y = \pm\frac{5}{2}x$

$a^2 = 25 \qquad b^2 = 4$

$a = 5 \qquad b = 2$

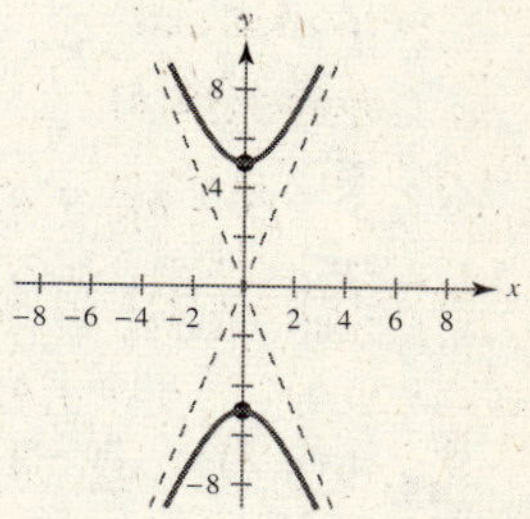

41. Vertices: $(\pm 2, 0)$

Asymptotes: $y = \pm\frac{3}{2}x$

Center: $(0, 0)$

$a = 2 \qquad \frac{b}{a} = \pm\frac{3}{2}$

$\frac{b}{2} = \pm\frac{3}{2}$

$b = 3$

$$\frac{x^2}{a^2} - \frac{y^2}{b^2} = 1$$

$$\frac{x^2}{2^2} - \frac{y^2}{3^2} = 1$$

$$\frac{x^2}{4} - \frac{y^2}{9} = 1$$

43. Center: $(0, 0)$

Vertices: $(0, -8)$, $(0, 8)$

Asymptotes: $y = \frac{4}{5}x, y = -\frac{4}{5}x$

Vertical transverse axis

$$\frac{y^2}{a^2} - \frac{x^2}{b^2} = 1$$

$$a = 8 \qquad \frac{a}{b} = \frac{4}{5}$$

$$\frac{8}{b} = \frac{4}{5}$$

$$b = 10$$

$$\frac{y^2}{8^2} - \frac{x^2}{10^2} = 1$$

$$\frac{y^2}{64} - \frac{x^2}{100} = 1$$

45. $\frac{(x-3)^2}{9} - \frac{(y+1)^2}{4} = 1$

Center: $(3, -1)$

Vertices: $(0, -1), (6, -1)$

$a^2 = 9 \qquad b^2 = 4$

$a = 3 \qquad b = 2$

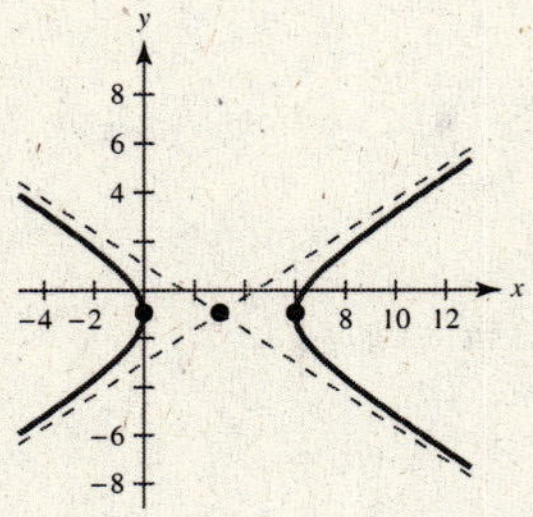

47.

$$8y^2 - 2x^2 + 48y + 16x + 8 = 0$$

$$(8y^2 + 48y) - (2x^2 - 16x) = -8$$

$$8(y^2 + 6y + 9) - 2(x^2 - 8x + 16) = -8 + 72 - 32$$

$$8(y+3)^2 - 2(x-4)^2 = 32$$

$$\frac{(y+3)^2}{4} - \frac{(x-4)^2}{16} = 1$$

Center: $(4, -3)$

Vertices: $(4, -1), (4, -5)$

$a^2 = 4 \qquad b^2 = 16$

$a = 2 \qquad b = 4$

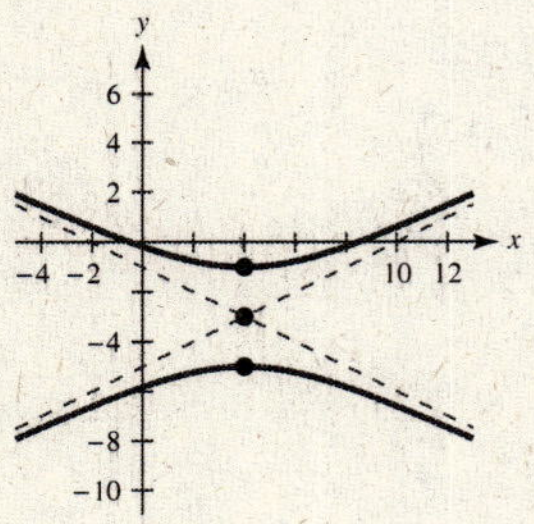

49. Center: $(-4, 6)$

Vertices: $(-6, 6), (-2, 6)$

Point: $(0, 12)$

Horizontal hyperbola

$$\frac{(x-h)^2}{a^2} - \frac{(y-k)^2}{b^2} = 1$$

$$a = 2$$

$$a^2 = 4$$

$$\frac{[0-(-4)]^2}{4} - \frac{(12-6)^2}{b^2} = 1$$

$$\frac{16}{4} - \frac{36}{b^2} = 1$$

$$4 - 1 = \frac{36}{b^2}$$

$$3 = \frac{36}{b^2}$$

$$b^2 = 12$$

$$\frac{(x+4)^2}{4} - \frac{(y-6)^2}{12} = 1$$

51. $\begin{cases} y = x^2 \\ y = 3x \end{cases}$

Keystrokes:

y_1: [Y=] [X,T,θ] [x^2] [ENTER]

y_2: 3 [X,T,θ] [GRAPH]

Solutions: $(0, 0), (3, 9)$

Check:	**Check:**
$0 = 0^2$	$9 \stackrel{?}{=} 3^2$
$0 = 3(0)$	$9 = 9$
	$9 \stackrel{?}{=} 3(3)$
	$9 = 9$

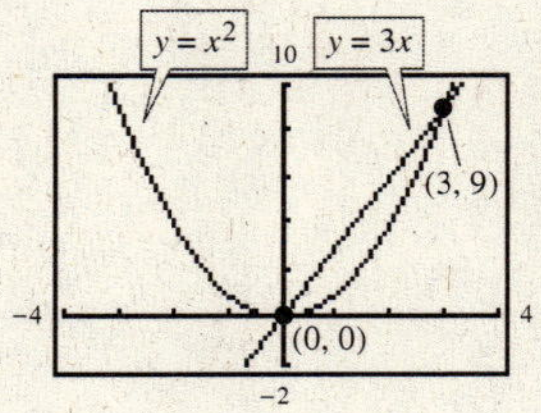

53. $\begin{cases} x^2 + y^2 = 16 \\ -x + y = 4 \end{cases} \Rightarrow \begin{array}{l} y = \pm\sqrt{16 - x^2} \\ y = x + 4 \end{array}$

Keystrokes:

y_1: [Y=] [√] 16 [−] [X,T,θ] [x²] [)] [ENTER]

y_2: [(−)] [√] 16 [−] [X,T,θ] [x²] [)] [ENTER]

y_3: [X,T,θ] [+] 4 [GRAPH]

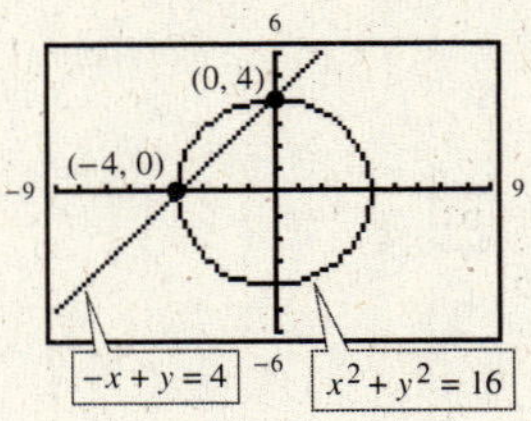

Solutions: $(-4, 0), (0, 4)$

Check:

$(-4)^2 + 0^2 \stackrel{?}{=} 16$

$16 = 16$

$-(-4) + 0 \stackrel{?}{=} 4$

$4 = 4$

Check:

$0^2 + 4^2 \stackrel{?}{=} 16$

$16 = 16$

$-0 + 4 \stackrel{?}{=} 4$

$4 = 4$

55. $\begin{cases} y = 5x^2 \\ y = -15x - 10 \end{cases}$

$$5x^2 = -15x - 10$$
$$5x^2 + 15x + 10 = 0$$
$$x^2 + 3x + 2 = 0$$
$$(x + 2)(x + 1) = 0$$

$x + 2 = 0$ $\qquad$ $x + 1 = 0$

$x = -2$ $\qquad$ $x = -1$

$y = 5(-2)^2 = 5(4) = 20$ $\qquad$ $y = 5(-1)^2 = 5(1) = 5$

$(-2, 20)$ $\qquad$ $(-1, 5)$

57. $\begin{cases} x^2 + y^2 = 1 \\ x + y = -1 \end{cases}$

$$y = -1 - x$$
$$x^2 + (-1 - x)^2 = 1$$
$$x^2 + 1 + 2x + x^2 - 1 = 0$$
$$2x^2 + 2x = 0$$
$$2x(x + 1) = 0$$

$2x = 0$ $\qquad$ $x + 1 = 0$

$x = 0$ $\qquad$ $x = -1$

$y = -1 - 0 = -1$ $\qquad$ $y = -1 - (-1) = 0$

$(0, -1)$ $\qquad$ $(-1, 0)$

59. $\begin{cases} 6x^2 - y^2 = 15 \\ x^2 + y^2 = 13 \end{cases}$

$$7x^2 = 28$$
$$x^2 = 4$$
$$x = \pm 2$$
$$(\pm 2)^2 + y^2 = 13$$
$$y^2 = 9$$
$$y = \pm 3$$

$(\pm 2, \pm 3)$

61. *Verbal Model:* 2 · [Length] + 2 · [Width] = [Perimeter]

$[\text{Length}]^2 + [\text{Width}]^2 = [\text{Diagonal}]^2$

Labels: Length $= x$

Width $= y$

System:

$$2x + 2y = 28$$
$$x^2 + y^2 = 10^2$$
$$x + y = 14$$
$$y = 14 - x$$
$$x^2 + (14 - x)^2 = 100$$
$$x^2 + 196 - 28x + x^2 - 100 = 0$$
$$2x^2 - 28x + 96 = 0$$
$$x^2 - 14x + 48 = 0$$
$$(x - 8)(x - 6) = 0$$

$x = 8$ $\qquad$ $x = 6$

$y = 6$ $\qquad$ $y = 8$

6 cm × 8 cm

63. *Verbal Model:* $\boxed{\text{Length of first piece}} + \boxed{\text{Length of second piece}} = 100$

$\boxed{\text{Area of Square 1}} = \boxed{\text{Area of Square 2}} + 144$

Labels: Length of first piece $= x$

Length of second piece $= y$

System:

$$x + y = 100$$
$$\left(\frac{x}{4}\right)^2 = \left(\frac{y}{4}\right)^2 + 144$$
$$x = 100 - y$$
$$\left(\frac{100 - y}{4}\right)^2 = \left(\frac{y}{4}\right)^2 + 144$$
$$\frac{10{,}000 - 200y + y^2}{16} = \frac{y^2}{16} + 144$$
$$10{,}000 - 200y + y^2 = y^2 + 2304$$
$$-200y = -7696$$
$$y \approx 38.48$$
$$x \approx 100 - 38.48$$
$$\approx 61.52$$

38.48 inches; 61.52 inches

Chapter Test for Chapter 10

1. Center: $(-2, -3)$

Radius: 4

$(x - h)^2 + (y - k)^2 = r^2$

$(x + 2)^2 + (y + 3)^2 = 16$

2.

$$x^2 + y^2 - 2x - 6y + 1 = 0$$
$$(x^2 - 2x) + (y^2 - 6y) = -1$$
$$(x^2 - 2x + 1) + (y^2 - 6y + 9) = -1 + 1 + 9$$
$$(x - 1)^2 + (y - 3)^2 = 9$$

Center: $(1, 3)$

Radius: 3

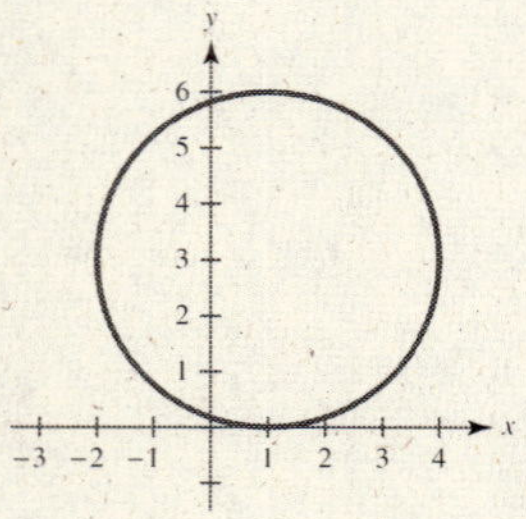

3.

$$x^2 + y^2 + 4x - 6y + 4 = 0$$
$$(x^2 + 4x) + (y^2 - 6y) = -4$$
$$(x^2 + 4x + 4) + (y^2 - 6y + 9) = -4 + 4 + 9$$
$$(x + 2)^2 + (y - 3)^2 = 9$$

Center: $(-2, 3)$

Radius: 3

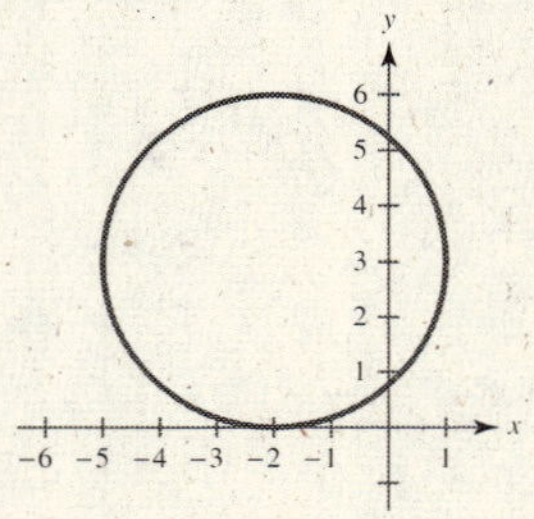

4.
$$x = -3y^2 + 12y - 8$$
$$x + 8 = -3(y^2 - 4y)$$
$$x + 8 - 12 = -3(y^2 - 4y + 4)$$
$$x - 4 = -3(y - 2)^2$$

Vertex: $(4, 2)$

$4p = -\frac{1}{3}$

$p = -\frac{1}{12}$

Focus: $\left(4 - \frac{1}{12}, 2\right) = \left(\frac{48}{12} - \frac{1}{12}, 2\right) = \left(\frac{47}{12}, 2\right)$

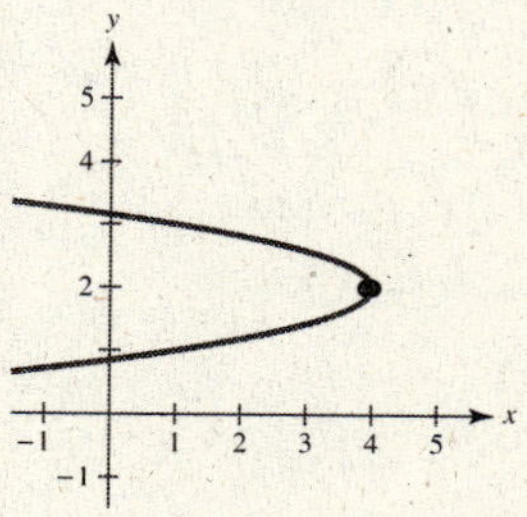

5. Vertex: $(7, -2)$

Focus: $(7, 0)$

Vertical parabola

$(x - h)^2 = 4p(y - k)$

$(x - 7)^2 = 4p[y - (-2)]$

$p = 2$

$(x - 7)^2 = 4(2)(y + 2)$

$(x - 7)^2 = 8(y + 2)$

6. Vertices: $(-3, 0), (7, 0)$

Co-vertices: $(2, 3), (2, -3)$

Horizontal axis

Center: $(2, 0)$

$a = 5 \qquad b = 3$

$a^2 = 25 \qquad b^2 = 9$

$$\frac{(x - h)^2}{a^2} + \frac{(y - k)^2}{b^2} = 1$$
$$\frac{(x - 2)^2}{25} + \frac{y^2}{9} = 1$$

7. $16x^2 + 4y^2 = 64$

$$\frac{x^2}{4} + \frac{y^2}{16} = 1$$

Center: $(0, 0)$

Vertices: $(0, \pm 4)$

$a^2 = 16 \qquad b^2 = 4$

$a = 4 \qquad b = 2$

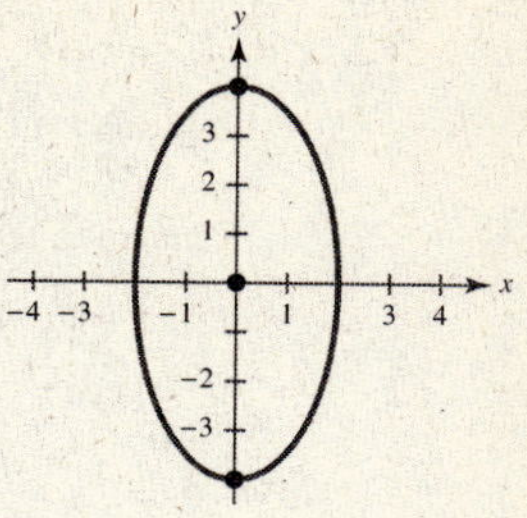

8.
$$25x^2 + 4y^2 - 50x - 24y - 39 = 0$$
$$(25x^2 - 50x) + (4y^2 - 24y) = 39$$
$$25(x^2 - 2x + 1) + 4(y^2 - 6y + 9) = 39 + 25 + 36$$
$$25(x - 1)^2 + 4(y - 3)^2 = 100$$
$$\frac{(x - 1)^2}{4} + \frac{(y - 3)^2}{25} = 1$$

Center: $(1, 3)$

$a^2 = 25$

$a = 5$

Vertices: $(1, -2), (1, 8)$

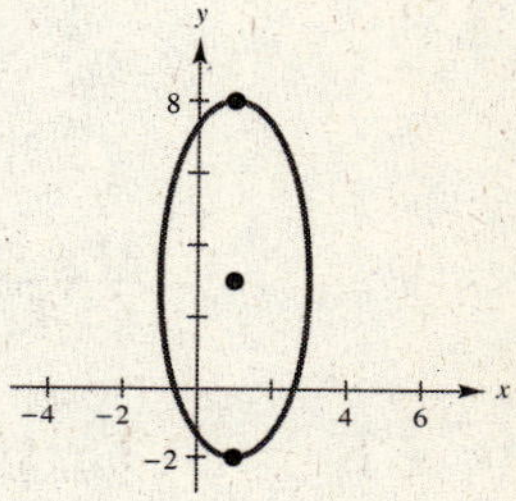

9. Vertices: $(\pm 3, 0)$

Asymptotes: $y = \pm\frac{2}{3}x$

$a = 3 \qquad \frac{b}{a} = \pm\frac{2}{3}$

$\frac{b}{3} = \pm\frac{2}{3}$

$b = 2$

$$\frac{x^2}{a^2} - \frac{y^2}{b^2} = 1$$
$$\frac{x^2}{3^2} - \frac{y^2}{2^2} = 1$$
$$\frac{x^2}{9} - \frac{y^2}{4} = 1$$

10. Vertices: $(0, -5), (0, 5)$

Asymptotes: $y = \pm\frac{5}{2}x$

Vertical transverse axis

$$\frac{y^2}{a^2} - \frac{x^2}{b^2} = 1$$

$$a = 5 \qquad \frac{a}{b} = \frac{5}{2}$$

$$\frac{5}{b} = \frac{5}{2}$$

$$b = 2$$

$$\frac{y^2}{25} - \frac{x^2}{4} = 1$$

11.
$$4x^2 - 2y^2 - 24x + 20 = 0$$
$$(4x^2 - 24x) - 2y^2 = -20$$
$$4(x^2 - 6x + 9) - 2y^2 = -20 + 36$$
$$4(x - 3)^2 - 2y^2 = 16$$
$$\frac{(x - 3)^2}{4} - \frac{y^2}{8} = 1$$

Center: $(3, 0)$

Vertices: $(1, 0), (5, 0)$

$$a^2 = 4$$
$$a = 2$$

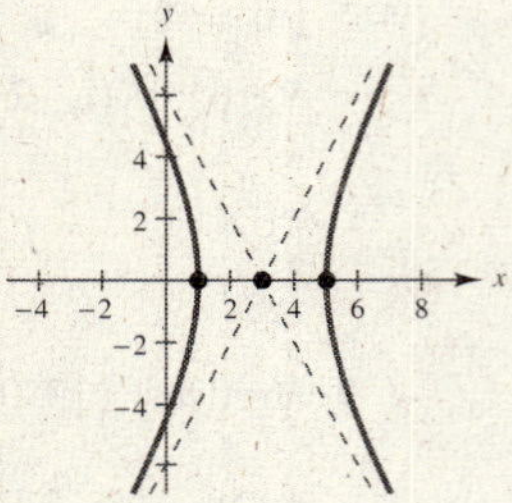

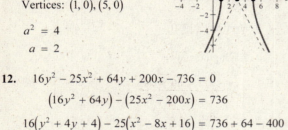

12.
$$16y^2 - 25x^2 + 64y + 200x - 736 = 0$$
$$(16y^2 + 64y) - (25x^2 - 200x) = 736$$
$$16(y^2 + 4y + 4) - 25(x^2 - 8x + 16) = 736 + 64 - 400$$
$$16(y + 2)^2 - 25(x - 4)^2 = 400$$
$$\frac{(y + 2)^2}{25} - \frac{(x - 4)^2}{16} = 1$$

Center: $(4, -2)$

Vertices: $(4, -7), (4, 3)$

$$a^2 = 25 \qquad b^2 = 16$$
$$a = 5 \qquad b = 4$$

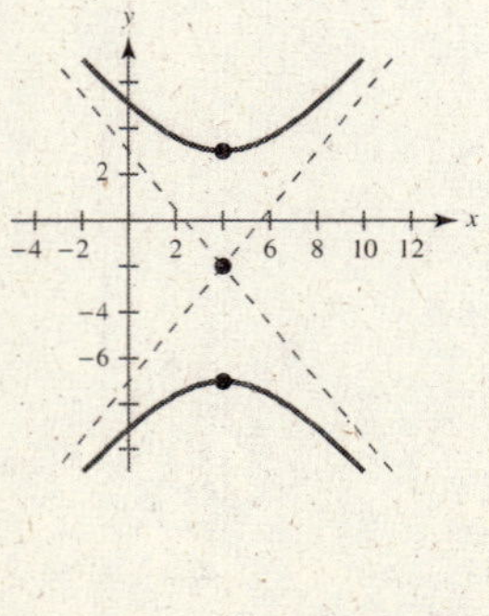

13.
$$\begin{cases} \frac{x^2}{16} + \frac{y^2}{9} = 1 \\ 3x + 4y = 12 \end{cases}$$

Solve for y in second equation.

$$4y = 3x + 12$$
$$y = -\frac{3}{4}x + 3$$

Substitute into first equation after multiplying by 144.

$$9x^2 + 16y^2 = 144$$
$$9x^2 + 16\left(-\frac{3}{4}x + 3\right)^2 = 144$$
$$9x^2 + 16\left(\frac{9}{16}x^2 - \frac{9}{2}x + 9\right) = 144$$
$$9x^2 + 9x^2 - 72x + 144 = 144$$
$$18x^2 - 72x = 0$$
$$18x(x - 4) = 0$$

$$18x = 0 \qquad x - 4 = 0$$
$$x = 0 \qquad x = 4$$
$$y = -\frac{3}{4}(0) + 3 \qquad y = -\frac{3}{4}(4) + 3$$
$$= 3 \qquad = 0$$

$(0, 3)$ $\quad$ $(4, 0)$

14.
$$\begin{cases} x^2 + y^2 = 16 \\ \frac{x^2}{16} - \frac{y^2}{9} = 1 \end{cases}$$

Multiply second equation by 144.

$$x^2 + y^2 = 16$$
$$9x^2 + 16y^2 = 144$$

Multiply first equation by 16.

$$16x^2 + 16y^2 = 256$$
$$9x^2 - 16y^2 = 144$$
$$25x^2 = 400$$
$$x^2 = 16$$
$$x = \pm 4$$

$$(\pm 4)^2 + y^2 = 16$$
$$16 + y^2 = 16$$
$$y^2 = 0$$
$$y = 0$$

$(4, 0), (-4, 0)$

15. $\begin{cases} x^2 + y^2 = 10 \\ x^2 = y^2 + 2 \end{cases}$

$$\begin{aligned} x^2 + y^2 &= 10 \\ x^2 - y^2 &= 2 \\ \hline 2x^2 &= 12 \\ x^2 &= 6 \\ x &= \pm\sqrt{6} \end{aligned}$$

$$\begin{aligned} \left(\pm\sqrt{6}\right)^2 + y^2 &= 10 \\ 6 + y^2 &= 10 \\ y^2 &= 4 \\ y &= \pm 2 \end{aligned}$$

$\left(\sqrt{6}, 2\right), \left(-\sqrt{6}, 2\right), \left(\sqrt{6}, -2\right), \left(-\sqrt{6}, -2\right)$

16. $r = 5000$

$$\begin{aligned} x^2 + y^2 &= r^2 \\ x^2 + y^2 &= 5000^2 \\ x^2 + y^2 &= 25{,}000{,}000 \end{aligned}$$

17. *Verbal Model:* $2 \cdot \boxed{\text{Length}} + 2 \cdot \boxed{\text{Width}} = \boxed{\text{Perimeter}}$

$\boxed{\text{Length}}^2 + \boxed{\text{Width}}^2 = \boxed{\text{Diagonal}}^2$

Labels: Length $= x$

Width $= y$

System:

$$\begin{aligned} 2x + 2y &= 56 \\ x^2 + y^2 &= 20^2 \\ x + y &= 28 \\ y &= 28 - x \\ x^2 + (28 - x)^2 &= 400 \\ x^2 + 784 - 56x + x^2 - 400 &= 0 \\ 2x^2 - 56x + 384 &= 0 \\ x^2 - 28x + 192 &= 0 \\ (x - 16)(x - 12) &= 0 \end{aligned}$$

$x = 16$, $y = 28 - 16 = 12$ $\qquad$ $x = 12$, $y = 28 - 12 = 16$

12 inches × 16 inches

Cumulative Test for Chapters 8–10

1. $4x^2 - 9x - 9 = 0$

$(4x + 3)(x - 3) = 0$

$4x + 3 = 0 \Rightarrow x = -\frac{3}{4}$ $\qquad$ $x - 3 = 0 \Rightarrow x = 3$

2. $(x - 5)^2 - 64 = 0$

$$\begin{aligned} (x - 5)^2 &= 64 \\ x - 5 &= \pm 8 \\ x &= 5 \pm 8 \\ x &= 13, -3 \end{aligned}$$

3. $x^2 - 10x - 25 = 0$

$$\begin{aligned} x^2 - 10x &= 25 \\ x^2 - 10x + 25 &= 25 + 25 \\ (x - 5)^2 &= 50 \\ x - 5 &= \pm\sqrt{50} \\ x &= 5 \pm 5\sqrt{2} \end{aligned}$$

4. $3x^2 + 6x + 2 = 0$

$a = 3, b = 6, c = 2$

$$\begin{aligned} x &= \frac{-6 \pm \sqrt{6^2 - 4(3)(2)}}{2(3)} \\ x &= \frac{-6 \pm \sqrt{36 - 24}}{6} \\ x &= \frac{-6 \pm \sqrt{12}}{6} \\ x &= \frac{-6}{6} \pm \frac{2\sqrt{3}}{6} \\ x &= -1 \pm \frac{\sqrt{3}}{3} \end{aligned}$$

5. $x^4 - 8x^2 + 15 = 0$

$(x^2 - 5)(x^2 - 3) = 0$

$x^2 = 5 \Rightarrow x = \pm\sqrt{5}$ $\qquad$ $x^2 = 3 \Rightarrow x = \pm\sqrt{3}$

6. $3x^2 + 8x \le 3$

$3x^2 + 8x - 3 \le 0$

$(3x - 1)(x + 3) \le 0$

Critical numbers: $-3, \frac{1}{3}$

Test intervals:

Positive: $(-\infty, -3)$

Negative: $\left(-3, \frac{1}{3}\right)$

Positive: $\left(\frac{1}{3}, \infty\right)$

Solution: $\left[-3, \frac{1}{3}\right]$

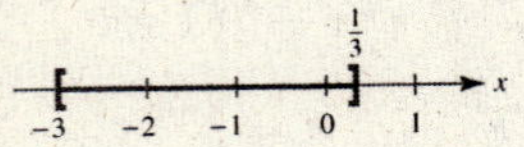

7. $\dfrac{3x + 4}{2x - 1} < 0$

Critical numbers: $x = -\frac{4}{3}, \frac{1}{2}$

Test intervals:

Positive: $\left(-\infty, -\frac{4}{3}\right)$

Negative: $\left(-\frac{4}{3}, \frac{1}{2}\right)$

Positive: $\left(\frac{1}{2}, \infty\right)$

Solution: $\left(-\frac{4}{3}, \frac{1}{2}\right)$

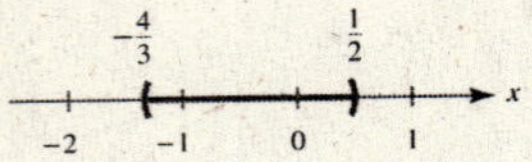

8. $x = -2$ and $x = 6$

$x + 2 = 0$ and $x - 6 = 0$

$(x + 2)(x - 6) = 0$

$x^2 - 4x - 12 = 0$

9. $f(x) = 2x^2 - 3, g(x) = 5x - 1$

(a) $$\begin{aligned}(f \circ g)(x) &= f[g(x)] \\ &= f[5x - 1] \\ &= 2(5x - 1)^2 - 3 \\ &= 2(25x^2 - 10x + 1) - 3 \\ &= 50x^2 - 20x + 2 - 3 \\ &= 50x^2 - 20x - 1\end{aligned}$$

Domain: $(-\infty, \infty)$

(b) $$\begin{aligned}(g \circ f)(x) &= g[f(x)] \\ &= g[2x^2 - 3] \\ &= 5(2x^2 - 3) - 1 \\ &= 10x^2 - 15 - 1 \\ &= 10x^2 - 16\end{aligned}$$

Domain: $(-\infty, \infty)$

10. $$\begin{aligned}f(x) &= \frac{5 - 3x}{4} \\ y &= \frac{5 - 3x}{4} \\ x &= \frac{5 - 3y}{4} \\ 4x &= 5 - 3y \\ 4x - 5 &= -3y \\ -\frac{4}{3}x + \frac{5}{3} &= y \\ -\frac{4}{3}x + \frac{5}{3} &= f^{-1}(x)\end{aligned}$$

11. $f(x) = 7 + 2^{-x}$

(a) $f(1) = 7 + 2^{-1} = 7 + \frac{1}{2} = \frac{15}{2}$

(b) $f(0.5) = 7 + 2^{-0.5}$

$= 7 + \dfrac{1}{\sqrt{2}} = \dfrac{14}{2} + \dfrac{\sqrt{2}}{2} = \dfrac{14 + \sqrt{2}}{2}$

(c) $f(3) = 7 + 2^{-3} = 7 + \frac{1}{8} = \frac{56}{8} + \frac{1}{8} = \frac{57}{8}$

12. $f(x) = 4^{x-1}$

Horizontal asymptote: $y = 0$

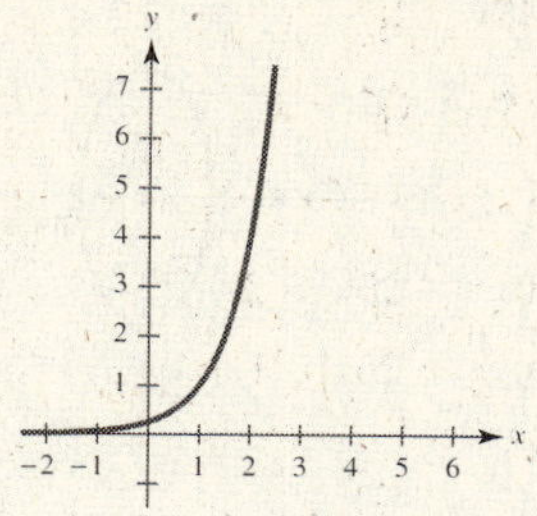

13.

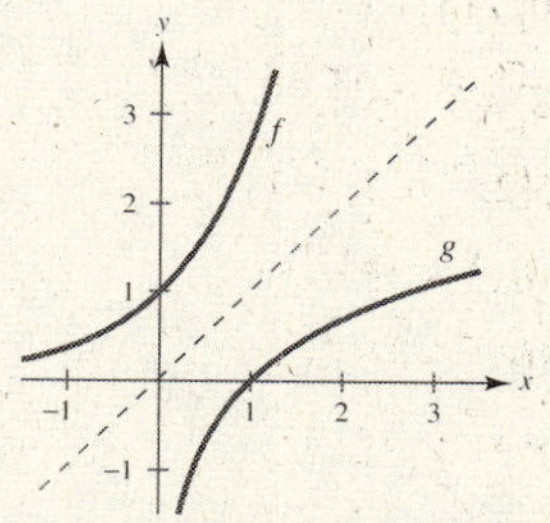

f and g are inverse functions, so the graphs are reflections in the line $y = x$.

14. $g(x) = \log_3(x - 1)$

Vertical asymptote: $x = 1$

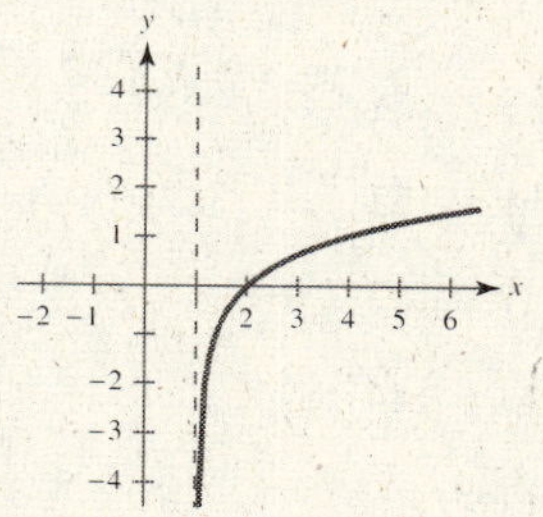

15. $\log_4 \frac{1}{16} = -2$ because $4^{-2} = \frac{1}{16}$.

16. $3(\log_2 x + \log_2 y) - \log_2 z = \log_2(xy)^3 - \log_2 z$

$$= \log_2 \frac{(xy)^3}{z} = \log_2 \frac{x^3y^3}{z}$$

17. $\log_{10} \frac{\sqrt{x+1}}{x^4} = \log_{10}\sqrt{x+1} - \log_{10} x^4$

$$= \log_{10}(x+1)^{1/2} - 4\log_{10} x$$

$$= \frac{1}{2}\log_{10}(x+1) - 4\log_{10} x$$

18. $\log_x\left(\frac{1}{9}\right) = -2$

$$\frac{1}{9} = x^{-2}$$

$$\frac{1}{9} = \frac{1}{x^2}$$

$$9 = x^2$$

$$3 = x$$

19. $4 \ln x = 10$

$$\ln x = \frac{10}{4}$$

$$\ln x = \frac{5}{2}$$

$$e^{\ln x} = e^{5/2}$$

$$x = e^{5/2}$$

$$x \approx 12.182$$

20. $500(1.08)^t = 2000$

$$1.08^t = \frac{2000}{500}$$

$$1.08^t = 4$$

$$\log_{1.08} 1.08^t = \log_{1.08} 4$$

$$t = \frac{\log 4}{\log 1.08}$$

$$t \approx 18.013$$

21. $3(1 + e^{2x}) = 20$

$$1 + e^{2x} = \frac{20}{3}$$

$$e^{2x} = \frac{17}{3}$$

$$\ln e^{2x} = \ln \frac{17}{3}$$

$$2x = \ln \frac{17}{3}$$

$$x = \frac{\ln(17/3)}{2}$$

$$x \approx 0.867$$

22. $C(t) = P(1.028)^t$

$$C(5) = 29.95(1.028)^5 \approx \$34.38$$

23. $A = Pe^{rt}$

$$A = 1000e^{0.08(1)}$$

$$A = \$1083.29$$

$$\text{Effective yield} = \frac{83.29}{1000} = 0.08329 \approx 8.33\%$$

24.
$$\begin{aligned} A &= Pe^{rt} \\ 6000 &= 1500e^{0.07t} \\ 4 &= e^{0.07t} \\ \ln 4 &= \ln e^{0.07t} \\ \ln 4 &= 0.07t \\ \frac{\ln 4}{0.07} &= t \\ 19.8 &\approx t \text{ years} \end{aligned}$$

25.
$$\begin{aligned} x^2 + y^2 - 6x + 14y - 6 &= 0 \\ (x^2 - 6x) + (y^2 + 14y) &= 6 \\ (x^2 - 6x + 9) + (y^2 + 14y + 49) &= 6 + 9 + 49 \\ (x - 3)^2 + (y + 7)^2 &= 64 \end{aligned}$$

Center: $(3, -7)$

Radius: 8

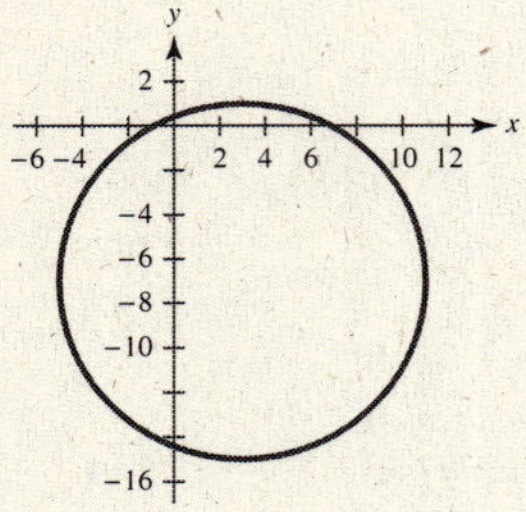

26.
$$\begin{aligned} y &= 2x^2 - 20x + 5 \\ y - 5 &= 2(x^2 - 10x) \\ y - 5 + 50 &= 2(x^2 - 10x + 25) \\ y + 45 &= 2(x - 5)^2 \\ \tfrac{1}{2}(y + 45) &= (x - 5)^2 \end{aligned}$$

Vertex: $(5, -45)$

$$4p = \tfrac{1}{2}$$
$$p = \tfrac{1}{8}$$

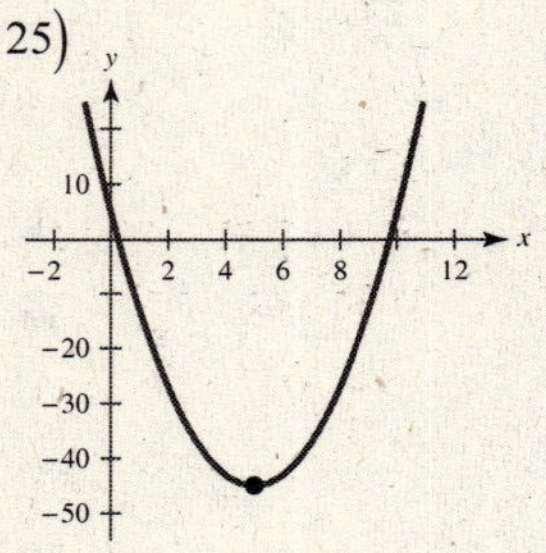

Focus: $\left(5, 45 + \frac{1}{8}\right) = \left(5, -\frac{360}{8} + \frac{1}{8}\right) = \left(5, -\frac{359}{8}\right)$

27.
$$\frac{(x - h)^2}{b^2} + \frac{(y - k)^2}{a^2} = 1$$
$$\frac{(x + 3)^2}{4} + \frac{(y - 2)^2}{25} = 1$$
$$b = 2 \qquad a = 5$$
$$b^2 = 4 \qquad a^2 = 25$$

28.
$$4x^2 + y^2 = 4$$
$$\frac{x^2}{1} + \frac{y^2}{4} = 1$$

Vertical ellipse

Center: $(0, 0)$

Vertices: $(0, \pm 2)$

$$a^2 = 4 \qquad b^2 = 1$$
$$a = 2 \qquad b = 1$$

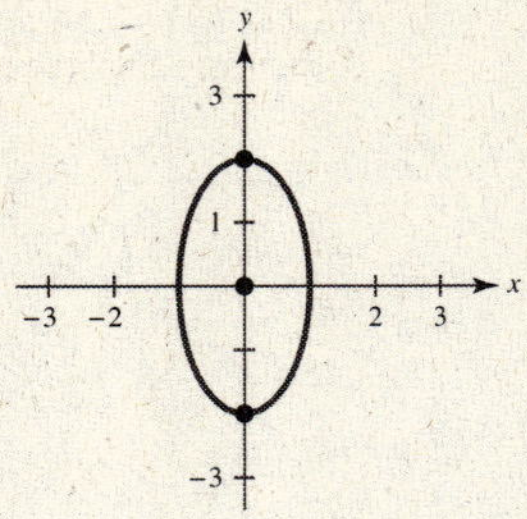

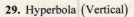

29. Hyperbola (Vertical)

Vertices: $(0, \pm 3)$

Asymptotes: $y = \pm 3x$

Center: $(0, 0)$

$$a = 3 \qquad \frac{a}{b} = \pm 3$$
$$a^2 = 9 \qquad \frac{3}{b} = \pm 3$$
$$b = 1$$

$$\frac{y^2}{a^2} - \frac{x^2}{b^2} = 1$$
$$\frac{y^2}{9} - \frac{x^2}{1} = 1$$

30.
$$\begin{aligned} x^2 - 9y^2 + 18y &= 153 \\ x^2 - (9y^2 - 18y) &= 153 \\ x^2 - 9(y^2 - 2y + 1) &= 153 - 9 \\ x^2 - 9(y - 1)^2 &= 144 \\ \frac{x^2}{144} - \frac{(y - 1)^2}{16} &= 1 \end{aligned}$$

Center: $(0, 1)$

Vertices: $(\pm 12, 1)$

$$a^2 = 144 \qquad b^2 = 16$$
$$a = 12 \qquad b = 4$$

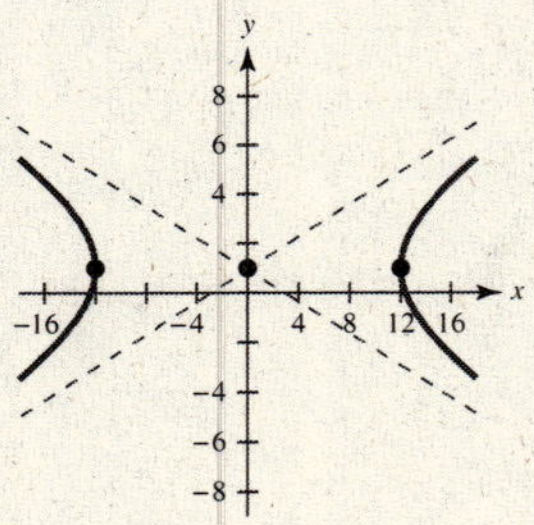

31. $\begin{cases} y = x^2 - x - 1 \\ 3x - y = 4 \end{cases}$

Substitute for y in second equation.

$$3x - (x^2 - x - 1) = 4$$
$$3x - x^2 + x + 1 - 4 = 0$$
$$-x^2 + 4x - 3 = 0$$
$$x^2 - 4x + 3 = 0$$
$$(x - 3)(x - 1) = 0$$

$x = 3$ | $x = 1$

$3(3) - y = 4$ | $3(1) - y = 4$

$-y = -5$ | $-y = 1$

$y = 5$ | $y = -1$

$(3, 5)$ | $(1, -1)$

32. $\begin{cases} x^2 + 5y^2 = 21 \\ -x + y^2 = 5 \end{cases}$

$$y^2 = 5 + x$$
$$x^2 + 5(5 + x) = 21$$
$$x^2 + 25 + 5x - 21 = 0$$
$$x^2 + 5x + 4 = 0$$
$$(x + 4)(x + 1) = 0$$

$x = -4$ | $x = -1$

$y^2 = 5 + (-4)$ | $y^2 = 5 + (-1)$

$y^2 = 1$ | $y^2 = 4$

$y = \pm 1$ | $y = \pm 2$

$(-4, \pm 1)$ | $(-1, \pm 2)$

33. *Verbal Model:* [Length] · [Width] = [Area]

2 · [Length] + 2 · [Width] = [Perimeter]

Labels: Length = x

Width = y

System:

$$xy = 32$$
$$2x + 2y = 24$$
$$x + y = 12$$
$$y = 12 - x$$
$$x(12 - x) = 32$$
$$12x - x^2 = 32$$
$$0 = x^2 - 12x + 32$$
$$0 = (x - 8)(x - 4)$$

$x = 8$ | $x = 4$

$y = 4$ | $y = 8$

4 feet × 8 feet

34. $y = -0.1x^2 + 3x + 6$

(a) *Keystrokes:*

[Y=] [(−)] .1 [X,T,θ] [x^2] [+] 3 [X,T,θ] [+] 6 [GRAPH]

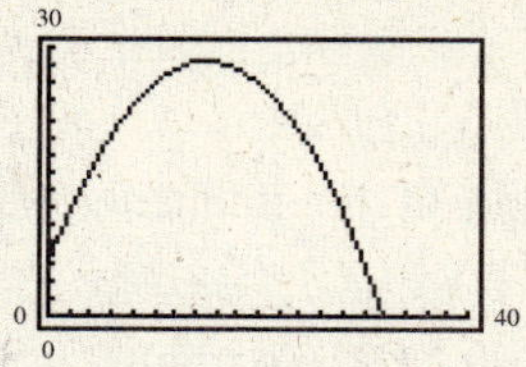

(b) Highest point: 28.5 feet

Range of path: $[0, 28.5]$

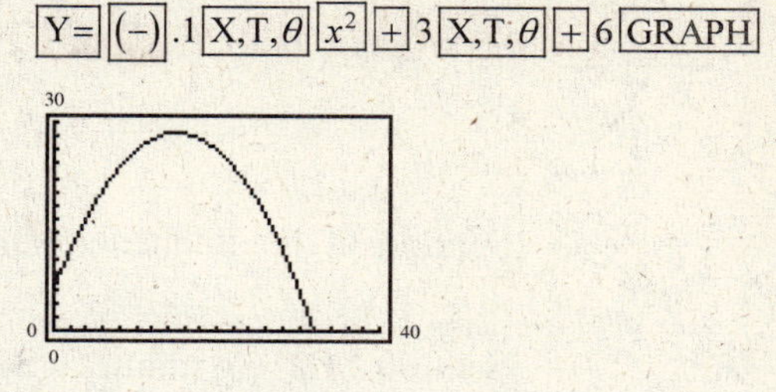

CHAPTER 11
Sequences, Series, and the Binomial Theorem

CHAPTER 11
Sequences, Series, and the Binomial Theorem

Section 11.1 Sequences and Series

1. $a_1 = 2(1) = 2$

$a_2 = 2(2) = 4$

$a_3 = 2(3) = 6$

$a_4 = 2(4) = 8$

$a_5 = 2(5) = 10$

$2, 4, 6, 8, 10, \ldots, 2n, \ldots$

3. $a_1 = \left(\frac{1}{4}\right)^1 = \frac{1}{4}$

$a_2 = \left(\frac{1}{4}\right)^2 = \frac{1}{16}$

$a_3 = \left(\frac{1}{4}\right)^3 = \frac{1}{64}$

$a_4 = \left(\frac{1}{4}\right)^4 = \frac{1}{256}$

$a_5 = \left(\frac{1}{4}\right)^5 = \frac{1}{1024}$

$\frac{1}{4}, \frac{1}{16}, \frac{1}{64}, \frac{1}{256}, \frac{1}{1024}, \ldots, \left(\frac{1}{4}\right)^n$

5. $a_1 = (-1)^1 2(1) = -2$

$a_2 = (-1)^2 2(2) = 4$

$a_3 = (-1)^3 2(3) = -6$

$a_4 = (-1)^4 2(4) = 8$

$a_5 = (-1)^5 2(5) = -10$

$-2, 4, -6, 8, -10, \ldots$

7. $a_1 = \left(-\frac{1}{2}\right)^2 = \frac{1}{4}$

$a_2 = \left(-\frac{1}{2}\right)^3 = -\frac{1}{8}$

$a_3 = \left(-\frac{1}{2}\right)^4 = \frac{1}{16}$

$a_4 = \left(-\frac{1}{2}\right)^5 = -\frac{1}{32}$

$a_5 = \left(-\frac{1}{2}\right)^6 = \frac{1}{64}$

$\frac{1}{4}, -\frac{1}{8}, \frac{1}{16}, -\frac{1}{32}, \frac{1}{64}, \ldots, \left(-\frac{1}{2}\right)^{n+1}, \ldots$

9. $a_1 = 5(1) - 2 = 3$

$a_2 = 5(2) - 2 = 8$

$a_3 = 5(3) - 2 = 13$

$a_4 = 5(4) - 2 = 18$

$a_5 = 5(5) - 2 = 23$

$3, 8, 13, 18, 23, \ldots, 5h - 2$

11. $a_1 = \frac{4}{1 + 3} = 1$

$a_2 = \frac{4}{2 + 3} = \frac{4}{5}$

$a_3 = \frac{4}{3 + 3} = \frac{4}{6} = \frac{2}{3}$

$a_4 = \frac{4}{4 + 3} = \frac{4}{7}$

$a_5 = \frac{4}{5 + 3} = \frac{4}{8} = \frac{1}{2}$

$1, \frac{4}{5}, \frac{2}{3}, \frac{4}{7}, \frac{1}{2}, \ldots, \frac{4}{n + 3}$

13. $a_1 = \frac{3(1)}{5(1) - 1} = \frac{3}{4}$

$a_2 = \frac{3(2)}{5(2) - 1} = \frac{6}{9} = \frac{2}{3}$

$a_3 = \frac{3(3)}{5(3) - 1} = \frac{9}{14}$

$a_4 = \frac{3(4)}{5(4) - 1} = \frac{12}{19}$

$a_5 = \frac{3(5)}{5(5) - 1} = \frac{15}{24} = \frac{5}{8}$

$\frac{3}{4}, \frac{2}{3}, \frac{9}{14}, \frac{12}{19}, \frac{15}{24}, \ldots, \frac{3n}{5n - 1}$

15. $a_1 = \frac{(1)^1}{1^2} = -1$

$a_2 = \frac{(-1)^2}{2^2} = \frac{1}{4}$

$a_3 = \frac{(-1)^3}{3^2} = -\frac{1}{9}$

$a_4 = \frac{(-1)^4}{4^2} = \frac{1}{16}$

$a_5 = \frac{(-1)^5}{5^2} = -\frac{1}{25}$

$-1, \frac{1}{4}, -\frac{1}{9}, \frac{1}{16}, -\frac{1}{25}, \ldots, \frac{(-1)^n}{n^2}, \ldots$

17. $a_1 = 2 + \frac{1}{4^1} = \frac{8}{4} + \frac{1}{4} = \frac{9}{4}$

$a_2 = 2 + \frac{1}{4^2} = \frac{32}{16} + \frac{1}{16} = \frac{33}{16}$

$a_3 = 2 + \frac{1}{4^3} = \frac{128}{64} + \frac{1}{64} = \frac{129}{64}$

$a_4 = 2 + \frac{1}{4^4} = \frac{512}{256} + \frac{1}{256} = \frac{513}{256}$

$a_5 = 2 + \frac{1}{4^5} = \frac{2048}{1024} + \frac{1}{1024} = \frac{2049}{1024}$

$\frac{9}{4}, \frac{33}{16}, \frac{129}{64}, \frac{513}{256}, \frac{2049}{1024}, \ldots, 2 + \frac{1}{4^n}$

19. $a_1 = \frac{(1+1)!}{1!} = \frac{2!}{1!} = \frac{2 \cdot 1}{1} = 2$

$a_2 = \frac{(2+1)!}{2!} = \frac{3!}{2!} = \frac{3 \cdot 2!}{2!} = 3$

$a_3 = \frac{(3+1)!}{3!} = \frac{4!}{3!} = \frac{4 \cdot 3!}{3!} = 4$

$a_4 = \frac{(4+1)!}{4!} = \frac{5!}{4!} = \frac{5 \cdot 4!}{4!} = 5$

$a_5 = \frac{(5+1)!}{5!} = \frac{6!}{5!} = \frac{6 \cdot 5!}{5!} = 6$

$2, 3, 4, 5, 6 \ldots, \frac{(n+1)!}{n!}$

21. $a_1 = \frac{2 + (-2)^1}{1!} = 0$

$a_2 = \frac{2 + (-2)^2}{2!} = \frac{6}{2 \cdot 1} = 3$

$a_3 = \frac{2 + (-2)^3}{3!} = \frac{-6}{3 \cdot 2 \cdot 1} = -1$

$a_4 = \frac{2 + (-2)^4}{4!} = \frac{18}{4 \cdot 3 \cdot 2 \cdot 1} = \frac{3}{4}$

$a_5 = \frac{2 + (-2)^5}{5!} = \frac{-30}{5 \cdot 4 \cdot 3 \cdot 2 \cdot 1} = -\frac{1}{4}$

$0, 3, -1, \frac{3}{4}, -\frac{1}{4} \ldots, \frac{1 + (-1)^n}{n^2}$

23. $a_{15} = (-1)^{15}[5(15) - 3] = -1[72] = -72$

25. $a_8 = \frac{8^2 - 2}{(8-1)!} = \frac{62}{7!} = \frac{62}{7 \cdot 6 \cdot 5 \cdot 4 \cdot 3 \cdot 2 \cdot 1} = \frac{31}{2520}$

27. $\frac{5!}{4!} = \frac{5 \cdot 4 \cdot 3 \cdot 2 \cdot 1}{4 \cdot 3 \cdot 2 \cdot 1} = 5$

29. $\frac{10!}{12!} = \frac{10!}{12 \cdot 11 \cdot 10!} = \frac{1}{132}$

31. $\frac{25!}{20!5!} = \frac{25 \cdot 24 \cdot 23 \cdot 22 \cdot 21 \cdot 20!}{20!5!}$

$= \frac{25 \cdot 24 \cdot 23 \cdot 22 \cdot 21}{5 \cdot 4 \cdot 3 \cdot 2 \cdot 1}$

$= 5 \cdot 6 \cdot 23 \cdot 11 \cdot 7$

$= 53{,}130$

33. $\frac{n!}{(n+1)!} = \frac{n \cdot 1}{(n+1)n \cdot 1} = \frac{1}{n+1}$

35. $\frac{(n+1)!}{(n-1)!} = \frac{(n+1)n(n-1)!}{(n-1)!} = (n+1)n$

37. $\frac{(2n)!}{(2n-1)!} = \frac{(2n)(2n-1)!}{(2n-1)!} = 2n$

39. (b); $a_n = \dfrac{6}{n+1}$

$a_1 = \dfrac{6}{2} = 3$

$a_2 = \dfrac{6}{3} = 2$

$a_3 = \dfrac{6}{4} = \dfrac{3}{2}$

$a_4 = \dfrac{6}{5}$

$a_5 = \dfrac{6}{6} = 1$

$a_6 = \dfrac{6}{7}$

$a_7 = \dfrac{6}{8} = \dfrac{3}{4}$

$a_8 = \dfrac{6}{9} = \dfrac{2}{3}$

$a_9 = \dfrac{6}{10} = \dfrac{3}{5}$

$a_{10} = \dfrac{6}{11}$

41. (c); $a_n = (0.6)^{n-1}$

$a_1 = (0.6)^0 = 1$

$a_2 = (0.6)^1 = 0.6$

$a_3 = (0.6)^2 = 0.36$

$a_4 = (0.6)^3 \approx 0.22$

$a_5 = (0.6)^4 \approx 0.13$

$a_6 = (0.6)^5 \approx 0.08$

$a_7 = (0.6)^6 \approx 0.05$

$a_8 = (0.6)^7 \approx 0.03$

$a_9 = (0.6)^8 \approx 0.02$

$a_{10} = (0.8)^9 \approx 0.01$

43. *Keystrokes* (calculator in sequence and dot mode):

[Y=] 4 [n] [x^2] [÷] [(] [n] [x^2] [−] 2 [)] [TRACE]

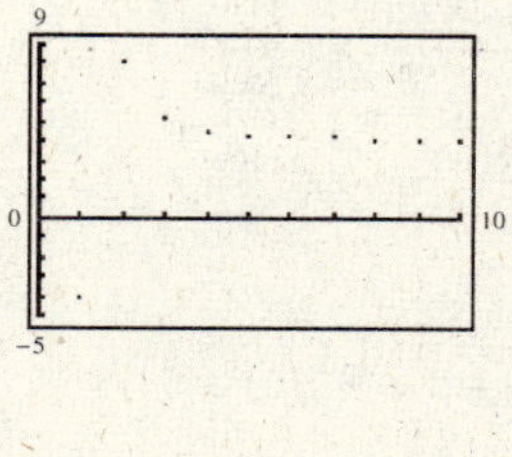

45. *Keystrokes* (calculator in sequence and dot mode):

[Y=] 3 [−] [(] 4 [÷] [n] [)] [TRACE]

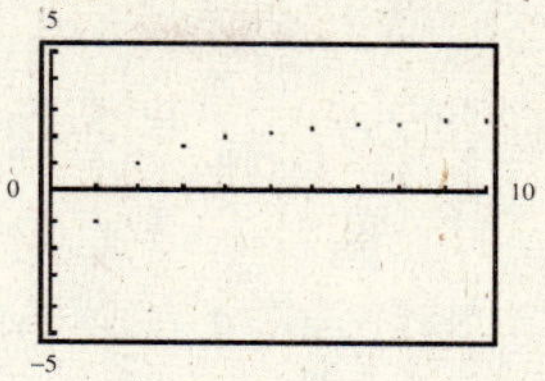

47. *Keystrokes* (calculator in sequence and dot mode):

[Y=] 100 [(] [(−)] 0.4 [)] [^] [(] [X, T, θ, n] [+] 1 [)] [TRACE]

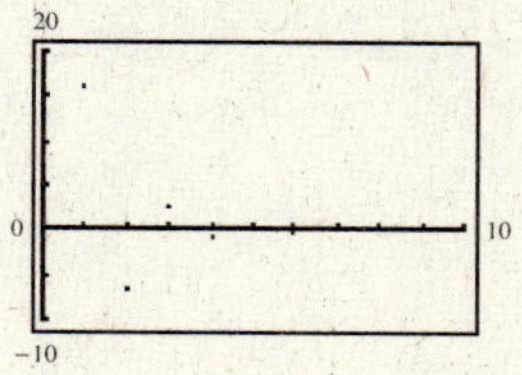

49. *n*: 1 2 3 4 5

Terms: 1 3 5 7 9

Apparent pattern: Each term is twice *n* minus 1.

$a_n - 2n - 1$

51. *n*: 1 2 3 4 5

Terms: 2 6 10 14 18

Apparent pattern: Each term is four times *n* minus 2.

$a_n = 4n - 2$

53. *n*: 1 2 3 4 5

Terms: 0 3 8 15 24

Apparent pattern: Each term is the square of *n* minus 1.

$a_n = n^2 - 1$

55. *n*: 1 2 3 4 5

Terms: 2 −4 6 −8 10

Apparent pattern: The terms have alternating signs with those in the even position being negative. Each term is double *n*.

$a_n = (-1)^{n+1}2n$

57. *n*: 1 2 3 4 5

Terms: $\dfrac{2}{3}$ $\dfrac{3}{4}$ $\dfrac{4}{5}$ $\dfrac{5}{6}$ $\dfrac{6}{7}$

Apparent pattern: The numerator is 1 more than *n* and the denominator is 2 more than *n*.

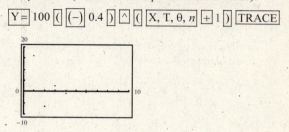

$a_n = \dfrac{n+1}{n+2}$

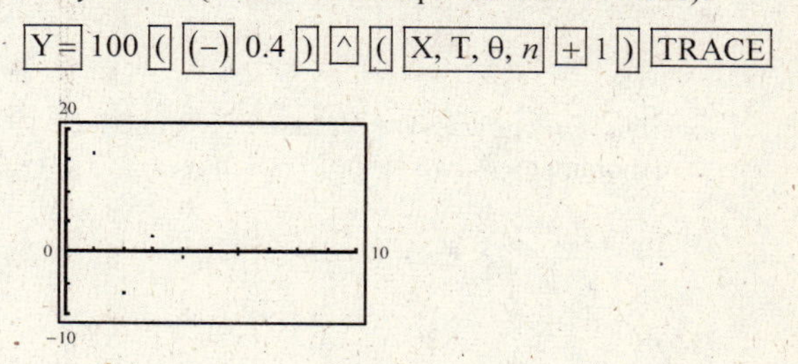

59. n: 1 2 3 4 5

Terms: $\frac{-1}{5}$ $\frac{1}{25}$ $\frac{-1}{125}$ $\frac{1}{625}$ $\frac{-1}{3125}$

Apparent pattern: The fraction $-\frac{1}{5}$ is raised to the n power.

$$a_n = \left(-\frac{1}{5}\right)^n$$

61. n: 1 2 3 4

Terms: 1 $\frac{1}{2}$ $\frac{1}{4}$ $\frac{1}{8}$

Apparent pattern: The numerator is 1 and the denominator is two to the $n - 1$ power.

$$a_n = \frac{1}{2^{n-1}}$$

63. n: 1 2 3 4 5

Terms: $1 + \frac{1}{1}$ $1 + \frac{1}{2}$ $1 + \frac{1}{3}$ $1 + \frac{1}{4}$ $1 + \frac{1}{5}$

Apparent pattern: The sum of one and $\frac{1}{n}$

$$a_n = 1 + \frac{1}{n}$$

65. n: 1 2 3 4 5

Terms: $-\frac{1}{2}$ $\frac{1}{6}$ $-\frac{1}{24}$ $\frac{1}{120}$ $-\frac{1}{720}$

Apparent pattern: The numerator is -1 to the n power and the denominator is $n + 1$ factorial.

$$a_n = \frac{(-1)^n}{(n+1)!}$$

67.
$$\begin{aligned}
a_1 &= 2(1) + 5 = 7\\
a_2 &= 2(2) + 5 = 9\\
a_3 &= 2(3) + 5 = 11\\
a_4 &= 2(4) + 5 = 13\\
a_5 &= 2(5) + 5 = 15\\
a_6 &= 2(6) + 5 = 17\\
s_1 &= a_1 = 7\\
s_2 &= a_1 + a_2 = 7 + 9 = 16\\
s_6 &= a_1 + a_2 + a_3 + a_4 + a_5 + a_6\\
&= 7 + 9 + 11 + 13 + 15 + 17\\
&= 72
\end{aligned}$$

69.
$$\begin{aligned}
a_1 &= \frac{1}{1} = 1\\
a_2 &= \frac{1}{2}\\
a_3 &= \frac{1}{3}\\
a_4 &= \frac{1}{4}\\
a_5 &= \frac{1}{5}\\
a_6 &= \frac{1}{6}\\
a_7 &= \frac{1}{7}\\
a_8 &= \frac{1}{8}\\
a_9 &= \frac{1}{9}
\end{aligned}$$

$$\begin{aligned}
s_2 &= a_1 + a_2 = 1 + \frac{1}{2} = \frac{3}{2}\\
s_3 &= a_1 + a_2 + a_3 = 1 + \frac{1}{2} + \frac{1}{3} = \frac{6 + 3 + 2}{6} = \frac{11}{6}\\
s_9 &= a_1 + a_2 + a_3 + a_4 + a_5 + a_6 + a_7 + a_8 + a_9\\
&= 1 + \frac{1}{2} + \frac{1}{3} + \frac{1}{4} + \frac{1}{5} + \frac{1}{6} + \frac{1}{7} + \frac{1}{8} + \frac{1}{9}\\
&= \frac{2520 + 1260 + 840 + 630 + 504 + 420 + 360 + 315 + 280}{2520} = \frac{7129}{2520}
\end{aligned}$$

71. $\sum_{k=1}^{5} 6k = 6(1) + 6(2) + 6(3) + 6(4) + 6(5) = 6 + 12 + 18 + 24 + 30 = 90$

73.
$$\begin{aligned}
\sum_{i=0}^{6} (2i + 5) &= [2(0) + 5] + [2(1) + 5] + [2(2) + 5] + [2(3) + 5] + [2(4) + 5] + [2(5) + 5] + [2(6) + 5]\\
&= 5 + 7 + 9 + 11 + 13 + 15 + 17 = 77
\end{aligned}$$

75. $\sum_{j=3}^{7}(6j-10) = (6\cdot 3-10)+(6\cdot 4-10)+(6\cdot 5-10)+(6\cdot 6-10)+(6\cdot 7-10)$

$= (18-10)+(24-10)+(30-10)+(36-10)+(42-10)$

$= 8+14+20+26+32 = 100$

77. $\sum_{j=1}^{5}\frac{(-1)^{j+1}}{j^2} = \frac{(-1)^{1+1}}{1^2}+\frac{(-1)^{2+1}}{2^2}+\frac{(-1)^{3+1}}{3^2}+\frac{(-1)^{4+1}}{4^2}+\frac{(-1)^{5+1}}{5^2}$

$= 1-\frac{1}{4}+\frac{1}{9}-\frac{1}{16}+\frac{1}{25} = \frac{3600}{3600}-\frac{900}{3600}+\frac{400}{3600}-\frac{225}{3600}+\frac{144}{3600} = \frac{3019}{3600}$

79. $\sum_{m=1}^{8}\frac{m}{m+1} = \frac{1}{1+1}+\frac{2}{2+1}+\frac{3}{3+1}+\frac{4}{4+1}+\frac{5}{5+1}+\frac{6}{6+1}+\frac{7}{7+1}+\frac{8}{8+1}$

$= \frac{1}{2}+\frac{2}{3}+\frac{3}{4}+\frac{4}{5}+\frac{5}{6}+\frac{6}{7}+\frac{7}{8}+\frac{8}{9} = \frac{15{,}551}{2520}$

81. $\sum_{k=1}^{6}(-8) = (-8)+(-8)+(-8)+(-8)+(-8)+(-8) = -48$

83. $\sum_{i=1}^{8}\left(\frac{1}{i}-\frac{1}{i+1}\right) = \left[\frac{1}{1}-\frac{1}{1+1}\right]+\left[\frac{1}{2}-\frac{1}{2+1}\right]+\left[\frac{1}{3}-\frac{1}{3+1}\right]+\left[\frac{1}{4}-\frac{1}{4+1}\right]+\left[\frac{1}{5}-\frac{1}{5+1}\right]+\left[\frac{1}{6}-\frac{1}{6+1}\right]$

$+\left[\frac{1}{7}-\frac{1}{7+1}\right]+\left[\frac{1}{8}-\frac{1}{8+1}\right]$

$= 1+\left(-\frac{1}{2}+\frac{1}{2}\right)+\left(-\frac{1}{3}+\frac{1}{3}\right)+\left(-\frac{1}{4}+\frac{1}{4}\right)+\left(-\frac{1}{5}+\frac{1}{5}\right)+\left(-\frac{1}{6}+\frac{1}{6}\right)+\left(-\frac{1}{7}+\frac{1}{7}\right)+\left(-\frac{1}{8}+\frac{1}{8}\right)-\frac{1}{9}$

$= 1-\frac{1}{9} = \frac{8}{9}$

85. $\sum_{n=0}^{5}\left(-\frac{1}{3}\right)^n = \left(-\frac{1}{3}\right)^0+\left(-\frac{1}{3}\right)^1+\left(-\frac{1}{3}\right)^2+\left(-\frac{1}{3}\right)^3+\left(-\frac{1}{3}\right)^4+\left(-\frac{1}{3}\right)^5$

$= 1+\left(-\frac{1}{3}\right)+\frac{1}{9}+\left(-\frac{1}{27}\right)+\frac{1}{81}+\left(-\frac{1}{243}\right) = \frac{243-81+27-9+3-1}{243} = \frac{182}{243}$

87. *Keystrokes:*

[LIST] [MATH] [5:SUM] [List] [OPS 5] [5:SEQ] 10 [X, T, θ] [x^2] [,] [X, T, θ] [,]

1 [,] 8 [,] 1 [)] [)] [ENTER]

$\sum_{n=1}^{8}10a^2 = 2040$

89. *Keystrokes:* [LIST] [MATH 5] [LIST] [OPS 5] [X, T, θ] [MATH] [PRB 4] [–] [X, T, θ] [,] [X, T, θ] [,] [2] [,] 6 [,] 1 [)] [ENTER]

$\sum_{j=2}^{6}(j!-j) = 852$

91. *Keystrokes:* [LIST] [MATH 5] [LIST] [OPS 5] [X, T, θ] 6 [÷] [MATH] [PRB 4] [,] [X, T, θ] [,] 0 [,] 4 [,] 1 [)] [ENTER]

$\sum_{j=0}^{4}\frac{6}{j!} = 16.25$

93. *Keystrokes:* [LIST] [MATH 5] [LIST] [OPS 5] [LN] [X, T, θ] [,] [X, T, θ] [,] 0 [,] 6 [,] 1 [)] [ENTER]

$\sum_{k=1}^{6}\ln k \approx 6.5793$

95. $\sum_{k=1}^{5} k$

97. $\sum_{k=1}^{6} 5k$

99. $\sum_{k=1}^{50} \frac{3}{1+k}$

101. $\sum_{k=1}^{10} \frac{1}{2k}$

103. $\sum_{k=0}^{12} \frac{1}{2^k}$

105. $\sum_{k=1}^{20} \frac{1}{k(k+1)}$

107. $\sum_{k=1}^{11} \frac{k}{k+1}$

109. $\sum_{k=1}^{20} \frac{2k}{k+3}$

111. $\sum_{k=0}^{6} k!$

113. (a) $A_1 = 500(1 + 0.07)^1 = \$535.00$

$A_2 = 500(1 + 0.07)^2 = \$572.45$

$A_3 = 500(1 + 0.07)^3 = \$612.52$

$A_4 = 500(1 + 0.07)^4 = \$655.40$

$A_5 = 500(1 + 0.07)^5 = \$701.28$

$A_6 = 500(1 + 0.07)^6 = \$750.37$

$A_7 = 500(1 + 0.07)^7 = \$802.89$

$A_8 = 500(1 + 0.07)^8 = \$859.09$

(b) $A_{40} = 500(1 + 0.07)^{40} = \$7487.23$

(c) *Keystrokes* (calculator in sequence and dot mode):

[Y=] 500 [(] 1 [+] 0.07 [)] [^] [n] [TRACE]

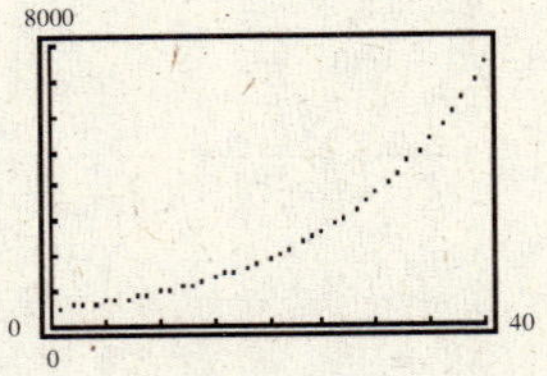

(d) Yes. Investment earning compound interest increases at an increasing rate.

115. $d_5 = \frac{180(5-4)}{5} = \frac{180(1)}{5} = \frac{180}{5} = 36°$

$d_6 = \frac{180(6-4)}{6} = \frac{180(2)}{6} = \frac{360}{6} = 60°$

$d_7 = \frac{180(7-4)}{7} = \frac{180(3)}{7} = \frac{540}{7} \approx 77.1°$

$d_8 = \frac{180(8-4)}{8} = \frac{180(4)}{8} = \frac{720}{8} = 90°$

$d_9 = \frac{180(9-4)}{9} = \frac{180(5)}{9} = \frac{900}{9} = 100°$

$d_{10} = \frac{180(10-4)}{10} = \frac{180(6)}{10} = \frac{1080}{10} = 108°$

117. (a) $a_3 = 32{,}000\left(\frac{3}{4}\right)^3 = 32{,}000\left(\frac{27}{64}\right) \approx 13{,}500$

Thus, after 3 years, the car is worth \$13,500.

(b) $a_6 = 32{,}000\left(\frac{3}{4}\right)^6 = 32{,}000\left(\frac{729}{4096}\right) \approx 5695$

Thus, after 6 years, the car is worth approximately \$5695. This value is less than half the value after 3 years. The value is decreasing at an increasing rate.

119. A sequence is a function because there is only one value for each term of the sequence.

121. $a_n = 4n! = 4(1 \cdot 2 \cdot 3 \cdot 3 \cdot \ldots \cdot (n-1) \cdot n)$

$a_n = (4n)! = (4 \cdot 1) \cdot (4 \cdot 2) \cdot (4 \cdot 3) \cdot (4 \cdot 4) \cdot \ldots \cdot (4n-1) \cdot (4n)$

123. True.

$$\sum_{k=1}^{4} 3k = 3 + 6 + 9 + 12 = 3(1 + 2 + 3 + 4) = 3\sum_{k=1}^{4} k$$

125. $-2n + 15$ for $n = 3$: $-2(3) + 15 = 9$

127. $25 - 3(n + 4)$ for $n = 8$: $25 - 3(8 + 4) = 25 - 3(12) = 25 - 36 = -11$

129. $x^2 + y^2 = 36$

Center: $(0, 0)$

Radius: 6

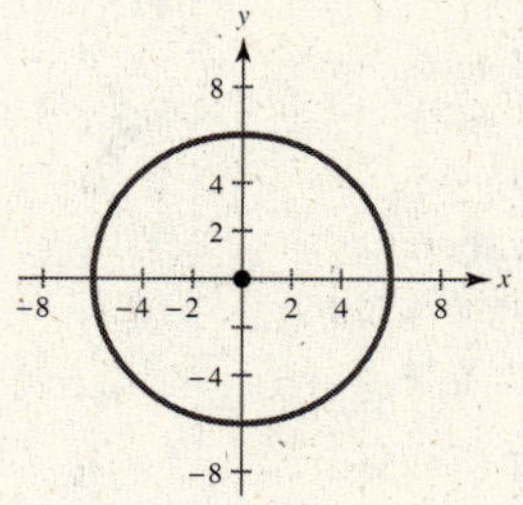

131. $x^2 + y^2 + 4x - 12 = 0$

$$(x^2 + 4x + 4) + y^2 = 12 + 4$$

$$(x + 2)^2 + y^2 = 16$$

Center: $(-2, 0)$

Radius: 4

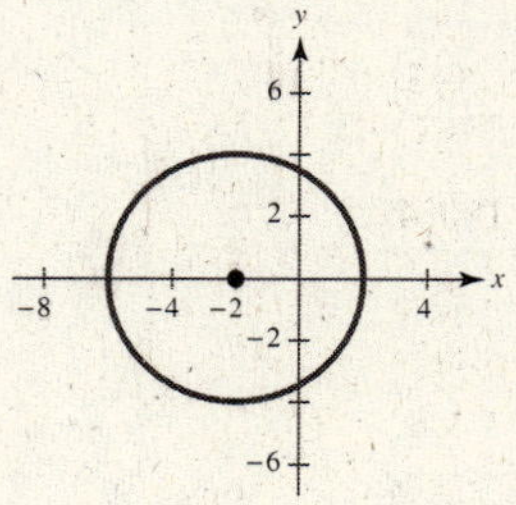

133. $x^2 = 6y$

Vertex: $(0, 0)$

$$4p = 6$$

$$p = \frac{6}{4} = \frac{3}{2}$$

Focus: $\left(0, \frac{3}{2}\right)$

135. $x^2 + 8y + 32 = 0$

$$x^2 + 8(y + 4) = 0$$

$$x^2 = -8(y + 4)$$

Vertex: $(0, -4)$

$$4p = -8$$

$$p = -\frac{8}{4} = -2$$

Focus: $(0, -6)$

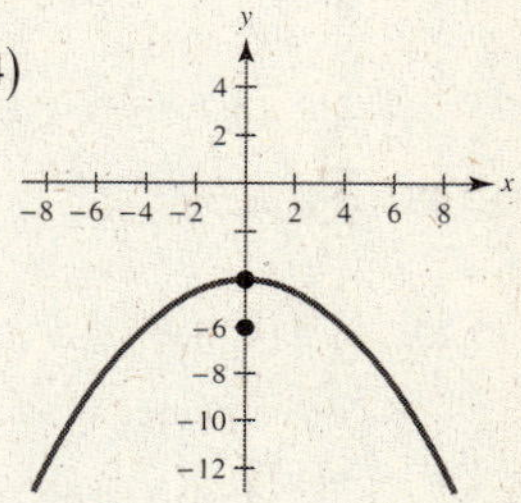

Section 11.2 Arithmetic Sequences

1. 2, 5, 8, 11, …

$d = 3$

$5 - 2 = 3, 8 - 5 = 3, 11 - 8 = 3$

3. 100, 94, 88, 82, …

$d = -6$

$94 - 100 = -6, 88 - 94 = -6, 82 - 88 = -6$

5. 10, −2, −14, −26, −38, …

$d = -12$

$-2 - 10 = -12, -14 - (-2) = -12,$

$-26 - (-14) = -12, -38 - (-26) = -12$

7. $1, \frac{5}{3}, \frac{7}{3}, 3, \ldots$

$d = \frac{2}{3}$

$\frac{5}{3} - 1 = \frac{2}{3}, \frac{7}{3} - \frac{5}{3} = \frac{2}{3}, 3 - \frac{7}{3} = \frac{2}{3}$

9. $\frac{7}{2}, \frac{9}{4}, 1, -\frac{1}{4}, -\frac{3}{2}, \ldots$

$d = -\frac{5}{4}$

$\frac{9}{4} - \frac{7}{2} = -\frac{5}{4}, 1 - \frac{9}{4} = -\frac{5}{4}, -\frac{1}{4} - 1 = -\frac{5}{4}$

11. $a_n = 4n + 5$

$a_1 = 9$

$a_2 = 13$

$a_3 = 17$

$a_4 = 21$

$d = 4$

13. $a_n = 8 - 3n$

$a_1 = 5$

$a_2 = 2$

$a_3 = -1$

$a_4 = -4$

$d = -3$

15. $a_n = \frac{1}{2}(n + 1)$

$a_1 = 1$

$a_2 = \frac{3}{2}$

$a_3 = 2$

$a_4 = \frac{5}{2}$

$d = \frac{1}{2}$

17. 2, 4, 6, 8, …

The sequence is arithmetic.

$d = 2$

$4 - 2 = 2, 6 - 4 = 2, 8 - 6 = 2$

19. 10, 8, 6, 4, 2, …

arithmetic; $d = -2$

$8 - 10 = -2, 6 - 8 = -2, 4 - 6 = -2, 2 - 4 = -2$

21. 32, 16, 0, –16, …

The sequence is arithmetic.

$d = -16$

$16 - 32 = -16, 0 - 16 = -16, -16 - 0 = -16$

23. 3.2, 4, 4.8, 5.6, …

The sequence is arithmetic.

$d = 0.8$

$4 - 3.2 = 0.8, 4.8 - 4 = 0.8, 5.6 - 4.8 = 0.8$

25. $2, \frac{7}{2}, 5, \frac{13}{2}, \ldots$

The sequence is arithmetic.

$d = \frac{3}{2}$

$\frac{7}{2} - 2 = \frac{3}{2}, 5 - \frac{7}{2} = \frac{3}{2}, \frac{13}{2} - 5 = \frac{3}{2}$

27. $\frac{1}{3}, \frac{2}{3}, \frac{4}{3}, \frac{8}{3}, \frac{16}{3}, \ldots$

The sequence is not arithmetic.

$\frac{2}{3} - \frac{1}{3} = \frac{1}{3}$

$\frac{4}{3} - \frac{2}{3} = \frac{2}{3}$

The difference is NOT the same.

29. $1, \sqrt{2}, \sqrt{3}, 2, \sqrt{5}$

The sequence is not arithmetic.

$\sqrt{2} - 1 = \sqrt{2} - 1 \approx 0.41$

$\sqrt{3} - \sqrt{2} = \sqrt{3} - \sqrt{2} \approx 0.32$

The difference is NOT the same.

31. ln 4, ln 8, ln 12, ln 16, …

The sequence is not arithmetic.

$\ln 8 - \ln 4 = \ln 8 - \ln 4 \approx 0.69$

$\ln 12 - \ln 8 = \ln 12 - \ln 8 \approx 0.41$

The difference is NOT the same.

33. $a_1 = 7, d = 5$

$a_1 = 7$

$a_2 = 7 + 5 = 12$

$a_3 = 12 + 5 = 17$

$a_4 = 17 + 5 = 22$

$a_5 = 22 + 5 = 27$

35. $a_1 = 11, d = 4$

$a_1 = 11$

$a_2 = 11 + 4 = 15$

$a_3 = 15 + 4 = 19$

$a_4 = 19 + 4 = 23$

$a_5 = 23 + 4 = 27$

37. $a_1 = 20, d = -4$

$a_1 = 20$

$a_2 = 20 - 4 = 16$

$a_3 = 16 - 4 = 12$

$a_4 = 12 - 4 = 8$

$a_5 = 8 - 4 = 4$

39. $a_1 = 6, a_2 = 11$

$a_1 = 6$

$a_2 = 11, d = 5$

$a_3 = 11 + 5 = 16$

$a_4 = 16 + 5 = 21$

$a_5 = 21 + 5 = 26$

41. $a_1 = 22, a_2 = 18$

$a_1 = 22$

$a_2 = 18, \ d = -4$

$a_3 = 18 - 4 = 14$

$a_4 = 14 - 4 = 10$

$a_5 = 10 - 4 = 6$

43. $a_1 = 3(1) + 4 = 7$

$a_2 = 3(2) + 4 = 10$

$a_3 = 3(3) + 4 = 13$

$a_4 = 3(4) + 4 = 16$

$a_5 = 3(5) + 4 = 19$

45. $a_1 = -2(1) + 8 = 6$

$a_2 = -2(2) + 8 = 4$

$a_3 = -2(3) + 8 = 2$

$a_4 = -2(4) + 8 = 0$

$a_5 = -2(5) + 8 = -2$

47. $a_1 = \frac{5}{2}(1) - 1 = \frac{3}{2}$

$a_2 = \frac{5}{2}(2) - 1 = 4$

$a_3 = \frac{5}{2}(3) - 1 = \frac{13}{2}$

$a_4 = \frac{5}{2}(4) - 1 = 9$

$a_5 = \frac{5}{2}(5) - 1 = \frac{23}{2}$

49. $a_1 = \frac{3}{5}(1) + 1 = \frac{8}{5}$

$a_2 = \frac{3}{5}(2) + 1 = \frac{11}{5}$

$a_3 = \frac{3}{5}(3) + 1 = \frac{14}{5}$

$a_4 = \frac{3}{5}(4) + 1 = \frac{17}{5}$

$a_5 = \frac{3}{5}(5) + 1 = \frac{20}{5} = 4$

51. $a_1 = -\frac{1}{4}(1 - 1) + 4 = 4$

$a_2 = -\frac{1}{4}(2 - 1) + 4 = \frac{15}{4}$

$a_3 = -\frac{1}{4}(3 - 1) + 4 = \frac{7}{2}$

$a_4 = -\frac{1}{4}(4 - 1) + 4 = \frac{13}{4}$

$a_5 = -\frac{1}{4}(5 - 1) + 4 = 3$

53. $a_1 = 4, d = 3$

$a_n = a_1 + (n - 1)d$

$a_n = 4 + (n - 1)3$

$a_n = 4 + 3n - 3$

$a_n = 3n + 1$

55. $a_1 = \frac{1}{2}, d = \frac{3}{2}$

$a_n = a_1 + (n - 1)d$

$a_n = \frac{1}{2} + (n - 1)\frac{3}{2}$

$a_n = \frac{1}{2} + \frac{3}{2}n - \frac{3}{2}$

$a_n = \frac{3}{2}n - \frac{2}{2}$

$a_n = \frac{3}{2}n - 1$

57. $a_1 = 100, d = -5$

$a_n = a_1 + (n - 1)d$

$a_n = 100 + (n - 1)(-5)$

$a_n = 100 - 5n + 5$

$a_n = -5n + 105$

59. $a_3 = 6, d = \frac{3}{2}$

$a_n = a_1 + (n - 1)d$

$6 = a_1 + (3 - 1)\frac{3}{2}$

$6 = a_1 + 3$

$3 = a_1$

$a_n = 3 + (n - 1)\frac{3}{2} = 3 + \frac{3}{2}n - \frac{3}{2}$

$a_n = \frac{3}{2}n + \frac{3}{2}$

61. $a_n = a_1 + (n - 1)d$

$15 = 5 + (5 - 1)d$

$15 = 5 + 4d$

$10 = 4d$

$\frac{5}{2} = \frac{10}{4} = d$

So,

$a_n = 5 + (n - 1)\frac{5}{2}$

$a_n = 5 + \frac{5}{2}n - \frac{5}{2}$

$a_n = \frac{5}{2}n + \frac{5}{2}$

63. $a_n = a_1 + (n - 1)d$

$16 = a_1 + (3 - 1)4$

$8 = a_1$

$a_n = 8 + (n - 1)(4)$

$a_n = 4n + 4$

65. $a_n = a_1 + (n - 1)d$

$30 = 50 + (3 - 1)d$

$-20 = 2d$

$-10 = d$

$a_n = 50 + (n - 1)(-10)$

$a_n = -10n + 60$

67. $d = \frac{8 - 10}{4} = -\frac{1}{2}$

$a_n = a_1 + (n - 1)d$

$10 = a_1 + (2 - 1)\left(-\frac{1}{2}\right)$

$10 = a_1 - \frac{1}{2}$

$\frac{21}{2} = a_1$

$a_n = \frac{21}{2} + (n - 1)\left(-\frac{1}{2}\right)$

$a_n = \frac{21}{2} - \frac{1}{2}n + \frac{1}{2}$

$a_n = -\frac{1}{2}n + 11$

69. $d = \frac{0.30 - 0.35}{1} = -0.05$

$a_n = a_1 + (n - 1)d$

$a_n = 0.35 + (n - 1)(-0.05)$

$a_n = 0.35 - 0.05n + 0.05$

$a_n = -0.05n + 0.40$

71. $a_1 = 14, a_{k+1} = a_k + 6$ so, $d = 6.$

$a_2 = a_{1+1} = a_1 + 6 = 14 + 6 = 20$

$a_3 = a_{2+1} = a_2 + 6 = 20 + 6 = 26$

$a_4 = a_{3+1} = a_3 + 6 = 26 + 6 = 32$

$a_5 = a_{4+1} = a_4 + 6 = 32 + 6 = 38$

73. $a_1 = 23, a_{k+1} = a_k - 5$ so, $d = -5.$

$a_2 = a_{1+1} = a_1 - 5 = 23 - 5 = 18$

$a_3 = a_{2+1} = a_2 - 5 = 18 - 5 = 13$

$a_4 = a_{3+1} = a_3 - 5 = 13 - 5 = 8$

$a_5 = a_{4+1} = a_4 - 5 = 8 - 5 = 3$

75. $a_1 = -16, a_{k+1} = a_k + 5$ so, $d = 5.$

$a_2 = a_{1+1} = a_1 + 5 = -16 + 5 = -11$

$a_3 = a_{2+1} = a_2 + 5 = -11 + 5 = -6$

$a_4 = a_{3+1} = a_3 + 5 = -6 + 5 = -1$

$a_5 = a_{4+1} = a_4 + 5 = -1 + 5 = 4$

77. $a_1 = 3.4, a_{k+1} = a_k - 1.1$ so, $d = -1.1.$

$a_2 = a_{1+1} = a_1 - 1.1 = 3.4 - 1.1 = 2.3$

$a_3 = a_{2+1} = a_2 - 1.1 = 2.3 - 1.1 = 1.2$

$a_4 = a_{3+1} = a_3 - 1.1 = 1.2 - 1.1 = 0.1$

$a_5 = a_{4+1} = a_4 - 1.1 = 0.1 - 1.1 = -1$

79. $\sum_{k=1}^{20} k = \frac{20}{2}(1 + 20) = 210$

81. $\sum_{k=1}^{50} (k + 3) = \frac{50}{2}(4 + 53) = 1425$

83. $\sum_{k=1}^{10} (5k - 2) = \frac{10}{2}(3 + 48) = 255$

85. $\sum_{n=1}^{500} \frac{n}{2} = \frac{500}{2}\left(\frac{1}{2} + 250\right) = 62{,}625$

87. $\sum_{n=1}^{30} \left(\frac{1}{3}n - 4\right) = \frac{30}{2}\left(-\frac{11}{3} + 6\right) = 35$

89. $\sum_{n=1}^{12} (7n - 2) = \frac{12}{2}(5 + 82) = 522$

91. $\sum_{n=1}^{25} (6n - 4) = \frac{25}{2}(2 + 146) = 1850$

93. $\sum_{n=1}^{8} (225 - 25n) = \frac{8}{2}(200 + 25) = 900$

95. $\sum_{n=1}^{50} (12n - 62) = \frac{50}{2}(-50 + 538) = 12{,}200$

97. $\sum_{n=1}^{12} (3.5n - 2.5) = \frac{12}{2}(1 + 39.5) = 243$

99. $\sum_{n=1}^{10} (0.4n + 0.1) = \frac{10}{2}(0.5 + 4.1) = 23$

101. (b)

102. (f)

103. (e)

104. (a)

105. (c)

106. (d)

107. *Keystrokes* (calculator in sequence and dot mode):

[Y=] [(−)] 2 [n] [+] 21 [TRACE]

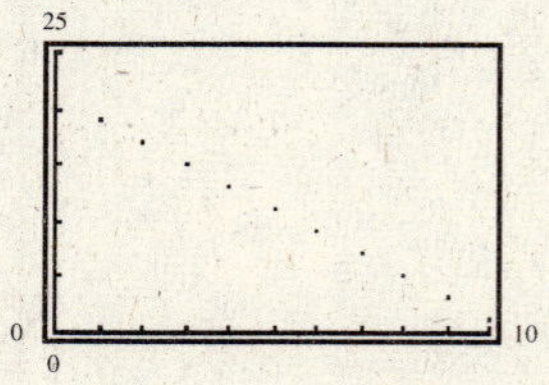

109. *Keystrokes* (calculator in sequence and dot mode):

[Y=] 0.6 [n] + 1.5 [TRACE]

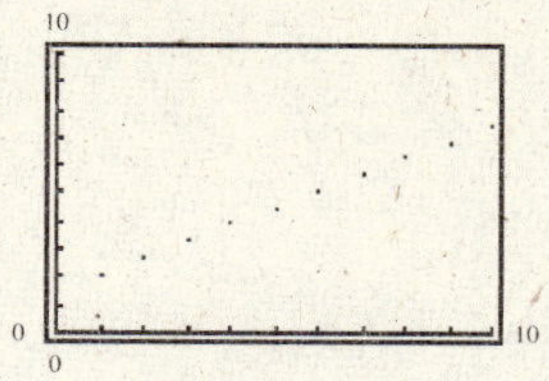

111. *Keystrokes* (calculator in sequence and dot mode):

[Y=] 2.5 [n] [−] 8 [TRACE]

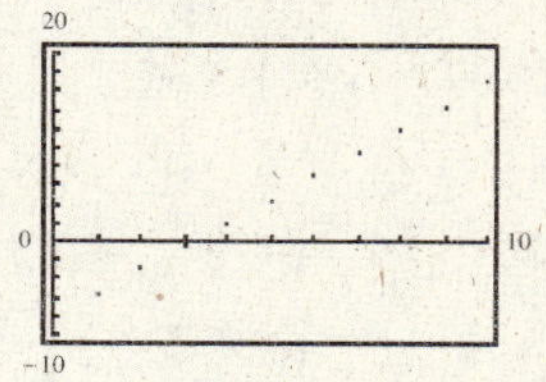

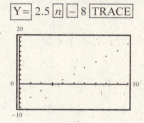

113. *Keystrokes:*

[LIST] [MATH 5] [LIST] [OPS 5] 750 [−] 30 [X, T, θ] [,]

[X, T, θ] [,] 1 [,] 25 [,] 1 [)] [ENTER]

$$\sum_{j=1}^{25}(750 - 30j) = 9000$$

115. *Keystrokes:*

[LIST] [MATH 5] [LIST] [OPS 5] 300 [−] 8 [X, T, θ]

[÷] 3 [,] [X, T, θ] [,] [1] [,] 60 [,] 1 [)] [ENTER]

$$\sum_{i=1}^{60}\left(300 - \tfrac{8}{3}i\right) = 13{,}120$$

117. Keystrokes:

[LIST] [MATH 5] [LIST] [OPS 5] 2.15 [X, T, θ]

[+] 5.4 [,] [X, T, θ] [,] 1 [,] 50 [,] 1 [)] [ENTER]

$$\sum_{n=1}^{50}(2.15n + 5.4) = 3011.25$$

119. $\sum_{n=1}^{75} n = \frac{75}{2}(1 + 75) = 2850$

121. $\sum_{n=1}^{50}(2n - 1) = \frac{50}{2}(1 + 99) = 2500$

123. $a_1 = \$54{,}000$

$a_2 = \$57{,}000$

$a_3 = \$60{,}000$

$a_4 = \$63{,}000$

$a_5 = \$66{,}000$

$a_6 = \$69{,}000$

Total salary $= \frac{6}{2}(54{,}000 + 69{,}000)$

$= 3(123{,}000) = \$369{,}000$

125. Sequence $= 20, 21, 22, \ldots, n = 20, d = 1$

$a_n = a_1 + (n - 1)d$

$a_n = 20 + (n - 1)1$

$a_n = 19 + n$

$$\sum_{n=1}^{20}(19 + n) = \frac{20}{2}(20 + 39) = 590 \text{ seats}$$

$$\frac{\text{Total cost}}{\text{Total seats}} = \text{Cost per ticket}$$

$$\frac{15{,}000}{590} = 25.43$$

Charge $25.43 to make at least $15,000.

127. Sequence $= 93, 89, 85, 81, \ldots$

$$\sum_{n=1}^{8}(97 - 4n) = \frac{8}{2}(93 + 65) = 632 \text{ bales}$$

129. Sequence $= 1, 2, 3, 4, \ldots$

$a_n = a_1 + (n - 1)d$

$a_n = 1 + (n - 1)(1)$

$a_n = 1 + n - 1$

$a_n = n$

$$\sum_{n=1}^{12} n = \frac{12}{2}(1 + 12) = 78 \text{ chimes}$$

3 chimes each hour × 12 hours = 36 chimes

Total chimes = 78 + 36 = 114 chimes

131. Sequence $= 16, 48, 80, \ldots, n = 8, d = 32$

$a_n = a_1 + (n - 1)d$

$a_n = 16 + (n - 1)32 = 16 + 32n - 32$

$a_n = 32n - 16$

$a_n = 16 + (8 - 1)32$

$a_n = 16 + 224$

$a_n = 240$

$$\sum_{n=1}^{8}(32n - 16) = \frac{8}{2}(16 + 240) = 1024 \text{ feet}$$

133. $\sum_{i=1}^{n} a_i = \frac{n}{2}(a_1 + a_n)$

$a_1 = 4.41 + 0.59(1) = 5$

$a_5 = 4.41 + 0.59(5) = 7.36$

$\sum_{i=1}^{5} 4.41 + 0.59n = \frac{5}{2}(5 + 7.36) = 30.9$ barrels

135. A recursion formula gives the relationship between the terms a_{n+1} and a_n.

137. Yes. Because a_{2n} is n terms away from a_n, add n times the difference d to a_n.

$a_{2n} = a_n + nd$

139. (a) $1 + 3 = 4$

$1 + 3 + 5 = 9$

$1 + 3 + 5 + 7 = 16$

$1 + 3 + 5 + 7 + 9 = 25$

$1 + 3 + 5 + 7 + 9 + 11 = 36$

(b) No. There is no common difference between consecutive terms of the sequence.

(c) $1 + 3 + 5 + 7 + 9 + 11 + 13 = 49$

$\sum_{k=1}^{n} (2k - 1) = n^2$

(d) Looking at the sums in part (a), each sum is the square of a consecutive positive integer.

$1 + 3 = 2^2$

$1 + 3 + 5 = 3^2$

$1 + 3 + 5 + 7 = 4^2$

$1 + 3 + 5 + 7 + 9 = 5^2$

$1 + 3 + 5 + 7 + 9 + 11 = 6^2$

So the sum of n odd integers is n^2. An odd integer is represented by $2k - 1$ which gives the formula

$\sum_{k=1}^{n} (2k - 1) = n^2.$

141. $\frac{(x-4)^2}{25} + \frac{(y+5)^2}{9} = 1$

Center: $(4, -5)$

$a^2 = 25$

$a = 5$, horizontal axis

Vertices: $(4 - 5, -5), (4 + 5, -5)$

$(-1, -5), (9, -5)$

143.

$$\begin{aligned} 9x^2 + 4y^2 - 18x + 24y + 9 &= 0 \\ 9x^2 - 18x + 4y^2 + 24y &= -9 \\ 9(x^2 - 2x) + 4(y^2 + 6y) &= -9 \\ 9(x^2 - 2x + 1) + 4(y^2 + 6y + 9) &= -9 + 9 + 36 \\ 9(x-1)^2 + 4(y+3)^2 &= 36 \\ \frac{(x-1)^2}{4} + \frac{(y+3)^2}{9} &= 1 \end{aligned}$$

Center: $(1, -3)$

$a^2 = 9$

$a = 3$, vertical axis

Vertices: $(1, -3 + 3), (1, -3 - 3)$

$(1, 0), (1, -6)$

145. $3 + 4 + 5 + 6 + 7 + 8 + 9 = \sum_{k=1}^{7} (k + 2)$

147. $12 + 15 + 18 + 21 + 24 = \sum_{k=1}^{5} (3k + 9)$

Mid-Chapter Quiz for Chapter 11

1. $a_n = 4n$

$a_1 = 4(1) = 4$

$a_2 = 4(2) = 8$

$a_3 = 4(3) = 12$

$a_4 = 4(4) = 16$

$a_5 = 4(5) = 20$

2. $a_n = 2n + 5$

$a_1 = 2(1) + 5 = 7$

$a_2 = 2(2) + 5 = 9$

$a_3 = 2(3) + 5 = 11$

$a_4 = 2(4) + 5 = 13$

$a_5 = 2(5) + 5 = 15$

3. $a_n = 32\left(\frac{1}{4}\right)^{n-1}$

$a_1 = 32\left(\frac{1}{4}\right)^{1-1} = 32$

$a_2 = 32\left(\frac{1}{4}\right)^{2-1} = 8$

$a_3 = 32\left(\frac{1}{4}\right)^{3-1} = 2$

$a_4 = 32\left(\frac{1}{4}\right)^{4-1} = \frac{1}{2}$

$a_5 = 32\left(\frac{1}{4}\right)^{5-1} = \frac{1}{8}$

4. $a_n = \dfrac{(-3)^n n}{n+4}$

$a_1 = \dfrac{(-3)^1 \cdot 1}{1+4} = -\dfrac{3}{5}$

$a_2 = \dfrac{(-3)^2 \cdot 2}{2+4} = 3$

$a_3 = \dfrac{(-3)^3 \cdot 3}{3+4} = -\dfrac{81}{7}$

$a_4 = \dfrac{(-3)^4 \cdot 4}{4+4} = \dfrac{81}{2}$

$a_5 = \dfrac{(-3)^5 \cdot 5}{5+4} = -135$

5. $\sum_{k=1}^{4} 10k = \frac{4}{2}(10 + 40) = 100$

6. $\sum_{i=1}^{10} 4 = \frac{10}{2}(4 + 4) = 40$

7. $\sum_{j=1}^{5} \dfrac{60}{j+1} = \dfrac{60}{2} + \dfrac{60}{3} + \dfrac{60}{4} + \dfrac{60}{5} + \dfrac{60}{6}$

$= 30 + 20 + 15 + 12 + 10$

$= 87$

8. $\sum_{n=1}^{4} \dfrac{12}{n} = 12 + 6 + 4 + 3 = 25$

9. $\sum_{n=1}^{5} (3n - 1) = \frac{5}{2}(2 + 14) = 40$

10. $\sum_{k=1}^{4} \left(k^2 - 1\right) = 0 + 3 + 8 + 15 = 26$

11. $\sum_{k=1}^{20} \dfrac{2}{3k}$

12. $\sum_{k=1}^{25} \dfrac{(-1)^{k-1}}{k^3}$

13. $\sum_{k=1}^{20} \dfrac{k-1}{k}$

14. $\sum_{k=1}^{10} \dfrac{k^2}{2}$

15. $d = \frac{1}{2}$

16. $d = -6$

17. $a_n = a_1 + (n-1)d$ $\quad a_n = 20 + (n-1)(-3)$

$11 = 20 + (4-1)d$ $\quad a_n = 20 - 3n + 3$

$-9 = 3d$ $\quad a_n = -3n + 23$

$-3 = d$

18. $a_1 = 32, d = -4$

$a_n = a_1 + (n-1)d$

$a_n = 32 + (n-1)(-4)$

$a_n = 32 - 4n + 4$

$a_n = -4n + 36$

19. $\sum_{n=1}^{200} 2n = \frac{200}{2}(2 + 400) = 100(402) = 40{,}200$

20. $a_n = a_1 + (n-1)d$

$= 0.50 + (n-1)(0.50)$

$= 0.50 + 0.50n - 0.50$

$a_n = 0.50n$

$\sum_{n=1}^{365} 0.50n = \frac{365}{2}(0.50 + 182.5) = \$33{,}397.50$

Section 11.3 Geometric Sequences and Series

1. 3, 6, 12, 24, ⋯

$r = 2$ since

$\frac{6}{3} = 2, \frac{12}{6} = 2, \frac{24}{12} = 2.$

3. 5, −5, 5, −5, …

$r = -1$ since

$\dfrac{-5}{5} = -1, \dfrac{5}{-5} = -1, \dfrac{-5}{5} = -1.$

5. $\frac{1}{2}, -\frac{1}{4}, \frac{1}{8}, -\frac{1}{16}, \ldots$

$r = -\frac{1}{2}$ since

$$\frac{-\frac{1}{4}}{\frac{1}{2}} = -\frac{1}{2}, \frac{\frac{1}{8}}{-\frac{1}{4}} = -\frac{1}{2},$$

$$\frac{-\frac{1}{16}}{\frac{1}{8}} = -\frac{1}{2}.$$

7. $75, 15, 3, \frac{3}{5}, \ldots$

$r = \frac{1}{5}$ since

$$\frac{15}{75} = \frac{1}{5}, \frac{3}{15} = \frac{1}{5}, \frac{\frac{3}{5}}{3} = \frac{1}{5}.$$

9. $1, \pi, \pi^2, \pi^3, \ldots$

$r = \pi$ since

$$\frac{\pi}{1} = \pi, \frac{\pi^2}{\pi} = \pi, \frac{\pi^3}{\pi} = \pi.$$

11. $50(1.04), 50(1.04)^2, 50(1.04)^3, 50(1.04)^4, \ldots$

$r = 1.04$ since

$$\frac{50(1.04)^2}{50(1.04)} = 1.04, \frac{50(1.04)^3}{50(1.04)^2} = 1.04.$$

13. The sequence is geometric.

$r = \frac{1}{2}$ since

$\frac{32}{64} = \frac{1}{2}, \frac{16}{32} = \frac{1}{2}, \frac{8}{16} = \frac{1}{2}.$

15. The sequence is not geometric, because

$\frac{15}{10} = \frac{3}{2}$ and $\frac{20}{15} = \frac{4}{3}.$

17. The sequence is geometric.

$r = 2$ since $\frac{10}{5} = 2, \frac{20}{10} = 2, \frac{40}{20} = 2.$

19. The sequence is not geometric, because

$\frac{8}{1} = 8$ and $\frac{27}{8} = \frac{27}{8}.$

21. The sequence is geometric.

$r = -\frac{2}{3}$ since

$$\frac{-\frac{2}{3}}{1} = -\frac{2}{3}, \frac{\frac{4}{9}}{-\frac{2}{3}} = -\frac{2}{3}, \frac{-\frac{8}{27}}{\frac{4}{9}} = -\frac{2}{3}.$$

23. The sequence is geometric.

$r = 1.1$ since

$$\frac{1.1}{1} = 1.1, \frac{1.21}{1.1} = 1.1, \frac{1.331}{1.21} = 1.1.$$

25.
$$\begin{aligned} a_n &= a_1 r^{n-1} \\ a_n &= 4(2)^{n-1} \\ a_1 &= 4(2)^{1-1} = 4 \\ a_2 &= 4(2)^{2-1} = 8 \\ a_3 &= 4(2)^{3-1} = 16 \\ a_4 &= 4(2)^{4-1} = 32 \\ a_5 &= 4(2)^{5-1} = 64 \end{aligned}$$

27.
$$\begin{aligned} a_n &= a_1 r^{n-1} \\ a_n &= 6\left(\tfrac{1}{2}\right)^{n-1} \\ a_1 &= 6\left(\tfrac{1}{2}\right)^{1-1} = 6\left(\tfrac{1}{2}\right)^0 = 6(1) = 6 \\ a_2 &= 6\left(\tfrac{1}{2}\right)^{2-1} = 6\left(\tfrac{1}{2}\right)^1 = 6\left(\tfrac{1}{2}\right) = 3 \\ a_3 &= 6\left(\tfrac{1}{2}\right)^{3-1} = 6\left(\tfrac{1}{2}\right)^2 = 6\left(\tfrac{1}{4}\right) = \tfrac{6}{4} = \tfrac{3}{2} \\ a_4 &= 6\left(\tfrac{1}{2}\right)^{4-1} = 6\left(\tfrac{1}{2}\right)^3 = 6\left(\tfrac{1}{8}\right) = \tfrac{6}{8} = \tfrac{3}{4} \\ a_5 &= 6\left(\tfrac{1}{2}\right)^{5-1} = 6\left(\tfrac{1}{2}\right)^4 = 6\left(\tfrac{1}{16}\right) = \tfrac{6}{16} = \tfrac{3}{8} \end{aligned}$$

29.
$$\begin{aligned} a_n &= a_1 r^{n-1} \\ a_n &= 5(-2)^{n-1} \\ a_1 &= 5(-2)^{1-1} = 5(-2)^0 = 5(1) = 5 \\ a_2 &= 5(-2)^{2-1} = 5(-2)^1 = 5(-2) = -10 \\ a_3 &= 5(-2)^{3-1} = 5(-2)^2 = 5(4) = 20 \\ a_4 &= 5(-2)^{4-1} = 5(-2)^3 = 5(-8) = -40 \\ a_5 &= 5(-2)^{5-1} = 5(-2)^4 = 5(16) = 80 \end{aligned}$$

31.
$$\begin{aligned} a_n &= a_1 r^{n-1} \\ a_n &= (-4)\left(-\tfrac{1}{2}\right)^{n-1} \\ a_1 &= (-4)\left(-\tfrac{1}{2}\right)^{1-1} = -4 \\ a_2 &= (-4)\left(-\tfrac{1}{2}\right)^{2-1} = (-4)\left(-\tfrac{1}{2}\right) = 2 \\ a_3 &= (-4)\left(-\tfrac{1}{2}\right)^{3-1} = (-4)\left(\tfrac{1}{4}\right) = -1 \\ a_4 &= (-4)\left(-\tfrac{1}{2}\right)^{4-1} = (-4)\left(-\tfrac{1}{8}\right) = \tfrac{1}{2} \\ a_5 &= (-4)\left(-\tfrac{1}{2}\right)^{5-1} = (-4)\left(\tfrac{1}{16}\right) = -\tfrac{1}{4} \end{aligned}$$

33. $a_n = a_1 r^{n-1}$

$a_n = 10(1.02)^{n-1}$

$a_1 = 10(1.02)^{1-1} = 10$

$a_2 = 10(1.02)^{2-1} = 10.2$

$a_3 = 10(1.02)^{3-1} = 10(1.040) = 10.40$

$a_4 = 10(1.02)^{4-1} = 10(1.061) = 10.61$

$a_5 = 10(1.02)^{5-1} = 10(1.082) = 10.82$

35. $a_n = a_1 r^{n-1}$

$a_n = 10\left(\frac{3}{5}\right)^{n-1}$

$a_1 = 10\left(\frac{3}{5}\right)^{1-1} = 10\left(\frac{3}{5}\right)^0 = 10(1) = 10$

$a_2 = 10\left(\frac{3}{5}\right)^{2-1} = 10\left(\frac{3}{5}\right)^1 = 6$

$a_3 = 10\left(\frac{3}{5}\right)^{3-1} = 10\left(\frac{3}{5}\right)^2 = 10\left(\frac{9}{25}\right) = \frac{18}{5}$

$a_4 = 10\left(\frac{3}{5}\right)^{4-1} = 10\left(\frac{3}{5}\right)^3 = 10\left(\frac{27}{125}\right) = \frac{54}{25}$

$a_5 = 10\left(\frac{3}{5}\right)^{5-1} = 10\left(\frac{3}{5}\right)^4 = 10\left(\frac{81}{625}\right) = \frac{162}{125}$

37. $a_n = a_1 r^{n-1}$

$a_n = \frac{3}{2}\left(\frac{2}{3}\right)^{n-1}$

$a_1 = \frac{3}{2}\left(\frac{2}{3}\right)^{1-1} = \frac{3}{2}$

$a_2 = \frac{3}{2}\left(\frac{2}{3}\right)^{2-1} = \frac{3}{2}\left(\frac{2}{3}\right) = 1$

$a_3 = \frac{3}{2}\left(\frac{2}{3}\right)^{3-1} = \frac{3}{2}\left(\frac{4}{9}\right) = \frac{2}{3}$

$a_4 = \frac{3}{2}\left(\frac{2}{3}\right)^{4-1} = \left(\frac{3}{2}\right)\left(\frac{8}{27}\right) = \frac{4}{9}$

$a_5 = \frac{3}{2}\left(\frac{2}{3}\right)^{5-1} = \left(\frac{3}{2}\right)\left(\frac{16}{81}\right) = \frac{8}{27}$

39. $a_n = a_1 r^{n-1}$

$a_n = 1(2)^{n-1}$

41. $a_n = a_1 r^{n-1}$

$a_n = 2(2)^{n-1}$

43. $a_n = a_1 r^{n-1}$

$a_n = 10\left(-\frac{1}{5}\right)^{n-1}$

45. $a_n = a_1 r^{n-1}$

$a_n = 4\left(-\frac{1}{2}\right)^{n-1}$

47. $a_n = a_1 r^{n-1}$

$a_n = 8\left(\frac{1}{4}\right)^{n-1}$

49. $r = \dfrac{a_2}{a_1} = \dfrac{\frac{21}{2}}{14} = \dfrac{3}{4}$

$a_n = ar^{n-1}$

$a_n = 14\left(\dfrac{3}{4}\right)^{n-1}$

51. $4r = -6$

$r = -\frac{6}{4} = -\frac{3}{2}$

$a_n = a_1 r^{n-1}$

$a_n = 4\left(-\frac{3}{2}\right)^{n-1}$

53. $a_n = a_1 r^{n-1}$

$a_{10} = 6\left(\frac{1}{2}\right)^{10-1} = \frac{3}{256}$

55. $a_n = a_1 r^{n-1}$

$a_{10} = 3\left(\sqrt{2}\right)^{10-1} = 48\sqrt{2}$

57. $a_n = a_1 r^{n-1}$

$a_{12} = 200(1.2)^{12-1}$

$a_{12} \approx 1486.02$

59. $a_n = a_1 r^{n-1}$

$a_{10} \approx 120\left(-\frac{1}{3}\right)^{10-1}$

$a_{10} \approx -0.00610$

61. $a_n = a_1 r^{n-1}$

$a_5 = 4\left(\frac{3}{4}\right)^{5-1} = \frac{81}{64}$

63. $a_n = a_1 r^{n-1}$

$3 = a_1(2)^{3-1}$

$3 = a_1(4)$

$\dfrac{3}{4} = a_1$

$a_n = \dfrac{3}{4}(2)^{n-1}$

$a_5 = \dfrac{3}{4}(2)^{5-1}$

$a_5 = \dfrac{3}{4}(16)$

$a_5 = 12$

65. $a_n = a_1r^{n-1}$

$12 = a_1\left(\frac{4}{3}\right)^{2-1}$

$12 = a_1\left(\frac{4}{3}\right)$

$9 = a_1$

$a_n = 9\left(\frac{4}{3}\right)^{n-1}$

$a_4 = 9\left(\frac{4}{3}\right)^{4-1}$

$a_4 = 9\left(\frac{4}{3}\right)^3 = 9\left(\frac{64}{27}\right) = \frac{64}{3}$

$a_5 = 9\left(\frac{4}{3}\right)^{5-1}$

$a_5 = 9\left(\frac{4}{3}\right)^4 = 9\left(\frac{256}{81}\right) = \frac{256}{9}$

67. (b)

68. (d)

69. (c)

70. (a)

71. $\sum_{i=1}^{10} 2^{i-1} = 1\left(\frac{2^{10}-1}{2-1}\right) = \frac{1024-1}{1} = 1023$

73. $\sum_{i=1}^{12} 3\left(\frac{3}{2}\right)^{i-1} = 3\left(\frac{\left(\frac{3}{2}\right)^{12}-1}{\frac{3}{2}-1}\right) = 3\left(\frac{128.74634}{0.5}\right) \approx 772.48$

75. $\sum_{i=1}^{15} 3\left(-\frac{1}{3}\right)^{i-1} = 3\left(\frac{\left(-\frac{1}{3}\right)^{15}-1}{-\frac{1}{3}-1}\right) = 3\left(\frac{-1.0000001}{-1.3333333}\right) \approx 2.25$

77. $\sum_{i=1}^{6} \left(\frac{3}{4}\right)^i = \frac{3}{4}\left[\frac{\left(\frac{3}{4}\right)^6-1}{\frac{3}{4}-1}\right]$

$= \frac{3}{4}\left[-\frac{3367}{4096} \div -\frac{1}{4}\right]$

$= \frac{3}{4}\left(\frac{13,468}{4096}\right)$

≈ 2.47

79. $\sum_{i=1}^{8} 6(0.1)^{i-1} = 6\left(\frac{(0.1)^8-1}{0.1-1}\right)$

$= 6\left(\frac{-0.999}{-0.9}\right)$

$= 6(1.\bar{1})$

≈ 6.67

81. $\sum_{i=1}^{10} 1(-3)^{i-1} = \left(\frac{(-3)^{10}-1}{-3-1}\right) = -14,762$

83. $\sum_{i=1}^{15} 8\left(\frac{1}{2}\right)^{i-1} = 8\left(\frac{\left(\frac{1}{2}\right)^{15}-1}{\frac{1}{2}-1}\right) \approx 16$

85. $\sum_{i=1}^{8} 4(3)^{i-1} = 4\left(\frac{3^8-1}{3-1}\right) = 4\left(\frac{6560}{2}\right) = 13,120$

87. $\sum_{i=1}^{12} 60\left(-\frac{1}{4}\right)^{i-1} = 60\left(\frac{\left(-\frac{1}{4}\right)^{12}-1}{\left(-\frac{1}{4}\right)-1}\right) \approx 60\left(\frac{-1.000}{-1.25}\right) \approx 48$

89. $\sum_{i=1}^{20} 30(1.06)^{i-1} = 30\left(\frac{1.06^{20}-1}{1.06-1}\right) \approx 1103.57$

91. $\sum_{i=1}^{18} \left(\sqrt{3}\right)^{i-1} = 1\left[\frac{\left(\sqrt{3}\right)^{18}-1}{\sqrt{3}-1}\right]$

$\approx \frac{19,682}{0.732050807569}$

$\approx 26,886.11$

93. $\sum_{n=0}^{\infty} \left(\frac{1}{2}\right)^n = \frac{1}{1-\frac{1}{2}} = \frac{1}{\frac{1}{2}} = 2$

95. $\sum_{n=0}^{\infty} 2\left(-\frac{2}{3}\right)^n = \frac{2}{1-\left(-\frac{2}{3}\right)} = \frac{2}{\frac{5}{3}} = \frac{6}{5}$

97. $\sum_{n=0}^{\infty} \left(\frac{1}{10}\right)^n = \frac{1}{1-\frac{1}{10}} = \frac{1}{\frac{9}{10}} = \frac{10}{9}$

99. $\sum_{n=0}^{\infty} 8\left(\frac{3}{4}\right)^n = \frac{8}{1-\frac{3}{4}} = \frac{8}{\frac{1}{4}} = 32$

101. $a_n = 20(-0.6)^{n-1}$

Keystrokes (calculator in sequence and dot mode):

[Y=] 20 [(] [(−)] 0.6 [)] [^] [(] [n] [−] 1 [)] [TRACE]

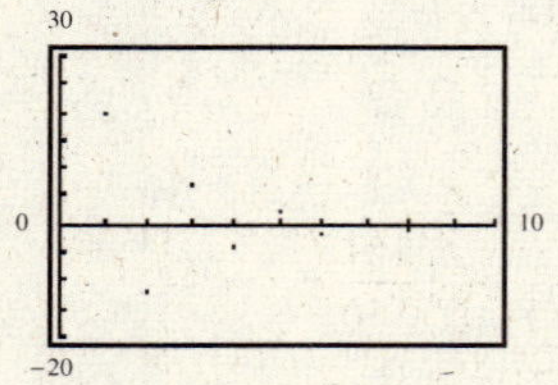

103. $a_n = 15(0.6)^{n-1}$

Keystrokes (calculator in sequence and dot mode):

[Y=] 15 [(] 0.6 [)] [^] [(] [n] [−] 1 [)] [TRACE]

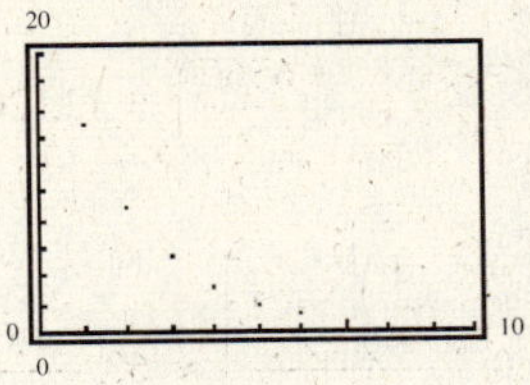

105. $a_0 = 250{,}000$

$a_1 = 250{,}000(0.75)$

$a_2 = 250{,}000(0.75)^2$

$a_3 = 250{,}000(0.75)^3$

$a_4 = 250{,}000(0.75)^4$

(a) $a_n = 250{,}000(0.75)^n$ or $187{,}500(0.75)^{n-1}$

(b) $a_5 = 250{,}000(0.75)^5 \approx \$59{,}326.17$

(c) the first year

107. $\text{Total salary} = \sum_{n=1}^{40} 30{,}000(1.05)^n$

$= 30{,}000\left(\dfrac{1.05^{40} - 1}{1.05 - 1}\right) \approx \$3{,}623{,}993.23$

109. $A = P\left(1 + \dfrac{r}{n}\right)^{nt}$

$a_{120} = 50\left(1 + \dfrac{0.09}{12}\right)^{12(10)} = 50(1.0075)^{120}$

$a_1 = 50(1.0075)^1$

$A = \left[50(1.0075)\right]\left[\dfrac{1.0075^{120} - 1}{1.0075 - 1}\right]$

$\approx \$9748.28$

111. $A = P\left(1 + \dfrac{r}{n}\right)^{nt}$

$a_{480} = 30\left(1 + \dfrac{0.08}{12}\right)^{12(40)} = 30\left(\dfrac{151}{150}\right)^{480}$

$a_1 = 30\left(\dfrac{151}{150}\right)^1$

$\text{Balance} = \left[30\left(\dfrac{151}{150}\right)\right]\left[\dfrac{\left(\dfrac{151}{150}\right)^{480} - 1}{\left(\dfrac{151}{150}\right) - 1}\right] \approx \$105{,}428.44$

113. $A = P\left(1 + \dfrac{r}{n}\right)^{nt}$

$a_{360} = 100\left(1 + \dfrac{0.06}{12}\right)^{12(30)} = 100(1.005)^{360}$

$a_1 = 100(1.005)^1$

$A = \left[100(1.005)\right]\left[\dfrac{1.005^{360} - 1}{1.005 - 1}\right] \approx \$100{,}953.76$

115. $a_n = 0.01(2)^{n-1}$

(a) $\text{Total income} = \sum_{n=1}^{29} 0.01(2)^{n-1}$

$= 0.01\left[\dfrac{2^{29} - 1}{2 - 1}\right]$

$\approx \$5{,}368{,}709.11$

(b) $\text{Total income} = \sum_{n=1}^{30} 0.01(2)^{n-1}$

$= 0.01\left[\dfrac{2^{30} - 1}{2 - 1}\right]$

$\approx \$10{,}737{,}418.23$

117. (a) $P = (0.999)^n$

(b) $P = (0.999)^{365} = 0.694069887 \approx 69.4\%$

119. Total area of shaded region

$= \dfrac{1}{4} + \dfrac{3}{4} \cdot \dfrac{1}{4} + \dfrac{9}{16} \cdot \dfrac{1}{4} = \dfrac{16}{64} + \dfrac{12}{64} + \dfrac{9}{64} = \dfrac{37}{64}$

or

Total area of shaded region

$= \sum_{i=1}^{3} \dfrac{1}{4}\left(\dfrac{3}{4}\right)^{n-1} = \dfrac{1}{4}\left[\dfrac{\left(\dfrac{3}{4}\right)^3 - 1}{\dfrac{3}{4} - 1}\right] = \dfrac{37}{64} \approx 0.578 \text{ square unit}$

121. $\sum_{i=1}^{10} 2(100)(0.75)^n = 2(100)(0.75)\left[\dfrac{0.75^{10} - 1}{0.75 - 1}\right]$

$\approx 150(3.774745941) \approx 566.21$

$\text{Total distance} = 100 + \sum_{i=1}^{10} 2(100)(0.75)^n$

$= 100 + 566.21 \approx 666.21 \text{ feet}$

123. The second and third terms of a geometric sequence are 6 and 3, respectively. The first term is 12 because the ratio of

$\dfrac{a_3}{a_2} = \dfrac{3}{6} = \dfrac{1}{2},$

so the ratio of $\dfrac{a_2}{a_1} = \dfrac{1}{2}$ and $\dfrac{6}{a_1} = \dfrac{1}{2}$ and $a_1 = 12$.

125. The terms of a geometric sequence decrease when $a_1 > 0$ and $a < r < 1$ because raising a real number between 0 and 1 to higher powers yields smaller numbers.

127. The nth partial sum of a sequence is the sum of the first n terms of a sequence.

129. $\begin{cases} y = 2x^2 \\ y = 2x + 4 \end{cases}$

$$2x^2 = 2x + 4$$
$$2x^2 - 2x - 4 = 0$$
$$x^2 - x - 2 = 0$$
$$(x - 2)(x + 1) = 0$$
$$x = 2 \qquad x = -1$$
$$y = 2(2)^2 = 8 \qquad y = 2(-1)^2 = 2$$
$$(2, 8), (-1, 2)$$

131. $A = P\left(1 + \frac{r}{n}\right)^{nt}$

$$2219.64 = 1000\left(1 + \frac{r}{12}\right)^{12(10)}$$
$$2.21964 = \left(1 + \frac{r}{12}\right)^{120}$$
$$2.21964^{1/120} = 1 + \frac{r}{12}$$
$$12\left(2.21964^{1/120} - 1\right) = r$$
$$0.08 \approx r$$
$$8\% \approx r$$

133. $A = P\left(1 + \frac{r}{n}\right)^{nt}$

$$10{,}619.63 = 2500\left(1 + \frac{r}{1}\right)^{1(20)}$$
$$4.247852 = (1 + r)^{20}$$
$$4.247852^{1/20} = 1 + r$$
$$4.247852^{1/20} - 1 = r$$
$$0.075 \approx r$$
$$7.5\% \approx r$$

135. $\frac{x^2}{16} - \frac{y^2}{9} = 1$

Horizontal axis

Vertices: $(\pm 4, 0)$

Asymptotes: $y = \pm\frac{3}{4}x$

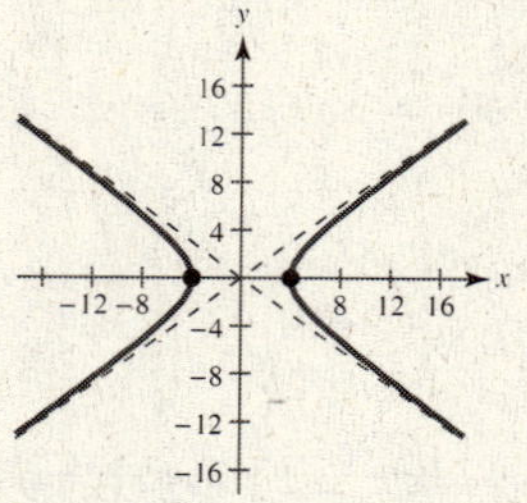

Section 11.4 The Binomial Theorem

1. ${}_6C_4 = {}_6C_2 = \frac{6 \cdot 5}{2 \cdot 1} = 15$

3. ${}_{10}C_5 = \frac{10 \cdot 9 \cdot 8 \cdot 7 \cdot 6}{5 \cdot 4 \cdot 3 \cdot 2 \cdot 1} = 252$

5. ${}_{12}C_{12} = \frac{12!}{0!12!} = 1$

7. ${}_{20}C_6 = \frac{20!}{14!6!} = \frac{20 \cdot 19 \cdot 18 \cdot 17 \cdot 16 \cdot 15}{6 \cdot 5 \cdot 4 \cdot 3 \cdot 2 \cdot 1} = 38{,}760$

9. ${}_{20}C_{14} = \frac{20!}{6!14!} = \frac{20 \cdot 19 \cdot 18 \cdot 17 \cdot 16 \cdot 15}{6 \cdot 5 \cdot 4 \cdot 3 \cdot 2 \cdot 1} = 38{,}760$

11. *Keystrokes:*

30 [MATH] [PRB 3] 6 [ENTER] ${}_{30}C_6 = 593{,}775$

13. *Keystrokes:*

52 [MATH] [PRB 3] 5 [ENTER] ${}_{52}C_5 = 2{,}598{,}960$

15. *Keystrokes:*

800 [MATH] [PRB 3] 797 [ENTER]

${}_{800}C_{797} = 85{,}013{,}600$

17. ${}_6C_2 = 15$

Row 6: 1 6 15 20 15 6 1

↑ entry 2

19. ${}_7C_3 = 35$

Row 7: 1 7 21 35 35 21 7 1

↑ entry 3

21. ${}_8C_4 = 70$

Row 8: 1 8 28 56 70 56 28 8 1

↑ entry 4

23. $(t + 5)^3 = (1)t^3 + (3)t^2(5) + (3)t(5)^2 + (1)5^3$

$= t^3 + 15t^2 + 75t + 125$

25. $(m - n)^5 = (1)m^5 + (5)m^4(-n) + (10)m^3(-n)^2 + 10m^2(-n)^3 + 5m(-n)^4 + 1(-n)^5$

$= m^5 - 5m^4n + 10m^3n^2 - 10m^2n^3 + 5mn^4 - n^5$

27. $(3a - 1)^5 = (1)(3a)^5 + (5)(3a)^4(-1) + (10)(3a)^3(-1)^2 + (10)(3a)^2(-1)^3 + (5)(3a)(-1)^4 + (1)(-1)^5$

$= 243a^5 - 405a^4 + 270a^3 - 90a^2 + 15a - 1$

29. $(2y + z)^6 = (1)(2y)^6 + 6(2y)^5z + 15(y)^4z^2 + 20(2y)^3z^3 + 15(2y)^2z^4 + 6(2y)z^5 + 1z^6$

$= 64y^6 + 192y^5z + 240y^4z^2 + 160y^3z^3 + 60y^2z^4 + 12yz^5 + z^6$

31. $(x^2 + 2)^4 = 1(x^2)^4 + 4(x^2)^3(2) + 6(x^2)^2(2)^2 + 4(x^2)(2)^3 + 1(2)^4$

$= x^8 + 8x^6 + 24x^4 + 32x^2 + 16$

33. $(x + 3)^6 = 1x^6 + 6x^5(3) + 15x^4(3)^2 + 20x^3(3)^3 + 15x^2(3)^4 + 6x(3)^5 + 1(3)^6$

$= x^6 + 18x^5 + 135x^4 + 540x^3 + 1215x^2 + 1458x + 729$

35. $(u - 2v)^3 = 1u^3 - 3u^2(2v) + 3u(2v)^2 - 1(2v)^3$

$= u^3 - 6u^2v + 12uv^2 - 8v^3$

37. $(3a + 2b)^4 = 1(3a)^4 + 4(3a)^3(2b) + 6(3a)^2(2b)^2 + 4(3a)(2b)^3 + 1(2b)^4$

$= 81a^4 + 216a^3b + 216a^2b^2 + 96ab^3 + 16b^4$

39. $\left(x + \dfrac{2}{y}\right)^4 = 1(x)^4 + 4(x)^3\left(\dfrac{2}{y}\right) + 6(x)^2\left(\dfrac{2}{y}\right)^2 + 4(x)\left(\dfrac{2}{y}\right)^3 + 1\left(\dfrac{2}{y}\right)^4$

$= x^4 + \dfrac{8x^3}{y} + \dfrac{24x^2}{y^2} + \dfrac{32x}{y^3} + \dfrac{16}{y^4}$

41. $(2x^2 - y)^5 = 1(2x^2)^5 + 5(2x^2)^4(-y) + 10(2x^2)^3(-y)^2 + 10(2x^2)^2(-y)^3 + 5(2x^2)(-y)^4 + 1(-y)^5$

$= 32x^{10} - 80x^8y + 80x^6y^2 - 40x^4y^3 + 10x^2y^4 - y^5$

43. $(x + y)^{10}$, 4th term

$(r + 1)^{\text{th}}$ term $= {}_nC_rx^{n-r}y^r$

$r + 1 = 4 \quad n = 10 \quad x = x \quad y = y$

$r = 3$

4th term $= {}_{10}C_3x^{10-3}y^3 = 120x^7y^3$

45. $(a + 6b)^9$, 5th term

$(r + 1)^{\text{th}}$ term $= {}_nC_rx^{n-r}y^r$

$r + 1 = 5 \quad n = 9 \quad x = a \quad y = 6b$

$r = 4$

5th term $= {}_9C_4a^{9-4}(6b)^4$

$= 126a^5(1296b^4)$

$= 163{,}296a^5b^4$

47. ${}_nC_r x^{n-r} y^r$

$n = 10 \quad n - r = 7 \quad r = 3 \quad x = x \quad y = 1$

${}_{10}C_3 x^7 1^3$

$${}_{10}C_3 = \frac{10 \cdot 9 \cdot 8}{3 \cdot 2 \cdot 1} = 120$$

49. ${}_nC_r x^{n-r} y^r$

$n = 4 \quad n - r = 2 \quad r = 2 \quad x = x^2 \quad y = (-3)$

${}_4C_2 (x^2)^2 (-3)^2$

$${}_4C_2 = \frac{4 \cdot 3}{(2 \cdot 1)} = 6$$

$(-3)^2 {}_4C_2 = 9(6) = 54$

51. $$\begin{aligned}(1.02)^8 &= (1 + 0.02)^8 \\ &= (1)^8 + 8(1)^7(0.02) + 28(1)^6(0.02)^2 + 56(1)^5(0.02)^3 + \dots \\ &\approx 1 + 0.16 + 0.0112 + 0.000448 \approx 1.172\end{aligned}$$

53. $$\begin{aligned}(2.99)^{12} &= (3 - 0.01)^{12} \\ &= 1(3)^{12} - 12(3)^{11}(0.01) + 66(3)^{10}(0.01)^2 - 220(3)^9(0.01)^3 + 495(3)^8(0.01)^4 - 792(3)^7(0.01)^5 + \dots \\ &\approx 531{,}441 - 21{,}257.64 + 389.7234 - 4.33026 + 0.03247695 - 0.0001732104 \\ &\approx 510{,}568.785\end{aligned}$$

55. $$\begin{aligned}\left(\tfrac{1}{2} + \tfrac{1}{2}\right)^5 &= 1\left(\tfrac{1}{2}\right)^5 + 5\left(\tfrac{1}{2}\right)^4\left(\tfrac{1}{2}\right) + 10\left(\tfrac{1}{2}\right)^3\left(\tfrac{1}{2}\right)^2 + 10\left(\tfrac{1}{2}\right)^2\left(\tfrac{1}{2}\right)^3 + 5\left(\tfrac{1}{2}\right)\left(\tfrac{1}{2}\right)^4 + 1\left(\tfrac{1}{2}\right)^5 \\ &= \tfrac{1}{32} + \tfrac{5}{32} + \tfrac{10}{32} + \tfrac{10}{32} + \tfrac{5}{32} + \tfrac{1}{32}\end{aligned}$$

57. $$\left(\tfrac{1}{4} + \tfrac{3}{4}\right)^4 = 1\left(\tfrac{1}{4}\right)^4 + 4\left(\tfrac{1}{4}\right)^3\left(\tfrac{3}{4}\right) + 6\left(\tfrac{1}{4}\right)^2\left(\tfrac{3}{4}\right)^2 + 4\left(\tfrac{1}{4}\right)\left(\tfrac{3}{4}\right)^3 + 1\left(\tfrac{3}{4}\right)^4 = \tfrac{1}{256} + \tfrac{12}{256} + \tfrac{54}{256} + \tfrac{108}{256} + \tfrac{81}{256}$$

59. The sum of the numbers in each row is a power of 2. Because Row 2 is $1 + 2 + 1 = 4 = 2^2$, Row n is 2^n.

61. The signs in the expansion of $(x + y)^n$ are all positive.

The signs in the expansion of $(x - y)^n$ alternate.

63. Pascal's Triangle is formed by making the first and last numbers in each row 1. Every other number in the row is formed by adding the two numbers immediately above the number.

65. $$\sum_{i=1}^{15} (2 + 3i) = \tfrac{15}{2}(5 + 47) = \tfrac{15}{2}(52) = 390$$

$a_1 = 2 + 3(1) = 5, \quad a_n = 2 + 3(15) = 47$

67. $$\sum_{k=1}^{8} 5^{k-1} = 1\left[\frac{5^8 - 1}{5 - 1}\right] = 97{,}656$$

Review Exercises for Chapter 11

1. $a_n = 3n + 5$

$a_1 = 3(1) + 5 = 8$

$a_2 = 3(2) + 5 = 11$

$a_3 = 3(3) + 5 = 14$

$a_4 = 3(4) + 5 = 17$

$a_5 = 3(5) + 5 = 20$

3. $$a_n = \frac{n}{3n - 1}$$

$$a_1 = \frac{1}{3(1) - 1} = \frac{1}{2}$$

$$a_2 = \frac{2}{3(2) - 1} = \frac{2}{5}$$

$$a_3 = \frac{3}{3(3) - 1} = \frac{3}{8}$$

$$a_4 = \frac{4}{3(4) - 1} = \frac{4}{11}$$

$$a_5 = \frac{5}{3(5) - 1} = \frac{5}{14}$$

5. $a_n = (n + 1)!$

$a_1 = (1 + 1)! = 2! = 2 \cdot 1 = 2$

$a_2 = (2 + 1)! = 3! = 3 \cdot 2 \cdot 1 = 6$

$a_3 = (3 + 1)! = 4! = 4 \cdot 3 \cdot 2 \cdot 1 = 24$

$a_4 = (4 + 1)! = 5! = 5 \cdot 4 \cdot 3 \cdot 2 \cdot 1 = 120$

$a_5 = (5 + 1)! = 6! = 6 \cdot 5 \cdot 4 \cdot 3 \cdot 2 \cdot 1 = 720$

7. $a_n = \frac{n!}{2n}$

$a_1 = \frac{1!}{2 \cdot 1} = \frac{1}{2}$

$a_2 = \frac{2!}{2 \cdot 2} = \frac{2}{4} = \frac{1}{2}$

$a_3 = \frac{3!}{2 \cdot 3} = \frac{3 \cdot 2 \cdot 1}{2 \cdot 3} = 1$

$a_4 = \frac{4!}{2 \cdot 4} = \frac{4 \cdot 3 \cdot 2 \cdot 1}{2 \cdot 4} = 3$

$a_5 = \frac{5!}{2 \cdot 5} = \frac{5 \cdot 4 \cdot 3 \cdot 2 \cdot 1}{2 \cdot 5} = 12$

9. $a_n = 3n + 1$

11. $a_n = \frac{1}{n^2 + 1}$

13. $a_n = -2n + 5$

15. $a_n = \frac{3n^2}{n^2 + 1}$

17. $\sum_{k=1}^{4} 7 = 7 + 7 + 7 + 7 = 28$

19. $\sum_{i=1}^{5} \frac{i-2}{i+1} = -\frac{1}{2} + 0 + \frac{1}{4} + \frac{2}{5} + \frac{1}{2}$

$= \frac{1}{4} + \frac{2}{5}$

$= \frac{5}{20} + \frac{8}{20}$

$= \frac{13}{20}$

21. $\sum_{k=1}^{4} (5k - 3)$

23. $\sum_{k=1}^{6} \frac{1}{3k}$

25. 50, 44.5, 39, 33.5, 28, …

$44.5 - 50 = -5.5$

$39 - 44.5 = -5.5$

$33.5 - 39 = -5.5$

$28 - 33.5 = -5.5$

$d = -5.5$

27. $a_1 = 132 - 5(1) = 127$

$a_2 = 132 - 5(2) = 122$

$a_3 = 132 - 5(3) = 117$

$a_4 = 132 - 5(4) = 112$

$a_5 = 132 - 5(5) = 107$

29. $a_1 = \frac{1}{3}(1) + \frac{5}{3} = \frac{6}{3} = 2$

$a_2 = \frac{1}{3}(2) + \frac{5}{3} = \frac{7}{3}$

$a_3 = \frac{1}{3}(3) + \frac{5}{3} = \frac{8}{3}$

$a_4 = \frac{1}{3}(4) + \frac{5}{3} = \frac{9}{3} = 3$

$a_5 = \frac{1}{3}(5) + \frac{5}{3} = \frac{10}{3}$

31. $a_1 = 80$

$a_2 = 80 - \frac{5}{2} = \frac{160}{2} - \frac{5}{2} = \frac{155}{2}$

$a_3 = \frac{155}{2} - \frac{5}{2} = \frac{150}{2} = 75$

$a_4 = \frac{150}{2} - \frac{5}{2} = \frac{145}{2}$

$a_5 = \frac{145}{2} - \frac{5}{2} = \frac{140}{2} = 70$

33. $a_n = dn + c$

$10 = 4(1) + c$

$6 = c$

$a_n = 4n + 6$

35. $a_n = dn + c$

$1000 = -50(1) + c$

$1050 = c$

$a_n = -50n + 1050$

37. $\sum_{k=1}^{12} (7k - 5) = \frac{12}{2}(2 + 79) = 486$

39. $\sum_{j=1}^{120} \left(\frac{1}{4}j + 1\right) = \frac{120}{2}\left(\frac{5}{4} + 31\right) = 60\left(\frac{129}{4}\right) = 1935$

41. *Keystrokes:*

[LIST] [MATH 5] [LIST] [OPS 5] 1.25

[X, T, θ] [+] 4 [,] [X, T, θ] [,] 1 [,] 60 [,] 1 [)] [ENTER]

$\sum_{i=1}^{60} (125i + 4) = 2527.5$

43. $\sum_{n=1}^{50} 4n = \frac{50}{2}(4 + 200) = 5100$

45. $\sum_{n=1}^{12} (3n + 19) = \frac{12}{2}(22 + 55) = 462$ seats

47. $8, 20, 50, 125, \frac{625}{2}, \ldots$

$r = \frac{5}{2}$ since

$\frac{20}{8} = \frac{5}{2}, \frac{50}{20} = \frac{5}{2}, \frac{125}{50} = \frac{5}{2}, \frac{\frac{625}{2}}{125} = \frac{5}{2}.$

49. $a_n = a_1 r^{n-1}$

$a_n = 10(3)^{n-1}$

$a_1 = 10(3)^{1-1} = 10$

$a_2 = 10(3)^{2-1} = 30$

$a_3 = 10(3)^{3-1} = 90$

$a_4 = 10(3)^{4-1} = 270$

$a_5 = 10(3)^{5-1} = 810$

51. $a_n = a_1 r^{n-1}$

$a_n = 100\left(-\frac{1}{2}\right)^{n-1}$

$a_1 = 100\left(-\frac{1}{2}\right)^{1-1} = 100$

$a_2 = 100\left(-\frac{1}{2}\right)^{2-1} = -50$

$a_3 = 100\left(-\frac{1}{2}\right)^{3-1} = 25$

$a_4 = 100\left(-\frac{1}{2}\right)^{4-1} = -12.5$

$a_5 = 100\left(-\frac{1}{2}\right)^{5-1} = 6.25$

53. $a_n = a_1 r^{n-1}$

$a_n = 4\left(\frac{3}{2}\right)^{n-1}$

$a_1 = 4\left(\frac{3}{2}\right)^{1-1} = 4\left(\frac{3}{2}\right)^{0} = 4(1) = 4$

$a_2 = 4\left(\frac{3}{2}\right)^{2-1} = 4\left(\frac{3}{2}\right)^{1} = 4\left(\frac{3}{2}\right) = 6$

$a_3 = 4\left(\frac{3}{2}\right)^{3-1} = 4\left(\frac{3}{2}\right)^{2} = 4\left(\frac{9}{4}\right) = 9$

$a_4 = 4\left(\frac{3}{2}\right)^{4-1} = 4\left(\frac{3}{2}\right)^{3} = 4\left(\frac{27}{8}\right) = \frac{27}{2}$

$a_5 = 5\left(\frac{3}{2}\right)^{5-1} = 4\left(\frac{3}{2}\right)^{4} = 4\left(\frac{81}{16}\right) = \frac{81}{4}$

55. $a_n = a_1 r^{n-1}$

$a_n = 1\left(-\frac{2}{3}\right)^{n-1}$

57. $a_n = a_1 r^{n-1}$

$a_n = 24(3)^{n-1}$

$r = \frac{a_2}{a_1} = \frac{72}{24} = 3$

59. $a_n = a_1 r^{n-1}$

$a_n = 12\left(-\frac{1}{2}\right)^{n-1}$

61. $\sum_{n=1}^{12} 2^n = 2\left(\frac{2^{12} - 1}{2 - 1}\right) = 8190$

63. $\sum_{k=1}^{8} 5\left(-\frac{3}{4}\right)^k = -\frac{15}{4}\left(\frac{\left(-\frac{3}{4}\right)^8 - 1}{-\frac{3}{4} - 1}\right) \approx -1.928$

65. $\sum_{n=1}^{120} 500(1.01)^n = 505\left(\frac{1.01^{120} - 1}{1.01 - 1}\right) \approx 116{,}169.538$

67. *Keystrokes:* [LIST] [MATH] [5:SUM] [LIST] [OPS] [5:SEQ] 200 [(] 1.4 [)] [^] [(] [X, T, θ] [−] 1 [)] [,] [X, T, θ] [,] 1 [,] 75 [,] 1 [)] [ENTER]

$\sum_{k=1}^{75} 200(1.4)^{k-1} \approx 4.556 \times 10^{13}$

69. $\sum_{i=1}^{\infty} \left(\frac{7}{8}\right)^{i-1} = \frac{1}{1 - \frac{7}{8}} = \frac{1}{\frac{1}{8}} = 8$

71. $\sum_{k=1}^{\infty} 4\left(\frac{2}{3}\right)^{k-1} = \frac{4}{1 - \frac{2}{3}} = \frac{4}{\frac{1}{3}} = 12$

73. (a) $a_n = a_1 r^n$

$a_n = 120{,}000(0.7)^n$

(b) $a_5 = 120{,}000(0.7)^5 = \$20{,}168.40$

75. $\sum_{i=1}^{90} 1000(1.125)^i = 1000\left[\frac{1.125^{90} - 1}{1.125 - 1}\right]$

$\approx 321{,}222{,}672$ visitors

77. ${}_8C_3 = \frac{8!}{3!5!} = \frac{8 \cdot 7 \cdot 6 \cdot 5!}{3 \cdot 2 \cdot 5!} = 56$

79. ${}_{15}C_4 = \frac{15!}{11!4!} = \frac{15 \cdot 14 \cdot 13 \cdot 12}{4 \cdot 3 \cdot 2 \cdot 1} = 1365$

81. *Keystrokes:*

40 [MATH] [PRB 3] 4 [ENTER] ${}_{40}C_4 = 91{,}390$

83. *Keystrokes:*

25 [MATH] [PRB 3] 6 [ENTER] $_{25}C_6 = 177{,}100$

85. $_5C_3 = 10$

Fifth row of Pascal's Triangle:

1 5 10 10 5 1

$\uparrow$ $_5C_3$

87. $_8C_4 = 70$

Eighth row of Pascal's Triangle:

1 8 28 56 70 56 28 8 1

$\uparrow$ $_8C_4$

89. $(x-5)^4 = 1(x)^4 + 4(x)^3(-5) + 6(x)^2(-5)^2 + 4(x)(-5)^3 + 1(-5)^4 = x^4 - 20x^3 + 150x^2 - 500x + 625$

91. $(5x+2)^3 = (1)(5x)^3 + (3)(5x)^2(2) + (3)(5x)(2)^2 + 1(2)^3 = 125x^3 + 150x^2 + 60x + 8$

93. $(x+1)^{10} = 1x^{10} + 10x^9(1) + 45x^8(1)^2 + 120x^7(1)^3 + 210x^6(1)^4 + 252x^5(1)^5 + 210x^4(1)^6 + 120x^3(1)^7 + 45x^2(1)^8$
$+ 10x(1)^9 + 1(1)^{10}$
$= x^{10} + 10x^9 + 45x^8 + 120x^7 + 210x^6 + 252x^5 + 210x^4 + 120x^3 + 45x^2 + 10x + 1$

95. $(3x-2y)^4 = 1(3x)^4 + 4(3x)^3(-2y) + 6(3x)^2(-2y)^2 + 4(3x)(-2y)^3 + (-2y)^4$
$= 81x^4 - 216x^3y + 216x^2y^2 - 96xy^3 + 16y^4$

97. $(u^2+v^3)^5 = (1)(u^2)^5 + (5)(u^2)^4(v^3) + (10)(u^2)^3(v^3)^2 + (10)(u^2)^2(v^3)^3 + (5)(u^2)(v^3)^4 + (1)(v^3)^5$
$= u^{10} + 5u^8v^3 + 10u^6v^6 + 10u^4v^9 + 5u^2v^{12} + v^{15}$

99. $(x+2)^{10}$, 7th term

$(r+1)^{th}$ term $= {}_nC_r x^{n-r} y^r$

$r + 1 = 7,\ n = 10,\ x = x,\ y = 2$

$r = 6$

7th term $= {}_{10}C_6 x^{10-6}(2)^6 = 210x^4(64) = 13{,}440x^4$

101. $_nC_r x^{n-r} y^r$

$n = 10,\ n - r = 5,\ r = 5,\ x = x,\ y = (-3)$

$_{10}C_5 = 252 \cdot (-3)^5 = -61{,}236$

Chapter Test for Chapter 11

1. $a_n = \left(-\frac{3}{5}\right)^{n-1}$

$a_1 = \left(-\frac{3}{5}\right)^{1-1} = 1$

$a_2 = \left(-\frac{3}{5}\right)^{2-1} = -\frac{3}{5}$

$a_3 = \left(-\frac{3}{5}\right)^{3-1} = \frac{9}{25}$

$a_4 = \left(-\frac{3}{5}\right)^{4-1} = -\frac{27}{125}$

$a_5 = \left(-\frac{3}{5}\right)^{5-1} = \frac{81}{625}$

2. $a_n = 3n^2 - n$

$a_1 = 3(1)^2 - 1 = 2$

$a_2 = 3(2)^2 - 2 = 12 - 2 = 10$

$a_3 = 3(3)^2 - 3 = 27 - 3 = 24$

$a_4 = 3(4)^2 - 4 = 48 - 4 = 44$

$a_5 = 3(5)^2 - 5 = 75 - 5 = 70$

3. $\sum_{n=1}^{12} 5 = 12(5) = 60$

4. $\sum_{k=0}^{8} (2k - 3) = -3 + (-1) + 1 + 3 + 5 + 7 + 9 + 11 + 13$

$= 45$

5. $\sum_{n=1}^{5} (3 - 4n) = \frac{5}{2}[-1 + (-17)] = -45$

6. $\sum_{n=1}^{12} \frac{2}{3n + 1}$

7. $\sum_{k=1}^{6} \left(\frac{1}{2}\right)^{2k-2}$

8. $a_n = a_1 + (n - 1)d$

$a_n = 12 + (n - 1)4 = 12 + 4n - 4 = 4n + 8$

$a_1 = 4(1) + 8 = 12$

$a_2 = 4(2) + 8 = 16$

$a_3 = 4(3) + 8 = 20$

$a_4 = 4(4) + 8 = 24$

$a_5 = 4(5) + 8 = 28$

9. $a_n = a_1 + (n - 1)d$

$a_n = 5000 + (n - 1)(-100)$

$a_n = 50000 - 100n + 100$

$a_n = -100n + 5100$

10. $\sum_{n=1}^{50} 3n = \frac{50}{2}(3 + 150) = 3825$

11. $-4, 3, -\frac{9}{4}, \frac{27}{16}, \ldots$

$r = -\frac{3}{4}$ since

$$\frac{3}{-4} = -\frac{3}{4}, \frac{-\frac{9}{4}}{3} = -\frac{3}{4}, \frac{\frac{27}{16}}{-\frac{9}{4}} = -\frac{3}{4}.$$

12. $a_n = a_1 r^{n-1}$

$a_n = 4\left(\frac{1}{2}\right)^{n-1}$

13. $\sum_{n=1}^{8} 2(2^n) = 4\left(\frac{2^8 - 1}{2 - 1}\right) = 1020$

14. $\sum_{n=1}^{10} 3\left(\frac{1}{2}\right)^n = \frac{3}{2}\left(\frac{\frac{1^{10}}{2} - 1}{\frac{1}{2} - 1}\right) = \frac{3069}{1024}$

15. $\sum_{i=1}^{\infty} \left(\frac{1}{2}\right)^i = \frac{\frac{1}{2}}{1 - \frac{1}{2}} = \frac{\frac{1}{2}}{\frac{1}{2}} = 1$

16. $\sum_{i=1}^{\infty} 10(0.4)^{i-1} = \frac{10}{1 - 0.4} = \frac{10}{0.6} = \frac{100}{6} = \frac{50}{3}$

17. ${}_{20}C_3 = \frac{20 \cdot 19 \cdot 18}{3 \cdot 2 \cdot 1} = 1140$

18. $(x - 2)^5 = 1(x^5) - 5x^4(2) + 10x^3(2)^2 - 10x^2(2)^3 + 5x(2)^4 - 1(2)^5$

$= x^5 - 10x^4 + 40x^3 - 80x^2 + 80x - 32$

19. The coefficient of x^3y^5 in expansion of $(x + y)^8$ is 56, since ${}_8C_3 = 56$.

20. $14.7 - 4.9 = 9.8$

$24.5 - 14.7 = 9.8$

So, $d = 9.8$.

$a_n = a_1 + (n - 1)d$

$a_n = 4.9 + (n - 1)9.8$

$a_n = 4.9 + 9.8n - 9.8$

$a_n = 9.8n - 4.9$

$\sum_{n=1}^{10} 9.8n - 4.9 = \frac{10}{2}(4.9 + 93.1) = 490$ meters

21. Balance $= 80\left(1 + \frac{0.048}{12}\right)^1 + \cdots + 80\left(1 + \frac{0.048}{12}\right)^{540} = 80(1.004)\left(\frac{1.004^{540} - 1}{1.004 - 1}\right) = \$153,287.87$

Appendix A Introduction to Graphing Calculators

1. $y = -3x$

Keystrokes:

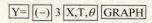

[Y=] [(−)] 3 [X,T,θ] [GRAPH]

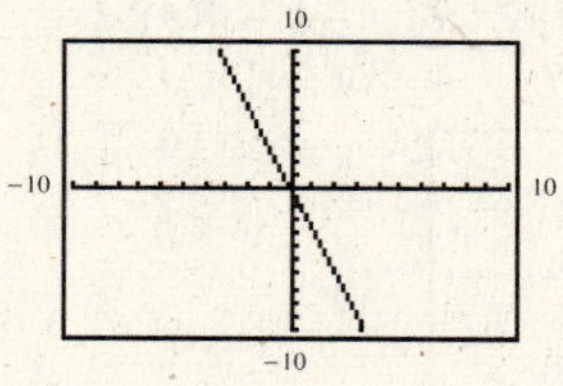

3. $y = \frac{3}{4}x - 6$

Keystrokes:

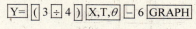

[Y=] [(] 3 [÷] 4 [)] [X,T,θ] [−] 6 [GRAPH]

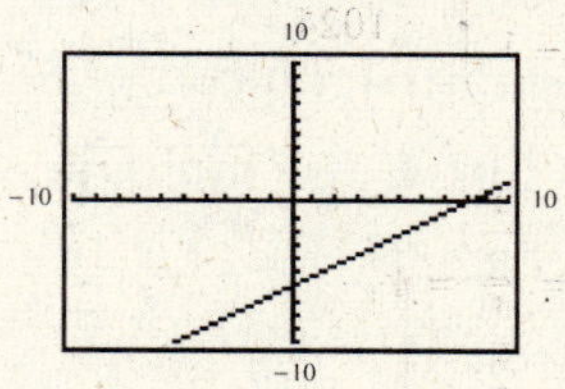

5. $y = \frac{1}{2}x^2$

Keystrokes:

[Y=] [(] 1 [÷] 2 [)] [X,T,θ] [x^2] [GRAPH]

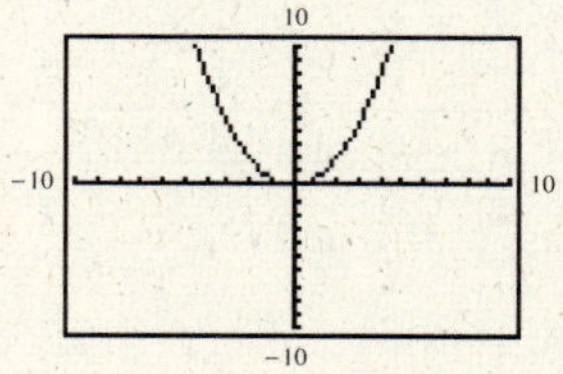

7. $y = x^2 - 4x + 2$

Keystrokes:

[Y=] [X,T,θ] [x^2] [−] 4 [X,T,θ] [+] 2 [GRAPH]

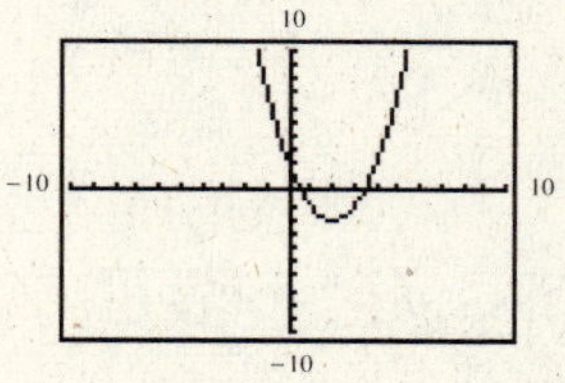

9. $y = |x - 5|$

Keystrokes:

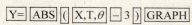

[Y=] [ABS] [(] [X,T,θ] [−] 3 [)] [GRAPH]

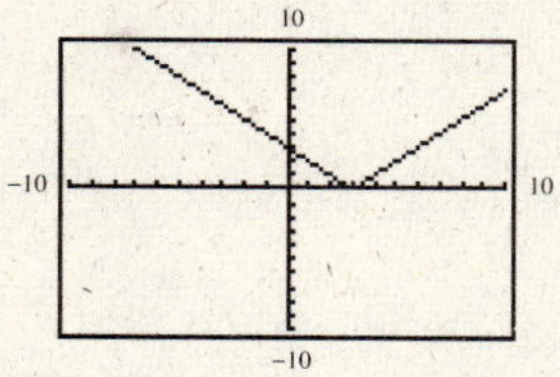

11. $y = |x^2 - 4|$

Keystrokes:

[Y=] [ABS] [(] [X,T,θ] [x^2] [−] 4 [)] [GRAPH]

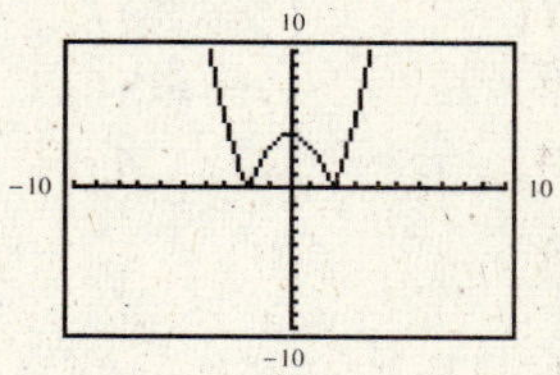

13. $y = 27x + 100$

Keystrokes:

[Y=] 27 [X,T,θ] [+] 100 [GRAPH]

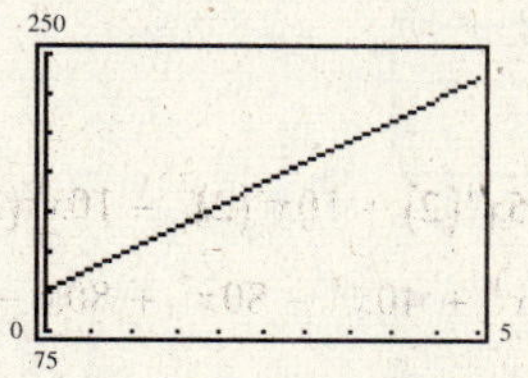

15. $y = 0.001x^2 + 0.5x$

Keystrokes:

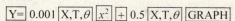

[Y=] 0.001 [X,T,θ] [x^2] [+] 0.5 [X,T,θ] [GRAPH]

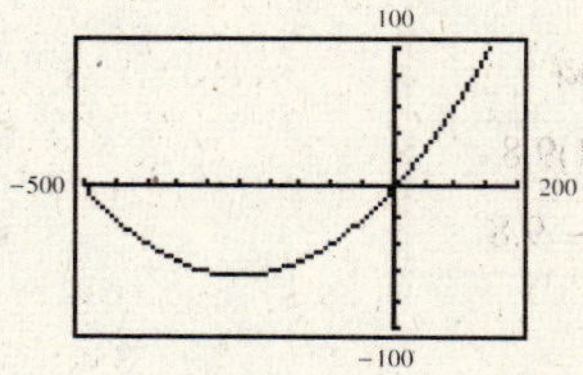

17. $y = 15 + |x - 12|$

Keystrokes:

[Y=] 15 [+] [ABS] [(] [X,T,θ] [−] 12 [)] [GRAPH]

Xmin = 4
Xmax = 20
Xscl = 1
Ymin = 14
Ymax = 22
Yscl = 1

19. $y = -15 + |x + 12|$

Keystrokes:

[Y=] [(−)] 15 [+] [ABS] [(] [X,T,θ] [+] 12 [)] [GRAPH]

Xmin = −20
Xmax = −4
Xscl = 1
Ymin = −16
Ymax = −8
Yscl = 1

21. $y = -4, y = -|x|$

Keystrokes: y_1: [Y=] [(−)] 4 [ENTER]

y_2: [(−)] [ABS] [X,T,θ] [GRAPH]

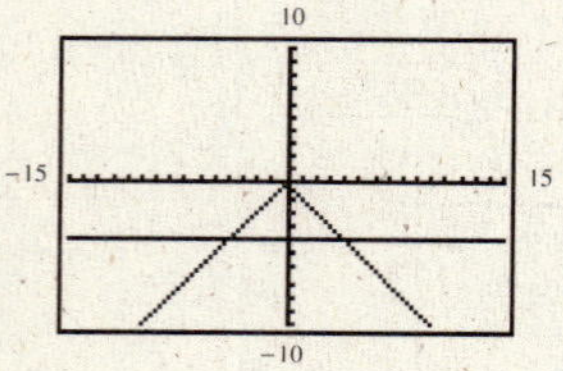

Triangle

23. $y = |x| - 8, y = -|x| + 8$

Keystrokes: y_1: [Y=] [ABS] [X,T,θ] [−] 8 [ENTER]

y_2: [(−)] [ABS] [X,T,θ] [+] 8 [GRAPH]

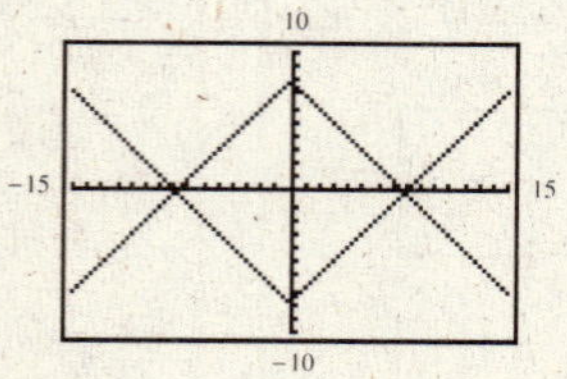

Square

25. $y_1 = 2x + (x + 1)$

$y_2 = (2x + x) + 1$

Keystrokes:

y_1: [Y=] 2 [X,T,θ] [+] [(] [X,T,θ] [+] 1 [)] [ENTER]

y_2: [(] 2 [X,T,θ] [+] [X,T,θ] [)] [+] 1 [GRAPH]

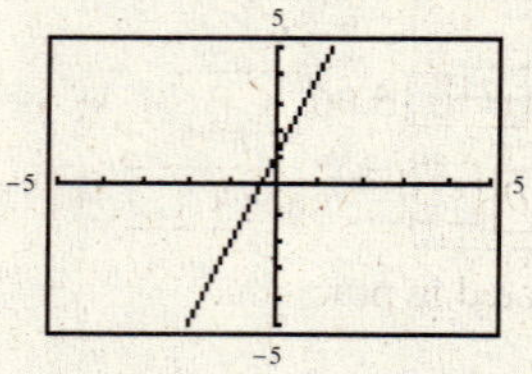

Associative Property of Addition

27. $y_1 = 2\left(\frac{1}{2}\right)$

$y_2 = 2$

Keystrokes: y_1: [Y=] 2 [(] 1 [÷] 2 [)] [ENTER]

y_2: 1 [GRAPH]

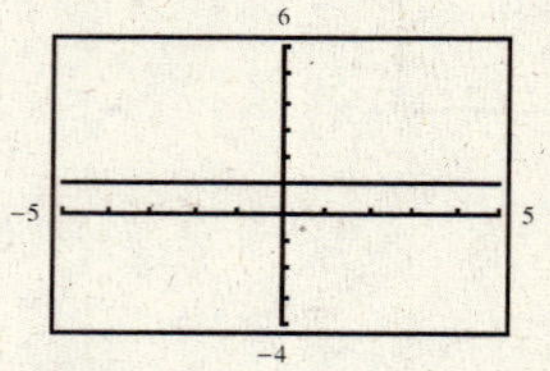

Multiplicative Inverse Property

29. $y = 9 - x^2$

Keystrokes: [Y=] 9 [−] [X,T,θ] [x^2] [GRAPH]

Trace to x-intercepts: $(-3, 0)$ and $(3, 0)$

Trace to y-intercept: $(0, 9)$

31. $y = 6 - |x + 2|$

Keystrokes:

[Y=] 6 [−] [ABS] [(] [X,T,θ] [+] 2 [)] [GRAPH]

Trace to x-intercepts: $(-8, 0)$ and $(4, 0)$

Trace to y-intercept: $(0, 4)$

33. $y = 2x - 5$

Keystrokes: [Y=] 2 [X,T,θ] [−] 5 [GRAPH]

Trace to x-intercept: $\left(\frac{5}{2}, 0\right)$

Trace to y-intercept: $(0, -5)$

35. $y = x^2 + 1.5x - 1$

Keystrokes:

[Y=] [X,T,θ] [x^2] [+] 1.5 [X,T,θ] [−] 1 [GRAPH]

Trace to x-intercepts: $(-2, 0)$ and $\left(\frac{1}{2}, 0\right)$

Trace to y-intercept: $(0, -1)$

37. *Keystrokes:*

y_1: [Y=] 0.5 [X,T,θ] [x^2] [−] 5.06 [X,T,θ] [+] 110.3 [ENTER]

y_2: [(−)] 0.221 [X,T,θ] [x^2] [+] 5.88 [X,T,θ] [+] 75.8 [GRAPH]

Set window as indicated in problem.

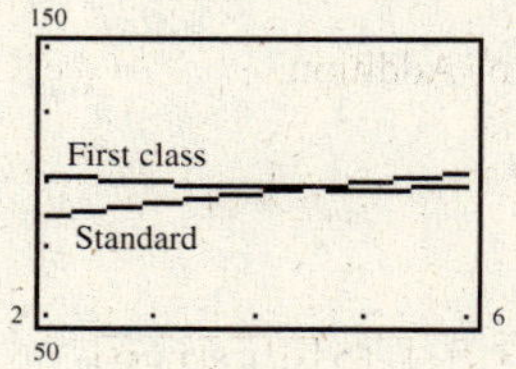